HANDBOOK OF AMPHIBIANS AND REPTILES OF NORTH-EAST AFRICA

HANDBOOK OF AMPHIBIANS AND REPTILES OF NORTH-EAST AFRICA

STEPHEN SPAWLS
TOMÁŠ MAZUCH
ABUBAKR MOHAMMAD

BLOOMSBURY WILDLIFE
LONDON • OXFORD • NEW YORK • NEW DELHI • SYDNEY

Front cover, main: Ethiopian Tree Agama.
Front cover, left to right: Osgood's Alpine Toad; Speckled Sand Snake; Somali Semaphore Gecko; Leopard Tortoise.
Spine: Kim Howell's Reed Frog.
Back cover: Nile Monitor eating Olive Sand Snake.
Page 2: Lort Phillips's Semaphore Gecko.

BLOOMSBURY WILDLIFE
Bloomsbury Publishing Plc
50 Bedford Square, London, WC1B 3DP, UK
29 Earlsfort Terrace, Dublin 2, Ireland

BLOOMSBURY, BLOOMSBURY WILDLIFE and the Diana logo are trademarks of Bloomsbury Publishing Plc

First published in the United Kingdom 2023

A catalogue record for this book is available from the British Library.

Library of Congress Cataloguing-in-Publication data has been applied for.

ISBN: HB: 978-1-4729-9144-7
ePub: 978-1-4729-9143-0
ePDF: 978-1-4729-9142-3

10 9 8 7 6 5 4 3 2 1

Design by Ginny Zeal
Maps by Julian Baker, Stephen Spawls, Tomáš Mazuch and Abubakr Mohammad

Printed and bound in India by Replika Press Pvt. Ltd.

FSC www.fsc.org MIX Paper from responsible sources FSC® C016779

To find out more about our authors and books visit www.bloomsbury.com and sign up for our newsletters.

Contents

Introduction

This is the first comprehensive guide to the herpetofauna of the seven countries (Egypt, Sudan, South Sudan, Ethiopia, Eritrea, Djibouti, Somalia) of north-east Africa. We have described the 114 species of amphibian and 460+ species of reptile known from the region. It is not the 'last word' in north-east African herpetology. There will never be a last word. There is a vast amount to be discovered. New discoveries, range extensions and taxonomic changes are always occurring. Biological data on the animals themselves is continuously being updated. Huge areas of north-east Africa have yet to be visited by herpetologists. And African herpetology is desperately short of workers. There is much research to be done. Be part of it. The active enthusiast, be they scientist or amateur, can make a big difference. You don't need to be an expert. So get into the field. Observe, document, photograph and collect; keep records. The distributional data for even the commonest of our amphibians and reptiles are patchy. Many map records result from a single observation by an enthusiast in a herpetologically unexplored area. Even if you object to collecting live animals you may find specimens dead; preserve them or freeze them and take them to a museum. Get in contact with zoologists at local institutions. Publish your data. In the wise words of Harry Greene, eminent American herpetologist, 'good natural history is a source of timeless, priceless information for the biological sciences'.

How to use this book; the text and the maps

This book is meant for use in the laboratory, in the home and in the field. Put it in your field bag or your car. When you see an amphibian or a reptile, leaf through the book and try to match it with a picture. If it matches, then check the distribution map; is it in the right area? Then look at the text; we have tried to emphasise key points of identification. Check the maximum size. In the laboratory you can use our keys (these work in couplets; there are two mutually exclusive statements numbered 1a and 1b, you choose the one that fits your animal, this takes you to another couplet, and so on, until eventually you reach a name). If you don't have the book with you, or the animal doesn't appear to be shown, take a photograph and make notes. Particularly useful identification pointers are the colours and patterns of the animal, its location, its size and any behaviour it displays. With frogs the shape, size and colour are useful, as is the location (on the ground, in a bush, etc.). With turtles and tortoises, the size, colour and shape of the shell is useful; as is the locality (in or near water, or not). With lizards the head and body shape and any head ornamentation (spikes, horns, visible ear holes) matter. With snakes, the colour, pattern, size, shape of the head, the type of pupil and the behaviour may give clues to identity. However, remember to be cautious when approaching certain reptiles to check identification, particularly snakes; see the notes on safety and reptiles towards the end of this section.

The maps show the distribution of each species, as known at present; red shading indicates where the species in question is known to occur. A question mark indicates an unconfirmed record, for reasons usually explained in the text. If you have identified a species that seems to be well outside its known range, check the habitat; is it suitable? A forest chameleon will not be in semi-desert. However, reptile distributions are poorly known in north-east Africa, and big range extensions can await discovery. If your record represents a range extension and you have a dead specimen or a good photograph, contact an expert or a museum, noting exactly where the record was made.

LEFT: Female Eritrean Rock Agama.

What are amphibians and reptiles?

The name 'amphibians' means 'both lives', as they live on land and in water. The first amphibians crawled from the lakes and rivers onto land about 370 million years ago. Some ancient species were huge, >8m long. Those giant amphibians are now extinct; we are left with three orders: the frogs and toads (Anura), the salamanders and newts (Caudata), and the poorly known, worm-like caecilians (Gymnophiona). There are no salamanders or newts in sub-Saharan Africa. North-east Africa has more than 110 species of frog and one caecilian. Impressive although this sounds, in reality the region has a very impoverished amphibian fauna; Kenya alone has as many species of frog as all seven countries of north-east Africa combined. This is a consequence of the relative aridity of the latter region, its geological history and, perhaps, the lack of herpetological research, both in the field and laboratory. There are almost certainly a number of undocumented species of amphibian in the wetter regions of South Sudan and in the Ethiopian highlands.

Amphibians need water; their skins are not waterproof (with the exception of foam-nest frogs of the genus *Chiromantis*). Water is lost by evaporation via the skin; if it is not replaced the frog will die by desiccation. Frogs need moisture to breed, for eggs to develop, and for their larvae (tadpoles) to exist in the aquatic stage, prior to metamorphosis. Within these constraints, Africa's frogs have adapted to exist in difficult situations. A few live on the shores of lakes and rivers, but most live in areas without permanent water; they have evolved lifestyles to avoid dehydration and to breed. In the dry season, many retreat underground where the soil remains moist, and some form sealed cocoons. When rain falls and creates temporary pools, frogs emerge in huge numbers (giving rise to local legends that they have dropped from the sky). Activity begins after dark; frogs are rarely active by day. The males usually call at the edge of pools to attract the females, which find a male by the species-specific sound he makes. Mating may follow. Fertilisation is nearly always external, the male climbs onto the female's back, in various positions; usually he grasps the female under the armpits in a position known as amplexus. He produces a stream of sperm that cascades over the eggs as the female releases them. The eggs are usually laid in or near the water. They hatch to produce a larva, the tadpole, that is strikingly unlike the adult. Tadpoles have gills like a fish; they eat algae and detritus, grow rapidly, and lose their tail and gills to become little frogs. These froglets eat insects and grow very rapidly in the remaining water. When the pool dries up, the frogs either burrow deep in the mud, seeking patches that will remain damp, or find some other moist retreat, such as a hole or deep fissure in a big log. There they remain, awaiting the arrival of the next rainy season.

Frogs are 'cold-blooded' ectotherms, meaning that the heat for life processes must come from the environment. Presence or absence of frogs can be an excellent indicator of environmental health. Because of their permeable (leaky) skins, frogs are highly sensitive to environmental conditions and pollution; they are rapidly affected by toxins that enter through their skin. Although frogs have functional paired lungs, they can also absorb oxygen through their skin and the membranes in their mouths.

Vannutelli's Tree Frog.

South-west forest, Bonga, Ethiopia.

Adar, West Darfur.

Reptiles are a diverse class of tetrapod ('four-footed') vertebrates, although some have lost their limbs. The group includes living snakes, lizards, turtles and tortoises, crocodiles and tuatara, and also included the extinct dinosaurs. Some scientists include birds in the group, as they are closely related to modern crocodiles, but the following brief discussion of reptilian characters does not apply to birds.

Although diverse, reptiles share many features. They have a dry waterproof skin and most reproduce by internal fertilisation, followed by the laying of amniotic, self-contained eggs, which develop without parental involvement. There is no larval stage, so the offspring look like their parents. Most reptiles are ectotherms, meaning that they derive their body heat from external sources; this explains why there are rarely nocturnal lizards at high altitude, where night-time temperatures are too cold. The lack of internal heating means that reptile energy budgets are low, and thus they consume relatively less food than mammals; some snakes survive on 6–8 meals a year. Although snakes are largely solitary, social behaviour is shown by some lizards that live in structured colonies. Many reptiles exhibit ritual courtship behaviour and some males fight over females.

Reptiles are an old group; the oldest clearly reptilian fossil is >300 million years old. There was a major radiation of reptiles between 250 and 65 million years ago, including many species of dinosaur, followed by the Cretaceous–Palaeogene mass extinction. Reptilian evolution continues today, with a radiation of smaller forms. Modern reptiles are well represented in our area, with >470 species and this total will rise as the north-east African reptile fauna becomes better known. It is to be hoped that at the same time the exploitative effects of humanity upon the environment will not cause further extinctions.

Beautiful Cat Snake.

North-east Africa: the landscape and zoogeography of the herpetofauna

The seven countries of north-east Africa have a remarkably diverse landscape. Elevation ranges from 133m below sea level in the Qattara Depression in Egypt, to 4,530m on the peak of Ras Dashen in mountainous north-west Ethiopia. The vegetation varies from flat desert through semi-desert, arid and moist savanna to woodland and high forest. The landscape is dominated by a number of major

A MAP OF NORTH-EAST AFRICA SHOWING THE TOPOGRAPHY OF THE REGION

Dissei Island, Eritrea.

geographical features: true desert, extensive high volcanic mountains and plateaux, extensive flood plains, large rivers flowing through arid lands, and a long sea coast with a dry interior. This variety, and its development over time and across a wide latitudinal range, has to a large extent shaped the flora and fauna.

Some brief notes on the dramatic geological and vegetational developments in our region may be illuminating. During the last 100 million years, continental drift has moved Africa about 15° northwards, gradually closing a massive seaway, the Tethys Ocean, that lay between Africa and Europe and covered parts of the Sahara (huge areas of Egypt are underlain by marine limestone). Between 75 and 55 million years ago, forest covered much of North Africa, and Central Africa was under savanna woodland. The situation then reversed. By the late Pliocene and early Miocene (15 million to seven million years ago), the rainforest was more or less where it is now. Savanna woodland covered most of the Sahara and north-east and East Africa, interspersed by stretches of semi-desert vegetation.

In addition, major uplift has occurred. Seventy million years ago below what is now Ethiopia and parts of south-west Arabia a mass of hot rock (a mantle plume) began to well up. Driven by convection, the hot rock hit the base of the lithosphere, the strong, rigid and cool continental crust. As it spread, it forced the land above up, to form a very large upland, the Ethiopian dome, roughly 30 million years ago. Shortly afterwards, movement of the Earth's tectonic plates caused a crack across the centre of the Ethiopian plateau, forming the Rift Valley. The Red Sea may have formed around this time (there is debate as to the dates), but between ten and five million years ago a land bridge connected Africa and Arabia, allowing faunal exchange. Over the last few million years the orographic rainfall created by the Ethiopian highlands drained away to form extremely deep dissected river gorges. Over the same time period, the River Nile came into existence as the Mediterranean Sea level sank, the Ethiopian highlands tilted and an east–west river system in what is now Egypt and the Sudan was captured. The continuously flowing mighty River Nile that we know today has only existed for the last 12,000 years.

Some particularly significant climate changes have occurred during the last 150,000 years. This variation

Simien Mountains, Ethiopia.

A MAP OF NORTH-EAST AFRICA SHOWING THE VEGETATION OF THE REGION

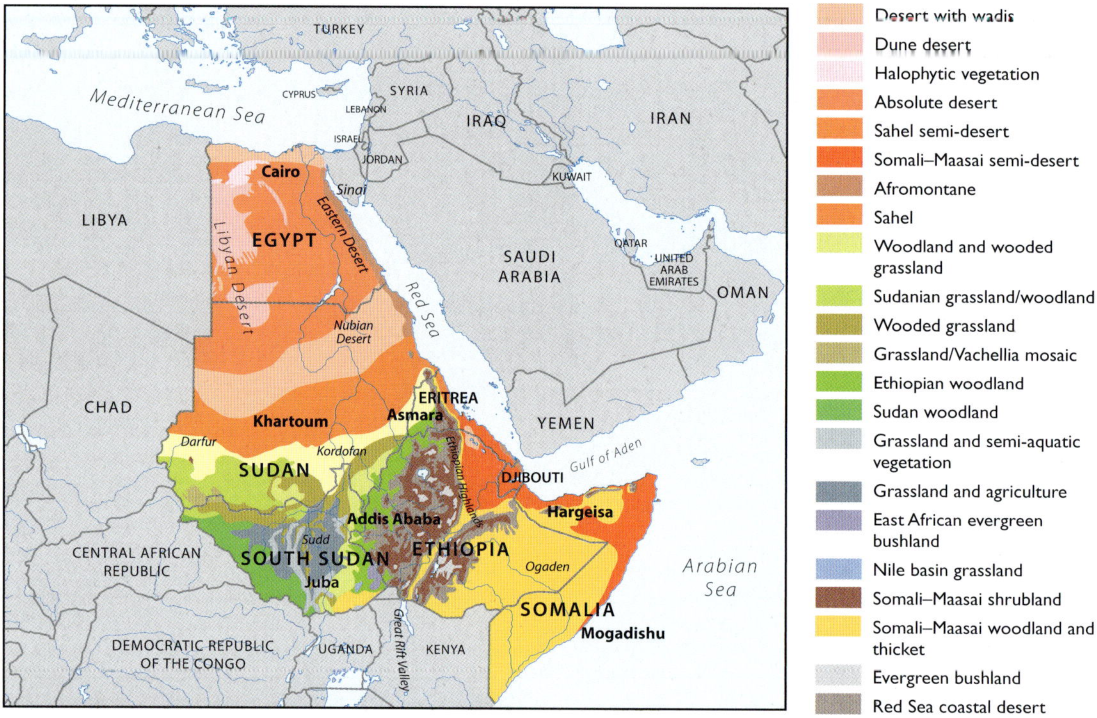

Near Kismayu, Somalia.

Imatong Mountains, South Sudan.

is global, and a consequence of changes in Earth's orbit. Between 150,000 and 8,000 years ago, Africa experienced several cycles of dry/cold and wet/warm periods, during which the extent of forest cover decreased and increased. Around 8,000 years ago, the Central African forest had greatly expanded, Lake Chad was gigantic and surrounded by wetlands, and the Sahara was largely covered by savanna woodland. After further brief arid/moist cycles, over the last 3,000 years North Africa has gradually become drier, with rainfall decreasing and the desert pushing south.

How are these changes of significance to our herpetofauna? More than 65 per cent of north-east Africa now receives on average less than 200mm annual rainfall, which means it is technically desert, and a further 15 per cent of the region has less than 400mm annual rainfall. Thus the herpetofauna of north-east Africa is largely dominated by arid-adapted species. In north-east Africa there are >40 species of the little half-toed *Hemidactylus* geckos, a genus that favours dry, rocky country, whereas in the five countries of East Africa, which are more tropical and more biodiverse, only 18 species of *Hemidactylus* occur. Similar totals apply to the arid-country loving lacertid lizards (of which there are more than 40 in our area,

A MAP OF NORTH-EAST AFRICA HIGHLIGHTING PROMINENT PLACE NAMES

compared to 23 in East Africa); likewise the agamas, fast-moving lizards that live in open country (37 species occur in north-east Africa, just 16 in East Africa). Conversely, north-east Africa has only 114 species of frog (including some 35 species endemic to the Ethiopian highlands) and a single species of caecilian. In moister, more tropical East Africa >200 species of frog and 12 species of caecilian occur. There are also more than 60 species of chameleons in East Africa, whereas only 16 chameleon species are known from north-east Africa.

Thus a picture emerges of a herpetofauna dominated by what are zoogeographically known as Saharo-Sindian species, animals associated with the great deserts and their surroundings that extend from Mauritania in the west, right across North Africa, to the dry lands of the Middle East. Typical examples of this fauna include desert vipers of the genus *Cerastes*. Another group of largely arid-adapted species occurs in the block of land hemmed in between the Ethiopian mountains and the sea, extending through eastern Ethiopia and Somalia to northern and eastern Kenya and parts of northern Tanzania. This is the Somali-Maasai fauna, named for the peoples of that arid region. Emblematic of this habitat is

Southern Sinai, Egypt.

Fayum, Egypt.

Red Sea Hills, Sudan.

the Somali-Maasai Clawed Gecko *Holodactylus africanus*. This area is also home to some unusual species and genera that are strictly endemic to arid habitats, often with very small ranges, for example the strange agamas, *Xenagama*, and the short snakes, *Brachyophis*.

Several other faunal groups are also dominant in north-east Africa. A number of species associated with the Sahel and Sudan Savanna of West and Central Africa reach our region on the great plains of Sudan and South Sudan, to the west of the western base of the Ethiopian Plateau, e.g., African Spurred Tortoise *Centrochelys sulcata* and Hook-nosed Snake *Scaphiophis albopunctatus*. A number of species of the southern and south-eastern savannas also reach our area, such as Leopard Tortoise *Stigmochelys pardalis*. A small handful of species of the great forests of Central Africa, like Eastern Gaboon Viper *Bitis gabonica*, occur in the dense woodlands, forests and hill forests along South Sudan's southern border. In Egypt, particularly along the north coast and in the Sinai, there are some species associated with the Mediterranean region, for example Javelin Sand Boa *Eryx jaculus*, and the Irano-Turanian fauna of Eurasia, for example Variable Green Toad *Bufotes sitibundus*. Also in Egypt there are some species of essentially Arabian origin that arrived via a land bridge, for example Burton's Carpet Viper *Echis coloratus*. Further south, species of the *Hemidactylus robustus* group and *Pristurus* geckos entered Africa via the Bab Al Mandab land bridge. There are also a handful of adaptable pan-African savanna species that occur widely in southern parts of north-east Africa,

Imatong Mountains, South Sudan.

such as Common Egg-eater *Dasypeltis scabra* and Puff Adder *Bitis arietans*. A few of these have penetrated along the Nile into Egypt, like Five-lined Skink *Trachylepis quinquetaeniata* and Nile Monitor *Varanus niloticus*. And finally, there is the unique herpetofauna of montane Ethiopia.

The rising Ethiopian dome created a most unusual highland region. Half of all the land in Africa above 2,000m, and 80 per cent of the land in excess of 3,000m, is in Ethiopia. On these highlands, especially in the grassland areas, there are a number of savanna species of East African affinities that reached the highlands from the south before northern Kenya became as arid as it is now. In the Ethiopian highlands, particularly in the south, mean annual rainfall is similar to that of the great Central African forest on the Congo River. And yet not a single snake species associated with the forests of Central Africa, for example

Rhinoceros Viper *Bitis nasicornis*, bush vipers of the genus *Atheris*, black tree snakes of the genus *Thrasops*, nor birds like the forest hornbills, occur in the forests of Ethiopia. This indicates that the great forests of Central Africa were never in faunal contact with the Ethiopian forests. The Central African forest had retreated from the region before the present Ethiopian forests came into existence. Consequently, highland Ethiopia has an unusual and impoverished herpetofauna, with a number of pan-African and savanna species that have penetrated to unusually high elevation, and a handful of endemic species formed by speciation after uplift. The number of endemic amphibians is relatively large, but numbers of endemic snakes and lizards are low, although they include two spectacular big vipers of the genus *Bitis* and the unusual high-altitude genus *Pseudoboodon* (the Ethiopian mountain snakes). We have expanded further on this unusual fauna in some of the family and genera introductions.

Observing, gathering data, and collecting reptiles and amphibians

In general, amphibians and reptiles are best left in the wild; that is where they live and they are an important part of the ecosystem. In addition, some areas in north-east Africa are protected, and animals should not be collected. It is our hope that users of this book enjoy watching reptiles in wild places, where successful identification adds to the enjoyment. However, in certain circumstances, collecting reptiles for scientific purposes is justified. Consequently, we give some advice on observing, recording, finding and collecting amphibians and reptiles, which are aimed at those with appropriate permission from local authorities.

A camera is very useful. The rise of the small versatile digital camera and the camera phone has made a huge difference to the gathering of field data. Many of us now take such handy portable devices into the field. With them, if you see an amphibian or a reptile, you can photograph it, and the image can be sent worldwide. More often than not, the species can be identified from the picture. Contact your local natural history museum. As authors we are always pleased to receive these data; many range extensions and new records that appear in this book are the result of an enthusiast photographing a wild reptile or amphibian and sending us the image. We don't intend to give detailed photographic advice here, but the following pointers may be useful. Firstly, get as close as safely possible to your subject without scaring it (but be careful with snakes, unless you know your subject is harmless, then stay at least twice the snake's length away). If light is low, use flash. Take a few shots, and keep the camera as steady as possible (important with small cameras and phone cameras). Make sure the head is clearly visible and in sharp focus. Activate the GPS function if you have one.

Weather affects fieldwork. In southern parts of north-east Africa, the broad climatic picture is of dry and wet seasons, and as one travels further north the temperate seasons of spring, summer, autumn and winter become increasingly obvious. Wet seasons, or spring/early summer, are the best times for observing reptile and amphibian activity. Their activity is greatly reduced during the dry season and winter, although some lizards and diurnal snakes, like sand snakes, may still be active. The field herpetologist will see little in Egypt in January or in highland Ethiopia in March.

Most amphibians and reptiles are secretive. They spend much of their time hidden, often in very specific refuges like rock crevices within a small range. If approached in the open, they tend to freeze or move away unobtrusively. The casual observer may see few

Finch's Agama.

in the wild, but a little expert hunting will increase your chances of seeing these creatures. There are various useful techniques. Turning over ground cover such as rocks, logs and vegetation heaps will expose hiding animals, but take care where you put your fingers, and replace the cover afterwards. Lifting up rock flakes, breaking up fallen logs, and removing dead bark from trees may reveal animals, but bear in mind that this destroys hiding places (and may also expose you to a dangerous snake!). Areas around rural housing, or abandoned buildings are often prime reptile habitat, particularly for *Hemidactylus* geckos. In forest and woodland, raking through leaf litter will often expose concealed amphibians or reptiles. When walking in the countryside, keep your eyes open. Move slowly, and look around. You can also look for tracks in sandy areas and follow them to where the animal is hiding; it may be buried, or in a hole. If you sit and watch a termite mound on a warm day animals may emerge to bask. In open areas, lizards are often active. Rocky hills, especially those with sloping sheet rock areas and cracks, are usually excellent places to see lizards. In open country, isolated big trees are important refuges.

Lowland savanna, near Turmi, Ethiopia.

Water sources are good places to see amphibians and reptiles. By day, frogs will hide under waterside vegetation or ground cover, between reed stems or between the leaf stem and trunk of fleshy plants, especially bananas. Frog-eating snakes may be in waterside vegetation, and frogs are sometimes observed in water during the day. Birds and some mammals, especially squirrels, may give away the presence of a snake. If you hear birds or animals mobbing something, then investigate.

Because of the unique nature of their skin, most of our amphibian species are nocturnal. Looking for amphibians (and reptiles) at night can be very rewarding, although the risks of any sort of night work must always be carefully assessed in north-east Africa. You will need a good lamp or torch; a head torch is essential if you want to catch or photograph amphibians at night. At the start of the rainy season, amphibian activity increases dramatically, with frogs rushing to temporary pools to breed; they can be located by their calls. A group of frogs around a pool can make a noise audible hundreds of metres away; the males producing their advertisement calls to attract females of their species. Each frog species has its own call. Where we are familiar with them, we have described these calls onomatopoeically.

Frog activity may also attract snakes. Many types of snakes are active at night, as are most geckos. Some reptiles, especially chameleons and green snakes, show up clearly at night, when they look pale against dark vegetation. However, apart from crocodiles,

Puff Adder.

reptile eyes do not reflect light well, thus it is no good looking for reptilian eyeshine. In urban areas, dripping taps and open water containers will attract frogs, which can smell water. Walls and drains act as traps and snakes may be found crawling along them. Storm drains often trap animals. Geckos may hunt around lamps at night. Diurnal lizards may sleep on vertical rock faces. Reptiles moving at night often make a surprising amount of noise, especially in well-vegetated areas, once in a while turn off your lamp and listen. Bear in mind that if you want to look for amphibians and reptiles at night, you should check before darkness falls if a watchman or security person will be present. Never walk at night in areas you haven't surveyed by day; you might get lost, or wander into danger (a cliff, a swamp, or into the presence of dangerous animals). Wear stout footwear and take care not to tread on snakes.

Driving slowly (20–30km/h) at night with your headlights dipped can also be rewarding, especially on quiet tarred roads. You may encounter frogs, geckos and snakes either lying on the surface to absorb heat, or crossing the road, especially if the latter is wet. Bear in mind that night driving in parts of Africa can be extremely dangerous. Hazards include unlit and unroadworthy vehicles, roadblocks, people and livestock on the road, and low standards of driving; you may encounter drivers with no concept of roadcraft, defensive driving, speed awareness or consideration for others. And at night you cannot see hazardous situations developing. Never be tempted to road cruise on a busy African road at night.

Useful equipment for the reptile and amphibian collector may include cloth bags with sewn-in drawstrings, stout plastic bags with rubber bands for frogs, and screw-topped plastic jars for little animals. Remember to keep containers with live animals in them away from sunlight or excessive heat, and take care not to squash them. Some frogs produce toxic skin secretions; you should never put more than one species of frog in a bag or container, and the water in any container with live frogs should be changed a couple of times a day. A hooked stick is useful for turning over ground cover, removing tree bark, poking about in holes and pressing down snakes.

For dangerous snakes, a grab stick, pair of tongs and a pair of goggles (for spitting cobras) are needed. A pair of industrial leather gloves is also useful, for protecting the hands when digging. Other useful equipment might include rubber boots and a net for frog collecting, a small shovel or trowel for digging out holes, and some plastic tubing, for sticking down holes so you don't lose your way when digging them out. Lizards may permit close approach and can be noosed using fishing line on a light rod (e.g. an old radio aerial or a long thin stick). Animals can sometimes be flushed out of holes with chloroform, petrol, mothballs or smoke, but such techniques may permanently damage the refuge, kill the animal in the hole or start a bush fire; thus their use should be carefully considered. Tortoise enthusiasts have developed a useful technique for locating an animal in a hole, using a snooker or pool cue; a tortoise poked with a pool cue makes a most distinctive sound.

Wadi Gindali, Egypt.

Striped Ethiopian Mountain Snake.

A small pair of binoculars is useful. You don't always have to walk to observe the smaller wildlife; the quiet watcher with binoculars in an elevated spot can often see a surprising amount. Local help is often available, people will offer their services, and they often possess a lot of local knowledge. If you have a book with you and speak the local language, you can show pictures of what you want. But never ask or allow people to catch snakes for you; if someone gets bitten there will be major problems.

Serious scientific collectors will need preservative and containers. Valuable specimens are often found dead. A 10 per cent formalin solution or 70 per cent alcohol solution can be used to preserve specimens. In general, formalin is a more efficient preservative, but pure alcohol preserves long strands of DNA necessary for molecular taxonomy, so if you have both, preserve some parts of the animal (a bit of skin or muscle will do) in a small container of alcohol (96 per cent clear ethanol is best) and put the rest in formalin (or in a freezer, which prevents decomposition until alcohol is available). All specimens should be labelled, mark the label with soft pencil. Locality is the most important piece of data; a specimen without a location is virtually useless. Larger animals should be cut open along the abdomen, and/or preservative injected into the soft body parts. In an emergency, you can preserve specimens with methylated spirit, alcohol (neat vodka, gin or whisky), petrol or concentrated salt solution. Specimens should be deposited in a public-access collection, such as a museum.

Always consider safety in the field. Anyone intending to collect dangerous snakes should be thoroughly experienced, and be part of a team with adequate medical support. Snakebite treatment may involve blood transfusions. The fieldworker should also have suitable protective clothing, food, drink and medical supplies; consider taking a supply of antivenom.

Amphibian and reptile conservation

Reptiles and amphibians have an image problem. Many people dislike them, and snakes and certain lizards are feared and killed if encountered. Environmental education may help with this, but even in highly developed countries, many people do not like reptiles and amphibians. It is for this reason that places such as zoos or reptile parks serve such an important function, in bringing the public face to face with reptiles and amphibians, and giving them a chance to appreciate their beauty and usefulness. For the same reason, any enthusiast who keeps local amphibians and reptiles (with appropriate permission, if needed) and shows them to their friends is benefitting conservation. In addition, amphibians and reptiles are an important part of ecosystems. They eat pests, they are useful environmental indicators (particularly frogs), and they possess both known and untapped potential in culinary, economic and medical terms. The benefits of amphibians, reptiles and biodiversity need to be emphasised in schools and colleges.

In many ways, the future is bright for the herpetofauna of Africa. Largely as a result of the electronic age, there has been a huge rise in awareness of the beauty and charms of our fauna, particularly among young people. There has also been a significant increase in interest in all aspects of natural history and biodiversity, and social media has enabled heightened awareness and a rapid response to events of environmental significance. Some of the countries of north-east Africa have tourist industries that employ people and bring in outside money. Thus their governments have a financial interest in protecting wild places and their inhabitants. There is a constantly improving national park and reserve system protecting biodiversity. In addition, many amphibians and reptiles are relatively small and secretive, they are not as vulnerable to habitat loss and exploitation as larger wild animals and can often survive on farmland and even in suburban areas; the hill suburbs of Addis Ababa still have chameleons, the parks of Cairo still yield Ocellated Skinks. In many cities, stone buildings serve as substitute cliffs and bluffs for populations of geckos.

But at the same time, threats exist. Few species have been thoroughly assessed in terms of current distribution and habitat requirements. The expanding human population and rapid development in our region means that habitats are severely under threat, from farming, logging, urbanisation, draining of wetlands and mineral exploitation. Even protected areas can be at risk from government action if they are deemed to stand in the way of progress. Some reptile and amphibian species are hugely in demand for the pet trade, or for food. Others are unable to exist in harmony with humanity; large reptiles (especially snakes) tend not to be tolerated by people, no matter how innocuous they are. And, of course, occasional invasive species, even small lizards, can threaten and displace the local fauna; see the accounts of some *Hemidactylus* geckos. In the species accounts, we have given details if a taxon has been assessed by the IUCN (International Union for the Conservation of Nature), which publishes the Red Lists of Threatened Species; definitions of threat levels can be found online. We have added our own comments where pertinent. If there is no comment on conservation status, it means the species has not been assessed and no major threats have been identified.

Harenna Forest, Ethiopia.

Without a doubt, some species are at risk. The vulnerable reptiles and amphibians are the endemic species, especially those with limited ranges that live in areas suitable for agriculture or exploitation. Many occur in restricted habitats. With population increases, these forests and wild lands are vulnerable, liable to be exploited for their timber and potential use for farming. For example, many Ethiopian endemic amphibians and chameleons live in tiny forested areas; they are greatly at risk from deforestation, and possibly climatic change as well. Consider the Bale Mountains Heather Frog *Ericabatrachus baleensis*, which was described in 1991. Since then, the type locality where it was found has been degraded by cattle, and it occurs there no longer. We have no idea how many are left and, if it still exists, how viable are the remaining populations. At present, our conservation priorities should be to identify vulnerable habitats, especially those with no protection, document their fauna and flora, and develop and implement protective strategies. Often a single small well-protected reserve will suffice to conserve a rare reptile and its habitat. But the crucial point is 'well protected', the Bale Mountains Heather Frog occurs within a national park but that has not helped it.

We would also like to draw attention to the importance of museums in conservation activities. No conservation strategy can succeed unless we know what organisms exist where. The ranges of almost all species described in this book are largely based upon museum specimens. Without those specimens, ranges would be unknown and taxonomic status impossible to clarify; thus no conservation strategy would be possible. There is a modern school of thought that decries collecting, that states that rare creatures should be always left alive; that range mapping should be based on field observations. This may be possible with larger mammal and bird species; it is not always so with amphibians and reptiles. There are few observers and few specimens, reptiles and

Finch's Agama.

amphibians are hard to see and find, the difference between many similar species is not obvious in the field, and field identification is difficult. We fully recognise the benefits of such forums as iNaturalist, Facebook and the Virtual Museum, but where reptile and amphibian range mapping is concerned, these are an adjunct to museum specimens, not a substitute. Many species can only be identified with certainty in the hand, with a key and often a binocular microscope, particularly the smaller ones. Specimens are necessary, and it is crucially important that they are deposited in local museum collections, to assist researchers.

Djibouti.

Brief notes on amphibian and reptile taxonomy

Taxonomy is that branch of science concerned with classification; working out the relationships between organisms, and providing a means of distinguishing between them. The most useful taxonomic unit is the species; it is important that we know what species are, and that they are robustly defined. Hence books like this. No faunal list can be drawn up unless we can say what species occur where; no researcher can say what organism they worked on unless the species can be unequivocally identified, no new locality records can be added if the specimen cannot be identified to species, and no conservation measures can succeed unless species are robustly identified.

Species have been defined in many ways, and the debate continues, but essentially a species is a group of organisms that are each other's closest relatives, can interbreed and produce fertile offspring. Us humans are a single species. The species is a dynamic unit, for obvious reasons. Life on earth started once and diversified; species change over time, new species are constantly being formed; organisms, and groups of organisms, evolve. All species are given unique two-part scientific names; the first, or generic, name applies to a genus, or group of very similar organisms, the second to the actual group of interbreeding organisms. For example, Sahara Horned Viper has the scientific name *Cerastes cerastes*. No other species can share this name. There are three other species of *Cerastes* and all have different specific names; for example Sahara Sand Viper has the scientific name *Cerastes vipera*. For some species, a third name, the subspecies may be added, where there are two (or more) subgroups within the population that somehow differ; for example there are two subspecies of Speckled Sand Snake *Psammophis punctulatus*; one has a single dorsal stripe, *P. p. punctulatus*, the other has three dorsal stripes, *P. p. trivirgatus*. In this book, we have occasionally mentioned subspecies in the identification section, where relevant; the reader should be aware that, among some zoologists, the status of subspecies is controversial.

Reptile taxonomy is made easier by the diversity of reptiles themselves and the fact that most are covered with scales, which can be counted. Amphibian taxonomy is not always as easy; as all frogs are relatively small and have a similar shape, although there are clues; especially with living frogs; their colours, body shape, digits and limbs, behaviour and calls are often diagnostic. Tortoises and turtles are identified by their size, shell shape and scale counts. Lizards are often identified by a mixture of scale counts, body shape and colour, snakes by scale counts, colour, teeth and body shape. These characters have proved useful for >250 years of herpetological taxonomy. Recently, a new set of tools has become available to laboratory-based taxonomists; the techniques that enable the isolation of standardised bits of DNA, and the computing power that permits the analysis of that DNA. With these tools, new forms can be defined. New evolutionary lineages and relationships can be discovered and old ones modified, through the differences in matching sections of DNA. Although traditional reptile and amphibian taxonomy, involving the detailed descriptions of animals in terms of morphology, appearance and measurements (and scale counts for reptiles) is still useful, especially with new forms that differ clearly from other members of the genus, the analysis of DNA has enabled researchers to fine-tune their identification, and often split what seemed in the

Djibouti.

past to be a single species into several species, evolving separately. Even the approximate date at which the two forms began to diverge, and accumulate differences, can be ascertained. This is a very useful development. But it has certain disadvantages. Often, those separate lineages cannot be told apart in the hand, or even under the microscope, and this can lead to taxonomic confusion. For example, Helmeted Terrapin *Pelomedusa subrufa*, a water tortoise widespread throughout sub-Saharan Africa and not particularly variable in appearance, has recently been split into ten named species. Useful though this is in a broad sense, the paper's authors made no attempt to construct a key that would enable any researcher to distinguish between them in the hand or even in the laboratory; individuals can only be assigned to species by analysis of their DNA. This creates difficulties, particularly in Africa where few scientists have access to a molecular lab. But, like its subjects, taxonomy is evolving. A visual key may be possible. We have constructed one for *Pelomedusa*. And in Africa, there is much to be discovered.

We should briefly say something about common names. English and vernacular names have value, particularly at present, where relationships revealed by biological molecules have often caused species to be reassigned to different genera. However, it should be remembered that local names are often imprecise, and applied to several different but similar-looking animals; in parts of Africa, the same local name is given to both worms and blind snakes).

Identifying amphibians and reptiles

Some species are easy to identify, others not. We have provided keys, although they have their limitations. Have a look at the pictures, become familiar with the variety of reptile and amphibian life in your area. We offer below some general rules for identification, but they are applicable to north-east African animals only; and you should use them with caution, particularly where snakes are concerned. See the note at the end of this section. Our diagrams (Figs. 1–25) show how to identify crucial features and count important scales.

Amphibians can often be identified fairly easily to genus (but beyond that, it can be hard!). Check the pictures. Most toads, are warty, and shades of brown; with squat bodies and big golden eyes; they hop and walk, rather than jump. The rocket or ridged frogs, *Ptychadena*, have long pointed noses and longitudinal glandular ridges and stripes; they can jump great distances. The clawed frogs, *Xenopus*, are unmistakable, flat and dark, with claws, and little eyes on top of their heads. A little, brown warty frog on the ground will probably be a puddle frog, *Phrynobatrachus*. The reed frogs, *Hyperolius*, are tiny climbers, often with vivid colours, or green, on waterside vegetation; a larger equally colourful frog also found climbing, will be a *Leptopelis*, tree frog. The piglet frogs, *Hemisus*, are globular with tiny eyes and pointed snouts. Rubber frogs, *Phrynomantis*, are black and red. Foam-nest frogs, *Chiromantis*, are large, grey or white, and usually found in trees in arid country. A little warty frog without a tympanum, in dry country, will be a sand frog, *Tomopterna*.

Most chameleons can be readily identified by their swivelling eyes set in turrets, their grasping feet and prehensile tail. Monitor lizards are large, most are >70cm long, and no other African lizard grows so big. Agamas have rounded heads and thin necks; large male agamas have broad, bright heads, usually vividly coloured. The smaller agamas often have blue on the throat.

Geckos are the only lizards active at night. Most have vertical eye pupils and soft skin. Skinks have shiny bodies, are often striped and/or brightly coloured, with small limbs and no distinct narrowing of the neck. Lacertid lizards are small, fast-moving, often striped and most common in arid country.

In north-east Africa, there is no single way to tell a harmless snake from a dangerous one. Any fat snake over 4m long must be a python. Any snake with more than one conspicuous stripe running along the body is probably back-fanged and not dangerous. Any snake >2m long is probably dangerous. Any grey, green or greenish tree snake more than 1.3m long is almost certainly dangerous, it will probably be a mamba or Boomslang. If it is >1.3m long and inflates the front half of its body in anger, it will be a dangerous back-fanged tree snake (Boomslang or vine snake).

Any snake that spreads a hood, flattens its neck or raises the forepart of the body off the ground when threatened is almost certainly dangerous, and will probably be an elapid (other possibilities include night adders, Rufous Beaked Snake and Semi-ornate Snake). Any snake with conspicuous bars, cross-bands, rings or V-shapes, especially on the neck or the front half of its body, on the back or on the belly is probably dangerous. Most cobras have dark bars or blotches on the underside of their necks. A fat-bodied snake, with a sub-triangular head, that lies quietly when approached, is probably a viper or adder. Small black, dark grey or brown snakes with tiny eyes, no obvious neck and a short fat tail that ends in a spike will almost certainly be burrowing asps.

Any snake with roughly rectangular or triangular markings on its back or sides, or rows of semi-circular markings along the flanks, is probably a viper. A snake that forms C-shaped coils and rubs them together, making a noise like water falling on a hot plate, will be either a carpet viper (dangerous) or an egg-eater (harmless). Any small snake with a blunt rounded head *and* a blunt rounded tail, with eyes either invisible or visible as little dark dots under the skin, and tiny scales that are the same size all the way around the body will be either a blind snake or a worm-snake, and harmless.

A final note of caution. There are some dangerous myths about how to identify venomous and non-venomous snakes. These circulate on the internet, often with misleading endorsements of their benefits. Sometimes they are relevant to one area, but are incorrectly assumed to work elsewhere. A common myth, wholly incorrect, is that dangerous snakes have vertical pupils, and harmless ones have round pupils; this is untrue (black mambas have round pupils; harmless house snakes vertical pupils). Another incorrect idea is that harmless snakes have smooth scales, and dangerous ones have keeled scales. Yet another fallacy is that dangerous snakes have broad heads,

Asmara, Eritrea.

harmless ones narrow heads; or that dangerous snakes have single subcaudal scales and harmless ones double subcaudal scales. Bear in mind the apposite saying: the internet exposes us to the stupidity of others even more efficiently.

Safety; observing and handling amphibians and reptiles; snakebite

Amphibians and reptiles in north-east Africa can usually be safely observed with enjoyment, sometimes from quite close quarters, provided a little care is taken to avoid disturbing the animal and to minimise the risk to yourself. There is a pleasure in watching the activities of a calling frog, or a little agama colony, the slow progress of a tortoise through the savanna, or the probing search by a sand snake of holes that may conceal prey, which too few people pause to enjoy. Take time to observe reptiles. However, a few important points should be noted.

Although most frogs are innocuous, a handful have toxic skin secretions, which can cause severe local irritation, so if you handle a frog, wash your hands afterwards and don't touch your eyes, mouth or an open wound or graze. No turtle, tortoise or lizard in Africa is venomous, despite local legends that indicate some (especially agamas and chameleons) have a venomous bite. In addition, no species has poisonous excreta. Thus frogs, chelonians, lizards and harmless snakes may be safely approached. But do not go too close to an unidentified snake, as it may be dangerous. A distance of twice the snake's length is as close as you should get, unless you suspect the snake is a spitting cobra, in which case do not go closer than 5m. Beware of approaching any snake, particularly a large one, that appears agitated (spread hood, raised head or neck, hissing loudly, opening its mouth) as it may strike or rush forward if cornered.

Beware of crocodiles, they are dangerous. Nile Crocodile is the only reptile in north-east Africa that may attack humans to eat them. Even quite small crocodiles may attack people. You should never wade or swim anywhere that Nile Crocodiles might live (basically, any waterbody, up to about 1,800m altitude), including lakes, rivers and even quite small dams. It is advisable to stay at least 8–10m from the water's edge.

Care must be taken if you wish to catch or handle some reptiles. Most will bite if cornered or seized. Some chelonians can deliver a powerful bite, handlers have lost fingers. A bite from a smaller lizard can hurt, a big lizard like a large agama or a monitor can deliver a severe, skin-breaking bite, and may scratch with their claws or lash with their tails. We don't advise handling big lizards without experience.

Unless you are an expert, you should never handle snakes. This includes apparently dead ones; they might not be dead. Be prudent if offered the opportunity to handle a snake at a snake park or by a snake charmer; do the handlers seem professional? Remember also that reptiles usually have bacteria on their skin and in their mouths. These are unlikely to cause any problems, provided you are not bitten, that you thoroughly wash your hands after holding amphibians and reptiles, and avoid touching your eyes and mouth while doing so. If you demonstrate reptiles to people, ensure that they wash their hands after handling an animal.

A number of snakes in our region are dangerously venomous and can kill. So a few rules for avoiding snakebite are given here. To avoid being bitten, always treat snakes (and other reptiles) with respect; avoid teasing or molesting a snake. Do not try to catch or kill them unless you are an expert. Snakes do not make unprovoked attacks. If confronted by a snake, move backwards slowly. To avoid being bitten, always look where you are going. Wear adequate shoes; don't put your hands or feet in places you cannot see. Use a lamp or torch at night. Don't sleep on the ground, or gather firewood after dark. Clear suitable snake hiding places from near your home, especially if you have small children. If someone gets venom in their eyes after an encounter with a spitting cobra, the eyes should be gently washed with large quantities of water (or other bland fluids such as milk), and the victim taken for a medical check-up. If someone has been bitten by a snake, treat it as a medical emergency, abandon other plans and organise rapid safe transport to a large hospital (clinics and health posts are unlikely to be able to adequately treat a major snakebite case). Do not try to catch or kill the snake. Leave the wound alone, don't start cutting it, giving shocks, using ice, etc. Immobilisation therapy (keeping the bitten area, especially limbs, as still as possible, often by splinting) may be useful in some cases, as detailed in our other publications. A medical professional may consider use of antivenom before arrival at hospital in an extreme case. But don't let fears about snakebite put you off going into wild places; the risks are very small.

Acknowledgements

Many people have helped with this book; directly and indirectly. We are profoundly grateful to them all, and sincerely apologise if we have missed anyone in what follows.

To the museums and institutions with north-east African collections and their curators (a museum list is given below) and the field herpetologists who have worked with and assisted us, we offer our thanks; Arthur Tiutenko, Wolfgang Wüster for his elapid and viper expertise, Sandra Goutte for expert frog advice, Dr Afework Bekelle at Addis Ababa University and Katama Amberbir and Goitam Redda at the Natural History Museum in Addis Ababa, Jean-Francois Trape, for his wide-ranging expertise, Alan Resetar at the Field Museum, Chicago, and Wolfgang Böhme and Morris Flecks, at the Alexander Koenig Zoological Research Museum in Bonn. Patrick Malonza and Victor Wasonga, at the Herpetology Section, National Museums Kenya in Nairobi. Patrick Campbell, David Gower, Simon Loader and Mark Wilkinson at the Natural History Museum, South Kensington. Tom thanks Vladimir Trailin and Pavel Novák for sharing info on Ethiopian herps, help in the field and general support. František Kovařík, Hassan Sh Abdirahman Elmi (Amoud University, Somaliland), Abdiqaadir Abdilahi, Abdisalaan Shabele, Ahmed Ibrahim Awale, Omar Shucayb and Shukri Haji Ismail Bandar (Ministry of Environment and Rural Development) from Somaliland team. Kateřina Kadlečková, Olga Mazuchová, Jiří Šmíd, Pavel Konečný, Thomas B. Price, Kamiel Hamers, Peter Uetz, Salvador Carranza and Werner Conradie offered general support in many ways. Bin Shinaa, Mahad Jebiye and Bashir Abdisamad shared info on the Somali herpetofauna, Mohamed Salah, Mohammed Adam Babayaro, Harin Mani Bakheet, Mohammed Adan Abdelrahman, Telal Ahmad and Scott Larkan for info on Sudan and South Sudan herps, and Mathieu Mahamoud-Issa for info on Djibouti herps. Patrick Campbell at NHMUK, Annamaria Nistri at MZUF, Giuliano Doria at MSNG, Giuseppe Bardelli at MSNM, and Silke Schweiger at NMW (for acronyms, see Literature and resources, below). Our acknowledgements also extend to the Natural History Museum in Khartoum, Sudan, the University of Sennar (Sudanese Wildlife Research Centre), Sudanese Wildlife Society, Sudanese Natural Heritage Society, and Wildlife Administration in Khartoum, Sudan.

Next; the photographers. Many herpetological photographers kindly allowed us to use their pictures; including top professionals who can command high fees for their pictures, yet generously allowed us to use their super images for the reward of a copy of this book. We offer our heartfelt thanks to them all, and apologise for any omissions; we had so many fine pictures, we couldn't use them all. In alphabetical order: Madad Alashaa, Abdelbagi Alsayed Ali, Saulo Bambi, Huda Ali Banfas, the Bio-Ken archive, Jonas Arvidsson, Robin James Backhouse, Sherif Baha El Din, Aviad Bar, Margaret Bartkowiak, Roger Barbour, David Blackburn, Nik Borrow, Bill Branch, Bob Drewes, Alan Channing, Anthony Childs, Laurent Chirio, Jaromír Chvástek, Tim Colston, Pierre-André Crochet, Davide Dal Largo, George Dannhauser, Max Dehling, Massimo Delfino, Gerald Dunger, Euan Edwards, Arwa Elaagib, Ismail Elsadig, Elfatih Eltagi, Dietmar Emmrich, Ian Engelbrecht, Jesse Erens, Yannick Francioli, Paul Freed, Philippe Geniez, Sandra Goutte, Eli Greenbaum, Ron Gutberlet, Václav Gvoždík, Thomas Hakannson, David Hegner, Harald Hinkel, Daniel Hollands, Barry Hughes, Ivan Ineich, Caspian Johnson, Kate Jackson, Jaroslav Karhánek, Tony Kamphorst, Abeje Kassie, Sebastian Kirchhof, Bertrand LaFrance, Malcolm Largen, Luca Luiselli, Quentin Luke, John Lyakurwa, Johan Marais, Sam Marshall, Justin Mathews, Mike McLaren, Konrad Mebert, Michele Menegon, David Modrý, Jiří Moravec, Petr Nečas, Stu Nielsen, Annamaria Nistri, Pavel Novák, Torsten Panner, Jean-Luc Perrett, Martin Pickersgill, Mike Poole, Kyle Ray, Eduardo Razzetti, Mike Roberts, Mark-Oliver Rödel, Abubakr Ringim, Steve Russell, Mohamed Salah, Entesar Salih, Bill Roland Schroeder, Khalid Shamboul, Siddig Awad Shimaila, Nuh Ismail Ahmed Shire, Robert Sindaco, Jiří Šmíd, Tim Spawls, Sisay Taye, Royjan Taylor, Colin Tilbury, Vladimir Trailin, Jean-Francois Trape, Marc Vaillant, Casper Van de Geer and the Watamu Turtle Watch, Lorenzo Vinciguerra, Vitězslav Vlk, Philipp Wagner, Van Wallach, Victor Wasonga, Andre Van Hecke, Scott Weinstein, and Elvira Wolfer.

We also remember, with gratitude, the pioneers whose assiduous efforts laid the foundations, in particular Wilhelm Peters, Giuseppe Scortecci, Arthur Loveridge, Benedetto Lanza, H. W. Parker, Malcolm Largen, Mohammed A El-rayyah, George E. Boulenger, John Anderson, Stanley S. Flower and Franz Werner; it is on their shoulders we stand.

And, finally; Steve thanks his wife Laura, for her endless support and kindly tolerance, and we thank the team at Bloomsbury; Jim Martin, who championed our proposal, and Jenny Campbell, who has expertly guided our efforts to fruition.

Literature And Resources

The internet has smoothed our path regarding herpetological resources. We have made constant use of a crucial online resource, the Reptile Database, available at reptile-database.org. It is run by Peter Uetz and Jiří Hošek, who we salute. Their efforts, aided by an army of volunteers, have made our work far easier. On the database all species are listed, with pertinent literature. Of equal importance, the team there scan journals, critically read the revisionary material and accord it the appropriate significance. In the past, it was often a struggle to get up-to-date taxonomic information. Now that information is available instantly. Equally useful websites are Darrel Frost's Amphibian Species of the World (amphibiansoftheworld.amnh.org), Vlada Trailin's summative website for the Ethiopian herpetofauna (reptiles-of-ethiopia-and-eritrea.com) and Amphibiaweb (amphiaweb.org). Also useful are several Facebook fora, including Nature of the Horn of Africa, East African Snakes and other reptiles, Reptiles and Amphibians of the Middle East and North Africa.

Via the internet you will also find references to the 'classic' works of north-east African herpetological literature, including older material by authors such as John Anderson, Donald G. Broadley, Benedetto Lanza, Giuseppe Scortecci, Arthur Loveridge and H. W. Parker; Parker's two major works (*The Lizards of British Somaliland*, and *The Snakes of Somaliland and the Socotra Islands*) have proved endlessly useful, and free downloads are available. Recent literature with pan-African or broad relevant regional coverage include Bill Branch's *Tortoise, Terrapins and Turtles of Africa*, Channing and Rödel's *Field Guide to the Frogs & Other Amphibians of Africa*, Chippaux and Jackson's *Snakes of Central and Western Africa*, Sherif Baha El Din's *A Guide to the Reptiles and Amphibians of Egypt*, Geniez's *Snakes of Europe, North Africa and the Middle East*, Largen and Spawls' *The Amphibians and Reptiles of Ethiopia and Eritrea*, Mazuch's *Amphibians and Reptiles of Somaliland and Eastern Ethiopia*, Sindaco and Jeremcenko's *The Reptiles of the Western Palearctic, Vol 1* , Sindaco, Venchi and Grieco's *The Reptiles of the Western Palearctic, Vol 2*, Tilbury's *Chameleons of Africa*, Trape and Mane's *Guide des Serpents d'Afrique Occidentale*, Trape, Trape and Chirio's *Lézards, Crocodiles et Tortues d'Afrique Occidentale et du Sahara*.

Museums with extensive collections relevant to our area include NHMUK – Natural History Museum, London (general), MNHN – Museum national d'Histoire naturelle, Paris (general), MZUF – Museo Zoologico dell'Università, Florence (Somalia), MSNG – Museo Civico di Storia Naturale, Genova (Somalia), MSNM – Museo Civico di Storia Naturale, Milan (Somalia), CAS – California Academy of Sciences, San Francisco (Somalia), MCZ – Museum of Comparative Zoology, Cambridge, MA (Somalia), MVZ – Museum of Vertebrate Zoology, Berkeley, CA (Somalia), NMW – Naturhistorisches Museum, Wien (Sudan herps), ZMB – Museum für Naturkunde, Berlin (general), ZFMK – Zoologisches Forschungsinstitut und Museum Alexander Koenig, Bonn (Sudan, Ethiopia), and WML – World Museum, Liverpool (Eritrea, Sudan).

For any health professional involved or interested in snakebite treatment, if you Google 'Guidelines for the prevention and clinical management of snakebite in Africa', you can download for free a superb book by Dr David Warrell, a pioneer of tropical snakebite treatment. Several other websites have proved very useful, especially the Global Biodiversity Information Facility (gbif.org), the Biodiversity Heritage Library, iNaturalist, and the IUCN Red List. Several scientific journals regularly publish African herpetological material, including *Zootaxa*, and the *African Journal of Ecology*. The only organisation solely devoted to African herpetology is the Herpetological Association of Africa, which produces a scholarly journal and a regular newsletter. Details of all these can be sourced from the internet; and they welcome contributions.

Finally; we are also always pleased to receive information and pictures relating to any aspect of north-east African herpetology; our contacts are as follows Abubakr Mohammad at justanaconda@yahoo.com, Tomáš Mazuch at somalia@seznam.cz and Stephen Spawls at stevespawls@hotmail.com.

CLASS AMPHIBIA

KEY TO THE ORDERS OF AMPHIBIAN IN NORTH-EAST AFRICA

1a: Limbless, resembling worms	caecilians, Order Gymnophiona
1b: Adults have four limbs	frogs and toads, Order Anura

ORDER GYMNOPHIONA; CAECILIANS

Limbless amphibians that look like earthworms, with distinct body rings. They burrow into leaf litter and soft soil. Their eyes are covered by bone or skin. About 200 species are known, in moist tropical regions; more than 20 species occur in tropical Africa, with a single species in our area, in south-west Ethiopia.

FAMILY INDOTYPHLIDAE INDO-AFRICAN CAECILIANS

A family of 24 species, patchily distributed in India, the Seychelles and Africa, mostly in wet forests. Their eyes are visible under the skin, and not covered by bone. Two species in Africa.

AFRICAN FOREST CAECILIANS *SYLVACAECILIA*

A genus with, at present, a single Ethiopian species. However, an as-yet undescribed form has been reported from the forests of south-east Ethiopia, south of the Bale Mountains.

ETHIOPIAN CAECILIAN, ALEKU CAECILIAN

Sylvacaecilia grandisonae
(Taylor, 1970)

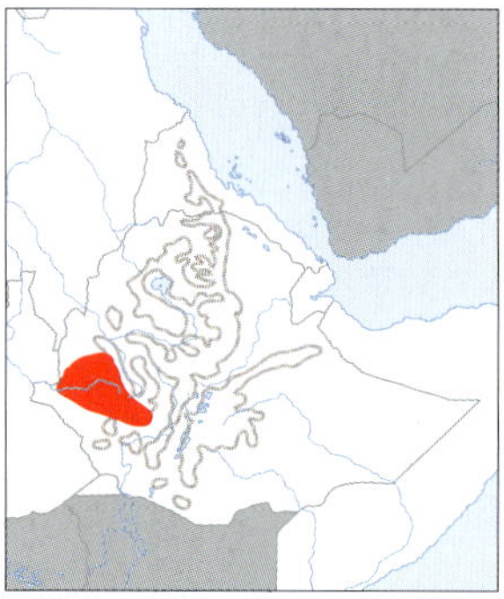

Sylvacaecilia grandisonae, near Jimma.

IDENTIFICATION: Unmistakable, the only caecilian in our area; a purplish-brown, worm-like caecilian of Ethiopia's south-west forest. The tiny dark eyes are visible in a pale patch, in the middle of the elongate snout. The skin glands are visible as tiny pale dots. It has 84–95 primary (all complete) body rings, more closely spaced posteriorly, and 69–72 secondary rings. Largest known adult was 26cm long. The aquatic larvae have spiracles to allow escape of water past the internal gills and sensory pits on the sides of the head and body; these disappear as they metamorphose into adults when 11–12cm long. Adults purplish-brown above, paler and pinkish below.

HABITAT AND DISTRIBUTION: Endemic to south-west Ethiopia, where known from a handful of localities, all within or near the edge of broadleaf forest, at 1,500–2,200m.

CONSERVATION STATUS: Data Deficient. However, it has a limited range in an area where forest is heavily exploited, so its ability to cope with deforestation will be significant in its survival, and should be investigated.

NATURAL HISTORY: Specimens have been found in leaf litter and humus on the forest floor, where they feed on termites and earthworms. Probably nocturnal, but poorly known. They evidently migrate to breed at small streams and springs, where it is assumed females deposit their eggs in burrows close to water. The larvae that emerge from the eggs burrow in soft mud by day, and probably emerge in darkness to feed on aquatic invertebrates. If handled they release copious amounts of slime while squirming vigorously in an attempt to escape.

ORDER ANURA; FROGS

Four-limbed amphibians, most have a larval stage; a tadpole with gills. Adult frogs have no tail. All are carnivores, eating mostly arthropods, so we have not mentioned diet in the species accounts, unless it is unusual. Almost all have a unique male advertisement call that attracts the female; this can often be used to identify the species. About 6,800 species of frog are known worldwide; almost 800 species in Africa. Slightly more than 110 species occur in our area. This is an impoverished fauna, as a consequence of the aridity of large areas of north-east Africa, the relatively late appearance of Ethiopian forests and lack of research in South Sudan.

KEY TO THE GENERA OF FROGS IN NORTH-EAST AFRICA

We have tried to make this key user-friendly, using a mixture of easily visible characters and locality. However, amphibians are often difficult to key. It has been jocularly said that no African frog is difficult to identify to the level of the genus, and no African frog is easy to identify to species. Occasionally we have had to refer to some rather obscure and subjective characters, particularly the webs and tubercles on the arms and feet. It may help to look at the illustrations for the various genera, to get a feel for their general appearance. Note: genera in the same family are not necessarily near each other in this key. The term length as used here always refers to snout–urostyle length (the urostyle is the long fused bone at the base of the spine, the 'back end' of the body); see Fig 25; not the stretched length that would include the legs.

1a: A totally aquatic frog, body very flat, never sits with head up and legs tucked underneath, with very small upward-pointing eyes right on top of the head, immovable eyelids, claws on three or more toes, a row of stitch-like lateral line organs on head and abdomen — *Xenopus*, Clawed frogs

1b: Not wholly aquatic, body not flat, sits with head up and legs tucked under the body, eyes not very small, on side of head or at junction between side and top of head, eyes can be closed, toes without claws, no lateral line organs — 2

2a: Head small and sharply pointed, snout hardened for digging, an obvious transverse groove behind small eyes, limbs short and powerful — *Hemisus*, Piglet frogs or Snout-burrowers

2b: Not having the above combination of characters — 3

3a: Size moderate (length <54mm), body robust, no tympanum, foot very broad, unwebbed, with a much-reduced outer toe high on the side of its neighbour, known only at high altitude in Bale Mountains, Ethiopia — *Balebreviceps*, Bale Mountains Rain Frog

3b: Not having the above combination of characters — 4

4a: Terrestrial frogs with fairly small dark eyes, dorsum and upper thigh always some mixture of red and/or pink and black, no other colours, skin smooth, upper jaw toothless — *Phrynomantis*, Rubber frogs

4b: Not having the above combination of characters — 5

5a: Tips of toes and fingers not always expanded, and terminal section of digits usually straight — 13
(N.B. The exceptions here are Forest White-lipped Frog *Amnirana albolabris*, which does have expanded digits, and Bale Mountains Heather Frog *Ericabatrachus baleensis*, where the digit tips are expanded into a pair of flaps)

5b: Tips of toes and fingers usually expanded slightly, and terminal section of digits kinked (out of line) 6

6a: Hand with fingers arranged in two opposable pairs, relatively large (up to 9cm long), usually mottled white, brown or grey, no clear pattern *Chiromantis*, Foam-nest frogs

6b: Hand with fingers all in same plane, relatively small (usually <6cm long), colours often vivid, sometimes green or yellow, often with clear spots or stripes 7

7a: Eye with horizontal or rounded pupil 8

7b: Eye with vertical pupil 10

8a: Only in northern Egypt, dorsum uniformly coloured, usually greenish *Hyla savignyi*, Savigny's Tree Frog

8b: Always in countries south of Egypt, dorsum not necessarily uniformly coloured 9

9a: Large, up to 5cm, terrestrial, tympanum clearly visible, only Somalia *Lanzarana largeni*, Lanza's Frog

9b: Small, rarely more than 3.5cm long, arboreal, tympanum usually invisible *Hyperolius*, Reed frogs

10a: Dorsum prominently patterned with relatively large round blotches or broad stripes, never arboreal, adult males have heavily pigmented vocal sac which, when expanded, balloons either side of large gular flap 11

10b: Dorsum not prominently patterned with relatively large rounded blotches or broad stripes, sometimes arboreal, adult males lack external modification of throat or have a gular flap that allows the unpigmented vocal sac to expand without restriction 12

11a: Known only in highland Ethiopia above 2,000m, chest and abdomen usually patterned, with a bright yellow patch at femur/abdominal junction, tympanum small and hard to see *Paracassina*, Mountain Kassinas

11b: Widespread throughout region, rarely above 2,000m, chest and abdomen without dark markings, no bright yellow patch at femur/abdominal junction, tympanum usually well developed *Kassina*, Running frogs or Kassinas

12a: Relatively large (usually 4–6cm long), vomerine teeth present in front centre of roof of mouth, never have small spines on skin, often green, rarely with distinct striping *Leptopelis*, Tree frogs

12b: Relatively small (usually 2–3cm long), vomerine teeth absent, always have small spines on skin, often brown and with distinct striping *Afrixalus*, Leaf-folding or Spiny reed frogs

13a: Skin not conspicuously thick, warty or spinose, no parotid gland present 14

13b: Skin thick, warty and sometimes spinose, a parotid gland (often large, elongate and prominent) present immediately behind each eye, upper jaw toothless 26

14a: A fine skin ridge (raphe) down centre of dorsum, no tarsal tubercles, males have an unusually elongate third finger, in our area known only in south-west South Sudan *Arthroleptis poecilonotus*, Mottled Squeaker

14b: No fine skin ridge down centre of dorsum, tarsal tubercles present, males without elongate third finger, widespread in our area 15

15a: Inner metatarsal tubercle massive and strongly compressed for burrowing 16

15b: Inner metatarsal tubercle not exceptionally large or strongly compressed 18

16a: Huge and squat when adult, >10cm long, two sharp tooth-like outgrowths on lower jaw *Pyxicephalus*, African bullfrogs

16b: Smaller, chunky but not squat, adults 4–7cm long, no sharp tooth-like outgrowths on lower jaw 17

17a: Small and toad-like, 4–5cm long, usually brown or grey, dorsal surface with many small rounded warts *Tomopterna*, Sand frogs

17b: Larger and frog-like, 5–7cm long, usually rufous or with rufous stripes, dorsal surface smooth or with a few warts *Hildebrandtia*, Ornate frogs

18a: Tips of digits divided into pair of flaps, appearing heart-shaped, known only from southern slopes of Bale Mountains, south-east Ethiopia — *Ericabatrachus baleensis*, Bale Mountains Heather Frog

18b: Tips of digits not divided, not confined to Bale Mountains — 19

19a: Small, usually less than 3cm long, dorsal surface warty, but without skin ridges, usually mottled brown, sometimes green — 20

19b: Large, usually more than 5cm long, dorsal surface usually smooth or with skin ridges, rarely warty, a range of colours — 21

20a: Underside distinctly spotted, dorsal warts not arranged in pairs — *Cacosternum*, Dainty frogs

20b: Underside not spotted, dorsal warts often arranged in one or two chevron-shaped pairs between shoulders — *Phrynobatrachus*, Puddle Frogs

21a: Adults huge, 12–16cm long, a transverse groove across head behind eyes — 22

21b: Adults smaller, usually <10cm long, no transverse groove across head behind eyes — 23

22a: Tympanum totally hidden, eyes on side of head, head enormously broad in large adults — *Conraua beccarii*, Beccari's Giant Frog

22b: Tympanum clearly visible, eyes on junction of side and top of head, head not unusually broad in large adults — *Hoplobatrachus occipitalis*, Eastern Groove-crowned Bullfrog

23a: Dorsum usually with several pairs of fine, conspicuous, typically continuous longitudinal skin folds — *Ptychadena*, Rocket or Grass frogs

23b: Dorsum without several pairs of fine folds, although it may have a pair of broad or narrow, partial or elongate dorsolateral skin folds — 24

24a: A golden, brown, green, largely uniform (sometimes heavily mottled) dorsal band, as wide as back, from snout to vent, within two prominent dorsolateral folds, but no smaller folds within them — *Amnirana*, White-lipped frogs

24b: No broad dorsal band as wide as back; if two prominent dorsolateral folds exist then usually there are smaller folds either side — 25

25a: Known only from Egypt (and a specimen from Eritrea), with a well-developed metatarsal tubercle; dorsum usually vivid green — *Pelophylax*, Green frogs

25b: Known only from countries south of Egypt, without metatarsal tubercle, not usually vivid green — *Amietia*, River frogs

26a: In our area confined to Egypt, with green dorsal blotches — *Bufotes*, Green toads

26b: Does not have above combination of characters — 27

27a: Tympanum absent, known only from highlands of south-east Ethiopia — *Altiphrynoides*, Ethiopian alpine toads

27b: Tympanum present, widespread in our area — 28

28a: Small, adults <5cm long, not in Egypt, flat-backed, with stout limbs — *Mertensophryne lughensis*, Lugh Pygmy Toad

28b: Does not have above combination of characters — *Sclerophrys*, typical African toads

FAMILY PIPIDAE CLAWED FROGS

A family of primitive, tongue-less frogs adapted to spend most of their life in water, having totally webbed feet and a series of lateral line organs (fish also have this) to detect pressure fluctuations. Four genera; one in South America and the other three in Africa, suggesting a Gondwanan origin.

CLAWED FROGS *XENOPUS*

An African genus of smooth, flat, streamlined frogs, with powerful hindlimbs and large, fully webbed feet, with claws on 3–4 toes. They all look similar. Many have multiple sets of chromosomes, instead of the usual pair. About 30 species are known, four occur in our area, but their taxonomy is undergoing revision and this number will almost certainly increase. Clawed frogs live, feed and mate in the water, although occasionally large groups will move across country (usually at night) from one waterbody to another. They are both predators and scavengers. Their little eyes on top of the head enable them to look outside of the water without exposing their bodies. Their use between 1930 and 1960 for pregnancy testing (if a female *Xenopus* was injected with urine from a pregnant woman, the frog began spawning) has led to vagrant introduced colonies of *Xenopus* worldwide. Clawed frogs call underwater, they have no vocal cords and their characteristic clinking, grunting or clicking sounds are produced by modified laryngeal cartilage elements rubbing against each other.

KEY TO THE NORTH-EAST AFRICAN MEMBERS OF THE GENUS *XENOPUS*

1a: Occurs only in Ethiopia or Eritrea	2
1b: Not in Ethiopia or Eritrea	3
2a: No subocular tentacle, three claws on foot	*Xenopus largeni*, Largen's Clawed Frog
2b: Prominent subocular tentacle, four claws on foot	*Xenopus clivii*, Peracca's Clawed Frog
3a: Subocular tentacle present	*Xenopus fischbergi*, Fischberg's Clawed Frog
3b: No subocular tentacle	*Xenopus victorianus*, Lake Victoria Clawed Frog

LAKE VICTORIA CLAWED FROG

Xenopus victorianus
Ahl, 1924

Xenopus victorianus, Yambio.

IDENTIFICATION: A medium-sized aquatic frog, in our area only in South Sudan. The snout and the back are smooth. There are three claws on the back foot. No subocular tentacles, but has 13 sensory organs around the eye. Average length 5–6cm male, 7–8cm female. Colour very variable, from dark grey or black to olive-green or yellow, immaculate or with many small and a few irregular larger spots. Taxonomic Notes: specimens from Garamba National Park and South Sudan originally described as *X. laevis*.

HABITAT AND DISTRIBUTION: In a wide range of habitats, including dry and moist savanna, woodland and forest, even farmland and peri-urban areas; where it occupies marshes, slow-flowing rivers, lakes, pools, in the rainy season may enter temporary pools. Quite tolerant of polluted and stagnant water. Generally at low altitude in South Sudan, although it ranges up to 2,500m elsewhere. In our area, known only from southern South Sudan, but also in Garamba National Park, in the DR Congo, across the border from Western and Central Equatoria. Elsewhere along the Albertine Rift, east to western Kenya, and south to north-west Tanzania.

CONSERVATION STATUS: Least Concern.

NATURAL HISTORY: Usually in deep water by day but may move into shallow water at night, where they hang, head up, watching for prey, with the head at the surface. The advertisement call, made underwater, sounds like someone tapping briskly on a porcelain plate with a coin; it has also been described as a series of rapid trills. They mate in the water, but few details known. A large group was observed crossing a road at night in western Kenya in July.

FISCHBERG'S CLAWED FROG

Xenopus fischbergi
Evans, Carter, Greenbaum, Gvoždík, Kelley, McLaughlin, Pauwels, Portik, Stanley, Tinsley, Tobias & Blackburn, 2015

IDENTIFICATION: A small aquatic frog, recently described, in our area only in South Sudan, (possibly Sudan). The snout and back are smooth. There are three claws on the back foot. A subocular tentacle is present. There are 12–15 sensory organs around the eye. Average length around 5cm male, 5–6.3cm female. Colour pale to olive or greenish-grey, faintly mottled.

HABITAT AND DISTRIBUTION: Occurs in a range of habitats, including savanna, farmland and peri-urban areas; where it lives in temporary and permanent pools, ponds and rivers. During the dry season it

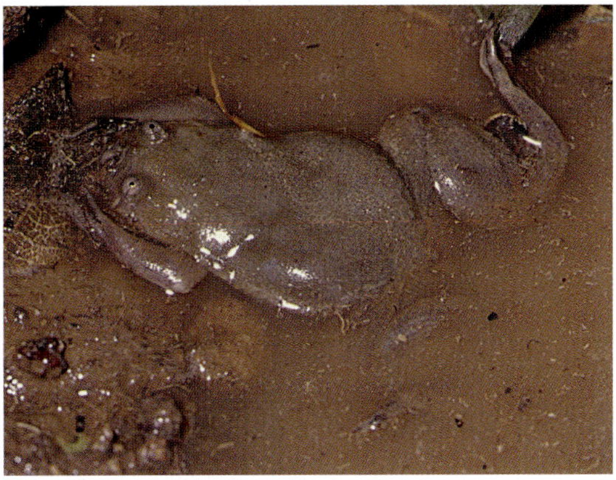

Xenopus fischbergi, Benin.

shelters under stones or bark on the banks of large rivers and ponds. Quite tolerant of polluted and stagnant water. In South Sudan its range is below 1,000m, although occurs at higher elevations elsewhere. At present known only in Eastern and Central Equatoria, but probably occurs in Western Equatoria and Western Bahr El Ghazal, and possibly southern South Darfur in Sudan. Elsewhere, west to Burkina Faso and north-east Ivory Coast.

CONSERVATION STATUS: Least Concern.

NATURAL HISTORY: Poorly known, but presumably similar to other Clawed frogs, in deeper water by day, shallow water by night. The advertisement call, made underwater, is a single loud click. The tadpoles are gregarious.

PERACCA'S CLAWED FROG, CLIVI'S CLAWED FROG

Xenopus clivii
Peracca, 1898

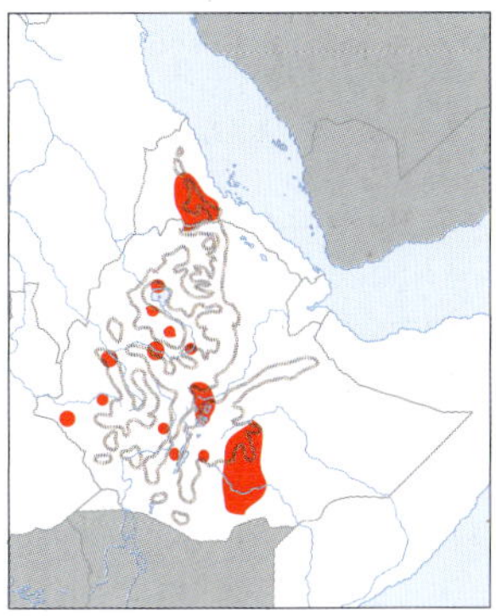

Xenopus clivii, Metu.

IDENTIFICATION: A medium-sized aquatic frog, in our area only in Ethiopia and Eritrea. Snout and back smooth. Four claws on back foot. Has a prominent subocular tentacle. Average length 6cm male, 7–8cm female. Colour very variable, from almost white to grey, olive and dull brown, sometimes with large irregular, vaguely paired dorsal blotches, or with random patches of black pigment. Underside cream to yellow or orangey-brown, immaculate or spotted pale to dark grey.

HABITAT AND DISTRIBUTION: A wide range of habitats, including grassland at high altitude, moist savanna, woodland and forest, where it occupies marshes, slow-flowing rivers, lakes and pools. Quite tolerant of polluted water in and around Addis Ababa and other towns, provided it is flowing. Occurs between 820 and 2,750m. In our area, in the central highlands of Eritrea (Asmara, Keren, Adi Keyh, Saganeiti, Adi Ugri), and Ethiopia (Adua, Lake Tana, Bahar Dar, south to Negele, and south-west to Jimma, Welkite and Godare). Informally reported from Jonglei in South Sudan, but records lacking.

CONSERVATION STATUS: Least Concern.

NATURAL HISTORY: Seems to be active day and night, around Bahar Dar it can be observed in the lake and adjacent marshes resting near the surface. The advertisement call, made underwater, is a long trill, repeated. Reported to breed in permanent, stagnant and slow-flowing water.

LARGEN'S CLAWED FROG

Xenopus largeni
Tinsley, 1995

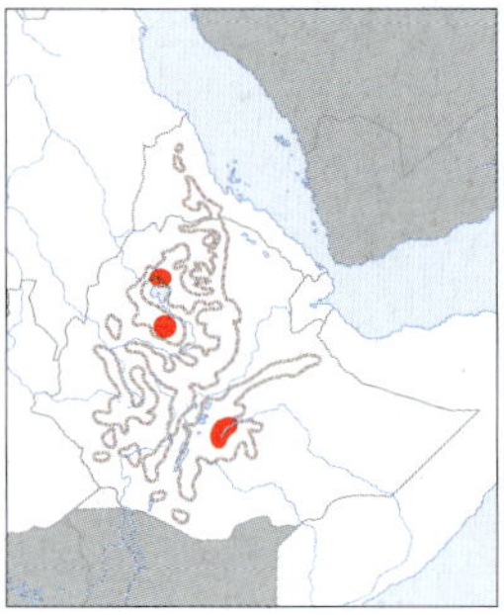

IDENTIFICATION: A small aquatic frog, found only in Ethiopia. Snout and back smooth. Three claws on back foot. No subocular tentacle. Small for a *Xenopus*; average length 4.5cm male, 5.5cm female. Colour dark brown or dull olive above, with small darker spots that are hard to see. The underside is dull yellow, mottled grey to a greater or lesser extent.

HABITAT AND DISTRIBUTION: Originally known from two localities either side of the Bale Mountains in south-west Ethiopia, in a single stream flowing through agricultural land at 2,500m in the Arussi Mountains, and in pools, streams and marshes in grassy clearings in an area denuded of forest south-west of Bale. The species was recently found north-west of the Rift Valley in streams in the agricultural highlands of Gojjam, in the mountains north of Debre Markos, and near Amba Giorgis, north of Gondar; between 2,300 and 2,700m.

CONSERVATION STATUS: Endangered.

NATURAL HISTORY: Virtually nothing known, but presumably similar to other clawed frogs. The advertisement call, made underwater, is a long trill, repeated.

Xenopus largeni, Amhara.

FAMILY HYLIDAE TREE FROGS AND ALLIES

A huge family of more than 1,000 species in at least 50 genera, found worldwide except sub-Saharan Africa. A single genus reaches North Africa, with two species, one in our area.

NORTHERN TREE FROGS *HYLA*

There is some debate about the limits of this genus, which at one point contained more than 300 species, but it is now taken to comprise about 17 species of largely green, arboreal frogs found in southern Europe, Asia and North Africa. Three species reach Africa, one in our area.

MIDDLE EAST TREE FROG

Hyla savignyi
Audouin, 1827

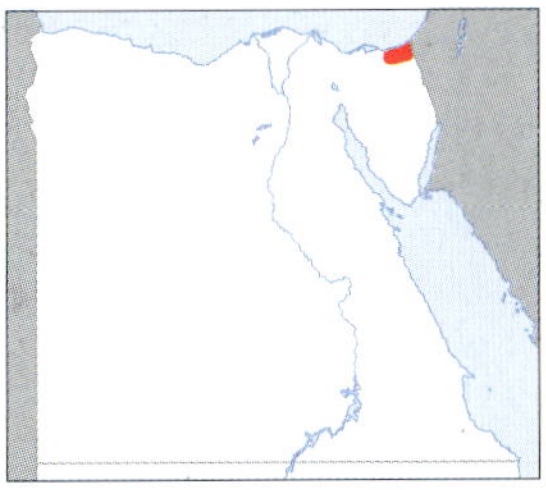

IDENTIFICATION: A green tree frog of northern Sinai. The body is fairly squat; the head short, with big eyes and horizontal pupils. Dorsal skin smooth but granular. The tympanum is distinct, about half the size of the eye. Limbs slender, toe tips enlarged. In Egypt, males reach a maximum size of about 3.5cm, females 4.6cm. Usually green, occasionally brown, olive or yellow, with a pair of brown dorso-lateral stripes through the snout, which may reach the hindlimbs or break up. The underside is white and granular.

HABITAT AND DISTRIBUTION: In our area, found in vegetated areas, agriculture and gardens, near water sources, both natural and artificial, in desert, near sea level. Known only from the extreme north-east Sinai Peninsula, between Al Arish, Sheikh Zowayid and Rafah. Elsewhere, north-east to Iran and Turkey, also parts of the south-west Arabian coast.

CONSERVATION STATUS: Least Concern, although within Egypt the tiny range and limited habitat available indicates it may be Vulnerable nationally.

NATURAL HISTORY: A nocturnal climbing species, occasionally active by day on overcast days and in winter. The advertisement call is a strident, rapid, rasping *craak, craak, craak*. Usually spends day well hidden in trees and bushes. Males call near dusk. In Jordan, breeding occurs between mid-December and late January; amplexus may occur in trees or on ground, but eggs are glued in clumps of *c.*30 to submerged vegetation; tadpoles metamorphose in 2–3 months.

Hyla savignyi, Israel/Palestine.

FAMILY BUFONIDAE TOADS

A toothless family of about 650 species in more than 50 genera, and occurring worldwide apart from Australia (although an introduced problem species, Cane Toad *Rhinella marina*, occurs there). About 17 genera occur in Africa, four in our area. The name 'toad' originates in the UK, and originally distinguished the warty species from British 'frogs', which are smooth. The name is now used loosely to describe stout, squat, short-limbed, often large, warty frogs.

AFRICAN TOADS *SCLEROPHRYS*

An African genus of warty frogs, with relatively short stout limbs, short snouts, huge prominent eyes, often the tympanum is big and visible. Many are large, up to 9–10cm long or more. They usually live in savanna or semi-desert. The pupil is horizontal, and immediately behind each eye is an elongate, often swollen parotid gland, which exudes sticky white toxin (bufotoxin) if the toad is molested. Toads are often common and have loud calls, and are thus annoying in urban areas. Mating usually occurs in shallow water, and toads will utilise temporary pools, some are tolerant of urbanisation and polluted water. Large numbers of pigmented eggs are laid in long strings of jelly. Around 40 species are known in Africa, 14 in our area. Originally in the worldwide genus *Bufo*, the genus *Amietophrynus* was coined for the African species, but subsequently the name *Sclerophrys* was found to have precedence.

KEY TO THE NORTH-EAST AFRICAN MEMBERS OF THE GENUS *SCLEROPHRYS*

Note: Some of these toads are very hard to identify, even by experts. Some species cannot be distinguished with certainty other than by DNA and/or call analysis. Some species also hybridise; to the extent of forming new species, sometimes with multiple sets of chromosomes, leading to considerable variation in the appearance of animals of the same species. Some species result from a single taxon being split; expert opinion is divided upon their validity. However, several species can be relatively easily identified, and location is often useful here. We have tried to make this key user-friendly, but, as before, in some cases we have had to revert to using obscure and subjective characters.

1a: A squat chunky toad with a very short snout, the upper surface of which is swollen and often horny between the eye and nostril, a large dark tubercle under the tarsus, back with many rounded hemispherical red and yellow warts — *Sclerophrys pentoni*, Penton's Toad

1b: Does not have the above combination of characters — 2

2a: Small, adults <5.5cm long, no distinct tarsal ridge on the back foot — 3

2b: Adults usually larger than 6cm long (except *Sclerophrys blanfordii*), have a distinct longitudinal ridge on the inner margin of the tarsus — 5

3a: Never larger than 4cm long, in our area confined to Egypt — *Sclerophrys kassasii*, Nile Valley Toad

3b: Adults up to 6.5cm long, in the south of our area, below 20°N, not in Egypt — 4

4a: Parotid glands indistinct, in savanna — *Sclerophrys steindachneri*, Steindachner's Toad

4b: Parotid glands obvious, in forest in South Sudan — *Sclerophrys funerea*, Sombre Toad

5a: Limbs relatively thin and spindly, a huge tympanum, throat of adult males lacks black pigment, known only from arid regions in the east of our area — *Sclerophrys dodsoni*, Dodson's Toad

5b: Does not have the above combination of characters — 6

6a: Small, <5cm long, tympanum very small, 15–50 per cent diameter of the eye, dorsal colour subtle browns and rufous, sometimes with small rufous spots, only in dry lowlands in east of our area — *Sclerophrys blanfordii*, Blanford's Toad

6b: Does not have the above combination of characters — 7

7a: Back of thigh usually distinctly marked with red pigment, parotid glands huge, twice as long as wide — 8

7b: Back of thigh without red markings, parotid glands usually less than twice as long as wide — 9

8a: Dorsal surface virtually always marked by distinctive, paired, maroon blotches, high contrast, parotid glands regular in shape — *Sclerophrys garmani*, Garman's Toad

8b: Dorsal surface with paired drab grey-brown blotches, low contrast, parotid glands often irregular in shape, broadening at one end — *Sclerophrys xeros*, Desert Toad

9a: Vividly marked with three or four paired blotches, often bordered yellow, throat of adult males not suffused with dark pigment, in our area confined to highland Ethiopia *Sclerophrys kerinyagae*, Kerinyaga Toad

9b: Does not have the above combination of characters 10

10a: Parotid glands low and flattened, or indistinct 11

10b: Parotid glands usually prominent 13

11a: Back of thighs and flanks vividly marked, with orange, yellow or green spots and blotches, back not notably flat, known only from central Ethiopian Rift Valley and Awash River *Sclerophrys langanoensis*, Lake Langano Toad

11b: Back of thighs and flanks marked in combination of dull browns and greys, back notably flat, not known from central Ethiopian Rift Valley 12

12a: Stocky, with relatively wide parotid glands, lower-pitched mating call, with longer notes *Sclerophrys pusilla*, Southern Flat-backed Toad

12b: Slenderer, with elongate parotid glands, high-pitched mating call of short notes *Sclerophrys maculata*, Northern Flat-backed Toad

13a: Dorsal surface usually with huge warts, sometimes as few as 12 between back of parotid gland and vent, never has white spots; highland Ethiopia and Eritrea, above 1,800m *Sclerophrys asmarae*, Asmara Toad

13b: Warts on dorsal surface large but not very large, sometimes spotted with white, not confined to highland Ethiopia and Eritrea *Sclerophrys regularis*, Common African Toad

ASMARA TOAD

Sclerophrys asmarae
(Tandy, Bogart, Largen & Feener, 1982)

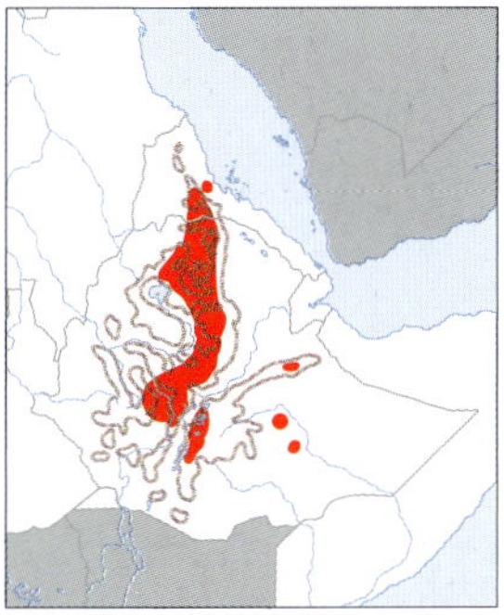

IDENTIFICATION: A big, chunky toad, in highland Ethiopia and Eritrea. The back is rough, with large, hemispherical warts, which sometimes have spines. Tympanum fairly large and prominent. Parotid glands prominent, three times longer than wide, and deeply pitted. The average male is 7–8cm long, females are slightly larger, reaching just over 9cm. The glands under the forearm are not fused. Dorsal colour usually yellowish-brown, olive-grey or greenish-brown, with paired darker blotches, including a pair on the snout. A yellow vertebral stripe is often present. Underside whitish or cream, sometimes weakly mottled grey. Males dark grey or blackish on throat. Taxonomic Notes: These toads have 20 pairs of chromosomes in their cell nuclei, suggesting they are a hybrid species involving *Sclerophrys kerinyagae* × *S. regularis*.

Sclerophrys asmarae, Asmara.

HABITAT AND DISTRIBUTION: An endemic of montane grassland and moist savanna on the Ethiopian and Eritrean plateau, from just north of Asmara (Medpi Saul) south to the highlands east of Lake Awassa, between 1,800 and 3,000m. There are two curious low-altitude records, from Saati, on the

Sclerophrys asmarae, Simien Mountains.

Eritrean maritime plain at 200m, and another west of Imi at 600m in eastern Ethiopia.

CONSERVATION STATUS: Least Concern.

NATURAL HISTORY: Terrestrial, and largely nocturnal. Advertisement call a low-pitched snore, lasting more than a second. Breeds in temporary pools in the rainy season; when males call after dark at the water's edge. Sometimes active by day when heavily overcast in the Entoto Hills above Addis Ababa.

BLANFORD'S TOAD

Sclerophrys blanfordii
(Boulenger, 1882)

IDENTIFICATION: An attractive little rufous toad of low country in the Horn of Africa. The back is rough, with small, hemispherical warts, which sometimes have spines. Tympanum relatively small, less than half the diameter of the eye. The parotid glands are indistinct; this, the tympanum and the small size are good field characters. Both sexes are usually 4–5cm long, females larger than males. Dorsal colour usually rufous-brown, or warm brown, either unmarked or with an irregular pattern of brownish blotches or small maroon spots. Underside whitish or cream, sometimes mottled darker; the back of the thigh is uniform. Males have a grey-brown or yellow throat.

HABITAT AND DISTRIBUTION: Near-desert and semi-desert country of north-east Africa, where associated with permanent or near-permanent water sources in arid rocky country; an unusual mixture of habitats that may explain its rarity, between sea level and 1,750m. Known from Eritrea, Djibouti, northern Somalia and north-east Ethiopia, including east of Afabet and on Dahlak Kebir Island in Eritrea, the Tonsa River, the lower Awash River in Ethiopia, and a handful of localities across northern Somalia, to Eyl on coast.

CONSERVATION STATUS: Least Concern.

NATURAL HISTORY: Terrestrial, breeds in shallow pools in the rainy season. The advertisement call is a medium-pitched, rapid snore, which lasts about half a second. Active at night, by day they shelter among aquatic vegetation or under stones near the water's edge. They may aestivate during the dry season, but where there is abundant permanent water throughout the year (for example large springs, such as Fil Fil in Eritrea and Medisho in Somaliland), these frogs are active and occur in large numbers for much of year, although numbers are low in the coolest months, December to February). In small permanent springs, where water dries to a trickle at times, they are less common, but solitary individuals may be encountered. Their ecological requirements indicate that they exist as a series of small isolated populations, each confined to a particular spring, well or oasis.

Sclerophrys blanfordii, Hudisso, Goolis Mountains.

Sclerophrys blanfordii, Fil-fil.

DODSON'S TOAD

Sclerophrys dodsoni
(Boulenger, 1895)

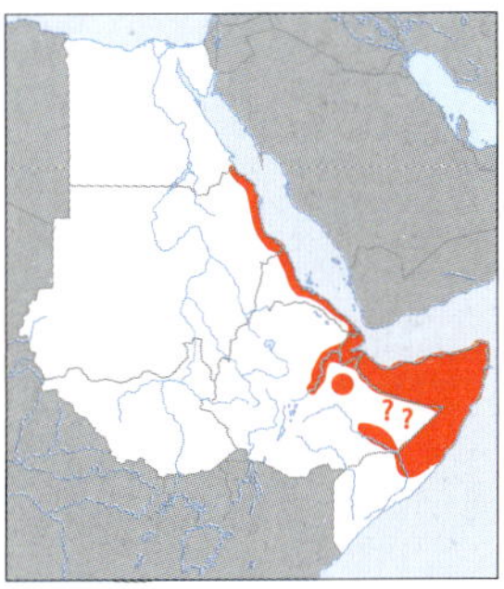

IDENTIFICATION: A small, relatively thin-limbed toad with a big tympanum, of dry country in the Horn of Africa. The back may be smooth, or have small, hemispherical warts, which sometimes have spines and extend onto the flanks. Tympanum large, almost the size of the eye and close to it. The parotid glands are low and indistinct. Adult males usually just over 5cm, females 6.5cm. Dorsal colour usually brown, olive or grey, faintly mottled, sometimes with darker spots. Underside whitish or cream, and smooth. Juveniles often distinctively spotted red. Some males have yellow or orange throats.

HABITAT AND DISTRIBUTION: Near-desert and semi-desert country of the low north-east; endemic to our region. In Egypt associated with rocky wadis in dry near-desert, in Sudan, Eritrea, Ethiopia and Somalia in dry savanna and semi-desert, at altitudes between sea level and 1,800m, but mostly fairly close to the coast. Occurs from Gebel Elba in Egypt south through Erkowit in Sudan, to Eritrea and dry eastern Ethiopia, as far south as Missarole, in Hiran, south-west of Bulo Barde in Somalia.

CONSERVATION STATUS: Least Concern.

NATURAL HISTORY: Terrestrial, nocturnal. The advertisement call is a brief yap, resembling a dog's bark. They breed in temporary pools in the rainy season and aestivate during the dry season; in Egypt this may be in deep rock fissures. In Egypt sometimes active in winter and spring. They seem extremely tolerant of dry conditions and may be found (always in small

Sclerophrys dodsoni, Ghinda.

Sclerophrys dodsoni, Las Anod.

Sclerophrys dodsoni, juvenile, Suakin.

numbers, maximum 2–4, but usually solitary) far from permanent water sources (compare *Sclerophrys blanfordii*). Observed in Gebel Elba perched on boulders, watching for prey below. The males call at the edges of temporary pools, and in Ethiopia have been observed calling by day. In Somaliland often enters buildings, looking for water.

SOMBRE TOAD

Sclerophrys funerea
(Bocage, 1866)

IDENTIFICATION: A medium-sized toad, in our area known only from Talanga Forest in South Sudan. Back broad, flat and rough, with numerous rounded warts, which on females may have spines. Tympanum fairly large, 75 per cent size of the eye. Parotid glands large and oval, twice as long as they are wide. There is a curious cluster of conical glands behind the gape. Males reach nearly 6cm, females 6.5cm long. Dorsal colour usually rufous-brown (often more rufous towards snout), with series of often irregular, poorly defined blotches on back. Often a pale vertebral stripe may be present, usually with a pale or rufous bar between eyes. Underside off-white.

Sclerophrys funerea, Congo.

HABITAT AND DISTRIBUTION: A forest species. In our area only in South Sudan; at Talanga Forest, 950m altitude, near Katire, Eastern Equatoria, but also known from Garamba National Park in DR Congo, so probably in Central and Western Equatoria. Elsewhere, occurs west to Gabon, and south to north-west Angola.

CONSERVATION STATUS: Least Concern.

NATURAL HISTORY: Unknown in South Sudan and little known elsewhere, but presumably terrestrial and nocturnal. Hides in leaf litter in the forest. Males have no vocal sacs, so they might not call. Breeds throughout the year in Central Africa; eggs deposited in small streams.

GARMAN'S TOAD

Sclerophrys garmani
(Meek, 1897)

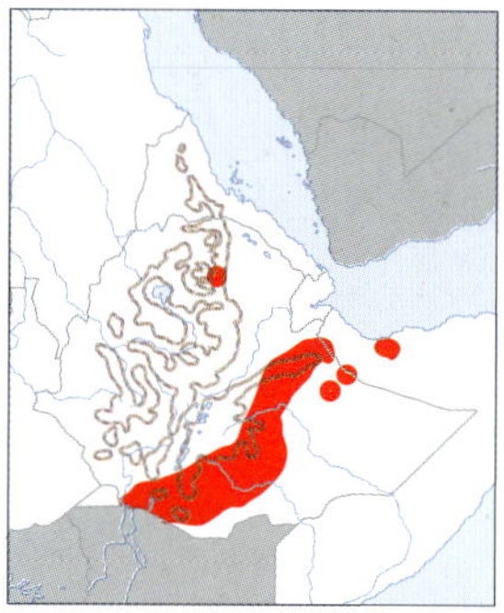

IDENTIFICATION: A big toad with huge parotid glands and paired maroon blotches on the back, in our area in southern Ethiopia, with an isolated population in northern Somalia. The back is flat and rough, with numerous, mainly small but some large hemispherical warts, which sometimes have spines. Tympanum fairly large, almost as big as the eye and usually distinct. The parotid glands are huge, more than twice as long as wide and almost touch the eye, this and the maroon blotches are good field characters. Males up to 10.5cm, females to 11.5cm long. Dorsal colour usually some shade of brown, dark, cream or rufous, with one to three paired maroon blotches on the back, sometimes indistinct but often with clear edges. Often a pale vertebral stripe is present. Juveniles often have vivid rufous parotid glands on a brown back. There are red patches at the back of

Sclerophrys garmani, Negele.

Sclerophrys garmani, Tog Wajaale.

the thighs. Underside whitish or cream. Males have a dark throat.

HABITAT AND DISTRIBUTION: Mid- to high-altitude savanna in much of our area, but also semi-desert; distribution curiously disjunct. Records in northern Somalia include Berbera (the type locality) and Tog Wajaale, on the Ethiopia–Somali border south-east of Borama and it is probably widespread in the Goolis and Ogo Mountains, but never common. Also recorded at Lake Hashenge in northern Ethiopia at 2,500m. Widespread in southern and eastern Ethiopia, between 400 and 2,000m. Elsewhere south to Botswana and South Africa, although these may prove to be a different species.

CONSERVATION STATUS: Least Concern.

NATURAL HISTORY: Terrestrial, breeds in shallow pools in the rainy season. Aestivates in the dry season. The advertisement call is a loud *kwaak*, lasting up to a second. Active at night, by day they shelter among aquatic vegetation or under cover near the water's edge. They lay the usual long string of toad eggs, up to 2,000 eggs may be produced per clutch.

NILE VALLEY TOAD

Sclerophrys kassasii
(Baha El Din, 1993)

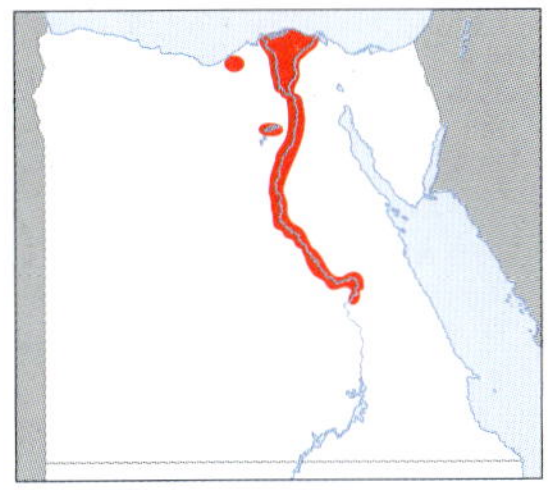

IDENTIFICATION: A warty, little toad of Egypt's Nile Valley, found nowhere else. Snout short. The back is rough, with small hemispherical warts that sometimes have spines. Tympanum relatively large, usually about three-quarters the diameter of the eye. Parotid glands indistinct, spinose and oval-shaped. Males reach a maximum of 3.5cm, females almost 4cm. Dorsal colour usually light olive-grey, rufous or pink, with 3–4 pairs of vague, irregular darker markings; a whitish vertebral line is often present. Red patches usually present on backs of thighs. Adult males have a noticeable yellowish wash, with a reduced dorsal pattern. The underside is white, either unmarked or with some dark flecks. Throat of male pale orange.

HABITAT AND DISTRIBUTION: Marginal vegetation along watercourses, associated irrigation canals and flooded fields. Found in reedbeds, rice fields, overgrown canals and waterways, sometimes in floating clumps of vegetation. Known only from the Fayum and the Nile Valley in Egypt, from the Delta south to Luxor.

CONSERVATION STATUS: Least Concern.

NATURAL HISTORY: A highly aquatic species, virtually always near or in water. Terrestrial. Mostly active at night (although males may call at any time), by day they shelter among marginal vegetation or other cover near the water's edge. The advertisement call is a rattling, slow snore. Often common, even in the heart of Cairo.

Sclerophrys kassasii, Nile Delta.

KERINYAGA TOAD

Sclerophrys kerinyagae
(Keith, 1968)

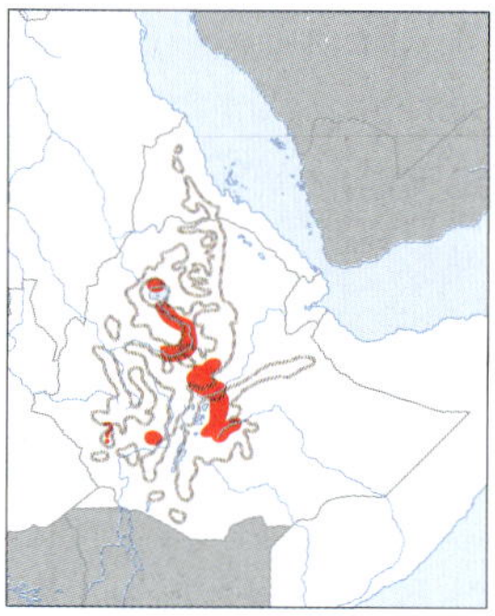

IDENTIFICATION: A fairly big toad, often vividly marked, with elongate parotid glands and paired markings, in our area confined to medium and high altitudes in Ethiopia. The back is flat and rough, with numerous warts that sometimes have spines. Tympanum fairly large, at least half the size of the eye. Parotid glands large and elongate , twice as long as wide and deeply pitted. Males and females reach a maximum size of about 7.5cm in Ethiopia (slightly larger in Kenya). Dorsal colour often vivid, a mixture of warm brown, yellow, green, olive and rufous, with a series of (usually four) paired blotches. Usually has a yellow vertebral stripe. Underside whitish or cream. Breeding males have a pale throat. Taxonomic Notes: Ethiopian and East African populations are separated by several hundred kilometres of low arid land, leading to suggestions that they might be specifically distinct.

HABITAT AND DISTRIBUTION: A species of mid- to high-altitude grassland, savanna and forest margins in Ethiopia, between 1,400 and 3,300m. Known from north of Lake Tana, south through highlands both east and south-west of the Rift Valley to the northern foothills of the Bale Mountains.

CONSERVATION STATUS: Least Concern.

NATURAL HISTORY: Terrestrial, males call at night in the rainy season from the flooded margins of temporary pools and flooded hollows. Juveniles may be active on overcast days in the rainy season. Aestivates in the dry season. They lay the usual long string of toad eggs. The advertisement call is a rapidly pulsed snore that drops in pitch, lasting about two seconds.

Sclerophrys kerinyagae, Sodere.

LAKE LANGANO TOAD

Sclerophrys langanoensis
(Largen, Tandy & Tandy, 1978)

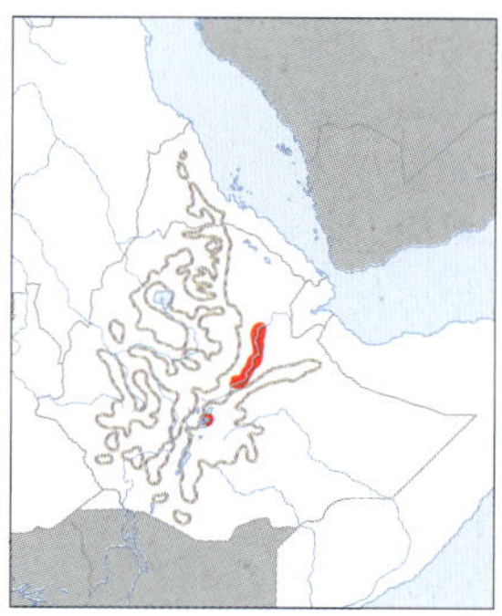

IDENTIFICATION: A medium-sized, warty, strikingly marked toad, endemic to Ethiopia's north-central Rift Valley and Awash River. Snout short. Back rough, with small hemispherical warts that sometimes have spines. Tympanum relatively large, usually about ¾ the diameter of the eye. Parotid glands reduced and indistinct. Males reach a maximum of 6cm, females slightly larger. Dorsal colour varies; a mix of rufous, brown-olive and grey, sometimes with large, paired, rufous blotches. Backs of thighs and the flanks conspicuously marked with circular or subcircular orange or bright yellow blotches. Underside whitish or cream, often with darker mottling. Throat of adult males suffused yellow, orange or grey-brown.

Sclerophrys langanoensis, Lake Langano.

HABITAT AND DISTRIBUTION: Degraded mid-altitude savanna, around Lake Langano, in central Rift Valley, and in low-altitude semi-desert and dry grassland along Awash River. The altitude at Lake Langano is just below 1,600m, on the lower Awash River 600–1,000m. Awash River localities range from Metahara and Awash National Park north to Lake Hertale. It may occur between the two localities (at Lake Zwai, for example).

CONSERVATION STATUS: Least Concern.

NATURAL HISTORY: Nocturnal and terrestrial. Unusually for an African savanna amphibian, it lives and breeds beside permanent water. The advertisement call is a low-pitched, grating snore, lasting less than a second. In Lake Langano, males call in shallow water at night while supporting themselves on submerged vegetation with the head above water level. By day, they shelter amongst the dense lakeside vegetation. Awash specimens are found on banks of rivers, streams and pools. Unlike other toads in the area, which breed explosively in temporary pools, this species appears to breed throughout the year.

NORTHERN FLAT-BACKED TOAD

Sclerophrys maculata
(Hallowell, 1855)

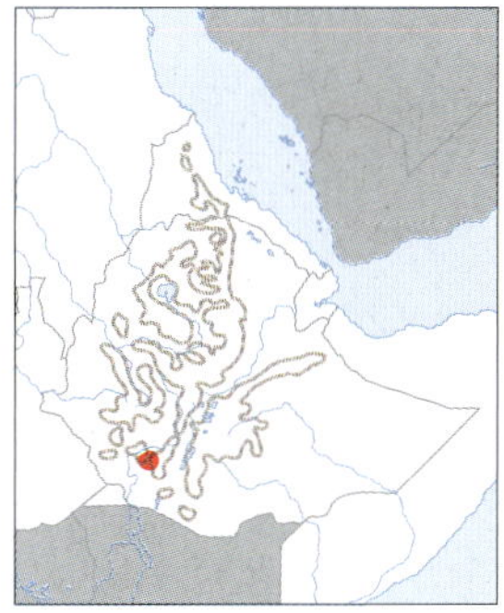

IDENTIFICATION: A medium-sized flat-backed toad of south-west Ethiopia. The back is broad and flat, and varies from rough, with numerous spiny warts, to smooth with flat warts. Tympanum fairly large, half size of eye in males, maybe larger in females. The parotid glands are long and flat with spiny warts. Males reach a maximum size of about 6cm, females 7cm. Dorsal colour very variable; from light beige to brown, rufous or almost black, sometimes with paired, low-contrast darker dorsal blotches. A light narrow vertebral line is sometimes present, and breeding males may turn bright yellow. Underside white to grey. Breeding males have a black throat. Taxonomic Notes: This species and the next (*Sclerophrys pusilla*) were long regarded as conspecific. A recent attempt to separate the two has caused confusion; no specimens from South Sudan were examined (probably *S. pusilla*), and those from the central Blue Nile drainage in Ethiopia were tentatively assigned to *S. pusilla*.

HABITAT AND DISTRIBUTION: Rainforest and moist savanna. The solitary record from our area is in south-west Ethiopia, near Lote, at 2,000m, north of Jinka in mid-altitude savanna. Elsewhere, widespread across West Africa, from Sierra Leone to Cameroon, possibly further east.

CONSERVATION STATUS: Least Concern.

NATURAL HISTORY: Unknown in our area, but elsewhere terrestrial and nocturnal. The advertisement call is a short deep, repetitive seesawing *oarr aarr oarr aarr*. Males call at night in the rainy season, often from a concealed position at the edge of temporary pools in depressions. In some areas, there is rapid breeding at the start of the rains, in others reproductive behaviour is more prolonged, even in the dry season if water is available. The usual long string of up to 8,000 eggs is produced. Aestivates in the dry season.

Sclerophrys maculata, Yambio.

SOUTHERN FLAT-BACKED TOAD

Sclerophrys pusilla
(Mertens, 1937)

IDENTIFICATION: A medium-sized flat-backed toad known from South Sudan and possibly Ethiopia. Back broad and flat, with distinct often large warts with small black spines; seen from the side it looks quite spiny. Tympanum large, 0.5–0.9 times the size of eye and very close to it or almost touching; a useful field character. Parotid glands more than twice as long as wide, often deeply pitted. Males reach a maximum size of about 6.5cm, females 8cm. Dorsal colour usually some shade of brown, from light yellowish-brown to warm brown or almost black; uniform or with paired blotches, which may be regular and well defined, or vague, irregular and hardly contrasting. Yellow vertebral line often present. Underside dirty white with dark flecking. Breeding males have a black throat, and may turn yellow when breeding. Taxonomic Notes: See the previous species.

HABITAT AND DISTRIBUTION: Forest margins, moist and dry savanna, often near water sources, at altitudes between 400 and 1,200m to sea level elsewhere. Specimens from South Sudan are assigned to this species; and it is widespread across the country south of 7°N. Some Ethiopian specimens from Debra Libanos, the Muger River and the Blue Nile gorge west of the Shafartak Bridge tentatively assigned to this species. Elsewhere; west to the Central African Republic and Cameroon, and south to northern South Africa.

CONSERVATION STATUS: Least Concern.

NATURAL HISTORY: Terrestrial. Advertisement call a low-frequency, drawn-out *quaaaak*. Males call at night in rainy season, often from a concealed position at edge of temporary pools in depressions. Explosive breeding at the start of the rains, with hundreds of pairs breeding simultaneously in shallow water. The males may fight over the females. The usual long string of eggs is produced. Aestivates in the dry season. In Zambia, a breeding aggregation was attacked by crabs, which tore apart the females while the males were still clasping them.

Sclerophrys pusilla, Congo.

PENTON'S TOAD

Sclerophrys pentoni
(Anderson, 1893)

IDENTIFICATION: A fairly large, squat toad with a very blunt head and many red warts. The back and sides have large, rounded, reddish warts, often spiny. Tympanum large, half the size of the eye or more. Parotid glands usually large and protruding, can look almost circular, width more than half the length (but in some males are much more indistinct). Adult males 5–7cm, females 6–9.5cm. Dorsal colour variable; fawn, shades of yellow, brown or grey, but the prominent rufous and yellow warts give the impression of a rufous animal. Underside whitish or cream. Breeding males have dark or black throats.

HABITAT AND DISTRIBUTION: Moist to dry savanna, semi-desert and near-desert, often in agricultural areas, between sea level and 1,100m. Occurs west from Eritrea across Sudan and Chad to Senegal and Mauritania. However, despite its huge range and relative abundance in some areas, there are few records from our region. Three records in Eritrea (Ghinda, Ailet, Barentu); the type specimens were from Suakin, in Sudan, and there are a few records from that area just north of the Eritrean border. It is also known from Zalingai, in Central Darfur, Neem and En Nahud, Northern Kordofan, in Sudan, and northernmost South Sudan; populations probably interconnected.

CONSERVATION STATUS: Least Concern.

NATURAL HISTORY: Terrestrial, largely nocturnal, but sometimes active by day, especially if it is overcast. Advertisement call a loud, low-frequency snore, lasting about half a second. May be present in huge numbers, breeding explosively in temporary pools in the rainy season, particularly those without marginal vegetation. Males call from the edges of such pools. A female was found to contain 2,600 eggs. A Ghana specimen was regurgitated by a West African Carpet Viper *Echis ocellatus*. Aestivates in the dry season. Often in very arid areas.

Sclerophrys pentoni, Neem.

COMMON AFRICAN TOAD, EGYPTIAN SQUARE-MARKED TOAD

Sclerophrys regularis
(Reuss, 1834)

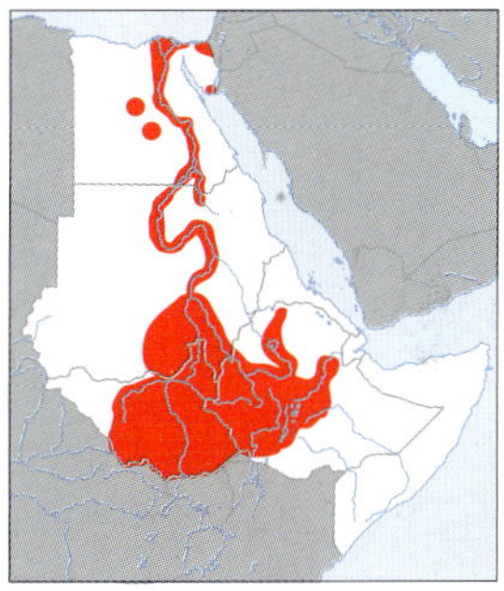

IDENTIFICATION: A big, usually brown toad, often common, and widespread in our area. The back is broad and flat, with distinct often large warts with small black spines. Tympanum close to eye, large, roughly the size of the eye (or even larger in females). The parotid glands are prominent, more than twice as long as wide, sometimes kidney-shaped, often deeply pitted. Males reach a maximum size of about 9cm, females 13cm. Dorsal colour very variable, usually some shade of brown, from tan to warm brown, sometimes rufous or grey, olive, almost black, or yellow and black; with both paired and random brown blotches, which may be regular and well defined, or vague and irregular, large or small. Yellow vertebral line often present. Some Egyptian males are distinctly spotted white. Underside dirty white; breeding males have a dark throat.

HABITAT AND DISTRIBUTION: Occupies a wide range of habitats, from forest edges to moist and dry savanna, grassland, semi-desert and near-desert, agricultural areas, very tolerant of urbanisation (common in the heart of Cairo) and arid areas (although, oddly, not found in dry eastern Ethiopia or Somalia). From sea level to above 2,500m in highland Ethiopia. In our area, sporadically recorded in the Afar Triangle, widespread in highland and western Ethiopia, widely known from the south and the Nile basin in South Sudan, thence along the Nile and associated tributaries through Sudan to Egypt, to the Delta. Isolated populations occur in Egypt in Kharga and Bahariya oases, and north Sinai. Elsewhere; west to Senegal and Mauritania, and on the west-central African coast, to northern Angola and Zambia.

CONSERVATION STATUS: Least Concern.

Sclerophrys regularis, Uganda.

Sclerophrys regularis, Fayum.

NATURAL HISTORY: Terrestrial. Advertisement call a slow snore, with audible pulses; males call at night in the rainy season (or during most of the year in highland Ethiopia), at the edge of both permanent water and temporary pools. The usual long string of eggs is produced. Aestivates in the dry season.

STEINDACHNER'S TOAD

Sclerophrys steindachneri
(Pfeffer, 1893)

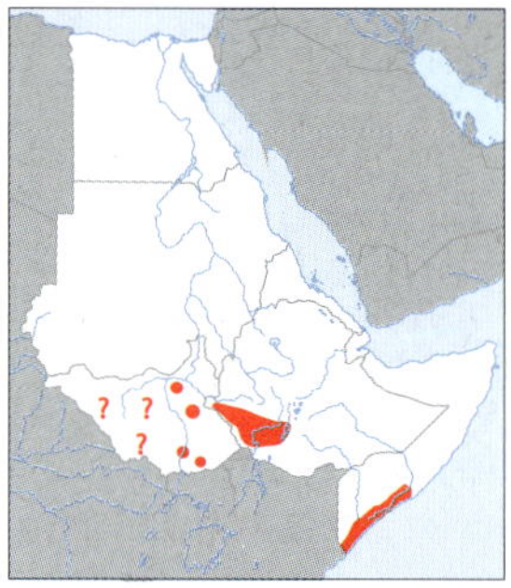

IDENTIFICATION: A distinctive small warty toad, with red patches on the limbs and their junctions. Back flat and rough, with numerous small rounded spiny warts. Tympanum slightly more than half the size of the eye. Parotid glands indistinct but more than twice as long as wide. Has a somewhat vacant look. Males reach just over 5cm, females a couple of millimetres larger. Dorsal colour shades of brown (tan, olive, medium, olive), a row of darker, paired markings may be clear or indistinct. Often a pale vertebral stripe is present, and there is usually a pale bar between the eyes. Red patches on the limbs and limb/body junctions. Underside granular and off white, breeding males have a yellow throat.

HABITAT AND DISTRIBUTION: Woodland, moist and dry savanna, often near permanent water, from sea level to 1,400m. In our area in coastal Somalia (south of 03°N), south-west Ethiopia (Lake Abaya, lower Omo River, Gambela area), sporadically recorded across South Sudan (localities include Juba, Ulang in Jonglei state, Bahr el Gebel, 50km south-east of Adok in Unity State, Talanga Forest); probably widespread in upper Nile basin, but records lacking. Elsewhere, west to Nigeria, south to north-east Tanzania.

CONSERVATION STATUS: Least Concern.

NATURAL HISTORY: Nocturnal and terrestrial. Advertisement call a series of low-pitched, metallic knocks, 3–4 times per second. Males emerge in rainy season, gather in large numbers and call from concealed sites at edges of rivers, marshes and temporary pools, or on floating vegetation.

Sclerophrys steindachneri, near Gambela.

DESERT TOAD

Sclerophrys xeros
(Tandy, Tandy, Keith & Duff-MacKay, 1976)

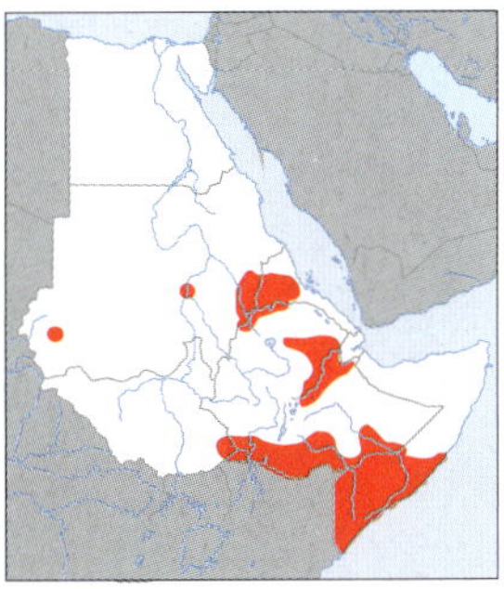

IDENTIFICATION: A big toad of dry country, with red thighs and low-contrast back markings, and an owl-like call. The back is broad and flat, with distinct, often large and somewhat flattened warts with small black spines. Tympanum close to eye, sometimes oval, large, roughly the size of the eye. Parotid glands prominent, more than twice as long as wide, often deeply pitted, and they may broaden considerably at either end, be hourglass-shaped or virtually circular; this is a useful field character. Males reach a maximum size of about 8cm, females almost 10cm. Dorsal colour very variable; usually some shade of brown, tan, grey or olive; dark or light, sometimes with a series of darker paired brown or olive blotches, which are usually hardly contrasting. A paler vertebral line is often present. The backs of the thighs, and sometimes the forelimbs and posterior flanks, are distinctly marked with red blotches. Underside granular and white; breeding males have a dark throat.

HABITAT AND DISTRIBUTION: Dry savanna, semi-desert and edges of desert. From sea level, to above 1,500m in Ethiopia and 2,000m in Eritrea, south-west of Asmara. In our area, known from Somalia south of 06°N, south-eastern and southern Ethiopia and the Afar Triangle, north-east of Awash National Park. Occurs in central and western Eritrea, extreme south-east South Sudan (Kapoeta, and the border with north-west Kenya). Probably occurs in a broad belt across central Sudan, but few records save a cluster in Kassala State (Aroma, New Halfa, Kassala town), 30km north of Khartoum, and Jebel Marra, in Darfur. Elsewhere, west to Mauritania, and north into mountains of Sahara in Algeria and Chad.

CONSERVATION STATUS: Least Concern.

Sclerophrys xeros, Hamashkoraib.

Sclerophrys xeros, Metahara.

NATURAL HISTORY: Terrestrial and nocturnal. Advertisement call a low-pitched hoot, like the call of a large owl. In semi-desert, these toads emerge in huge numbers following the first big storms of the rainy season, and explosive breeding occurs. Males call at edge of large and small temporary pools, both night and day; fertilised eggs are laid and develop rapidly. Tadpole mortality in Ethiopia was observed to be high, as smaller pools dry out. Adults aestivate in the dry season, underground or below large rocks.

FOREST TOADS *MERTENSOPHRYNE*

An African genus of about 15 species of small toad. The solitary member in our area, Lugh Toad *Mertensophryne lughensis*, was originally placed in genus *Poyntonophrynus*, pygmy toads, but a recent phylogenetic analysis indicates that it clusters in *Mertensophryne*. However, other works retain the species in *Poyntonophrynus*; all other members of *Mertensophryne* lack a tympanum.

LUGH PYGMY TOAD

Mertensophryne lughensis
(Loveridge, 1932)

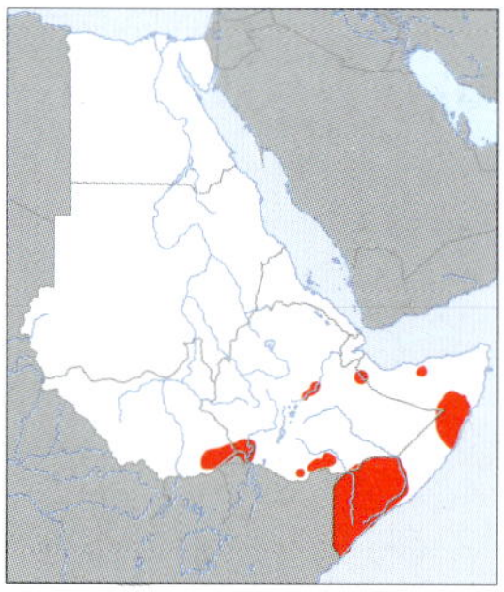

Mertensophryne lughensis, North of Erigavo.

Mertensophryne lughensis, North of Erigavo.

IDENTIFICATION: A distinctive small, flat toad of the Horn of Africa. Back flat and rough, with numerous small, usually flattened warts. Tympanum small, about half size of eye. Parotid glands large and elongate, but often near-invisible. Males reach about 3.5cm, females 4.5cm or slightly larger. Dorsal colour light or dark brown, sometimes pinkish-brown, mottled, and warts may be tipped rufous or pink. Rarely has any clear paired dorsal blotches, but some have a pair of dark markings on head. Breeding males in eastern Kenya turn vivid pink, with blue upper limbs, but this isn't noted elsewhere. Underside pale with black speckling, breeding males have a yellow throat. Taxonomic Notes: Usually placed in *Poyntonophrynus*, most members of which, like this species, have tympani, and inhabit dry areas.

HABITAT AND DISTRIBUTION: Dry savanna and semi-desert, from sea level to 1,400m (but found at higher altitudes and more well-watered areas at Laikipia, in Kenya). Two records in northern Somalia (north of Erigavo and Borama). Widespread in eastern Somalia, south of Gardo, southern Ethiopia (also Awash National Park), and the Ilemi Triangle in Eastern Equatoria, south-east South Sudan. Elsewhere, northern and eastern Kenya. Records very sporadic, and probably much more widespread than indicated but overlooked.

CONSERVATION STATUS: Least Concern.

NATURAL HISTORY: Nocturnal and terrestrial. Advertisement call a series of short, goat-like bleats. Aestivates in the dry season, buried in soil or sheltering under rocks. Males emerge in the rainy season, immediately gather in large numbers and call among sparse vegetation at edges of temporary pools. In northern Somalia specimens were found in the rainy season under rocks by day; but could not be found in the same places in the dry season. At Kokuro, South Sudan, several were collected emerging from burrows after dusk, and it is speculated that they excavated the burrows (which were up to 1m long, and 5cm deep) themselves. Freshly metamorphosed specimens were found in August at Borama, northern Somalia.

ETHIOPIAN ALPINE TOADS *ALTIPHRYNOIDES*

A genus of two species of toad, without tympani, endemic to the mountains and hills of southern Ethiopia, largely west of the Rift Valley.

KEY TO THE NORTH-EAST AFRICAN MEMBERS OF THE GENUS *ALTIPHRYNOIDES*

1a: Small, <32mm long, only in Bale Mountains — *Altiphrynoides malcolmi*, Malcolm's Alpine Toad

1b: Large, up to 62mm long, on Mt Guge and highlands east of Lake Awassa, including Bale Mountains — *Altiphrynoides osgoodi*, Osgood's Alpine Toad

MALCOLM'S ALPINE TOAD

Altiphrynoides malcolmi
(Grandison, 1978)

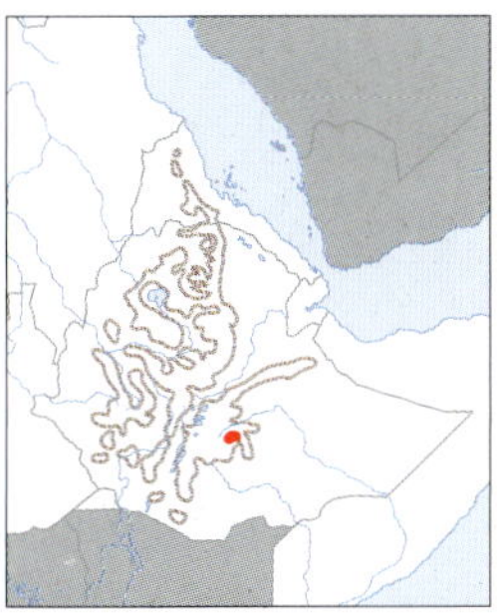

Altiphrynoides malcolmi, Bale Mountains.

IDENTIFICATION: A small brown, rufous or green striped frog of Bale Mountains in Ethiopia. Back smooth. No tympanum. Parotid glands elongate but indistinct, and become a series of elongate, narrow warts along the back. Males reach just over 2cm, females 3cm. Dorsal colour pink, rufous, grey, brown, yellowish or green, with up to five longitudinal stripes, which may be pale green, white, yellow or red, sometimes with dark edges. Underside white or cream, speckled or suffused with grey, rufous or purple.

HABITAT AND DISTRIBUTION: Montane moorland, low heather scrub, giant heather forest, *Hagenia*, *Podocarpus* and high-altitude forest, between 2,500 and 4,000m. Known only from Bale Mountains in south-east Ethiopia.

CONSERVATION STATUS: Least Concern, but recent research indicates that it is substantially declining, probably due to human encroachment into and degradation of its habitat.

Altiphrynoides malcolmi, Bale Mountains.

NATURAL HISTORY: Nocturnal and terrestrial. No vocal sac but males can produce a quiet creaking advertisement call, audible only 2–3m. They shelter under rocks, in vegetation and under fallen timber, emerging after dark. Huge numbers have been found under large rotting logs. A known breeding site was on the grassy bank of a small stream, at 3,200m, where males were found wandering in search of females. Unusually, these frogs have internal fertilisation, and females then spawn communally, creating a thick gelatinous blob in which the tadpoles hatch.

OSGOOD'S ALPINE TOAD

Altiphrynoides osgoodi
(Loveridge, 1932)

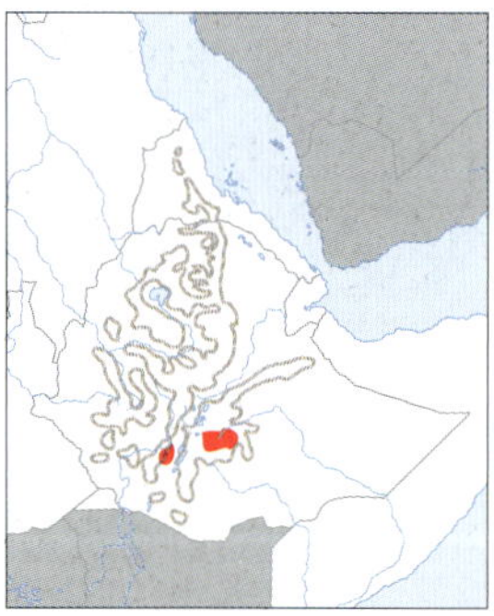

IDENTIFICATION: A medium-sized, often warty frog, of the highlands of southern Ethiopia. Back can be smooth, have elongate knobbed ridges or many hemispherical warts. No tympanum. Parotid glands large, flattened and thus indistinct. Males reach just over 5cm, females 6cm. Dorsum very variable; yellow, tan, green, brown with black blotching or olive, sometimes with a fine pale vertebral line. Flanks often curiously granular. Underside dirty white or yellow, speckled or suffused with grey, brown or purple blotches. Taxonomic Notes: placed for a while in genus *Spinophrynoides*, but synonymised with *Altiphrynoides* in 2006.

HABITAT AND DISTRIBUTION: Ethiopian endemic. High grassland, montane moorland, *Hagenia*, *Podocarpus* and high-altitude forest, between 1,950 and 4,000m. Known from the high country east of Lake Awassa, from Kofele east to the Bale Mountains; also Mt Guge, west of Lake Abaya.

CONSERVATION STATUS: Critically Endangered. Regularly seen in the Bale area in the 1980s and 1990s, it has hardly ever been seen since, its decline probably due to human encroachment into and degradation of its habitat.

Altiphrynoides osgoodi, Bale Mountains.

Altiphrynoides osgoodi, Bale Mountains.

NATURAL HISTORY: Nocturnal and terrestrial. Although they have no vocal sac, males do call, producing a succession of low creaks, audible only over 6–8m. They shelter under rocks, in vegetation and under fallen timber, emerging after dark. One was found under a rock on the desolate Sanetti Plateau, Bale Mountains, at almost 4,000m. A breeding site was found in a grassy glade surrounded by *Hypericum* woodland at 3,200m; the frogs congregated around a small pool and a long thin string of pigmented eggs was deposited in the pool.

EURASIAN TOADS *BUFOTES*

A Palearctic genus of about 16 species of small to medium-sized toad, often with a dorsal pattern of unpaired, green blotches, in Europe, Asia and parts of North Africa. Two species in our area. Taxonomy complicated; many were regarded as colour variants of European Green Toad *B. viridis*, and can be distinguished (if at all) only by mix of locality and molecular analysis.

KEY TO THE NORTH-EAST AFRICAN MEMBERS OF THE GENUS *BUFOTES*

1a: In our area, known only in northern Sinai; iris usually dark (with strong vermiculations), blackish-yellow or green; mid-dorsal stripe rare; often with lateral red spots *Bufotes sitibundus*, Variable Green Toad

1b: In our area, widespread in Egypt; iris usually very pale, greenish-yellow; quite often with mid-dorsal stripe; usually no lateral spots *Bufotes boulengeri*, African Green Toad

VARIABLE GREEN TOAD

Bufotes sitibundus
(Pallas, 1771)

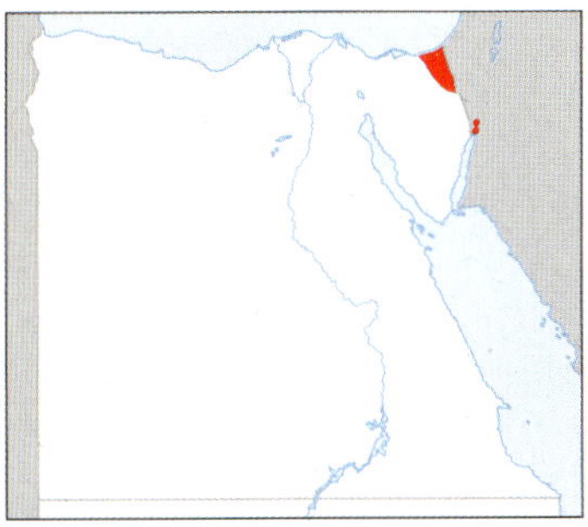

IDENTIFICATION: A medium-sized, usually brown-cream toad with larger greenish blotches, known from northern Sinai. Head large, back broad and flat. Skin very rough with small to medium-sized tubercles. Tympanum fairly large, about half size of the eye. Parotid glands prominent, large, parallel and elongate, about twice as long as wide, deeply pitted. Maximum size 10cm, average about 8–9cm, males smaller. Colour and pattern variable; dorsally whitish to greyish, with well-defined large green spots or patches, often combined with red or orange dots on upperparts; vertebral stripe rare. Iris usually dark (strongly vermiculated), blackish-yellow or green. Underside creamy-white, often with small darker spots. Taxonomic Notes: Long known as *Bufo viridis*, Green Toad. The name *Bufotes variabilis*, to which Middle Eastern populations were traditionally referred, is a junior synonym of *B. sitibundus*.

HABITAT AND DISTRIBUTION: Desert to semi-desert in our area, from sea level to about 400m. Found at suitable mesic microhabitats in dunes and in small waterbodies, ponds and freshwater irrigation creeks. Known only from north-east Sinai in Egypt, in neighbourhood of Al Arish and Rafah, then Sheikh Zowayid and El Magdhaba south of Al Arish and south from Wadi el Gedeirat near the north-east border. Elsewhere; east to Iran, north to south-west Russia and Kazakhstan.

Bufotes sitibundus, Israel/Palestine.

CONSERVATION STATUS: Least Concern, often locally common, but possibly at risk from urbanisation and loss of wetlands for breeding.

NATURAL HISTORY: Terrestrial. Mostly nocturnal. Advertisement call a soft, bird-like liquid trill, increasing in volume and lasting up to ten seconds. Found in water only during the short mating season. During dry periods they hide in deep, cool and mesic burrows, holes or underneath objects such as logs and rocks. They breed in a variety of waterbodies, mostly in anthropogenic habitats. Active year-round, breeds in spring. Females lay egg-strands up to 4m long, containing up to 15,000 eggs. Diet various invertebrates and molluscs.

AFRICAN GREEN TOAD

Bufotes boulengeri
(Lataste, 1879)

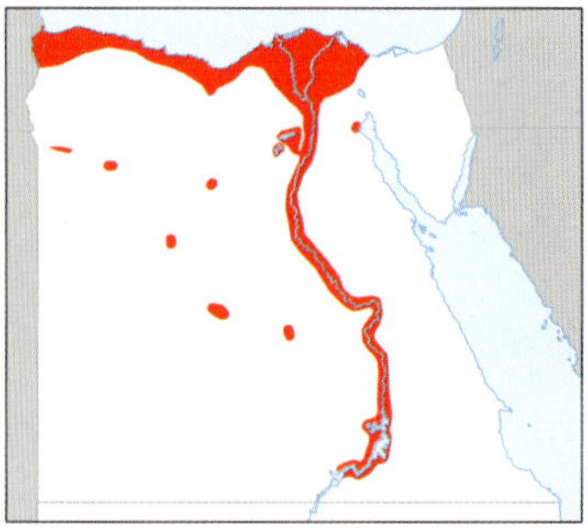

IDENTIFICATION: A medium-sized, usually brown-cream toad with large greenish blotches, widespread in Egypt but not Sinai. Head large, back flat and broad. Skin very rough with small to medium-sized tubercles. Tympanum fairly large, about 45 per cent size of the eye. Parotid glands large, parallel and elongate, almost twice as long as wide, deeply pitted. Males smaller and slenderer than females. Maximum size 10cm, average about 8cm, pygmy-sized populations reported in Egypt and Algeria at the edge of Sahara. Colour and pattern very variable; dorsally reddish-brown, olive-green or creamy with superimposed green spots, sometimes fused into large patches; a pale vertebral stripe (or a hint of it) is usually present. Iris pale, with fewer vermiculations than previous species, and greenish-yellow. Underside whitish or yellowish, sometimes with small green or dark spots. Taxonomic Notes: Two subspecies recognised, in our area nominate *B. b. boulengeri*.

Bufotes boulengeri, Tunisia.

HABITAT AND DISTRIBUTION: Semi-desert with scattered mesic microhabitats, wadis, oases, cultivation on desert margins, from sea level to about 2,700m (in Morocco). Widespread across most of Egypt west of Suez Canal, in the Delta, the Nile Valley south to Abu Simbel, on Mediterranean coastal plains and at most oases in Western Desert. Elsewhere; along coast west to Morocco, south in Sahara to Haggar Mountains in Algeria.

CONSERVATION STATUS: Least Concern.

NATURAL HISTORY: Active year-round. Terrestrial and nocturnal, but males may call also during daytime. Advertisement call a soft, bird-like trill. Tolerates very dry conditions, but only survive at places with permanent or at least seasonal water. Found in wells, cisterns and brackish marshes. Breeding occurs in spring, from February and March to May. Mating can occur on land or in water. Embryonic development lasts 3–5 days and larvae complete their development within one or two months. Newly metamorphosed toads measure about 15–20mm. Adults feed on invertebrates, mainly on beetles.

FAMILY MICROHYLIDAE NARROW-MOUTHED FROGS

A family of relatively small but diverse frogs, with nearly 700 species, found on every continent except Europe. However, revisionary studies have removed several African species from the group, and only a single genus now occurs in our area, rubber frogs, *Phrynomantis*.

RUBBER FROGS *PHRYNOMANTIS*

An African genus of attractive red-and-black, medium-sized frogs, none larger than 7cm in our area. All have small heads, fairly blunt snouts and small dark eyes; often the tympanum is hard to see or absent. They are sometimes called 'snake-necked frogs', as the neck is unusually long, enabling them to move their heads. Inhabitants of savanna and semi-desert; their specialised diet consists of ants and termites. They walk slowly and climb well. They often hide in termitaria, burrows or tree holes and become active after heavy rain. The male call is a long, far-reaching, melodious electronic-sounding trill. If threatened, they may make small jumps, or arch the body upward, while hiding their head between their limbs; further stress causes the red or orange to fade. All secrete strong skin toxins. If these toxins enter a cut, or touch a mucous membrane, symptoms such as headache, painful swellings, breathing difficulties, nausea and an elevated heart rate may appear, although they usually disappear within a few hours. Six species are known, three in our area. Older texts may use the name *Phrynomerus* for this genus.

KEY TO THE NORTH-EAST AFRICAN MEMBERS OF THE GENUS *PHRYNOMANTIS*

1a: Almost the entire dorsal surface red or orange, sometimes with a fine black vertebral stripe, in our area confined to western Ethiopia and South Sudan — *Phrynomantis microps*, West African Rubber Frog

1b: Dorsal surface red and/or pink, striped or blotched, with extensive black areas, in our area confined to southern Ethiopia and Somalia — 2

2a: Dorsal surface usually with two unbroken red, orange or pink dorsolateral stripes, known only from extreme southern Somalia, tympanum present but hard to see — *Phrynomantis bifasciatus*, Red-banded Rubber Frog

2b: Dorsal surface usually with broad, elongate dorsolateral blotches which are pink in juveniles, pink or orange in adults, interspersed with black; known from southern Ethiopia and Somalia, tympanum absent — *Phrynomantis somalicus*, Somali Rubber Frog

WEST AFRICAN RUBBER FROG

Phrynomantis microps
Peters, 1875

IDENTIFICATION: A small, flattish, red- or orange-backed frog of South Sudan or lowlands of western Ethiopia. Body pear-shaped, back smooth. Tympanum tiny and hard to see. The limbs are thin; the digit tips enlarged to form discs. Males reach a maximum of about 4.5cm, females 6cm. Dorsal colour usually vivid pink, red or orange, sometimes with a fine black vertebral; limbs banded or spotted red and black. Juveniles may be more grey than red. Underside black or dark grey with white spots, and males have a black throat, brown in females.

HABITAT AND DISTRIBUTION: Moist and dry savanna, between sea level and around 600m in Ethiopia and South Sudan; higher elsewhere. In our area, there are only a few confirmed records; Gambela in Ethiopia, Rumbek (in Lakes State) and between Juba and Torit in South Sudan. Records from northern Uganda and Garamba National Park indicate that it is probably widespread in southern South Sudan, and a record from Zakouma National Park in Chad suggests it might occur in extreme southwest South Sudan. Elsewhere, west across the savanna to Senegal.

CONSERVATION STATUS: Least Concern.

NATURAL HISTORY: Nocturnal and largely terrestrial, but will climb trees, hiding in holes and leaf axils. Advertisement call a long (1–3 seconds) loud,

Phrynomantis microps, Gambela.

Phrynomantis microps, in amplexus, Benin.

melodious trill. They shelter in holes, in vegetation, under rocks, in burrows and tree holes, and emerge after dark. Mating follows large rainstorms. Males gather around temporary pools, both large and small. They call from within cavities, but also on the ground and on grass tufts. Eggs are laid in discrete batches, and breeding continues throughout the wet season. Tadpoles hatch within a day and are gregarious.

SOMALI RUBBER FROG

Phrynomantis somalicus
(Scortecci, 1941)

IDENTIFICATION: A rare, small, flattish, red, pink and black frog of Ethiopia and the southern third of Somalia. Body pear-shaped (rectangular when young), the back is smooth. Tympanum absent. The limbs are thin; the digit tips enlarged to form discs. Males reach a maximum of about 3.5cm, females 4.5cm, possibly more. Dorsal colour of adults black, with red, pink or orange markings thus: an interrupted V-shape on the head and neck, two paired but non-matching irregular blotches at mid-body and another above the vent. The legs are banded or spotted red/pink/orange and black. Juveniles have a fine red vertebral line. Underside grey with fine white mottling.

HABITAT AND DISTRIBUTION: Known only from savanna habitats; the Ethiopian specimen was at 1,200m altitude, all others below 300m. Only seven specimens are known. The sole Ethiopian example was from the Genale River, north of Negele, the Somali ones from Dinsor, Mogadishu and the vicinity of Afgoi, in the south; the only other known specimen was a juvenile from Kulal Camp, Tsavo East National

Phrynomantis somalicus, preserved, Afgoi.

Phrynomantis somalicus, juvenile, Genale River.

Park in Kenya. Probably occurs in the intervening areas, but the lack of records indicate a very secretive species.

CONSERVATION STATUS: Least Concern.

NATURAL HISTORY: Poorly known; no adult has been knowingly photographed alive. Presumably nocturnal, partially arboreal and emerges following heavy rain, like other members of the genus. One Somali specimen entered a house during torrential rain; two were calling at night in a flooded grassy depression, emitting a 'high, whining call'. The Ethiopian example was in a hole at the base of a tree, the Kenyan animal under a log.

Phrynomantis bifasciatus, Kenya.

RED-BANDED RUBBER FROG

Phrynomantis bifasciatus
(Smith, 1847)

IDENTIFICATION: A red-striped black frog, in our area only in southern Somalia. The body is pear-shaped, the back is smooth. Tympanum tiny and hard to see. Limbs thin; digit tips enlarged to form discs. Males reach a maximum of about 5.3cm, females 6.5cm. Dorsal colour black, with two broad dorsolateral vivid orange or red stripes and a red rump; the limbs banded or spotted red and black. Sometimes the dorsolateral stripes split posteriorly and the inner stripe joins with the red rump; this often coincides with diffuse broad grey stripes in the centre of the back. Underside black or dark grey with extensive white spots; males have a black throat, grey in females.

HABITAT AND DISTRIBUTION: In moist and dry savanna and woodland, near sea level in Somalia, higher elsewhere. A single record in our area, on the Somali coast just north of the Kenya border at Ras Kamboni. Elsewhere, south into Kenya, to northern South Africa, south-west to Namibia and Angola.

CONSERVATION STATUS: Least Concern.

NATURAL HISTORY: Nocturnal and largely terrestrial, but will climb trees, hiding in holes and leaf axils. Advertisement call a long (1–7 seconds) loud, melodious, electronic-sounding trill, audible over some distance. They shelter in holes, in vegetation, under rocks, in burrows and tree holes, and emerge after dark. Mating follows heavy rain. Males gather around temporary pools, both large and small, even around water-filled animal footprints; calling from cavities, but also in the open and on grass tufts. Eggs laid in discrete batches, and breeding continues throughout the rainy season. Tadpoles hatch within four days and are gregarious.

FAMILY HEMISOTIDAE PIGLET FROGS OR SNOUT-BURROWERS

A family erected for a single genus of snout-burrowing frogs, *Hemisus*. Originally called Shovel-nosed frogs, then Snout-burrowers, the evocative new name Piglet frogs has now been proposed. Recent molecular work indicates they are a true monophyletic family that diversified >60 million years ago, and are members of a clade of frogs confined to Africa and Madagascar, the Afrobatrachia.

PIGLET FROGS OR SNOUT-BURROWERS *HEMISUS*

An unusual sub-Saharan African genus of small to medium-sized, egg-shaped frogs with tiny heads and small bead-like eyes; nine species are known, three occur in our area. They use their hard pointed snouts and short, powerful limbs to burrow rapidly into soil, snout first; no other African frog does this. The diet appears to be only ants and termites. They have very smooth skin and tiny eyes, with a distinct transverse skin fold behind the eye. Their eggs are laid in burrows and the mother tends them, protecting them and keeping them moist. When the tadpoles have hatched, she may create a slide that the tadpoles can move along, or carry them to the water on her back and stretched-out legs.

KEY TO THE NORTH-EAST AFRICAN MEMBERS OF THE GENUS *HEMISUS*

Note: some individuals are very hard or even impossible to identify in the hand to species level. Even experts struggle (particularly with *Hemisus marmoratus* and *H. guineensis*), although locality may help.

1a: Dorsal surface usually looks light, or tan, with dark markings or spots, not yellow, in our area not confined to South Sudan or Ethiopia, usually in savanna	*Hemisus marmoratus*, Marbled Piglet Frog
1b: Dorsal surface looks dark, often with yellow markings or spots, in our area confined to either South Sudan or Ethiopia, often in woodland	2
2a: Only in highland Ethiopia	*Hemisus microscaphus*, Ethiopian Piglet Frog
2b: In our area, not in Ethiopia	*Hemisus guineensis*, Guinea Piglet Frog

MARBLED PIGLET FROG

Hemisus marmoratus
Peters, 1854

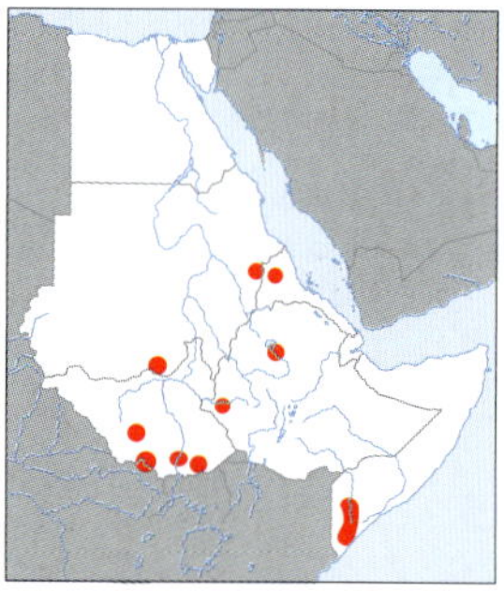

IDENTIFICATION: A small globular frog with a tiny head, which is sharply pointed, and a hard snout. The body is broad and squat, the back may be smooth, or covered with small warts. Tympanum invisible. The limbs are short and powerful. Males reach a maximum of about 3.5cm, females 5.5cm. Dorsal colour usually a coarsely mottled mixture of tan, brown, grey, olive and sometimes black on the flanks. A thin vertebral stripe can be present. Underside smooth and white; males have a dark purple to black throat.

HABITAT AND DISTRIBUTION: Woodland, moist and dry savanna, from sea level up to c.600m (although also known from Bahar Dar, in Ethiopia, at 1,850m). Records sporadic in our area; known from southern Somalia, south of 03°N, the Gambela area and Bahar Dar in Ethiopia, Agordat in Eritrea, Kassala and Southern Kordofan in Sudan, and Torit, Juba, Mongalla and Southern National Park in South Sudan; also known from just across the border in Garamba National Park in DR Congo. It probably occurs much more widely in lowlands of the South Sudan Nile basin, but is unrecorded there. Elsewhere, west to Senegal, south to northern South Africa.

CONSERVATION STATUS: Least Concern.

NATURAL HISTORY: Nocturnal and terrestrial. Advertisement call a long, high-pitched, cricket-like

Hemisus marmoratus, Gambela.

buzz, lasting several seconds. They live close to water, in burrows or cavities they have dug under rocks and fallen logs, using the hardened tip of their snout as a shovel; they may dig down half a metre or more. It is not known if they forage underground, but they are only found above ground at times of high humidity, associated with heavy rain. Males call from below ground or on the surface. The male grasps the female around the hind legs (an unusual position, called inguinal amplexus). The parents dig a pit near the water and the eggs are laid there. Females guard the eggs, and after hatching the tadpoles adhere to the mother; when rain is falling she carries the tadpoles on her back and legs to the water, or creates a slick slide that the tadpoles move down.

GUINEA PIGLET FROG

Hemisus guineensis
Cope, 1865

IDENTIFICATION: A small globular frog with a tiny head, which is sharply pointed, with a hard snout. The body is broad and squat, the back may be smooth, or covered to a greater or lesser extent with small hemispherical warts. Tympanum invisible. Limbs short and powerful. Males reach a maximum of about 3.7cm, females 5.5cm. Dorsal colour variable, either a finely mottled mixture of tan, brown and grey, or uniform brown or grey with white, yellow or orange spots. A thin yellow vertebral stripe may be present. Underside smooth and pinkish or white; males have a dark purple to black throat.

HABITAT AND DISTRIBUTION: Wooded savanna and forest, from sea level up to around 600m or more. Confusion between this species and *H. marmoratus* means its distribution in our area is poorly understood (there are no formal records) but it probably occurs in parts of southern South Sudan, for example Western Equatoria, as it is recorded from Garamba National Park in DR Congo. Elsewhere, west to Senegal, in forest, south to Botswana and South Africa.

CONSERVATION STATUS: Least Concern.

NATURAL HISTORY: Very similar to the previous species; i.e. nocturnal and terrestrial, burrows with the snout, emerges when rain is falling, females guard the eggs. Advertisement call a long, high-pitched, trill lasting 2+ seconds.

Hemisus guineensis, Tanzania.

ETHIOPIAN PIGLET FROG

Hemisus microscaphus
Laurent, 1972

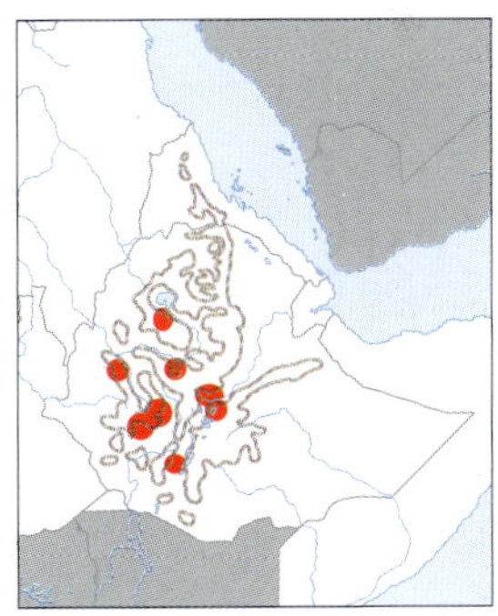

IDENTIFICATION: A small globular frog with a tiny head and pointed snout, endemic to the western highlands of Ethiopia. The body is broader than most *Hemisus*, and squat, the back is smooth. Tympanum invisible. Limbs short and powerful. Males reach a maximum of about 3.5cm, females 4.5cm. Dorsal colour dark grey, brown or purplish-brown, mottled, sometimes finely spotted yellow, the size of the spots increasing on the flanks. The hard snout tip is pinkish-white. Underside smooth and dirty yellow, blotched purplish-brown; males have a dark purple-brown to black throat.

HABITAT AND DISTRIBUTION: High savanna, grassland and forest in high rainfall areas, at altitudes between 1,500 and 2,700m. Endemic to Ethiopia, and known from a handful of localities west of the Rift Valley; from Injibara, near Bahar Dar, south to around Bonga and Mizan Teferi in the south-west forest, west to Abiyu. Also known from Chencha in the highlands west of Lake Abaya. Probably much more widespread in the western highlands but unreported. The only specimen from an area with annual rainfall less than 800mm is the type, collected on the west shore of Lake Zwai in the Rift Valley.

CONSERVATION STATUS: Least Concern.

NATURAL HISTORY: Poorly known. Presumably similar to other *Hemisus*, i.e. nocturnal and terrestrial, burrows with the snout, emerges when rain is falling, females guard the eggs. The advertisement call is unknown. Specimens in Addis Ababa were found in burrows beneath logs at the edge of an artificial dam. In January, at Abiyu, west of Gore, 16 specimens were found at a depth of 15–30cm below the soil surface, in forest 20m from a stream. In October, advanced tadpoles were found in a reed-fringed pool in grassland near Injibara.

Hemisus microscaphus, Addis Ababa

FAMILY BREVICIPITIDAE RAIN FROGS

A family of five genera and c.35 species of unusual, small, blunt-snouted, globular frogs, originally in the family Microhylidae. They are members of an African/Malagasy clade of frogs, the Afrobatrachia, and sister to the Hemisotidae. Most occur in moist savanna, woodland and forest in south-east Africa (a couple are found in the deserts of the south-west corner of Africa). However, a single species, Bale Mountains Rain Frog *Balebreviceps hillmani*, is known from Ethiopia, north of arid northern Kenya, and recent research indicates it has been isolated from its relatives for about 43 million years.

BALE MOUNTAINS RAIN FROG *BALEBREVICEPS*

A monotypic genus; the single species is an Ethiopian endemic.

BALE MOUNTAINS RAIN FROG

Balebreviceps hillmani
(Largen & Drewes, 1989)

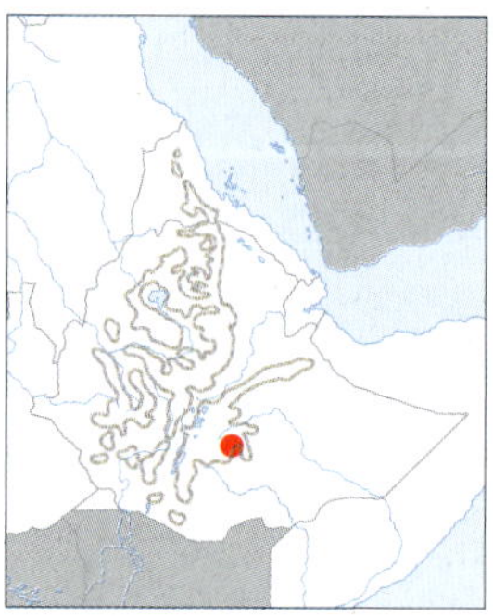

IDENTIFICATION: Unmistakable; a small globular black-and-orange frog, with a blunt snout, endemic to the Bale Mountains in Ethiopia. Body is broad, the back rough and pitted but without conspicuous warts. No tympanum. The limbs are short and stout; no spade-like metatarsal tubercle. Males reach a maximum of about 3.9cm, females 5.3cm. Dorsal colour rich brown, with two broad orange or tan dorsolateral stripes, which extend down behind the eye, and always fuse posteriorly; these stripes may expand towards the midline so the whole back looks yellow. Underside dirty yellow blotched purplish-brown.

HABITAT AND DISTRIBUTION: Giant heather and *Podocarpus*/*Hagenia* forest at altitudes between 2,800 and 3,200m. Endemic to south-east Ethiopia, where known only from Harenna Forest, on the southern slopes of the Bale Mountains.

CONSERVATION STATUS: Critically Endangered. Although its entire range lies within a national park that is nominally protected, many people live in the park; they harvest firewood and building material, and graze stock in its habitat. Recent research indicates that the type locality is being rapidly degraded. However, only a single road traverses the locality, and specimens and suitable habitat can be found away from the road at similar altitude.

NATURAL HISTORY: Poorly known and no advertisement call has been noted. Most specimens have been found under logs, rocks and live moss by day, so presumably it is nocturnal, despite the cold. The absence of the spade-like metatarsal tubercle possessed by its near relatives (rain frogs, *Breviceps*) may be an indicator of the softness and thickness of the substrate it lives in, or perhaps the lack of predators. When threatened, these frogs inflate the body, stand up, stiff-limbed, head lowered, and exude a thick, white sticky and possibly toxic fluid from the dorsal pores. Breeding details unknown, but mature females contain large eggs, suggesting that, like *Breviceps*, the larvae perhaps develop within the eggs and hatch as froglets, so no free-swimming tadpoles exist.

Balebreviceps hillmani, Bale Mountains.

FAMILY HYPEROLIIDAE CLIMBING FROGS AND KASSINAS

A very diverse family of 17 genera and about 220 species of small, delicate and attractive frogs that were recently split into two subfamilies: Hyperoliinae and Kassininae. One genus occurs on the Seychelles, one on Madagascar, the remaining 15 are mainland African; four occur in our area; *Kassina and Hyperolius* (in the Hyperoliinae) and *Kassina* and *Paracassina* (in the Kassininae). They are members of the Afrobatrachia.

RUNNING FROGS OR KASSINAS *KASSINA*

A sub-Saharan African genus of attractive spotted and striped, medium-sized frogs, none much larger than 5cm, with beautiful calls. They have huge eyes with vertical pupils, slender limbs and usually walk instead of jumping (although they can jump). Most are terrestrial, although some will climb. The taxonomy of the genus and closely associated genera has changed a lot recently. At present, there are about 15 species, four occur in our area.

KEY TO THE NORTH-EAST AFRICAN MEMBERS OF THE GENUS *KASSINA*

1a: Dorsal surface and flanks uniformly covered with 20+ relatively large, sub-circular dark spots, usually with fine light edging, yellow patches in groin, in forest — *Kassina maculosa*, Central African Running Frog

1b: Dorsal surface and flanks often with a few spots and/or stripes, but rarely finely edged, no yellow patches in groin, not in forest — 2

2a: Grey or golden-brown above with a dumbbell or X-shaped blotch between the shoulders, tips of digit visibly broad, in very dry country — *Kassina maculifer*, Spotted Running Frog

2b: Yellow or tan above, with dorsal spots or stripes or a mix of both, but no dumbbell or X-shapes, digit tips not notably broad, in a range of habitats — 3

3a: Dorsal surface sometimes immaculate or with a few dark spots, in lines or random — *Kassina somalica*, Somali Running Frog

3b: Dorsal surface never immaculate, usually with an unbroken or weakly interrupted vertebral stripe, often with two dorsolateral stripes, or a series of spots in a line — *Kassina senegalensis*, Senegal Running Frog

CENTRAL AFRICAN RUNNING FROG

Kassina maculosa
(Sternfeld, 1917)

IDENTIFICATION: A small slender frog, the back covered with sub-circular dark spots; in our area only in South Sudan. The body is slender, the head short and rounded, with big eyes and vertical pupils, and the back is smooth. Tympanum is about half the size of the eye or less, and may be hard to see. Limbs slender, toe tips slightly enlarged. Males and females reach a maximum of about 3.8cm Dorsal colour brown, grey or olive, the back covered with dark, light-edged subcircular spots (rarely fused). Underside greyish-white, but limb-body junctions have yellow patches.

HABITAT AND DISTRIBUTION: Various types of for est and savanna. In our area known only from moist savanna between Juba and Torit in South Sudan, at an altitude of c.600m, but also from Garamba National Park, in DR Congo, so probably in Western Equatoria. Elsewhere, west to Cameroon, between sea level and 2,000m.

Kassina maculosa, Cameroon.

CONSERVATION STATUS: Least Concern.

NATURAL HISTORY: Poorly known, but similar to other members of the genus, i.e. nocturnal and terrestrial; emerges following heavy rains and breeds in stagnant pools. Advertisement call a brief rising whistle. Males may climb into waterside vegetation to call, or from floating vegetation.

SPOTTED RUNNING FROG, SPOTTED KASSINA

Kassina maculifer
(Ahl, 1924)

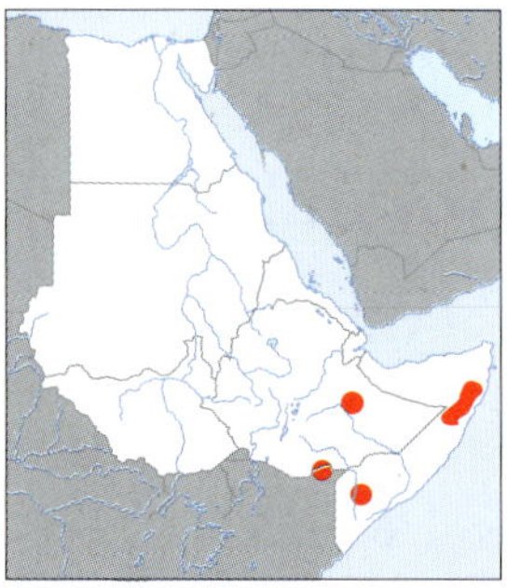

IDENTIFICATION: A small slim frog with a dumbbell back marking, in the dry east. The body is slender, the head short and rounded, with big eyes and vertical pupils, and the back is smooth. Tympanum about half the size of the eye, and may be hard to see. The limbs are slender, toe tips enlarged. Males reach a maximum of about 4.1cm, females 4.3cm. Dorsal colour grey or goldish-tan, with dark blotches, paired or irregular, or absent, and a dumbbell or X-shape between the shoulders (an excellent aid to identification). Underside yellowish-white; males have a dark throat. Taxonomic Note: Long known as *K. parkeri*, but *K. maculifer* has priority.

HABITAT AND DISTRIBUTION: Dry savanna and semi-desert, between sea level and 1,200m. Endemic to the Horn of Africa, but records very sporadic. Known from Nugaal (around Garowe) and Bari (Wadi Dudo environs) in north-east Somalia; Xuddur and Dinsor in south-central Somalia, a couple of imprecise localities in Ethiopia (north-west of Fik, in the northern Ogaden, and near Chelago, just west of Malka

Kassina maculifer, Kenya.

Murri in the south); also Wajir and Banisa in north-east Kenya. Probably widespread between these localities, but obviously very rare or very secretive.

CONSERVATION STATUS: Least Concern.

NATURAL HISTORY: Essentially unknown, presumably like other members of the genus that live in arid habitats; i.e. nocturnal and terrestrial; emerging after heavy rains and breeding rapidly in temporary pools. Advertisement call a brief rising note of very short duration. In Wajir, in Kenya, these frogs lived in concrete rainwater tanks at the school; here they perched by day on the walls above water level, emerging at night; in northern Somalia specimens were also found in a shallow tank near a waterhole, and emerged at night, hiding in vegetation. Described as being capable of 'prodigious jumps', unlike other members of the genus.

SENEGAL RUNNING FROG, SENEGAL KASSINA

Kassina senegalensis
(Duméril & Bibron, 1841)

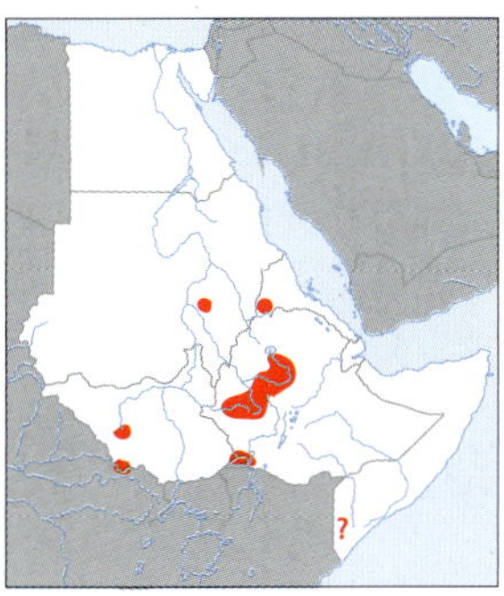

IDENTIFICATION: A small slim frog, yellow and black, with big eyes. The body is slender, the head short and rounded, with big eyes and vertical pupils, the back is smooth. Tympanum is less than half the size of the eye, and is hard to see, being pigmented and slightly below and behind the eye. The limbs are slender, toe tips very slightly enlarged. Males and females reach a maximum of about 5cm. Dorsal colour usually yellowish-brown (sometimes grey, brown or pinkish-brown), usually with up to three dark, longitudinal stripes, although the outer stripes often break up into a line of dark spots. Underside white, cream or greyish-white; males have a dark throat. Taxonomic Note: *K. somalica* (the next species) was long regarded as a subspecies of this frog, and all records from eastern Ethiopia and Somalia have been transferred to that species. Not all authorities accept the split.

Kassina senegalensis, Yambio.

Kassina senegalensis, Bahar Dar.

HABITAT AND DISTRIBUTION: Forest edges, woodland, moist and dry savanna and semi-desert, from sea level to above 2,500m. In our area, a confirmed record from Eritrea, at Barentu, and a provisional record from the south-west. Sporadically recorded in western Ethiopia, between 1,000 and 2,000m, and a handful of records from South Sudan (Wau and environs, Yambio, Ilemi Triangle) and Sudan (Wad Madani, Southern Darfur and Kordofan). A specimen from southern Somalia might also be this species but needs confirmation. Elsewhere, west to Senegal and south to South Africa; huge range, probably the largest of any African frog; future research will probably split the species due to the amount of variation.

CONSERVATION STATUS: Least Concern.

NATURAL HISTORY: Nocturnal and largely terrestrial (sometimes climbs into plant axils). The advertisement call is a beautiful, short *quoip*, like a water drop falling into a bell-shaped bucket. In the rainy season males start to call from mid-afternoon onwards. They hide in burrows, termitaria and under ground cover. They usually walk, even when stressed, although they can make small jumps. They emerge following rain and call as they approach breeding sites, which may be permanent water, large and small temporary pools. Males often call from a place of concealment, but can also be found calling on open ground far from water; when the male has grasped the female, she may carry him a relatively long distance (200m or more) to the water. Eggs are laid in water, and may sink or stick to vegetation.

SOMALI KASSINA

Kassina somalica
Scortecci, 1932

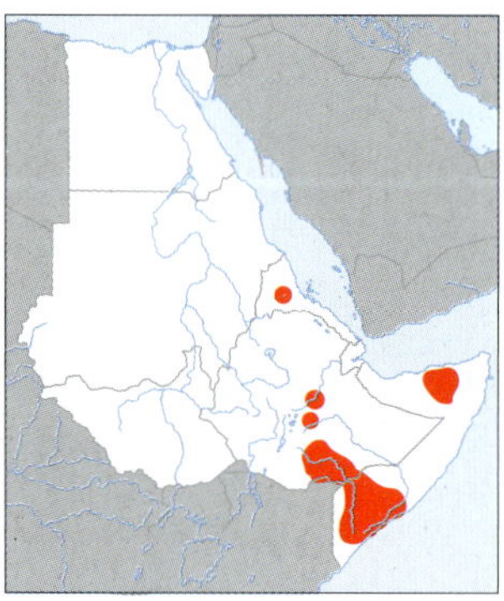

IDENTIFICATION: A small slim frog of dry country, golden with black spots, with big eyes. The body is slender, the head short and rounded, with big eyes and vertical pupils, and the back is smooth. Tympanum small. Limbs slender, toe tips very slightly enlarged. Males and females reach a maximum of about 4cm. Dorsal colour usually yellow-brown, golden or yellowish-grey, with widely separated black spots, either random or forming one or three lines. Underside grey-white. Taxonomic Note: See previous species.

HABITAT AND DISTRIBUTION: Dry savanna and semi-desert, from sea level to 100m. In our area, sporadically recorded from Eritrea (Agordat), Sool (south of Taleh and at Halin) and Bari (Galgalo Oasis) regions in northern Somalia. Occurs on Awash River, fairly widespread in lowlands of south-east Ethiopia, with a cluster of records in southern Somalia, south of 03°N. Tadpoles collected at Buq west of Erigavo in northern Somalia probably also belong to this species. Elsewhere, south to eastern Kenya and north-east Tanzania.

CONSERVATION STATUS: Least Concern.

NATURAL HISTORY: Poorly known, but probably similar to the previous species, i.e. nocturnal and terrestrial, hide in burrows and under ground cover. Advertisement call a beautiful, short *quoip*, near-identical to the previous species. They usually walk, even when stressed, although they can make small jumps. They emerge following rain; mating was observed in Awash National Park in May and at Afgoi, Somalia in August. More than 100 were found under a large rock adjacent to a drying pool in Tsavo East National Park in July.

Kassina somalica, Ginir.

MOUNTAIN KASSINAS OR SNAIL-EATING FROGS *PARACASSINA*

An Ethiopian genus comprising two species of attractive, spotted, medium-sized frogs that live at high altitude. They are very similar to running frogs but differ in having unique modifications to the skull, jaw muscles and tongue that enable them to eat snails.

KEY TO THE NORTH-EAST AFRICAN MEMBERS OF THE GENUS *PARACASSINA*

1a: Field impression of a brown or grey frog, with many smallish, light-edged black blotches, known only from west of the Ethiopian Rift Valley — *Paracassina obscura*, Western Mountain Kassina

1b: Field impression of a green or yellow frog, with a few large black or brown blotches, known only from east of the Ethiopian Rift Valley — *Paracassina kounhiensis*, Eastern Mountain Kassina

WESTERN MOUNTAIN KASSINA

Paracassina obscura
(Boulenger, 1895)

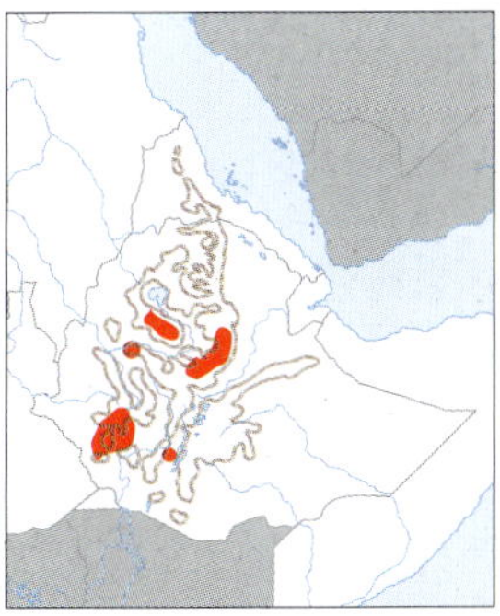

IDENTIFICATION: A chunky spotted, walking frog of western Ethiopia. Body short and compact, head short and rounded, with big eyes and a vertical pupil (hard to see); back smooth. Tympanum small and hard to see. The limbs are stout, toe tips very slightly enlarged. Males and females reach a maximum of about 4.5cm. Dorsal colour shades of brown or grey, the back has many dark, light-edged spots (although some specimens from the forested south-west have relatively few spots). There is a bright yellow or orange patch at the leg/groin junction (sometimes absent in the west), hidden unless the frog is walking. Underside grey, brown or purplish with dirty white or yellowish marbling.

HABITAT AND DISTRIBUTION: Ethiopian endemic. Mostly in grassland at high elevation on the western plateau of Ethiopia, between 1,800 and 3,000m, but also known from the forested south-west, where recent research indicates it is quite widespread. It occurs from Injibara south through the highlands to Jimma, east to Debre Berhan, and on the hills around Chencha, west of Lake Abaya; records from forests of the south-west include Bibita Mountain at 1,500m, south-west of Gore, in a clearing in the forest at Godere Mission, at 820m, and at the forest edge at Jimma, 1,750m.

Paracassina obscura, Bonga

CONSERVATION STATUS: Least Concern. However, it is not known how this species copes with urbanisation. In the early 1970s, this frog was common around Addis Ababa, but it was not encountered anywhere near the city in the 1990s, despite extensive searching.

NATURAL HISTORY: Poorly known. Nocturnal, and usually terrestrial, but one was found in a tree, 1.75m above ground level. Advertisement call similar to

Kassina senegalensis, but less melodious. Usually walk, although they can make small jumps. If stressed, tend to curl into a ball with the head down. Emerge following rainstorms; and breeding continues throughout the long rainy season of western Ethiopia, from April to September. Males call from water, by day as well as at night. They use permanent and temporary ponds, pools, swamps and ditches, while clinging to submerged vegetation. Unusually, they feed largely on snails.

EASTERN MOUNTAIN KASSINA

Paracassina kounhiensis
(Mocquard, 1905)

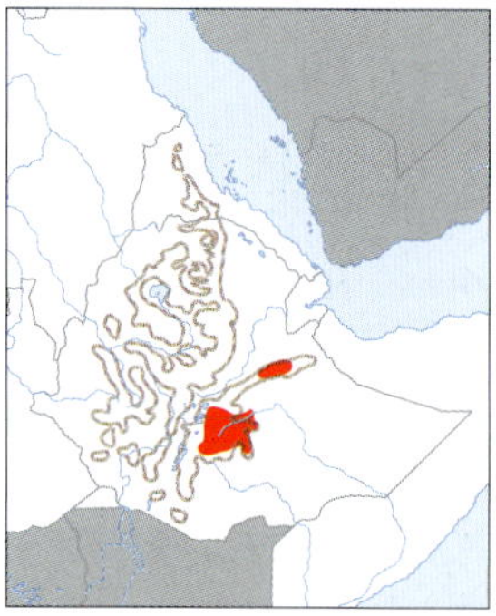

IDENTIFICATION: A chunky, spotted, walking frog of high eastern and south-eastern Ethiopia. Body short and compact, head short and rounded, with big eyes and a vertical pupil; back smooth. Tympanum small and hard to see. The limbs are stout, toe tips very slightly enlarged. Males and females reach a maximum of about 5cm. Dorsal colour yellow, cream or green, with large black or brown spots or blotches, which may coalesce. There is a bright yellow, orange or greenish-yellow patch at the leg/groin junction, hidden unless the frog is walking, although this isn't so obvious on yellow-and-black individuals. Underside yellow or cream, sometimes immaculate, or with darker spots or blotches.

HABITAT AND DISTRIBUTION: An Ethiopian endemic, known from high savanna, grassland and in forest clearings, also in agricultural land, on the eastern and south-eastern plateau of Ethiopia, at altitudes between 1,980 and 3,200m. A handful of records from the vicinity of the Kounhi Valley (the type locality) and further east, in the Ahmar Mountains, but the bulk of records are in the highlands east of a line between Lakes Langano and Chamo in the Rift Valley, extending to the Bale Mountains.

CONSERVATION STATUS: Vulnerable. However, this frog is capable of living in agricultural areas; it was common in the large-scale wheat farms around Dodolla.

NATURAL HISTORY: Poorly known. Nocturnal and terrestrial. Advertisement call similar to *Kassina senegalensis*, but higher-pitched and tinnier. They usually walk, but can make small jumps. If stressed, they tend to curl into a ball with the head down. They emerge following rainstorms; and breeding activity continues throughout the long rainy season, from April to December. Males form noisy congregations among vegetation in marshes, pools and along rivers, both permanent and temporary. In the western foothills of the Bale Mountains, calling and copulation was observed in April–June. By day they can be found under rocks and other ground cover, both on the high plateau and in the Bale forest. Unusually, they feed entirely on land, exclusively on slugs and snails.

Paracassina kounhiensis, Bale Mountains.

Paracassina kounhiensis, Koffelle.

SPINY REED FROGS OR LEAF-FOLDING FROGS *AFRIXALUS*

A sub-Saharan African genus of attractive, alert-looking, often striped, climbing frogs. The back is covered with tiny spines (asperities), which are not always visible but can usually be felt if the back is gently touched. They are often vividly coloured, but their colours may fade in the dark. *Afrixalus* are small, none much larger than 3cm. Their eyes are large, with vertical, diamond-shaped pupils, and they have slender limbs. Calls are buzzes or clicks. At present, some 30+ species are known, five in our area, with fairly disjunct distributions, which we have used in the key. The alternative name of leaf-folding frog refers to the behaviour of several species; they lay eggs on a leaf and fold the leaf over, gluing the edges together to protect the eggs. They are sometimes called banana frogs, as they are often found between the leaves and stem of banana plants.

KEY TO THE NORTH-EAST AFRICAN MEMBERS OF THE GENUS *AFRIXALUS*

1a: A dark patch between eyes; only in southern Somalia	*Afrixalus delicatus*, Delicate Spiny Reed Frog
1b: No dark patch between the eyes, in our area not in Somalia	2
2a: Dorsal surface usually green, confined to south-west Ethiopia	*Afrixalus clarkei*, Clarke's Spiny Reed Frog
2b: Dorsal surface never green, not confined to south-west Ethiopia	3
3a: Dorsal surface usually yellow or orange, never with broad dorsal stripes; confined to South Sudan	*Afrixalus weidholzi*, Weidholz's Spiny Reed Frog
3b: Dorsal surface not yellow or orange, has broad dorsal stripes, in our area not confined to South Sudan	4
4a: Dorsal pattern two pale dorsolateral stripes that meet on snout, confined to highland Ethiopia above 1,400m.	*Afrixalus enseticola*, Oromia Spiny Reed Frog
4b: Dorsally, four long dark bands separate three pale stripes of similar width, in lowland savanna below 1,000m in western Ethiopia and South Sudan	*Afrixalus quadrivittatus*, Four-lined Spiny Reed Frog

DELICATE SPINY REED FROG

Afrixalus delicatus
Pickersgill, 1984

IDENTIFICATION: A small slender frog of southern Somalia. The body is slim, the head short, with big eyes and diamond-shaped pupils, the back covered with very tiny spines, which may be hard to see. Tympanum invisible. The limbs are slender, toe tips enlarged. Males reach a maximum of about 2cm, females 2.4cm. Dorsal colour various shades of brown (golden, orangey) or greyish-white, the flanks brown often with white speckling, and there is usually a dark triangular blotch between the eyes. Underside pale.

HABITAT AND DISTRIBUTION: Coastal woodland, bush, thicket and semi-desert, at low altitudes. In our area, known only from around Gelib, on the lower Juba River in Somalia, at 25m, but is also known from the Tana Delta in Kenya, so probably extends along the Somali coastal plain to the mouth of the Juba River, possibly further north. Elsewhere, south to north-east South Africa.

CONSERVATION STATUS: Least Concern.

NATURAL HISTORY: A nocturnal climbing species. The high-pitched advertisement call is in two parts; a brief *zip* followed by a long trill, lasting up to 6 seconds; the *zip* informs other males how close they are. Emerge following rain; a small group of males will climb onto low vegetation and call; other males wait silently nearby to intercept approaching females. Eggs are deposited in leaves at the water's edge, which are then folded and glued closed.

Afrixalus delicatus, Kenya.

WEIDHOLZ'S SPINY REED FROG

Afrixalus weidholzi
(Mertens, 1938)

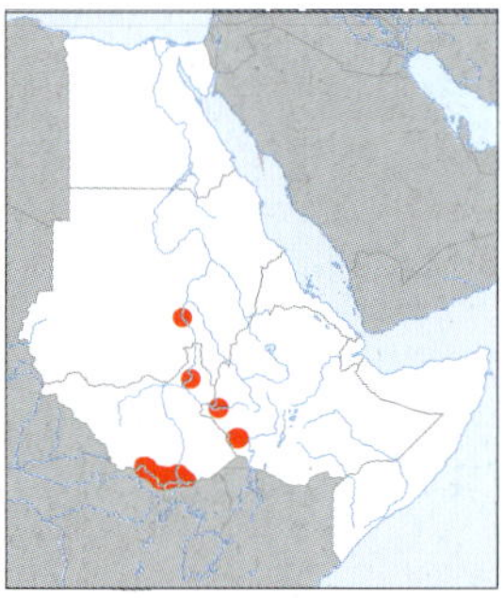

IDENTIFICATION: A small slender yellow frog, with large protruding eyes and diamond-shaped pupils, which may just reach South Sudan. The body is slim, the head short, and the back is covered with very tiny spines. Tympanum invisible. The limbs are slender, toe tips enlarged. Males reach a maximum of about 2.3cm, females 2.5cm. Dorsal colour by day yellow, usually with a black vertebral line; flanks black or brown. At night, becomes orangey-brown. Underside white and can seem almost translucent posteriorly, in gravid females the eggs can be seen through the skin.

HABITAT AND DISTRIBUTION: Moist and dry savanna, from sea level to >1,000m. No records for our area, but it occurs just across the DR Congo border in Garamba National Park, so probably present in Central and West Equatoria, in South Sudan. Elsewhere, west across the savanna to The Gambia, but apparently rare.

CONSERVATION STATUS: Least Concern.

NATURAL HISTORY: A nocturnal climbing species. The high-pitched advertisement call is a long insect-like quiet buzz lasting five seconds or more. It is not known where they aestivate, but they have been observed in West Africa on palm leaves during the dry season, and in the wet season they are visible by day sitting on leaves of waterside vegetation, unlike other *Afrixalus*. At night they call from tall grass, often some distance from water, and also from low grass in shallow pools. Calling males are territorial. Breeding occurs in temporary pools, large and small; in Garamba National Park (DR Congo) breeding recorded in May.

Afrixalus weidholzi, Benin.

FOUR-LINED SPINY REED FROG

Afrixalus quadrivittatus
(Werner, 1908)

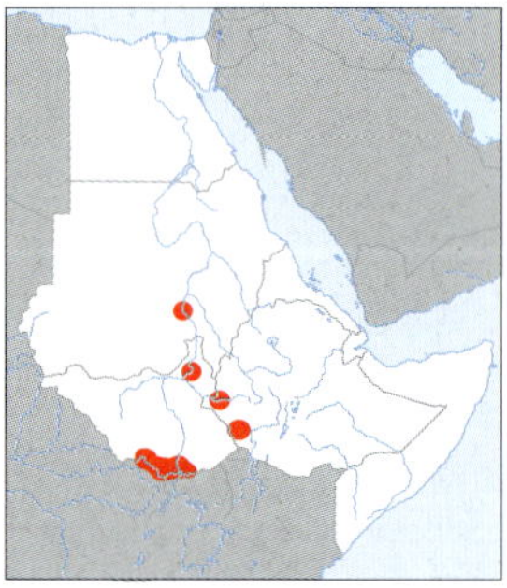

IDENTIFICATION: A small slender frog with big eyes and three pale prominent stripes that meet on the nose. The body is slim, the head short, with big eyes and diamond-shaped pupils, the back of the male is covered with well-developed tiny spines, and those of the females are smaller. Tympanum invisible. Limbs slender, toe tips enlarged. Males reach a maximum of about 2.7cm, females 2.8cm. The back is brown, yellow-brown or fawn, with three broad pale stripes that meet on the nose, although some in Ethiopia have very subdued stripes. The flanks may be brown, or brown above, white below; underside white, the male's throat vivid yellow. Taxonomic Note: Status of this species is problematic, because of uncertainties concerning its relationship with the similar-looking *A. vittiger* and *A. fulvovittatus*, consequently its exact distribution is unclear.

Afrixalus quadrivittatus, Gambela.

HABITAT AND DISTRIBUTION: Forest fringes, woodland and moist and dry savanna, from sea level to above 1,500m. Distribution poorly understood. Known localities include the vicinity of Mt Bibita and Gambela in south-west Ethiopia, Kosti in Sudan, Tonga and Khor Attar (near Malakal), Juba, Talanga Forest, Kagelu and Yei in South Sudan, and just across the DR Congo border in Garamba National Park. Probably occurs throughout the Nile basin south of 10°N, but records lacking. Elsewhere, south to Tanzania and northern Zambia.

CONSERVATION STATUS: Least Concern.

NATURAL HISTORY: A nocturnal climbing species. The high-pitched advertisement call is in two parts; a brief buzz followed by 6–18 rapid clicks. In dry areas they emerge following rain, although near permanent water (e.g. Lake Victoria) they can be found throughout the year. Males call from tall grasses and reeds. Eggs are deposited in leaves, usually grasses growing in water, which are then folded and glued closed.

CLARKE'S SPINY REED FROG

Afrixalus clarkei
Largen, 1974

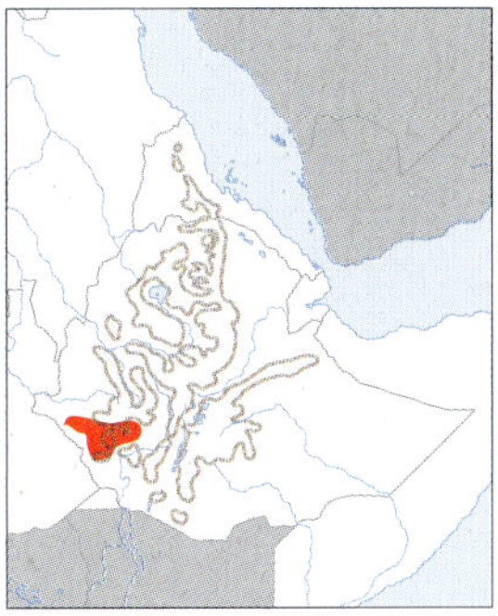

IDENTIFICATION: A small green tree frog, endemic to forests of south-west Ethiopia. Body slim, head short, with big eyes and diamond-shaped pupils. The back of the male is covered with very tiny spines, which may be hard to see. Tympanum invisible. Limbs slender, toe tips enlarged. Males and females reach a maximum of about 2.4cm The back is usually green, sometimes yellow-green, turquoise, olive or dark brown; immaculate, or with random brown speckling, or with two dark stripes that usually don't reach the head. The flanks are brown above, often with white speckling. Underside cream; male's throat vivid yellow.

HABITAT AND DISTRIBUTION: Forested and deforested areas (including agricultural regions) between

Afrixalus clarkei, Sheka.

Afrixalus clarkei, Jimma.

820 and 2,030m. Endemic to Ethiopia, the distribution extends north-west from Jimma to Bedelle, west to Chira, Godare and the environs of Bonga; also reported from the vicinity of Mt Bibita, although the identity of these specimens is uncertain.

CONSERVATION STATUS: Endangered. However, recent research indicates it can tolerate agriculturally degraded landscapes; it was originally believed to be dependent upon pristine forest.

NATURAL HISTORY: A nocturnal climbing species. Advertisement call a long buzz (sometimes described as a 'creak'), followed by a number of metallic clicks. Have been found in the leaf axils of false banana plants, and (rather surprisingly) under rocks; in the breeding season males call from tall grasses and other waterside vegetation; in marshes, riverine floodplains and the vicinity of forest pools. Little else is known of their biology; presumably the eggs are glued onto leaves which are then folded.

OROMIA SPINY REED FROG, GRASSLAND SPINY REED FROG

Afrixalus enseticola
Largen, 1974

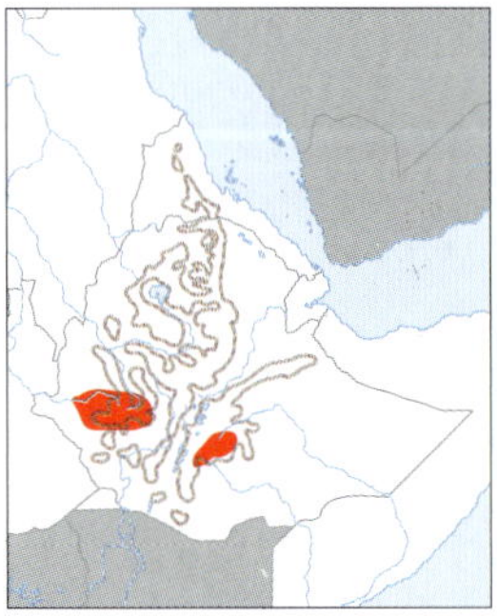

IDENTIFICATION: A small brown, striped tree frog, endemic to southern Ethiopia. The body is slim, the head short, with big eyes and diamond shaped pupils. Spines on the back of the male are tiny and feeble, and often absent in females. Tympanum invisible. Limbs slender, toe tips enlarged. Males reach a maximum of about 2.4cm, females 2.8cm. The back is usually a shade of brown or rufous, with two light dorsolateral stripes that start on the head but don't reach the hind limbs. Some lack the stripes but have a few light snout markings; in others the stripes broaden to cover the whole back. Flanks are brown above, shading to white. Underside cream; male's throat vivid yellow.

HABITAT AND DISTRIBUTION: Forest, deforested areas, grassland in the vicinity of forest and farmland, particularly with wild bananas. Endemic to southern

Afrixalus enseticola, Sheka.

Afrixalus enseticola, Bonga.

Ethiopia, on both sides of the Rift Valley, at altitudes between 1,400 and 2,750m. East of the Rift Valley it is known from Kofele, forest on the southern slopes of the Bale Mountains and Kebri Mangest, presumably it occurs in the intervening country. Appears fairly widespread in the south-west forest; in the triangle bounded by Jimma, Mettu and Mizan Teferi.

CONSERVATION STATUS: Vulnerable.

NATURAL HISTORY: A nocturnal climbing species, but poorly known. Advertisement call apparently unknown. Have been found in the leaf axils of false banana plants. In the breeding season males call from tall grasses and other waterside vegetation; at pools and marshes in forest glades. Presumably, like other *Afrixalus*, eggs are glued onto leaves which are then folded.

REED FROGS *HYPEROLIUS*

A sub-Saharan African genus of small, attractive, brightly coloured frogs, none much larger than 3.5cm; and like the previous three genera, part of the Afrobatrachia. They have large eyes with horizontal pupils. The tympanum is invisible. They have smooth skin, slender limbs, gripping discs on their digits, and are skilled climbers and jumpers, living mostly in waterside vegetation. Often difficult to identify. They display a wide variety of colour patterns; sometimes the male, female and juveniles look quite different. The lack of measurable features on their body also creates difficulties for identification. In Channing and Howell's authoritative 2006 book *Amphibians of East Africa*, this is the only genus with no key. The number of known species listed varies from 111 to 145, depending upon the authority (even so, this is the most speciose genus of African frogs). Nine species occur in our area; and this indicates an impoverished fauna (there are 14 species in Kenya alone, not including the varieties of *Hyperolius viridiflavus*), as a consequence of the aridity of our area. However, undocumented species will probably be found in the upper Nile basin in South Sudan. In addition, the status of some populations of tiny green *Hyperolius* in the *H. nasutus* complex is unclear; we have assigned those from South Sudan and low western Ethiopia to *H. papyri*, and, more tentatively, those from high Ethiopia and southern Somalia to East African *H. howelli*.

KEY TO THE NORTH-EAST AFRICAN MEMBERS OF THE GENUS *HYPEROLIUS*

This is a difficult genus to key. We have made use of colour and locality, which should help identification. In the field call is also extremely useful. Occasional specimens may evade identification. Use your judgement.

1a: Large, up to 4.3cm, never has large dorsal spots, or broad dorsolateral or dorsal stripes, only lateral if striped at all; usually green, grey or brown, digit tips usually red	*Hyperolius viridiflavus*, Common Reed Frog
1b: Does not have the above combination of characters	2
2a: Tiny, slim, not squat, <2.5cm, predominantly green, sharp-nosed	3
2b: Doesn't have the above full combination of characters	4
3a: Found in South Sudan and low-altitude western Ethiopia	*Hyperolius papyri*, Nile Reed Frog
3b: In highland Ethiopia and possibly southernmost Somalia	*Hyperolius howelli*, Kim Howell's Reed Frog
4a: In our area, only in southern Somalia	5
4b: In our area, not in Somalia	7
5a: Green or blue, squat and robust, small, <2.3cm, virtually always has black speckling on head and sometimes on body	*Hyperolius pusillus*, Water Lily Reed Frog

5b: Usually green, sometimes yellow, sometimes spotted, large, up to 3.8cm — 6

6a: Females brown, pink or rufous, with big dark-edged high-contrast spots, males green or yellow with fine discrete black speckles, iris brown with fine yellow edging near pupil — *Hyperolius argus*, Argus Reed Frog

6b: Both sexes green, yellow or ivory, never with large high-contrast spots, if speckling present lacks isolated discrete spots, iris usually fairly uniform yellow — *Hyperolius tuberilinguis*, Tinker Reed Frog

7a: Small, <3.2cm, usually has a broad light dorsolateral stripe almost to the urostyle — *Hyperolius cinnamomeoventris*, Cinnamon-bellied Reed Frog

7b: Large, up to 4.2cm, lacks a broad light dorsolateral stripe — 8

8a: A broad dark line or row of dark spots that extends from the snout through eye and to a greater or lesser extent on the flanks, and a light line above that — *Hyperolius kivuensis*, Kivu Reed Frog

8b: May have a thin green or dark dorsolateral stripe or row of dots that extends through the eye, but without a light line above it — *Hyperolius balfouri*, Balfour's Reed Frog

COMMON REED FROG

Hyperolius viridiflavus
(Duméril & Bibron, 1841)

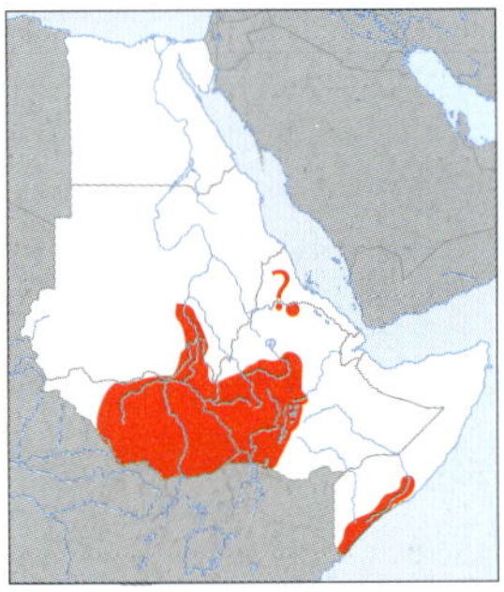

IDENTIFICATION: A small squat frog, widespread, usually has red digit tips and the unexposed leg skin is translucent and pink. Body oval; head short, with big eyes and horizontal pupils. The skin is smooth (although it may be rugose in aestivating animals). Limbs slender, toe tips enlarged. Males reach a maximum of about 3.2cm, females 4.3cm. A very wide range of colours is shown; females often green, immaculate or with yellow spots, males and juveniles whitish-grey, brown or yellow, immaculate, spotted black, or with a broad dark lateral stripe. Toes and toe tips often red, and unexposed leg skin pink. White or yellow below, often with dark spots.

HABITAT AND DISTRIBUTION: Forest fringes, woodland, moist and dry savanna and semi-desert, tolerant of agriculture. Occurs from sea level to above 2,400m. In our area occurs in southern Somalia, south from 04°N along rivers and on the coastal plain; in Ethiopia the type was from Adua (on the border with Eritrea, where can be expected), and it is widespread in the Rift Valley south of Awash National Park, and across the west. Occurs in and around the

Hyperolius viridiflavus, female, Jimma.

Hyperolius viridiflavus, male, Jimma.

Nile basin in South Sudan. Northernmost record appears to be from Kosti, on the White Nile in Sudan. Elsewhere, south to South Africa and Angola.

CONSERVATION STATUS: Least Concern.

NATURAL HISTORY: A nocturnal climbing species. High-pitched advertisement call a brief, melodic and distinctive reverberating *tink*, like a metal rod or bottle being struck repeatedly, with slightly less than a second between each *tink*. They emerge in the wet season, the males calling from any tall plant near the water or growing in it. On overcast days males may call during the day. During the day they shelter in leaf axils, thick vegetation, flowers, etc., but may also sit quite exposed. They will also enter buildings, or forage for insects under outside lights. Unlike other reed frogs, they deposit their eggs in water, utilising temporary pools, larger ponds and stagnant water. They have a startling adaptation to dry conditions; the number of iridophores (reflective cells) in the skin greatly increases, reflecting radiation, and any water landing on the frog's back runs onto the thin pink unexposed skin on the limbs and is rapidly absorbed.

WATER LILY REED FROG

Hyperolius pusillus
(Cope, 1862)

IDENTIFICATION: A tiny squat frog, green, almost translucent, with small but distinct black spots, of southern Somalia. Body oval, head short, with big eyes and horizontal pupils. The skin is smooth. Limbs slender, toe tips enlarged. Males reach a maximum of about 2.1cm, females 2.3cm. Usually green, sometimes olive or blue-white, usually has a black stripe or line of dots from the snout through the eye, and small, distinct black dots on the head and body. Sometimes has a narrow pale dorsolateral stripe. Underside white or greenish, the viscera are often visible through the skin, as are the eggs of gravid females.

HABITAT AND DISTRIBUTION: Forest fringes, woodland, coastal thicket and moist savanna. Found at low elevation in our area, but above 1,000m elsewhere. Recorded in southern Somalia, in the Juba Sugar Project at Gelib and Ola Uager, on the Kenyan border, so presumably occurs along the coastal plain

Hyperolius pusillus, Kenya.

Hyperolius pusillus, Kenya.

Hyperolius pusillus, Kenya.

between them. Elsewhere, south through coastal Kenya to South Africa, inland to the Caprivi Strip.

CONSERVATION STATUS: Least Concern.

NATURAL HISTORY: A nocturnal climbing species. Advertisement call a series of somewhat blurred, high-pitched clicks. Active in the wet season, the males calling from floating vegetation in temporary and permanent ponds, swamps and artificial water sources. The name stems from their habit of sitting on water lily leaves, and they lay their eggs between the leaves. During the day they shelter in leaf axils and thick vegetation.

NILE REED FROG

Hyperolius papyri
(Werner, 1908)

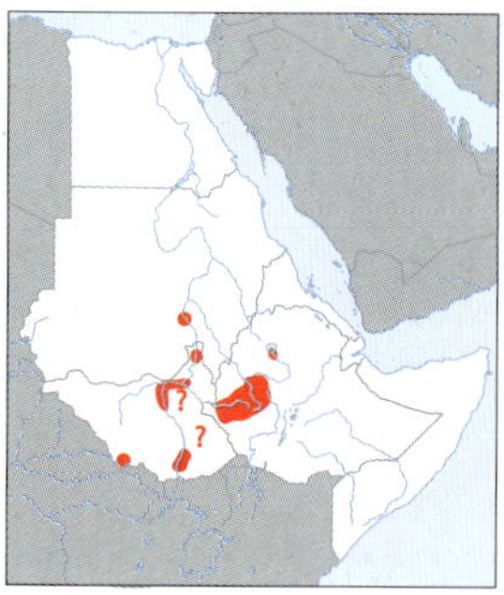

IDENTIFICATION: A tiny slim frog, green, with white dorsolateral stripes, from western Ethiopia and South Sudan. The body is elongate; the head short, with big eyes and horizontal pupils. The skin is smooth. Limbs slender, toe tips enlarged. Males and females reach a maximum of about 3cm, less in highland Ethiopia. Usually green, bright or dull, or rufous-brown, usually with a pair of white dorsolateral stripes. A fine dark vertebral line may be present. There is often extensive fine dark dorsal speckling, especially around the dorsolateral stripes. Underside translucent silvery greenish-white or blue-greenish. Taxonomic Note: A major 2013 revision of the *H. nasutus* superspecies (of which this frog and the next species are members) did not consider South Sudanese or Ethiopian examples, so their status is unclear. This species was subsequently elevated from the *H. nasutus* group in 2019, using South Sudanese examples. Provisional molecular data indicate that it occurs west of the Ethiopian Rift Valley, the following species on the valley floor and further east.

HABITAT AND DISTRIBUTION: Woodland, moist and dry savanna, from 200m to around 2,000m. A cluster of Ethiopian records west of the Rift Valley (Bahar Dar, Nekemte, Bedelle, Metu, Jimma, Gambela). Records from Sudan and South Sudan are sparse, but it probably occurs throughout most of eastern South Sudan, as it is known from Renk, Upper Nile, Khor Attar, Yambio, Juba and environs, in the far south of Central Equatoria. In Sudan, known from Kosti, and presumably from there into South Sudan.

CONSERVATION STATUS: Data Deficient.

NATURAL HISTORY: A nocturnal climbing species. Advertisement call a series of brief high-pitched buzzes. Active in the wet season, the males calling from tall grasses, sedges and reedbeds; in the long rainy season they often commence calling well before dark. In Ethiopia usually at the fringes of permanent or near-permanent swamps, streams, pools and lakes.

Hyperolius papyri, Yambio.

KIM HOWELL'S REED FROG

Hyperolius howelli
Du Preez & Channing, 2013

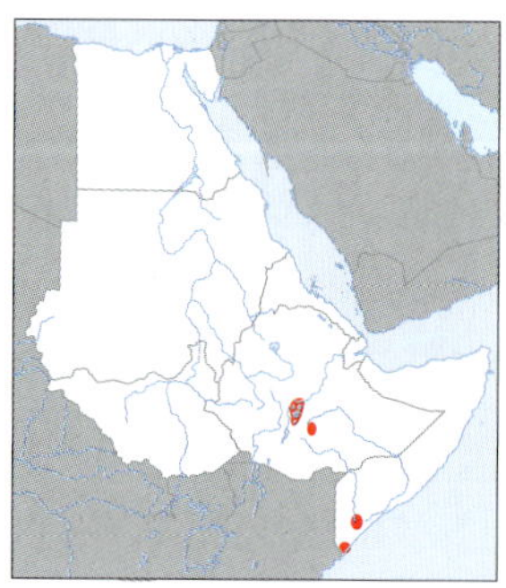

IDENTIFICATION: A tiny slim frog, green, usually with white dorsolateral stripes, from highland Ethiopia and southern Somalia. Body elongate; head short, with a shark-like snout, big eyes and horizontal pupils. Skin smooth. Limbs slender, toe tips enlarged. Males and females reach a maximum of about 2cm. Usually green, bright or dull, occasionally rufous-brown, usually with a pair of white dorsolateral stripes. There is often extensive fine dark dorsal speckling, especially around the dorsolateral stripes. Underside translucent silvery greenish-white or blue-greenish.

HABITAT AND DISTRIBUTION: Woodland, moist and dry savanna, from sea level to above 2,000m in Ethiopia, where it occurs in the central Rift Valley and eastern highlands, e.g. recently recorded in Bale Mountains National Park, on the lower slopes of Harenna Forest. In Somalia, known from Ola Uager, on the border with Kenya, and the Juba Sugar Project at Gelib, so probably in the intervening areas. Elsewhere, to Tanzania.

CONSERVATION STATUS: Data Deficient.

NATURAL HISTORY: A nocturnal climbing species. Advertisement call a short chirp that starts with a brief buzz, followed by three notes, in <0.2 second. Active in the wet season, males calling from tall grasses, sedges and reedbeds; in the long rainy season they often commence calling well before dark. In Ethiopia usually at the fringes of permanent or near-permanent swamps, streams, pools and lakes. Huge numbers occur at Lake Awassa, in Ethiopia, and have been found at night moving across country, several hundred metres from the lakeshore.

Hyperolius howelli, Harenna Forest.

ARGUS REED FROG

Hyperolius argus
Peters, 1854

IDENTIFICATION: A small squat frog, females beautifully spotted, males green or yellow, in southern Somalia. The body is oval, the head short, with big eyes and horizontal pupils. The skin is smooth. Limbs slender, toe tips enlarged. Males reach a maximum of about 3.4cm, females 3.8cm. Females are rufous, brown or pinkish; the dorsal surface has several big, dark-edged yellow or white spots, a V-shape of the same colour connects the snout and the eyes. Underside pink or rufous. Males green or greenish-yellow, often with a pale dorsolateral stripe, speckled black to a greater or lesser extent.

Hyperolius argus, Kenya.

HABITAT AND DISTRIBUTION: Woodland, coastal thicket and moist savanna. Found at low elevations in our area, but higher elsewhere. In our area in southern Somalia, at the Juba Sugar Project at Gelib, and Ola Uager on the border with Kenya, so presumably extends along the coastal plain. Elsewhere, south through coastal Kenya to South Africa, never far from the coast.

CONSERVATION STATUS: Least Concern.

NATURAL HISTORY: A nocturnal climbing species. Advertisement call a series of thin, rapid high-pitched clicks or *chik* sounds. Active in the wet season, males calling from vegetation above water. Often around permanent water, but known from temporary ponds in Kenya, provided they are reasonably large. Eggs are attached to vegetation underwater. A Somali specimen was found in the jaws of a Puff Adder *Bitis arietans*.

TINKER REED FROG

Hyperolius tuberilinguis
Smith, 1849

IDENTIFICATION: A relatively large green reed frog of southern Somalia. The body is oval, the head short, with big eyes and horizontal pupils. Skin smooth. Limbs slender, toe tips enlarged. Males reach a maximum of about 3.4cm, females 3.8cm. Usually a virulent shade of green or yellow, immaculate, but may be olive, light brown or ivory, non-green individuals sometime haphazardly speckled black, juveniles may have a dark lateral stripe. Underside cream, unexposed limbs pink.

HABITAT AND DISTRIBUTION: Woodland, coastal thicket and moist savanna. Found at low elevation in Somalia but above 1,000m elsewhere. In our area in southern Somalia, at from the Juba Sugar Project at Gelib, so presumably extends along the coastal plain south of there. Elsewhere, south through coastal Kenya to Mozambique and South Africa.

CONSERVATION STATUS: Least Concern.

NATURAL HISTORY: A nocturnal climbing species. Advertisement call a short series of high-pitched clicks or taps (usually 2–3, but up to six), sometimes preceded by a growl-like croak to deter other males. Active in the wet season; males call from vegetation just above water level. They utilise temporary and permanent ponds, swamps and artificial water sources, provided marginal vegetation is fairly dense. By day they shelter in leaf axils and thick vegetation. Eggs are attached to vegetation just above water level.

Hyperolius tuberilinguis, Kenya.

Hyperolius tuberilinguis, Kenya.

KIVU REED FROG

Hyperolius kivuensis
Ahl, 1931

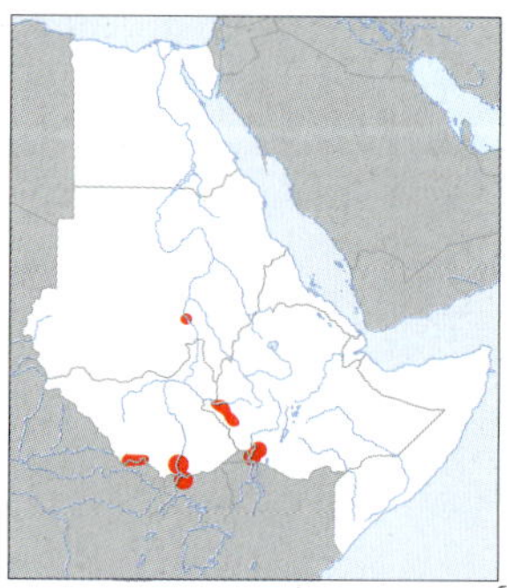

IDENTIFICATION: A relatively large reed frog with a dark eye-stripe, in lowlands of western Ethiopia and South Sudan. Body oval, head short, with big eyes and horizontal pupils. The skin is smooth. Limbs slender, toe tips enlarged. Males reach a maximum of about 3cm, females 3.8cm. Dorsal colour usually green, olive, brown or white, but has a dark stripe or line of large dots through the eye, which may taper off or extend the length of the body. Underside cream or dirty white, unexposed limbs pink or yellow.

HABITAT AND DISTRIBUTION: Forest, woodland and moist savanna, at low altitudes in our area but above 2 ,000m elsewhere. A handful of records from extreme south-west Ethiopia, in the Gambela area south-east to Godare and eastern slopes of Mt Bibita, Maji and Turmi. In South Sudan known from the area around Yambio, and recorded in Garamba National Park, DR Congo, so probably just across the border in West and Central Equatoria. Also reported from Kosti on the White Nile, way north in Sudan, so presumably occurs in the intervening Nile basin, but unrecorded. Elsewhere, south on the east side of the Central African forest to Malawi, Zambia and Angola.

CONSERVATION STATUS: Least Concern.

NATURAL HISTORY: A nocturnal climbing species. Advertisement call a short harsh chirp; males also make a spacing call to keep other males away, six evenly spaced clicks in 0.5 second. Active in the wet season, males calling while concealed in vegetation just above the water level in temporary and permanent ponds, swamps and artificial water sources. Often shelter in leaf axils. Eggs are attached to vegetation.

Hyperolius kivuensis, Gambela.

BALFOUR'S REED FROG

Hyperolius balfouri
(Werner, 1908)

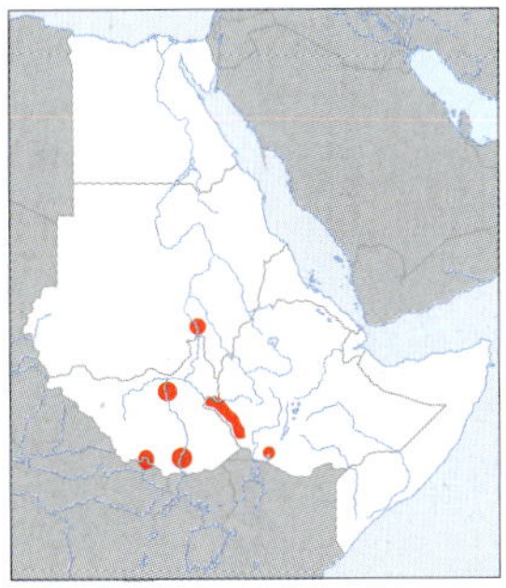

IDENTIFICATION: A relatively large reed frog, very variable in colour but often has a green dorsolateral stripe. The body is oval, the head short, with big eyes, a golden iris and horizontal pupils. Skin smooth. Limbs slender, toe tips enlarged. Males reach a maximum of about 3.4cm, females 4.2cm. Dorsal colour usually brown, olive, grey-brown or pale grey. Often has a fairly narrow green dorsolateral stripe that doesn't reach the rump. Underside translucent cream or yellow, unexposed limbs pink.

HABITAT AND DISTRIBUTION: Known to enter degraded forest, but usually in woodland and moist savanna, at low altitudes in our area but higher in the Albertine Rift and Cameroon. A handful of records from south-west Ethiopia (Gambela, Godare, Bench Maji, north shore of Lake Chew Bahir). In South Sudan known from the vicinity of Juba, and the heart of the Sudd swamp in Unity State, 80km south of Adok. Also Garamba National Park, DR Congo, so probably just across the border in West and Central Equatoria. Also reported from Kosti on the White Nile, way north in Sudan, so presumably occurs in the intervening Nile basin. Elsewhere, west to Cameroon, south to Rwanda.

CONSERVATION STATUS: Least Concern.

NATURAL HISTORY: A nocturnal climbing species. Advertisement call described as a deep, coarse, slow unmelodic creak. Active in the wet season, males call from the water surface or while perched on grass or small shrubs at the water's edge and as high as 45cm above ground. Breeding occurs in swamps, permanent and temporary pools, and quiet but flowing water; in Garamba it commenced in April and frogs preferred larger streams. Several specimens entered houses, presumably in search of insects.

Hyperolius balfouri, Gambela.

CINNAMON-BELLIED REED FROG

Hyperolius cinnamomeoventris
Bocage, 1866

IDENTIFICATION: A medium-sized reed frog, very variable in colour, often with a broad, short dorsolateral stripe. The body is oval, the head short, with big eyes and horizontal pupils. The skin is smooth. Limbs slender, toe tips enlarged. Males reach a maximum of about 2.8cm, females 3.3cm. Dorsal colour various shades of brown, rufous, olive, yellow or green. Males often have a broad, yellow dorsolateral stripe that doesn't reach the rump. Green females may have a thin black lateral line separating the white ventral colour from the green. Underside translucent white, thighs pink.

HABITAT AND DISTRIBUTION: Forest and moist savanna. The only record in our area was on the east slope of Mt Bibita, in south-west Ethiopia, at 1,400m, and there seem to be no specimens from South Sudan, but it is recorded in Garamba National Park, DR Congo, so almost certainly occurs just across the border in West and Central Equatoria, South Sudan. Elsewhere, west to Cameroon and south to Angola, so presumably occurs in Central Africa's forests, although records are lacking.

CONSERVATION STATUS: Least Concern.

NATURAL HISTORY: A nocturnal climbing species. Advertisement call a series of brief clicks or double clicks. Active in the wet season, males call from low down on plants in shallow water. Breeding occurs in streams, marshes, ponds and seeps. In Garamba breeding commenced in March. Nearly 13,000 specimens were collected there, most at edges of streams, >90 per cent close to the water's edge. About 250 eggs are laid in shallow water.

Hyperolius cinnamomeoventris, Congo.

Hyperolius cinnamomeoventris, Congo.

FAMILY ARTHROLEPTIDAE SQUEAKERS, AFRICAN TREE FROGS AND FOREST FROGS

A diverse family of eight genera and about 150 species of tropical African frogs; their distribution centred on the forests, woodland and moist savanna of West and Central Africa. There is no common name for the whole family, which includes some very unusual frogs, for example Hairy Frog *Trichobatrachus robustus* and Cat-eyed Frog *Nyctibates corrugatus*, both of which are the only species in their genera. The two most speciose genera are *Leptopelis*, tree frogs, with nearly 60 species and *Arthroleptis*, the squeakers, with c.50 species.

TREE FROGS (OR AFRICAN TREE FROGS) *LEPTOPELIS*

A sub-Saharan African genus of relatively large, chunky, attractive frogs, with huge eyes, vertical pupils and a look of wide-eyed innocence. Many are green. They mostly live in forest, woodland and wooded savanna, and deposit their eggs in nests in soil, near water. Originally in the family Hyperoliidae, recent molecular evidence has shifted them to the Arthroleptidae. Eleven species occur in our area, seven of which are endemic to Ethiopia and all seven descended from a common ancestor less than 10 million years ago. One Ethiopian endemic, *L. gramineus*, was recently split into five separate species, there may be more; the taxonomy is not fully resolved (see couplets 3 and 4 in the key). Despite the name, at least four of the 11 in our area live on the ground.

KEY TO THE NORTH-EAST AFRICAN MEMBERS OF THE GENUS *LEPTOPELIS*

1a: Tips of digits not noticeably rounded, and same width or slightly broader than preceding joint, body squat, usually on ground — 2

1b: Tips of digits noticeably rounded, and, in adults, usually broader than preceding joint, typically in trees or bushes — 5

2a: Dorsal colour usually light brown, occasionally green, with a large darker brown or darker green dorsal marking, sometimes with a pale centre — *Leptopelis bocagii*, Bocage's Tree Frog

2b: Dorsal colour usually some shade of green, sometimes greenish-brown, blue-brown or yellow — 3

3a: Digital discs sometimes distinct, very slightly wider (up to 30 per cent) than preceding joint — 4

3b: Digital discs indistinct, no wider than preceding joint — *Leptopelis montanus*, Ethiopian Montane Tree Frog

4a: Dorsal skin obviously smooth, with occasional small individual warts; east of Ethiopian Rift Valley — *Leptopelis diffidens*, Harenna Burrowing Tree Frog

4b: Dorsal skin rough, very warty; west of Ethiopian Rift Valley — *Leptopelis gramineus*, Ethiopian Burrowing Tree Frog

5a: Brown or pink, white below, a vague dark mark between the eyes, in our area only in southern Somalia — *Leptopelis concolor*, Witu Tree Frog

5b: Does not have the above combination of characters, not in Somalia — 6

6a: In our area, only in Ethiopia — 7

6b: In our area only in South Sudan — 10

7a: Often has three dark stripes on dorsal surface, and back and thighs suffused to a greater or lesser extent brick-red; only in highlands around Mt Gughe, west of Lake Abaya — *Leptopelis susanae*, Susan's Tree Frog

7b: Usually lacks stripes, never brick-red on dorsal surface, not confined to Mt Gughe area — 8

8a: Slightly squat in appearance, dorsal surface often uniform green, or brown; only in highlands within Ethiopian loop of Blue Nile — *Leptopelis yaldeni*, Yalden's Tree Frog

8b: Slender in appearance, not found in the Ethiopian loop of Blue Nile — 9

9a: Known only from forested or deforested areas of south-west Ethiopia below 2,200m, limb junctions often blue-green, and rarely a dark stripe through eye — *Leptopelis vannutellii*, Vannutelli's Tree Frog

9b: Montane forests of Ethiopian plateau at altitudes above 1,900m, usually in south-east, limb junctions not blue-green, often a dark stripe through eye *Leptopelis ragazzii*, Ragazzi's Tree Frog

10a: Large frog, reaching 5.8cm, back shades of light brown with three narrow dark longitudinal dorsal stripes *Leptopelis oryi*, Ory's Tree Frog

10b: Smaller, up to 4.8cm, adults brown, juveniles green, without three narrow dorsal stripes *Leptopelis viridis*, Green Tree Frog

ETHIOPIAN BURROWING TREE FROG

Leptopelis gramineus
(Boulenger, 1898)

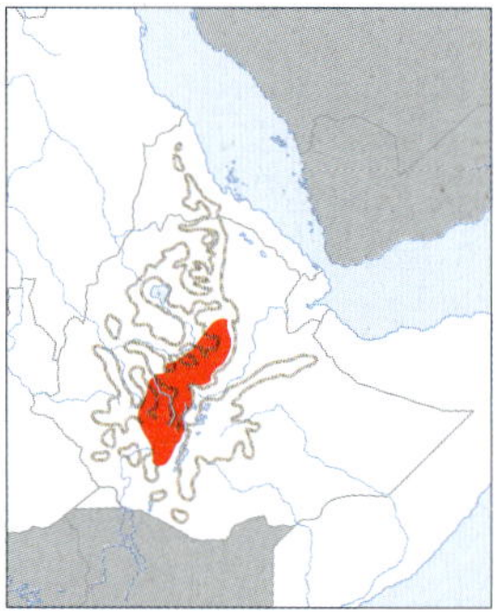

IDENTIFICATION: A big squat terrestrial frog, green, in highland Ethiopia. Body oval, head short, with big eyes and vertical pupils. Tympanum visible, about 2⁄3 the size of the eye. The dorsal skin is rough, with small flattened lumps. Limbs stocky, digit tips not enlarged. Males reach a maximum of about 4.5cm, females 6.3cm. Usually green, sometimes brown, blue-grey or yellow. A dark line extends from the snout through the eye, and curves down towards the forelimbs. The flanks and dorsal surface often have big subcircular dark blotches. Underside may be white or orange, immaculate or marbled to a greater or lesser extent purplish-brown. Taxonomic Notes: Originally a widespread, variable highland species, this frog was recently split into three species, then 2 more; their status vigorously disputed. For the time being, we accept three: this one and the following two. See also our notes on *Leptopelis montanus*.

HABITAT AND DISTRIBUTION: Endemic to Ethiopia, in grassland, moist savanna, woodland, forest fringes, agriculture and urban areas, between 1,900 and 3,500m. Occurs in the highlands along the western wall of the Ethiopian Rift Valley, from Dessie south to west of Lakes Abaya and Chamo, and west to around Nekemte.

CONSERVATION STATUS: Not Assessed. The recent taxonomic reorganisation means its new distribution is not clear, but it is a tolerant species found in areas of intense agriculture and even within the heart of Addis Ababa, so probably not under any threat.

NATURAL HISTORY: A nocturnal, terrestrial species (although newly metamorphosed juveniles will climb grass). Advertisement call a quiet *clack*, sometimes preceded by a short groan. Active in both wet seasons, males calling from a concealed position near temporary pools and ditches, usually by night, but sometimes active and calling by day, especially on overcast days in June. The eggs are laid in soil. In long rainy season (May–September), there seems to be little breeding activity after July. In dry season they aestivate underground, but also under fallen logs or large rocks, and in Addis Ababa may use compost heaps and mounds of rotting vegetation.

Leptopelis gramineus, Addis Ababa.

HARENNA BURROWING TREE FROG

Leptopelis diffidens
Tiutenko & Zinenko, 2021

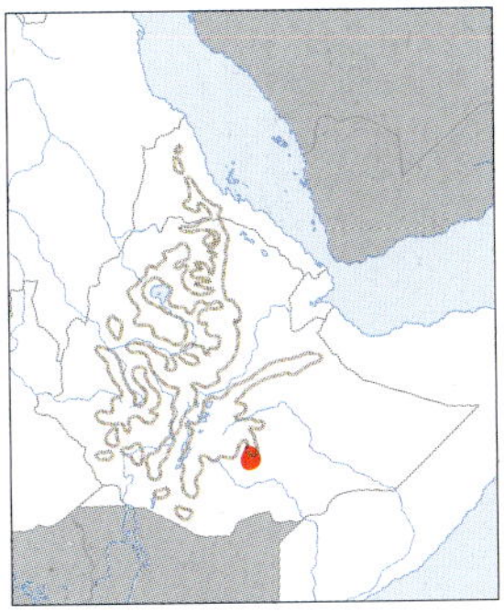

IDENTIFICATION: A squat terrestrial frog, green and smooth, of Bale Mountains in Ethiopia. Body oval, but flattened; head short, with big eyes and vertical pupils. Dorsal skin smooth. Tympanum visible, 50 per cent or more the diameter of eye. Limbs stocky, digit tips slightly enlarged. Males reach a maximum of about 3cm, females 4cm. Dorsal colour light to very dark green, or brown; uniform or with three longitudinal dorsal stripes, slightly lighter than ground colour. A dark line extends from snout through eye, and breaks into dots. At limb junctions dorsal colour changes to blue and then pink. Underside pale and pattern-less. Taxonomic Notes: Elevated out of synonymy of *L. gramineus*; see previous species.

HABITAT AND DISTRIBUTION: Forest glades. Endemic to Ethiopia, known only from Harenna Forest, on southern slopes of Bale Mountains, between 1,400 and 2,300m.

CONSERVATION STATUS: newly described and not evaluated, but its habitat (although within a national park) is under threat of development, and it does not appear to tolerate any urbanisation.

NATURAL HISTORY: Active by day and night in wet season, and by night in dry season. A terrestrial species (although newly metamorphosed juveniles will climb grass). Advertisement call a very quiet quack, inaudible from more than a few metres away; a single note, repeated roughly every 20 seconds, although disturbed males immediately ceased calling and took >10 minutes before recommencing. Breeding males call from tussocks in water and from the ground in dense grass at edges of pools and puddles. Appears to be confined to forest glades that become flooded in the rainy season. The eggs are deposited in soil. Metamorphosing juveniles found in May and June. In the dry season found under logs and large rocks.

Leptopelis diffidens, Harenna Forest.

OCELLATED BURROWING TREE FROG

Leptopelis montanus
Tiutenko & Zinenko, 2021

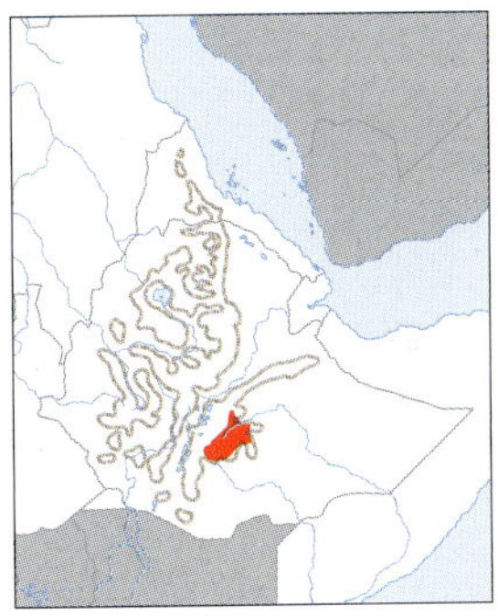

IDENTIFICATION: A squat terrestrial frog, green and warty, often with ocellate markings, of eastern highland Ethiopia. Body oval, head short, with big eyes and vertical pupils. The dorsal skin is warty. Tympanum visible but small, about half the size of the eye. Limbs stocky, digit tips not wider than the terminal joint. Males reach a maximum of about 3.5cm, females 5cm. Usually green, sometimes brown or almost black, uniform or with three broad brown longitudinal dorsal stripes; flanks and back marked to a greater or lesser extent with vaguely circular brown blotches. A dark line extends from snout through eye, and breaks into dots or subcircular blotches, sometimes bordered white. Underside pale, white or yellow. Taxonomic Notes: formerly subsumed in *L. gramineus*, recently resurrected. An immediate contentious response was that *Leptopelis montanus* is invalid, and southern specimens of this taxon should be called *Leptopelis xeniae*, northern ones *Leptopelis shebellensis*. We have not adopted this proposal.

HABITAT AND DISTRIBUTION: Open grassland, high savanna, woodland, forest and agriculture. Endemic to Ethiopia, known only from high country east of the Ethiopian Rift, from Asela south-east to Goba, south to lower southern slopes of Bale Mountains, between 1,800 and 3,400m.

CONSERVATION STATUS: Newly described and not evaluated, but seems to be tolerant of agriculture and urbanisation.

NATURAL HISTORY: Active by night in wet season, sometimes by day, individuals moving in flooded ditches near Dodolla on an overcast day. A terrestrial species. Advertisement call a quiet clack. Breeding males seen calling at the margins of flooded ditches by day. The eggs are buried in soil. In the dry season it has been found under large logs and rocks.

Leptopelis montanus, Goba.

BOCAGE'S TREE FROG

Leptopelis bocagii
(Günther, 1865)

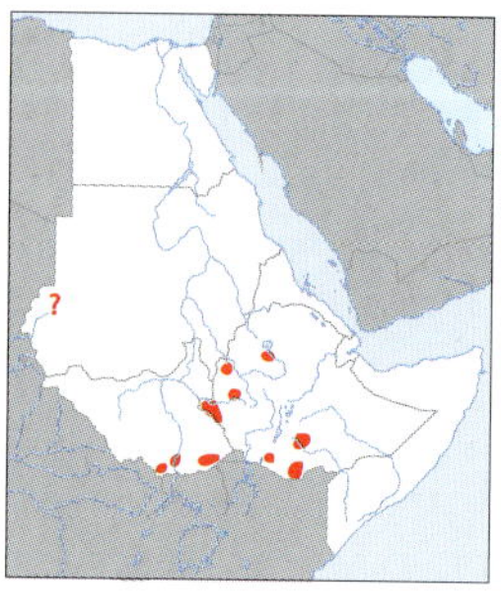

Leptopelis bocagii, Negele.

IDENTIFICATION: A squat terrestrial frog, brown with a dark saddle, from medium to low savanna in Ethiopia and South Sudan. Body oval; head short, with big eyes and vertical pupils. The dorsal skin has small, smooth warts. Tympanum visible, 50 per cent or more the diameter of the eye. Limbs stocky and, unlike most *Leptopelis*, digit tips are not wider than the terminal joint. Males reach a maximum of about 5.6cm, females 6.5cm. Usually brown, occasionally almost cream, with a large dark dorsal saddle, uniform or infilled to a greater or lesser extent paler brown. A broad dark stripe through eye may extend onto flanks. Underside cream or white, sometimes with mottling. Juveniles are green. Taxonomic Notes: Specimens from the moist lowlands (both savanna and forested areas) in Ethiopia are different to those from Sudan, being slenderer and more arboreal, and might prove to be a different species.

HABITAT AND DISTRIBUTION: Mid- to low-altitude savanna, woodland and agriculture, between 400 and 1,900m in our area, to sea level elsewhere. Fairly widespread in south, south-west and west Ethiopia, also known from Lake Tana and upper reaches of Blue Nile. In South Sudan known only from Central and Eastern Equatoria (Juba, Yei, Katire, Talanga Forest), but probably more widespread, as it is known from just across the border from Jonglei at Gambela, in Ethiopia, and also at Faradje, in DR Congo, near border with Central Equatoria. There is a curious record from Jebel Marra, in Darfur, Sudan. Elsewhere, south along eastern side of Central African forest to Zimbabwe and Angola.

CONSERVATION STATUS: Least Concern.

NATURAL HISTORY: Terrestrial (although it occasionally climbs low shrubs, and shelters in plant axils), active by night in wet season; and usually only emerges following heavy storms. Advertisement call a low-pitched double croak. Breeding males call from burrows on open ground or in low vegetation. Eggs are deposited in deep holes during rainstorms. In the dry season they burrow underground and secrete a thin cocoon to prevent desiccation.

WITU TREE FROG, UNIFORM TREE FROG

Leptopelis concolor
Ahl, 1929

IDENTIFICATION: A small, squat, almost globular frog, brown or pink, of southern Somalia. Body oval, head very short, with big eyes and vertical pupils. The dorsal skin is smooth. Tympanum visible, about half the size of the eye. Limbs fairly stocky, digit tips spherical, wider than the terminal joint. Males reach a maximum of about 3.4cm, females 5cm. Usually brown or pinkish-brown, with a vague dark mark between eyes and irregular (sometimes Y-shaped) darker dorsal blotches. A dark line through the eyes breaks up behind tympanum. Underside white or pinkish.

HABITAT AND DISTRIBUTION: In our area, only in riverine woodland of the southern Somali coastal plain, at altitudes below 50m, but higher elsewhere. The only record in our area is from the Juba Sugar Project, at Gelib, so presumably extends along the coastal plain south of there, as it occurs in the Tana Delta in Kenya. Elsewhere, south to northern coastal Tanzania.

CONSERVATION STATUS: Least Concern.

NATURAL HISTORY: Terrestrial (but occasionally climbs low shrubs), active by night in wet season. Advertisement call a sharp clack, sometimes followed by two or three very different screams. Breeding males call from widely spaced sites in thick bushes or on tall grass; in Somalia males were calling high up on sugar cane in October. Mating occurs near temporary ponds, and eggs are laid in cavities in mud surrounding ponds; tadpoles hatch when the nest floods. Adults will move a long way across country to find a suitable pond, Kenyan specimens were found crossing roads hundreds of metres from the nearest pool. They aestivate in the dry season, presumably buried in soil.

Leptopelis concolor, Kenya.

ORY'S TREE FROG

Leptopelis oryi
Inger, 1968

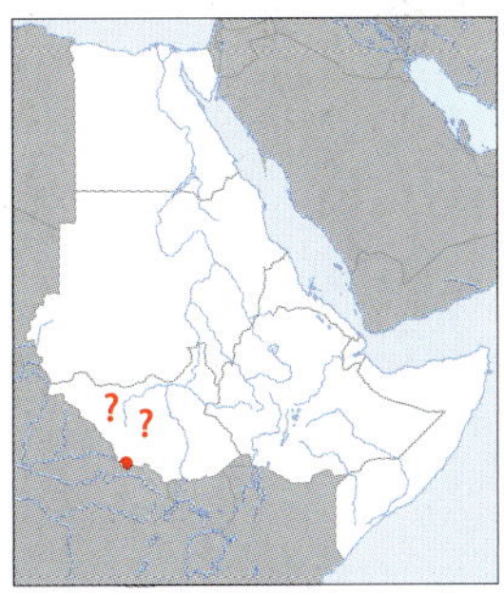

IDENTIFICATION: A small squat frog, finely striped, of south-west South Sudan. Body oval, head very short, with big eyes and vertical pupils. Dorsal skin smooth. Tympanum visible, about half the size of the eye. Limbs slim, digit tips spherical, slightly wider than the terminal joint Males reach a maximum of about 4.3cm, females 5.8cm. Usually brown or pinkish-brown, with three fine longitudinal stripes on back. A dark line through the eyes may broaden behind the tympanum or fade into the dorsal colour. The upper lip is white. Crossbars on the limbs are very fine or absent. Underside white or pinkish.

HABITAT AND DISTRIBUTION: Moist savanna and forest margins, at around 700–900m altitude. In our area, known only from the vicinity of Yambio, in South Sudan, but also in Garamba National Park in DR Congo, so probably widespread in Western Equatoria. Elsewhere, to north-west Uganda, at Murchison Falls and Budongo Forest.

CONSERVATION STATUS: Least Concern.

NATURAL HISTORY: A nocturnal species, climbing into vegetation. Advertisement call a single, loud, low-pitched long clack, lasting up to 0.15 second. Breeding males call from tall grass and reeds. Mating occurs near ponds, rivers and around swamps, in DR Congo males commenced calling in March, and tadpoles were found from May to September. In the dry season, specimens were found in rodent burrows or under rocks.

Leptopelis oryi, Yambio.

GREEN TREE FROG

Leptopelis viridis
(Günther, 1869)

IDENTIFICATION: A small squat frog, brown (despite the name), of South Sudan. Body oval, head very short, with big eyes and vertical pupils. Dorsal skin smooth. Tympanum visible, about half the size of the eye. The limbs are slim, the digit tips spherical, slightly wider than the terminal joint. Males reach a maximum of about 3.5cm, females 4.8cm. Dorsal surface usually light brown, often with a darker brown horseshoe-shaped or irregular asymmetrical blotch, sometimes N-shaped. A dark line through the eyes may broaden behind the tympanum or fade. Underside may be white or greyish-white, sometimes with brown, green or purple mottling. Juveniles are bright green.

HABITAT AND DISTRIBUTION: Moist and dry savanna, tolerant of agriculture, in our area at altitudes of 600–1,000m, to sea level elsewhere. In our area, only in vicinity of Yambio, and Lui, Western Equatoria, South Sudan. However, it is known from Garamba National Park in DR Congo and in eastern Central African Republic, so probably more widespread in Western Equatoria and western Bahar El Ghazal. Elsewhere, west across savanna to Senegal.

CONSERVATION STATUS: Least Concern.

NATURAL HISTORY: A nocturnal species that climbs into vegetation, during day they bury themselves. Advertisement call a loud clack, sometimes followed by a long buzz. Breeding males call from bushes, tall grass and reeds, occasionally from ground; in Ghana there was never more than one male per bush around a swamp. Mating occurs near ponds, temporary pools, rivers and around swamps. In DR Congo activity commenced in April and extended until July. Eggs laid in moist soil. They aestivate in the soil in the dry season.

Leptopelis viridis, Yambio.

SUSAN'S TREE FROG

Leptopelis susanae
Largen, 1977

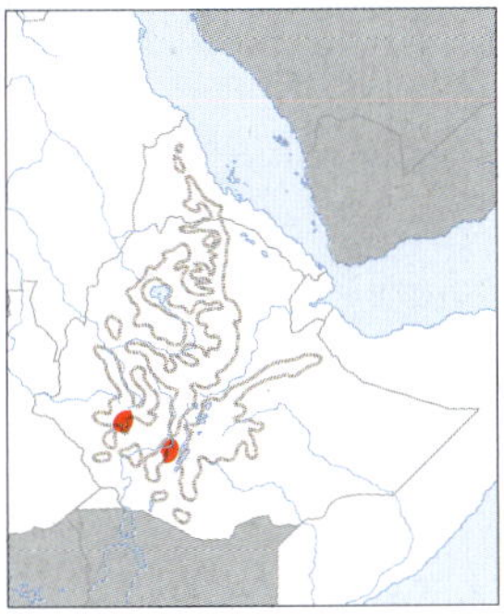

IDENTIFICATION: A medium-sized tree frog, often brick-red on back or limbs, endemic to Gughe Mountains in Ethiopia. Body oval, head very short, with big eyes and vertical pupils. Dorsal skin has many small warts. Tympanum visible, slightly over half the size of the eye. Limbs slim, digit tips spherical, wider than the terminal joint. Males reach a maximum of about 3.6cm, females 5.3cm. Dorsal colour very variable: green, brick-red, yellow-green, grey or brown; immaculate or spotted or striped. There is a series of large dark subcircular marks on flanks. A dark line through the eyes may extend behind the tympanum to the front limb junction. Underside may be white or greyish-white, sometimes with brown mottling. Taxonomic Notes: A recent phylogenetic analysis placed this distinctive climbing species, unexpectedly, within the ground-dwelling *L. gramineus* group, indicating a recent re-evolution of arboreality.

Leptopelis susanae, Dorze.

Leptopelis susanae, Dorze.

HABITAT AND DISTRIBUTION: An Ethiopian endemic of high-altitude forest and forest remnants. Known from a handful of sites in Gughe Mountains, west of Lake Abaya, at 2,200–2,700m; recently reported further west, at Saja Forest, 60km north of Bonga, at 1,950m.

CONSERVATION STATUS: Endangered.

NATURAL HISTORY: Poorly known. A nocturnal species, climbing into vegetation. Advertisement call a quiet click, sometimes followed by a low creaking sound. Breeding males call from bushes near or overhanging water at edges of stream, in late April. No further breeding details known. Named for Susan Largen, the zoologist wife of Malcolm Largen, who did major herpetological work in Ethiopia.

YALDEN'S TREE FROG

Leptopelis yaldeni
Largen, 1977

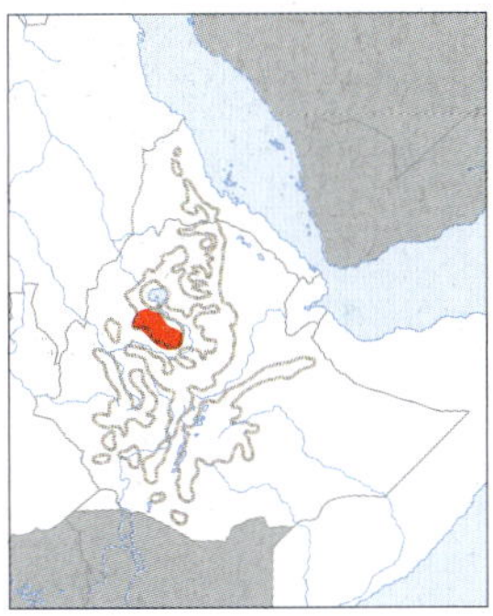

IDENTIFICATION: A medium-sized tree frog, known only from the high land enclosed by the loop of the Blue Nile in Ethiopia. Body oval; head short, with big eyes and vertical pupils. Dorsal skin smooth. Tympanum visible, about ⅔ the size of the eye. Limbs

Leptopelis yaldeni, Chagni.

Leptopelis yaldeni, Injibara.

Leptopelis yaldeni, Debre Marcos.

slim, digit tips spherical, wider than the terminal joint. Males reach a maximum of about 3.4cm, females 4.3cm. Dorsal colour variable: green, light brown, rufous or rich purple-brown, immaculate, marbled or with vague irregular darker markings. Flanks are often marbled black and some shade of white. Underside white or greyish-white, but underside of limbs distinctly pink.

HABITAT AND DISTRIBUTION: An Ethiopian endemic, of high grassland and wooded streams, at 2,000–2,700m. Known only from old Gojjam Province, at Dangila, south-west of Lake Tana, through Injibara south to Debre Marcos, all within land enclosed by the loop of the Ethiopian Blue Nile.

CONSERVATION STATUS: Vulnerable.

NATURAL HISTORY: Poorly known. A nocturnal species, terrestrial but will climb. Advertisement call a long scream, followed by a click. Breeding males call from exposed positions, near pools or slow-flowing streams, on ground or on grasses, reeds and sometimes bushes, up to 2m above water. Breeding occurs throughout the seven-month rainy season, from April to October. Sex ratio of specimens collected in September curiously skewed 60:1 in favour of males. Green individuals commonest in high-altitude open country and pastures, brown ones at lower altitudes near well-vegetated streams. Presumably, like other *Leptopelis*, they conceal their eggs in moist soil.

RAGAZZI'S TREE FROG

Leptopelis ragazzii
Boulenger, 1896

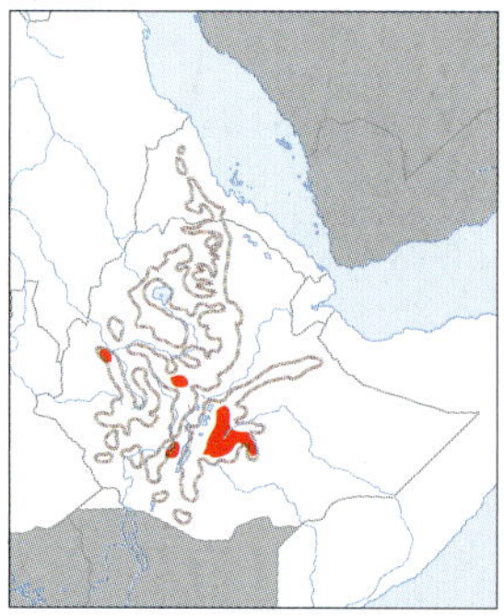

IDENTIFICATION: A fairly large tree frog, often vivid green, confined to Ethiopian montane forests. Body oval; head short, with big eyes and vertical pupils. Dorsal skin smooth, sometimes with fine irregular warts. Tympanum visible but small, less than half the diameter of the eye. Limbs slim, digit tips spherical, wider than the terminal joint. Males reach a maximum of about 4.2cm, females 5cm. Dorsal colour variable: usually green or brown, immaculate or with darker, irregular meandering blotches. A dark stripe through eye often becomes a marbling of black and dorsal colour on flanks, or series of brown blotches. The flanks are often marbled black and some shade of white. Underside white, cream or yellow, often mottled to some extent with grey or brown. Taxonomic Notes: This species and the next (*L. vannutellii*) are virtually indistinguishable, particularly when preserved, and identification has often rested on locality; molecular analysis indicates a recent common ancestor.

HABITAT AND DISTRIBUTION: An Ethiopian endemic of highland forest and forest remnants, at altitudes between 1,900 and 3,100m, although commonest above 2,300m. Curious distribution, with a handful of very widely scattered records west of Ethiopian Rift Valley (Didessa River Valley, Wonchi Crater Lake, Upper Gibe River), also known from highlands around Mt Gughe, west of Lakes Abaya and Chamo, and appears widespread in high country east of Ethiopian Rift Valley, east of a line between Shashamene and Yirga Chefe, to Bale Mountains and Harenna Forest.

CONSERVATION STATUS: Vulnerable.

NATURAL HISTORY: Poorly known. A nocturnal species, terrestrial but will climb. Advertisement call a sharp high-pitched click, followed by a low creak. Breeding males call from low vegetation, from 25cm to 2m above ground, in vicinity of streams. Calling heard between April and September. A copulating pair was observed on ground in long grass near a stream. Presumably, like other *Leptopelis*, they conceal their eggs in moist soil. Specimens were found underneath logs and rocks in the dry season.

Leptopelis ragazzii, Dorze.

Leptopelis ragazzii, Bale.

Leptopelis ragazzii, Harenna Forest.

Leptopelis ragazzii, Harenna Forest.

VANNUTELLI'S TREE FROG

Leptopelis vannutellii
Boulenger, 1898

IDENTIFICATION: A medium-sized tree frog, in forests of south-west Ethiopia, showing a wide range of colours. Body oval; head short, with big eyes and vertical pupils. Dorsal skin smooth, sometimes with fine irregular warts. Tympanum visible, about half the diameter of the eye. Limbs slim, digit tips spherical, wider than the terminal joint. Males reach a maximum of about 4cm, females 4.5cm. Dorsal colour extremely variable: light to dark brown, light and deep green, yellow or grey; immaculate, spotted yellow or with darker, irregular meandering blotches. A dark stripe through eye occasionally present. Flanks often marbled or spotted, with range of colours; brown, green-yellow, or turquoise. Underside white, cream, yellow or blue-green, and skin in limb junctions often blue-green, which is a useful field character.

HABITAT AND DISTRIBUTION: An Ethiopian endemic, from broadleaf forest of the south-west, although it tolerates tree agriculture, particularly banana and coffee plantations. Known from between 1,500 and 2,000m; its range extends from forests between Metu and Bedelle south through Jimma, Bonga and Mizan Teferi to Mt Bibita, east across Omo River to hills near Bulki.

Leptopelis vannutellii, Bonga.

Leptopelis vannutellii, Bonga.

CONSERVATION STATUS: Least Concern.

NATURAL HISTORY: Poorly known. A nocturnal climbing species. Advertisement call a quiet but high-pitched click, followed by a low creak. Breeding males call from branches of trees and bushes near or overhanging pools, small streams and marshes, at heights between 1 and 2.5m. Often in trees, although a mating pair was found on ground and specimens were found on wet roads and paths at night around Jimma. Shelters in arboreal vegetation tangles. They appear to breed throughout much of year, even between November and February, which are the months with least rainfall in its forest habitat. Presumably, like other *Leptopelis*, they conceal their eggs in moist soil.

SQUEAKERS *ARTHROLEPTIS*

A tropical African genus of small, chunky toad-like frogs, rarely more than 3cm long. Most are brown or rufous, often with a darker hourglass shape on the back. Most live in leaf litter, in forests, woodland and wooded savanna. They have a fine skin ridge, a raphe, on the centre of the back, and many males possess an elongate third finger. Their eggs develop directly into froglets, with no tadpole stage. Around 50 species are known, but only a single species occurs in our area. Several species of *Arthroleptis* were previously placed in genus *Schoutedenella*, which is no longer used.

PSEUDO-MOTTLED SQUEAKER

Arthroleptis poecilonotus
Peters, 1863

IDENTIFICATION: A small squat frog of South Sudan. Body sub-rectangular, head short, eyes fairly large, pupil horizontal. Dorsal skin smooth, occasionally with small warts, and faintly granular. Tympanum visible, about half the size of the eye. Limbs slim, digits do not have discs; males have along third finger. Males reach a maximum of about 2.7cm, females 3.4cm. Colour varies a lot but dorsal surface usually brown, grey or rufous, often with a vague darker, or dark-edged hourglass-shaped dorsal blotch. A dark line through the eyes becomes a black spot just above the tympanum. Underside white or grey, often with darker marbling laterally; male's throat dark. Taxonomic Notes: This little frog is part of a species complex, which may be split in the future; South Sudanese examples are probably going to be recognised as a new species. Details here apply to that form.

HABITAT AND DISTRIBUTION: Forest and moist savanna; tolerant of agriculture and degraded forest. In our area, known only at Katire, in Eastern Equatoria, the vicinity of Yei, Central Equatoria, and Bengangai Game Reserve, north-west of Yambio in Western Equatoria, in South Sudan, at around 800m. Also Garamba National Park, in DR Congo, just cross the border from Western Equatoria. Records from Central African Republic suggest it may occur in Western Bahr el Ghazal. Elsewhere, as presently defined, west to Nigeria and Gabon.

CONSERVATION STATUS: Not evaluated.

NATURAL HISTORY: Terrestrial and nocturnal. Advertisement call a long whistle, with rising frequency, likened to chirp of a cricket. Usually found in leaf litter. Breeding males call from the ground. Mating occurs on ground, the eggs are laid in a nest, a cavity at the base of the leaf litter, and after 15–20 days hatch into tiny froglets.

Arthroleptis poecilonotus, Ghana.

FAMILY PHRYNOBATRACHIDAE PUDDLE FROGS

Previously in family Ranidae ('Common frogs'), now elevated to a family containing a single sub-Saharan African genus of about 90 species. Name 'puddle frog' derives from their habit of calling and breeding even in the tiniest of puddles and pools. The majority are small (<3cm) energetic brown frogs, in most species there is more than one colour phase, with or without a vertebral line. In the rainy season they may be active and heard calling by day as well as at night. Their backs are usually warty, often with pairs of elongate warts arranged in a chevron shape or in non-parallel ridges. Eight species occur in our area, an impoverished fauna (Kenya alone has 11 species). However, undiscovered species are undoubtedly present, particularly in tropical South Sudan. At present, an undescribed form is known from south-west Ethiopia, and two species (Parker's Puddle Frog *Phrynobatrachus parkeri*, and Buta Puddle Frog *Phrynobatrachus scapularis*), not described here, are known from Garamba National Park; their status is uncertain but if valid, they might occur in South Sudan. Neither species has ever been photographed in colour.

KEY TO THE NORTH-EAST AFRICAN MEMBERS OF THE GENUS *PHRYNOBATRACHUS*

These small frogs are not easy to identify in the hand, even experts struggle, and some zoogeographic patterns may not make 'sense'. However, the locality may provide clues, several species have tiny ranges.

1a: Large, adults up to 3.8cm, two prominent pairs of chevron-shaped warts between shoulders, often with a red, green or yellow vertebral stripe, tympanum faintly visible — *Phrynobatrachus natalensis*, Natal Puddle Frog

1b: Small, adults up to 3cm, does not have combination of characters above — 2

2a: In our area, only Somalia — 3

2b: In our area, not in Somalia — 4

3a: Maximum size 2.1cm, tympanum invisible — *Phrynobatrachus mababiensis*, Mababe Puddle Frog

3b: Maximum size 3cm, tympanum clearly visible — *Phrynobatrachus acridoides*, East African Puddle Frog

4a: Sudan, up to 3cm long, often with distinct dark flank stripe — *Phrynobatrachus perpalmatus*, Webbed Puddle Frog

4b: Ethiopia, usually <2.5cm, without a dark flank stripe — 5

5a: Digits tips slightly expanded — 6

5b: Digit tips not expanded — 7

6a: Maximum size 2.1cm, dark stripe between nostril and eye, vertebral line rare and narrow if present, known only on Mt Bibita in extreme south-west Ethiopia — *Phrynobatrachus bibita*, Mount Bibita Puddle Frog

6b: Maximum size 1.7cm, no dark stripe between nostril and eye, vertebral line often present and broad; Mt Gughe and highlands south-east of Rift Valley in Ethiopia — *Phrynobatrachus inexpectatus*, Largen's Dwarf Puddle Frog

7a: Maximum size 2.7cm, dorsal warts evenly scattered or absent, only Lake Awassa — *Phrynobatrachus bullans*, Bubbling Puddle Frog

7b: Maximum size 2.3cm, dorsal warts absent, scattered or form vague lines, widespread in highland Ethiopia — *Phrynobatrachus minutus*, Ethiopian Dwarf Puddle Frog

NATAL PUDDLE FROG, SNORING PUDDLE FROG

Phrynobatrachus natalensis
(Smith, 1849)

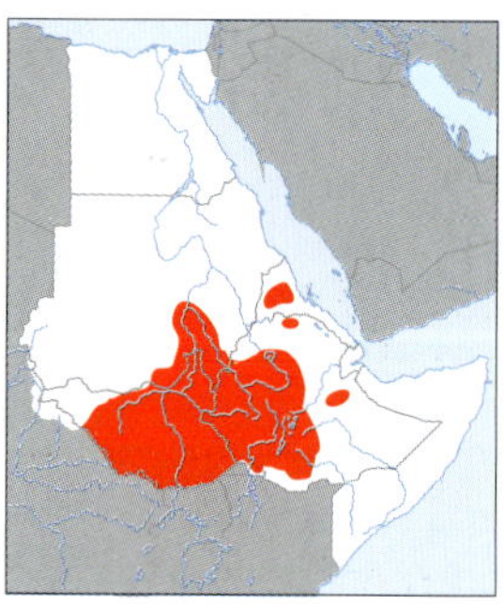

IDENTIFICATION: A relatively large, warty and very variable frog, widespread in the west of our area. Body squat, quite deep, head short, the eyes fairly large, pupil horizontal. Dorsal skin usually warty (rarely smooth). Tympanum faintly visible, about half the size of the eye. Limbs quite stout; digits lack discs. Males reach a maximum of about 3.6cm, females 3.8cm. Colour very variable; pink, grey or brown, sometimes with a broad or narrow red, green or yellow vertebral stripe, sometimes the entire dorsal surface is red or green (a survey in Sudan found the proportion of unstriped to striped individuals was 5:1). Underside white or pink, sometimes with fine spotting and brown freckling on throat; in mature males this is almost black. Taxonomic Notes: Molecular analysis indicates that, as currently defined, this taxon is a complex of cryptic species; true *P. natalensis* occurs in southern Africa, and those from our area are an undescribed form.

HABITAT AND DISTRIBUTION: In wide range of habitats, from semi-desert through dry and moist

Phrynobatrachus natalensis, Fil-fil.

Phrynobatrachus natalensis, south of Turmi.

savanna, woodland and lowland forest; tolerant of agriculture and degraded land; from 300 to above 2,200m, elsewhere to sea level. Occurs in central Eritrea (Asmara, Keren, Fil-fil and adjacent areas), most of western Ethiopia (and a couple of localities in Ahmar Mountains). Probably occurs virtually throughout South Sudan, although records sporadic, but it occurs as far north as Renk in Upper Nile State. In Sudan, probably occurs as far north as the 400mm rainfall isohyet, possibly further north; localities include Khartoum and environs (Jebel Awlia, Soba, Al Sunut Forest), Sennar, Rabak and Kosti, Dilling, Kadugli, Dinder National Park and the Nuba Mountains. It is feasible that increased flooding is carrying them further north along the Nile. Elsewhere, south to South Africa, west to Senegal; a little frog with a huge range. See taxonomic notes above.

CONSERVATION STATUS: Least Concern.

NATURAL HISTORY: Terrestrial and largely nocturnal, but active by day in many places, especially on overcast days in rainy season. Advertisement call a distinctive croak, sometimes likened to a snore, rapidly repeated, with many males joining in together, to produce a far-carrying see-sawing buzz that can often be heard a long way away. Calling may commence in late morning, and go on all afternoon and into the night. Breeding males call from the ground, usually tucked under a grass clump, but may call from floating vegetation, around temporary pools (even quite small puddles), swamps, lakes, dams and slow-moving rivers. Mating occurs on the ground, with eggs laid in shallow water where they float as a single layer on the surface.

MABABE PUDDLE FROG

Phrynobatrachus mababiensis
FitzSimons, 1932

IDENTIFICATION: A small puddle frog, just reaches southern Somalia in our area. Body slightly elongate. Head short, eyes fairly large, pupil horizontal. Dorsal skin warty, sometimes with very large flat warts; often 1–2 pairs of chevron-shaped warts between the shoulders. Tympanum not visible. Limbs quite stout; digits very slightly expanded at the tips, but without discs. Males reach a maximum of about 1.9cm, females 2.1cm. Colour usually brown or grey, with paired or unpaired, irregular darker blotches. Very rarely a rufous vertebral stripe. Dark marbling on flanks. Underside white or grey.

HABITAT AND DISTRIBUTION: In our area, only around Gelib, on the lower Juba River in Somalia, at an altitude of around 20m, but also known from the Tana Delta in Kenya, so probably extends along the Somali coastal plain to the mouth of the Juba River, possibly further north. Elsewhere, south to north-east South Africa, west to Angola.

CONSERVATION STATUS: Least Concern.

NATURAL HISTORY: Terrestrial and largely nocturnal, but sometimes active on overcast days in the rainy season. Advertisement call a short insect-like buzz, followed by a rapid series of individual clicks. Breeding males call from ground, usually under a grass clump, around temporary pools (even very small ones), swamps, lakes, dams and slow-moving rivers. Calling may commence by day, and sometimes diminishes after nightfall. Mating occurs on ground, and eggs are laid in shallow water where they float as a single layer on surface.

Phrynobatrachus mababiensis, Kenya.

EAST AFRICAN PUDDLE FROG

Phrynobatrachus acridoides
(Cope, 1867)

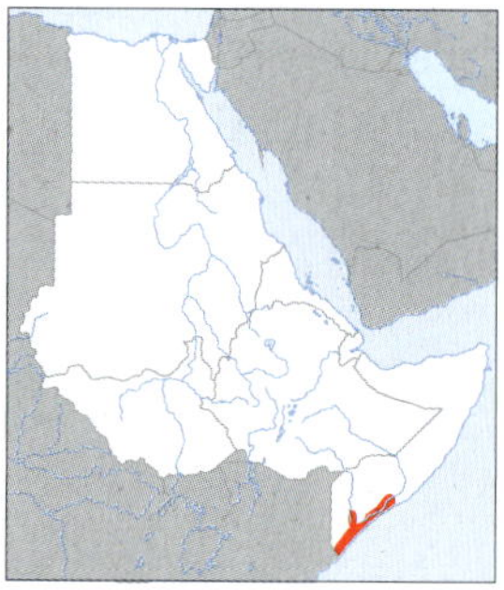

IDENTIFICATION: A small puddle frog, just reaches southern Somalia in our area. Body slightly elongate. Head short, eyes fairly large, pupil horizontal. The dorsal skin is warty, often a pair of long, skewed chevron-shaped warts between the shoulders. Tympanum visible. The limbs are quite stout; digits very slightly expanded to discs at the tips. Males reach a maximum of about 2.8cm, females 3cm. Dorsal colour usually brown, greyish-brown or beige; sometimes with a vertebral stripe (thin or broad), with paired or unpaired, irregular darker or lighter blotches. Underside white or grey, with darker mottling.

HABITAT AND DISTRIBUTION: In our area on the Somali coast, south of 01°N, also at Gelib, on the lower Juba River, at an altitude of around 20m, so presumably extends along the Juba to the sea. Elsewhere, south to north-east South Africa.

CONSERVATION STATUS: Least Concern.

NATURAL HISTORY: Terrestrial; both diurnal and nocturnal. Advertisement call a distinctive croak, *kwaak, kwaak, kwaak*, sometimes likened to a snore, or a drill cutting into a thick bit of wood, repeated several times. Almost semi-aquatic in rainy season, rarely leaving the water, breeding males call day and night from the ground, well hidden on vegetation at water surface. They breed around puddles, temporary pools, swamps, lakes and dams. Mating occurs on the ground, and eggs are laid in shallow water where they float as a single layer on the surface.

Phrynobatrachus acridoides, Kenya.

WEBBED PUDDLE FROG

Phrynobatrachus perpalmatus
Boulenger, 1898

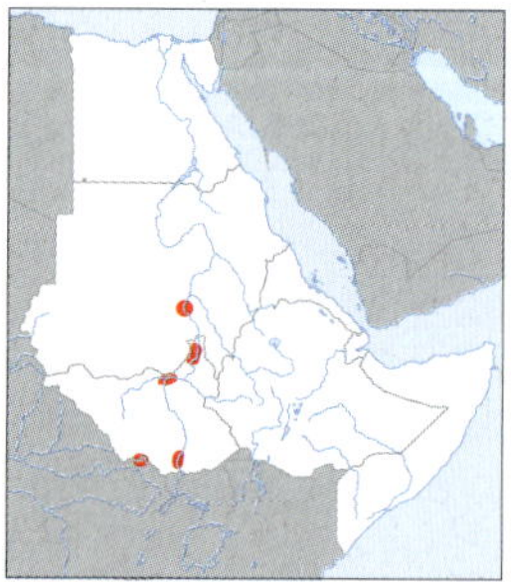

Phrynobatrachus perpalmatus, Congo.

IDENTIFICATION: A small puddle frog, in South Sudan and Sudan. Body slightly elongate. Head short, the eyes fairly large, pupil horizontal. Dorsal skin smooth to granular, sometimes with a few warts. Tympanum visible. Limbs quite stout; digits very slightly expanded to discs at tips. Males reach a maximum of about 2.5cm, females 3cm. Dorsal colour usually brown or greyish-brown, with irregular unbounded dark spots and blotches, sometimes with a vertebral stripe. A dark line through the eye may tail off or extend the length of the flank.

HABITAT AND DISTRIBUTION: Semi-desert, moist and dry savanna, in our area between 400 and 900m. Records in north-east Africa very sporadic. There appears to be only a single one in Sudan (Al Qutaynah, 80km south of Khartoum). Records from South Sudan include Renk and Tonga, in Upper Nile State, a sight record from Juba in Central Equatoria, also known from Garamba in DRC, so probably in Western Equatoria. Elsewhere, south to Zambia and northern Mozambique.

CONSERVATION STATUS: Least Concern.

NATURAL HISTORY: Poorly known. Terrestrial and nocturnal. Advertisement call a metallic clicking or coarse rasping croak. Most specimens in Garamba National Park were near permanent water sources, and breeding occurred May–August, with tadpoles metamorphosing September–October. Presumably, like other puddle frogs, mating occurs on the ground and eggs are laid in shallow water where they float as a single layer on the surface.

MOUNT BIBITA PUDDLE FROG

Phrynobatrachus bibita
Goutte, Reyes-Velasco & Boissinot, 2019

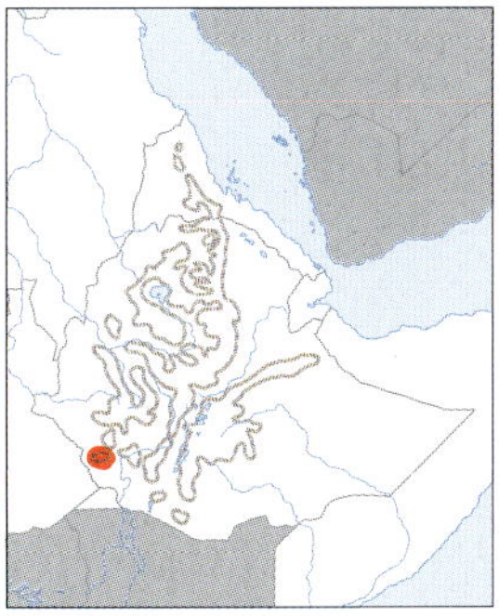

IDENTIFICATION: A tiny gracile golden-brown frog, known only from Mt Bibita in south-west Ethiopia. Body slender and elongate. Head pointed, eyes fairly large, pupil horizontal. Dorsal skin has small warts, and elongate warts form two short oblique ridges behind the eyes and ridges between the shoulders, which may form a chevron. Tympanum not visible. Limbs slim; digits slightly expanded to discs at the tips. Males reach a maximum of about 1.7cm, females 2.1cm. Dorsal colour usually golden to light brown, with a dark chevron between the shoulders. A dark line usually runs from the snout through the eye, to the arm junction. Some individuals have a pale vertebral line. Underside white or grey.

HABITAT AND DISTRIBUTION: An Ethiopian endemic, recently described. All individuals were found around at an overgrown pond in primary forest, at 1,970m altitude, on the eastern slopes of Mt Bibita, in south-west Ethiopia.

Phrynobatrachus bibita, Mount Bibita.

CONSERVATION STATUS: Not Assessed.

NATURAL HISTORY: Poorly known. Advertisement call a series of pulsed notes, said to be at higher frequency than *P. natalensis*. All specimens collected in June, at night; the females and one mating pair were perched on vegetation around 30cm above the water in an overgrown pond; one male was in the water. The females were all gravid. Unusually, clutches, each of roughly 30 eggs, were found on leaves overhanging the water.

LARGEN'S DWARF PUDDLE FROG

Phrynobatrachus inexpectatus
Largen, 2001

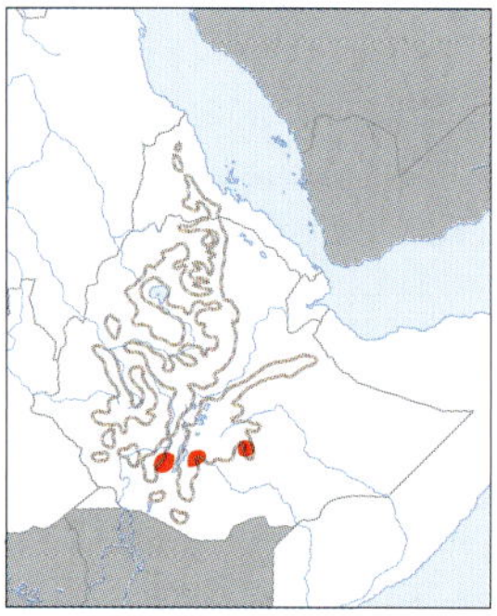

IDENTIFICATION: A small brown puddle frog, endemic to forests of south-central Ethiopia. Body oval and squat. Head pointed, eyes large and prominent, pupil horizontal. The dorsal skin is smooth with some enlarged warts. Tympanum indistinct. Limbs fairly stout; digits slightly expanded to discs at the tips. Males reach a maximum of about 1.9cm, females 2.1cm. Dorsal colour usually some shade of brown (rufous or coppery), with some irregular darker mottling. Some individuals have a yellow or green vertebral line; females may have black flank patches. Underside silvery-white or cream, sometimes with light grey mottling.

HABITAT AND DISTRIBUTION: An Ethiopian endemic. Forest glades, remnant forest and agriculture, at 1,400–2,650m. Known from three localities; Harenna Forest in the Bale Mountains, around the town of Bore, east of Lake Abaya in the Bale foothills, and at Dorze in the Gughe Mountains, west of Lake Abaya.

CONSERVATION STATUS: Data Deficient.

NATURAL HISTORY: Poorly known. Terrestrial and active day and night. Advertisement call described as a high-pitched rattle, followed by several clicks. Specimens were found on the forest floor, under trees, in puddles, and amongst vegetation around pools. Males were observed calling from the water and from low grass at pool margins day and night in Harenna Forest, but on Mt Gughe males called from water at night in a short-grazed, heavily cultivated area. In Harenna Forest mating was observed in puddles during the day following rainstorms. Eggs are attached to vegetation in puddles.

Phrynobatrachus inexpectatus, Harenna Forest.

ETHIOPIA DWARF PUDDLE FROG

Phrynobatrachus minutus
(Boulenger, 1895)

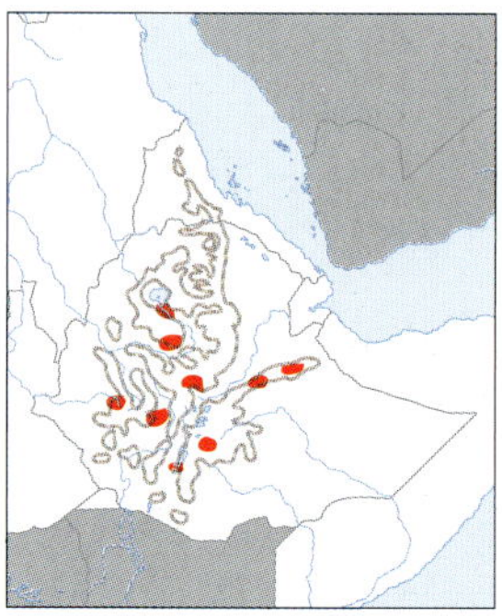

IDENTIFICATION: A small grey or brown puddle frog, endemic but widespread in central Ethiopia. Body oval and squat. The snout is rounded; eyes large and prominent, pupil horizontal. Dorsal skin smooth in some warty to a greater or lesser extent in others; sometimes there are elongate warts between the shoulders. Tympanum present, but indistinct in some individuals. Limbs fairly stout; digits not expanded. Males reach a maximum of about 2cm, females 2.3cm. Dorsal colour usually some shade of brown, with some irregular darker or lighter mottling. Some individuals have light vertebral line. Underside silvery-white or cream. Males have a yellow throat in life.

HABITAT AND DISTRIBUTION: An Ethiopian endemic; in forest glades, high savanna and grassland, at altitudes largely between 1,300 and 2,800m. Rather sporadically recorded from largely highland areas; from Bahar Dar and Debre Markos in north-west, Addis Ababa, a cluster of records in the forests of the south-west in the Bedelle, Jimma and Bonga area. Records east of the Rift Valley include the Ahmar Mountains, Bore, and a surprising low-altitude record at Arba Minch. All records in the Bale Mountains area have been shown to be another species, *P. inexpectatus*.

CONSERVATION STATUS: Least Concern.

NATURAL HISTORY: Poorly known. Terrestrial and nocturnal. Advertisement call described as a series of pulsed notes, similar to *P. bibita*. Specimens were found in vegetation at the margins of waterbodies, and under rocks near small streams. Presumably, like other puddle frogs, mating occurs on the ground and eggs are laid in shallow water where they float as a single layer on the surface.

Phrynobatrachus minutus, Bonga.

BUBBLING PUDDLE FROG

Phrynobatrachus bullans
Crutsinger, Pickersgill, Channing & Moyer, 2004

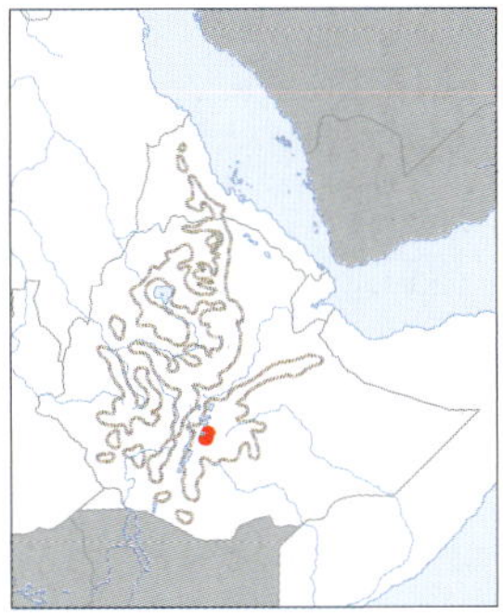

IDENTIFICATION: A small warty puddle frog, with a single record in our area. Body ovoid. Head short, eyes fairly large, pupil horizontal. Dorsal skin covered with regularly spaced, hemispherical warts. Tympanum just visible. The limbs are quite stout; digit tips not expanded. Males reach a maximum of about 2.4cm, females 2.7cm. Dorsal colour usually light or dark brown, with irregular brown or green blotches, sometimes a pale vertebral line is present. Underside white.

Phrynobatrachus bullans, Tanzania.

HABITAT AND DISTRIBUTION: Savanna and grassland. In our area a single, curiously disjunct record on east shore of Lake Awassa, in Ethiopia's Rift Valley, at 1,700m. The nearest record is from Homa Bay, on Kenya's Lake Victoria, >900km away. Elsewhere, northern Tanzania.

CONSERVATION STATUS: Least Concern.

NATURAL HISTORY: Terrestrial; both diurnal and nocturnal. Advertisement call a long series of pulsed notes, starting softly and gradually increasing in volume over 16 seconds. Found around shallow temporary pools, rice fields and seasonal rivers. Males call from concealed positions, partially submerged. The eggs are laid in shallow water where they float as a single layer on the surface.

FAMILY PETROPEDETIDAE TORRENT, WATER AND BALE MOUNTAINS HEATHER FROGS

A small family of three genera and 12 species of tropical African frogs, found in woodland, forest and mountains, and largely associated with flowing streams. There is no common name for the whole family. A single monotypic genus occurs in our area.

BALE MOUNTAINS HEATHER FROGS *ERICABATRACHUS*

A monotypic Ethiopian genus; a small montane frog, endangered, with a tiny range.

BALE MOUNTAINS HEATHER FROG

Ericabatrachus baleensis
Largen, 1991

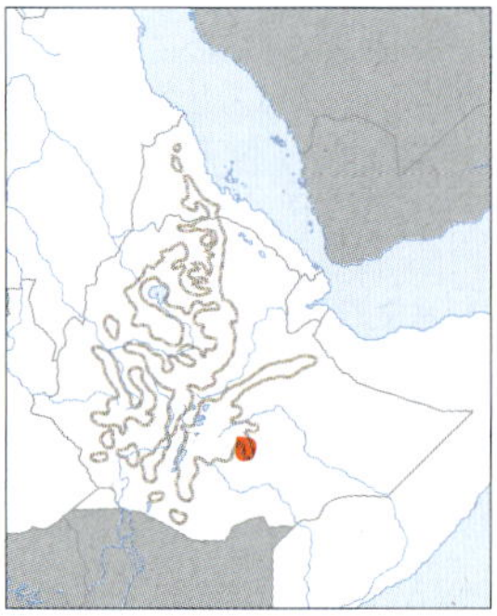

IDENTIFICATION: A tiny frog, resembling a puddle frog, with heart-shaped digit tips; known only from Ethiopia's Bale Mountains. Body ovoid. Head short, eyes fairly large, pupil horizontal. The dorsal skin is warty, sometimes forming elongate dorsal ridges. Tympanum invisible. Limbs thin; digit tips split into two flaps, which appear heart-shaped from above and spherical from below. Males reach a maximum of about 2.2cm, females 2.7cm. Dorsal colour variable; cream, shades of brown or dull green, mottled or marbled white or darker brown; there is usually a dark bar between the eyes and a dark stripe from behind the eye to the forelimb junction. Underside black, deep pink or purple, with large white spots, and bright yellow at the hindlimb junction; throat sometimes suffused rufous-brown.

HABITAT AND DISTRIBUTION: Forest. Ethiopian endemic; known only from Harenna Forest on the southern slopes of Bale Mountains, where it occurs from giant heather forest at 3,200m to broadleaf forest at 2,400m.

CONSERVATION STATUS: Critically Endangered. Its habitat in Bale Mountains National Park is nominally protected, but many people live in the park, taking firewood, farming and grazing stock. Recent research found the type locality degraded and no frogs. However, only a single road traverses the forest;

Ericabatrachus baleensis, Harenna Forest.

Ericabatrachus baleensis, Harenna Forest.

specimens and suitable habitat can be found away from the road at similar altitude.

NATURAL HISTORY: Poorly known. Terrestrial. Advertisement call unknown. Most specimens were found in vegetation at the edge of a mountain stream, they attempted to escape by leaping into the water and swimming to and clinging onto large boulders. Females contained large unpigmented eggs, which might indicate either direct development, without a tadpole stage, or a terrestrial nest.

Ericabatrachus baleensis, underside, Harenna Forest.

FAMILY PTYCHADENIDAE GRASSLAND FROGS OR ROCKET FROGS AND ALLIES

Previously placed in family Ranidae ('Common frogs'), now elevated to a small family of three genera and about 55 species of African frogs. These frogs are relatively large and share common features of the skull and clavicle. All three genera occur in our area. The largest group, the rocket frogs, *Ptychadena*, are very dissimilar to the other two.

ORNATE FROGS *HILDEBRANDTIA*

A sub-Saharan Africa genus of three species of big, attractive, stocky, brightly coloured frogs, with a somewhat supercilious look. They inhabit savanna and semi-desert. All have a large inner metatarsal tubercle, and a hard protrusion on the foot used for digging in the hard soil. They are rarely seen above ground, spending most of their lives buried, but emerge after heavy rain. Three species, of which two occur in our area.

KEY TO THE NORTH-EAST AFRICAN MEMBERS OF THE GENUS *HILDEBRANDTIA*

1a: In our area only Somalia and eastern Ethiopia, dorsal surface rufous, immaculate, throat white
Hildebrandtia macrotympanum, Northern Ornate Frog

1b: In our area only South Sudan, dorsal surface striped and green or brown, throat dark with highly contrasting white bars
Hildebrandtia ornata, Common Ornate Frog

NORTHERN ORNATE FROG

Hildebrandtia macrotympanum
(Boulenger, 1912)

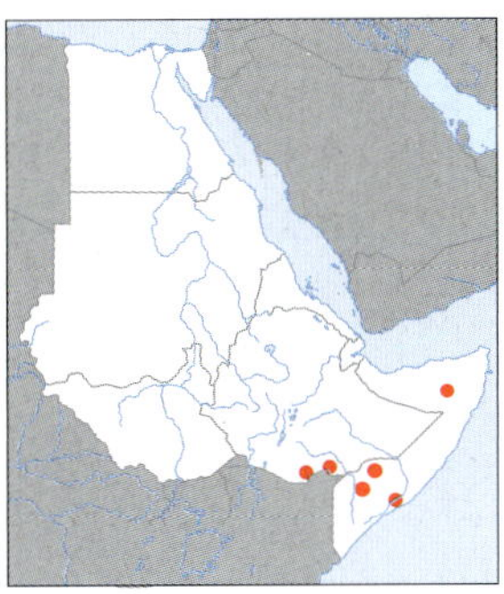

IDENTIFICATION: A big rufous frog of dry country in Horn of Africa. Body ovoid. Head short, eyes fairly large, pupil horizontal. Dorsal skin smooth. Males have curious black vocal sacs just in front of the armpits. Tympanum large, often wider than eye. Limbs stout; digit tips not widened. Males reach a maximum of about 4.8cm, females 5.3cm. Dorsal colour rufous-brown to almost red. A rich brown stripe from the snout passes through the eye, narrows, then broadens on flanks. A fine black dot between the eyes. Underside white, with some black speckling on the throat.

HABITAT AND DISTRIBUTION: Semi-desert and dry savanna, between sea level and 1,100m. Sporadic records include Gardo in north-east Somalia, a cluster of records from between the lower Juba and Webi Shebelle Rivers in southern Somalia, a couple of records from extreme southern Ethiopia in the vicinity of the Kenya border near Malka Murri and in savanna south-east of Filtu. It probably occurs throughout this landscape. Elsewhere, known from Mt Ololokwe, Tsavo East National Park and the vicinity of Mt Kasigau in Kenya.

CONSERVATION STATUS: Least Concern.

NATURAL HISTORY: Poorly known. Terrestrial and nocturnal. Advertisement call a deep grunt-like hoot, with several uttered in fairly rapid succession. Spend most of their lives buried deep underground, but emerge following large storms and breed in temporary pools and dams. Males call from shallow water, commencing after dark; Kenyan males were calling in November following a huge rainstorm. The Filtu specimens were found in the rainy season, in late October, after recent rain; local trees were in leaf.

Hildebrandtia macrotympanum, near Filtu.

COMMON ORNATE FROG

Hildebrandtia ornata
(Peters, 1878)

Hildebrandtia ornata, Gambela.

IDENTIFICATION: A big attractive striped frog of western Ethiopia and South Sudan. Body ovoid. Head short, eyes fairly large, pupil horizontal. The dorsal skin is smooth. Males have curious slate-grey vocal sacs just in front of the armpits. Tympanum large, almost the size of the eye. The limbs are stout; digit tips not widened. Males reach a maximum of about 6cm, females 7cm. Dorsal colour very variable, but those in our area are usually brown, a broad rufous or brown stripe from the snout splits into a dorsolateral stripe and a broad descending stripe onto the flanks. A green vertebral stripe may be present; occasional all-green animals occur. Underside white; on the dark chest and throat are two highly contrasting white bars, which sometimes divide.

HABITAT AND DISTRIBUTION: Dry and moist savanna. Rather sporadically recorded in our area; localities include Gambela in Ethiopia, North Kordofan, Sudan, Torit, and between Juba and Torit in South Sudan. Records from Garamba National Park in DR Congo, northern Uganda and eastern Central African Republic indicate it might be widespread in southern South Sudan. Elsewhere, west to Senegal, south to South Africa and Angola.

CONSERVATION STATUS: Least Concern.

NATURAL HISTORY: Terrestrial and nocturnal. Advertisement call a short, loud nasal bellow, *wheee-iik*. They spend much of their lives buried underground, but emerge following large storms and breed in temporary pools and dams. Males call from shallow water or semi-concealed positions near the water's edge, their huge paired vocal sacs ballooning out from under the jaw. Mating was noted in April in Garamba National Park. The female releases eggs singly, in shallow water, moving after each release.

LANZA'S FROG *LANZARANA*

A genus endemic to the Horn of Africa, with a single species, known only from Somalia. Originally placed in the genus *Hildebrandtia*, later moved to its own genus, its DNA has never been analysed, but general consensus is that *Hildebrandtia*, then *Ptychadena*, are its closest relatives.

LANZA'S FROG

Lanzarana largeni
(Lanza, 1978)

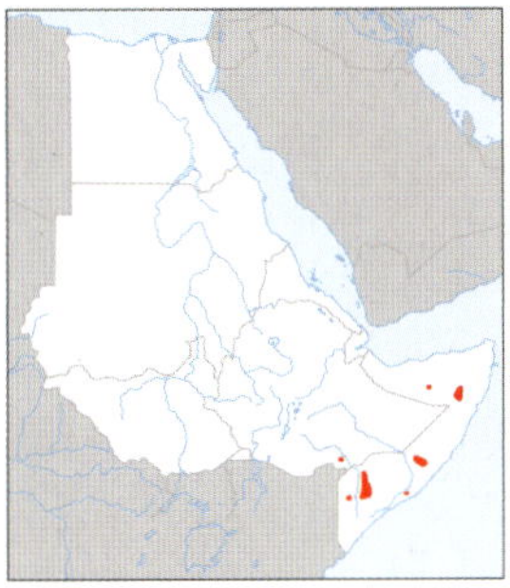

IDENTIFICATION: A chunky, spotted terrestrial frog of Somalia. Stocky, with a short, slightly flattened head and rounded snout, moderately prominent eyes and a horizontal pupil. Dorsal surface smooth with a few low warts. A skin fold runs from eye around top of the tympanum to the arm. Tympanum about half the eye diameter. Limbs moderately long with slightly expanded tips on fingers; unusually the tips of the second to fourth finger dilate and become discoidal in correlation with the rainy season. Males with external vocal sacs in front of the armpits. Males reach a maximum of about 4.5cm, females 4.9cm. Dorsal surface hazel-grey with more or less extensive light brown marbling. There is usually a dark butterfly-shaped transverse spot between the eyes, two pairs of larger irregular blotches between shoulders and a dark spot above the rump. A dark stripe extends from the snout through the eye and tympanum to the flanks; a side branch may extend to the shoulder. A pale vertebral stripe is usually present. Underside white, breeding males have a white throat. Almost pattern-less specimens are known.

HABITAT AND DISTRIBUTION: Semi-desert, dry savanna and xerophilous open woodland, at low elevations between 80 and 630m. Endemic to Somalia. Only records from north are historical, in Nogal Valley, at Halin and Garowe, with recent records near God Heeli, in upper Nogal Valley. Also Mahas and Bud Bud, in Hiraan, and a cluster of records south of 04°N, in more mesic habitat between the Juba and Shebelle Rivers (Dinsor, Afgoi, El Ure). Probably more widespread, but overlooked; may occur in extreme eastern Kenya and Ethiopia.

CONSERVATION STATUS: Least Concern; widespread and apparently common.

NATURAL HISTORY: Poorly known. Terrestrial and nocturnal. Advertisement call unknown. Largely lives underground, especially in central and northern regions, in burrows and crevices, under boulders or in root tangles. Active after sporadic heavy rainstorms, or during wet season, when the water creates temporary pools and fills wells and dams, stimulating males

Lanzarana largeni, Nogal Valley.

Lanzarana largeni, Nogal Valley.

to call from shallow water; recently observed by day following rain, moving between clumps of vegetation. Specimens from less arid areas in the south were found on the surface under piles of rotting leaves around wells, on the walls of water reservoirs or underground on the walls of deep limestone wells. Probably not a true burrowing frog, unlike *Hildebrandtia*. The specific name honours Malcolm Largen (1941–2017), British biologist, who worked at Liverpool Museum and Haile Selassie University, Addis Ababa in Ethiopia, his main interests were the amphibians and reptiles of Ethiopia and Eritrea, he authored many scientific articles and in 2010 co-authored the first book on the herpetofauna of Ethiopia and Eritrea.

ROCKET, RIDGED, GRASS OR SHARP-NOSED FROGS *PTYCHADENA*

A successful African genus (has also dispersed to Madagascar and other islands in western Indian Ocean) of small to medium-sized frogs, characterised by long snouts, long slender hind legs and a series of dorsal and lateral skin folds or ridges. They have broad, muscular back legs and the ability to make huge jumps; one species holds the world record for the longest jump by any frog. This leaping ability means they often forage a long way away from standing water, returning to a suitable refuge near dawn. They hide by day among plants near water, and when disturbed they leap either into the water or into vegetation, where they then burrow down and hide. They do not climb (although they may shuffle up among vegetation), and their toe tips are not noticeably enlarged, so these are not referred to in the species accounts. The male vocal sacs expand through slits either side of the jaw. Breeding usually occurs at night, often in temporary pools. Some species are very tolerant of urbanisation, and can exist in relatively polluted water.

The conservative morphology of rocket frogs has created taxonomic uncertainties; they all look rather similar. About 60 species are known, their number recently boosted by revisions of the Ethiopian highland forms, which have added several new species. About 23 species are known from our area (the reason for this uncertainty is given in the taxonomic notes for the relevant species). Rocket frogs are often difficult to identify to species level, even by experts. Taxonomists have sometimes shuffled the same animal between three or four species; separate species have been combined, single species split, species redefined. For field identification, habitat may give clues. In our key, especially for the couplets from 11 onwards, we have had to revert to using some webbing details and very precise measurements, impractical for use with live animals. This is a consequence of a 2021 revision by Sandra Goutte and her team of the Ethiopian highland taxa originally lumped under *P. neumanni*, which have a recent common ancestor; many appear very similar, and they hybridise. We suggest using a combination of the keys, the pictures and habitat to identify these frogs, but certain identification in the field may not be possible.

KEY TO THE NORTH-EAST AFRICAN MEMBERS OF THE GENUS *PTYCHADENA*

1a: Not usually in the Ethiopian highlands, found elsewhere below 1,500m	2
1b: Usually in Ethiopian highlands, above 1,500m	11
2a: Large, adults 6cm long or more	3
2b: Small, adults <6cm long	6
3a: Never has vertebral stripe, dorsal surface brown, rufous or tan, usually a pale triangle on snout, up to 6.2cm long, widespread in our area	*Ptychadena anchietae*, Anchieta's Rocket Frog
3b: Lacks above combination of characters, in our area only in South Sudan	4
4a: Heavily built, body deep, three pairs of dorsal skin folds, maximum length of females 6cm, iris usually uniform brown or slightly more yellow at top	*Ptychadena trinodis*, Dakar Rocket Frog

4b: Body elongate, not deep, 3–4 pairs of dorsal skin folds, maximum length of female >6cm, iris has a highly contrasting yellow band across top, darker below 5

5a: Up to 8cm, rarely has light vertebral stripe *Ptychadena oxyrhynchus*, Sharp-nosed Rocket Frog

5b: Up to 6.5cm, usually has light vertebral stripe *Ptychadena perreti*, Perret's Rocket Frog

6a: Dorsal skin folds weak and often indistinct, dorsal surface often rufous, with or without darker mottling, dorsal surface not striped *Ptychadena tellinii*, Tellini's Rocket Frog

6b: Dorsal skin folds prominent and distinct, dorsal surface often striped, rarely rufous 7

7a: Small, compact, maximum length 3.6cm *Ptychadena pumilio*, Western Dwarf Rocket Frog

7b: Larger, elongate, maximum length more than 3.6cm 8

8a: Endemic to southern Somalia, maximum length 4cm *Ptychadena gansi*, Somali Rocket Frog

8b: Not endemic to Somalia, maximum length 5.8cm 9

9a: Dorsal folds often not continuous, back looks lumpy, often with red spotting *Ptychadena schillukorum*, Schilluk Rocket Frog

9b: Dorsal folds usually continuous, back looks ridged but not lumpy, rarely with red spots 10

10a: In our area, only South Sudan, 2–4 pairs of dorsal skin folds, dorsolateral fold often short or incomplete, rarely with green wash on thighs or leg-body junction *Ptychadena bibronii*, Bibron's Rocket Frog

10b: In our area widespread, South Sudan to Ethiopia and along Nile to Egypt, four pairs of dorsal skin folds, dorsolateral fold usually elongate, often with green wash on thighs and leg-body junction *Ptychadena nilotica*, Nile Rocket Frog

11a: Large (male >4.5cm, female >5.5cm), dorsal folds formed by rows of short ridges, foot webbing extends beyond second phalange of fourth and fifth toe, dark grey vocal sacs *Ptychadena cooperi*, Dark-bellied Rocket Frog

11b: Small, male <4.7cm, female <5.6cm 12

12a: Very small (male <3.2cm, female <3.45cm), short hind limbs (male thigh length/length 0.37–0.4, female thigh length/length 0.33–0.47), foot webbing extends at most to second phalange of fourth toe, vocal sacs at least partially pigmented grey 13

12b: Small to medium body length (male 3.1–4.7cm, female 3.3–5.6cm), foot webbing extends at most to second phalange of fourth toe 15

13a. Outer metatarsal tubercle present, dorsal skin folds uninterrupted except in sacral region; back of thigh with pattern of dark streaks on pale background, confined to area immediately south of Bahar Dar *Ptychadena wadei*, Wade's Rocket Frog

13b. Outer metatarsal tubercle absent, back of thighs mottled or marbled dark and light brown or cream, not confined to Bahar Dar 14

14a: No stripe or blotch on the tympanum, small warts almost always present in adult males during the breeding season, known only from the highlands east of Lake Zwai *Ptychadena nana*, Smallest Rocket Frog

14b. Light blotch on tympanum, no warts, only Bale Mountains *Ptychadena robeensis*, Robe Rocket Frog

15a. No stripe or blotch on tympanum, vocal sacs yellow or cream, adult male with thickened forearms *Ptychadena erlangeri*, Erlanger's Rocket Frog

15b. Lacks above combination of characters 16

16a: Long hind limbs (male tibia length/length 0.52–0.77, female tibia length/length 0.55–0.85; male thigh length/length 0.42–0.64, female thigh length/length 0.43–0.80) 17

16b: Short hind limbs (male tibia length/length 0.37–0.56, female tibia length/length 0.44–0.61; male thigh length/length 0.33–0.5, female thigh length/length 0.40–0.52) 20

17a. Large body length (male 36.6–47mm, female 46.4–55.4mm), long hind limbs (male tibia length/length 0.57–0.77, female tibia length/length 0.57–0.62), well-developed metatarsal tubercles and relatively large toe discs, light vocal sacs *Ptychadena goweri*, Gower's Rocket Frog

17b: Lacks above combination of characters 18

18a: Vertebral stripe absent and only few faint dark markings on back, dark grey vocal sacs *Ptychadena harenna*, Harenna Forest Rocket Frog

18b: Lacks above combination of characters 19

19a: Tympanum small (male tympanum diameter/eye diameter 0.44–0.81, females tympanum diameter/eye diameter 0.49–0.8), eyes close to one another (male: distance between nostrils/between eyes 0.91–1.43, female: distance between nostrils/between eyes 0.95–1.48), vocal sacs dark grey or cream and grey *Ptychadena doro*, Chicken Rocket Frog

19b: Tympanum larger (male tympanum diameter/eye diameter 0.56–0.88, female tympanum diameter/eye diameter 0.60–1.0), eyes further apart (male: distance between nostrils/between eyes 0.84–1.16, female: distance between nostrils/between eyes 0.85–1.31), vocal sacs dark grey *Ptychadena neumanni*, Neumann's Rocket Frog

20a. Body length small (male 31–38mm, female 34.5–40.5mm), snout short (male snout length/length 0.09–0.15, female snout length/length 0.12–0.14), vocal sacs most often bicoloured, cream and grey *Ptychadena levenorum*, Levene's Rocket Frog

20b. Body length medium to large (male 33.8–44.2mm, female 38.4–49.2mm), snout longer (male snout length/length 0.14–0.17, female snout length/length 0.13–0.17) 21

21a. Eyes close to one another (male distance between eyes/eye diameter 0.42–0.61, female distance between eyes/eye diameter 0.43–0.69), vocal sacs light grey or cream, males covered with small warts during breeding season *Ptychadena amharensis*, Amhara Rocket Frog

21b: Eyes further apart (male distance between eyes/eye diameter 0.51–0.86, female distance between eyes/eye diameter 0.52–1.0), vocal sacs at least partially dark grey 22

22a. Forearm moderately long (forearm length/length 0.16–0.20) and not thickened in adult males, vocal sacs bicoloured cream and grey, warts almost always absent in males *Ptychadena beka*, Enough Rocket Frog

22b. Forearms relatively long (forearm length/length 0.19–0.22), vocal sacs dark grey or bicoloured, cream and dark grey *Ptychadena delphina*, Clicking Rocket Frog

ANCHIETA'S ROCKET FROG

Ptychadena anchietae
(Bocage, 1868)

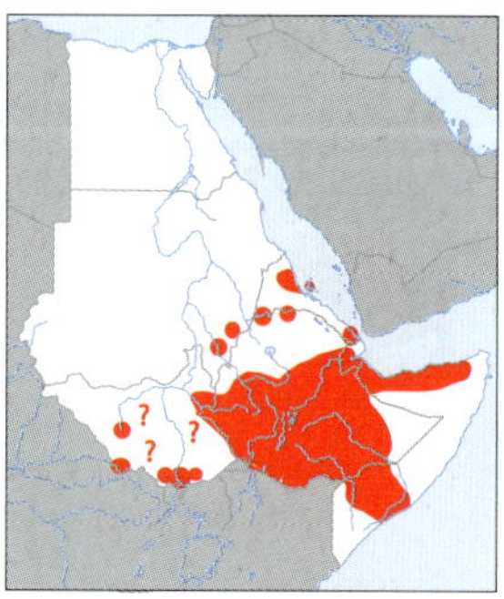

IDENTIFICATION: A big, unstriped sharp-nosed frog, with a pale triangle on the snout, widespread and often common in lowlands of the Horn of Africa. Body compact. Head relatively large and the snout is long and pointed; the eyes fairly large, pupil horizontal but looks oval. Dorsal skin has series of four pairs of long folds, although these may be flat and indistinct. Tympanum has a pale border, and slightly smaller than the eye diameter. Males reach a maximum of about 5.3cm, females 6.2cm. Dorsal colour various shades of brown, rufous or tan. Unlike most *Ptychadena*, there is never a vertebral stripe. Dorsal surface varies from immaculate to extensively blotched. Flanks marbled black, forming a diffuse stripe to a greater or lesser extent. Underside white, sometimes suffused yellow posteriorly. Taxonomic Note: The species *P. baroensis* was recently described from the Baro River, in westernmost Ethiopia. Its formal description was invalid, but specimens of *P. anchietae* from lowlands west of the Ethiopian plateau may belong to this smaller form, which has a less prominent pale snout triangle.

HABITAT AND DISTRIBUTION: Semi-desert, dry and moist savanna, tolerant of agriculture and urbanisation, between sea level and 1,500m, sometimes higher. Northern and southern Somalia (absence from the centre probably due to under-collecting), Eritrea (quite widespread in centre, also Dahlak Kebir Island and Gaarre, in the south), Djibouti, most of lowland Ethiopia, several localities in South Sudan (Yambio, Juba, Torit, Katire, Talanga Forest, Wau, Malakal and environs). In Sudan, seemingly confined to low states near the Ethiopian border (eastern Gadarif, eastern Blue Nile), but probably more widespread. Elsewhere, extends south to South Africa and southwest to Angola.

CONSERVATION STATUS: Least Concern.

NATURAL HISTORY: Terrestrial and nocturnal. Advertisement call a distinctive, see-sawing high-pitched trill, repeated rapidly; males also produce a harsh territorial call. Usually found fairly near a water source (both seasonal and permanent), and are active as long as there is standing water. In the dry season shelter in holes, under big logs and rocks. Males call, partially concealed or in open, from shallow water, commencing after dark. If disturbed, they jump away from the water. Eggs are laid in temporary pools, and float in a single layer on the water surface; up to 300 eggs recorded in a single raft.

Ptychadena anchietae, Dire Dawa.

Ptychadena anchietae, Cal Madow Mountains.

Ptychadena anchietae, Yambio.

Ptychadena anchietae, Sodere.

BIBRON'S ROCKET FROG

Ptychadena bibronii
(Hallowell, 1845)

IDENTIFICATION: A fairly large, striped, sharp-nosed frog, in our area just reaches South Sudan. Body compact. Head relatively large and snout long and pointed; eyes fairly large, pupil horizontal but looks oval. Dorsal skin has a series of 2–4 pairs of long folds. Tympanum has a pale border, and is slightly smaller than the eye diameter. Males reach a maximum of about 4.8cm, females 5.6cm. Dorsal colour varies; grey, brown or green, with large black spots; a vertebral stripe of rufous, brown or yellow is often present. Flanks granular, sometimes with darker marbling. Underside white to grey.

HABITAT AND DISTRIBUTION: Moist and dry savanna, sometimes degraded forest, between sea level and 1,000m. In our area, the only definite records are from Torit and Talanga Forest, Eastern Equatoria, in South Sudan, but known also from Garamba National Park in DR Congo, so probably in Central and Western Equatoria. Elsewhere, west to Senegal. Conservation Status; Least Concern.

Ptychadena bibronii, Ivory Coast.

NATURAL HISTORY: Terrestrial and nocturnal. Advertisement call a distinctive series of quacks, like a little duck. Breeds in temporary pools and ponds. Males call from vegetation or otherwise concealed. Aestivates in the dry season. Eggs are laid in temporary pools, and float in a single layer on the water surface; one raft contained 300 eggs.

SOMALI ROCKET FROG

Ptychadena gansi
Laurent, 1965

IDENTIFICATION: A fairly small sharp-nosed frog, with low-contrast striping; endemic to Somali. Body compact. Head relatively large and snout long and pointed; eyes fairly large, pupil horizontal but looks oval. Dorsal skin has four pairs of long ridges, often interrupted. Tympanum lacks a pale border, and is slightly smaller than the eye diameter. Males and females reach a maximum of about 4cm. Dorsal colour usually brown or grey-brown, with scattered small black spots; a pale vertebral stripe usually present. A dark stripe from the nostril breaks up behind the eye. Flanks granular, sometimes with darker marbling. Taxonomic Notes: Similar in appearance to *P. mossambica*; some authorities consider the two to be conspecific; a molecular analysis would be interesting.

HABITAT AND DISTRIBUTION: Dry savanna and semi-desert, below 200m. Endemic to Somalia, from south of 04°N, east of 42°E, on the coastal plain, most records on floodplains close to the lower reaches of the Juba and Shebelle Rivers.

CONSERVATION STATUS: Least Concern.

NATURAL HISTORY: Poorly known. Presumably terrestrial and nocturnal. Advertisement call unknown. Males observed calling from flooded fields; breeding noted in temporary pools, sometimes in places where water runs off rock faces. Presumably aestivates in the dry season.

Ptychadena gansi, preserved.

NILE ROCKET FROG

Ptychadena nilotica
(Seetzen, 1855)

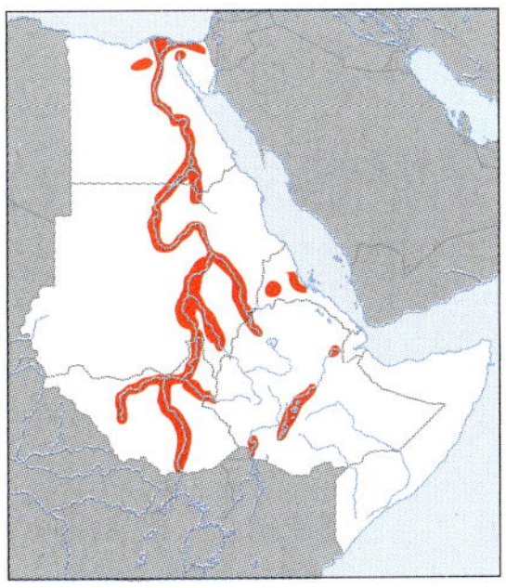

IDENTIFICATION: A fairly large, common, striped, sharp-nosed frog, often with green on body; widespread in our area, from South Sudan north to the Egyptian coast. Body compact. Head relatively large and snout long and pointed; eyes fairly large, pupil horizontal but looks oval. The skin surrounding the eye is sometimes blue-green. Dorsal skin has a series of four sharply defined pairs of long folds. Tympanum has a pale border and is slightly smaller than the eye diameter. Males reach a maximum of about 4.7cm, females 5.8cm. Dorsal colour variable; black, brown, grey or green, with large black spots; a vertebral stripe, usually broad but sometimes narrow, of green, brown or yellow often present. Flanks granular, sometimes with darker marbling. Underside white to cream. Taxonomic Notes: Resurrected from the pan-African *P. mascareniensis* complex, which contains at least 12 independently evolving lineages. Those in our area assigned to *P. nilotica*, including *P. filwoha*, originally described as an Ethiopian endemic. Some examples of *P. mascareniensis* might occur in South Sudan, pending clarification we include them here. Note; *P. nuerensis* was described recently from lowlands of western Ethiopia. Formal description incomplete; for now we leave it in *P. nilotica*.

HABITAT AND DISTRIBUTION: Woodland, moist and dry savanna, semi-desert, even desert near waterbodies; sea level to above 2,000m. In our area, essentially known from the two Niles and their tributaries and surroundings, from headwaters in Rwanda to Uganda, and thence through South Sudan, Sudan and Egypt. Away from Nile catchment, also known from the Ethiopian Rift Valley (Awash National Park south to Lake Chamo), several localities in eastern and central Eritrea, the Gendoa River in north-west Ethiopia, from Asaiyta near Djibouti border, and the Faiyum, southern Suez Canal and north-east Sinai in Egypt. Probably more widespread, but confusion with

Ptychadena nilotica, Asayita.

Ptychadena nilotica, Aswan.

Ptychadena nilotica, Fayum.

records of *P. mascareniensis* needs resolving. Elsewhere, south to northern South Africa.

CONSERVATION STATUS: Least Concern.

NATURAL HISTORY: Terrestrial, active by day (usually in afternoon) and night. Advertisement call a distinctive series of nasal quacks, followed by clucking sounds. By day, often concealed among waterside vegetation; if disturbed it jumps. Breeds in temporary pools, ponds, rice fields, slow streams, dams; tolerant of agriculture and breeds in ditches and flooded fields, in Egypt often in canals. Males call from the edges of water. Aestivates in the dry season. Eggs are laid in temporary pools. If held, this frog becomes rigid, secretes foam through the skin and utters a moan that may distract predators.

SHARP-NOSED ROCKET FROG

Ptychadena oxyrhynchus
(Smith, 1849)

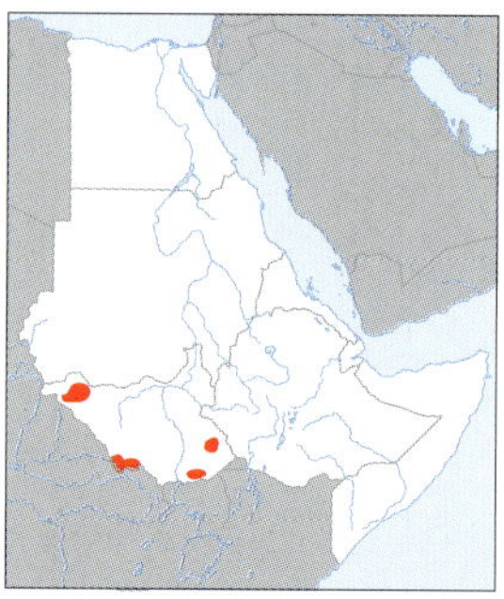

IDENTIFICATION: A big rocket frog, with an acutely pointed snout, in our area only in South Sudan. Body squat and compact, described as 'gnome-like'. Head relatively large and snout long and sharply pointed; eyes fairly large, pupil horizontal but looks oval. Dorsal skin has 3–4 pairs of long, narrow, distinct ridges. Tympanum has a pale border and is almost as big as the eye. Males reach a maximum of 6cm, females 8cm, perhaps more. Dorsal colour usually brown, grey-brown or rufous, with scattered small black spots. Appears to lack a vertebral stripe. Underside white or grey.

HABITAT AND DISTRIBUTION: Moist and dry savanna and semi-desert, at 500–1,000m in our area, but to sea level and higher elsewhere. Only a handful of records in South Sudan (Boma, in Jonglei State, the Boro River in Western Bahar El Ghazal State, Gilo in Eastern Equatoria, and Yambio, Western Equatoria), but probably more widespread; known from northern Uganda and Garamba National Park in DR Congo, so probably in Central Equatoria. Elsewhere, west to Senegal, south to South Africa, outside of the rainforest.

CONSERVATION STATUS: Least Concern.

NATURAL HISTORY: Terrestrial and nocturnal. Advertisement call a series of loud, low-pitched screeches, about one per second, *craaa…craaa… craaa*. May forage some distance from standing water. Males call from partially concealed sites in shallow water or at water's edge, usually at night but also on overcast days. Breeds in temporary pools of any size (even little ones on rock outcrops), flooded grasslands, pans and ditches. In Garamba National Park, breeding commenced in March, at start of rainy season. They aestivate underground in the dry season. The usual raft of eggs is laid. This frog holds the record for the longest jump of any frog; 9.83m in three successive leaps.

Ptychadena oxyrhynchus, Cameroon.

PERRET'S ROCKET FROG

Ptychadena perreti
Guibé & Lamotte, 1958

IDENTIFICATION: A medium-sized rocket frog, with an acutely pointed snout, in our area only in South Sudan. Body squat and compact; head relatively large and snout long and sharply pointed; eyes fairly large, pupil horizontal but looks oval. The dorsal skin has four pairs of distinct ridges, interrupted in places. Tympanum has a pale vertical bar and is almost as big as the eye. Males reach a maximum of 5.0cm, females 6.5cm. Dorsal colour usually brown, or grey with numerous dark spots. There is a beige to red vertebral band, sometimes with a fine pale line in the centre. Underside white or grey.

HABITAT AND DISTRIBUTION: Forest, degraded forest and moist savanna, tolerates agriculture. Found at around 900m in our area, to sea level elsewhere. Known from the Inimvua River, near Maridi in Western Equatoria, South Sudan, but might be more widespread, it is known from just across the border in Garamba National Park in DR Congo. Elsewhere, west to Cameroon and Gabon.

Ptychadena perreti, Cameroon.

CONSERVATION STATUS: Least Concern.

NATURAL HISTORY: Terrestrial and nocturnal. Advertisement call a gabbling series of goose-like calls, and a sonorous clucking. Seems to be associated with marshes and ponds with nearby tree cover in Garamba; elsewhere fond of pools on forest roads. Breeding in temporary pools recorded. In Garamba National Park, male calling and breeding commenced in March, at start of rainy season.

WESTERN DWARF ROCKET FROG

Ptychadena pumilio
(Boulenger, 1920)

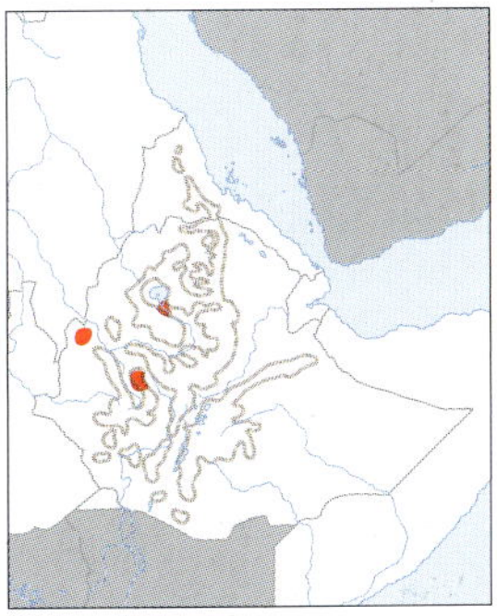

IDENTIFICATION: A small, slender rocket frog; in our area only in western Ethiopia. Body slender to compact. Head relatively small for a rocket frog. The eyes are fairly large, pupil horizontal but looks oval. The dorsal skin has four pairs of long, narrow, distinct ridges. Tympanum has a pale border and is almost as big as the eye. Males reach a maximum of 3.2cm, females 3.6cm. Dorsal colour usually brown or grey-brown, occasionally greenish, with scattered but largish dark spots, sometimes arranged in a regular pattern. Often has a fine or broad vertebral stripe, usually rufous or yellow. Underside white or grey.

HABITAT AND DISTRIBUTION: In Ethiopia, in moist savanna at medium to high altitudes, between 500 and 1,850m. Elsewhere, in degraded forest and dry savanna, to sea level. In our area there are only a handful of records, all in western Ethiopia (Lake Tana, Dabus River near Mendi, Didessa River). No South

Sudan records, but occurs Central African Republic, and extends west to Senegal.

CONSERVATION STATUS: Least Concern.

NATURAL HISTORY: Terrestrial and nocturnal. Appears to be two advertisement calls; one a nasal quack, the other a rattling bird-like screech. Males call while floating or from partially concealed positions. They breed in small to large temporary pools; eggs small and they float.

Ptychadena pumilio, Cameroon.

SCHILLUK ROCKET FROG

Ptychadena schillukorum
(Werner, 1908)

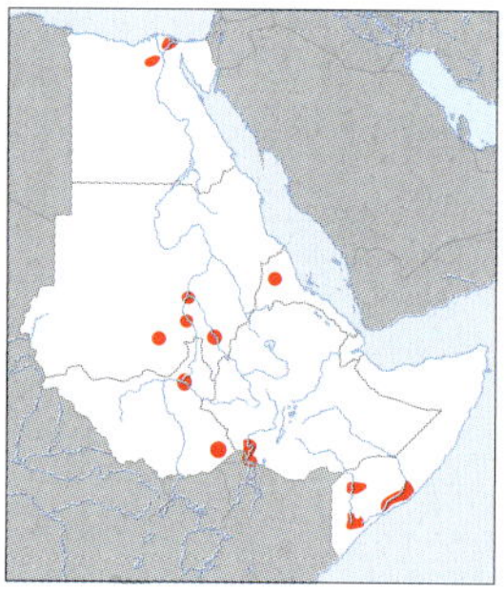

IDENTIFICATION: A medium-sized, squat rocket frog of low savanna, the numerous short dorsal folds make its back look lumpy. Body compact. Head large. Eyes fairly large, pupil horizontal but looks oval. Dorsal skin has 3–4 pairs of fragmented skin folds. Tympanum has a broad and sometimes incomplete pale border, and is almost as big as the eye. Males reach a maximum of 4.8cm, females 5.1cm. Dorsal colour usually brown or grey, sometime rufous, with scattered small to medium dark spots, often with some red on flanks. Sometimes has a broad, light-coloured vertebral stripe. Underside white.

HABITAT AND DISTRIBUTION: Moist and dry savanna, semi-desert (and desert watercourses); between sea level and 1,000m, or more. Although it has a huge range, records are somewhat sporadic in our area. Southern Somalia, south of 4°N, Agordat in western Eritrea, Turmi, the lower Omo River and Chew Bahir in southern Ethiopia, near Malakal (Upper Nile State) and north-west of Kapoeta (Eastern Equatoria State) in South Sudan, Ed Damazin, the Ingessana Hills, Roseires Dam, North Kordofan, Kosti and Khartoum in Sudan, and the Faiyum and Lake Manzala in the north-east Nile Delta, in Egypt. Probably much more widespread in north-east Africa, but under-collected. Elsewhere, west to Senegal, south to Mozambique.

CONSERVATION STATUS: Least Concern.

NATURAL HISTORY: Terrestrial and nocturnal. Advertisement call a loud rapid burst of *shuk…shuk.. shuk* sounds, sometimes likened to rapid drumming. Breeding occurs in swamps, flooded grasslands, slow-moving streams and shallow, temporary pools; males may call from the ground at the water's edge, in shallow water, while floating, or from vegetation in the water. They aestivate in the dry season. The diet, unusually, includes small vertebrates, an Egyptian specimen was found eating a young Nile Rocket Frog, *Ptychadena nilotica*. Presumably lays a floating raft of eggs, like other *Ptychadena*.

Ptychadena schillukorum, Turmi.

RED ROCKET FROG, TELLINI'S ROCKET FROG

Ptychadena tellinii
(Peracca, 1904)

IDENTIFICATION: A small, squat, attractive rocket frog; dorsal surface always rufous. Body stocky and looks somewhat stunted. Eyes fairly large, pupil horizontal but looks oval; lower half of iris very dark. The dorsal skin has several pairs of low ridges, which may be so low as to be invisible and the back looks smooth. Tympanum about two-thirds the size of the eye. Males reach a maximum of 3.9cm, females 4.7cm. Dorsal colour usually rufous, rufous-brown or orangey-brown, sometimes with a poorly defined darker central blotch with diffuse edges. A dark stripe from the snout passes through the eye and tympanum, becoming a broad wavy dark flank stripe. Underside white or grey. Taxonomic Notes: Long known, and still listed in some catalogues, as *P. schubotzi*.

HABITAT AND DISTRIBUTION: Gallery forest, moist and dry savanna; tolerant of agriculture; between 600 and 2,500m in our area. Eritrean records mostly in low-altitude areas of the south-west, south of a line joining Tesenay and Barentu, and on Ethiopian border near Senafe. In Ethiopia, at Kessa, Aykel, west of Gondar, and several localities on the Blue Nile between the Shafartak Bridge and Sudan border. There appear to be no records in South Sudan, but the species occurs in Garamba National Pak in DR Congo, so almost certainly occurs in West and Central Equatoria. Elsewhere, west to Sierra Leone.

CONSERVATION STATUS: Least Concern.

NATURAL HISTORY: Terrestrial and nocturnal. Advertisement call a high-pitched, metallic *kik…kik…kik*, rapidly repeated. They emerge at the start of the rainy season, males call from pond edges, hidden among vegetation at the water's edge or on open ground. They breed in temporary pools; females deposit up to 100 floating eggs at a time.

Ptychadena tellinii, west of Aykel, near Gondar.

DAKAR ROCKET FROG

Ptychadena trinodis
(Boettger, 1881)

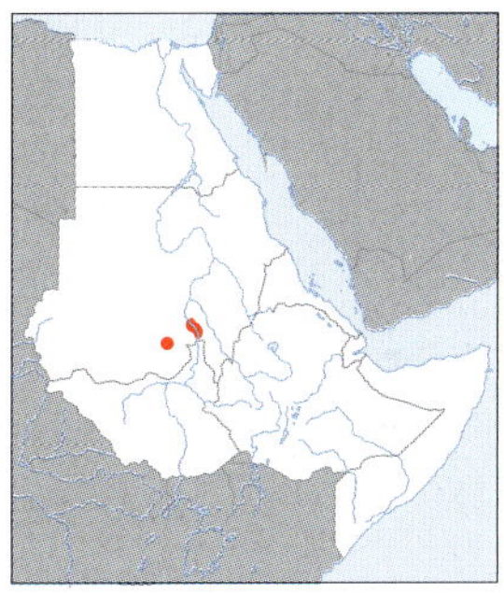

IDENTIFICATION: A big, heavily built rocket frog; in our area only in Sudan. Body stocky. Head relatively small for a rocket frog. Eyes fairly large, pupil horizontal but looks oval. The dorsal skin has three pairs of long, narrow, distinct ridges. Tympanum about two-thirds the size of the eye. Males reach a maximum of 5.2cm, females 6cm. Dorsal colour usually brown, sometimes grey-brown, with scattered but largish dark spots, sometimes arranged in a regular pattern, especially in the centre of the back. Virtually always has fine yellow or cream vertebral stripe, and a yellow or cream dorsolateral stripe, often tinged red anteriorly Underside white or grey.

HABITAT AND DISTRIBUTION: Dry savanna and semi-desert at 350–1,000m, but to sea level elsewhere. The only definite records in our area are North Kordofan (Sudan) and from Asalaya (near Kosti), on the White Nile in Sudan, but records from eastern Central African Republic (Abeche, Gongo) and Delimbe in northern Central African Republic suggest it probably occurs in Darfur in Sudan. Known from Garamba National Park in DR Congo, so probably in Western and Central Equatoria, South Sudan. Elsewhere, west to Mauritania and Senegal.

CONSERVATION STATUS: Least Concern.

NATURAL HISTORY: Terrestrial and nocturnal. Advertisement call a series (about one per second) of rather bird-like *cra…cra…cra…* sounds. They emerge at the start of the rainy season, and males call from among vegetation at the water's edge or in shallow water. They breed in temporary pools; eggs deposited in shallow water. They have a big metatarsal tubercle, suggesting they bury themselves in the dry season.

Ptychadena trinodis, Ghana.

DARK-BELLIED ROCKET FROG

Ptychadena cooperi
Parker, 1930

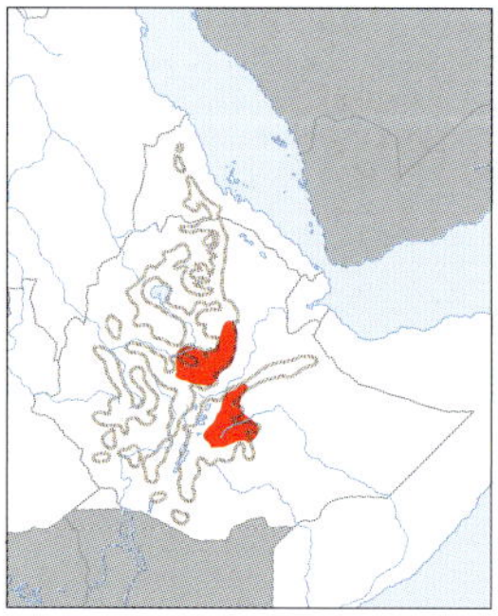

Ptychadena cooperi, Dodolla.

IDENTIFICATION: A big, squat rocket frog; with a very lumpy back, of highland Ethiopia. Body stocky. Eyes fairly large, pupil horizontal but looks oval. Dorsal skin has at least four pairs of long ridges, but all are interrupted, so the dorsal surface looks very warty. Tympanum about two-thirds the size of the eye. Males reach a maximum of 6cm, females 7.2cm. Dorsal colour usually a rich brown, but can be grey, or even black; brown and grey ones have scattered small dark spots. Virtually always has a fine yellow or cream vertebral stripe. Underside white or grey.

HABITAT AND DISTRIBUTION: Endemic to Ethiopia, in high grassland and savanna, tolerant of agriculture; at altitudes between 2,500 and 3,100m. Two populations occur. West of the Ethiopian Rift Valley, it occurs south along the escarpment from near Dessie to Addis Ababa, Debre Libanos and Guder. East of the Rift Valley, occurs south from the high country of Arsi Region, east of Koka Dam, to the northern slopes of the Bale Mountains.

CONSERVATION STATUS: Least Concern.

NATURAL HISTORY: Terrestrial and nocturnal, but sometimes active by day in the rainy season. Advertisement call a strident squeak, sometimes heard by day. These frogs emerge at the start of the rainy season, in May and are active until November; around Addis Ababa a few seem to remain active in the dry season near small streams. Males call from among vegetation or on ground. They may be seen sitting on grass tufts at edges of streams during the day; breed in temporary pools in grassland. The author of this species was Hampton Wildman Parker (1897–1968), British herpetologist and systematist, keeper of Zoology at the Natural History Museum. Some of his most important works were two monographs on the lizards and snakes of Somalia (1942, 1949). Parker described several northern Somali species, including *Xenagama taylori*, *Uromastyx macfadyeni*, *Hemidactylus curlei*, *Atractaspis scorteccii* and *Platyceps taylori*.

WADE'S ROCKET FROG

Ptychadena wadei
Largen, 2000

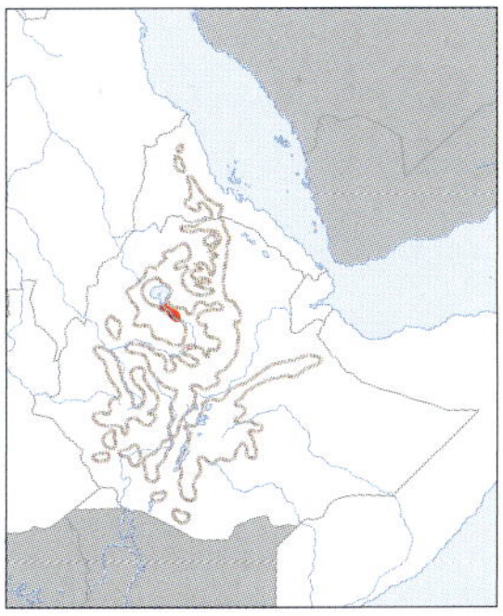

IDENTIFICATION: A small rocket frog; with prominent unbroken dorsal skin folds, known only from vicinity of Tissisat Falls on the Blue Nile, highland Ethiopia. Body elongate. Eyes fairly large, pupil horizontal but looks oval. Dorsal skin has at least five pairs of long ridges. Tympanum about two-thirds the size of the eye. Males reach a maximum of 3cm, females 3.5cm. Dorsal colour usually a shade of brown or grey-brown, with or without scattered dark spots A broad rufous, light brown or cream vertebral stripe seemingly always present, sometimes with a fine paler vertebral line. Underside cream, immaculate. Taxonomic Notes: Two specimens of Grassland Rocket Frog *P. porosissima* were collected in 1902 at Bahar Dar. No other specimens of this form are known nearer than Uganda; recent research suggests that these are not *P. porosissima*, but *P. wadei*.

HABITAT AND DISTRIBUTION: Endemic to Ethiopia, in high grassland and agricultural land of the north-west, along a 30km stretch of the Blue Nile between Bahar Dar and Tissisat Falls, at just below 1,800m.

CONSERVATION STATUS: Data Deficient.

NATURAL HISTORY: Unknown, but presumably similar to other rocket frogs, i.e. terrestrial and nocturnal. Advertisement call unknown. Juveniles collected between August and October, in damp grass and under rocks during the day.

Ptychadena wadei, south-east of Bahar Dar.

SMALLEST ROCKET FROG

Ptychadena nana
Perret, 1980

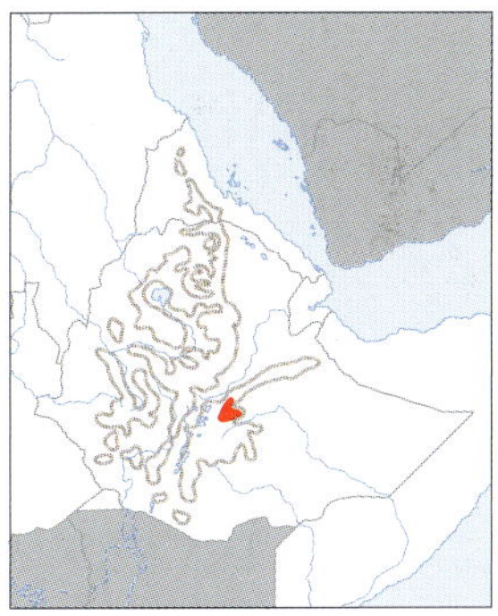

IDENTIFICATION: A very small, chunky rocket frog with prominent extensively broken dorsal skin folds, only in Arussi highlands, Ethiopia. Body squat. Eyes fairly large, pupil horizontal but looks oval. The dorsal skin has several pairs of long ridges but these are extensively interrupted, so the back just looks warty. Tympanum about two-thirds the size of the eye. Males reach a maximum of 2.7cm, females 3cm. Dorsal colour usually a shade of brown or grey-brown, with or without large scattered dark spots A broad or narrow, green, rufous or yellow vertebral stripe is usually present. Underside white or grey.

HABITAT AND DISTRIBUTION: Endemic to Ethiopia, in high grassland in eastern half of the Arussi plateau, east of Koka Dam and Lake Zwai, roughly between 07.50 and 08.30°N, 39.40 to 40°E, between 2,400 and 2,850m; range limited in south by the Shebelle River, and in the north-east by the Great Rift Valley.

CONSERVATION STATUS: Endangered.

NATURAL HISTORY: Poorly known. Terrestrial and nocturnal. Advertisement call a series of single pulsed notes about a quarter of a second in duration with a short gap between them. Males were calling in large numbers in late June in shallow, roadside puddles overgrown by grass.

Ptychadena nana, near Asela.

ROBE ROCKET FROG

Ptychadena robeensis
Goutte, Reyes-Velasco, Freilich, Abeje Kassie & Boissinot, 2021

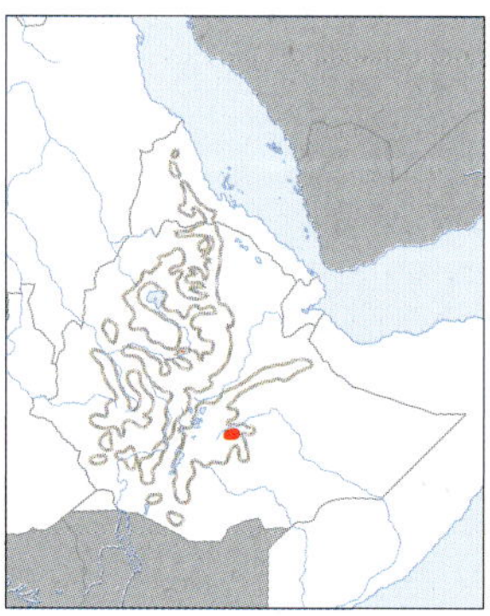

IDENTIFICATION: A small, wart-less rocket frog; known only in north-central Bale Mountains, Ethiopia. Body squat. Eyes fairly large, pupil horizontal but looks oval. Dorsal skin has four pairs of long ridges. Tympanum about two-thirds the size of the eye. Males reach a maximum of 3.1cm, females 3.4cm. Dorsal colour varies; brown, grey or green, usually with scattered dark spots A broad or narrow, green, rufous or yellow vertebral stripe is usually present. Underside uniformly white.

HABITAT AND DISTRIBUTION: Endemic to Ethiopia, in high grassland and at edges of coniferous forest, between 2,375 and 3,120m, and between the towns of Dinsho, Ali, Robe and Goba, in the northern foot-hills of the Bale Mountains.

CONSERVATION STATUS: Not Assessed.

NATURAL HISTORY: Poorly known. Terrestrial and nocturnal. Advertisement call a single pulsed note, rapidly fluctuating. Males found calling after 20.00 h in April, July and August (all in the rainy season), from shallow grassy puddles or flooded fields.

Ptychadena robeensis, Robe.

Ptychadena robeensis, Robe.

ENOUGH ROCKET FROG

Ptychadena beka
Goutte, Reyes-Velasco, Freilich, Abeje Kassie & Boissinot, 2021

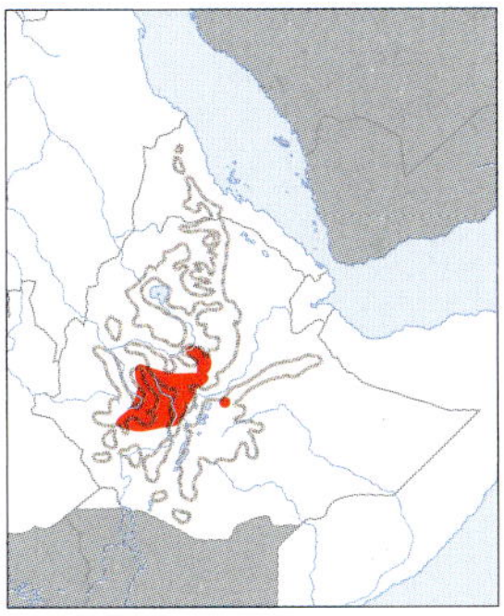

IDENTIFICATION: A medium-sized rocket frog; largely in the western plateau of Ethiopia. Body squat, but has a somewhat alert look. Eyes fairly large, pupil horizontal but looks oval. The dorsal skin has four pairs of long ridges, the central two on each side interrupted. Tympanum about two-thirds the size of the eye and usually has a light vertical bar. Males reach a maximum of 4.1cm, females 4.6cm. Dorsal colour brown or grey brown, usually with a few scattered dark spots. A broad or narrow cream-rufous or yellow vertebral stripe is usually present. Underside uniform white to light yellow. Taxonomic Notes: The curious specific name *beka* is a transliteration of an Amharic word; '*beka*', meaning 'enough' or 'that's it'; representing the hope, of those who described it, that this group's taxonomy is now settled.

HABITAT AND DISTRIBUTION: Endemic to Ethiopia, in high grassland, savanna and forest clearings, between 1,700 and 2,600m. West of the Rift Valley, it occurs south from the vicinity of the Shafartak Bridge on the Blue Nile, west to Nekemte, south-west to Bonga, south-east to Hoseana; east of the Ethiopian Rift Valley known from the vicinity of Asela.

Ptychadena beka, Addis Ababa.

CONSERVATION STATUS: Not Assessed.

NATURAL HISTORY: Poorly known. Terrestrial and nocturnal. Advertisement call a single pulsed note, rapidly fluctuating. Males found calling in roadside pools and flooded agricultural land, in large numbers, usually after 20.00 h (sometimes as late as 02.00 h), between June and August.

CLICKING ROCKET FROG

Ptychadena delphina
Goutte, Reyes-Velasco, Freilich, Abeje Kassie & Boissinot, 2021

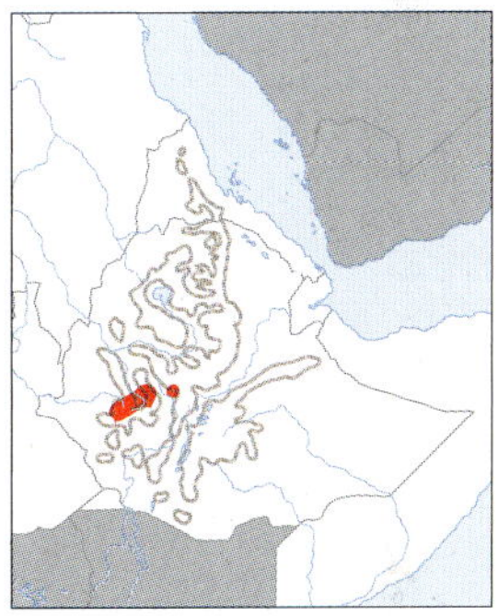

IDENTIFICATION: A fairly large rocket frog; most records from south-western forests of Ethiopia. Body squat. Eyes fairly large, pupil horizontal but looks oval. Dorsal ridges vary in number and definition. Tympanum about two-thirds the size of the eye, sometimes with a light mark. Males reach a maximum of 4.3cm, females 5.1cm. Dorsal colour light to dark brown, usually with elongate brown blotches symmetrically positioned either side of the vertebral line. A broad or narrow rufous, brown or yellow vertebral stripe is usually present. Underside uniform white to light yellow.

HABITAT AND DISTRIBUTION: Endemic to Ethiopia, in forest between 1,600 and 2,100m. Recorded in broadleaf forests of the south-west, between Bedelle, Metu, Gore and Dembi. However, there is also an unusual high-altitude record just east of Ambo, west of Addis Ababa, at 2,370m.

CONSERVATION STATUS: Not Assessed.

NATURAL HISTORY: Poorly known. Terrestrial and nocturnal. Advertisement call a series of clicks, resembling a dolphin, hence the specific name. Males were found calling in flooded grassland ponds or puddles, or in rain-fed puddles on the road, in June.

Ptychadena delphina, Bedelle.

CHICKEN ROCKET FROG

Ptychadena doro
Goutte, Reyes-Velasco, Freilich, Abeje Kassie & Boissinot, 2021

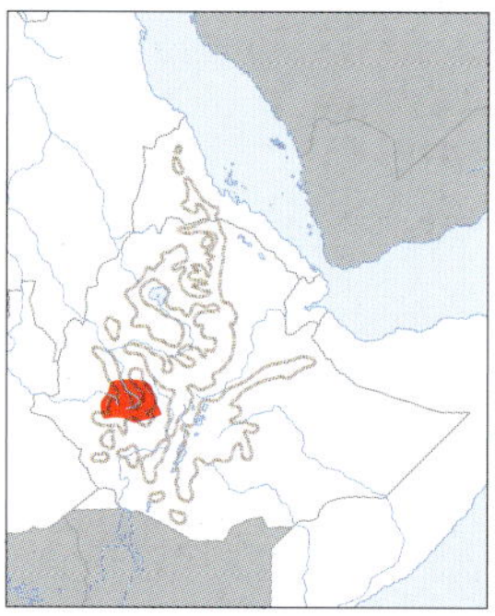

IDENTIFICATION: A medium-sized rocket frog; largely in south-western forests of Ethiopia. Body squat. Eyes fairly large, pupil horizontal but looks oval. Dorsal surface has three pairs of skin ridges, one pair continuous, two pairs interrupted. Tympanum about two-thirds the size of the eye, with a light vertical bar. Males reach a maximum of 3.8cm, females 4.8cm. Dorsal colour yellowish to dark brown, usually with elongate brown blotches symmetrically positioned either side of the vertebral line. A broad or narrow light brown or yellow vertebral stripe is usually present. Underside light yellow, sometimes with a hint of grey on the throat.

HABITAT AND DISTRIBUTION: Endemic to Ethiopia, in forest, between 1,650 and 2,300m. Recorded in broadleaf forests of south-west, roughly in a square between Jimma, Bedelle, Metu and Teppi.

CONSERVATION STATUS: Not Assessed.

NATURAL HISTORY: Poorly known. Terrestrial and nocturnal. The advertisement call resembles a chicken, hence the specific name; '*doro*' is Amharic for chicken. Males were found calling in flooded grassland ponds or puddles, or in rain-fed puddles on the road, in June.

Ptychadena doro, Jimma.

GOWER'S ROCKET FROG

Ptychadena goweri
Smith, Noonan & Colston, 2017

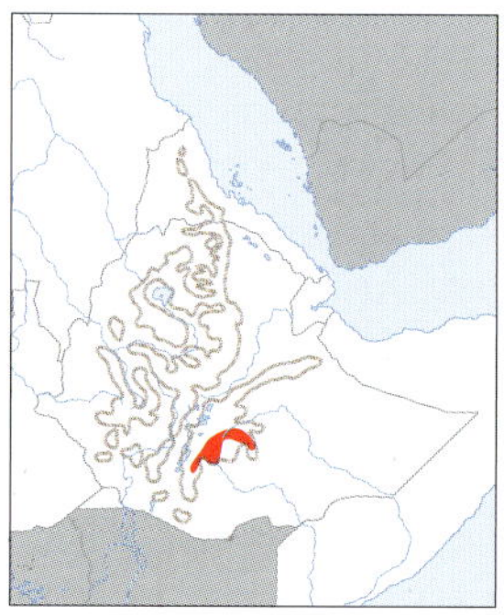

IDENTIFICATION: A fairly large rocket frog of high south-east Ethiopia. Body squat. Eyes fairly large, pupil horizontal but looks oval. Dorsal skin has three or four pairs of long ridges, the middle three usually interrupted. Tympanum about three-quarters the size of the eye or more. Males reach a maximum of 4.8cm, females 6cm. Dorsal colour shades of brown or grey, usually with scattered dark spots that may lie on the

Ptychadena goweri, Harenna Forest.

Ptychadena goweri, Harenna Forest.

interrupted ridges. A broad or narrow rufous, cream or yellow vertebral stripe is usually present. Underside uniform white or yellow.

HABITAT AND DISTRIBUTION: Endemic to Ethiopia, in forest and high savanna, between 1,700 and 2,600m. Lives in high country east of a line connecting Lakes Awassa and Chamo, but not beyond 40°E; localities include Harenna Forest, Hagere Maryam and around Kebri Mangest.

CONSERVATION STATUS: Not Assessed.

NATURAL HISTORY: Poorly known. Terrestrial; nocturnal and diurnal. Advertisement call a single pulsed note, 0.5 second in duration. Males found calling at night between April and September (in the rainy season), from shallow grassy puddles in clearings, or grassy meadows, may well be active most of year as specimens in Harenna Forest were active by day around flooded pools and in damp vegetation in December.

LEVENE'S ROCKET FROG

Ptychadena levenorum
Smith, Noonan & Colston, 2017

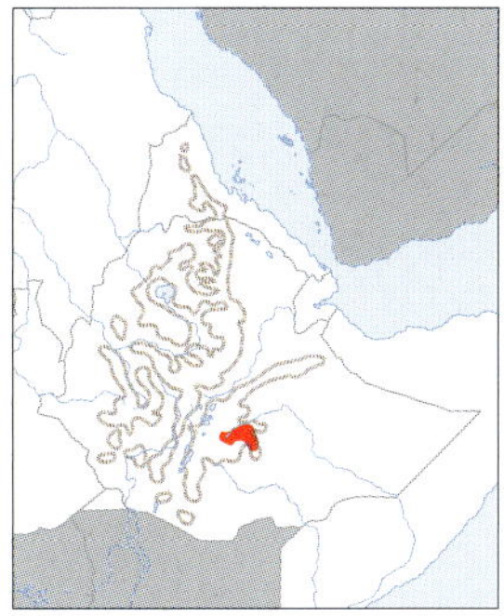

IDENTIFICATION: A small rocket frog of the Bale Mountains, Ethiopia. Body squat. The eyes are fairly large, pupil horizontal but looks oval. Dorsal skin has four pairs of long ridges, the middle two usually interrupted. Tympanum about three-quarters the size of the eye or more. Males reach a maximum of 3.6cm, females 4.0cm. Dorsal colour shades of brown, rufous or grey, usually with scattered dark spots, often large, which may lie on the interrupted ridges. A broad or narrow brown, green, cream or yellow vertebral stripe is usually present. Underside uniform white or light yellow.

HABITAT AND DISTRIBUTION: Endemic to Ethiopia, in forest and high savanna, between 3,000 and 3,400m on north side of the Bale Mountains, and lower (2,300–2,500m) on south side. Known from a handful of localities, between Dodolla and Dinsho on the north side and in Harenna Forest on the south side.

CONSERVATION STATUS: Not Assessed.

Ptychadena levenorum, Harenna Forest.

NATURAL HISTORY: Poorly known. Terrestrial; nocturnal and diurnal. Advertisement call a single pulsed note, 0.5 second in duration. Males were calling at night between April and August (during the rainy season), from shallow puddles in forest clearings or grassy meadows; males and females were found inside and under logs in Harenna Forest in December.

AMHARA ROCKET FROG

Ptychadena amharensis
Smith, Noonan & Colston, 2017

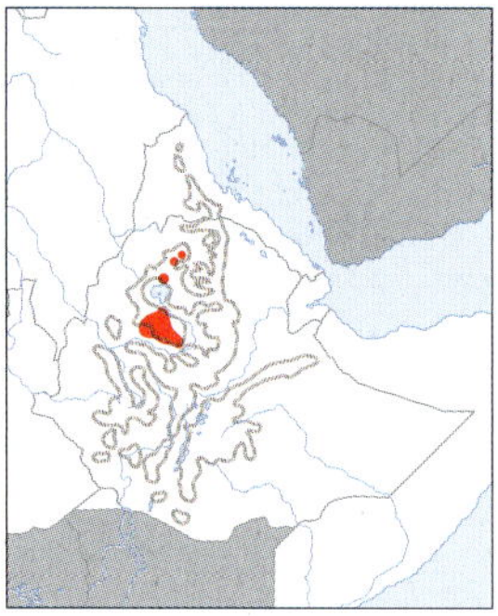

IDENTIFICATION: A small medium-sized frog of Gojjam province, in the loop of the Blue Nile, Ethiopia. Body squat. Eyes are fairly large, pupil horizontal but looks oval. Dorsal skin has four pairs of long ridges, all may be interrupted. Tympanum about three-quarters the size of the eye or more, usually with a light vertical bar. Males reach a maximum of 3.9cm, females 4.5cm. Dorsal colour shades of olive, orange, brown or yellowish brown, occasionally green, usually with small scattered dark spots, which may lie on the interrupted ridges A broad or narrow cream or sand-coloured vertebral stripe is usually present. Underside uniform white, cream or light yellow.

Ptychadena amharensis, near Debre Marcos.

Ptychadena amharensis, Debre Marcos.

HABITAT AND DISTRIBUTION: Endemic to Ethiopia, in high grassland and moist savanna, 1,800 and 2,700m, often in agriculture. Known from the old Gojjam province, inside the loop of the Blue Nile, with a few records north of Lake Tana; localities include a string of records along the road from Dejen to Debre Markos, Injibara, Bahar Dar, Gondar and Debarq, the northernmost record.

CONSERVATION STATUS: Not Assessed.

NATURAL HISTORY: Poorly known. Terrestrial; nocturnal and diurnal. Advertisement call a single pulsed note. Males were found calling at night in July and August, in flooded fields. Specimens near Debre Markos were in dense forest at small clearings along streams, active by day on rocks in and by the river.

HARENNA FOREST ROCKET FROG

Ptychadena harenna
Largen, 1997

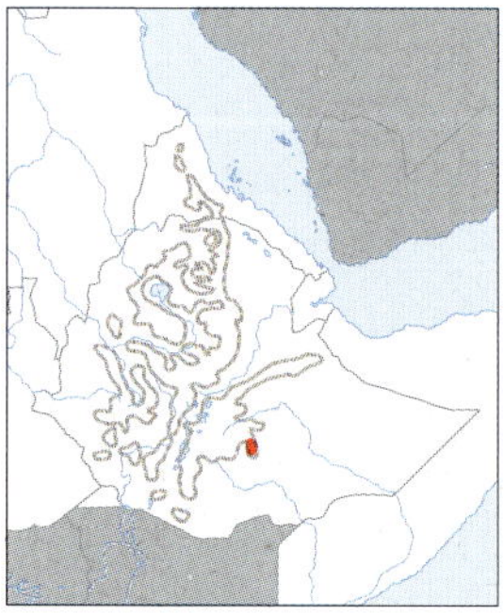

IDENTIFICATION: A small rocket frog of Harenna Forest, Ethiopia; it resembles a small *P. anchietae*. Body squat. Eyes fairly large, pupil horizontal but looks oval. The dorsal skin has four pairs of long ridges, often indistinct and always irregularly fragmented. Tympanum about three-quarters the size of the eye or more. Males reach a maximum of 4cm, females 4.5cm. Dorsal colour shades of brown, sometimes immaculate, or with scattered dark spots that may form patterns. A pale vertebral stripe sometimes present. There is usually a light, immaculate patch on the snout. Underside cream to vivid yellow.

HABITAT AND DISTRIBUTION: Endemic to Ethiopia, known only from clearings in Harenna Forest, in the Bale Mountains, at 1,400–1,550m.

CONSERVATION STATUS: Data Deficient.

NATURAL HISTORY: Poorly known. Terrestrial and nocturnal. The types were found in a shallow stagnant pool in *Podocarpus* forest.

Ptychadena harenna, Harenna Forest.

Ptychadena harenna, Harenna Forest.

ERLANGER'S ROCKET FROG

Ptychadena erlangeri
(Ahl, 1924)

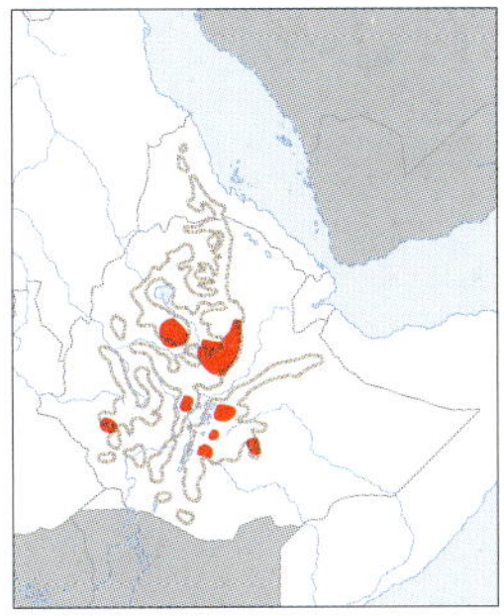

IDENTIFICATION: A medium-sized rocket frog of highland Ethiopia. Body stocky. Eyes fairly large, pupil horizontal but looks oval. The dorsal skin has at least four pairs of long ridges; the median pair usually unbroken. Tympanum about two-thirds the size of the eye. Males reach a maximum of 4.7cm, females 5.4cm. Dorsal colour brown, pinkish or olive, with scattered dark spots which may be small, large and/or elongate. A fine or broad white, yellow or cream vertebral stripe is usually present. Underside cream to yellow. Taxonomic Notes: A recent revision of this genus in Ethiopia by Sandra Goutte and her team suggested that many specimens formerly assigned to this species are in fact *P. neumanni*, and vice versa. Not all authorities agree. Until this is clarified, the notes here largely refer to *P. erlangeri* as originally defined.

HABITAT AND DISTRIBUTION: Endemic to Ethiopia, in high grassland, moist savanna and clearings in forest; and savanna, tolerant of agriculture; between 1,300 and 3,400m. As originally described, this frog was

Ptychadena erlangeri, north of Debre Berhan.

known west of the Ethiopian Rift Valley, from Debra Markos and Injibara in the loop of the Blue Nile, and near Bonga. Records east of the Rift Valley include Wondo Genet, Harenna Forest and near Hagere Maryam. The type was from Lake Abaya. New records west of the Rift Valley indicate it is widespread in the highland between Dessie, Debre Libanos and Addis Ababa, and is also known east of the Rift Valley in the highlands around Asela and Kofele.

CONSERVATION STATUS: Near Threatened.

NATURAL HISTORY: Terrestrial and largely nocturnal. Advertisement call unknown. Emerges at the start of the rainy season, in April or May and active until November; around Addis Ababa a few seem to remain active even later. Males call from shallow pools and flooded fields.

Ptychadena erlangeri, Wonchi Crater Lake.

NEUMANN'S ROCKET FROG

Ptychadena neumanni
(Ahl, 1924)

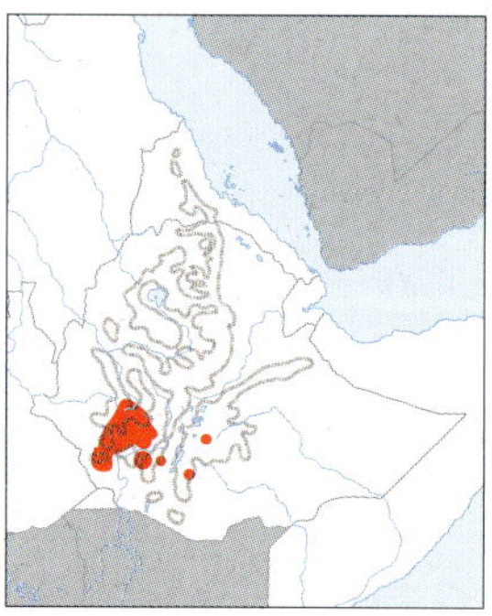

IDENTIFICATION: A medium-sized rocket frog of highland Ethiopia. Body stocky. Eyes fairly large, pupil horizontal but looks oval. The dorsal skin has at least four pairs of long ridges, mostly unbroken. Tympanum about two-thirds the size of the eye. Males reach a maximum of 4cm, females 4.7cm. Dorsal colour light to dark brown or grey-brown, with scattered dark spots which may be small, large and/or elongate. A fine or broad white, yellow or cream vertebral stripe is usually present. Underside cream to yellow. Taxonomic Notes: See the previous species. Whether the redefinition and restriction of this species to a much smaller range is justified remains to be seen.

HABITAT AND DISTRIBUTION: Endemic to Ethiopia, in high grassland, moist savanna, and clearings in forest; tolerant of agriculture. As originally defined, extremely widespread throughout high and mid-altitude Ethiopia west of the Rift Valley. East of the Rift Valley, it was also known from the Ahmar Mountains, and highlands south and west from the Bale Mountains to the eastern escarpment, largely between 1,800 and 3,800m. The aforementioned revision by Goutte *et al.* (2021) restricted this species largely to forested south-west Ethiopia, in a triangle between Gore, Maji and Sodo; with a handful of other isolated records, at Gadat (the type locality), Mount Gughe, Wondo Genet and Kebri Mangest.

CONSERVATION STATUS: Least Concern.

NATURAL HISTORY: Terrestrial and largely nocturnal. Advertisement call unknown. These frogs emerge at the start of the rainy season, in April or May and are active until November. Males call from and breed in all sorts of standing water, including puddles, swamps and slow-moving streams.

Ptychadena neumanni, Bonga.

FAMILY CONRAUIDAE GIANT FROGS

A family erected for a single genus of frogs, *Conraua*. Recent molecular work indicates they are a true monophyletic family that diversified >50 million years ago.

GIANT FROGS *CONRAUA*

A genus of seven currently recognised species (there are probably more cryptic forms) of medium-sized to giant frogs, one species, Goliath Frog *Conraua goliath*, is the largest in the world, reaching 34cm length and 3.25kg. Most occur in the forests of western and west-central Africa, with a single species in our area.

BECCARI'S GIANT FROG

Conraua beccarii
Boulenger, 1911

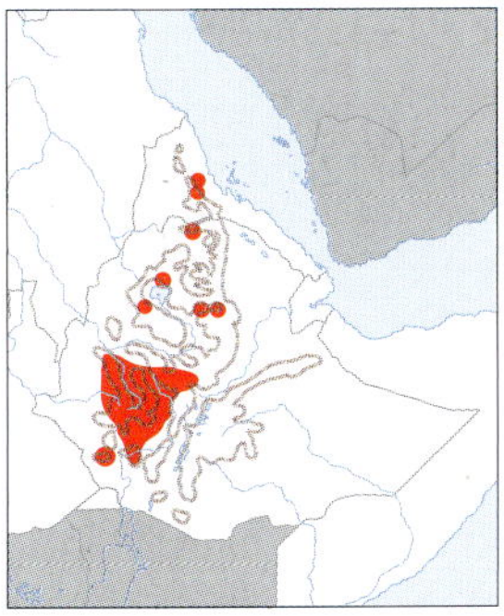

IDENTIFICATION: Large adults unmistakable, a huge, sinister-looking, broad-headed deep-water frog of highlands in western Ethiopia and Eritrea. Juveniles have an oval or sub-rectangular body and a short head, and look like a typical frog, but as they reach adulthood the head becomes astonishingly broad and flat, and the body is almost teardrop-shaped. Eyes large, with vertical pupils, and change position in very large adults, from the side of the head to more upward-pointing. Dorsal skin warty, with dorsolateral folds, and a thick obvious skin fold extends from behind the eye to the forelimb insertion. Tympanum invisible. Limbs stout, toes totally webbed. Males and females reach of maximum of about 17cm. Dorsal colour of juveniles olive-green or yellowish, adults tan, dirty brown, rufous or grey, sometimes with yellow spotting at limb junctions and on lower flanks. Underside dirty white.

Conraua beccarii, sub-adult, Sheka.

HABITAT AND DISTRIBUTION: Forest, moist savanna and high grassland, between 800 and 2,900m. Endemic to Eritrea and Ethiopia, from west of the Ethiopian Rift Valley. Known from Fil-fil (the type locality) and Asmara in Eritrea, although recent research suggests that it may no longer occur at the former, as the stream where the type was collected seems to have virtually dried up. In Ethiopia, sporadically recorded at Adua in the far north south to Mt Bibita in

Conraua beccari, juvenile, Bonga.

Conraua beccarii, adult, Bonga.

the south-west, including Gondar, Dessie, Addis Ababa, Gore, Jimma and Bonga.

CONSERVATION STATUS: Least Concern.

NATURAL HISTORY: Terrestrial. Active largely by night, although sometimes found by day sheltering in waterside vegetation, or resting near the bank of steep-shelving pools. Usually in vicinity of deep water in highland rivers or swamps. Quick to escape into water; it swims fast and dives if threatened. Advertisement call unknown, although these frogs may produce a deep grunt, almost like a growl. No breeding details known.

FAMILY DICROGLOSSIDAE FORK-TONGUED FROGS

A family of about 14 genera and >200 species; most are Asian but one is African.

GROOVE-CROWNED BULLFROGS OR TIGER FROGS *HOPLOBATRACHUS*

A genus of five species; one large species is widespread in West Africa and savannas surrounding the Central African forest. The other four all occur in Asia.

EASTERN GROOVE-CROWNED BULLFROG, AFRICAN TIGER FROG

Hoplobatrachus occipitalis
Günther, 1858

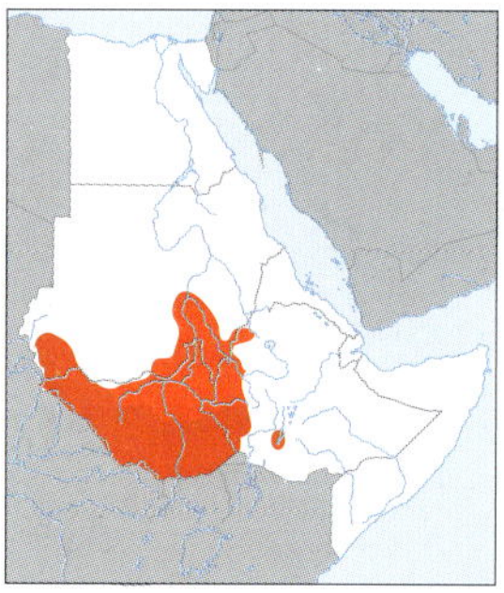

IDENTIFICATION: Big adults unmistakable; a huge, green frog, with a deep groove behind the large protruding eyes, which are on top of the head. Pupil horizontal, but looks diamond-shaped, or circular in bright light. Lower jaw has three sharp, tooth-like projections. The dorsal skin is warty; and the warts may form vague lines. A thick obvious skin fold extends from behind the eye above the tympanum and down to the jawline. Tympanum large, the same size as the eye, and visible. Limbs thick and stout, toes totally webbed. Males reach of maximum of about 11cm, females 16cm. Dorsal colour usually shades of green, brown, olive, or grey-green, with vague darker blotches. Flanks often marbled, white, or yellow and dark green; underside dirty white or yellow.

HABITAT AND DISTRIBUTION: A frog of forest fringes, moist and dry savanna, and semi-desert, between 200 and 1,500m, elsewhere to sea level. Sporadically recorded in western Ethiopia (Metemma, lower Blue Nile Gorge, Gambela, Arba Minch, Bench Maji). Probably occurs across most of South Sudan and the Nile basin in Sudan, but records sporadic; South Sudan localities include the Boro River in Western Bahr El Ghazal, Sudd in Unity State, Malakal

Hoplobatrachus occipitalis, south of Metemma.

Hoplobatrachus occipitalis, South Kordofan.

in Upper Nile State, Marial Bay in Warrab, Yambio, Maridi and Nzara in Western Equatoria, Kagelu, Yei and Juba in Central Equatoria, and Gilo and Katire in the Imatong Mountains. Known from Garamba National Park in DR Congo. In Sudan, known from the following states: Blue Nile (Damazin), Sennar, White Nile (from Kosti south to Joda on the South Sudan border), South Kordofan, West Kordofan, South Darfur, West Darfur, Gezira (Wad Madani) and Khartoum. Elsewhere, west to Senegal, south around the eastern edge of the great forests to Angola.

CONSERVATION STATUS: Least Concern.

NATURAL HISTORY: Terrestrial, and active largely by night, but in the rainy season can be seen by day in pools, their eyes above the surface. In the dry season tend to be near permanent water, hidden in vegetation by day, or aestivating, buried underground, but as the rainy season commences they disperse, moving >1km in a night, and occupying all sorts of temporary pools, including very small ones. Advertisement call a series of low-pitched, snorts, like someone saying 'hear' in a deep croaky voice. Males call while floating or sitting in shallow water. A mating pair move from pool to pool, depositing a few eggs, apparently gauging the capabilities of the pool in terms of tadpole numbers. They can be seen at night, floating spreadeagled in pools; their eyes reflect torchlight. They move so fast they can skitter across the water surface, like a bouncing stone.

FAMILY PYXICEPHALIDAE

An African family with no common name, of about 12 genera and around 90 species, some very large, for example the African bullfrogs, others very small, for example the Dainty frogs. Four genera occur in our area; *Amietia*, *Cacosternum*, *Pyxicephalus* and *Tomopterna*.

RIVER FROGS *AMIETIA*

A genus of about 15 species, all in southern and eastern Africa. They look like the 'typical' frogs of European folk tales, large with big eyes. Most live near permanent water where they breed.

NUTT'S RIVER FROG

Amietia nutti
Boulenger, 1896

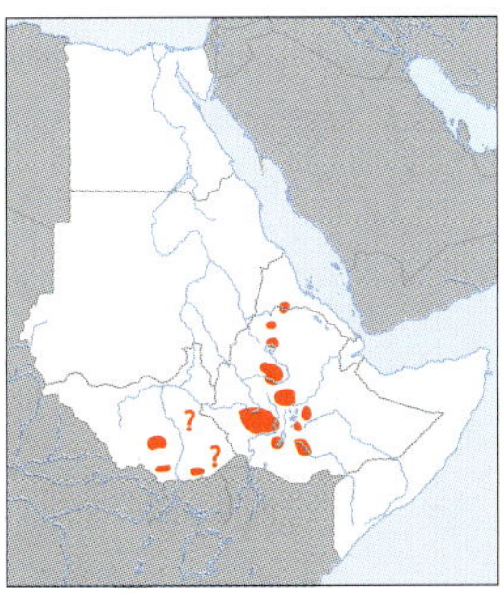

IDENTIFICATION: A fairly large, stereotypical frog, in our area usually associated with high and mid-altitude rivers in Ethiopia. Eyes large, with horizontal pupils. The number of dorsal skin ridges varies, from none to several pairs, but dorsolateral ridges are usually present and prominent. The ridges are often interrupted, making the back look warty. Tympanum about half the eye's diameter or slightly larger. Limbs fairly stout, toes totally webbed. Males reach about 7cm, females 9cm. Dorsal colour very variable; shades of brown, dark green, tan or rufous, with or without irregular darker blotches, and sometimes a pale vertebral line. A fine line runs from the nostril to the eye and a broad dark patch behind it. Underside white, although the throat may be mottled. Taxonomic Notes: Has been called *Rana angolensis*, *Afrana angolensis* and *Amietia angolensis*.

HABITAT AND DISTRIBUTION: Forest, moist savanna and high grassland, very tolerant of agriculture so long as permanent water remains, between 1,000 and 3,100m in our area, to sea level elsewhere. Known from Eritrea, Ethiopia and South Sudan. A single Eritrean record; Mendefera; widespread in Ethiopia, from Adua south through Gondar, Debre Marcos, Dejen, the highlands around Addis Ababa, to Metu, Bonga and Arba Minch, a few records east of the Rift Valley (Asela, Shashamene, Kibre Mengistu). South Sudan records largely in the south; they include Gilo and Katire in the Imatong Mountains (Eastern Equatoria), Kagelu near Yei (Central Equatoria) and

Amietia nutti, Addis Ababa.

Amietia nutti, Omo River.

Rumbek (Lakes State), probably widespread in the upper Nile basin (Blue Nile State) but under-collected. Elsewhere, south to Tanzania.

CONSERVATION STATUS: Not Evaluated.

NATURAL HISTORY: Terrestrial, and largely nocturnal. Advertisement call a series of brief knocks, like someone rapping briskly on wood, followed by a brief groan-like croak or two. Usually found in vicinity of permanent water (streams, swamps, lakes) with waterside vegetation, and often abundant. Sometimes visible by day, and often by night, near the water's edge, and if threatened jumps into the water. Males call from shallow water or under overhanging vegetation near the water's edge.

DAINTY FROGS OR CACOS *CACOSTERNUM*

A genus of about 16 species of tiny, attractive frogs, most occur in southern Africa. One species occurs in our area.

PLIMPTON'S DAINTY FROG

Cacosternum plimptoni
Channing, Brun, Burger, Febvre & Moyer, 2005

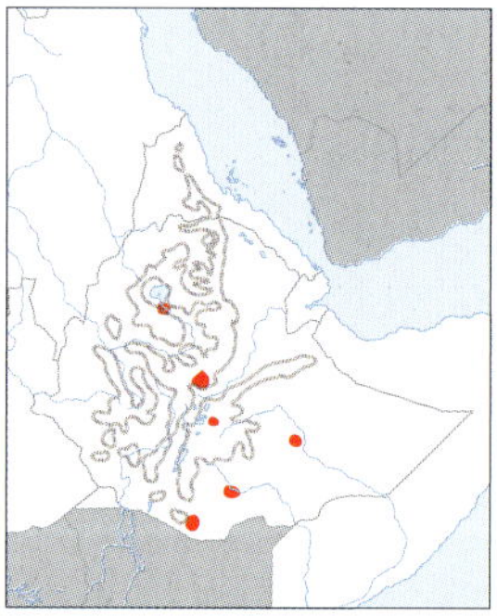

IDENTIFICATION: A tiny terrestrial green or brown frog, smaller than the top joint of a thumb, with a spotted underside, in our area only in Ethiopia. Eyes medium-sized, with horizontal pupils. Dorsal skin can be smooth or lumpy, sometimes with a few paired warts, especially towards the vent. Tympanum invisible. Limbs slender. Males reach about 1.8cm, females 2cm. Dorsal colour usually green or brown or a mixture (there is some evidence that males are usually brown, females green), sometimes with a vertebral line. A dark line from the nostril to the eye broadens on the flanks. Underside white with dark spots; adult males have an orange-brown throat. Taxonomic Notes: Described from Kenya; recent work suggests the Ethiopian examples are conspecific, but it is not certain.

Cacosternum plimptoni, Sebeta.

HABITAT AND DISTRIBUTION: Moist savanna and high grassland, tolerant of agriculture, at 800–2,500m in Ethiopia. In our area, known from the vicinity of Bahar Dar, quite widespread in the highlands south of Addis Ababa, and other localities include Asela, around Negele, Mega in southern Ethiopia, and the Durro River east of Goba. Elsewhere, from northern Tanzania and south-central Kenya (but see above).

Cacosternum plimptoni, Sebeta.

CONSERVATION STATUS: Least Concern.

NATURAL HISTORY: Terrestrial, and largely nocturnal. Advertisement call a series of brief harsh clicks, like a steel nail tapping a plate. Aestivates in the dry season in burrows, under large rocks, in deep cracks, etc., but emerges in numbers in both rainy seasons, and breeds in any available temporary waterbodies, including ditches, where their calls can be heard throughout day and night. Males call from cover on the shore, in the water or on submerged vegetation. Mating occurs in the water.

AFRICAN BULLFROGS *PYXICEPHALUS*

An African genus of huge, squat, dangerous-looking burrowing frogs, which includes the largest frogs in north-east Africa, up to 12cm length (23cm outside our area). They are stout, with short limbs, large wide heads, relatively small eyes, a horizontal pupil, somewhat obtusely wedged snout, and dorsal skin with ridges or warts. They live in savanna, wooded grasslands, floodplains, riverine woodlands and remain underground until heavy rains produce suitable temporary breeding pools. Males undertake parental care of tadpoles. Four species in Africa, but taxonomy not fully resolved, and it seems likely that the West and Central African form of *Pyxicephalus edulis* is a separate species. Two species in our area, one endemic.

KEY TO THE NORTH-EAST AFRICAN MEMBERS OF THE GENUS *PYXICEPHALUS*

1a: In our area, only in Sudan, South Sudan and southern Somalia (inter-riverine regions close to Kenya border), width of tympanum less than or equal to eye diameter, and positioned more than one half-eye diameter from the eye — *Pyxicephalus edulis*, Edible Bullfrog

1b: In our area, only in central and northern Somalia (Puntland), width of tympanum greater than eye diameter, and positioned less than one half-eye diameter from the eye — *Pyxicephalus obbianus*, Obbia Bullfrog

EDIBLE BULLFROG

Pyxicephalus edulis
Peters, 1854

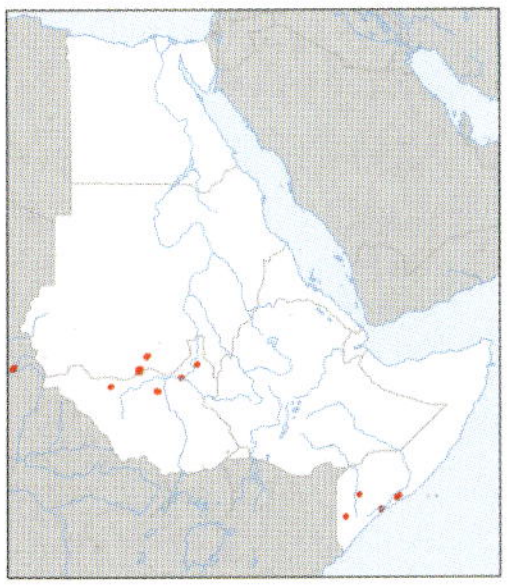

IDENTIFICATION: A huge, squat frog of dry country in southern Somalia and the border between Sudan and South Sudan. Head moderately wide. Tympanum less than or equal to eye diameter, less than a tympanum width from eye. Dorsal skin ridges thin and prominent. Very little webbing between toes. Unusually, males larger than females, reaching 14cm, females about 10cm. Tadpoles large, up to 4.6cm. Breeding males bright green, females duller and brownish; back pale yellow to cream, with darker olive-green spots, a pale vertebral stripe and crossbar between the eyes, and pale bars between upper lip and eye. Juveniles more greenish. Taxonomic Notes: Somali examples long called *P. flavigula*. The taxonomy

Pyxicephalus edulis, Ghana.

Pyxicephalus edulis, sub-adult, Ghana.

of animals from Sudan and South Sudan is problematic; some authors regard them as a separate species, informally known as *Pyxicephalus* sp, 'edulis west'.

HABITAT AND DISTRIBUTION: Dry savanna to open woodland on consolidated coastal dunes, riverine woodland, at elevations from near sea level to about 500m. In our area, two widely dispersed populations, one in southern Somalia, the other in border regions of Sudan and South Sudan. In Somalia, found on the Benadir coast, from Afgoi towards the Kenya border, inland on the Juba river as far north as Oroflillo (east of Bardera) and west of the river in Golgia (Afmadu district). In South Sudan, originally recorded at Paloich, Thar Jath Wetlands, and Khor Atar (near Malakal) in Upper Nile State, and Aweil in Northern Bahar El Ghazal, there are more recent records from along the Sudan/South Sudan border near

Pyxicephalus edulis, juvenile, Kenya.

the Ethiopian border west to the Central African Republic; localities include Unity State, Heglig and Kadugli. Probably more widespread, but under-collected. Elsewhere: the typical form extends south to South Africa, and 'edulis West' west to Senegal.

CONSERVATION STATUS: Least Concern.

NATURAL HISTORY: Terrestrial. Active at night, or by day after very heavy rains. Advertisement call a short *wapp-wapp*, like a small dog yapping. Breeds in shallow water early in the rainy season, or once sufficient rain has fallen. Aggregates in large numbers, the males fighting over calling sites and females. During the dry season they aestivate, buried inactive deep in sandy soil; they may remain underground for years. Although the diet is largely invertebrates, they will take small vertebrates, including lizards, other frogs, and even smaller individuals of their own species.

OBBIA BULLFROG

Pyxicephalus obbianus
Calabresi, 1927

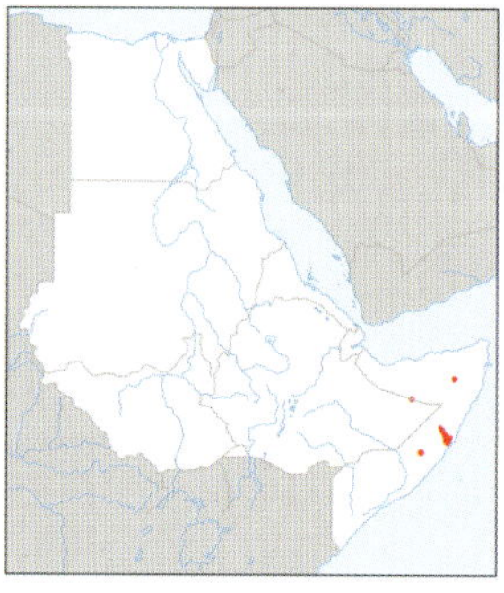

IDENTIFICATION: A huge squat frog, endemic to central Somalia. Dorsal skin ridges absent, but isolated warts are present. The head is short and very wide. Tympanum much larger than the eye diameter, less than a tympanum width from eye. Some webbing is present between toes. Males reach about 12cm, maximum size of adult females unknown, and tadpoles large, up to 5.5cm. Dorsum olive-yellowish or bluish-grey, with dark brown small or larger and fairly irregular markings, which are finer and denser in younger specimens. The pale vertebral stripe is confined to the snout; there is a crossbar between the eyes, pale bars between upper lip and eye, and banded limbs. Underside whitish, throat mottled yellow and brown. Taxonomic Notes: Described twice, so known as *P. cimmarutai* in older literature.

HABITAT AND DISTRIBUTION: Low-altitude, flat xerophilous open woodland, subdesert shrub and grassland with sparse low vegetation and scattered trees, on red or sandy loams, between sea level and 750m. Endemic to central Somalia and adjacent regions. Most records are from the plains between Hobyo and Galkayo in Mudug region and El Bur in Galguduud; isolated records further north; Gardo in Puntland and Haud region west of Bohodle on the border of Somaliland and Ethiopia. Probably occurs in similar habitats in adjacent Ogaden, eastern Ethiopia.

CONSERVATION STATUS: Least Concern.

NATURAL HISTORY: Poorly known. A terrestrial, fossorial species, probably nocturnal. They aestivate in the dry season, buried in sandy soil, becoming active and breeding after isolated heavy storms, or in wet season. Most specimens were collected between October and December. An adult from near Galkayo had a *Holodactylus* gecko in its stomach.

Pyxicephalus obbianus.

SAND FROGS *TOMOPTERNA*

An African genus of medium-sized, attractive, robust-bodied (toad-like), burrowing frogs, sometimes warty, but the parotid gland diagnostic of toads is always absent. Occur in drier parts of sub-Saharan Africa, largely avoiding forested and extremely dry areas. They breed after isolated heavy rainstorms or during the wet season, in any standing water. Sand frogs hibernate underground, buried in soil or hidden under rocks. About 17 species are known; three in north-east Africa (one endemic). An unidentified form (tentatively identified as *T. milletihorsini*) is known from Kordofan and Darfur, in western Sudan. Taxonomy not yet totally clarified, with situation complicated because several taxa are polyploid, leading to considerable variation in discrete species. New species are constantly being described. Several cannot be separated by morphology; only by advertisement calls and DNA analysis.

KEY TO THE NORTH-EAST AFRICAN MEMBERS OF THE GENUS *TOMOPTERNA*

1a: Tympanum indistinct or invisible; outer metatarsal tubercle distinctly elevated; from south-east Sudan, through extreme southern Ethiopia to southern Somalia — *Tomopterna tandyi*, Tandy´s Sand Frog

1b: Tympanum visible; outer metatarsal tubercle small; absent from are occupied by previous species — 2

2a: Infra-tympanic gland a discontinuous row of small glands, absent or present; 2–3 phalanges free of webbing on fourth toe; eastern Somaliland, Puntland to northern parts of central Somalia — *Tomopterna elegans*, Elegant Sand Frog

2b: Infra-tympanic gland a discontinuous row of small glands or a continuous ridge; 3–3.5 phalanges free of webbing on fourth toe; drier mountainous areas from Sudan to Ethiopia and central-west Somaliland — *Tomopterna kachowskii*, Kachowski's Sand Frog

ELEGANT SAND FROG

Tomopterna elegans
(Calabresi, 1927)

Tomopterna elegans, Daallo Forest.

IDENTIFICATION: A medium-sized toad-like frog, only in very dry country of Puntland, Somalia and adjacent areas. The body is stout and toad-like; head short and deep; snout rounded, moderately long, with big eyes, a cream-brown iris and almost triangular pupils; dorsal skin fairly smooth, with few enlarged warts. Tympanum visible or indistinct, round or slightly oval, about ⅓ to ⅔ the diameter of the eye. A row of glands, not fused into a ridge, runs from below the eye to the arm; 2–3 phalanges of fourth toe free of webbing. Males reach of maximum of about 4.1cm, females 4.8cm. Dorsal surface usually light cream-brown or grey, with darker irregular markings, often an H-shaped mark in shoulder region, and dorsal warts reddish with black margins; a pale vertebral stripe and a light bar between the eyes may be present. Throat margins dark in males, uniformly pale or slightly mottled in females; limbs barred. Underside whitish. Taxonomic Notes: Formerly included in the *T. cryptotis* complex; a valid species of arid environments at tip of Horn of Africa.

HABITAT AND DISTRIBUTION: In low dry country, rocky hilly landscapes and on sandy or gypseous plains, but also in montane forest, from near sea level to 2,155m. Endemic to north-east tip of Somalia (Puntland, eastern Somaliland and probably also north-central Somalia); localities include Erigavo and Mount Daallo (in montane dry evergreen forest). In north-east Somaliland in the Daror and Nogal Valleys, Puntland. Southern limit not well known, but probably also occurs in northern parts of central Somalia (Hobyo and Galkayo); may also occur in extreme eastern Ethiopia, in Ogaden.

Tomopterna elegans, Erigavo.

CONSERVATION STATUS: Least Concern.

NATURAL HISTORY: Poorly known. Males recently observed in Erigavo were calling at night around puddles after a short intense rainstorm. In Daallo Forest, adults were found at night around smaller pools at edge of *Juniperus procera* forest, during the short rainy season; at the same time the following year no rain had fallen, there were no pools and no active frogs. Biology presumably similar to other *Tomopterna*.

KACHOWSKI'S SAND FROG

Tomopterna kachowskii
(Nikolskii, 1900)

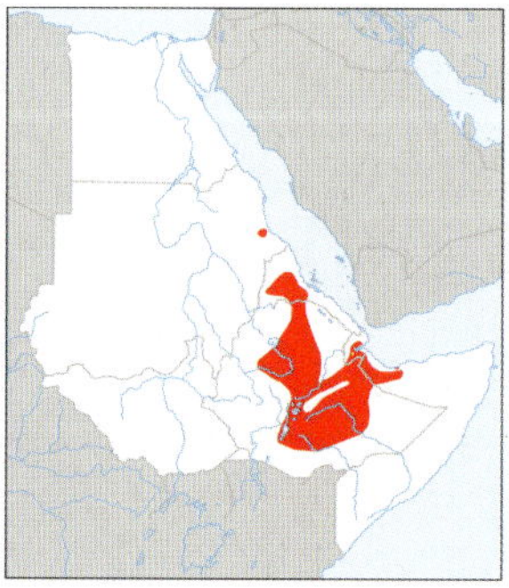

Tomopterna kachowskii, Red Sea Hills near Suakin.

IDENTIFICATION: A medium-sized toad-like frog, in dry mountainous areas of Sudan, Ethiopia and west-central Somaliland. The body is stout; head short, snout rounded, with big eyes, a brownish-yellow iris and horizontally oval pupils; skin of back fairly warty with a few enlarged warts. Tympanum visible, round, about ½ to ¾ the diameter of eye. A row of glands, not fused or fused into a ridge, runs from below the eye to the arm; 3–3.5 phalanges of fourth toe are free of webbing. Males reach a maximum of about 4.3cm, females 5.6cm. Dorsal colour light grey, pale brown or occasionally orange-brown with darker irregular markings, several warts on back reddish; a pale vertebral stripe and a bar between the eyes may be present; throat margins dark in males; limbs barred. Underside whitish. Taxonomic Notes: Long known as *T. cryptotis*, but *T. kachowskii* has recently been resurrected for montane populations in the Horn of Africa.

Tomopterna kachowskii, Fil-fil.

HABITAT AND DISTRIBUTION: Occupies a wide range of habitats, mostly in dry and moist savanna, semi-desert scrub and *Vachellia* woodland, at elevations of 200–2,500m. In our area absent only from high altitude, humid forested areas and desert environments. Known from Red Sea hills in Sudan, where recorded at Sinkat, Erkowit and west of Suakin, thence south to central mountains and western lowlands of Eritrea, over much of Ethiopia and Djibouti, and western Somalia along the Djibouti border. Also north-central Somalia (Ogo and Goolis Mountains), a handful of records from north-west Ethiopia (mostly in the Abay and Tekeze Valleys), and the Ogaden. Absent from humid south-west Ethiopia and the very arid Danakil Depression of north-east Ethiopia and adjacent Eritrea.

CONSERVATION STATUS: Least Concern; one the commonest frogs in drier environments of Ethiopia and Somaliland.

NATURAL HISTORY: Terrestrial and nocturnal. Spends day buried in soil or beneath stones, rocks and logs. Active after dusk, when emerges from cover and hunts insect prey. The advertisement call is not recorded. Breeds in the wet season or after a heavy storm; males form noisy congregations around temporarily flooded hollows, pools, streams and ditches. They call, almost invariably at night, from exposed positions close to the edge of the shallow water in which the eggs will later be deposited. Often found in livestock water reservoirs.

TANDY'S SAND FROG

Tomopterna tandyi
Channing & Bogart, 1996

IDENTIFICATION: A medium-sized toad-like frog with an invisible tympanum, only in the extreme south

Tomopterna tandyi, Turmi.

of our area. The body is stout; head short and fairly deep, snout rounded, with big eyes, a cream to brownish-yellow iris and horizontally ovoid pupils; skin of back fairly warty with a several enlarged warts. Tympanum indistinct or invisible. A row of glands, sometimes fused into a ridge, runs from below eye to arm; three phalanges of fourth toe free of webbing. Males reach a maximum of about 3.8cm, females 5.4cm. Dorsal colour light greyish-brown, with irregular darker markings and sometimes a pale scapular patch or darker H-shaped mark, many dorsal warts reddish with blackish margins; a pale vertebral stripe sometimes present. Often has a light bar between eyes. Throat margins dark in males; limbs barred. Underside white. Taxonomic Notes: Recently resurrected from the *T. cryptotis* complex; known from northern Tanzania, we tentatively refer the sand frogs of extreme southern Ethiopia and south Somalia to *T. tandyi*.

HABITAT AND DISTRIBUTION: Mostly savanna or open woodland with sandy soil at lower elevations, in our area at 500–1,000m. Known from south-eastern South Sudan, between Kapoeta and the Kenya border, also the Ilemi Triangle. Ethiopian records include the area between Lakes Turkana and Chew-Bahr in the south-west, and between Moyale and Dolo Odo in the far south. In Somalia, widely recorded between the Juba and Webe Shebelle Rivers, south of El Bur and Bud Bud. Elsewhere, from Kenya south to South Africa and Angola.

CONSERVATION STATUS: Least Concern.

NATURAL HISTORY: A burrowing frog in areas of soft and sandy soils. Breeds after heavy rains when temporary pools form, as well as in pans and small streams. Call a series of high-pitched metallic but melodious notes. A Turmi (Ethiopia) specimen was active at night in a small pool on the edge of drying seasonal river.

FAMILY RANIDAE TRUE OR 'TYPICAL' FROGS

A large family of about 26 genera and more than 420 species found worldwide, although a number of genera and species originally in Ranidae (including many African taxa) have been moved into new families. Two genera occur in our area.

GREEN FROGS *PELOPHYLAX*

Fairly large aquatic frogs with webbed toes. Fingers and toe tips not enlarged. They usually live near permanent water and breed throughout much of the year. More than 20 species are known, largely as a result of the splitting three variable Eurasian forms. The validity of some species is debated, particularly as they interbreed in zones of overlap. Most in Europe and northern Asia. Three species in our area, two in Egypt and an enigmatic one in Eritrea.

KEY TO THE NORTH-EAST AFRICAN MEMBERS OF THE GENUS *PELOPHYLAX*

1a: In our area, only in Eritrea	*Pelophylax demarchii*, Eritrean Green Frog
1b: In our area, only in Egypt	2
2a: Dorsal surface with numerous dark spots, north-west Egypt	*Pelophylax saharicus*, Sahara Green Frog
2b: Dorsal surface with a few dark spots; Nile and north-east Egypt	*Pelophylax bedriagae*, Levantine Green Frog

LEVANTINE GREEN FROG

Pelophylax bedriagae
(Camerano, 1882)

IDENTIFICATION: A big green or brown frog, in our area only in lower Egypt. The body is elongate; the head large, with big eyes and horizontal ovoid pupils. Dorsal skin granular, but without warts. Tympanum visible, almost the diameter of the eye. Limbs stocky, finger and toe tips not enlarged. Males reach a maximum of about 8cm, females 10cm. Dorsal surface usually green or brown, with scattered dark blotches, sometimes with a light vertebral stripe. Underside cream or white, sometimes with mottling. Taxonomic Notes: Egyptian specimens were called Marsh Frog *Rana ridibunda* in older literature.

Pelophylax bedriagae, south-west of Cairo.

HABITAT AND DISTRIBUTION: Associated with marginal vegetation and agriculture beside canals, swamps, rivers, ditches and other watercourses in the desert, between sea level and 100m in our area. From Luxor north along the Nile to the Delta, also the Faiyum, Suez Canal (vicinity of Ismailia) and Rafah in north-east Sinai. Elsewhere; around eastern Mediterranean to Turkey. Some authors believe this species was introduced to Egypt and is expanding its range; until recently it wasn't known south of Cairo.

CONSERVATION STATUS: Least Concern.

NATURAL HISTORY: Terrestrial, active by day and night in water, forages at night on land. Advertisement call a distinctive, high-pitched quack, somewhat like an angry squirrel. Breeding males call in the water, on aquatic vegetation, or on land, and breeding occurs almost year-round. The eggs are deposited in shallow waterbodies. Many adults are harvested for the trade in edible frogs legs.

SAHARA GREEN FROG

Pelophylax saharicus
(Boulenger, 1913)

IDENTIFICATION: A big green or brown frog, in our area only at Siwa Oasis, north-west Egypt. Body elongate; head large, with big eyes and horizontal ovoid pupils. Dorsal skin granular, without warts, dorsolateral ridges prominent. Tympanum visible, almost the diameter of the eye. Limbs stocky, finger and toe tips not enlarged. Males reach a maximum of about 5cm, females 6cm in Egypt, but much larger animals are reported elsewhere. Dorsal surface usually green or brown, with a few scattered dark blotches, sometimes immaculate, occasionally with a light vertebral stripe. Underside cream or white. Taxonomic Notes: Separated from *P. bedriagae* in Egypt by its call and distribution.

Pelophylax saharicus, Tunisia.

Pelophylax saharicus, Tunisia.

HABITAT AND DISTRIBUTION: Associated with various waterbodies (anthropogenic or natural) in desert or near-desert. In our area, only Siwa Oasis, at or below sea level! Elsewhere; west across the Maghreb to Morocco, with a limited penetration of the Sahara, and introduced in France.

CONSERVATION STATUS: Least Concern.

NATURAL HISTORY: Terrestrial, active day and night in water, forages at night on land. Advertisement call a distinctive, high-speed rattle, fastest when a group of males calls together. Breeding males call from shallow water, on aquatic vegetation, or on land, and breeding occurs almost throughout the year. The eggs are deposited in shallow waterbodies.

ERITREAN GREEN FROG

Pelophylax demarchii
(Scortecci, 1929)

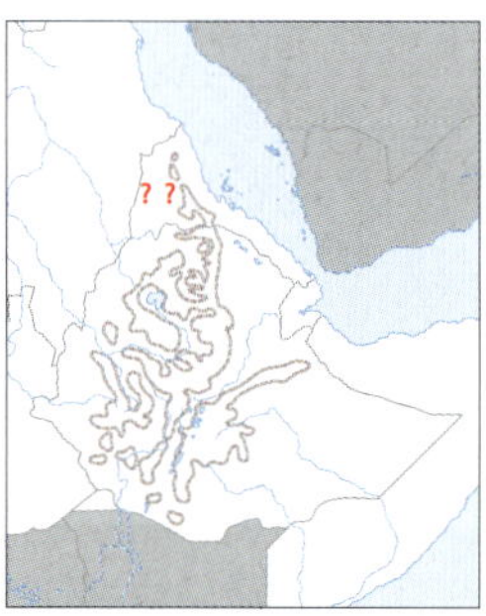

IDENTIFICATION: Known from just two specimens; a spotted Eritrean frog, presumably similar to other species of *Pelophylax*. Body elongate; head medium-sized, with big eyes and horizontal ovoid pupils. Dorsal skin granular, without warts, dorsolateral ridges prominent. Tympanum visible, almost the diameter of the eye. Limbs stocky, finger and toe tips not enlarged. The male was 11cm long, the female 10.8cm. Dorsal surface greyish to olive-green, with scattered dark blotches. Taxonomic Notes: Described in 1929 by Giuseppe Scortecci from 'Eritrea' (but without further data). One of the types is lost. No more specimens have been found, leading to speculation that it might be a species known from elsewhere, that it may not have originated in Eritrea and that Scortecci was misinformed, or was simply mislabelled.

HABITAT AND DISTRIBUTION: Known only from Eritrea. Never knowingly found alive to date.

CONSERVATION STATUS: Data Deficient.

NATURAL HISTORY: Unknown, presumably similar to other green frogs.

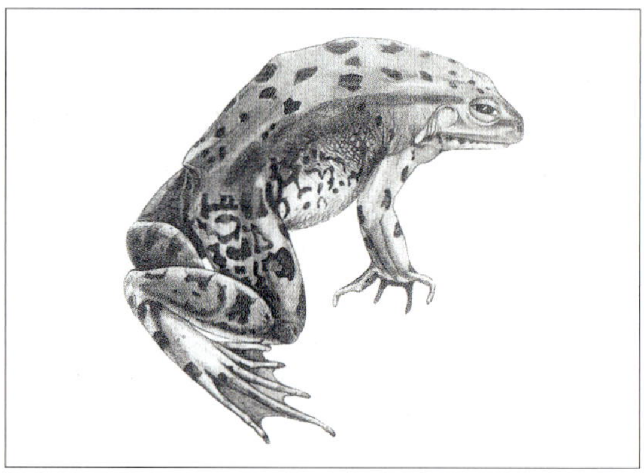

Pelophylax demarchii.

WHITE-LIPPED FROGS *AMNIRANA*

A tropical African genus of fairly large, mostly forest-dwelling, attractive frogs, some have expanded toe tips and climb, others do not. Males have large glands on the upper arm and prominent dorsolateral skin folds. About 12 species, but further species within the group await description. Two species in our area. Originally placed in genus *Rana*, then *Hylarana*.

KEY TO THE NORTH-EAST AFRICAN MEMBERS OF THE GENUS *AMNIRANA*

1a: Expanded discs on toes	*Amnirana albolabris*, Forest White-lipped Frog
1b: Toe tips not expanded	*Amnirana galamensis*, Galam White-lipped Frog

GALAM WHITE-LIPPED FROG, GOLDEN WHITE-LIPPED FROG

Amnirana galamensis
(Duméril & Bibron, 1841)

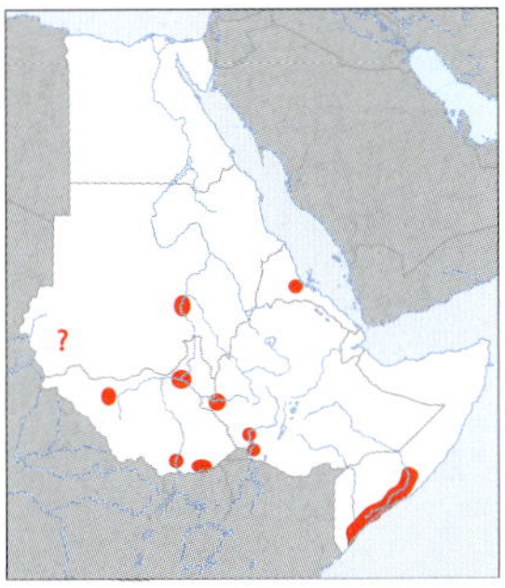

IDENTIFICATION: A big, golden-striped frog of the lowlands. Body robust to elongate; head large, with big eyes and horizontal ovoid pupils. Dorsal skin granular, largely smooth with occasional flat warts towards rump; dorsolateral skin ridges quite broad and flat. Tympanum visible, almost the diameter of the eye. The limbs are muscular, finger and toe tips not enlarged. Males reach a maximum of about 8cm, females 8.5cm. Dorsal surface usually gold, warm brown or rufous, marbled to a greater or lesser extent with black, with a dark flank stripe, often spotted same colour as back. Underside dirty white. Taxonomic Notes: The type of this species is from Senegal. Widespread across Africa north of the forest. Those east of Chad might be a separate species, but for the time being, in lieu of molecular investigation, we leave them in *A. galamensis*; if they prove to be separate, the name *A. bravana* is available for Somali specimens.

HABITAT AND DISTRIBUTION: Dry and moist savanna and semi-desert between sea level and 1,000m. In our area, known from the lower Webe Shebelle River in Somalia, south of 04°N, along the coastal plain to the Kenya border. A single record from Eritrea (Ghinda) and three from Ethiopia (Gambela, Mago National Park, lower Omo River). Sporadically recorded in South Sudan (Gilo, Torit, 60km west of Torit, and Juba, Aweil in Northern Bahar El Ghazal, Malakal in Upper Nile); in Sudan, from between Kosti and Khartoum, possibly further north; probably occurs in south-west, as it is known from northern Central African Republic. Elsewhere, the nominate extends west to Senegal, the eastern subspecies south to Mozambique.

Amnirana galamensis, Gambela.

Amnirana galamensis, Keilak.

Amnirana galamensis, Kenya.

CONSERVATION STATUS: Least Concern.

NATURAL HISTORY: Terrestrial, nocturnal, but occasionally forages by day. Advertisement call a rather ghostly *hoooh*, followed by a deep grunt. Breeding males call, widely spaced, from well-vegetated spots at water's edge. Usually associated with permanent water in dry season, but may aestivate. Appears in numbers and spreads out in wet season, utilising larger temporary ponds and swamps. Up to 4,000 eggs may be laid in a single layer. The skin contains bradykinins, toxins that affect the heart rate.

FOREST WHITE-LIPPED FROG

Amnirana albolabris
(Hallowell, 1856)

IDENTIFICATION: A big white-lipped frog of South Sudan. Body robust to elongate; head large, with big eyes and horizontal ovoid pupils. Dorsal skin granular, largely smooth with occasional flat warts towards rump and on flanks. Dorsolateral skin ridges fragment posteriorly in some specimens. Tympanum visible, almost the diameter of the eye. Limbs muscular, finger and toe tips enlarged into discs. Males reach a maximum of about 5.7cm, females 7.5cm. Dorsal surface usually shades of brown or rufous, occasionally green, the flank stripe sometimes darker than the dorsal colour, or the same; occasionally marbled black and white. Underside white or yellow, sometimes the yellow extends up the flanks.

HABITAT AND DISTRIBUTION: Primary and secondary forest, forest fringes and farmland at edges of forest; in our area at 800–1,000m, to sea level, and higher, elsewhere. In north-east Africa, known from a couple of records in South Sudan; at Kagelu, near Yei, Central Equatoria, and the Inimvua River, on the Sudan side of the border with Garamba National Park, on the Western/Central Equatoria border; also Garamba National Park in DR Congo. Possibly more widespread in south-west South Sudan. Elsewhere, west to Sierra Leone, though animals west of Nigeria may prove to be a different species.

CONSERVATION STATUS: Least Concern.

NATURAL HISTORY: Terrestrial and arboreal, often climbs low vegetation and small trees; active by day and night. Advertisement call a quiet *ouic…ouic*; males tend to call in choruses and it is hard to distinguish individuals. Breeding males call from edges of slow-flowing streams, swamps and temporary pools; eggs are deposited on the surface of stagnant water. Like the previous species, skin contains toxins that deter predators like dwarf crocodiles and catfish.

Amnirana albolabris, Kenya.

FAMILY RHACOPHORIDAE AFRO-ASIAN TREE FROGS

A large family of over 20 genera and about 450 species, including the famous flying frogs of South-East Asia. All genera are Asian save one, African foam-nest frogs *Chiromantis*.

FOAM-NEST FROGS *CHIROMANTIS*

A sub-Saharan African genus of fairly large frogs. At present, four species known, two in our area, but there is debate over their taxonomy; see *C. kelleri*. Foam-nest frogs are unusual in several ways. Savanna species are adapted for dry times; they secrete fluid under the skin to waterproof it, and change colour, becoming white to reflect heat. Consequently, they are sometimes seen sitting in the open on branches in dry country by day, where no other frog could survive. Their hands and feet are webbed but the fingers are arranged in opposing pairs, enabling them to grasp and climb; if threatened they make big ponderous jumps, limbs outstretched to grab a branch. When breeding, females secrete a fluid that is whipped by their back legs into a white foam nest; this is often clearly visible at lowland pools. The eggs are laid in the foam and hatch into tadpoles. After a few days the base of the foam nest breaks away and the tadpoles drop into a pool below.

KEY TO THE NORTH-EAST AFRICAN MEMBERS OF THE GENUS *CHIROMANTIS*

1a: In our area, confined to southern Somalia, one joint or less on fourth finger webless
Chiromantis xerampelina, Southern Foam-nest Frog

1b: In our area, not confined to southern Somalia, more than one joint of fourth finger webless
Chiromantis kelleri, Keller's Foam-nest Frog

SOUTHERN FOAM-NEST FROG

Chiromantis xerampelina
Peters, 1854

Chiromantis xerampelina, Kenya.

Chiromantis xerampelina, Kenya.

IDENTIFICATION: A lanky, elongate climbing frog of southernmost Somalia. Body elongate and head large, with very large eyes and horizontal, ovoid pupils. Dorsal skin rough, with small to medium warts. Tympanum visible, 60 per cent or more the diameter of the eye. Limbs quite stout for a tree frog, and digit tips expanded. Males reach a maximum of about 7.5cm, females 9cm. Dorsal surface usually brown, grey or rufous-brown, mottled to a greater or lesser extent with darker blotches, which may or may not be symmetrical on the midline, but provide superb camouflage. By day, turns chalky white. Areas of underside covered by folded limbs pink, unpigmented, the remainder mottled white.

HABITAT AND DISTRIBUTION: Low-altitude savanna and coastal woodland, near sea level. No definite records from our area, but recorded from Boni National Reserve in Kenya, which abuts the Somali border, so almost certainly present on the Somali side. Elsewhere, south to South Africa and Angola.

CONSERVATION STATUS: Least Concern.

NATURAL HISTORY: Arboreal and nocturnal. Largely active by night in wet season. In dry season, unusually, these frogs remain dormant on branches, their skin waterproofed and white. Advertisement call a series of irregularly pulsed nasal grunts (sometimes described as creaks). Breeding males call from an elevated position (branch, rock, vegetation, but usually over water). Females arrive, moving through the vegetation. Male grasps the female; she produces a secretion which she beats into foam by thrashing her legs. Often, several females group together. The female then releases the eggs, they are fertilised and drop into the foam, where they become tadpoles. The outer foam hardens and forms a waterproof cocoon. The lower part of the nest detaches after a few days, dumping the tadpoles in the pool below. Some monkeys eat the foam nests, as does Fornasini's Spiny Reed Frog *Afrixalus fornasinii*, which creeps close and snatches eggs from the foam before it has hardened.

KELLER'S FOAM-NEST FROG

Chiromantis kelleri
Boettger, 1893

IDENTIFICATION: A lanky, elongate climbing frog of semi-desert and dry savanna in our area. Body elongate and head large, with very large eyes and horizontal, ovoid pupils. Dorsal skin rough, with small to medium warts. Tympanum visible, 60 per cent or more the diameter of the eye. Limbs quite stout for a tree frog, and digit tips expanded. Males reach a maximum of about 5.6cm, females 9.2cm. Dorsal surface usually brown, grey or rufous-brown, mottled to a greater or lesser extent with darker blotches, which may or may not be symmetrical on the midline. By day, turns chalky white. The areas of the underside covered by the folded limbs is pink, unpigmented, the remainder mottled white. Taxonomic Notes: in East Africa, the dry-country inland foam-nest frogs are assigned to two species, *C. kelleri* and *C. petersi*; *C. xerampelina* is largely coastal. In some areas, intermediate forms occur, which display a mosaic of features from two or three forms, particularly in northern Somalia and on Kenya/Ethiopia border. For now, we regard all inland animals as *C. kelleri*; this might prove to be an over-simplification.

HABITAT AND DISTRIBUTION: Low-altitude savanna and semi-desert, between sea level and about 1,600m. Occurs over most of Somalia (although records sporadic at 04–08°N, and in Somaliland and Puntland). In Ethiopia, a handful of records from the Ogaden, Sof Omar, at Malka Murri on Kenya border, and the lower Omo River and Lake Chew Bahir. In South Sudan known from Juba, in Central Equatoria, and 50km north of Kapoeta, in Eastern Equatoria; its presence in north-west Kenya, at Kakuma and Lokichogio, indicate it is probably widespread in south-east South Sudan. Elsewhere, south to Kenya, possibly Tanzania.

CONSERVATION STATUS: Least Concern.

NATURAL HISTORY: Arboreal and nocturnal, rather similar to previous species in terms of activity patterns, dry-season strategy, breeding etc. Advertisement call a slow creak. In central Somaliland foam cocoons observed on branches of *Vachellia* trees above water, on rocks above water and at edges of small pools, in the October short rains.

Chiromantis kelleri, Sof Omar.

Chiromantis kelleri, Sof Omar.

CLASS REPTILIA

ORDER TESTUDINES; CHELONIANS (TURTLES AND TORTOISES)

A distinctive order of reptiles, with a protective shell, and hard outer scales covering an inner, bony layer. Inside, the ribs are fused to the shell. We use the useful collective term chelonian for all members of the order, tortoise for the land-dwelling chelonians, and terrapins and turtles for, respectively, small and large water-dwelling chelonians. Worldwide, c.360 species are known; in north-east Africa there are six land tortoises, five sea turtles, three big freshwater, soft-shelled turtles and eight terrapins. A small American species, Red-eared Terrapin *Trachemys scripta*, a popular pet, has been introduced into Egypt, in the Nile and associated waterways around Cairo; we have not described it or placed it in the key.

Many turtles and tortoises can totally or partially withdraw their head and limbs inside the shell. They are long-lived and may take several years to reach maturity. In many species, adult males have a concave plastron (underside of the shell), the male goes on top while mating and the concavity prevents him slipping off. Males often have longer and thicker tails than females. All chelonians lay eggs; females excavate a flask-shaped pit in soil or sand. In some species sex of the young is temperature-dependent, with cooler nests producing mostly male offspring. Land tortoises largely eat plants (some take invertebrates and carrion); most water-dwelling species are carnivorous. Due to their size and shape, they take a long time to gain heat, so are rarely found above 1,500m in north-east Africa. Chelonians have a sense of direction; sea turtles migrate across thousands of kilometres of ocean to lay eggs on their 'home' beach, and translocated land tortoises have found their way home over many kilometres.

Small land chelonians are fairly defenceless and have many enemies. Land chelonians are at risk from bush fires and are killed on roads. Large ones have one main enemy, humans. Although the situation is improving, risks include harvesting for the pet trade, harvesting of eggs, accidental and deliberate trapping in fish nets, and habitat loss and disorientation of hatchlings due to coastal tourist development. A disturbing development is the harvesting across the tropics of chelonians for the South-East Asian food trade, fuelled by the superstitious (and totally irrational) belief that eating the meat of a long-lived and hardy animal will impart those qualities to the consumer. Education is needed.

The order Testudines is split into two suborders, Cryptodira and Pleurodira. In north-east Africa, all land tortoises, sea turtles and the three large soft-shells are in the Cryptodira. The name means 'hidden neck', as they all withdraw their head into the shell by flexing the neck into an S-shape, in a vertical plane, and at the same time, in terrestrial species, the legs are withdrawn and the front legs protect the head. The freshwater terrapins of the genera *Pelomedusa* and *Pelusios* are in the suborder Pleurodira, which name means 'side-neck' as these terrapins withdraw the head sideways in defence, leaving one eye exposed. They have webbed hind feet.

KEY TO THE CHELONIAN GROUPS AND SOME SPECIES IN NORTH-EAST AFRICA

1a: Neck exposed (bent sideways) when head is withdrawn into shell. Suborder Pleurodira	2
1b: Neck hidden when head withdrawn into shell, or head cannot be withdrawn. Suborder Cryptodira	3
2a: Plastron has no hinge; head never reticulated or spotted, shell not conspicuously domed or deep	*Pelomedusa*, Helmeted terrapins
2b: Plastron with hinge at front, head often shows reticulations or spotting, shell often domed and deep	*Pelusios*, Hinged terrapins
3a: Living in or near water but rarely in sea, snout elongate and pig-like, shell leathery but never with seven equally spaced longitudinal ridges	12
3b: Does not have the above combination of characters	4

4a: Limbs modified as flippers, with 0–2 claws, in the sea or on beaches	5
4b: Limbs not modified as flippers, with 3–5 claws, not in the sea	9
5a: Carapace leathery, with seven longitudinal ridges, flippers clawless	*Dermochelys coriacea*, Leatherback Turtle
5b: Carapace not leathery, with symmetrical horny plates, flippers with 1–2 claws	6
6a: Carapace with four pairs of costals	7
6b: Carapace with five or more pairs of costals	8
7a: Hooked beak present, four pre-frontal head shields, usually two claws on each limb	*Eretmochelys imbricata*, Hawksbill Turtle
7b: Beak not hooked, two pre-frontal head shields, usually with single claw on each limb	*Chelonia mydas*, Green Turtle
8a: Carapace with five pairs of costals, head large, colour red-brown.	*Caretta caretta*, Loggerhead Turtle
8b: Carapace usually with 6–9 pairs of costals, adults olive, young olivaceous-black.	*Lepidochelys olivacea*, Olive Ridley Turtle
9a: Adult shell has hinge at rear of carapace	*Kinixys belliana*, Bell's Hinged Tortoise
9b: Adult shell without hinge	10
10a: In our area, confined to Egypt, small, adults never more than 15cm carapace length	*Testudo kleinmanni*, Egyptian Tortoise
10b: In our area, not found in Egypt, adults reaching 60–75cm carapace length	11
11a: In our area, in Ethiopia, South Sudan and Somalia, shell usually domed at top, and yellow with irregular black blotches (rarely uniform brown in very large animals)	*Stigmochelys pardalis*, Leopard Tortoise
11b: In our area, in Eritrea, Ethiopia and Sudan, shell flattened on top, uniform in colour, without dark blotches	*Centrochelys sulcata*, African Spurred Tortoise
12a: Plastron without skin flaps over rear limbs, carapace up to 1.2m, head, neck, limbs and shell often covered with small, light, dark-edged spots	*Trionyx triunguis*, Nile Soft-shelled Turtle
12b: Plastron with skin flaps (femoral flaps) that cover withdrawn rear limbs, carapace rarely larger than 70cm, limbs and shell never covered by small light, dark-edged spots	13
13a: Maximum size 50cm, ventral surface of adults with 5–9 skin calluses, several at front, 5–6 crescent-shaped skin folds on front of each forefoot	*Cyclanorbis senegalensis*, Senegal Flap-shelled Terrapin
13b: Maximum size 70cm, ventral surface of adults without calluses at front of plastron, four crescent-shaped skin folds on front of each forefoot	*Cyclanorbis elegans*, Nubian Flap-shelled Terrapin

SUBORDER CRYPTODIRA

SUPERFAMILY TESTUDINOIDEA

FAMILY TESTUDINIDAE TORTOISES

A family of land-dwelling tortoises, with about 60 species in the group, in 17 or so genera, eight of which occur in Africa, six in our area (but see our taxonomic notes on *Kinixys*). Recent taxonomic work has caused several name changes; in our area Leopard Tortoise, originally in the genus *Geochelone*, now has its own genus, *Stigmochelys*, and African Spurred Tortoise, for a long time *Geochelone sulcata*, is now *Centrochelys sulcata*. All land tortoises are on CITES Appendix II, meaning they can be legally exported only under permit.

LEOPARD TORTOISE *STIGMOCHELYS*

A monotypic genus.

LEOPARD TORTOISE

Stigmochelys pardalis
(Bell, 1828)

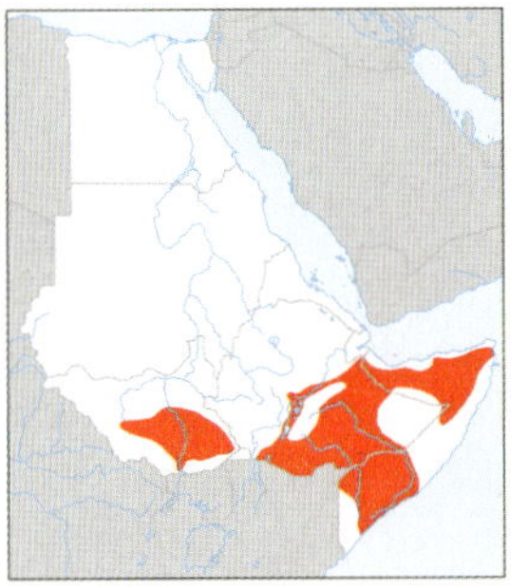

IDENTIFICATION: A big, yellow-and-black land tortoise with a domed, steep-sided shell, a blunt bullet-shaped head, and hooked beak. Head moderately large and fairly narrow, with large frontal scale on top of head usually absent. The costal and vertebral scales sometimes form blunt, tent-like peaks. Limbs stout, five claws on front feet, four on back. The anterior surface of the front legs is covered by 3–4 rows of large, non-overlapping, rather flat scales. Buttock tubercles present and well developed. Males have a longer thicker tail than females, with no terminal spike, and have a well-developed plastral concavity. Adult shells usually 50–65cm long (30–50cm outside our area, except in South Africa), but may be up to 78cm, hatchlings 4–5cm long and 23–50g; big adults usually female. Mass usually 20–35kg but up to 50kg. As illustrated, colour changes during life, the often dotted, black-edged scales of hatchlings changing to yellow or tan, with light or heavy black speckling, dots or radial patterns; with age the markings often fade to a dull uniform brown or grey. Animals from higher altitudes and more mesic (e.g., forested) habitats tend to be dark with heavy speckling; those in arid plains usually more or less uniform.

HABITAT AND DISTRIBUTION: Semi-desert, dry and moist savanna, rocky country, coastal woodland and thicket, marginally also in dry evergreen montane forest with *Juniperus* and *Buxus hildebrandtii*, from sea level to about 2,100m. Absent from dense forest and very arid habitats with sparse vegetation. Occurs widely in south-east South Sudan, in Eastern, Central and the east of Western Equatoria, and southern Lakes State, with a couple of isolated records in Western Bahar El Ghazal and Unity State. Also extreme southern Ethiopia and south-east into Somalia south of 04°N. Found virtually throughout the Ethiopian Rift Valley, at the southern edge of the Afar Triangle, into northern Somaliland and Puntland, as far south as north-west of Galkayo. Absent from desertic coastal plains and mesic and fairly cold high parts of the northern mountains (Cal Madow, Cal Miskaat and Bari Mountains), although recorded from near Shimbiris (2,460m), the highest point in Somalia. Absent from western Ethiopia, save a few records in the low Rift Valley foothills (Nazareth, Modjo). Not recorded in Djibouti, but nearby Somaliland records and similar habitat indicate it is probably present in the Ali Sabieh region. No records from much of central Somalia and eastern Ogaden, but this may be due to under-collecting; it probably occurs. Elsewhere, south to South Africa, Angola and Namibia.

Stigmochelys pardalis, hatchling, Sheikh.

Stigmochelys pardalis, Hargeisa.

Stigmochelys pardalis, west of Hargeisa.

Stigmochelys pardalis, old adult, Dilla.

CONSERVATION STATUS: Least Concern.

NATURAL HISTORY: Diurnal but shelters during heat of day. Active in rainy season, especially mornings, when mating occurs. Males fight during breeding season, butting and pushing their rivals. The males are equally rough with the females during mating, accompanied by asthmatic wheezing and hissing. The female digs a deep pit in an open area at dusk for her eggs, softening the soil with cloacal water or urine. On average 4–21 hard-shelled eggs (maximum 30), 3–5cm wide and 4–6cm long, depending on the female's size, are laid in darkness, and the female then depresses the soil over the pit, using her plastron. Several clutches may be laid each year. Incubation time 4–12 months, depending on the temperature. Nests may be excavated by small mammalian carnivores, and hatchlings are eaten by snakes, birds, monitor lizards and carnivores (even Lions, *Panthera leo*, will take them). Thus the young feed nervously from cover, but once they reach 20cm or so, they are safe from almost everything (except humans, occasionally Lions and hyenas) and can feed and move about more safely; at the same size they start to reproduce. They eat a wide variety of plants, and are recorded eating carrion, chewing bones and eating hyena or dog faeces to obtain calcium and other nutrients. When fresh green vegetation is unavailable and temperatures fall in the dry season, they hibernate, often with their heads rammed into vegetation, although they may emerge to bask. They are solitary and territorial, with home ranges up to 50km^2. However, in suitable areas with abundant food and water they aggregate in small temporary groups. They are at risk from fire, particularly in savanna areas where the vegetation is very thick, and from road-killing. They are not eaten by humans in our area, but are often kept as pets. Common in central and western Somaliland, even on the outskirts of cities like Hargeisa.

AFRICAN SPURRED TORTOISE *CENTROCHELYS*

A monotypic genus with a single species, the largest land tortoise in Africa.

AFRICAN SPURRED TORTOISE

Centrochelys sulcata
(Miller, 1779)

IDENTIFICATION: A huge brown tortoise with a flat-topped shell, of the Sahel. Head big and square, with a large frontal scale. The costal and vertebral scales sometimes form blunt, tent-like peaks, and the marginal scales at the front and back are often curved upwards; from above the back end often looks spiky. The limbs are stout; five claws are present on the front feet, four on the back. The anterior surface of the front legs is studded with big, non-overlapping scales, either flat or cuboid. The back of the thigh has 2–3 large conical tubercles. Males have a longer thicker tail than females, without a terminal scale. Males have a well-developed plastral concavity, and big males have large forked gular scales (the plastral scales under the neck). Adult shells usually 50–65cm long, but may be up to 80cm or slightly more; big adults usually males. hatchlings 4–6cm long and 40–50g. Adult mass typically 40–80kg but individuals >100kg are known. The black edging to juvenile scales may increase or decrease with age, but big adults are usually uniform brown, tan or grey.

HABITAT AND DISTRIBUTION: Semi-desert and near-desert, dry and moist savanna, from sea level possibly to above 2,000m, but usually below 1,000m. Oddly, for such a large and relatively easily identified species, its purported present distribution is clouded by misidentifications, translocated pets, subfossil material and possible local extinctions. In Ethiopia, known from a small cluster of north-western records near the border with Eritrea (Kafta Sheraro National Park, Yirga, Tekeze River), and recently reported from the Blue Nile gorge west of Dejen. Old records from central Eritrea (Mareb River, Keren, Godolefassi and Chenafena) might be accurate and the species may still occur, but recent searches failed to find it. More recent Eritrean records are from south-east of Barentu and on Ethiopian border west of Om Hajer. Occurs widely but sporadically across Sudan, between c.10°N and 16°N in the following states: Kassala, Gadarif, Sennar, Blue Nile, Northern and Southern Kordofan (where, in the Muglad Basin, at Neem and Baleela, breeds in seasonal wadis), Khartoum, Northern, Southern and Western Darfur. Older records from further north (Wadi Halfa, Dongola and Barbar) need reconfirmation. Other records from Ethiopia, Somalia and East Africa erroneous. Elsewhere, isolated populations west to Senegal and Mauritania, mostly at 12–20°N.

CONSERVATION STATUS: Vulnerable. The southward creep of the Sahara has greatly reduced its range; a near-complete carapace 5,000 years old is known from Abu Tabari, 17°34'N, 28°31'E, in northern Sudan, well north of the present range in central Sudan.

NATURAL HISTORY: Diurnal but often crepuscular, moving at dawn and dusk. Mostly active during rainy season, July–October; they drink deeply from rain pools when they first emerge. They sometimes make short movements at night in the dry season. Excavate deep, long burrows (up to 30m long, 15m deep), in sandy or gravelly soil. These cool moist refuges are

Centrochelys sulcata, Khartoum.

Centrochelys sulcata, juvenile, West Kordofan.

Centrochelys sulcata, Khartoum.

Centrochelys sulcata, captive juvenile.

used for aestivation in the dry season; other animals share the burrows. Males fight during the breeding season, butting, pushing and trying to flip their rivals. They are equally rough with females during mating; this is accompanied by asthmatic wheezing and hissing. Female digs a deep pit at dusk for her eggs, softening the soil with cloacal water or urine. On average 10–30 white, spherical hard-shelled eggs about 37–45mm in diameter and mass 30–70g are laid in darkness; the female then depresses the soil over the pit, using her plastron. Up to four clutches may be laid each year, at the end of the rains. Incubation times in Chad 144–149 days, but shorter times are reported for captive animals: 90–100 days. Hatchlings appear to time their emergence to coincide with the onset of the rains. They grow rapidly, feeding frantically on succulents and grasses in the wet season; a nine month juvenile can weigh 200g and a three-year juvenile >4kg. They may feed on carrion. Hatchlings are predated by monitor lizards, possibly large snakes and birds of prey, but adults are only at risk from humanity, fire and drought. Long-lived; some have reached 100 years. They are respected in parts of West Africa; Dogon communities in Mali believe the spirits of the ancestors live in them; old males may be 'pampered' in villages, and chiefs sit on them while dispensing judgements. In other areas, they may be eaten. Often kept as pets.

HINGED TORTOISES *KINIXYS*

A genus of attractive, omnivorous, small (maximum 40cm, usually 25cm or less) yellow, tan or brown tortoises, confined to sub-Saharan Africa and Madagascar. Eight species are known; three occur in our area. Largely active during the rains, activity is reduced in the dry season; some aestivate. Adults have a fibrous hinge on both sides near the back of the carapace (upper shell), enabling the rear shell to shut, protecting the hindquarters and back legs (they tend to hide head-first in holes). Taxonomy complicated due to the variability of diagnostic external characters (patterns, shell shapes etc).

KEY TO THE NORTH-EAST AFRICAN MEMBERS OF THE GENUS *KINIXYS*

1a: Rear of carapace strongly serrated, usually in forest	*Kinixys erosa*, Forest Hinged Tortoise
1b: Rear of carapace not strongly serrated, in savanna	2
2a: Widespread in our area, five claws on each forefoot, carapace often quite light, with little dark pigment	*Kinixys belliana*, Bell's Hinged Tortoise
2b: In our area confined to south-west South Sudan, four claws on each forefoot, carapace often darkly pigmented	*Kinixys nogueyi*, Western Hinged Tortoise

FOREST HINGED TORTOISE

Kinixys erosa
(Schweigger, 1812)

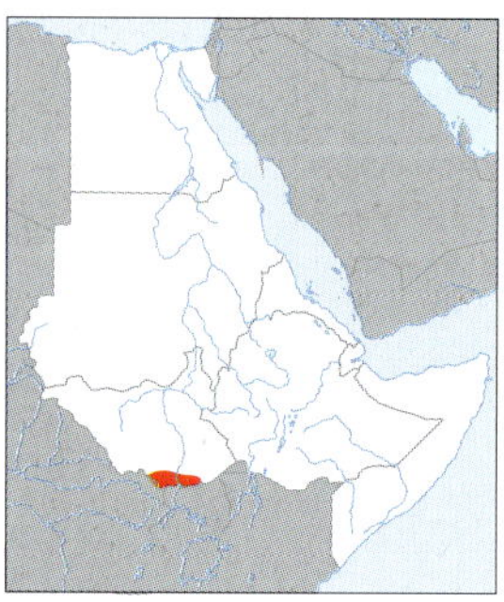

IDENTIFICATION: A big (maximum 37.5cm) hinged tortoise, in forest. Has a slightly concave shell, the scales are black, bright yellow and warm brown, head rounded with an elevated snout and beak has a single cusp. Marginal scales often upturned at the edges, especially at the rear, giving a bizarre serrated appearance to the back end. The front feet have five claws, the rear feet four. The anterior of the front foot is covered by 4–5 rows of large, overlapping scales. No buttock tubercles, the tail has a claw-like tubercle at the tip, males have longer and thicker tails than females. Average size 15–25cm but may reach 37cm or more, males larger than females. Hatchlings 4–6cm. Adult shell a vivid mix of bright yellow, brown and black, often the lower half of the pleural scales is bright yellow, giving the impression of a yellow side-stripe. Individual plastron scutes have black radial patches, the seams cream or yellow.

Kinixys erosa, Congo.

HABITAT AND DISTRIBUTION: Low- to mid-altitude forest, said to prefer moist areas. Recently reported from a handful of areas along the border between South Sudan and DR Congo; from near Nagishot in Eastern Equatoria west through Nimule, and in the vicinity of Yei, Central Equatoria; mostly at 700–900m altitude. Elsewhere, west to Sierra Leone, south-west to the Congo River mouth.

CONSERVATION STATUS: Data Deficient, although utilised for food and local medicine; some populations in West Africa have collapsed, indicating overexploitation.

NATURAL HISTORY: Sometimes diurnal but unusually for a tortoise may be active at night, and is known to travel long distances, up to 275m. Most active in rainy season in West Africa. Home range several hectares and usually includes a waterbody. Shelters under logs, in holes, leaf litter and vegetable debris. Presumably defends itself like other hinged tortoises, i.e. withdraws the limbs and closes the shell. Males fight albeit rarely. Mating occurs year-round in Central Africa, both day and night; the male ramming the female and even turning her over. Lays 1–10 large eggs (average 5–8), either buried or cached under vegetation, and more than one clutch may be laid in a season; females are sexually mature at 20cm length. Eggs weigh 17–28g, diameter roughly 4cm and can take 5–10 months to hatch. Omnivorous, eating fungi, plants, invertebrates and carrion. Prized as a food by some forest peoples, it is hunted with dogs and located by its distinctive smell. They swim well, and in swamp forest in West Africa may actually disperse by swimming with rising flood waters.

BELL'S HINGED TORTOISE

Kinixys belliana
Gray, 1831

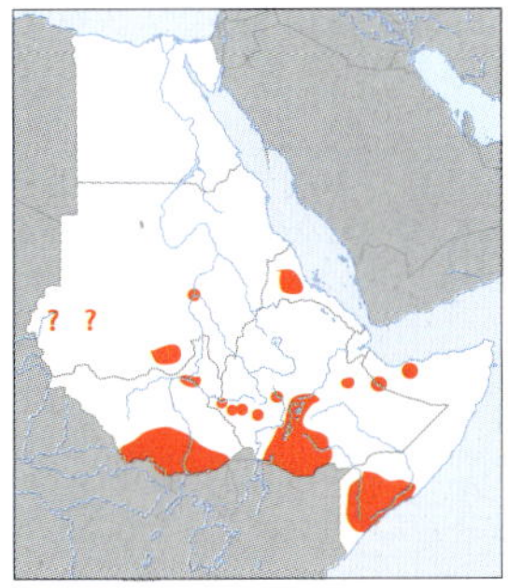

IDENTIFICATION: A small, brown, or yellow-and-brown hinged tortoise, in savanna, with a slightly domed shell (especially at the back), a pointed head and beak with a single cusp. A hinge is usually present at the rear of the shell in adults. Back end of shell steep and flattened in adults. The front feet have five claws, the rear feet four. The anterior of the front foot is covered by several rows of large, overlapping scales. No buttock tubercles, the tail has a terminal spine, males have longer and thicker tails than females. Average size 14–18cm, maximum about 22cm, hatchlings 3.5–4.8cm. Shell usually yellow or tan, with black, radial (spoke-shaped) patterns, which may be heavy or light on each scale. In large males the pattern usually fades to brown or grey. Plastron yellow. Taxonomic Notes: There is debate about whether the *Kinixys* in Central African Republic, and western Sudan are the West African *K. nogueyi* or *K. belliana*; we have treated them as *K. belliana*.

HABITAT AND DISTRIBUTION: Dry to moist savanna, woodland and coastal thicket, from sea level to nearly 2,000m. In Somalia, known only from the far south (south of 03°N) and a handful of records in western Somaliland (an old report from the environs of Berbera, a more recent one from Borama, where rare). Records from Eritrea, apart from the Anseba Valley, suggest translocated pets (Asmara, Keren). Quite widespread in eastern Ethiopia and Ethiopian Rift Valley, often to fairly high altitudes; localities include Modjo, Sodere, Lake Langano, Awash National

Park, Harar, Arba Minch, Turmi, Negele and Moyale. Absent from the Ogaden. A handful of records in western Ethiopia; Welkite, Jimma, Bedelle, Metu and Gambela. Sporadic in South Sudan; at Mongalla and Juba in Central Equatoria, Opari, 105km south-west of Torit, Tonj (midway between Rumbek and Wau) and Li Rangu, north of Yambio. Recently reported west of Malakal in Upper Nile State, with a cluster of records in southern Central and Western Equatoria, but no precise localities. Known from Garamba National Park in DR Congo, just across the border from Western Equatoria. Sudanese records include Khartoum, Kadugli and Talodi (in Nuba Mountains). Presence in north-east Central African Republic lends credence to unconfirmed sight records at Kutum, in Darfur, and Radom National Park, in the far west. Elsewhere, south to Tanzania and Angola, outside the forest. *K. nogueyi* (originally a subspecies of *K. belliana*) occurs further west; where they meet is not known.

Kinixys belliana, Wachile.

Kinixys belliana, Uganda.

Kinixys belliana, Welkite.

CONSERVATION STATUS: Not Assessed.

NATURAL HISTORY: Diurnal, usually moving during early to mid-morning and late afternoon, most active during rainy season. Usually aestivates during dry season in holes, under boulders or in rock cracks, where they also shelter at night. If approached, respond by withdrawing the limbs and closing the shell; if picked up they may urinate, defecate and hiss. Large home ranges but occurs at low densities, 2–3 per ha. Males fight in the breeding season. Lays 2–7 eggs in a pit in shade of a rock or bush. Several clutches may be laid in a season. Omnivorous, eating fungi, plants, small invertebrates (especially millipedes) and carrion. Seasonal shift in diet; mostly plants in wet season and animals in the dry. Captive specimens fond of red fruits and known to live up to 22 years.

WESTERN HINGED TORTOISE

Kinixys nogueyi
(Lataste, 1886)

IDENTIFICATION: A small, brown, black and yellow hinged tortoise, in south-west South Sudan. A domed shell which can look slightly tubular, a pointed head and beak with a single cusp. A hinge is usually present in adults at the rear of the shell, but seems often to be non-functional. The front and back feet have four claws. The anterior of the front foot is covered by several rows of large, overlapping scales. No buttock tubercles, the tail has a terminal spine, males have longer and thicker tails than females. Average size

Kinixys nogueyi, West Africa.

14–18cm in males, 20–24cm in females, maximum about 30cm, hatchlings 4–5cm. Shell usually yellow or tan, with heavy black, radial (spoke-shaped) patterns, sometimes the vertebral scales can be deep brown, maroon or purplish. In large individuals the pattern may fade. Plastron yellow or tan, often with large black or deep brown radial markings. Taxonomic Notes: The status of this taxon is debated, some authors give it full species status, others treat it as a subspecies (or even just a colour morph or regional variant) of *K. belliana*.

HABITAT AND DISTRIBUTION: Dry to moist savanna and woodland, from sea level to >2,000m. Recently reported (sight records) from vicinity of Tambura, at western edge of Western Equatoria, at around 700m. Elsewhere, west to Senegal.

CONSERVATION STATUS: Not Assessed.

NATURAL HISTORY: Presumably similar to *K. belliana*. Clutches of 2–5 eggs, roughly 3×3.5cm, hatching after 148–172 days in captivity. During flooding of Lake Volta in Ghana, many were collected floating on the water as their habitat was flooded.

MEDITERRANEAN TORTOISES *TESTUDO*

A genus of attractive, medium-sized tortoises associated with the Mediterranean Sea and lands east of there. Five species are known; one in our area, Egyptian Tortoise *T. kleinmanni*. Mediterranean Spur-thighed Tortoise *T. graeca* has occasionally been reported in Egypt, but never with credible supporting data.

EGYPTIAN TORTOISE

Testudo kleinmanni
Lortet, 1883

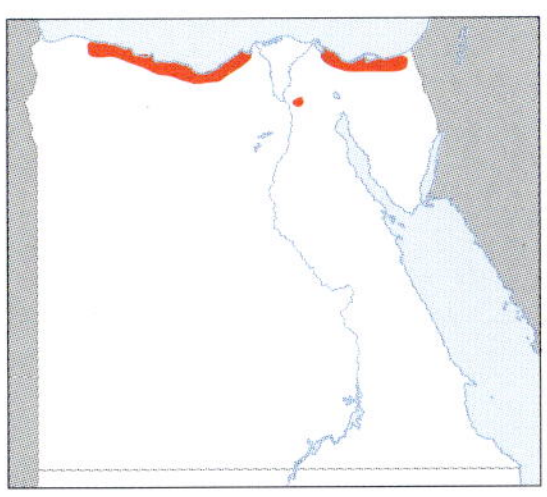

IDENTIFICATION: A very small land tortoise of the northern Egyptian coast. Eyes quite small. Shell domed and hemispherical. Front feet have five claws, rear feet four. Males have longer tails than females. Average size in Egypt 10–12cm, maximum about 14.5cm, females larger; larger animals reported elsewhere. Hatchlings about 3cm. Shell usually olive, vivid yellow or tan, sometimes pinkish or grey, each scale black-edged, to a greater or lesser extent the black may fade with age, so old adults look yellow. Plastron waxy yellow, often with 1–3 pairs of symmetrical black subtriangular patches.

HABITAT AND DISTRIBUTION: Arid semi-desert and near desert, in vegetated dunes, wadis and stony steppe, largely in areas with 50–150mm annual rainfall; mostly at sea level or slightly higher. Sporadically distributed along Egyptian coast, east and west of Nile Delta, extending inland 50km or more in places. Originally widespread but now limited to a few localities (Wadi Natrun, Wadi Degla, Lake Bardawil, Zaranik Protected Area, El Omayed Protected Area). Elsewhere, west along Libyan coast almost to Tunisian border, east into western Negev Desert.

Testudo kleinmanni, captive, juvenile.

Testudo kleinmanni, captive, Cairo.

Testudo kleinmanni, captive.

CONSERVATION STATUS: Critically Endangered, and in Egypt almost extirpated, where range ravaged by overgrazing, reclamation schemes, coastal developments associated with tourism and collection for the pet trade; its disappearance in Egypt means that it is now harvested in large numbers in Libya. Future looks bleak.

NATURAL HISTORY: Diurnal, active in middle of day in winter, mid-morning and afternoon in spring and at dawn (or inactive) during summer. Shelter under vegetation, but during hot summer may aestivate in rodent burrows. If approached, respond by withdrawing the limbs; if picked up they may urinate, defecate and hiss. Quite large home ranges; averaging 35ha in males, 16ha in females. Mating observed in captive Cairo animals in September–October, with males producing rhythmic bleats (sometimes likened to a dove's call) while copulating. Female digs a pit for eggs; in Cairo they were laid in June, and hatched September and October. Wild specimens noted as laying 1–3 oval eggs, roughly 3×2.5cm, although captive animals laid several clutches per year of up to five eggs. Incubation times 85–126 days. Captives are known to live up to 40 years. This tortoise is predated by crows and foxes; in the Negev Desert a fox control programme resulted in increasing numbers of this species.

SEA TURTLES

SUPERFAMILY CHELONIOIDEA

Marine turtles are remnants of a huge, ancient group, that entered the sea about 100 million years ago. Some were huge, *Archelon* was >3m long. Seven species of sea turtle in six genera remain. They are adapted to a life in the sea; they cannot withdraw their limbs, which are modified into flippers, and can excrete excess salt via the tear ducts. They come ashore only to lay eggs; up to 1,000 in a season, and the sex of the clutch is determined by incubation temperature; higher values produce all females. Officially, all are protected, but in reality they, and their eggs, are often harvested (they are fairly helpless on shore) and their nesting beaches and feeding areas threatened by development, artificial lighting, coastal armouring, pollution and beach erosion. On land, they look dull-coloured, clumsy and ridiculous, but a free turtle at sea is a creature of colour, grace and beauty; they should not be allowed to become extinct.

MODERN SEA TURTLES

FAMILY CHELONIIDAE

Advanced sea turtles with strengthened limbs to increase their swimming efficiency, and a hard shell. Fossils are known from the Cretaceous period. There are 33 genera, five of which are still living; four are represented in East Africa.

RIDLEY SEA TURTLES *LEPIDOCHELYS*

Ridleys are the smallest sea turtles, with broad shells. They are known for their 'arribadas', huge congregations (often many thousands) coming ashore on a single beach on one night to lay eggs simultaneously. Theoretically, there are so many eggs that the predators cannot cope, and no matter how many they eat, many remain to hatch safely. Sadly, humanity has found ways of dealing with them; Ridleys are now endangered, and few huge arribadas remain. Two species of Ridley turtle are known. Kemp's Ridley *L. kempi* is a Critically Endangered species found largely in the western Atlantic; it nests only on Mexican beaches. The Olive Ridley occurs in our area.

OLIVE RIDLEY TURTLE

Lepidochelys olivacea
(Eschscholtz, 1829)

IDENTIFICATION: The smallest sea turtle, with a smooth, round carapace, slightly wider than long. General impression is of a deep-bodied, dark, plate or saucer-shaped turtle. The head is sub-triangular, snout short and not compressed, beak slightly hooked and not bicuspid; the edges of the jaws are smooth. Two claws distinguishable by the time the animal reaches a carapace length of 28.5cm. Tail short in hatchlings and juveniles, longer in adult males. Maximum shell length about 70cm and mass up to 45kg, average 50–60cm, juvenile carapace length 4.3–4.5cm. Reach sexual maturity in 7–9 years. Carapace olive-grey, top of head brown with scales outlined yellow-white, dorsal skink surfaces and flippers dark grey, lower jaw yellow-brown, leading edges of flippers light yellow to brown. Hatchlings olivaceous-black, with plastral keels, lower jaw and cutting edges of upper jaw with white patches, throat grey. Trailing edges of flippers have white flecks. Skin of the ventrum greyish-white.

HABITAT AND DISTRIBUTION: Circumtropical, in warm waters of Atlantic, Pacific and Indian Oceans. Usually coastal, protected shallow waters, sometimes in open sea. In our area, not recorded on Egyptian Mediterranean coast (essentially absent from the Mediterranean, save odd vagrants at the western end), but a handful of Red Sea records. Sporadic records from the Eritrean coast, in small numbers, including from Massawa. Known from Djibouti in the Gulf of Aden. Reported from Somali coasts, although specific data lacking.

CONSERVATION STATUS: Vulnerable. Probably the commonest sea turtle worldwide, but endangered due to trade in its limb and neck leather, also often accidentally taken in shrimp trawls.

NATURAL HISTORY: Olive Ridleys favour shallow water but will forage in deep water, they have been taken in prawn trawls 110m deep. Sometimes in large 'herds', moving between nesting beaches and foraging areas, and sleeping communally on the surface. In some places, (not in our area, where there are no nesting records), hundreds of females simultaneously may come ashore to nest (an 'arribada'), usually on rising afternoon tide; in the past, arribadas of >46,000 turtles recorded. Mating occurs offshore from nesting beaches at time of nesting. Female lays two clutches (occasionally three), 17–60 days apart, the interval controlled by tides and weather. About 110 (range 30–168) white, soft-shelled round eggs laid, averaging 4×3.5cm. at a depth of about 50cm. Returns to nest

Lepidochelys olivacea, Kenya.

at one- to two-year intervals. Nest construction and egg-laying process rapid, usually taking less than an hour, the female finishing off by banging the sand down over the nest with the thickened sides of her plastron. Incubation short, 40–60 days. Olive Ridleys are carnivorous, feeding on marine invertebrates, squid and fish, and will take crabs in mangrove areas.

LOGGERHEAD TURTLES *CARETTA*

A genus with a single living species, although there are a number of fossil forms.

LOGGERHEAD TURTLE

Caretta caretta
(Linnaeus, 1758)

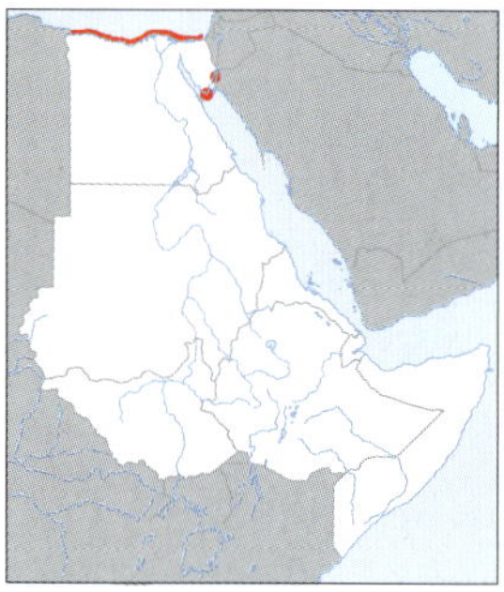

IDENTIFICATION: A big reddish-looking turtle with a huge, broad head. The snout is relatively short, the beak slightly hooked and unicuspid; edges of the jaws smooth but have broad surfaces which can crush hard-shelled prey. Two claws on all four flippers; over time the outer claw usually becomes recessed, at maturity barely visible. Tail longer in adult males than females. Maximum size in Egypt 1.24 m, average c.70–90cm, maximum mass 140kg. Hatchlings 4–5cm long. Carapace of hatchlings grey-brown when dry, pale red-brown when wet, becoming vividly streaked red-brown in adults, each scale dark-edged. Occasionally, very dark brown individuals occur, and rarely albino hatchlings. Underparts and skin dark brown to black in juveniles but centre of plastral shields light, beak and eyelids black or dark brown, becoming paler in adults.

HABITAT AND DISTRIBUTION: Worldwide, in tropical and temperate waters, usually fairly near coasts and in vicinity of reefs, but has been seen 240km from land. Sometimes enters estuaries. Common on Egypt's Mediterranean coast, where it sometimes nests, but few records from Red Sea. Reported on northern Somali coast. Generally rare; further south in Kenya, of 1,500 turtles recorded per year, about five are Loggerheads. The only breeding records from our areas are on the Egyptian Mediterranean Coast.

CONSERVATION STATUS: Vulnerable. Occasionally harvested for its meat.

NATURAL HISTORY: In southern Africa, juveniles spend first three years of life drifting in offshore waters before returning to the coast. Omnivorous, eating invertebrates (crabs, clams, jellyfish) and seagrass. Sexually mature when shell c.60–70cm long; at an age of 10–30 years. Females may breed just once in 2–3 years. Mate in summer in southern Africa, by day and night. Up to five clutches and a total of 500 eggs may be laid each season, the female emerging at night to dig a body pit and then a flask-shaped egg pit. Incubation 45–70 days. Individuals tagged in South Africa have been recaptured in Tanzania, having travelled >2,500km.

Caretta caretta, Kenya.

HAWKSBILL TURTLES *ERETMOCHELYS*

A genus with a single living species.

HAWKSBILL TURTLE

Eretmochelys imbricata
(Linnaeus, 1766)

IDENTIFICATION: A small sea turtle, with bird-like beaked jaws (hence the common name) and an oval carapace. The beak is not hooked and no cusps are present; its edge is smooth. The forelimbs have very long digits, each flipper two claws. Males distinguished from females by having a longer, narrower shell, longer tail and small plastron concavity. Maximum shell length about 90cm (occasionally larger), average 50–80cm; hatchlings have a carapace 4–5cm long. Juvenile carapace uniform brown, the plastron is dark, each scale has a large dark spot. Adults beautifully marked, the carapace scales are warm translucent amber, with radiating red, honey, yellow, black and brown streaks. Plastron yellow or tan, with brown markings, the flippers black, scales yellow-edged, and head yellow with black-centred scales.

HABITAT AND DISTRIBUTION: Circumtropical, in warm waters of the Atlantic, Pacific and Indian Oceans, sometimes in warmer temperate waters. Usually in shallow coastal waters, bays, areas with mangroves and estuaries, coral reefs and rocky areas with coral. One of the two most commonly seen turtles in north-east Africa (the other is Green Turtle), although it is a rare vagrant in the Mediterranean; not recorded on Egypt's north coast. Common throughout the Red Sea, Gulf of Aden and Somalia's Indian Ocean coast, where it nests frequently.

CONSERVATION STATUS: Critically Endangered. Heavily persecuted both in the past and present for its shell ('tortoiseshell').

NATURAL HISTORY: Forages in shallow water, especially around reefs, often nervous but can become habituated. Juveniles feed mostly on floating vegetation and scavenge around weed slicks, but adults carnivorous, favouring invertebrate prey such as sponges, jellyfish, molluscs and fish. Sexually mature at 8–10 years. Mates in shallow water off nesting beaches, by day and night. Two to four clutches laid at intervals of 15–19 days, between December and February in East Africa. Fifty to 200 eggs are laid each time, roughly 4cm in diameter, and take about two months to hatch. The flesh of some turtles may be toxic, due to feeding on poisonous corals and algae.

Eretmochelys imbricata, Kenya.

GREEN TURTLES *CHELONIA*

A genus with a single living species.

GREEN TURTLE

Chelonia mydas
(Linnaeus, 1758)

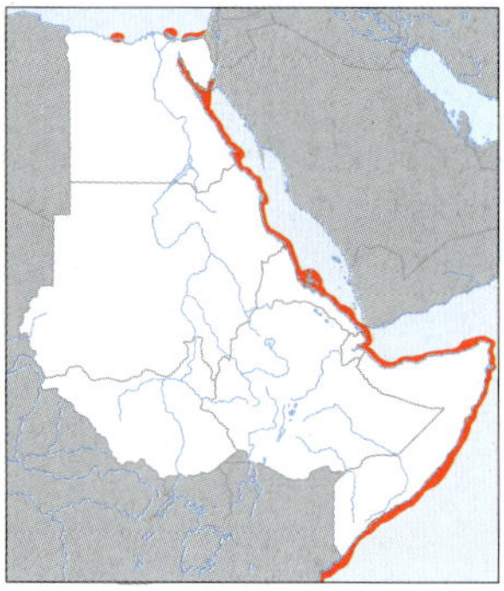

IDENTIFICATION: A fairly big, hard-shelled, deep-bodied sea turtle with a small blunt head, the beak is not hooked and lacks cusps. Younger individuals are flattish, big adults can develop huge domed shells, which look sub-circular in profile. The cutting edge of the horny jaw covering is serrated, which may be an adaptation to grazing. Shell ovate and smooth, the thin scales do not overlap. A single claw on each of the flippers; young have two claws on the front flippers. Males have a longer tail than females. Maximum carapace size about 1.4m, average 90cm to 1.2m, hatchlings 4–5cm, maximum mass in Indian Ocean 200kg (Atlantic specimens larger). Colour variable, olive-green, yellow-green, greenish-brown or black; vertebrals streaked or uniform. Hatchlings dark, with bronzing on vertebrals and a white border and plastron; juvenile carapace is dark grey at 20cm length.

HABITAT AND DISTRIBUTION: Tropical and warm temperate seas. Migrates across open sea, but forages in warm, shallow areas with abundant vegetation. One of the two commonest turtles on coasts of our area. Known from Egypt's Mediterranean coast, where it occasionally nests, but widespread along the Red Sea and entire Somali coastline, where it often nests and feeds. In the 1960s the abundance of Green Turtles on Somalia's Benadir coast prompted the building of a commercial turtle cannery, which was closed for political reasons. Most often observed around seagrass beds.

Chelonia mydas, Kenya.

CONSERVATION STATUS: Endangered, despite nominal protection. Sometimes harvested on north-east African coast, mostly for meat and oil, but decorated shells are also sold, and species under threat from coastal development, with altered beaches, night lights and associated scavengers.

NATURAL HISTORY: Often forages in relatively shallow water. Sometimes rests, basking on the surface and may bask on the shore, possibly to gain vitamin D. Can become habituated where not persecuted. Green turtles become sexually mature at 10–15 years. Mating occurs off the laying beaches; the male hooks the enlarged claws on their front flippers over the leading edge of the female's shell. Nesting occurs largely during the first six months of the year; adults may migrate huge distances to nest. A radio-tagged Green Turtle moved from the Chagos Islands to eastern Somalia, a distance of >6,000km. Females come ashore at night, moving up the beach using a curious 'humping' movement, moving both forelegs together (all other turtles crawl using the flippers alternately). Between 12 and 240 circular eggs (usually 100–150) are laid; the female returns to the beach several times, at intervals of 10–20 days, to lay further clutches. Incubation takes usually 45–60 days, and the hatchlings emerge simultaneously. Juveniles feed on small invertebrates, but adults largely herbivorous, grazing on algae, seaweed and seagrass, as well as taking marine invertebrates.

LEATHERBACK SEA TURTLES

FAMILY DERMOCHELYIDAE

A family of giant sea turtles, Leatherbacks are characterised by the lack of nasal bones, extremely reduced carapace bones and a unique internal shell of small polygonal bones. There are four fossil genera but only a single living species.

LEATHERBACK TURTLE

Dermochelys coriacea
(Vandelli, 1761)

IDENTIFICATION: Unmistakable; a huge, black or dark brown turtle without horny plates, covered by smooth leathery skin that resembles vulcanised rubber. On its back there are seven long ridges, one on each side and five on the plastron. Snout blunt, head huge and rounded. The beak is bicuspid, short and hooked, with sharp edges. Neck short and thick. The flippers are huge and lack claws. Males have concave plastra, rather flattened in profile, their tails are longer than the hind flippers; female tail is half as long. Maximum carapace length about 1.8 m, average size 1.3–1.7m, hatchlings about 6cm long; maximum mass about 850kg. Adults black or deep brown, spotted white, skin ridges paler, the plastron and underparts whitish, with usually some pink and grey-black coloration present. There is a curious distinct pink spot on the dorsal surface of the head, just above the pineal gland, in adults. In juveniles, the vertebral ridges resemble lines of pale beads and white edges to the flippers are conspicuous.

HABITAT AND DISTRIBUTION: Throughout the Atlantic, Pacific and Indian Oceans, ranging to high latitudes; tolerant of cold conditions. Essentially a deep-sea animal, but occasionally seen off coasts of north-east Africa. A handful of sight records from Mediterranean coast of Egypt, the Red Sea (Talla-talla Kabir Island, south-east of Suakin) and Somali coast, recently reported for the first time in Djibouti waters. Not known to nest on coasts of north-east Africa, but reported in Kenya.

CONSERVATION STATUS: Vulnerable. At risk from coastal development, loss of nesting beaches, etc., and they sometimes die after eating floating bits of plastic, mistaking these for jellyfish.

Dermochelys coriacea, Kenya.

NATURAL HISTORY: Highly adapted to life at sea, they can dive deep (up to 1,200m, but 50–60m more usual) and stay down for an hour or more. The circulatory systems in their flippers and oil content of their flesh are indicative of a warm-blooded animal, they can metabolise fat to generate internal heat and their huge size aids heat retention; one had a body temperature 18°C higher than the water, and such regulation is not possible by behavioural methods. Extremely rapid skeletal and body growth; it is suggested that they are sexually mature at 3–5 years. Undertake long journeys, and enter cold currents to feed. Courtship and mating have never been described. In South Africa, gravid females come ashore on moonless nights between December and January, preferably with high tide around midnight. They lay 50–170 (average c.100), billiard-ball sized spherical or ellipsoidal eggs. The female returns to the beach every 9–11 days, until she has deposited about 1,000 eggs. These take 53–75 days to hatch. Young emerge at night to avoid predation by ghost crabs, gulls and fish; the nests themselves are also at risk from jackals. Leatherbacks have weak jaws and adults feed largely on jellyfish; moving to high latitudes in summer to follow the jellyfish flotilla.

SOFT-SHELLED TURTLES

FAMILY TRIONYCHIDAE

The soft-shells are a family of freshwater turtles without hard scales, their big disc-shaped shells are covered by skin; the common name refers to the three claws on their forelimbs. More than 30 species are known, most in North America and South-East Asia. Three genera and five species occur in Africa, three in our area, the Nile Soft-shelled Turtle *Trionyx triunguis*, the only species in that genus, and two *Cyclanorbis*, spotted flap-shelled turtles. Soft-shells have long necks, long tubular snouts and prominent eyes on top of the head, enabling them to see above the surface and breathe easily without exposing much of the head. They are seriously exploited for food in parts of their range, as they have an oily but palatable flesh and edible eggs.

NILE SOFT-SHELLED TURTLE *TRIONYX*

A genus with a single living species, Nile Soft-shelled Turtle *T. triunguis*. Originally, >20 species in the New and Old World were assigned to this genus, but a 1987 review removed all living species bar one from the genus, although some 15 extinct forms remain.

NILE SOFT-SHELLED TURTLE

Trionyx triunguis
(Forskål, 1775)

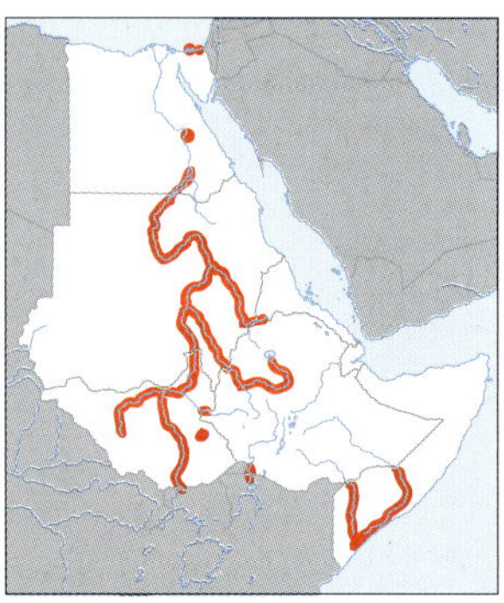

IDENTIFICATION: A large, flat, soft-shelled water turtle, without scales, body and shell covered with skin. It has a long, pointed tubular nose, prominent yellow eyes on top of the head and a long neck, which can be completely withdrawn under the shell. Carapace oval. The limbs are broad and webbed, with conspicuous nail-like claws on three of the four toes. Males have longer thicker tails than females. Shells of large females up to 1.2m long, average 50–80cm, hatchling size 3–4cm. Males usually smaller than females. Mass up to 25kg. Shell brown, olive or dull green, spotted to a greater or lesser extent, and edged yellow-white, limbs brown with yellow spots, whitish-yellow below.

HABITAT AND DISTRIBUTION: Permanent lakes, dams, large and small rivers, in deep and shallow water, known to enter the sea. Sea level to 1,800m. In our area, occurs along the White Nile below Murchison Falls in Uganda, north through South Sudan to Sudan and Egypt, although there are few recent records north of Lake Nasser in Egypt. Also known from Egypt's north-east Mediterranean coast. Other records include the Setit/Tekeze River on the border between Eritrea, Ethiopia and Sudan (thence down the Atbara River); the Blue Nile in Ethiopia, from Lake Tana downstream, the Ethiopian end of Lake Turkana (and possibly the lower Omo River), and the Juba and Webi Shebelle Rivers in Somali, from the Ethiopian border to the sea. Present in larger

Trionyx triunguis, Khartoum.

Trionyx triunguis, Lake Turkana.

tributaries of the White Nile, such as the Jur and Bahr El Ghazal, south almost to Yambio. Elsewhere; west to Senegal, south to Angola, outside Africa north to southern coastal Turkey.

CONSERVATION STATUS: Not Assessed, but rated Vulnerable in the 9th edition of the *Turtles of the World* catalogue. Widespread in West and Central African rivers and lakes, and thus probably under no threat, but eaten in parts of its range; almost extinct in Egypt.

NATURAL HISTORY: Long-lived, adults can survive 50 years in captivity. Spends most of its life in water, active by day, possibly by night. Shy, but may bask. Shuffles down in shallow water, partially concealing itself with sand or mud. Courtship details poorly known, but a captive male was observed on top of a female, biting her neck vigorously. Nests in the dry season to avoid flooding; several clutches of 25–60 eggs are laid in a 25cm deep pit in sand. In West Africa incubation is 70–80 days. Mostly carnivorous, eating fish which it ambushes by 'gulp feeding'; the neck is shot forward, the mouth opens and the throat expands, sucking water and prey into the vacuum. Other prey includes amphibians, reptiles, crabs, insects and floating plant material. Capable of aquatic respiration, absorbing up to 70 per cent of oxygen via the skin. Bites fiercely if handled incautiously. Occasionally eaten in parts of its range.

SPOTTED FLAP-SHELLED TURTLES *CYCLANORBIS*

A genus of two African species, both in our area. Like other soft-shells, their big disc-shaped shells are covered by skin. They have paddle-like limbs, and there is a curious large flexible skin flap either side of the rear plastron, which conceals the back legs when they are withdrawn. The two species are covered within the general key to the order (p. 132).

SENEGAL FLAP-SHELLED TURTLE

Cyclanorbis senegalensis
(Duméril & Bibron, 1835)

IDENTIFICATION: A fairly big, flat, soft-shelled water turtle, without scales, body and shell covered with skin. Has a pointed tubular nose, bright yellow eyes and spotted neck, which is partially withdrawn if threatened. Carapace rounded in juveniles, becoming more oval in adults. Limbs broad and webbed, with conspicuous long claw, and five or six crescent-shaped skin folds on front of each forefoot. The ventral surface of adults has five to nine skin calluses, several at

Cyclanorbis senegalensis, juvenile, Gambela.

front. Males have longer thicker tails than females. Shells of large adults up to 50cm long, average 30–40cm, hatchling size 4.5–5cm. Adults can weigh up to 12kg. Shell smooth in adults, but juvenile shells have a low keel and numerous wavy rows of tubercles. Carapace colour brownish with irregular, black, sub-rectangular mottling, often with yellow or tan edging, and mottling fades with age. Juvenile carapace dark, pinkish-grey or tan, finely spotted to a greater or lesser extent. Adult plastron brown with yellow edging in juveniles, in adults pale (pink, ivory, cream or white), the callosities bluish-grey or pinkish. Head and neck spotted in juveniles; tend to fade in adults.

Cyclanorbis senegalensis, Ivory Coast.

HABITAT AND DISTRIBUTION: In rivers (even quite small ones), lakes, large ponds and permanent marshes, in forest, savanna and semi-desert, between sea level and 600m. In our area, on the White Nile from Juba in South Sudan north to near Khartoum, and on the Sobat, extending upstream to the Baro River catchment, around Gambela in Ethiopia. Also recently reported (without exact localities or specimens) from parts of western and central Upper Nile State. Seems to be absent from western South Sudan. Elsewhere, there is a population centred on southern Chad, and a coherent population in West Africa, from Nigeria west to Senegambia.

CONSERVATION STATUS: Vulnerable, although reportedly still common in some areas; it is eaten by fishermen.

NATURAL HISTORY: Active by day and may bask at the surface or in floating vegetation, rarely on shore. Recent research in South Sudan indicates it is more common in smaller tributaries and backwaters, whilst the larger *C. elegans* is dominant in the main stream. Has a very long neck, nearly half the carapace length, and uses it to snatch prey. Courtship details unknown. Comes ashore at night to lay up to 12 eggs, roughly 25×45mm, in shallow nests that can be some distance from water. Several clutches may be laid in the rainy season. Juveniles hatch during the wet season (April–November) in Sudan; hatchlings found in August in Gambela, western Ethiopia. Omnivorous, eating aquatic animals and fruit; may scavenge carcasses.

NUBIAN FLAP-SHELLED TERRAPIN

Cyclanorbis elegans
(Gray, 1869)

IDENTIFICATION: A big, flat, soft-shelled water turtle, without scales, body and shell covered with skin, in our area only in the Nile system. Pointed, relatively broad and short tubular nose, yellow eyes and a long neck, which can be partially withdrawn when threatened. Carapace rounded in juveniles, more oval in adults. Limbs broad and webbed, with long conspicuous claws, and four crescent-shaped skin folds on front of each forefoot. The ventral surface of adults has up to four calluses, but towards the back. Males have longer, thicker tails than females. Shells of large adults up to 70cm, average 40–60cm, hatchling size unknown, but a juvenile had a shell 17cm long. Adults can weigh up to 20kg. Shell smooth in adults, but juvenile shells have a low keel and numerous rows of small tubercles. Adult carapace olive-green to light brown, with green spotting at edges, but fades with age; large adults can be almost black. Neck and head brownish, spotted yellow to a greater or lesser extent. Underside white or pinkish. Juvenile carapace brown with large yellowish blotches, and smaller yellow spots at margins; head and neck grey-brown, with yellow spotting.

HABITAT AND DISTRIBUTION: Lives in rivers at low altitude. In our area, on the White Nile in Uganda, north through Juba in South Sudan to Khartoum, and on the Sobat, extending upstream to Nasir, almost on Ethiopian border. Elsewhere, west to Ghana but range fragmented into three populations.

CONSERVATION STATUS: Critically Endangered. For a long time no specimens were reported in South Sudan, but recently rediscovered there; worryingly it is also being traded there at high prices for the Far Eastern market. Possibly extinct further west. Its eggs and flesh are eaten in South Sudan and elsewhere.

NATURAL HISTORY: Poorly known. Observed swimming and basking by day. The long neck, nearly half the carapace length, is probably used to snatch prey. Courtship and breeding details poorly known, but recent research in South Sudan found females using same site for several years, usually sandbanks in smaller rivers; several females at same site. On average 20–30 eggs (but up to 38) are laid in the early dry season (November–December), enabling them to hatch before rising water covers them. Diet unknown, but probably eats aquatic animals and possibly fruit.

Cyclanorbis elegans, Juba.

SUBORDER PLEURODIRA

A suborder of three families and just under 100 tropical species; one family in our area.

FAMILY PELOMEDUSIDAE AFRICAN SIDE-NECKED TERRAPINS

The suborder Pleurodira is an ancient group that occurs in South America, Africa, Madagascar and Australia; their ancestors probably lived on the supercontinent Gondwana. There are two genera in the primitive aquatic family Pelomedusidae; the hinged terrapins, *Pelusios*, and helmeted terrapins, *Pelomedusa*.

HELMETED OR MARSH TERRAPINS *PELOMEDUSA*

Originally, this African genus contained a single species, Helmeted Terrapin *P. subrufa*, widely distributed throughout sub-Saharan Africa and south-west Arabia, mostly in temporary waterbodies. A recent review based on DNA indicates that *Pelomedusa* is a species complex of at least 13 evolutionary lineages or clades; names have now been given to ten of these. At least four such species occur in north-east Africa. An additional genetically confirmed but undescribed species occurs in Kordofan, Sudan, and perhaps in adjacent Upper Nile state of South Sudan, and another species (of uncertain status) is known from Jebel Marra in Darfur, western Sudan.

Biology is similar in all species that occur in north-east Africa. They are aquatic, and bask on the water surface or on the shore. They bury themselves in mud when waterbodies dry up, and may stay there for several years. They emerge after rain (this sudden appearance of many individuals gives rise to the legend that they have dropped from the sky). They walk long distances across country looking for pools; they can smell water and may climb into tanks. Capable of running relatively fast. If picked up they scratch, bite and produce a foul-smelling fluid from glands on the flanks. Mating takes place in the water, the male rubs his snout against the female's hindquarters; if she is receptive he grasps her shell with his feet and rubs his barbels onto her head, and he may expel a steam of water from his nostrils over her head. Eggs are deposited in a pit dug by the female near the water. Clutch size usually 10–30 eggs, maximum 42, roughly 3.5×2.5cm, mass 10–15g. Helmeted terrapins are opportunistic carnivores, eating any small animal they can seize, depending upon how big they are; they also take carrion.

KEY TO NORTH-EAST AFRICAN MEMBERS OF THE GENUS *PELOMEDUSA*

The recent division of what was originally one species into 13 races was mainly based on DNA sequences, and no attempt was made to create a physical key. However, we have devised a key, based on carapace coloration, plastral coloration and pattern, coloration of the head and limbs, shape of pectoral scales and whether or not they are in contact.

1a: Pectoral scales triangular and widely separated from plastral midseam; plastron uniformly pale yellow, sometimes orangey; dorsal coloration usually vividly hazelnut-brown; Red Sea coast and hills of Sudan, eastern Sudan (east of Khartoum), Eritrea — *Pelomedusa gehafie*, Eritrean Helmeted Terrapin

1b: Pectoral scales usually rectangular with wide or narrow midseam contact; plastron pale yellow with or without pattern, or dark; dorsal coloration usually not hazelnut-brown; distribution not as above — 2

2a: Light-coloured dorsally; plastron pale yellow with more or less wide dark brown edges or several fine stripes along seams of all scales, giving a somewhat reticulate appearance; northern Somalia and adjacent low and dry central and eastern Ethiopia (northern Ogaden) — *Pelomedusa somalica*, Somali Helmeted Terrapin

2b: Usually dark-coloured dorsally; plastron uniformly pale yellow or dark; distribution not as above — 3

3a: Medium-sized, shell up to 22cm long; pectoral scales on plastron rectangular with wide midseam contact; plastral colour uniform pale yellow or blackish; dry to moist savanna in southern Ethiopia to south Somalia — *Pelomedusa neumanni*, Tanzanian Helmeted Terrapin

3b: Small to medium-sized, shell up to 16cm long; pectoral scales on plastron rectangular with wide midseam contact or triangular with narrow contact; plastral colour almost uniform dark brown or slightly suffused yellow; wetlands in savanna-forest mosaic in southern and western South Sudan — *Pelomedusa schweinfurthi*, Schweinfurth's Helmeted Terrapin

ERITREAN HELMETED TERRAPIN

Pelomedusa gehafie
(Rüppell, 1835)

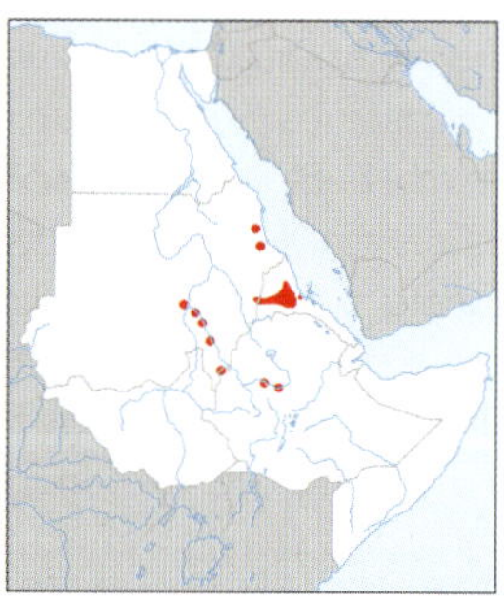

IDENTIFICATION: A medium-sized, light brownish-coloured helmeted terrapin, widespread in northern Eritrea and adjacent eastern and northern Sudan. The head is broad with a protruding snout and two small tentacles (barbels) under the chin, and a large undivided temporal scale on each side. The neck is long; limbs broad and webbed, with sharp claws. The shell is hard, fairly flat and wide with no plastron hinge. Pectoral scales triangular in adults and do not touch, an excellent aid to identification. Males have longer thicker tails than females. Shells of large adults up to 22cm long, average 15–18cm, hatchlings 2.5–3.5cm. Adult mass >1.3kg. Coloration distinctive; carapace vivid hazelnut-brown (but often coated with green algae). Plastron uniformly pale yellow, sometimes orangey, especially in younger animals, but can be brown in older, large animals. Head and limbs brown above, yellowish-white below. Iris pale brownish-grey. Juveniles resemble adults, without plastral patterns, although seams may be slightly suffused brown. Easily distinguished from all other *Pelomedusa* by its brownish coloration and lack of contact between pectoral scales.

HABITAT AND DISTRIBUTION: Semi-desert, dry savanna, dry submontane grassland and agriculture, from near sea level up to about 2,100m. Usually in temporary or small waterbodies, shallow streams or dams. Widespread in central and northern Eritrean highlands and eastern and western lowlands, notable localities include: near Massawa, Ghinda, Filfil, Mendefera, between Asmara and Keren, Afabet, Barentu and Agordat. Known also from Red Sea Hills in Sudan, where reported at Khor Arbaat north-west of Port Sudan and Erkowit; in eastern Sudan at Kassala and perhaps as far west as Jabal Karari, also Omdurman near Khartoum. Specimens from several localities (Jebel Aulia dam, Wad Madani, Gebel Moya in Sennar) in Blue Nile Valley, Sudan, probably of this species, as are populations from Ethiopian Blue Nile Valley (mouth of Guder River and 10km west of Mabil).

CONSERVATION STATUS: Least Concern; widespread and (in Eritrea) common.

NATURAL HISTORY: Biology probably similar to other helmeted terrapins, see generic introduction. A Ghinda adult was found in an almost dry concrete reservoir. A group seen floating on surface in a small dam in hilly landscape with large boulders and *Euphorbia* woodland were very shy, and dived deep when approached closer than 20m. Several dozen specimens from Khor Arbaat were in pools in rocky canyons, which are fed by fresh water from a spring.

Pelomedusa gehafie, Ghinda.

Pelomedusa gehafie, Ghinda.

NEUMANN'S HELMETED TERRAPIN

Pelomedusa neumanni

Petzold, Vargas-Ramírez, Kehlmaier, Vamberger, Branch, Du Preez, Hofmeyr, Meyer, Schleicher, Široký & Fritz, 2014

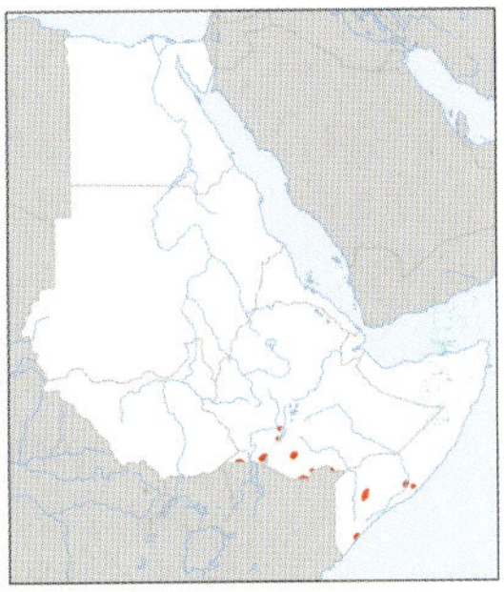

IDENTIFICATION: A medium-sized robust, helmeted terrapin, of variable coloration, known from south-west Ethiopia and possibly southern Somalia. Head comparatively broad and deep with a protruding snout and two little tentacles (barbels) under the chin. A large undivided temporal scale is present either side. Shell hard and moderately flat. Pectoral scales rectangular with wide midseam contact. Males have longer thicker tails than females, and have narrower, flatter shells and grow larger than females. Shells of large adults up to 22cm long, average 16–20cm, hatchlings 2.5–3.5cm. Coloration variable, the carapace mostly horn-coloured (dark greyish-brown), sometimes coated with green algae; plastron often completely yellow. Almost black specimens are known and appear to be associated with forest. Head and limbs dark brownish-grey above, lighter (greyish-yellow) below. Iris yellowish-grey with darker edging. Hatchlings generally dark grey or black above and below, the submarginal scales are pale horn-coloured with darker seams. The plastral fore-lobe may be suffused by yellow. Taxonomic Notes: One of several forms elevated to species status in a 2014 review.

HABITAT AND DISTRIBUTION: Dry and moist savanna (and forests outside our area), from near sea

Pelomedusa neumanni, Turmi.

Pelomedusa neumanni, Turmi.

level to about 1,300m (elsewhere up to 2,000m). Usually in temporary or small waterbodies. The only genetically confirmed specimens in our area are from Turmi, south-west Ethiopia. However, specimens probably of this species are known from the southern Ethiopian Rift Valley (near Lakes Abaya and Chamo), thence east and south towards Kenya, roughly south of Genale River. Also southern Somalia, between the Juba and Shebelle Rivers, and the Ilemi Triangle in extreme south-east South Sudan. Elsewhere: south-east Uganda, most of Kenya and northern Tanzania. Widespread and common.

CONSERVATION STATUS: Least Concern.

NATURAL HISTORY: Biology similar to other helmeted terrapins. Aquatic, they are found in any temporary waterbody. In the dry season they bury themselves in mud or soil, emerging after rain. At Turmi, an adult male was active during night in a small pond at the edge of a large wadi, feeding on the eggs of the toad *Sclerophrys regularis*.

SCHWEINFURTH'S HELMETED TERRAPIN

Pelomedusa schweinfurthi
Petzold, Vargas-Ramírez, Kehlmaier, Vamberger, Branch, Du Preez, Hofmeyr, Meyer, Schleicher, Široký & Fritz, 2014

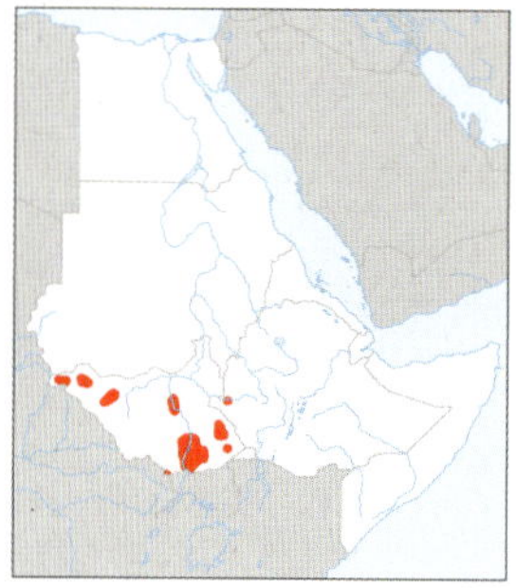

IDENTIFICATION: A small to medium-sized, darkish helmeted terrapin, of wetlands in southern and western South Sudan. Poorly known, its external anatomy is similar to other species of the genus (pectoral scales rectangular with wide midseam contact, or triangular with narrow midseam contact). Temporal scale large, mostly undivided; two small barbels under chin. Shell length up to 16cm. Carapace dark brown, plastron uniform dark brown or slightly suffused yellow, head and limbs light below (pale grey to yellowish), dark grey above, head sometimes orangey-brown. Some variation in colour; adults from Gambela (probably this species) in lowlands of western Ethiopia, have a centrally yellow, black-edged plastron, juveniles a black plastron; in southern Sudan the carapace is black. Taxonomic Notes: Like the previous species, elevated to species status in a 2014 review. Molecular analysis indicates it is closely related to an undescribed species of *Pelomedusa* from Kordofan in Sudan.

HABITAT AND DISTRIBUTION: Swamps, streams and ponds in grassland, savanna, shrubland and woodland of the Northern Congolian forest-savanna mosaic, at about 400–1,000m. Widespread in South Sudan, from Boma National Park and the extreme south-west in Jonglei State, across East and West Equatoria (Lohutok, Torit, Liria, between Juba and Mongalla and Gemmaiza), the southern Sudd swamps, west to Warrab and Bahr Al Ghazal States. Known from Wau, thence towards the border between Central African Republic and Sudan, west of Raga; and from Radom National Park in Southern Darfur. A couple of sight records from Upper Nile State. Specimens from Gambela, western Ethiopia (where habitat identical) probably this species. Elsewhere: north-east DR Congo (Garamba region), south-east Central African Republic and probably northern Uganda.

CONSERVATION STATUS: Data Deficient.

NATURAL HISTORY: Poorly known. This species, unlike other species from our area, is tied to wetlands. Specimens from DR Congo (near South Sudan border) were observed basking on tussocks of grass or lower parts of bushes above water surface in swampy grassland. At Gambela found in a small temporary pool beside the road in high grassy moist savanna.

Pelomedusa schweinfurthi, Gambela.

Pelomedusa schweinfurthi, Gambela.

SOMALI HELMETED TERRAPIN

Pelomedusa somalica

Petzold, Vargas-Ramírez, Kehlmaier, Vamberger, Branch, Du Preez, Hofmeyr, Meyer, Schleicher, Široký & Fritz, 2014

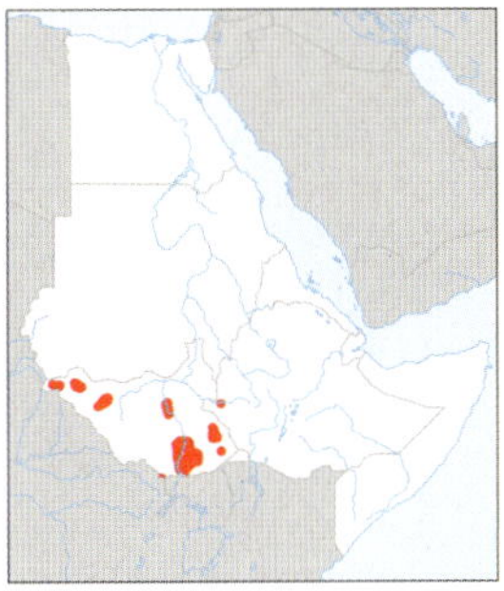

IDENTIFICATION: A small to medium-sized, light-coloured helmeted terrapin, widespread in northern Somalia, lowland and dry eastern and central Ethiopia. Head broad with protruding snout and two little tentacles (barbels) under the chin, and a large undivided temporal scale on each side. Appears to be smiling on account of the curved mouth. Pectoral scales triangular to rectangular with a wide or, rarely, narrow midseam contact. Males have longer thicker tails than females, have narrower, flatter shells, and grow larger than females. Shells of large adults up to 18cm long, average 14–16cm, hatchlings 2.5–3cm. Carapace pale brown (occasionally dark brown), often coated with green algae. Plastron distinctly coloured, usually pale yellow; seams have dark brown edges of varying thicknesses, or there are several fine stripes along seams of all scales, giving a somewhat reticulate appearance. Head and limbs lighter below, usually yellowish; brownish-grey or sometimes dark brown above. Juveniles darker with a dark mark in the centre of the plastron, radiating from the seams. Iris pale greenish-grey. Taxonomic Notes: Molecular analysis indicates this is the ancestral form of all *Pelomedusa* north of the equator, and consists of two distinct lineages, north and south of the eastern Ethiopian highlands, which meet in Borama district in Somaliland.

HABITAT AND DISTRIBUTION: Semi-desert, dry savanna or *Vachellia* woodland, at 500–1,800m. Known only from northern Somaliland and adjacent dry low eastern and central Ethiopia. Its range extends north and east from Koka Dam in Ethiopia, along the upper Awash River, to the border close to Djibouti (but not recorded in the latter) and south into the northern Ogaden. In Somalia, known from the Cal Miskaat Mountains in Puntland and several localities on the border with Somaliland (including El Donfar, Buran and Wadi Run), it is probably more widespread here but under-collected. Known from Erigavo district and Ogo Mountains in central and western Somaliland (where common), the environs of Hargeisa, Daarbuduq, Las Geel and around Borama.

CONSERVATION STATUS: Least Concern.

NATURAL HISTORY: Biology similar to other helmeted terrapins. Recent research in central and western Somaliland found this species in shallow streams, temporary or small waterbodies, even small rain pools on roads. Also found in 'berkads' (concrete reservoirs in arid country that collect rainwater for dry season use) during the wet season from April to September. In dry season seen only in small dams or wadis fed by permanent spring water, for example in Daarbuduq. Usually solitary in small waterbodies, but aggregations may bask in substantial water sources. An adult male at Toon in Somaliland was aestivating in dry mud in a berkad below *Vachellia* trees in February, in the dry season. A hatchling was found in grass close to a waterhole north-east of Hargeisa in July.

Pelomedusa somalica, Daarbuduq.

Pelomedusa somalica, Daarbuduq.

HINGED TERRAPINS *PELUSIOS*

A genus of medium to large hinged water terrapins, with dark shells, living mostly in permanent waterbodies, although a few inhabit temporary pools; often seen basking. At present, 17 species are recognised; mostly in sub-Saharan Africa, a couple on islands of the western Indian Ocean and Madagascar. Four species occur in north-east Africa. Adults can be identified to genus in the hand by the presence of a hinge on the plastron, enabling them to withdraw the head and front limbs and shut the front of the shell, if a hand-held specimen is touched on the nose or forelimbs it will shut as described (although this seems to be non-functional in *P. broadleyi*). The shell is thick and domed, often with a vertebral ridge or keel; it may change colour during life, and in big adults is often extensively cracked and chipped. The incubation temperature seems to affect the sex ratio of hatchlings, at least in captives.

KEY TO THE NORTH-EAST AFRICAN MEMBERS OF THE GENUS *PELUSIOS*

This is a difficult genus to key, but locality provides a clue. The key refers frequently to shell features, look at the accompanying diagrams.

1a: Plastral fore-lobe at least twice as long as length of inter-abdominal seam	2
1b: Plastral fore-lobe about 1.5 times as long as length of inter-abdominal seam	3
2a: Plastron centrally yellow, widespread in Sudan and South Sudan	*Pelusios adansonii*, Adanson's Hinged Terrapin
2b: Plastron and bridge dark brown, streaked and spotted, only in and around north end of Lake Turkana	*Pelusios broadleyi*, Lake Turkana Hinged Terrapin

3a: Posterior rim of carapace usually strongly serrated, sometimes sinuate, black border on both plastron lobes, in southern Ethiopia and Somalia *Pelusios sinuatus*, Serrated Hinged Terrapin

3b: Posterior rim of carapace smooth or slightly serrated; extreme south-west South Sudan *Pelusios chapini*, Congo Hinged Terrapin

ADANSON'S HINGED TERRAPIN

Pelusios adansonii
(Schweigger, 1812)

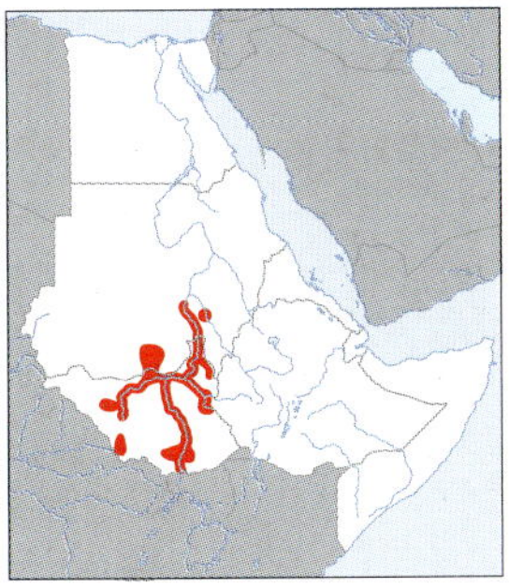

IDENTIFICATION: A medium-sized (up to 24cm) hinged terrapin with a yellow-brown, brown or black shell, a hinged plastron, broad head and slightly projecting snout, with two little tentacles (barbels) under chin. Shell hard, fairly deep and rounded. Limbs broad and webbed, with sharp claws. Males have longer thicker tails than females. Shells of large adults up to 22cm long, average 13–17cm, hatchling size unknown but probably 3–5cm. Adult mass unknown. Shell carapace usually blackish-brown, becoming brown laterally. Plastron yellow (both dull and bright) centrally, black at sides. Head grey-brown dorsally, with yellow vermiculations, yellow at back; sometimes there is a yellow stripe from the orbit to tympanum.

HABITAT AND DISTRIBUTION: Rivers, streams and backwaters in savanna of central and west Africa, north of the forest and south of the Sahara, between sea level and 1,000m. In our area, largely associated with the Nile basin. In South Sudan occurs along the Nile from near the Uganda border, north into Sudan, to Khartoum. Known from the Yubu (Yubo) River, near Tambura, so probably north on the Sue, Jur and Bahr El Ghazal to where it meets the White Nile. Known from Sennar on the Blue Nile, and from several localities in southern Kordofan (south of Kadugli). Found along the Sobat River, upstream to Ethiopia in the Gambela basin. Also reported at several localities west of Wau. Recently reported from eastern Upper Nile State, but without precise localities. Elsewhere, west to Senegal and the Cape Verde Islands.

CONSERVATION STATUS: Not Assessed.

NATURAL HISTORY: Little known. Presumably similar to other hinged terrapins, i.e. diurnal, basks in exposed sites, and lays hard-shelled eggs in a pit dug by the female near the water's edge. Several clutches of up to 15 eggs, roughly 3×2cm, may be laid several times per season. Feeds mostly on aquatic invertebrates, especially molluscs. If their pools dry up they bury themselves in the mud and aestivate.

Pelusios adansonii, Khartoum.

LAKE TURKANA HINGED TERRAPIN

Pelusios broadleyi
Bour, 1986

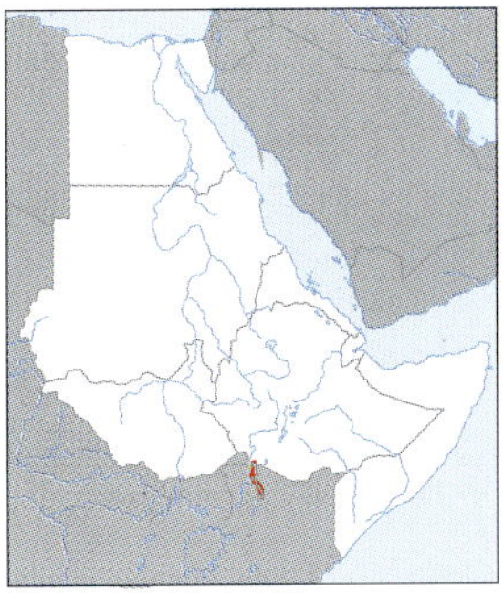

Pelusios broadleyi, Kenya.

Pelusios broadleyi, Kenya.

IDENTIFICATION: A small hinged terrapin, known from Lake Turkana and associated streams on the Kenya side. Shell flat (slightly domed in females) with a knobbly central keel, most pronounced in juveniles. Unusually, plastron small and somewhat rigid, and rarely closed. Limbs broad and webbed, with sharp claws, and big transverse scales on the anterior surface of the front legs. Males have longer thicker tails than females. Shells of large adults up to 18cm, possibly larger, average size unknown; hatchlings 1.7–1.9cm. Shell carapace grey or red-brown, each scale finely speckled, the lower third of the costals sometimes very light, giving the impression of a pale ring on the lower carapace. Plastron brown to black, sometimes with yellow blotches at centre, in juveniles the plastron and bridge may be extensively blotched yellow, orange or pink. Head dark, with light vermiculations, the chin and neck paler, the eyes yellow or green.

HABITAT AND DISTRIBUTION: Originally described from seasonal streams at Loyengalani, on south-east shore of Lake Turkana in Kenya. Reported in lower Omo Delta at 300–450m in Kenya, so probably extends to Ethiopia; informally reported near Chirim, in extreme south-west Ethiopia.

CONSERVATION STATUS: Vulnerable, due to small range, as known so far.

NATURAL HISTORY: Presumably similar to other hinged terrapins, i.e. diurnal, basks in exposed sites (although Loyengalani and Koobi Fora are hot and arid, so might be nocturnal, and the types were collected at night). Reported to bury themselves in mud surrounding seasonal pools in riverbed, where they aestivate if pools dry up; captives rapidly bury

Pelusios broadleyi, Kenya.

themselves if given the opportunity. They lay hard-shelled eggs in a pit dug by the female near the water's edge; captive specimens dug pits 10–13cm deep and laid clutches of 13–21 eggs, which at temperatures of 27–30°C hatched 46–53 days later. Captive animals are omnivorous.

CONGO HINGED TERRAPIN

Pelusios chapini
Laurent, 1965

IDENTIFICATION: A fairly large (up to 38cm) hinged terrapin with a brown shell, a hinged plastron and blunt head with two little tentacles (barbels) under chin. Beak strongly bicuspid. The shell is smooth and domed (keeled in juveniles). Limbs broad and webbed, with sharp black claws. Males have longer thicker tails than females. Shells of adults up to 38cm, average unknown, hatchlings 3cm. Adult mass unknown. Shell carapace usually grey, blackish-brown to brown or greenish-brown above; plastron black, edged or blotched yellow.

HABITAT AND DISTRIBUTION: Shallow, slow-flowing rivers and creeks, swamps, shallow pans and isolated temporary pools. No records from South Sudan, but known from Bagbele, a village in Garamba National Park, on DR Congo/South Sudan border. However, the border in that region follows the Nile/Congo watershed, all other records are from the Congo system, so it might not extend into South Sudan.

CONSERVATION STATUS: Not Assessed.

NATURAL HISTORY: Diurnal, may bask on shores and in shallow water. Known to leave rivers for flooded savanna during rainy season. Congo specimens were observed resting in water vegetation and debris, with head and shell partially out of water. Shy and nervous. Lays hard-shelled eggs in a pit dug by female, a clutch of seven eggs, roughly 3.5cm diameter, were laid in a hole 60cm from water's edge; a hatchling measuring 27×34mm in Gabon in February. Omnivorous, it eats plant material, fruit, fish and aquatic invertebrates, even carrion. Some specimens were caught in nets, fish traps or shallow pools.

Pelusios chapini, Gabon.

SERRATED HINGED TERRAPIN

Pelusios sinuatus
(Smith, 1838)

IDENTIFICATION: A huge dark hinged terrapin with a hinged plastron. Head very broad, snout elongated, with a weakly bicuspid beak and two little tentacles (barbels) under chin. Shell hard, domed and fairly deep, juvenile plastron distinctly keeled with tent-like peaks, which may vanish in adults or persist as a few raised bumps. Rear marginal scales distinctly serrated, thus spiky rear end is an excellent field character, even in large adults. Unusually for a *Pelusios*, plastron concavity almost absent. Limbs broad and webbed, with sharp claws. Males have longer thicker tails than females. Shells of large adults up to 55cm long in upland Kenya, but smaller elsewhere, most adults average 20–30cm; hatchlings 4–5cm or slightly larger. Adult mass >20kg. In juveniles shell grey or greenish, each scale with a paler centre and fine radial markings, but fades in adults to uniform dark grey or black (sometimes called 'black water turtle'), sometimes with pale edges to scales. In juveniles, top of head and limbs grey or greenish-grey with fine black speckling, and sides of head and chin are light, fading in adults to uniform grey or black. Plastron yellow or orange, with a black border. Taxonomic Notes: The subspecies in our area is *P. s. bottegi*, which is larger than the nominate.

HABITAT AND DISTRIBUTION: Rivers, lakes and waterholes. In our area mostly in rivers, between sea level and 1,550m, in Ethiopia and Somalia. Known from the Daua, Genale, Juba and Webi Shebelle Rivers. The uppermost records on those rivers are at Imi (Webi Shebelle, Ethiopia), Mandera (Daua; Kenya/Ethiopia), Dolo (Juba, Somalia) and north of Negele (Genale, Ethiopia). The species presumably extends downstream to the sea, although there are few

Pelusios sinuatus, Kenya.

Pelusios sinuatus, Kenya.

records; other known Somali localities are Lugh, Bardera, Alessandra Island near Jilib and Belet Amin on the Juba; and Bulo Barde, Balad and Havai on the Shebelle. Might be present on the lower Omo River, as it occurs in Lake Turkana. Elsewhere, south to South Africa.

CONSERVATION STATUS: Not Assessed.

NATURAL HISTORY: Diurnal, swimming in big rivers and lakes, often seen basking on shores, logs, exposed rocks and mudflats, with head and legs fully extended to catch the warmth. Juveniles sometimes active at night in shallow water, avoiding predation by fish eagles. Breeding details poorly known; clutches of 7–30 eggs recorded, a captive female laid a clutch of 27 eggs, roughly 3.5cm diameter, which hatched in two months. Kenya hatchlings 5–6cm long were found heading towards the river 300m away; 8cm juveniles in July in Kenya were active in an oxbow lake. Largely carnivorous, eating mostly molluscs and snails, but also fish, amphibians and insects, and will scavenge carrion, at which times they are often attacked by crocodiles. Occasionally take floating fruit. If handled, they bite and scratch, and eject a foul-smelling fluid from glands near the rear limbs. Nests sometimes excavated by Nile Monitors *Varanus niloticus*, and mongooses.

SCALED REPTILES

The taxonomy of lizards and snakes is changing, as molecular analysis throws fresh light on the group. At present, all snakes and lizards are placed in the order Squamata, the scaled reptiles, an extremely diverse group of tetrapods, with nearly 11,000 species (just over 7,000 lizards and almost 4,000 snakes); recent work indicates the oldest fossil squamates date back to the Permian, 260 million years ago. This division may change, although this is not of direct relevance to a field guide like this.

The squamate reptiles are distinctive animals; they have scales, diapsid skulls (which have two large holes on each side), and their upper jaw shows kinesis, i.e. it can move relative to the cranium. Males have a pair of copulatory organs (the hemipenes) and females largely lay eggs. Lizards and snakes occur worldwide, except in Antarctica, and there are virtually none in latitudes above the Arctic Circle. Their greatest diversity is in the tropics.

LIZARDS

Lizards are a diverse, abundant and visible group of reptiles; most visitors to north-east Africa observe some lizards. The world's largest lizard is the Komodo Dragon *Varanus komodoensis* of South-East Asia, a monitor lizard that can weigh >150kg and grows to more than 3m long. Our largest lizard, Nile Monitor *Varanus niloticus*, belongs to the same genus and can reach 2.5m, possibly more. The smallest lizards in our area are pigmy geckos of the genus *Tropiocolotes*, some of which are only 6cm long as adults.

At present there are eight superfamilies and just over 40 families of lizard worldwide. Six superfamilies, 13 families and around 250 species of lizard are recorded in north-east Africa; there are almost certainly many more undiscovered species. Lizards in north-east Africa are found from sea level to montane moorland at 4,000m; in general they are more abundant at low altitudes, especially in open country where there are rocks and opportunities to bask. The most numerous families in north-east Africa are the geckos, with 80 species, but there are also 50 species of skink, 41 species of lacertid lizard and 37 species of agama. No African lizards are venomous, despite legends to the contrary, although most will bite in defence and big species can scratch and lash their tails. All lizards have scales and most have four limbs; there are a few with reduced limbs, some have no legs. Most lizards have external ears, and most have eyelids and can close their eyes. Many can lose their tails in defence (chameleons and monitor lizards cannot) and the regenerated tail may have a different shape to the original. Lizards usually lay eggs; a few species retain the eggs internally until they are about to hatch. Unlike snakes, some lizards are social animals, living in structured colonies, observing hierarchies, defending territories and interacting with other members of the colony. Lizards largely eat invertebrates, although a few consume larger animals and some eat plant food; in our accounts we only mention diet if it is unusual. Their many enemies include snakes, small mammalian carnivores and predatory birds. Many can be identified to family quite quickly, see our key below, for which there are some useful clues. For example, any lizard active at night without obvious scales or eyelids is almost certainly a gecko, monitor lizards are huge, chameleons are slow-moving and often found in trees, skinks are shiny and the only lizards that burrow in soil or sand, striped fast-moving lizards in open county will be lacertids, and a lizard perched in a prominent place with a thin neck and vivid colours is probably an agama.

KEY TO THE FAMILIES OF LIZARD IN NORTH-EAST AFRICA

This key assigns all north-east African lizards to family level. Higher taxonomic orders (Infraorder and Superfamily) are listed in the relevant introductions.

1a: Body covered with rectangular segments of skin that form regular rings, without limbs — Worm lizards, family Amphisbaenidae

1b: Body covered with granules or scales, not forming regular rings, limbs present — 2

2a: Has both a vertical pupil and large eyes that can close, with eyelids — Eublepharid geckos, family Eublepharidae

2b: Does not have the above combination of characters — 3

3a: Top of head covered with granular or small, irregular scales — 4

3b: Top of head covered with large, symmetrical shields — 7

4a: Eyelids present, eye closes — 5

4b: No eyelids, eye can't close — 10

5a: Head much wider than neck, dorsal scales often keeled, overlapping and spiky (if dorsal scales not keeled then tail broad and spiky), tongue short and broad — Agamas, family Agamidae

5b: Head same width or slightly wider than neck, dorsal scales smooth or granular, not overlapping, never spiky, tongue long and slender, tail cannot be shed — 6

6a: Digits fused into opposed bundles for grasping, tongue very long and telescopic, eye very tiny, in revolving turret, tail cylindrical, no adult larger than 70cm — Chameleons, family Chamaeleonidae

6b: Digits separate, in a single plane, tongue long and forked, eye normal, not in revolving turret, tail laterally flattened, adults always larger than 70cm — Monitor lizards, family Varanidae

7a: Shiny, dorsal scales usually highly polished, overlapping, smooth or keeled, ventral scales similar but lack keels and slightly larger, femoral pores absent, eyelids present or absent — Skinks, family Scincidae

7b: Not shiny, dorsal scales usually matt, not overlapping, with a strong median keel or small and granular, ventral scales rectangular, larger than dorsals, femoral pores or a row of differentiated scales on posterior thigh, eyelids present 8

8a: Lateral granular fold absent, slim, often less than 30cm Lacertid or Typical lizards, family Lacertidae

8b: Lateral granular fold present, fairly stocky, often more than 30cm when adult 9

9a: Tongue long, scales on tail rectangular, not keeled or spiny Plated lizards, family Gerrhosauridae

9b: Tongue short, scales on tail keeled and/or spiny Girdled lizards, family Cordylidae

10a: Unregenerated tail of male has curious serrated crest, pupil round, digits thin with claws, toe tips not expanded Clawed geckos, family Sphaerodactylidae

10b: Does not have above combination of four characters 10

11a: Digits either dilated along entire length but with obvious claws only on third and fourth digits, or all toes widely divided at extreme tip into two fan-shaped pads Leaf-toed geckos and allies, family Phyllodactylidae

11b: Digits not as above 'Typical' or 'Common' geckos, family Gekkonidae

EUBLEPHARID GECKOS

FAMILY EUBLEPHARIDAE

Eublepharid geckos are small, slow-moving, nocturnal terrestrial lizards. Unlike most other geckos, they have claws rather than specialised toe pads. Most live in arid regions. Many are attractively marked with spots and bands, and are popular in the pet trade. There are 36 species in the family, in six genera; they are curiously widespread, with species in the south-west USA, Central America, East and West Africa, the Middle East and extreme South-East Asia. Four species are African, three occur in our area.

KEY TO THE NORTH-EAST AFRICAN GENERA AND SPECIES OF THE FAMILY EUBLEPHARIDAE

1a: Maximum size up to 18cm, dorsal surface granular, with many irregular rows of enlarged, weakly keeled tubercles *Hemitheconyx taylori*, Taylor's Fat-tailed Gecko

1b: Maximum size up to 15cm, dorsal surface granular but without enlarged tubercles 2

2a: Small, maximum length 11cm; upper surface of head convex; snout rounded; rostral separated from nasal by 1–3 scales; nasals separated by 3–5 scales; upper labials 8–12; lower labials 9–13; tail wide (diameter usually much greater than thigh diameter) *Holodactylus africanus*, Somali-Maasai Clawed Gecko

2b: Medium-sized, maximum length 15.3cm; upper surface of head flat; snout pointed; rostral separated from nasal by 5–7 scales; nasals separated by 8–14 scales; upper labials 13–18; lower labials 14–18; tail narrower (diameter usually equal or slightly greater than thigh diameter) *Holodactylus cornii*, Corni's Clawed Gecko

FAT-TAILED GECKOS *HEMITHECONYX*

An African genus of medium-sized, stocky, slow-moving terrestrial geckos with bluntly rounded snouts, vertical pupils, eyelids that can close, short weak limbs and thin digits with short claws, not expanded pads. Nocturnal, they occur in arid areas and feed largely on termites. The conspicuously short tail stores fat reserves in time of plenty, becoming swollen. They lay soft-shelled eggs. Fat-tailed geckos have a very peculiar distribution. One species is confined to West Africa, the other found only in Somalia and eastern Ethiopia. Their nearest relatives are the north-east African clawed ground geckos *Holodactylus*.

TAYLOR'S FAT-TAILED GECKO

Hemitheconyx taylori
Parker, 1930

IDENTIFICATION: A big, stocky, fat-tailed, vividly marked ground dwelling gecko of the Horn of Africa. The head is rounded; eye small with distinctive movable eyelids, a thin neck, and feet with short claws. Tail short and stumpy, constricted at base, about 25–35 per cent of total length; regenerated tail shorter, more globular and blunter, lacking constrictions and tubercles. The dorsal surface is covered with a mixture of small polygonal raised granular scales, intermixed with larger, bluntly keeled button-like tubercles. Males have larger heads than females, as well as post-cloacal white hemispherical tubercles and 21–28 precloacal pores. Maximum size about 18cm (maximum snout–vent length 12.2cm), average 14–16cm, hatchlings 7–8cm. The back has two distinct dark brown or black, broad, dark-edged crossbars, as does the tail; on a ground colour of warm brown, pink or pinkish-grey, with small dark spots. Pattern more vivid in juveniles, the crossbars may be higher contrast, the dorsal surface more yellow or orange, and front crossbar extends round onto sides of head. The iris is a startling bluish-pink. Ventral surface uniformly pale white or cream, except the tail, limbs and chin which are suffused with brown.

HABITAT AND DISTRIBUTION: Usually in stony semi-desert and dry savanna, between 490 and 1,400m. Eastern Ethiopia and northern Somalia. Ethiopian records include Degeh Bur and Aware, and Bohodle on Ethiopia/Somali border, thence north to Hargeisa, Oodweyne and plains south of Sheikh in Somaliland, but no further north, and not in northern mountain chain or north of there. No records from Sanaag (probably due to under-collecting) but it reappears in border country between Puntland and Somaliland, from Daror Valley south through Sol Haud (Gardo) to Nogal Valley. Probably more widespread in the Ogaden and the Haud, and central and eastern Puntland, but these areas are under-collected. Widespread and common, however quite large numbers are collected for the pet trade, although it is a delicate species and rarely thrives in captivity; it is a finicky feeder and often will only take termites.

Hemitheconyx taylori, juvenile, Oog, north-west of Las Anod.

Hemitheconyx taylori, male with regenerated tail, Las Anod.

CONSERVATION STATUS: Least Concern, but in high demand in the pet trade, so local protection is needed.

NATURAL HISTORY: A terrestrial, slow-moving and lethargic nocturnal gecko, active at night. Solitary; during the day hides under flat rocks, in termitaria and burrows, up to 1m deep; in the cold and dry seasons it remains torpid. Activity, feeding and reproduction triggered by rainfall and rising temperatures in April and May. Diet largely (perhaps exclusively) termites; faeces of wild-caught individuals contain termites, and captive individuals reluctant to eat much else. Resulting fat is stored in the tail and sustains the animal in the dry season, November to March. Captive specimens laid two soft-shelled eggs, roughly

28×18–19mm, probably 2–3 clutches in the rainy season. Incubation in captive animals at 28°C was about 80 days. Despite its lethargic mien, when threatened, this gecko stands tall on its legs, inflates and bobs the body, hisses loudly, lashes its tail, and may bite savagely if handled.

NORTH-EAST AFRICAN CLAWED GROUND GECKOS *HOLODACTYLUS*

An African genus of small, nocturnal, stocky, ground-dwelling lethargic geckos with rounded heads, thin necks, vertical pupils (although the eye just looks black) and movable eyelids with very pale vivid edges. Body elongate, tail short and swollen, not segmented, body scales homogenous, limbs short and slender, with claws. Males lack pre-anal or femoral pores. They are slow-moving, live in dry country and feed largely (possibly exclusively) on termites, gorging themselves and storing fat in their tails. Two species known, both endemic to the Horn of Africa, one appears endemic to Somalia. Included in the family key above.

SOMALI-MAASAI CLAWED GECKO

Holodactylus africanus
(Boettger, 1893)

IDENTIFICATION: Unmistakable; a small, stocky, ground-dwelling gecko with a stumpy tail, clawed feet, rounded head, dark eye, with a vertical pupil (hard to see) and distinctive yellow movable eyelids. Body cylindrical, the tail, which is constricted at the base, is short and stumpy, and about 24–28 per cent of total length; regenerated tails are more bluntly rounded. Toes slim, feet clawed. Body and limbs with fairly homogeneous scalation, covered by small, smooth or rough (occasionally slightly keeled), granular scales, very rarely intermixed with enlarged tubercles, giving it a velvety appearance. Both sexes lack pores, male with well-developed hemipenial bulge and two pairs of cloacal spurs. Maximum size 11cm (maximum snout–vent length 8.3cm, large animals usually females), average 7–10cm, hatchling size about 4.5–5.5cm. Dorsum pale sandy-grey, yellowish or tan, with a prominent series of 8–9 broad, red-brown crescent-shaped bands crossing back and tail, sometimes accompanied by a pale vertebral stripe, which extends to tail tip; sometimes there are less distinct lateral stripes. Underside pinkish-white. Juvenile colours usually more vivid than adults.

HABITAT AND DISTRIBUTION: Semi-desert, dry steppes or savanna in low country, from near sea level to approximately 1,400m in our area. Recorded at a string of localities from Zeila in extreme north-west Somalia, south along the Djibouti and Ethiopia/Somalia border, widely across Somaliland (apart from mountainous areas above 1,400m), east to Bosaso in Puntland, then south across the Ogaden and western Puntland (including Gardo and Garowe), and most of southern and central Somalia. Elsewhere, south through dry eastern and northern Kenya to Tanzania.

CONSERVATION STATUS: Least Concern, although exported from Tanzania for the pet trade, but doesn't do well in captivity.

NATURAL HISTORY: A nocturnal, solitary, slow-moving terrestrial animal; by day they shelter beneath stones, in termitaria or burrows, emerging after dark to crawl slowly about on the soil, in search of termites and possibly beetles. Known from both sandy and stony substrates. They are able to change colour, to match the soil, and if threatened they freeze, waiting motionless until the intruder disappears. Activity, feeding and reproduction triggered by rainfall and rising temperatures in April and May, on occasion several animals were observed in the vicinity of termite nests feeding on the winged insects as they emerge. The abundance of prey in the rainy season suggests the animal gorges itself and stores fat in the tail against hard times. Usually buries two soft-shelled eggs, 23–26×10.5–13mm, in a hole in the sand. Incubation in captivity 53–62 days. One was found in the stomach of an Obbia Bullfrog *Pyxicephalus obbianus*. This

Holodactylus africanus, Sheikh.

Holodactylus africanus, Geerissa.

charming little gecko is greatly feared in Somalia; it is erroneously believed to be venomous and the bite causes your skin to drop off.

CORNI'S CLAWED GECKO

Holodactylus cornii
Scortecci, 1931

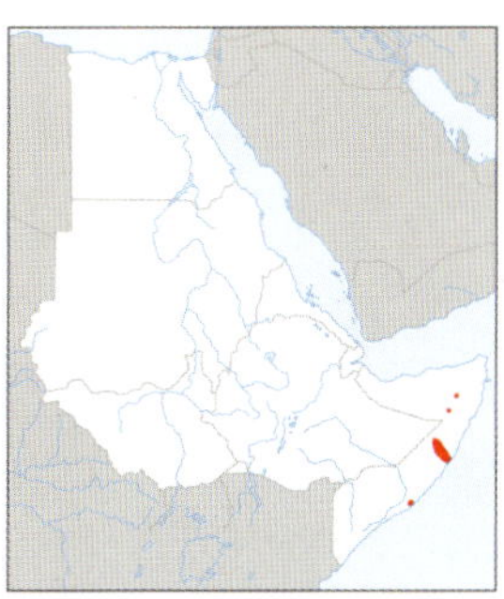

IDENTIFICATION: A rare, clawed gecko, terrestrial, with distinctive pale movable eyelids, of eastern Somalia. Massive triangular head and thin neck; eye large and black, with a hard-to-see vertical pupil, and distinctive white or yellow movable eyelids. The front edge of the ear-opening has a very small backwards-pointing spine. Body slim and cylindrical, limbs short and thin, toes not dilated but with short claws. Tail slightly constricted at base, swollen but depressed and narrow, c.27–29 per cent of total length. Dorsal scales granular and homogenous (with occasional slightly keeled scales), giving it a velvety appearance. Males and females lack pores, males have a well-developed hemipenial bulge and two pairs of cloacal spurs. Maximum size 15.3cm (maximum snout–vent length 10.9cm, large animals usually females), average 9–13cm, hatchling size unknown. Dorsum whitish, pale sandy-grey or yellowish, with a series of five broad, dull brown fairly irregular bands on back, usually accompanied by a wide pale vertebral stripe from snout to tip of tail. All paler parts of head, body, limbs and tail have scattered circular brown spots. Underside pinkish-white. Juvenile colours more vivid, and usually have four whitish crossbars on body.

HABITAT AND DISTRIBUTION: Semi-desert and subdesert with sandy or gypseous soils, low hills, grassland, open *Vachellia–Commiphora* savanna and flat xerophilous open woodland, between near sea level and 740m. Endemic to eastern regions of the Somali peninsula, known in the north from Gardo, from an imprecise locality in the Nogal Valley (at 420m) in Puntland and between Galkayo and Hobyo in central Somalia, where Scortecci collected a series along the road. Southernmost record 30km north of Mogadishu on coast.

CONSERVATION STATUS: Data Deficient.

NATURAL HISTORY: This species is known from fewer than 15 individuals. Biology unknown; probably similar to *H. africanus*, although its larger size, more slender habit, different shape and smaller tail, different head shape and more gracile fore legs may indicate a different way of life and diet. Both species are sympatric at several localities, notably Gardo, Galkayo and perhaps the Nogal Valley.

Holodactylus cornii, preserved, Warshiikh.

CLAWED GECKOS

FAMILY SPHAERODACTYLIDAE

A family of more than 230 species, in 12 genera, of diverse, relatively small clawed geckos, with a near worldwide distribution in tropical and temperate regions; absent only from Australasia and the southern half of Africa. In our area represented by a single genus, *Pristurus*.

SEMAPHORE GECKOS *PRISTURUS*

A Saharo-Arabian genus of small, active, diurnal geckos with curiously rounded heads (they resemble small agamas) and large eyes with round or oval pupils. Body cylindrical, covered by small, fairly homogenous granular scales without larger tubercles, and no pre-anal or femoral pores. Tail very long, the unregenerated tail (especially of males) often has a curious serrated crest; the feet are clawed, covered by a single row of sub-digital scales. They are tree- or ground-dwelling, tolerant of arid conditions and live in semi-desert and near-desert, feeding on small arthropods, mainly ants. They usually lay one hard-shelled egg and communicate using tail signals (particularly males); the tails are waved vertically and curled above the back, hence the common name. More than 25 species are known, mostly in north-east Africa, Socotra and the Middle East. Six species occur in our area, two of them endemic. Their taxonomy is not yet fully clarified, new species are constantly being discovered; at present several new species from Somalia are in the process of being described.

KEY TO THE NORTH-EAST AFRICAN MEMBERS OF THE GENUS *PRISTURUS*

1a: Dorsal and ventral scales keeled; only found on Benadir coast, southern Somalia
Pristurus simonettai, Simonetta's Semaphore Gecko

1b: Dorsal and ventral scales smooth; not known from Benadir coast, southern Somalia 2

2a: Nostril in contact with rostral shield 3

2b: Nostril separated from rostral shield 4

3a: Mainly tree-dwelling; tail (especially in males) strongly compressed, with dorsal crest of elongate, lanceolate scales that extends anteriorly beyond level of vent, at least as a row of enlarged scales; fold of upper eyelid usually weakly enlarged; maximum snout–vent length 4cm *Pristurus flavipunctatus*, Rüppell's Semaphore Gecko

3b: Rock-dwelling; tail less strongly compressed, that of males with a more feeble dorsal crest or row of enlarged scales that does not extend anteriorly beyond level of vent, mid-dorsals never enlarged, on contrary usually slightly smaller than scales on flanks; fold of upper eyelid moderately enlarged; maximum snout–vent length about 3cm *Pristurus migiurtinicus*, Migiurtine Semaphore Gecko

4a: Usually has highly contrasting dorsal pattern, with pale vertebral stripe, and short, narrow tooth-shaped crossbars; dorsal granules clearly smaller than ventral scales; 18–24 lamellae beneath fourth toe; claws longer than basal scales; generally on sand *Pristurus crucifer*, Cross-marked Semaphore Gecko

4b: Coloration and pattern not as in 4a; dorsal and ventral granules equal in size; usually 15–19 (occasionally up to 21) lamellae beneath fourth toe; claws equal to or shorter than basal scales; generally on stony ground 5

5a: Small, maximum size 6.7cm (snout–vent length 3.2cm); snout rounded, not depressed or beak-like; nostril not vertical; 15–17 (occasionally 14 or 18) lamellae beneath fourth toe; tail short, *c.*51–56 per cent of total length, with 10–12 tail bands; at higher altitudes (usually above 1,300m)
Pristurus phillipsii, Lort Phillips's Semaphore Gecko

5b: Comparatively larger, maximum size 8.2cm (snout–vent length 3.9cm); snout pointed, depressed and almost beak-like; nostril vertical; 18–19 (occasionally 16–17 or 20–21) lamellae beneath fourth toe; tail long, about 57–60 per cent of total length, with 13–16 tail bands; at low altitudes (usually below 1,100m)
Pristurus somalicus, Somali Semaphore Gecko

CROSS-MARKED SEMAPHORE GECKO

Pristurus crucifer
(Valenciennes, 1861)

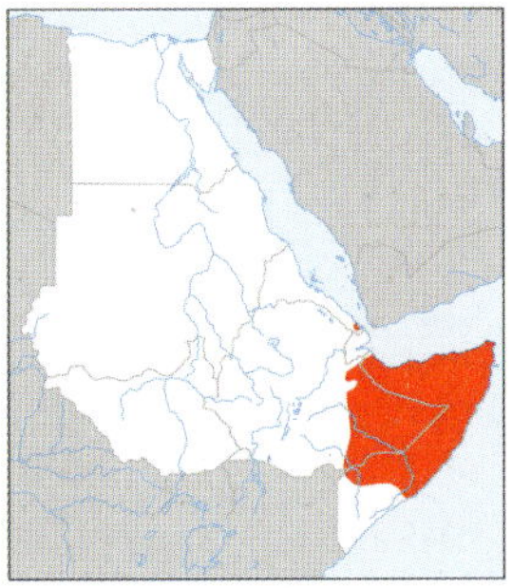

IDENTIFICATION: A little, ground-dwelling, grey or brownish gecko, usually strongly patterned with a distinct pale vertebral stripe. Body fairly slender, cylindrical or slightly depressed; legs short and slim, tail long, about 56–66 per cent of total length. Tail of male compressed laterally with a low, slightly serrated crest; tail of female has no crest and is slightly compressed. Dorsal scales homogenous, smooth, overlapping, smaller than ventrals. Maximum size about 10cm (maximum snout–vent length 4cm), average 6–9cm, hatchling size 2–2.5cm. Coloration distinctive, strongly patterned, dorsum grey, grey-brown, brown or rufous above, dependent on soil colour (e.g. grey in gypseous soil regions, red on red Haud sands), some have black spots on neck. There is a pale vertebral stripe, with short, narrow tooth-shaped crossbars and a dark dorsolateral stripe punctuated by white or cream spots; flanks may be speckled red. The stripe continues on tail, punctuated by pale blotches, although in some the back looks plain. Whitish below, chin yellow mottled grey. Juveniles similarly coloured. Taxonomic Notes: Length of tail, toes and claws varies, which has led to description of several different forms (*longipes*, *stefanii*, *laticephalus*) currently regarded as synonyms of *P. crucifer*. Molecular analysis indicates that *P. phillipsii* and *P. somalicus* are sister species.

HABITAT AND DISTRIBUTION: Semi-desert and dry savanna, both sandy and (more rarely) stony, usually with low sparse vegetation. Gravel plains, hard laterite on limestone, gypseous plains, but also in coastal dunes, relatively mesic *Vachellia* savanna and margins of cultivated fields from near sea level to around 1,650m. Known from Massawa and Assab in Eritrea and probably along coast in between, but records lacking. Also reported at Dire Dawa, Ethiopia. Extends from just south of Ethiopia/Somalia/Djibouti border, across the Ogaden and most of northern and central Somalia (away from the northern mountains), south to Rahanweyn country, Mogadishu, Bur Heybo and Doriyanle; south-west to Mandera in north-east Kenya. Probably in Djibouti but unrecorded. Elsewhere, south-west Arabia. Widespread and common.

CONSERVATION STATUS: Least Concern.

NATURAL HISTORY: Diurnal, occasionally active after dusk. Terrestrial, fast and very agile. Solitary, single individuals live in a small hole, dug in sand near base of a bush. Hunts by ambush, dashing out to snatch passing insects, particularly beetles and ants. Their strongly contrasting back pattern resembles the stalks of dry grass and leafless thorny branches of low *Vachellia*

Pristurus crucifer, Haud.

Pristurus crucifer, Dire Dawa.

bushes, under which they usually reside. Animals in very sandy habitats (e.g. the Haud) have longer tails and clawed toes. A single egg (sometimes two) are laid; in Somaliland a hatchling was observed in July. Diet beetles and ants. Sympatric in much of its range with *P. flavipunctatus*, at higher altitudes in Somaliland with *P. phillipsii*, at lower elevations of Somaliland, Puntland and central Somalia with *P. somalicus*.

RÜPPELL'S SEMAPHORE GECKO

Pristurus flavipunctatus
Rüppell, 1835

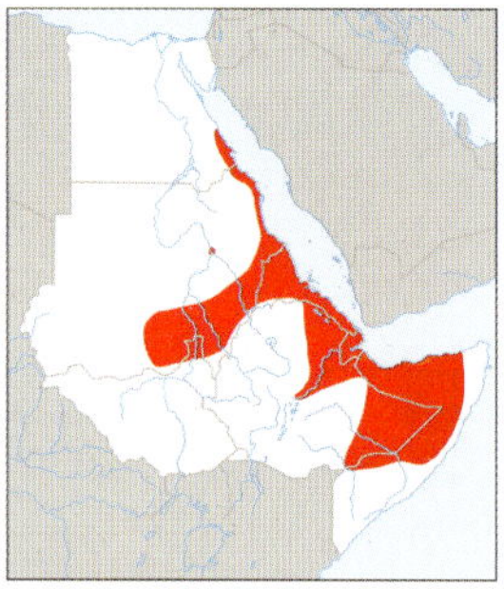

IDENTIFICATION: A small, generally tree-dwelling, greyish-brown lizard, in males the tail crest extends onto the body. Body slightly laterally compressed, stouter in females, legs long and slim, with short claws, tail long, about 53–63 per cent of total length. The male's tail is strongly compressed laterally with a high, serrated crest of lanceolate scales on the upper, and shorter on the lower, edge; tail of female has no crest and is scarcely compressed. Dorsal scales homogenous and granular (mid-dorsal scales sometimes lightly enlarged, forming a weak vertebral ridge in males). Maximum size about 10cm (maximum snout–vent length 3.8cm), average 7–9cm, hatchling size 2.5cm. Cryptically coloured. There is a dark line through the eye, and the lips and chin are pale. Dorsum pale or dark grey-brown, with a series of c.8 chevron-like dark marks or crossbars on mid-dorsum, which extends onto the tail; flanks may be spotted or streaked red or orange. There is often a yellow or orange vertebral stripe. A dark line extends from snout through eyes to above ears, with 1–3 dark short lines above shoulder. The side of the head is paler below, dark above. White below, immaculate or spotted brown, throat and belly sometimes spotted red. Juveniles similarly coloured, with a very pronounced yellow vertebral stripe and pinkish tail.

Pristurus flavipunctatus, Geerissa.

Pristurus flavipunctatus, Adi Teklezam.

Pristurus flavipunctatus, Cal Madow Mountains.

HABITAT AND DISTRIBUTION: In near-desert, semi-desert, woodland and urban areas, provided there are trees, from sea level to about 2,300m. Occurs along Red Sea coast from south-east Egypt (just north of Marsa Alam) south to coastal hills and plains of Sudan (Durru, Suakin, Port Sudan), Eritrea (Dissei Island) and Djibouti, thence into Somaliland, Puntland south through central Somalia to Dollo, Belet Uen and El Bur. Also the White Nile Valley, from Atbara south to Kordofan, notable records at Talodi, Bara, near Kosti, Singa and Sennar on the Blue Nile. It extends into the Afar Region of Ethiopia, thence south across the Ogaden. Elsewhere, in south-west Arabia (Saudi Arabia and Yemen).

CONSERVATION STATUS: Least Concern; uncommon and highly local in Egypt but widespread and common in Eritrea and Somaliland.

NATURAL HISTORY: Arboreal and diurnal; favouring *Vachellia* and *Balanites aegyptiaca* trees and *Tamarix* thickets, but will live on rocks. Even in apparently unsuitable habitats like the virtually treeless steppe of central Somalia, a single isolated tree may support a population of these geckos. In Eritrea north of Asmara, specimens lived on large quartz boulders among tall grass on a hill; in Somaliland and Awash in Ethiopia found on rocks. They appear to be social; in Somaliland true pairs (sometimes more) were observed sleeping together 2–3m above ground at tips of branches. Communicate by tail signals, waving the tail vertically or curling it. A single, tiny hard-shelled near-spherical egg, 6×7mm, is laid in a depression under a rock or in a deep tree crack; several clutches per season. In Ethiopia one was taken by a domestic cat, and most of the local population had lost their tails. These geckos seem to occupy much the same niche as dwarf geckoes, *Lygodactylus*, which replace them in the south of our area.

MIGIURTINE SEMAPHORE GECKO

Pristurus migiurtinicus
Scortecci, 1933

IDENTIFICATION: A small, rock-dwelling grey-brown gecko of the north Somali coast; the male's tail crest does not extend onto the body. Body cylindrical or very slightly laterally compressed. Legs fairly long and slim, toes clawed. Tail long, about 53–60 per cent of total length, and in males is slightly compressed laterally, with a moderately high, serrated crest of enlarged scales. Female's tail has no crest and is not compressed. Mid-dorsal scales not enlarged. Maximum size about 7cm (maximum snout–vent length 2.9cm), average 5–6cm, hatchling size unknown. Coloration cryptic; dorsum pale or dark grey-brown, with a series of *c.*8 dark, posteriorly white, square or round blotches along spine and onto tail. A pale vertebral stripe is sometimes present. Flanks marked with longitudinal rows of reddish, pale-bordered streaks. A dark line extends through eye to above ear-opening but can be indistinct in front of eye; side of head uniform in colour. Underside whitish, with some black or brown spots. Taxonomic Notes: Long regarded as single species, *P. rupestris*, from eastern Iran, throughout Arabia to Somalia and Eritrea. However, molecular studies indicate that populations in south-west Arabia, Jordan and probably Africa differ from true *P. rupestris*. Thus the oldest available name for African populations is *P. migiurtinicus*.

HABITAT AND DISTRIBUTION: Arid near-desert, semi-desert and dry savanna, always in stony or rocky terrain with sparse vegetation. Largely near coast, but occasionally inland and at higher altitudes; from sea level to approximately 2,000m. In Africa known from four disjunct regions (which might prove interconnected). Originally described from Bosaso, in Puntland, where recorded at Biyo Kulule, Galgalo, El Figra; nearby records in Cal Miskaat Mountains include Habeno, Abal, Bahaja, Warmadow and Guriasamo. Another population in western Somaliland, in hills between Dilla and Quljeet. Also Borama district and Djibouti, where reported from Obock and Gulf of Tadjoura. The fourth population is on Romia and Andeber Islands, Dahlak Archipelago, Eritrea.

CONSERVATION STATUS: Not Assessed.

NATURAL HISTORY: Diurnal, but sometimes active at dusk. Virtually always on rock, and although

Pristurus migiurtinicus, preserved type, Bosaso.

terrestrial it climbs well. Shelters in rock cracks and between boulders. Dahlak animals were under rocks by day, on open soil among seasonal vegetation at night. Often on walls in villages and towns. They are 'sit-and-wait' predators, favouring ants. Territorial, males wave and curl their tails to signal. A single hard-shelled egg is laid several times per season; the holotype had a single egg, 4.5×3.4mm, in her oviduct in September.

LORT PHILLIPS'S SEMAPHORE GECKO

Pristurus phillipsii
Boulenger, 1895

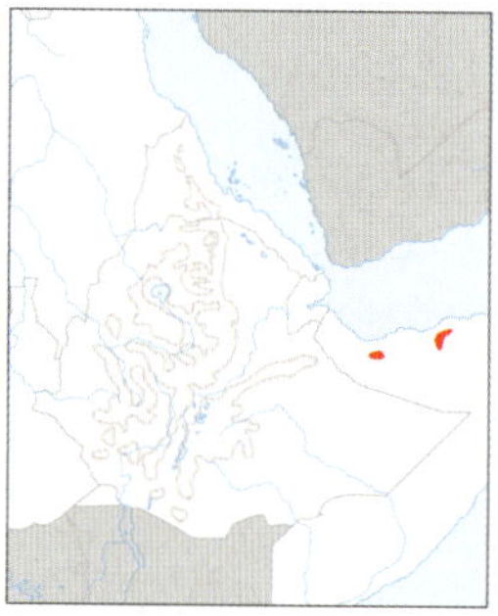

IDENTIFICATION: A small, stout, terrestrial semaphore gecko, with a rather bird-like head, endemic to northern mountains of Somaliland. Eye small with an (oddly) vertically elliptic pupil. Body cylindrical, legs stout with claws. Tail short, c.51–56 per cent of total length (shorter in females). Tail of male compressed laterally with a moderately serrated crest; tail of female has no crest and is cylindrical. Maximum size 6–7cm (maximum snout–vent length 3.2cm), average 5.5–6.3cm, hatchling size unknown. Strongly sexually dimorphic: males slimmer and larger with a more

Pristurus phillipsii, Cal Madow Mountains.

Pristurus phillipsii, Yuffleh.

distinct neck and longer tail with a bulbous base. Coloration highly cryptic; dorsum brown or grey-brown. Along spine is a series of 8–9 faint round botches, paler posteriorly. A pale pinkish-orange vertebral stripe is usually present; flanks have 3–4 longitudinal rows of dark and white streaks or spots, and may be stippled red. A distinct large black blotch is present in the groin and above the shoulder. Tail has 10–12 paler bands. Belly cream, undertail greyish or brownish, throat grey, marbled brown. In bright sunlight they fade to a brownish-grey, but pattern remains visible.

HABITAT AND DISTRIBUTION: Largely in mesic montane habitats and, occasionally, associated lower arid areas, with more than 500mm of annual rainfall, at 1,070–2,100m. In open woodland, densely wooded cedar and box forests, may descend to relatively mesic *Vachellia* savanna with sparse herbaceous cover, evergreen scrub and stony plains. Known from Goolis Mountains between Gaan Libaah and Sheikh, where it is found on the top and southern slopes, as far as Go'oo village, and also 16km south of Sheikh. Also at western edge of Cal Madow Mountains in Sanaag region, from Gal Sheikh north through the Yuffleh plains to hills north-west of Erigavo and Daallo forest; on the northern slopes it occurs along the road from Erigavo to Maydh on the coast. Probably in the Wagar Range, and the central and eastern Cal Madow mountains, but as yet unrecorded.

CONSERVATION STATUS: Least Concern. Although restricted, it is common and widespread in suitable habitats.

NATURAL HISTORY: Solitary, terrestrial and diurnal, living in small territories. Favours stony ground, where they bask, monitor their territory and opportunistically hunt while perched on the top of small rocks, 10–20cm in diameter. When disturbed they shelter beneath stones. In Daallo forest north of Erigavo, where clay soil is covered by limestone gravel and typical vegetation involves *Juniperus* and *Sideroxylon* trees, *Buxus hildebrandtii*, *Cadia purpurea* and *Dodonaea viscosa* shrubs; they favour open habitats like meadows or edges of rocky roads. Here they are especially numerous during the rainy season (April–September), as the temperature rises. During this season the females lay a single hard-shelled egg, buried in soil under rocks. Several eggs may be laid per season. Tail signals, like other members of genus. At lower elevations (below 1,000m) replaced by the similar and closely related *P. somalicus*. Observed in sympatry with *P. crucifer* and *P. flavipunctatus*.

SIMONETTA'S SEMAPHORE GECKO

Pristurus simonettai
(Lanza & Sassi, 1968)

IDENTIFICATION: A diminutive, slender, reddish, terrestrial gecko, endemic to Benadir coast of southern Somalia. The eye has a vertical subelliptical pupil. Body cylindrical or slightly depressed, elongate, limbs and toes long and slim. Tail long and thin, about 63–66 per cent of total length; sub-cylindrical, without any crest. Dorsal scales keeled and overlapping, fairly homogenous, although tend to be larger and more rectangular on back, smaller and rounded on neck and flanks. Maximum size about 7.5cm (maximum snout–vent length 2.7cm), average 6–7cm, hatchling size probably 2–2.5cm. Coloration distinctive, strongly patterned, dorsum hazel-brown to rufous with a broad pale greyish or pinkish vertebral stripe, with fairly short, wide zigzag tooth-shaped crossbars, from nape to tail base, where it becomes a line. A broad white stripe extends dorsolaterally from behind eye to tail, where it fades. Flanks whitish, spotted brown, merging into white of belly. A dull light stripe extends from nostril to upper edge of eye. Head and dorsal surface of limbs with several minute, scattered, irregular dark spots. Belly whitish, other underside parts yellowish to pinkish-red. Juveniles similarly coloured. Taxonomic Notes: A distinctive animal, originally placed in its own genus *Geisopristurus*, later synonymised with *Pristurus*, although its relationships within the genus are presently unknown.

Pristurus simonettai, preserved type.

HABITAT AND DISTRIBUTION: A southern Somali endemic, confined to static sand dunes and flat sandy areas, with open mixed bushland and scattered trees on southern Somali coast, between sea level and 100m; never more than 40km inland. Enters cities and has been collected in gardens. Endemic to coastal inter-riverine area between the Juba and Shebelle Rivers, at just three localities; Kismayu, Sar Uanle (about 20km south of Kismayu) and the environs of Mogadishu (extends 6km from Mogadishu along the Afgoi road). Probably widespread in the intervening country (all suitable habitat) but no records.

CONSERVATION STATUS: Least Concern.

NATURAL HISTORY: Poorly known. Terrestrial and diurnal, sometimes active at dusk. The juvenile paratype was collected on a consolidated light reddish, calcareous dune, in xerophilous woodland with shrubs or small trees. Several other psammophilous species of reptiles occur in the same habitat (Worm lizards *Agamodon anguliceps*, *A. compressus*, Révoil's Somali Short Snake *Brachyophis revoili*, Sand lizards *Pseuderemias brenneri*, *P. striata*) and several species of Writhing skinks (*Mochlus* spp.). A fast-moving species, which favours sandy habitats and shelters in vegetation. Often extremely abundant. Said to curl the tail upwards. Diet and reproductive details unknown, but probably similar to other *Pristurus*. A juvenile, <2 months old, was collected at end of August.

SOMALI SEMAPHORE GECKO

Pristurus somalicus
Parker, 1932

IDENTIFICATION: A smallish pale grey or pink diurnal Somali gecko. The snout is almost beak-like, depressed and pointed, the nostril vertical, separated from the rostral scale. Eye moderately sized, pupil vertically elliptic. Body cylindrical, legs long and slim with claws. Tail long, about 57–60 per cent of total length (shorter in females). Tail of male compressed laterally with a moderately serrated crest; tail of female has no crest and is cylindrical. Dorsal scales homogenous, smooth, slightly overlapping. Maximum size 8.2cm (maximum snout–vent length 3.9cm), average 7–8cm, hatchling size 2.5cm. Sexual dimorphism considerable, males slimmer and larger with longer tails, and tail base slightly more bulbous. Pale grey, grey-brown or pinkish above, with a series of 8–9 faint reddish square-shaped blotches on mid-dorsum, males may have a paler pinkish-orange vertebral stripe. Flanks have longitudinal rows of thin stippled streaks; red in males or brown in females. There is a black blotch above shoulder. Head with tiny dark spots; sometimes an indistinct stripe extends behind eye. Tail with 13–16 paler bands. Belly whitish-cream, throat with grey blotches, underside of tail greyish or pale brownish with darker bands. The scales of the male's crest are yellowish. Males generally bicoloured, the anterior half brown, posterior and tail grey. In bright sunlight the pattern disappears. Juveniles plain grey with a faint pattern. Taxonomic Notes: Originally lumped with *P. phillipsii*, but recent fieldwork and analysis indicate the two are not conspecific, although intermediate (and as yet undescribed) forms exist.

Pristurus somalicus, Maydh.

Pristurus somalicus, Inaafmadow.

Pristurus somalicus, Dagah Shebell.

HABITAT AND DISTRIBUTION: Favours arid rocky, unvegetated habitats, usually with >200mm of annual rainfall, between sea level and about 1,100m (but commonest below 800m). In near-desert, semi-desert, arid sub-shrub steppe and dry savanna, in flat or low hilly country. In Somaliland restricted to the Guban, from Berbera east along coast north of the mountain chain, to Heis, Maydh and Las Korey, into Puntland, where common and widespread everywhere in suitable habitats below 1,100m. In southern and eastern Somaliland known from the Darror, Nogal Valley and Haud region, south to Bohodle on the Ethiopian border. Central Somali records include Galkayo, near Hobyo and El Bur, south to Belet Uen, and environs of Jesomma, in southern Somalia to 04°N. Probably occurs in the eastern Ogaden in Ethiopia.

CONSERVATION STATUS: Least Concern. Widespread and very common in suitable habitats.

NATURAL HISTORY: Fast and agile, diurnal, active at dusk during hotter parts of year. Terrestrial, tolerant of heat and often seen basking on stones or running, even in middle of day; hides under stones if threatened. Becomes very pale, almost white, in summer, darker in winter. It often stands by ant trails, dipping its head rapidly to grab insects as they pass. A single hard-shelled egg is laid several times per season. In Somaliland it is one of the most common reptiles on hot stony coastal plains, and is often found only few metres away from the sea, on highly saline rocks, stones and gravel. In northern Somalia at higher elevations and in more mesic habitats it is replaced by its sister species, *P. phillipsii*. Sympatric with *P. flavipunctatus*, *P. migiurtinicus* and *P. crucifer*.

LEAF-TOED GECKOS

FAMILY PHYLLODACTYLIDAE

A family of more than 150 species, in ten genera, of diverse geckos, widespread in both the New and Old Worlds. In our area represented by two genera, *Ptyodactylus* and *Tarentola*. The notes on reproduction, and structure of the feet, of the following family, Gekkonidae, are relevant also to this family.

FAN-TOED GECKOS *PTYODACTYLUS*

A genus of relatively large geckos, unmistakable on account of their digits, adapted to grip rocks, being strongly dilated at the tip into a pair of pads that resemble a fan. They are slender, with huge eyes and long, thin limbs. They live on rocks and buildings and are territorial, the male producing a startlingly loud clicking call (resembling rocks being banged together) in defence of his territory. Twelve species occur across northern Africa and Arabia, four in our area, but there remains debate about the status of some taxa.

KEY TO THE NORTH-EAST AFRICAN MEMBERS OF THE GENUS *PTYODACTYLUS*

1a: Tail longer than or equal to snout–vent length	2
1b: Tail shorter than snout–vent length	3
2a: Stocky, terminal pads large, back mottled but without crossbars, in our area south from Gebel Elba in Egypt to Ethiopia	*Ptyodactylus ragazzii*, Ragazzi's Fan-toed Gecko
2b: Slim, terminal pads small, back mottled and has vaguely regular crossbars, in our area from southern Egyptian border region north	*Ptyodactylus hasselquistii*, Egyptian Fan-toed Gecko
3a: Nostril distinctly tubular, smaller (maximum snout–vent length 73mm), limbs shorter and stockier	*Ptyodactylus siphonorhina*, Saharan Fan-toed Gecko
3b: Nostril swollen but not tubular, larger (maximum snout–vent length 90mm), limbs longer and slenderer	*Ptyodactylus guttatus*, Spotted Fan-toed Gecko

RAGAZZI'S FAN-TOED GECKO

Ptyodactylus ragazzii
Anderson, 1898

IDENTIFICATION: A fan-footed rock-dwelling gecko of southern Red Sea coasts. Stocky, body depressed, limbs stout, digit tips expanded to fan shapes. Unregenerated tail thin, 55 per cent or more of total length and can be shed. Scales small, juxtaposed granules, interspersed by scattered, small tubercles, which may be slightly keeled. Maximum size about 20cm, average 14–18cm, hatchling size 5–6cm. Dorsal colour usually pinkish-grey or grey-brown, heavily mottled with darker blotches and sometimes little scatterings of white scales. Tail often banded; upper limbs blotched with same colours. Immaculate white or cream below.

HABITAT AND DISTRIBUTION: Dry savanna, semi-desert and near-desert, on rocky hills, large and small, gorges, or on buildings, from sea level to above 2,400m. Occurs from north-west Somaliland (east as far as Agabar, north of Hargeisa), on Ethiopian border, through Djibouti and Eritrea (mostly in lowlands, but inland, for example at Senafe, 2,400m), along coast of Sudan, just reaching south-east Egypt at Gebel Elba. Due to taxonomic debate about the identity of this species and similar taxa, its range elsewhere is uncertain, although spotted specimens from Kordofan and perhaps Kassala in Sudan probably belong here.

CONSERVATION STATUS: Least Concern.

NATURAL HISTORY: Rock-dwelling, will live on and in buildings. Doesn't usually venture onto soil or sand, but may cross roads at night. Active by day and night; by day in cooler periods, and nocturnal or crepuscular in the hot season; although it may utilise deep rock cracks as heat sources in cold periods, warming inside and emerging briefly to hunt. Lives in rock crevices, and similar refuges in buildings. At night may hunt insects around lamps. It can run very fast and also jump; if pursued it sprints a short distance, then leaps sideways while turning, to run in a different direction, to confuse predators. Males territorial and will fight intruding males, seizing each other's limbs, shuffling and twisting, trying to shake the opponent off the surface; their bones can be heard creaking. Males also produce a brief series of strident clicks, like two rocks being banged together, to alert rival males and possibly attract females. Two hard-shelled eggs are laid, glued to the rock in a recess; they often nest communally, with many eggs deposited in same deep crack. Up to six clutches per season recorded in captivity.

Ptyodactylus ragazzii, Jidhi.

Ptyodactylus ragazzii, Erkowit.

Ptyodactylus ragazzii, Fil-fil.

EGYPTIAN FAN-TOED GECKO

Ptyodactylus hasselquistii
(Dondorff, 1798)

IDENTIFICATION: A fairly big slender fan-toed gecko of the Nile and Egyptian Red Sea coast. Eye large, the pupil vertical with scalloped edges. Body slightly depressed, limbs stout, digit tips expanded to fan shapes. Unregenerated tail thin, 50 per cent or slightly more of total length and can be shed. The scales are small, juxtaposed granules, interspersed by scattered, small tubercles, which may be slightly keeled. Maximum size about 17cm, average 12–16cm, hatchling size probably 5–6cm. Dorsal colour quite variable, and may change, but usually shades of pinkish-grey or brown, heavily mottled darker brown, and with vague, broad dark-edged mottled brown crossbars. Sometimes there is a line of light narrow lozenges along spine. Occasional individuals appear pattern-less. Tail often banded; upper limbs blotched with same colours. Immaculate white or cream below. Taxonomic Notes: Recent research indicates this is a species complex, with deeply divergent clades that will probably be split.

HABITAT AND DISTRIBUTION: Dry savanna, semi-desert and near-desert, on rocks or buildings, from sea level to around 800m; in Sinai usually replaced by *P. guttatus* above that altitude. Occurs in and around Cairo, south Sinai, with a handful of records along Egyptian Red Sea Coast , on Egyptian Nile from near Edfu to Aswan, and in Sudan south as far as Khartoum; recorded from Wad Ben Naga, Karima, Jebel Barkal, Dongola, Abu Sari, Quikkah, bridge by Akasha, Wadi Halfa, Ferka East, Abri, Al-Khandaq, 15km south of Meroe, El Kadada and Khartoum. A presumably introduced population on the north Sinai coast. Elsewhere, around edges of Arabian Peninsula.

Ptyodactylus hasselquistii, Safaga.

Ptyodactylus hasselquistii, juvenile, Hamata.

Ptyodactylus hasselquistii, Safaga.

CONSERVATION STATUS: Least Concern.

NATURAL HISTORY: Similar to previous species, i.e. rock-dwelling, active by day and night, fast-moving, territorial, produces a series of clicks, two hard-shelled eggs laid communally. Often present in large numbers. In Sinai it sometimes ventures onto sand at night. Observed on hotel walls about 40km south of Quseir in Egypt side-by-side with *Hemidactylus granosus* and *H. flaviviridis*.

SPOTTED FAN-TOED GECKO

Ptyodactylus guttatus
Heyden, 1827

IDENTIFICATION: A fairly big, slender, fan-toed gecko of north-east Egypt. The eye is large, pupil vertical with scalloped edges. Body slightly depressed, limbs stout, digit tips expanded to fan shapes. Unregenerated tail thin, less than half total length and can be shed. The scales are small, juxtaposed granules with a few small tubercles, especially on neck. Maximum size about 17cm, average 11–15cm, hatchling size unknown. Dorsal colour quite variable, and may change, but usually shades of orange or brown with light spots all over, sometimes a series of light and dark, vague crossbars. Tail strongly banded, light and dark. Immaculate white or cream below.

Ptyodactylus guttatus, Sinai.

HABITAT AND DISTRIBUTION: Semi-desert and near-desert, on rocks or buildings, from sea level to around 2,000m or more. Occurs in Cairo, east to Gulf of Suez, south to Jebel El Shayeb near Hurghada, widespread in northern Sinai, in higher hills in south Sinai. Elsewhere, north to Syria and across Arabia.

CONSERVATION STATUS: Least Concern.

NATURAL HISTORY: Similar to previous species, i.e. rock dwelling, active day and night, (apparently more diurnal in Sinai where it overlaps with the previous species), fast-moving, territorial, male and female produce a series of clicks. They mate in spring, two hard-shelled eggs laid communally, 3–4 times a year.

SAHARAN FAN-TOED GECKO

Ptyodactylus siphonorhina
Anderson, 1896

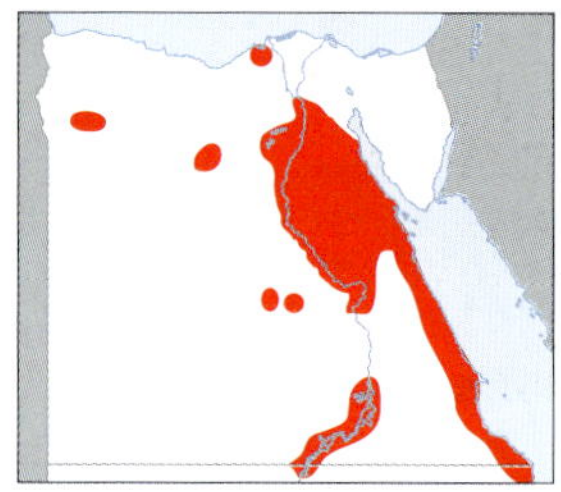

IDENTIFICATION: A fairly big, slender, fan-toed gecko of Egypt and possibly Libya (see distribution notes below). Eye large; pupil vertical with scalloped edges. Body slightly depressed, limbs stout, digit tips expanded to fan shapes. Unregenerated tail thin, less than half total length and can be shed. Scales are small, juxtaposed granules with a few small tubercles, especially on neck. Maximum size about 15cm, average 11–14cm, hatchling size unknown. Dorsal colour quite variable, and may change, but usually shades of grey, pinkish-grey or brown, with crossbars of white spots alternating with brown spots. Tail strongly banded, light and dark. Immaculate white or cream below. Taxonomic Notes: Status of this species debated, some authorities regard it as a subspecies or just a variant of *P. guttatus*, with which it forms a monophyletic group.

HABITAT AND DISTRIBUTION: Semi-desert and desert, on rocks, sometimes buildings, at low altitudes. Sporadically distributed in Egypt (Qattara Depression, and most of the Western Desert oases), also sporadically along Egyptian Red Sea coast, inland to highly arid hill country. Occurs in vicinity of the Nile around Lake Nasser, possibly just entering northern Sudan. Might extend into Libya, but status of extralimital animals there not clarified.

CONSERVATION STATUS: Not Assessed.

NATURAL HISTORY: Similar to previous species, i.e. rock-dwelling, on boulders, vertical rocks, in caves, also buildings. Active day and night, fast-moving, territorial, male and female produce a series of clicks, females lay two eggs which are glued to the rock. In Egypt, regularly seen on open rocks during day, waiting for prey, in winter suns itself near a suitable retreat. The author who discovered and described this species was John Anderson (1833–1900), Scottish anatomist and zoologist, who devoted the latter half of his life to the study of Egyptian amphibians and reptiles. He described many species and the comprehensive monograph he published in 1898 is still relevant.

Ptyodactylus siphonorhina, Wadi El Rada.

WALL GECKOS *TARENTOLA*

A genus of large chunky nocturnal geckos, characterised by dorsal rows of large tubercles, spiky tails and wide-splayed pads on the toes, with claws on the third and fourth digits. They are territorial and vocal, and live on trees, rocks and buildings. About 30 species known. They have crossed the Atlantic, being known from the Caribbean, and have radiated on the Canary Islands (four species) and Cape Verde Islands (ten species); *T. annularis* and *T. mauritanica* are invasive species in the USA. About ten species occur in the northern half of Africa, but the status of some taxa is debated. Four species (possibly five, see *T. ephippiata*) occur in our area.

KEY TO THE NORTH-EAST AFRICAN MEMBERS OF THE GENUS *TARENTOLA*

1a: Huge, adults up to 28cm long, dorsal tubercles not surrounded by rosettes of small scales, usually white spots or crossbars on shoulders, widespread — *Tarentola annularis*, White-spotted Wall Gecko

1b: Adults up to 25cm long; lacks above combination of characters — 2

2a: In our area not in Egypt, maximum size 25cm, usually grey or grey-brown with broad dark crossbars — *Tarentola ephippiata*, African Wall Gecko

2b: In our area only in Egypt, maximum size 17cm, brown or grey, with narrow crossbars — 3

3a: Tubercles on temporal (cheek) region clumped and often in contact; only in Qattara Depression and north-west oases of Egypt — *Tarentola mindiae*, Qattara Gecko

3b: Tubercles on temporal (cheek) region isolated, surrounded by smaller scales; only within 100km of Egyptian Mediterranean coast — *Tarentola fascicularis*, Eastern Moorish Gecko

WHITE-SPOTTED WALL GECKO

Tarentola annularis
(Geoffroy, 1827)

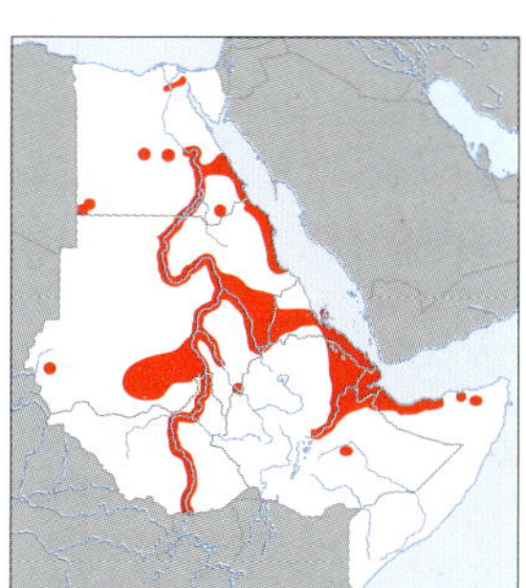

IDENTIFICATION: A huge, knobbly gecko, widespread in our area. Stocky, body depressed, limbs stout, digit tips expanded along entire toe, with claws on third and fourth digits. Unregenerated tail broad and tapering, 53–55 per cent or more of total length and can be shed; sharp-edged lateral tail scales make it look spiky. Dorsal surface covered with small scales, and regular rows of larger (albeit relatively small) unkeeled tubercles. Maximum size about 28cm, average 18–24cm, hatchling size oddly small, 3–4cm. Dorsal colour variable; shades of brown, grey, rufous, almost black, but usually 2–3 pairs of white dots, or one or two white or pale crossbars on neck and shoulders, and back darkly and irregularly mottled. Underside immaculate white or cream. Taxonomic Notes: Two

Tarentola annularis, Kassala.

Tarentola annularis, Dissei Island.

Tarentola annularis, Awash National Park.

subspecies recognised. *T. a. relicta* in Sudan, South Sudan and northern Uganda lacks the four white, dark-rimmed scapular dots and has multiple keeled scales and tubercles; the nominate subspecies has the white dots and a single keel on each scale and tubercle.

HABITAT AND DISTRIBUTION: Dry savanna, semi-desert and desert, on rocky hills and outcrops, sheet rock, on cliffs, in gorges or on stony ground, or on buildings, from sea level to above 1,600m in Somaliland. Very widespread in our area, from the southern Nile Delta right along the Nile to the Ugandan border, and from the southern Red Sea coast of Egypt through Eritrea and Djibouti to north-west Somaliland (records sporadic, but probably along entire coast). Extends west from Djibouti to Dire Dawa and lower Awash River in Ethiopia. Sporadically recorded from Eritrean lowlands, including several Dahlak Islands (Dissei, Andeber, Dahlak Kebir, Dehel, Durghella, Isratu, Harmil and Nokra). In Sudan, as well as right along Nile, also known from Kordofan, Darfur, Kassala and Sennar, probably occurs almost throughout the dry savanna and semi-desert where there are rocks. Recorded from oases in Egypt's Western Desert and Jebel Awaynat on the Egypt/Libya/Sudan border. Elsewhere, south to northern Uganda, west to Mauritania. An introduced population reported from Israel/Palestine.

CONSERVATION STATUS: Least Concern, although large numbers are exported for the pet trade.

NATURAL HISTORY: Usually on rocks, but will live on stony plains, where it shelters in burrows under rocks or in earth cracks. Often in culverts, and on stone buildings, where may hunt insects around lamps. Nocturnal, although it may emerge near dusk. Males territorial and aggressive, and will fight intruding males. A quick, agile species. The males also produce a brief series of croaks to alert rival males and possibly attract females, who may call back. Two hard-shelled near-spherical eggs are laid in a suitable refuge; 2–8 clutches per season. Often surprisingly common, seven were seen near a lamp at Khartoum airport; over 40 specimens were observed on the wall of a hut in Sudan. It is a well-adapted species, tolerant of extremely hot and inhospitable areas, in some harsh environments (e.g. the smaller Dahlak Islands) it is the only reptile. They will eat small vertebrates as well as arthropods.

AFRICA WALL GECKO, FIG TREE GECKO

Tarentola ephippiata
O'Shaughnessy, 1875

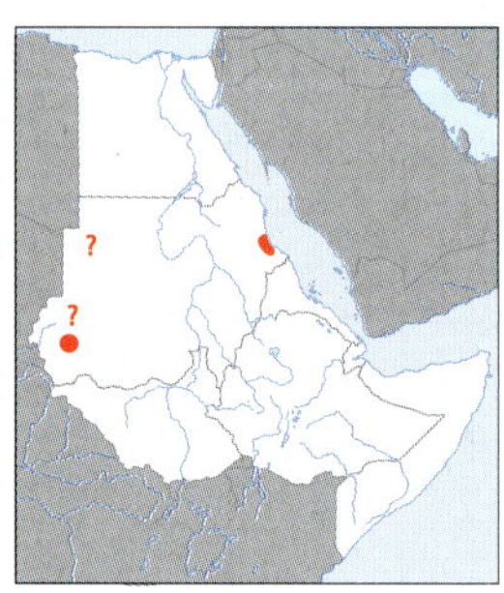

IDENTIFICATION: A big grey-and-black gecko, in our area in eastern and western Sudan. Stocky, body depressed, limbs stout, digit tips expanded along entire toe, with claws on third and fourth digits. The unregenerated tail is broad and tapering, 53–55 per cent or more of total length and can be shed. The dorsal surface is covered with small scales, and regular rows of larger (albeit relatively small) unkeeled tubercles. Maximum size about 18cm (although specimens up to 25cm reported), average 12–16cm, hatchling size unknown. Dorsal colour usually a shade of grey, a dark bar through eye broadens on shoulders and forms a crossbar, the rest of the back has broad dark and light grey crossbars, edged black, tail banded light and dark grey. Occasional individuals have a series of large pale blotches on spine. Underside white, cream or yellow. Specimens from north-east Sudan have seven narrow, X-shaped markings on back from nape to tail base, Y-shaped markings connect these to each other and to a dorsolateral stripe, forming a reticulated pattern. The space between the dorsolateral stripes is light brown (in alcohol), the sides of the body are white-grey. Taxonomic Notes: Specimens from semi-desert and desert across Africa, between 15 and 30°N, are regarded by some authorities as a separate species, *T. hoggarensis*, and that form might reach extreme north-west Sudan. The subspecies in north-east Sudan is *T. e. nikolausi*.

Tarentola ephippiata, Ghana.

Tarentola ephippiata, Chad.

HABITAT AND DISTRIBUTION: Most records in dry savanna and semi-desert, but moist savanna and rocky hills in desert in parts of its range; from sea level to above 2,000m. However, in our area known only from around Durrur, Erkowit and Suakin, on Red Sea coast of north-east Sudan, and Western Darfur; in western Sudan, although it might occur in the intervening area. Elsewhere, west to Senegal.

CONSERVATION STATUS: Least Concern.

NATURAL HISTORY: Usually on trees, under bark or in cracks, in Nigeria it favoured crevices in fig trees, but will also move on the ground at night, and will live on buildings. Reported on small rocky hills. Nocturnal, although it may emerge near dusk or even late afternoon. It might be territorial, although details unknown. A quick, agile species. Two hard-shelled near-spherical eggs are laid in a suitable refuge, said to bury its eggs in desert habitats.

QATTARA GECKO

Tarentola mindiae
Baha El Din, 1997

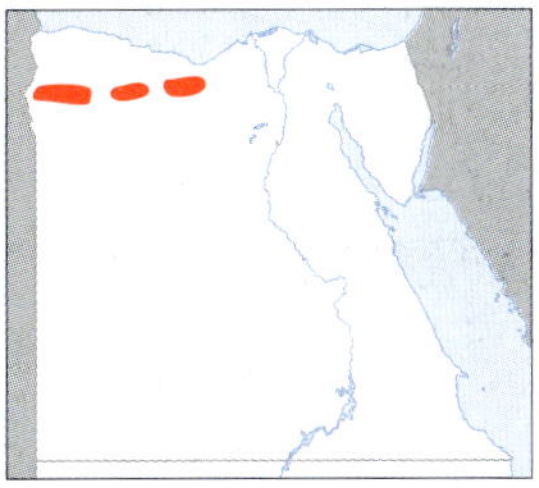

IDENTIFICATION: A big grey-and-black gecko, endemic to deserts in north-west Egypt. Unregenerated tail broad and tapering, usually >50 per cent of total length and can be shed. The dorsal surface is covered in regular bands of protruding tubercles, with a strong central keel, surrounded by rosettes of 5–8 smaller scales in a horseshoe. Maximum size about 17cm, hatchling size unknown. Dorsal colour usually brown or grey, with 5–6 dark irregular crossbars between the neck and tail; the tail is banded light and dark. Underside grey-white, each scale with one or more tiny dark spots.

Tarentola mindiae, Qattara Depression.

HABITAT AND DISTRIBUTION: In dense stands of vegetation in otherwise flat, barren sandy desert, below 200m. In our area, known from the Qattara Depression and Siwa Oasis in north-west Egypt. Elsewhere, just extends into eastern Libya west of Siwa Oasis, but might be more widespread.

CONSERVATION STATUS: Least Concern.

NATURAL HISTORY: Most specimens were on trees but some on the ground at night. Found in dense stands of vegetation, surrounded by sandy desert, but some found walking on open sand. Occurs at high densities in suitable habitat; five were found under the bark of a single dead *Vachellia* tree. Captive animals showed territorial behaviour. Up to four clutches of two eggs are laid per year, buried in shallow soil. They reach sexual maturity in less than a year.

EASTERN MOORISH GECKO

Tarentola fascicularis
(Daudin, 1802)

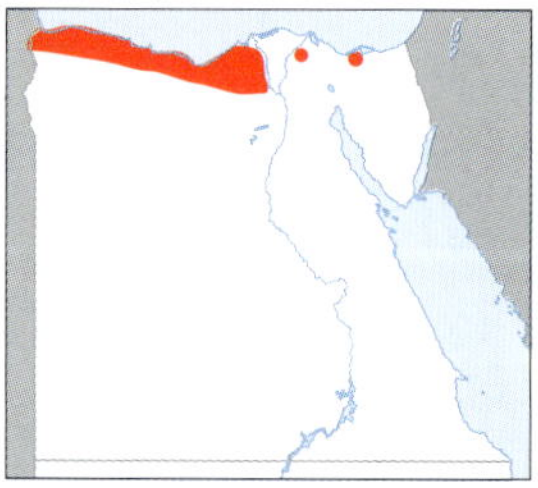

IDENTIFICATION: A big knobbly grey-and-black gecko of the north Egyptian coast and hinterland. Unregenerated tail broad and tapering, usually >50 per cent of total length and can be shed. Dorsal surface covered in regular bands of protruding tubercles, with multiple keels, surrounded by rosettes of 5–8 smaller scales in a horseshoe. Maximum size about 17cm, possibly larger, average 14–16cm, hatchling 5–6cm. Dorsal colour usually yellowish-brown or grey, with 4–5 dark irregular crossbars between the neck and tail. Tail banded light and dark; in occasional individuals the back looks almost uniform. Underside white. Limited ability to change colour, possibly for thermoregulation or camouflage. Taxonomic Notes: Egyptian animals elevated out of the synonymy of *T. mauritanica*, a species complex that occurs most of the way around the Mediterranean, although status of the various populations is not yet fully clarified. Two subspecies recognised, the nominate in our area.

HABITAT AND DISTRIBUTION: In sandy coastal desert at low elevations. In our area, known from the Egyptian Mediterranean coast, at the western edge of the Nile Delta (but doesn't reach Cairo), west to the Libyan border, usually within 80km of the coast. A few records in the Nile Delta and north-west Sinai. Elsewhere; west to Tunisia.

CONSERVATION STATUS: Not Assessed (as presently defined).

NATURAL HISTORY: Mostly on rocks, often on buildings and ruins, very occasionally trees, and sometimes on small isolated rock formations in the desert, indicating they must move across open country at night. Active by night in summer, crepuscular or diurnal in cooler months, and will bask near suitable refuges; often at high densities. Territorial, using clicks to establish their presence. Several clutches of two eggs are laid per year, in a suitable recess.

Tarentola fascicularis, Sinai.

GECKOS

FAMILY GEKKONIDAE

Geckos are small, plump, mostly nocturnal lizards, often grey, pinkish or brown, with large heads. Their eyes are large, as they hunt by sight in near darkness. Their eyelids are fused and immovable. Their eyes are covered by a transparent spectacle which they lick clean with their tongues. A number have vertical pupils that contract to a curious shape (usually three small holes) in bright light. They have soft, granular skin, often with scattered, enlarged scales known as tubercles. Their tails are easily shed, in some species just a single touch will cause the tail to fall off; it then twitches for some time. The regenerated tail may be quite a different shape and colour from the original, particularly if it is storing fat. Male geckos often have femoral and precloacal pores, the secretions from which may have a territorial or reproductive function. Unlike other lizards, some geckos can communicate by sound, a useful skill for a nocturnal animal. Nocturnal north-east African geckoes mostly just squeak, a sound likely to be heard when the animal is seized, but some Asian geckos utter a strident grunt, and some southern African species have a repertoire of clicks and barks. More than 1,400 gecko species are known worldwide, mostly in tropical and warmer temperate regions; around 63 are so far recorded in north-east Africa.

Geckos are superb climbers. The toes are usually dilated at the tips into pads, called scansors, on the lower surface of which is a mass of hair-like, often branched structures (setae). There is often a claw on the toe, which may be retractable. When the gecko is climbing, a combination of the setae tangling with minute irregularities, Van der Waals forces (tiny electrical forces) and a sharp claw allow it to ascend walls (and glass) and hang upside-down. Geckos often enter buildings and hunt on the walls and ceilings, catching insects attracted to lamps. Most eat arthropods exclusively, but a few are omnivorous, and dwarf geckos (*Lygodactylus*) eat arthropods, nectar and flowers. Some nocturnal geckos can change colour fairly rapidly, between light and dark. Most geckos lay two eggs, which are soft when laid, they stick to the surface (often a rock crevice or tree cavity) and harden, and are quite resistant to desiccation. Some are communal egg-layers, and agglomerations of many eggs may be found in suitable spots. Several clutches may be laid in a season, and females have curious neck pouches (endolymphatic sacs), where calcium for the egg-shells is stored.

Due to their habit of entering buildings, rock and timber piles, etc., they are often accidentally translocated; some species have spread far around the world in this fashion, and then spread beyond their point of introduction. There are some curious superstitions about geckos, for example they are thought to cause leprosy (a legend stemming from the way their skin sloughs in pale patches). And yet, geckos are beneficial animals, eating mosquitoes and other troublesome insects, for which they should be seen as the householder's friend.

KEY TO THE NORTH-EAST AFRICAN GECKO GENERA AND SOME SPECIES

1a:. Eye pupil vertical	2
1b: Eye pupil round	6
2a: Digits thin, not expanded at tips	3
2b: Digits expanded at tips	5
3a: Dorsal scales with clearly enlarged tubercles among smaller scales	*Cyrtopodion scabrum*, Keeled Rock Gecko
3b: Dorsal scales similar sizes	4
4a: Dorsal scales snake-like, imbricate (overlapping), limbs spindly, usually have flattened posture, head not noticeably deep, very small, rarely more than 6cm	*Tropiocolotes*, Pygmy geckos
4b: Dorsal scales tiny, circular, limbs stockier, with upright posture and a deep head, larger, adults 8+cm long	*Stenodactylus*, Desert clawed geckoes
5a: Claws retractable, back covered with smooth, non-overlapping scales of similar size, adults up to 16cm	*Homopholis fasciata*, Banded Velvet Gecko

5b. Claws not retractable, back usually with some keeled scales of varying sizes, adults of most species less than 14cm *Hemidactylus*, Half-toed geckos

6a: Five obvious thin toes, not expanded at tips and no digital pads *Ancylodactylus dickersonae*, Dickerson's Forest Gecko

6b: Four obvious large toes and one tiny toe, large toes dilated at tips and digital pads obvious *Lygodactylus*, dwarf geckos

CLAWED DESERT GECKOS *STENODACTYLUS*

A genus of c.10 species of slow-moving, terrestrial nocturnal geckos (none larger than 15cm), which occur in the flat and sandy deserts and near-deserts of the Middle East, Turkey, North and north-east Africa. Three species occur in our area, only one widespread. They have large heads and prominent eyes, with strange-shaped vertical pupils; body scales small and homogenous. Toes are finger-like, undilated.

KEY TO THE NORTH-EAST AFRICAN MEMBERS OF THE GENUS *STENODACTYLUS*

1a: Tail tapers abruptly, becoming very thin below cloacal tubercles, which are in several rows and very obvious, protruding at base of tail *Stenodactylus petrii*, Anderson's Short-fingered Gecko

1b: Tail not very thin, cloacal tubercles in single row and not protruding strongly 2

2a: Snout concave in profile, dorsal scales smooth and homogenous, adults up to 10cm, widespread in our area *Stenodactylus sthenodactylus*, Elegant Gecko

2b: Snout convex in profile, dorsal scales rough and heterogeneous, adults up to 8cm, in our area only northern Egypt *Stenodactylus mauritanicus*, Northern Elegant Gecko

ELEGANT GECKO

Stenodactylus sthenodactylus
(Lichtenstein, 1823)

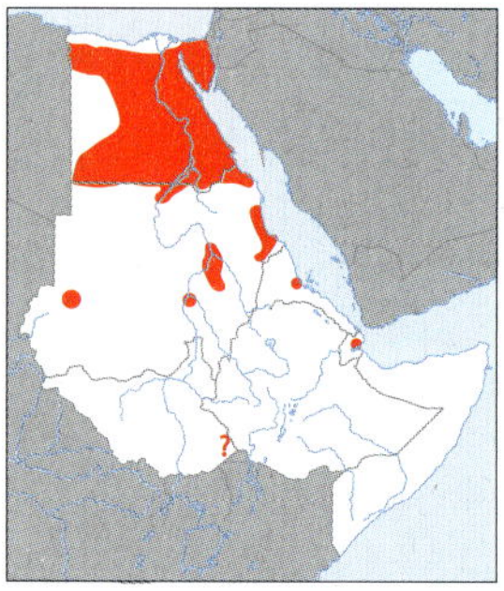

IDENTIFICATION: A small gecko with a big head and huge eye with a vertical pupil; iris often marked by a weird radiating pattern of fine lines. Body slightly depressed, limbs slender, with five thin clawed toes; toes have tiny lateral fringes. Tail not very thin, less than 50 per cent total length and can be shed. On either side of the tail base is a curious row of two to four white, tooth-like tubercles. Dorsal surface covered with small, uniform, juxtaposed circular scales. Maximum size about 11cm, average 7–9cm, hatchling size 5–6cm. Colour very variable, usually brown, grey, pinkish-grey or rufous, sometimes with dorsal crossbars and/or speckling, with a confused mixture of light ocelli and dark speckling, and a banded tail; immaculate white or cream below. Taxonomic Notes: Monotypic; a southern population described from north-west Kenya and adjacent areas as *S. s. zavattarii* is no longer regarded as valid.

HABITAT AND DISTRIBUTION: Gravel plains and small dunes in desert and semi-desert, mostly at low altitude, but up to 1,500m in parts of its range. In open desert, usually near wadis. Occurs almost

Stenodactylus sthenodactylus, Quseir.

Stenodactylus sthenodactylus, Quseir.

throughout Egypt, apart from Mediterranean coast west of the Nile Delta. Elsewhere, largely sporadic. Known from Obock in Djibouti, Massawa in Eritrea, sporadically along Sudan Red Sea Coast (Port Sudan, Durrur, Suakin, Tokar), other Sudanese records include Khartoum, Wadi Halfa, El Kadada, Wad Ben Nagar, Tabo, between Suakin and Kassala, Barbar, Merowe and El Fasher. Elsewhere; west to Mauritania, north and east to Jordan and Syria, and also occurs in north-west Kenya (and thus perhaps in south-east South Sudan).

CONSERVATION STATUS: Least Concern, although huge numbers are exported from Egypt for the pet trade.

NATURAL HISTORY: Terrestrial and nocturnal, emerges at dusk on warm nights and creeps around looking for food. In deserts of North Africa most active between April and September. Hides in holes and under ground cover. Captive males show territorial behaviour. It has a click-type call and if seized produces a range of calls of differing descending frequencies and duration, which may confuse predators. If threatened, may stand up high, twist its body into a circle, or run. Males approaching a possibly receptive female lift the tail and wave it horizontally. Tail also vigorously waved when threatened. One or two spherical eggs about 1cm diameter are laid, buried in sand. In North Africa the eggs hatch 65–80 days after laying. Diet insects, especially small desert beetles.

NORTHERN ELEGANT GECKO

Stenodactylus mauritanicus
Guichenot, 1850

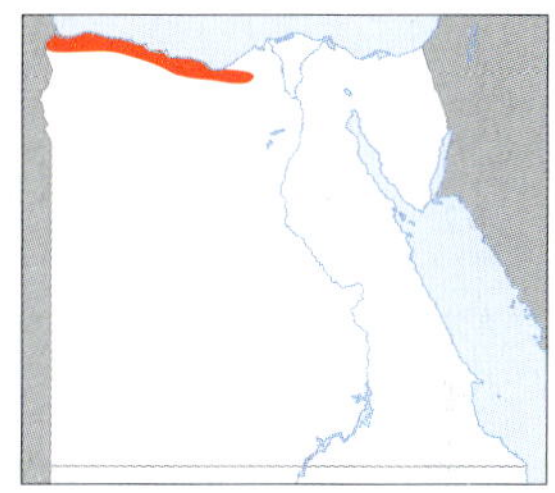

IDENTIFICATION: A small gecko with a big head and huge eye with a vertical pupil; iris often marked by a weird radiating pattern of fine lines. Body slightly depressed, limbs slender, with five thin clawed toes. Tail not very thin, less than 50 per cent total length and can be shed. On either side of the tail base is a curious row of two to four white, small tooth-like tubercles. Dorsal surface covered with small, uniform, juxtaposed circular scales. Maximum size about 8cm, average 6–7cm, hatchling unknown. Colour very variable, usually brown or grey, the surface with a mixture of large or small dark-edged blotches, ocelli, or sometimes sub-rectangular crossbars. Usually a brown line extends back from the eye and loops round the back of the occiput. Tail banded. White or cream below.

Stenodactylus mauritanicus, Tunisia.

HABITAT AND DISTRIBUTION: In Egypt on rock-strewn sand and gravel plains with some vegetation, at low altitudes. On Mediterranean coast west of Nile Delta to Libyan border, nearly always within 50km of coast. Elsewhere, west to Western Sahara and Morocco.

CONSERVATION STATUS: Not Evaluated.

NATURAL HISTORY: Very similar to previous species, i.e. terrestrial and nocturnal, hides in holes or under cover, mostly active in spring and summer, lays two eggs, diet small arthropods.

ANDERSON'S SHORT-FINGERED GECKO, PETRIE'S GECKO

Stenodactylus petrii
Anderson, 1896

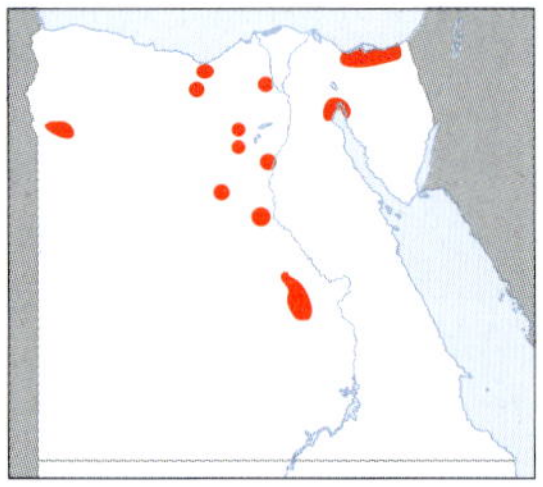

IDENTIFICATION: A small gecko with a big head and huge eye with a vertical pupil; iris often marked by a strange radiating pattern of fine lines. Body slightly depressed, limbs very slender, with five thin clawed toes. Tail less than 50 per cent of total length and tapers abruptly, becoming very thin. On either side of the tail base are several rows of prominent, white, tooth-like cloacal tubercles. Dorsal surface covered with small, smooth, uniform, juxtaposed circular scales. Maximum size about 12.5cm, average 8–11cm, hatchling 5cm. Colour variable; brown, grey, pinkish-grey or pinkish (especially in Egypt), the back with vague darker crossbars, light and dark mottling or ocellate spots, there is usually a broad diffuse brown bar on the flanks. Tail banded. Immaculate white or cream below.

HABITAT AND DISTRIBUTION: In true desert, near-desert and semi-desert, mostly at low altitudes but up to 1,500m in parts of range. Often in dunes, but also sandy plains and lightly vegetated wadis. In our area, sporadic in Egypt, in northern and eastern Sinai, along northern Nile Valley and oases in Western Desert. Seems absent from Eastern Desert. There are a few scattered older records from northern Sudan, but their identity has been questioned; pending re-evaluation or fresh records, it seems safest to regard the species as absent from Sudan. Elsewhere; west to Mauritania, largely on coast but also sporadically in Sahara. Just enters Israel/Palestine.

CONSERVATION STATUS: Least Concern.

NATURAL HISTORY: Terrestrial and nocturnal, emerges at dusk on warm nights and walks slowly, or waits in suitable spot, standing high, looking for food; known to wander 100m or more in a night. Sometimes locally common. Usually active March–September. Hides under large rocks or in burrows; reported to dig a new burrow daily, usually at base of a bush, and the tail (despite its narrowness) is coiled up to block the hole. Unlike other members of genus, it rarely sheds its tail. A fairly loud clicking call is made. Two spherical eggs are laid, from May, in captivity they hatched after 60–80 days. Diet arthropods, including scorpions; described as pausing, tail rapidly coiling and uncoiling, before leaping on prey.

Stenodactylus petrii, Al Arish.

Stenodactylus petrii, Tunisia.

AFRICAN FOREST GECKOS *ANCYLODACTYLUS*

A genus of poorly known, small slim forest geckos, they have large eyes with round pupils. Many have a chain of light, sub-triangular or rhomboidal blotches or chevrons on the spine; and several species (especially their juveniles) have a yellow underside. They do not have expanded digits and their thin elongate toes have sharp claws; rear toes often kinked. They climb superbly, living largely on trees and rocks, although they may descend to the ground. There is debate as to whether they are nocturnal, diurnal or crepuscular, although the yellow lenses of one species, *A. africana*, indicate it is diurnal. Most occur in forest and woodland. About 20 species occur in Africa, but there are probably a number of cryptic species within the group. One species occurs in north-east Africa.

DICKERSON'S FOREST GECKO

Ancylodactylus dickersonae
(Schmidt, 1919)

IDENTIFICATION: A relatively large slender forest gecko. The body is depressed; tail just over half total length. The scales are small, juxtaposed granules, interspersed by four or fewer rows of enlarged dorsolateral tubercles, those on flanks elongate and spiky; usually no tubercles on the tail. Toes long and slim. Males have 7–8, exceptionally nine precloacal pores. Maximum size about 9.1cm; average about 7cm; hatchling unknown. Colour very variable; in some the back is olive-grey, olive-green or brown, a series of light grey chevrons or rhombuses run from the back of the head to the tail; highlighted by surrounding darker brown or grey pigmentation. Others may be uniform pale or dark brown, and some have a fine or broad light vertebral stripe. Below, grey or yellow; throat with diffuse speckles, not a distinct pattern. Limbs are lightly banded darker brown or grey. Taxonomic Notes: As presently defined, this species is found in six East African countries. Its populations are largely isolated and will probably prove to be separate species.

HABITAT AND DISTRIBUTION: Mid-altitude hill and rainforest; in Ethiopia and South Sudan from 1,200m to over 2,300m; lower elsewhere. In our area, known only from the Imatong Mountains in South Sudan and at a handful of forested localities in south-west Ethiopia, in Jimma, Sheka, Kaffa and northern Bench Maji zones. Elsewhere, west to DR Congo, south to Uganda and Tanzania.

Ancylodactylus dickersonae, Illubabor.

Ancylodactylus dickersonae, Illubabor.

CONSERVATION STATUS: Not Evaluated.

NATURAL HISTORY: Poorly known. Secretive, diurnal and arboreal; waiting in ambush under edges of bark, in tree cracks, in deep shade between trunks, etc., dashing out for food. Will move onto buildings, especially wooden ones. They lay eggs, but no other breeding details known. Eat invertebrates. Often parasitised by tiny red mites.

VELVET GECKOS *HOMOPHOLIS*

A genus of medium to large arboreal and rock-dwelling geckoes, some nocturnal, others diurnal. Usually shades of grey, brown or black. Eye pupil vertical. They have soft, velvety skin, with small granular scales. The toes are dilated, with 8–12 V-shaped undivided scansors and a retractile claw. Tail thick and relatively short. Precloacal pores are present. Four species are known, three in southern Africa, one in our area. Most closely related to three genera not known from our region, i.e. *Afroedura* (Angola and Zimbabwe to South Africa), *Blaesodactylus* (Madagascar) and *Geckolepis* (Madagascar and Comoro Islands).

BANDED VELVET GECKO

Homopholis fasciata
(Boulenger, 1890)

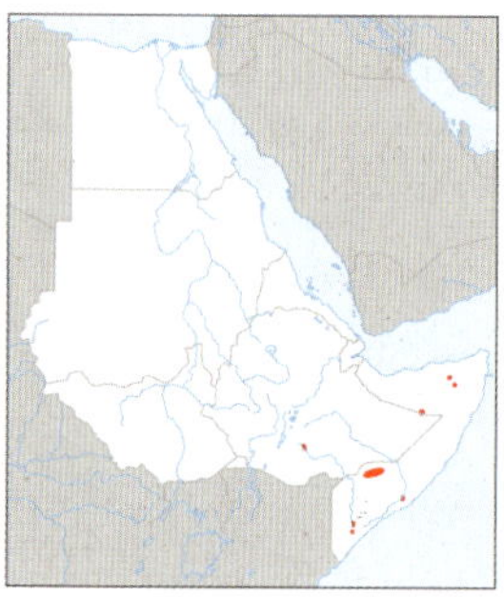

IDENTIFICATION: A fairly large, stocky gecko, head short and indistinct on a relatively long neck; snout short, wide and rounded; large pale eyes with vertical pupils. Toes broadly dilated, with big claws, though often concealed. Tail stout, cylindrical, not segmented and relatively short, about 34–41 per cent total length, it is easily shed. Scales moderate, smooth, roughly homogeneous, rounded and granular, not overlapping. Males have two precloacal pores. Maximum size about 16cm, average 9–13cm, hatchling 4.5cm. Colour grey, beige, olive or purplish-brown. A broad dark band forms a crescent on back of head, with 3–6 dark-edged, wavy brown crossbars on the back (in some cases, the light bands are very narrow), the tail has a series of 6–7 wide black, brown and grey crossbars. Lips and chin pale, often with fine brown reticulations. Underside whitish or yellow, sometimes marbled brown. Taxonomic Notes: Northern populations (from Juba River in southern Somalia north to Somaliland and Puntland) are regarded as a valid subspecies, *H. f. erlangeri*, by some authorities differing by more greyish coloration and a dorsal pattern consisting of crossbars with a pair of dorsolateral longitudinal stripes; there is no U-shaped crossband from eye to eye across nape and there are more than nine dark bands on the tail. Molecular analysis may clarify the situation.

HABITAT AND DISTRIBUTION: Dry and moist savanna with big trees, at elevations between 95 and 1,300m. All those in our area are *H. f. erlangeri*. Known from a handful of largely isolated records in our area; hampered by colonial names, they include a single imprecise southern Ethiopian record, the 'upper Ganale' (River), presumably north or north-east of Negele,

Homopholis fasciata, preserved, north of Gelib.

Homopholis fasciata, Tanzania.

'Goscia', a narrow band of riparian forest in southern Somalia along the Juba River between the sea and Bardera, 'Umfudu' (Munfuudhi), the type locality of *erlangeri*, also on the Juba, and Balad on the Shebelle River. Also in 'Rahanuin' country (the Rahanweyn are a people living largely between the Juba and Shebelle Rivers in southern Somalia), and the Haud in northern Somalia. The northernmost records, on the Somaliland/Puntland border, are Buran and an unnamed locality in Sol Haud, about 10km west of Samaysa Dheer village. Probably widespread in the intervening country, but under-collected due to its rarity and secretive habits. Elsewhere, the nominate subspecies occurs in Kenya and Tanzania.

CONSERVATION STATUS: Least Concern; not in the pet trade in our area.

NATURAL HISTORY: Poorly known and uncommon (but in Tanzania large numbers have been collected from dead branches of *Vachellia* trees, suggesting it is locally numerous). Lives mostly on trees, fond of baobabs, big fig trees and acacias, although sometimes on ground at night. Nocturnal, but might show some diurnal activity. Hides in holes and cracks in big old trees. Two hard-shelled eggs, roughly 1.5–2cm in diameter, are laid; several clutches per year. They have a powerful bite if handled, and eat insects and other arthropods. Adults produce distress calls.

BENT-TOED GECKOS *CYRTOPODION*

A Middle East genus of 24 species of small to medium-sized, nocturnal, ground or rock-dwelling geckos that favour arid environments. They have clawed, angular digits without adhesive pads. Head flattened, with prominent eyes with vertical pupils. The body scalation is heterogeneous, small scales intermixed with small or large trihedral keeled tubercles, with pre-anal pores in males. Various forms occur from Turkey across Arabia to north-west India and north Iran, with introduced populations in north-east and North Africa and USA. A single, probably introduced, species occurs in our area.

ROUGH BENT-TOED GECKO

Cyrtopodion scabrum
(Heyden, 1827)

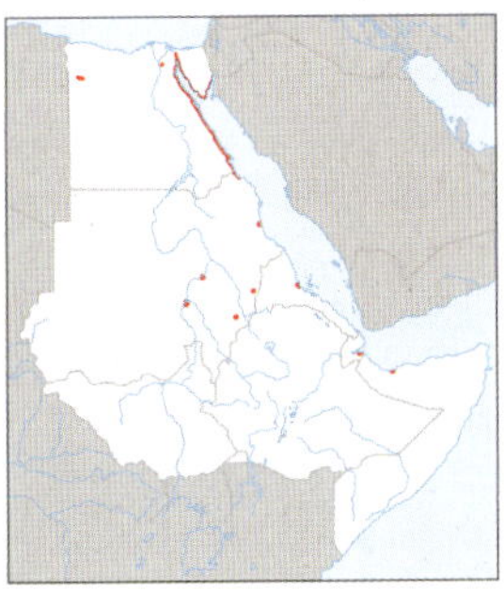

IDENTIFICATION: A terrestrial, medium-sized gecko (to 12cm) with claws but no pads, largely along the coast in our area. The neck is distinct, limbs long, digits long, slim and bent, with claws. Body slightly depressed; tail relatively slender, segmented, about 55–60 per cent of total length. Body scales heterogeneous, small flat scales intermixed with large, mostly trihedral, strongly keeled tubercles, in 12–16 more or less longitudinal series. Males have 4–7 pre-anal pores. Maximum size about 12cm average 8–10cm, hatchling about 3–3.5cm. Dorsum sandy-buff to pale grey, with conspicuous brownish markings that tend to form longitudinal rows of spots on back and 10–14 dark crossbars on tail. Regenerated tail uniform, greyish or pinkish-brown. Underside white.

HABITAT AND DISTRIBUTION: Deserts and semi-deserts, usually in rocky places, at low altitudes. In Africa there are few observations from unaltered habitats, it is a waif species and all populations are probably introduced. Often in urban habitats, on low external walls, occupied and abandoned houses, undersides of bridges, gardens and rubbish dumps.

Cyrtopodion scabrum, Quseir.

Cyrtopodion scabrum, Quseir.

Widespread in urban areas on Red Sea coast in Egypt, including Gulf of Suez and Aqaba in southern Sinai. Isolated records around Cairo and Siwa Oasis. Occurs along Sudanese Red Sea coast (an old record from Port Sudan); recent inland records include Khartoum, Shendi, Kassala and Gadarif. An 1827 coastal Eritrean record was probably from the vicinity of Massawa. Also recently reported from Djibouti City and Berbera; probably widespread along Red Sea coast and Gulf of Aden. As with other adaptable waif 'house' gecko species (e.g., *Hemidactylus robustus*, *H. flaviviridis*), its range can be expected to expand both on coasts and inland. Elsewhere, from southern Turkey and Arabia to north-west India and Afghanistan. Introduced to north-east Algeria and USA (Nevada, Texas).

CONSERVATION STATUS: Least Concern.

NATURAL HISTORY: Nocturnal, sometimes active before sunset. Mostly terrestrial but climbs well, using its claws. Usually solitary and highly territorial, but may aggregate in suitable places. Often synanthropic, seen hunting insects by 'sit-and-wait' methods on building walls around lights. In Suez, animals were mostly encountered on buildings and brick walls. A Quseir (Egypt) animal was hidden under a concrete block on the sandy bottom of man-made drainage canal in the middle of town. One or two hard-shelled eggs, roughly 7×10.5mm, are laid; a communal nesting site with eight eggs was found in Suez. Two Port Sudan females were found under boulders near two eggs, 10×8.5mm and 11×8mm, apparently belonging to this species. In Egypt reproductive activity starts early in spring (March) and continues in summer and autumn.

TROPICAL, HOUSE OR HALF-TOED GECKOS *HEMIDACTYLUS*

A genus of tropical or subtropical nocturnal geckos, widespread across the Americas, southern Europe, Asia and Africa, and the islands of the Pacific and Indian Oceans. With more than 190 species, it ranks among the top ten most species-rich genera of reptiles, it is the third-most speciose gecko genus globally, surpassed only by *Cyrtodactylus* and *Cnemaspis*. Adapted to semi-desert and dry savanna, *Hemidactylus* has radiated and the greatest species diversity occurs in our area, with more than 40 species; 15 are endemic. Some species (informally known as 'House' geckos) are commensal with humanity, and may be seen near lights at night. Most are excellent climbers, but several are strictly ground-dwelling, with often highly variable habits. They have broad toes with retractile claws and prominent scansors; usually in pairs. Some species can change colour to a limited extent. Many species can make sounds, clicks and grunts, and are territorial, living in structured colonies with a dominant male who keeps conspecific males away. They eat arthropods and usually lay two hard-shelled eggs. Taxonomy is not yet totally clarified; new species are constantly being discovered and named.

KEY TO THE NORTH-EAST AFRICAN MEMBERS OF THE GENUS *HEMIDACTYLUS*

1a: Tail more or less conspicuously constricted at base, depressed and often swollen or root-shaped — 2

1b: Tail lacks significant basal constriction, cylindrical or sometimes depressed, usually tapering gradually throughout its length — 9

2a: Back covered with more or less homogeneous scales — 3

2b: Back covered with heterogeneous scales — 4

3a: Back covered only with small, granular scales; large, maximum snout–vent length c.95mm — *Hemidactylus flaviviridis*, Yellow-bellied Gecko

3b: Back covered with slightly overlapping flat scales; small, maximum snout–vent length c.43mm — *Hemidactylus curlei*, Curlei's Gecko

4a: Back covered with a mix of granules and tubercles, latter only moderately large and either smooth or weakly keeled 5

4b: Back covered with granular scales among them rows of large, trihedral and strongly keeled tubercles 6

5a: Tail with uniformly small scales, males with two precloacal pores, endemic to Somaliland *Hemidactylus laevis*, Somaliland Mountain Gecko

5b: Tail with small scales and six large tubercles in each segment males with 12–27 preano-femoral pores, range Eritrea to Ethiopia *Hemidactylus laticaudatus*, Andersson's Gecko

6a: Original tail very short and wide, *c.*45 per cent of total length; divided by annular grooves into very narrow segments, covered by small scales with four slightly enlarged tubercles in each segment, endemic to small area around Gardo in Puntland *Hemidactylus taylori*, Taylor's Gecko

6b: Original tail short to moderately long and comparatively slim, slightly <50 per cent to *c.*53 per cent of total length; segmented to whorls with small scales and 6–10 large tubercles in each segment, from Sudan to Somalia (but not Gardo area in Puntland) 7

7a: Fewer than 15 precloacal pores *Hemidactylus jubensis*, Ethiopian Gecko

7b: More than 15 preano-femoral pores 8

8a: Upper surface of snout covered by large, strongly keeled tubercles; males with 28–36 preano-femoral pores in continuous series; regenerated tail grossly swollen and often leaf-shaped; maximum snout–vent length 57mm *Hemidactylus ruspolii*, Prince Ruspoli's Gecko

8b: Upper surface of snout covered by small granules (only rarely admixed keeled tubercles); males with a series of 20–46 preano-femoral pores interrupted mid-ventrally; regenerated tail no more than moderately swollen and root-shaped; maximum snout–vent length *c.*69mm *Hemidactylus angulatus*, East African House Gecko

9a: Innermost digit, particularly on front foot, very short, with short flat claw *Hemidactylus frenatus*, Pacific House Gecko

9b: Innermost digit normal in length compared to other digits, claw angled upwards 10

10a: Dorsal scalation of juxtaposed or weakly imbricating granules or tubercles 21

10b: Dorsal scalation of strongly imbricating scales 11

11a: Scales on back heterogeneous 12

11b: Scales on back homogeneous 14

12a: Both smaller and enlarged scales on back keeled or carinate, males with 6–10 preanal pores *Hemidactylus tropidolepis*, Ogaden Gecko

12b: Enlarged scales keeled, smaller scales usually smooth or rarely obtusely keeled, males with 6–20 preanal pores 13

13a: Head bulky, forehead flat or slightly convex; snout wide; upper nasals usually in contact; 6–8 upper labials; 27–35 scales between eyes; 24–30 gular scales in first row behind two pairs of chin shields; enlarged dorsal scales more or less rhomboid; not known from Sudan *Hemidactylus squamulatus*, Nyika Gecko

13b: Head long, narrow; forehead slightly concave; snout narrow; upper nasals not in contact; ten upper labials; about 18 scales between eyes; ten gular scales in first row behind two pairs of chin shields; enlarged dorsal scales more or less rounded; only Blue Nile Valley in Sudan *Hemidactylus floweri*, Flower's Sudan Gecko

14a: Dorsal scales keeled, only southern Somalia 15

14b: Dorsal scales smooth 16

15a: All scales, except subcaudals, strongly keeled; scales of back with 3–6 keels *Hemidactylus klauberi*, Klauber's Gecko

15b: Only scales on back keeled, with only one keel *Hemidactylus funaiolii*, Archer's Post Gecko

16a: Mid-body scale rows 50–59; scales on occiput and nape similar to those on back, smooth and imbricate 17

16b: Mid-body scale rows 59–102; scales on occiput and nape different from those on ack, granular and juxtaposed 18

17a: Males with eight preanal pores, body and tail usually with narrow white lines or stripes dorsally, range Ethiopia *Hemidactylus ophiolepis*, Snake-scaled Gecko

17b: Males with 22–25 preano-femoral pores, body and tail usually dorsally spotted or uniform; Somaliland *Hemidactylus ophiolepoides*, Somaliland Ground Gecko

18a: Hind part of thighs without enlarged tubercles 19

18b: Hind part of thighs with distinctly enlarged tubercles 20

19a: Larger size, maximum snout–vent length *c*.49mm; wide and higher bulky head; free distal joints of digits short; original tail slightly swollen behind cloaca; pale top of snout; northern to southern Somalia, through Ogaden *Hemidactylus isolepis*, Uniform-scaled Gecko

19b: Smaller size, maximum snout–vent length *c*.42mm; narrower, pointier and less bulky head; free distal joints of digits very short; original tail narrow and smoothly tapering; dark spot in middle of snout; south-west Ethiopia and adjacent Kenya *Hemidactylus lanzai*, Lanza's Gecko

20a: No enlarged tubercles on occiput, larger size, maximum snout–vent length *c*.61mm; eastern Ethiopia with Ogaden, Haud and adjacent Somaliland *Hemidactylus albopunctatus*, White-spotted Gecko

20b: Few enlarged tubercles scattered over occiput, smaller size, maximum snout–vent length *c*.48mm; Benadir coast of southern Somalia *Hemidactylus modestus*, Tana River Gecko

21a: Back with rows of greatly enlarged, more or less strongly keeled, sub- or trihedral tubercles 22

21b: Back with small uniform granules, sometimes with weakly enlarged tubercles that are no more than feebly keeled 36

22a: Free distal joints of digits distinctly very short, terrestrial 23

22b: Free distal joints of digits not very short, terrestrial or scansorial 24

23a: Upper nasals not in contact, fewer sub-digital lamellae, 4–6 beneath first toe, 7–11 beneath fourth toe, 4–6 precloacal pores; Djibouti to Somalia *Hemidactylus citernii*, Citerni's Gecko

23b: Upper nasals in contact, more sub-digital lamellae, 5–7 beneath first toe, 10–11 beneath fourth toe, 8–10 precloacal pores; Egypt to Eritrea *Hemidactylus foudaii*, Elba Gecko

24a: Back with vivid pattern of four black transverse bands between neck and tail base; 3–6 lamellae beneath first toe; 6–9 lamellae beneath fourth toe; maximum snout–vent length 40mm *Hemidactylus bavazzanoi*, Somali Banded Gecko

24b: No vivid pattern of dark transverse bands (or if such markings present then 7–10 lamellae beneath first toe; 9–14 lamellae beneath fourth toe and maximum snout–vent length 80mm or more) 25

25a: Proximal region of tail strongly depressed, outermost row of tubercles forming a distinct ventrolateral series; 6–8 lamellae beneath first toe; 9–11 lamellae beneath fourth toe; males with 6–11 preanal pores *Hemidactylus barodanus*, Enigmatic Gecko

25b: Proximal region of tail no more than moderately depressed, outermost row of tubercles not forming a ventrolateral series but lying close to lateral midline 26

26a: Large size, maximum snout–vent length 80mm or more; 7–10 lamellae beneath first toe; 9–14 lamellae beneath fourth toe; males with 6–13 preanal pores (in *H. macropholis*, adult males of *H. arnoldi* unknown); range Djibouti to Somalia 27

26b: Small to medium, maximum snout–vent length *c*.60mm; 5–8 lamellae beneath first toe; 8–13 lamellae beneath fourth toe; males with 3–11 preanal pores; Egypt to Somalia 28

27a: An enlarged scale, surrounded by granules of sole, situated between bases of first and fifth toes; prominent dorsal pattern of dark transverse bars, each widest at level of vertebral line; Djibouti and extreme north-west Somaliland *Hemidactylus arnoldi*, Arnold's Gecko

27b: No enlarged scale on sole of foot; no prominent dorsal pattern, except in some juveniles which have transverse bars widest at their lateral margins; not known from Djibouti and extreme north-west Somaliland *Hemidactylus macropholis*, Boulenger's Gecko

28a: Upper nasals in contact 29

28b: Upper nasals not in contact 30

29a: Subcaudal scales transversally enlarged; Afar Triangle in Ethiopia and adjacent Eritrea, northern Somaliland *Hemidactylus afarensis*, Afar Gecko

29b: Subcaudal scales uniformly small; south-east Egypt to central and northern Eritrea and Sudan *Hemidactylus sinaitus*, Speckled Gecko

30a: Larger size, maximum snout–vent length *c.*62mm; forehead distinctly concave; first upper labial widely excluded from nostril; back covered by distinctly large rounded, strongly keeled tubercles; northern and central Somalia *Hemidactylus granchii*, Granchi's Gecko

30b: Smaller size, maximum snout–vent length *c.*57mm (except *H. pauciporosus* 62mm); forehead flat or slightly concave; first upper labial usually in contact with nostril; back covered with round or oval comparatively smaller more or less keeled tubercles; outside Somalia, except *H. pauciporosus* and *H. robustus* 31

31a: Dorsal scales elongated, strongly keeled and arranged in very regular longitudinal rows, giving animal a striated appearance; lateral tail tubercles nail-like, distinctly protruding, stand out from dorsal tail outline; endemic to Awash region in central Ethiopia *Hemidactylus awashensis*, Awash Gecko

31b: Dorsal scales usually not much elongated, more or less keeled and arranged in less regular longitudinal rows, not giving animal a striated appearance, lateral tail tubercles not or only slightly protruding 32

32a: Head notably depressed and long; dorsal tubercles small, not strongly keeled; tail pattern strongly contrasting (black and white) in adults; high mountains of south Sinai, Egypt *Hemidactylus mindiae*, Mt Sinai Gecko

32b: Head not notably depressed and long; dorsal tubercles small to large, weakly to strongly keeled; tail pattern not strongly contrasting in adults; except *H. turcicus* not in high mountains of south Sinai, Egypt 33

33a: Tail almost smooth, with small dorsal tubercles restricted to few anterior segments; lateral tail tubercles not or only scarcely protrude from dorsal tail outline; dorsal body tubercles small, weakly keeled *Hemidactylus robustus*, Red Sea Gecko

33b: Tail with large tubercles extending almost entire length; lateral tubercles usually protruding from dorsal tail outline; dorsal body tubercles moderately large and usually distinctly keeled 34

34a: Larger, maximum snout–vent length *c.*62mm, 10–13 upper labials, 9–10 transverse rows of scales in each whorl at base of tail; northern Somaliland *Hemidactylus pauciporosus*, North Somali House Gecko

34b: Slightly smaller, maximum snout–vent length *c.*56mm, 7–11 upper labials, 7–8 transverse rows of scales in each whorl at base of tail; Egypt 35

35a: 9–11 upper labials, anterior shin shields in contact with second lower labial (in 100 per cent) and in wide mutual contact, 3–7 precloacal pores *Hemidactylus granosus*, Heyden's Gecko

35b: 7–10 upper labials, anterior shin shields usually not in contact with second lower labial (only in *c.*12 per cent) and in punctual mutual contact, 6–10 precloacal pores *Hemidactylus turcicus*, Turkish Gecko

36a: Dorsal scalation more or less homogeneous, consisting of small granular scales with or without a few that are only very slightly enlarged, especially on posterior half of back 37

36b: Dorsal scalation heterogeneous, both small granules and some clearly enlarged tubercles 39

37a: No enlarged scales at all *Hemidactylus megalops*, Parker's Gecko

37b: Few enlarged scales, especially on posterior half of back near groin 38

38a: Head long, narrow and flat; adhesive pads well developed; tail with enlarged tubercles; endemic to southern Somalia *Hemidactylus fragilis*, Rahanuin Gecko

38b: Head short, fairly high; adhesive pads moderately developed; tail without enlarged tubercles, endemic to northern Somalia *Hemidactylus somalicus*, Somali Gecko

39a: Ground-dwelling; adhesive pads under-developed; head oviform, short, relatively deep *Hemidactylus puccionii*, Puccioni's Gecko

39b: Usually arboreal; adhesive pads well developed; head oval, long and much depressed 40

40a: Large, maximum snout–vent length c.94mm; snout long; enlarged tubercles on back in 8–12 (usually ten) longitudinal rows; males with 45–57 precloacal-femoral pores in continuous row *Hemidactylus platycephalus*, Tree Gecko

40b: Smaller, maximum snout–vent length c.57mm; snout shorter; enlarged tubercles on back in 10–14 longitudinal rows; males with 28–33 precloacal-femoral pores in a sub- or continuous row 41

41a: Sub-digital lamellae extending onto sole of foot, 8–10 transverse rows of scales in each whorl at base of tail, 6–9 lamellae beneath fourth toe; only extreme south Somalia *Hemidactylus mabouia*, Tropical House Gecko

41b: Sub-digital lamellae extending onto sole of foot, 7–8 transverse rows of scales in each whorl at base of tail, 9–12 lamellae beneath fourth toe; not in extreme south Somalia *Hemidactylus smithi*, Smith's Tree Gecko

AFAR GECKO

Hemidactylus afarensis
Šmíd, Mazuch, Nováková, Modrý, Malonza, Abdirahman-Elmi, Carranza & Moravec, 2020

IDENTIFICATION: A small ground-dwelling gecko with heterogeneous dorsal scales and usually very contrasting dark and pale coloration, of the Afar Triangle and north-west Somaliland. Head oval and slightly depressed; snout moderately long. Body moderately long and stout, slightly depressed. Limbs somewhat short and slim; digits moderately long and slightly dilated. Tail not constricted at its base, weakly segmented, about 50–54 per cent of total length. The subcaudal scales are transversely enlarged. Dorsal scales heterogeneous, small, smooth or slightly keeled sub-imbricate scales interspersed by 11–14 rows of closely spaced, strongly keeled, trihedral, conical and more or less rounded fairly large tubercles. Enlarged slightly keeled tubercles on hind thighs; anterior chin shields large, in wide contact. Males have 4–6 precloacal pores. Maximum total length about 10cm (maximum snout–vent length 4.9cm), average 8–9cm. Dorsum light beige-brownish, with six wide crossbars (occasionally seven or exceptionally missing, sometimes interrupted); these are usually dark brown (rarely blackish or pale brown). Specimens from volcanic areas tend to be dark, those in sandy regions light. Tail with numerous (12–15) distinct dark brown or blackish bands, among which are narrow bands composed of a single row of yellowish tubercles. A dark stripe runs from posterior margin of eye to above ear. Limbs and top of head mottled dark brown and white or yellow. Iris sandy or brownish-grey. Regenerated tail yellowish, mottled dark brown, covered by larger and irregularly sized, keeled scales. Body pinkish-white below, tail underside yellowish, mottled brown. Taxonomic Notes: Recently described, originally lumped in Horn of Africa with *Hemidactylus sinaitus*. This taxon and the closely related and similar *H. citernii* and *H. foudaii* form part of a large African lineage of small terrestrial *Hemidactylus* with reduced adhesive pads on digits and often homogeneous dorsal scalation.

HABITAT AND DISTRIBUTION: On rocky (occasionally sandy) habitats in hot arid country, from

Hemidactylus afarensis, male, Awash.

Hemidactylus afarensis, gravid female, Geerissa.

Hemidactylus afarensis, gravid female, Geerissa.

near-desert to dry savanna, at elevations from sea level to about 1,400m. Most records in the inhospitably hot volcanic rubble and semi-desert of the Afar Triangle, from Awash National Park east into northern Somaliland, north towards Eritrea, sporadic records from latter (Keren, Assab); notable Ethiopian localities are near Berhale, Mile, Gewane, Metahara, and Dire Dawa; north-west Somaliland records at Geerissa, Artu and Berbera, in the Guban coastal strip. An old record from Las Korey in the Sanaag region of eastern Somaliland needs re-evaluation. Apparently absent from Djibouti, but records near the northern and southern border and suitable habitat indicate it probably occurs there.

CONSERVATION STATUS: Least Concern. In Ethiopia widespread and common.

NATURAL HISTORY: Largely terrestrial and nocturnal, but climbs low rocks and walls while hunting prey at night. Large numbers were found at night on cracked and fissured lava boulders in grassland with scattered *Vachellia* trees in Awash National Park; individuals of *H. awashensis* and *Tarentola annularis* were also present. Shelters by day under lava boulders. Specimens from Geerissa in Somaliland were collected in dry savanna with sandy soil, devoid of stones or rocks. In Gewane this gecko was found on the same walls of buildings as other geckos typical for this dry and hot area, *H.* aff. *awashensis*, *H. flaviviridis* and *Tarentola annularis*, the latter two were observed here preying on Afar Gecko. Lays hard-shelled eggs.

WHITE-SPOTTED GECKO

Hemidactylus albopunctatus
(Loveridge, 1947)

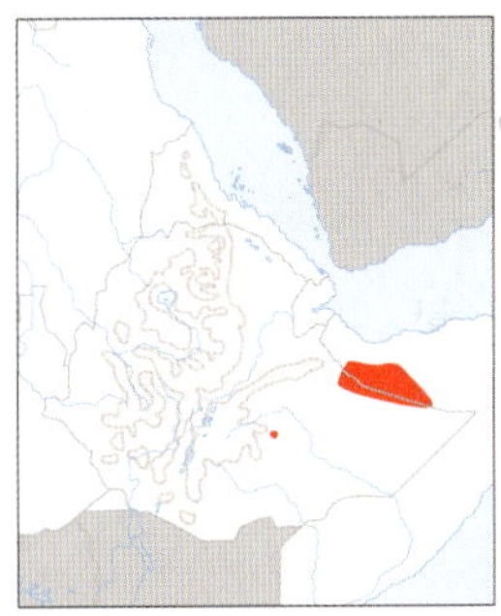

IDENTIFICATION: An attractive, chunky, little, purple-brown, white-blotched, deep-headed gecko of the Haud region and northern Ogaden. Head oval, fairly narrow; eyes medium-sized, iris dark; ear-opening is small. The limbs are stout and short. Digits short, stout, equally wide along entire length; adhesive pads very small, mostly divided. Tail not segmented, cylindrical, short, about 47–50 per cent of total length, tapering smoothly, subcaudal scales might be very slightly enlarged. Unusually, body and tail covered by small, overlapping scales of roughly equal size, arranged in 76–102 rows around mid-body. Head scales smaller without enlarged flat tubercles on occiput. Male has 12–19 precloacal pores. Maximum length about 10.2cm, (maximum snout–vent length 6.1cm),

Hemidactylus albopunctatus, female, Ogaden.

Hemidactylus albopunctatus, female, Burao.

average 8–9cm. Coloration distinctive, dorsum is brownish-grey to purple-brown with series of distinct, round or transversely expanded white blotches along vertebral midline (about five to seven on body), a black streak from nostril passes through eye and above ear-opening, and is bordered below by a white line, on the occiput is a whitish V-shaped marking. Underside white. Regenerated tail swollen, shorter, and without pale markings. Juveniles bicoloured, head, body and fore limbs black, hind limbs and tail pale greyish or orange brown, pale transverse markings preserved. Taxonomic Notes: Closely related to *H. modestus* with which it shares most characters.

HABITAT AND DISTRIBUTION: A typical species of flat arid savanna and bushland, at 600–1,750m. Favours inland sandy areas with patches of grass, scrubs or other succulent vegetation, especially aloes. Not in rocky areas. Most records on red sandy soil; known from several localities in the Haud along border between Ethiopia and Somaliland, from Bohodle to about 30km south-east of Salahle. Recently reported further north in the Haud at Shanshacade and known from around Hargeisa and Burao. Two Ethiopian records: Kebri Beyah in northern Ogaden and about 10km north-east of Sof Omar, further west. Probably much more widespread, and likely to occur further south in red sand areas of the Ogaden.

CONSERVATION STATUS: Least Concern.

NATURAL HISTORY: Nocturnal and terrestrial; active on ground at night; by day shelters in holes or under ground cover. A specimen found by day near Hargeisa was under a stone on a sandy plain with scattered *Vachellia* trees, shrubs and dense *Aloe megalacantha* thickets; another was under a stone at edge of a field. At Shanshacade, adults and juveniles were active at night on sandy ground with small shrubs, along with *H. tropidolepis,* which has similar habitat requirements. Two hard-shelled eggs are laid several times per year, during wet season.

Hemidactylus albopunctatus, female, Shanshacade.

Hemidactylus albopunctatus, juvenile, Shanshacade.

EAST AFRICAN HOUSE GECKO

Hemidactylus angulatus
Hallowell, 1854

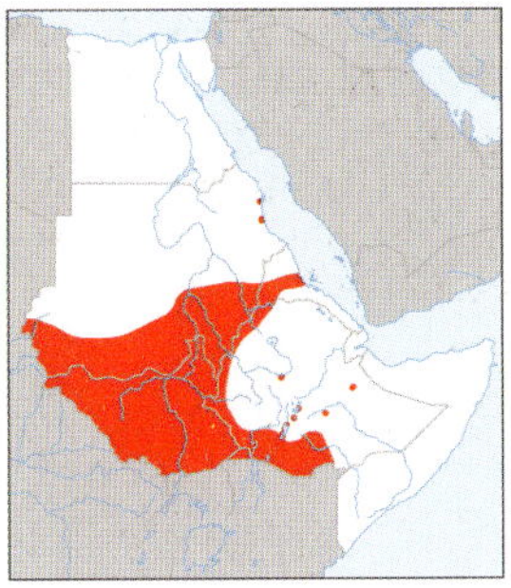

IDENTIFICATION: A fairly large gecko with a depressed, stocky body and many tubercles, often common and known from Sudan to southern Ethiopia. Head sub-triangular, snout fairly long, eyes large, prominent; ear-opening fairly large, elliptic. Limbs short, rather slim; digits long, moderately dilated; adhesive pads distally well developed, proximally much-reduced. Tail segmented, the swollen basal portion (especially in adults) marked by a faint constriction, moderately long, about 50–53 per cent of total length. Subcaudal scales transversely enlarged. Regenerated tail shorter and wider, carrot-shaped, depressed and significantly swollen; covered by small, uniform scales, pale brown and mottled dark brown. Dorsal scales heterogeneous; small granular scales interspersed by 14–25 rows of large, strongly keeled, sub-trihedral and mostly oval tubercles; enlarged tubercles on hind part of thighs. Anterior chin shields large, in wide contact. Males have a series of 20–46 precloacal pores, interrupted in the midline. Maximum total length about 15cm (maximum snout–vent length 6.9cm), average 10–12cm. Ground colour variable, may be grey-brown, red-brown or orange, pink, sometimes even bright red, with dark brown X-shaped blotches on back and neck that are usually arranged in four transverse rows; tail spotted with additional 8–11 dark bands; a distinct dark stripe from nostril across eye to above the ear-opening; top of head and limbs mottled dark brown, about half of tubercles on limbs, tail and especially flanks and neck are white. Underside white. Juveniles similar, but white tubercles on body more pronounced and tail is brightly orangey and more clearly banded. Taxonomic Notes: Formerly known in north-east Africa as Brook's Gecko *H. brooki*; that name is now restricted

Hemidactylus angulatus, male, south-west Ethiopia.

Hemidactylus angulatus, Negele.

Hemidactylus angulatus, Metemma.

to Asian specimens. The variation in colour and morphology, coupled with molecular analysis, indicate that this species, as presently defined, represents a complex of several cryptic undescribed forms, closely related to *H. ruspolii*.

HABITAT AND DISTRIBUTION: Moist and dry savanna, woodland, occasionally semi-desert but not desert. Occurs from sea level to about 2,000m. In our area widespread and common in southern half of Sudan, South Sudan, central Eritrea and southern Ethiopia. Northernmost records within continuous range are from Omdurman (Khartoum) and Kassala in Sudan, Keren in Eritrea, easternmost records from neighbourhood of Malka Mari and Sheik Mahomed, and between them in southern Ethiopia. A rather dubious record from Gildessa north-east of Dire Dawa might be due to confusion with *H. barodanus*. Two records from Port Sudan and Muhammad Qol on Sudan's Red Sea coast. Known from Eritrea, the Danakil region and most of the Ogaden in Ethiopia, but not Djibouti or Somalia. Quite widespread in southern Ethiopia (but not in montane or forested areas). Widespread, often locally common; the most common *H.* in low southern Ethiopia. Elsewhere, west to Senegal, south to north-east Tanzania.

CONSERVATION STATUS: Least Concern.

NATURAL HISTORY: Nocturnal, equally at home on ground, trees, rocks and buildings; often on walls at night hunting insects near lamps. One of a group loosely called 'house geckoes', with huge distributions due to dispersal by humanity and adaptability to urbanisation. Other such species in north-east Africa include *H. flaviviridis*, *H. frenatus*, *H. mabouia*, *H. robustus* and *Cyrtopodion scabrum*. In natural settings it shelters under ground cover, in abandoned termitaria, under bark and in cracks in trees and rocks. Lays two hard-shelled eggs, roughly 10–11mm in diameter.

ARNOLD'S GECKO

Hemidactylus arnoldi
Lanza, 1978

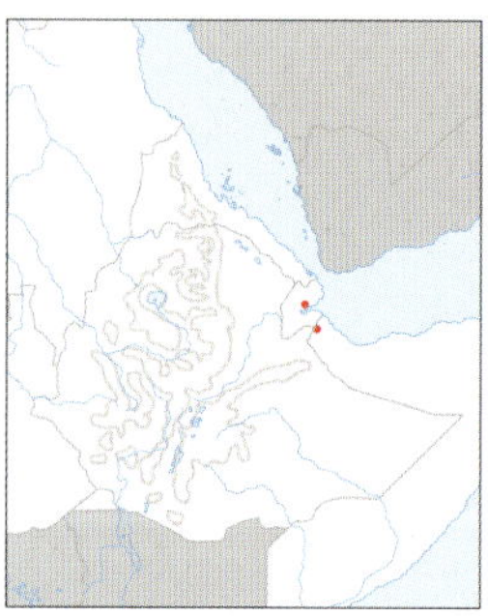

IDENTIFICATION: A fairly large, stocky gecko, distinctly banded, endemic to basaltic outcrops in Djibouti and adjacent Somaliland. Head big, broad and triangular, eyes large, prominent; ear-opening fairly large, elliptic. Body stout, sub-cylindrical, slightly depressed. Limbs short, adhesive pads well developed.

Hemidactylus arnoldi, Djibouti.

Uniquely among *Hemidactylus*, it has a conspicuously large rounded scale between small granules on sole, between bases of inner and outer toes. Tail segmented, tapering, not constricted at its base, subcaudal scales strongly enlarged, about 43–48 per cent of total length. Dorsal scales heterogeneous; small granular scales interspersed by 20 or more longitudinal rows of strongly keeled, sub-trihedral and mostly rounded or slightly oval tubercles, enlarged tubercles on hind thighs. No precloacal pores present in the only known two young males. Maximum total length 15.7cm (maximum snout–vent length 8.2cm). Coloration very distinctive, dorsum grey-brown, vividly marked by a series of broad, dark brown transverse bands with wavy blackish edges, one on neck and three on back, widest at level of the vertebral midline, with another 8–9 distinct crossbars on tail; head mottled darker greyish-brown, an irregular wide dark stripe runs from eye to above ear-opening, three dark brown spots on nape; iris light brownish-grey. Underside greyish-white, mottled grey on throat and tail, and a little less also on body and limbs.

HABITAT AND DISTRIBUTION: Near-desert volcanic areas in a harsh, hot and arid area, at 400–820m. Known from two locations. Originally described from the Guban in north-west Somaliland, very close (3.5km) to border with Djibouti and Ethiopia. Recently discovered near Dittilou village near the Day Forest National Park in Djibouti. Probably more widespread in Djibouti, and possibly adjacent Ethiopia.

CONSERVATION STATUS: Data Deficient, but probably under no threat; it occurs mostly in inaccessible areas with low human activity; in Djibouti it probably occurs in the protected Day Forest National Park.

NATURAL HISTORY: Virtually unknown, only four specimens found. The two types (an adult female and young male) were collected in a dry, stony region completely devoid of vegetation. Probably rupicolous. Biology presumably similar to other *Hemidactylus* of dry country, i.e. nocturnal, shelters under ground cover (stones or rocks) or in walls near human settlements; lays two eggs.

AWASH GECKO

Hemidactylus awashensis
Šmíd, Moravec, Kratochvíl, Nasher, Mazuch, Gvoždík & Carranza, 2015

IDENTIFICATION: A medium-sized brownish-grey gecko with strongly heterogeneous dorsal scales, endemic to area around Awash National Park in Ethiopia. Head pointed, relatively wide, eyes large, very prominent; ear-opening moderately large, crescent-shaped. The body is stout, slightly depressed. The tail is weakly segmented, fairly long, tapering smoothly, slightly depressed, especially at the base, with an indistinct median dorsal furrow, not constricted at its base, about 53–57 per cent of total length. Subcaudal scales transversally enlarged. Dorsal scales heterogeneous, small granular scales interspersed by 14 regular rows of strongly keeled, large, prominent, trihedral tubercles. These give the animal a striated appearance, an excellent aid to identification. Enlarged conical tubercles on back of thighs. Anterior chin shields large, in broad contact. Males have 4–5 precloacal pores. Maximum total length about 12cm (maximum snout–vent length 5.5cm), average 9–11cm, females smaller. Dorsum beige-brownish to grey, with four vertebral dark elongated blotches, which sometimes extend transversely into X-shapes. Dorsal tubercles usually dark, those on flanks pale; tail with about 12 dark brown and light bands, which increase in contrast distally, being blackish and white along its terminal portion, more pronounced in juveniles. A dark line runs from snout through eye. The back of the head has

Hemidactylus awashensis, male, Metahara.

Hemidactylus awashensis, juvenile, Metahara.

Hemidactylus awashensis, male, Metahara.

dark spots, labials whitish. Iris yellowish or greyish-brown. Pinkish-white below. Taxonomic Notes: Recently described, conflated in past with the unrelated *H. macropholis*. A member of the '*H. robustus*' group. Two similar-looking taxa from Gewane and Asaiyta (both in Afar region) are genetically distinct and in the process of being described.

HABITAT AND DISTRIBUTION: Dry open savanna with lava fields and outcrops, at about 1,000m. Endemic to Ethiopia, known only from Metahara town and Awash National Park, both south of the road to the Awash River and north to Fantalle Volcano. A similar form in Eritrea (Fil-fil and Gaarre) is probably an undescribed species.

CONSERVATION STATUS: Not Assessed, but fairly common.

NATURAL HISTORY: Strictly nocturnal and largely terrestrial, shy and fast-moving. At the type locality, Metahara town, and Kereyu Lodge in Awash National Park, they were on the ground, on low rocks and walls around lights, hunting insects, accompanied by the more dominant *H. flaviviridis*. In natural settings they are active at night on lava flows, or isolated volcanic rocks in grassland in *Vachellia tortilis* savanna; by day they shelter in fissures and caves, often in the company of the syntopic *H. afarensis* and *Tarentola annularis*. Many larger individuals have regenerated tails, suggesting they rely heavily on tail autotomy (losing their tails) as an escape mechanism.

ENIGMATIC GECKO

Hemidactylus barodanus
Boulenger, 1901

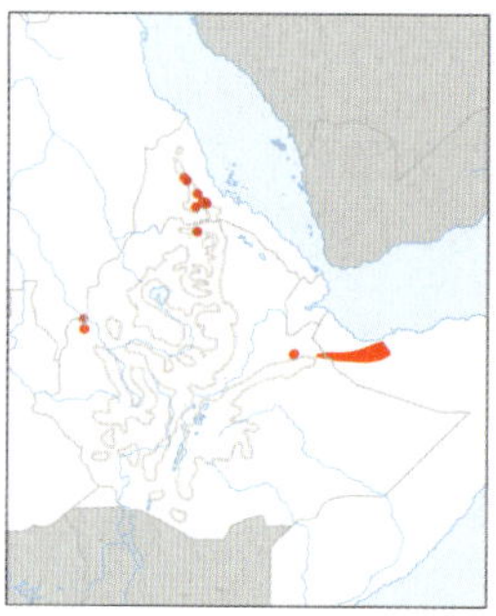

IDENTIFICATION: A fairly large stout rupicolous gecko with a distinct carrot-shaped tail with serrated edges. Head wide, eyes large, less prominent; ear-opening large, oval. Body fairly short, stout, depressed. Limbs short and robust; adhesive pads well developed. Tail segmented, carrot-shaped, the swollen basal portion marked by a faint constriction, distinctly flattened, with a slight median dorsal furrow, moderately long, about 52–54 per cent of total length. Subcaudal scales transversally enlarged. Regenerated tail similar in shape, but shorter and covered by uniform small scales. Dorsal scales heterogeneous, small granular scales interspersed by 12–16 rows of larger, obtusely keeled or sub-trihedral, usually oval-shaped tubercles. Anterior chin shields large, in broad contact. Male has 6–11 precloacal pores. Maximum total length about 16cm (maximum snout–vent length

Hemidactylus barodanus, Mendefera.

Hemidactylus barodanus, Shiikh Abdaal, north-east of Hargeisa.

Hemidactylus barodanus, Mendefera.

Hemidactylus barodanus, Las Geel.

7.8cm), average 13–14cm. Colour variable; dorsum brownish-grey to creamy-yellow, mottled darker brown on head and tail. On neck and back are five dark brown, broad, transverse X-shaped spots, interspersed by a single longitudinal series of pale vertebral pale spots that bifurcate on their edges (an excellent aid to identification). A dark stripe runs from the nostril across the eye to above the ear-opening; iris yellowish-brown. Underside whitish. Juveniles similar but the posterior 50 per cent of the tail is white with dark bands, and there are small white spots on back. Taxonomic Notes: Genetic analysis of this very varied species indicates that, as presently known, it is a complex of several cryptic and undescribed species. True *H. barodanus* is known with certainty only from Gaan Libaah, Somaliland, the type locality.

HABITAT AND DISTRIBUTION: From dry low to medium and high-altitude savanna, between 700 and 2 300m, usually on rocks in evergreen thicket and scrub. Usually at higher elevations. Occurs in three widely separated regions. In Somaliland known from Goolis Mountains and Gaan Libaah west across Ogo Mountains to environs of Borama, thence along border north-west to around Dire Dawa in eastern Ethiopia (localities include Sheikh, Mandheera and Las Geel). Also northern Ethiopia (Adua) and widespread in mountains of Eritrea, including Mendefera, Dekemhare, Saganeiti, Asmara, Elabered and Keren. Finally, in border country of western Ethiopia and eastern Sudan, south of Abbasi River, between Quezan and Belfodio, and near Shimaila River. Whether these populations are connected is unknown; widespread (as presently defined, see above); common in Somaliland and mountains south of Asmara in Eritrea.

CONSERVATION STATUS: Least Concern.

NATURAL HISTORY: A fairly big rupicolous species, often on large quartz and granitic boulders. Usually nocturnal but sometimes active by day in deep shade. Occasionally on walls of abandoned buildings or concrete bridges near settlements; juveniles shelter under rocks on ground. Often sympatric with other larger species of *Hemidactylus*, e.g. *H. laticaudatus* and *H. pauciporosus*. Two hard-shelled eggs, roughly 10–12mm diameter, are glued in rock crevices; communal clutches known from suitable places. Diet invertebrates.

SOMALI BANDED GECKO

Hemidactylus bavazzanoi
Lanza, 1978

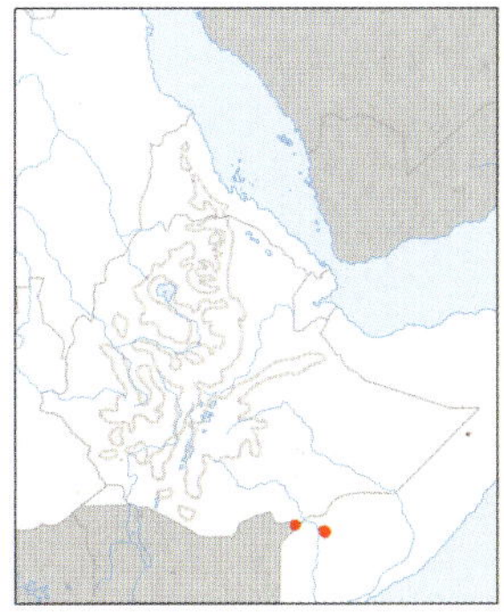

IDENTIFICATION: A small, vividly coloured terrestrial stocky gecko, from the upper Juba Valley in southern Somalia. Head oval; eyes fairly large, prominent; ear-opening small, oval. Body cylindrical, slightly depressed, rather slim. Tail segmented, tapering and not constricted at its base, covered with small smooth overlapping scales; original tail length unknown. Dorsal scales heterogeneous, small smooth scales interspersed by 16–18 rows of strongly keeled mostly oval tubercles. Enlarged tubercles on hind thighs. Anterior chin shields long and in broad contact. Both males have seven precloacal pores. Maximum total length about 8cm (maximum snout–vent length 4cm). Back orange or pinkish-brown, top of head russet, with a dark brown or blackish crescent-shaped neck bar that extends through eyes to snout, two wide dark blackish crossbars on body and several on tail; iris bicoloured, upper half yellowish-grey, lower blackish-brown. Regenerated tail not segmented, covered by small smooth scales, dark above, light grey below. Underside pinkish-white.

HABITAT AND DISTRIBUTION: Known only from two male specimens; one found 20km south-east of Lugh, towards Baidoa, on the Juba River in southern Somalia, the other from the banks of the Daua River, Mandera, in Kenya; both around 230m altitude. Both localities are in river valleys with *Vachellia–Commiphora* bushland, so the species might be associated with riparian gallery forest. Its broad habitat is semi-desert; unlikely to be developed, but riverine valleys are potential sites for large-scale agriculture.

CONSERVATION STATUS: Data Deficient.

NATURAL HISTORY: One of the rarest and least known geckoes in our area. The Kenyan specimen (collected in 1971) was under a palm log in riverine forest in the morning, the type specimen from Somalia (found in 1977, at night) was on a low bush from which it let itself fall to the ground, here it attempted to escape with rather slow movements. Biology presumably similar to other ground-dwelling *Hemidactylus* of dry country, i.e. nocturnal, shelters under ground cover or in holes, lays hard-shelled eggs and eats invertebrates.

Hemidactylus bavazzanoi, Kenya.

CITERNI'S GECKO

Hemidactylus citernii
Boulenger, 1912

IDENTIFICATION: A small ground-dwelling gecko with large dorsal tubercles, very short distal joints on fingers and distinct coloration, from Djibouti and much of Somalia. Head oval, eyes very large and prominent; ear-opening fairly small and round. Body stout to slim, sub-circular, slightly depressed. Free distal joints of digits very short (an excellent aid to identification). Tail segmented, slim, tapering, cylindrical, not constricted at its base; about 52–54 per cent of total length. Subcaudal scales transversely enlarged. Dorsal scales heterogeneous, small, smooth, rugose or striated granular scales interspersed by 14–16 rows of strongly keeled, sub-trihedral and oval-shaped tubercles, which are fairly large (but distinctly smaller in near-desertic coastal populations). Anterior chin shields large and in broad contact. Males have 4–6 precloacal pores. Maximum total length about 8.6cm (maximum snout–vent length 4.2cm), average 6–8cm. Dorsum sandy, pale yellowish-brown or grey, distinctly marked with about eight dark brown crossbars, each composed of three blotches, giving appearance of three longitudinal dashed stripes; this pattern is more pronounced anteriorly (and persists in preserved animals) but fades towards and onto tail base, although the tip is banded light and dark. The enlarged tubercles are yellow; a dark stripe from nostril through eye becomes a row of lateral spots. Some specimens have a dark arrow-shaped mark on snout. Iris yellow-brown. Underside pinkish-white. Juveniles similar, but tubercles more intense yellow. Taxonomic Notes: Ongoing genetic analysis of this very varied species indicates that, as presently known, it is a complex of several cryptic and as yet undescribed species. The molecular evidence also places it (along with *H. afarensis*, *H. foudaii*, *H. sinaitus*) in a clade of African terrestrial species

Hemidactylus citernii, Bawn.

Hemidactylus citernii, Maydh.

Hemidactylus citernii, Las Geel.

with greatly reduced adhesive pads and lacking enlarged dorsal tubercles, although it does not look like them.

HABITAT AND DISTRIBUTION: Semi-desert and dry savanna, from sea level to about 1,800m. Usually avoids sandy habitats. Widespread throughout Somaliland, western Puntland, and southern Somalia. A single Djibouti record (5km west of Obock); probably more widespread and overlooked there. In Somaliland known from the stony plains of the coastal Guban strip, widespread in dry stony and rocky terrain of Ogo, Goolis, Wagar and Cal Madow Mountains, south to the northern edge of the sandy Haud region (including Las Geel, Berbera, Hudisso, Yufle, Maydh, Buran and Halin). Not at high altitude. In Puntland, Candala and the Bosaso plain south along the border with Somaliland to the southern edges of the Nogal Valley, from west of Jurile, south and north of Xidda, Wadi Run and Garowe. Not known from central Somalia, but possibly due to under-collecting. Originally described from 'Rahanuin' country in southern Somalia, but few subsequent records, save Bud Bud and Jesomma. Historical records from Kenya and Tanzania are misidentifications.

CONSERVATION STATUS: Least Concern; widespread, in much of Somalia one of the most common *Hemidactylus*.

NATURAL HISTORY: Nocturnal and terrestrial. Favours dry stony terrain, hunts arthropods at night. Berbera specimens were found hunting low down on *Balanites* tree trunks. By day shelters under rocks, in holes and occasionally under the bark of dead trees. A Djibouti specimen was found under old clothing on a fossil reef, near the sea. When threatened they may raise their tail, and move slowly or in little jumps. Not usually in urban areas, and solitary. Sympatric with the terrestrial species *H. somalicus*, *H. megalops* and *H. robustus*. One or two small eggs, roughly 6–7mm in diameter, are laid.

CURLEI'S GECKO

Hemidactylus curlei
Parker, 1942

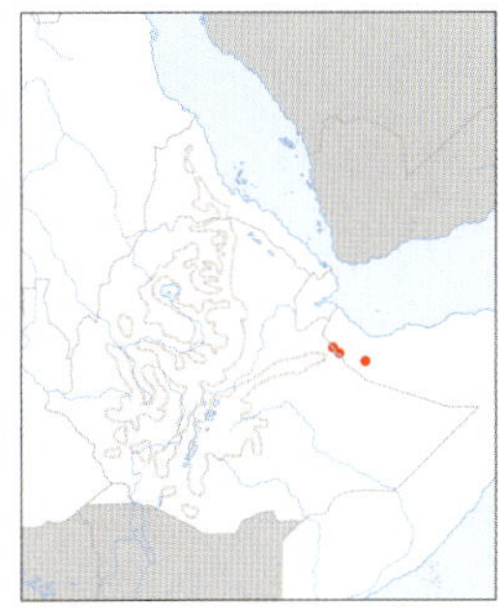

IDENTIFICATION: A small yellowish-brown flat-looking gecko with a carrot-shaped tail, endemic to western Somaliland. Head wide and long, eyes moderately large, not very prominent; ear-opening small, oval. The body is very depressed and stocky. Limbs short and stout; digits fairly short. This species is known from only eight specimens, all with missing or regenerated tails, with faded patterns, unsegmented and without enlarged subcaudal scales. Dorsal scales homogeneous, flat, more or less circular and slightly overlapping. Anterior chin shields in broad contact. Males have four pre-cloacal pores and inconspicuous hemipenial pockets. Maximum size about 9cm (maximum snout–vent length 4.3cm). Dorsum yellowish and back heavily marked with purplish-black spots and blotches, sometimes forming irregular transverse bars. A dark streak from nostril through eye and above the ear continues along necks and flanks as a row of spots to base of tail; flanks and limbs with scattered, circular, white spots; tail banded light and dark. Underside white. Taxonomic Notes: No molecular data, so its relationships are unknown.

HABITAT AND DISTRIBUTION: A poorly known endemic of Somaliland and adjacent Ethiopia. Inhabits dry hilly savanna on limestone or granite, with sandy or loamy soil at altitudes of 1,300–1,600m. Known from three localities; two in Somaliland, just west of Hargeisa, and Borama; the third in Ethiopia but only a few metres from the Somaliland border, between

Quljeet and Alaua villages. No recent records, but probably occurs in parts of Ogo Mountains.

CONSERVATION STATUS: Data Deficient.

NATURAL HISTORY: Virtually nothing known. All specimens were collected on stony ground, so presumably terrestrial. They must be very rare and secretive; one of the authors has looked for them in all known localities, several times in different seasons of the year, by day and night, over ten years, without success. The very depressed head and body, and relatively well-developed adhesive pads suggest that *H. curlei* lives on and hides in fissures of granite rocks characterising its habitat, but this needs confirmation. Biology presumably similar to other *Hemidactylus*, i.e. nocturnal, lays eggs, eats invertebrates.

Hemidactylus curlei, preserved.

YELLOW-BELLIED GECKO

Hemidactylus flaviviridis
Rüppell, 1835

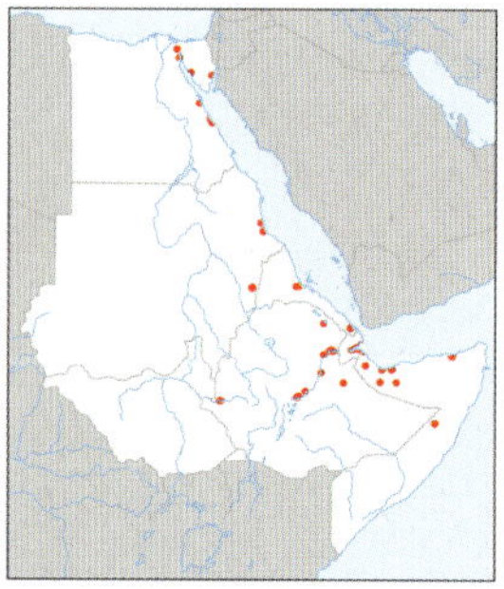

IDENTIFICATION: A large stocky yellowish or grey gecko with homogeneous dorsal scales and carrot-shaped tail, an urban invader along our coastline. Head oval; eyes large, prominent; ear-opening moderate, triangular. Body stout and depressed. Limbs fairly short and thick. Tail very distinctive, strongly segmented, usually quite swollen and carrot-shaped, more or less constricted at its base, about 50–55 per cent of total length. Subcaudal scales transversely enlarged. Regenerated tail not segmented and covered by uniformly small scales. Dorsal scales homogeneous, back covered by small, granular, smooth scales. Anterior chin shields in broad contact. Male has 4–14 pre-cloacal pores and conspicuous hemipenial pockets. Maximum known size about 19cm (maximum snout-vent length 9.5cm), average 12–17cm, hatchlings about 5cm. Back grey, pinkish-brown, pale yellow or yellowish-green, unmarked or with rather feeble dark crossbars (five on back) that have wavy margins and whitish posterior edges. Iris yellow or brown-grey. Underside pale to bright yellow. Taxonomic Notes: In our area closely related only to *H. frenatus*, another non-native species.

Hemidactylus flaviviridis, Gewane.

HABITAT AND DISTRIBUTION: Although the type is from Eritrea, this is an Asian species, from India, Nepal, Pakistan and eastern Afghanistan. Spread by human agency to ports on our coast, from where it has dispersed inland in cargoes, in north-east Africa known only in urban areas of Egypt, Sudan, Eritrea, Djibouti and Somalia, where it favours old buildings, from sea level to about 1,300m. Occurs along Egyptian coast south from Suez along Red Sea coast. In Sudan recorded in Port Sudan and Suakin (probably more widespread), and inland from Kassala; in Eritrea at Massawa, Ghinda and Assab, in Ethiopia over most of Danakil region, notably Afdera, Logiya, Asaiyta, Mile, Gewane, Metahara in Awash and Dire Dawa; there is a curious inland record from Gambela, in wet western Ethiopia. In Djibouti at Obock, Tadjoura and Djibouti City. Northern Somali records include Zeila, Geerissa, Bulhar, Hargeisa, Berbera, Burao and Bosaso; in central Somalia at Galkayo. Probably much more widespread, but overlooked. Elsewhere, from

Hemidactylus flaviviridis, Berbera.

northern India west along the coast to Iraq and around the margins of Arabia. Locally very common. Needs monitoring, it is an aggressive coloniser that displaces indigenous geckos, and is now the most common gecko in urban habitats of Ethiopia and Somalia (e.g. Logiya, Metahara, Gewane and Berbera).

CONSERVATION STATUS: Least Concern.

NATURAL HISTORY: Nocturnal, a fast-moving agile climber, rarely on ground. Commensal with humans, often on walls at night near lights hunting insects, and in our area not known from any undisturbed habitat. By day they shelter in furniture, structural cavities or crevices. They may congregate in suitable places, a group of ten was observed on a wall around a door in Gewane. Often sympatric with other geckos (e.g., *H. afarensis*, *H. awashensis*, *Ptyodactylus ragazzii* and *Tarentola annularis*). A pair of hard-shelled eggs, 10 × 13mm, laid about 20 days after copulation, in any suitable hole or crevice. Diet a wide range of invertebrates.

FLOWER'S SUDAN GECKO

Hemidactylus floweri
Werner, 1908

IDENTIFICATION: A small terrestrial gecko in the Blue Nile region of Sudan, known from a single specimen. Head long, eyes moderately large. Body sub-cylindrical, slightly depressed. Digits moderately long and thick, equally wide along entire length; adhesive pads moderately developed. Tail not segmented, sub-cylindrical, slightly depressed, about 50 per cent of total length, tapering smoothly. The scalation is unusual. Dorsal scales small, slightly conical and overlapping, smooth or obtusely keeled; becoming larger towards flanks; interspersed by ten longitudinal rows of larger keeled scales that are round or slightly oval with a round posterior edge. Enlarged conical tubercles present on hind thighs. The single male has 18 (not 14 as mentioned in the original description) precloacal pores. Size 8cm (snout–vent length 4.1cm). Colour pale greyish-brown; with a number of fine, vague darker crossbars or spots; a clear, narrow dark brown stripe runs from nostrils across eye to the temporal area; on head above this are two stripes, one light and one dark, from above eye towards rostral scale. Underside whitish. Taxonomic Notes: Previously regarded as a subspecies of both *H. tropidolepis* and *H. squamulatus*, but clearly differs from both. It bears some similarity to *H. squamulatus*, but differs from the latter, *inter alia*, in different head shape, dorsal scalation, more upper labials (ten, versus usually 6–8), no contact between uppermost nasals (versus usually wide), larger ventral and head scales (on throat, between eyes and between ear and corner of mouth) and longer and slenderer digits with long distal joints and claws.

HABITAT AND DISTRIBUTION: Presumably a Sudanese endemic, as known from a single adult male, described in 1907, from 'Blue Nile', later amended to 'a few miles north of Wad Medani'. This is now an area of intense agriculture on the banks of the Blue Nile, but was formerly semi-arid to dry open savanna, at an altitude of about 410m. No subsequent specimens found, despite intense research effort.

Hemidactylus floweri, preserved type.

CONSERVATION STATUS: Not evaluated.

NATURAL HISTORY: Nothing known. Presumably similar to other terrestrial *Hemidactylus* in the area. The specific name honours Major Stanley Smyth Flower (1871–1946), English zoologist and conservationist, who collected the type; his field was the herpetology of South-East Asia and the Middle East, and he made important collections in Sudan and Egypt.

ELBA GECKO

Hemidactylus foudaii
Baha El Din, 2003

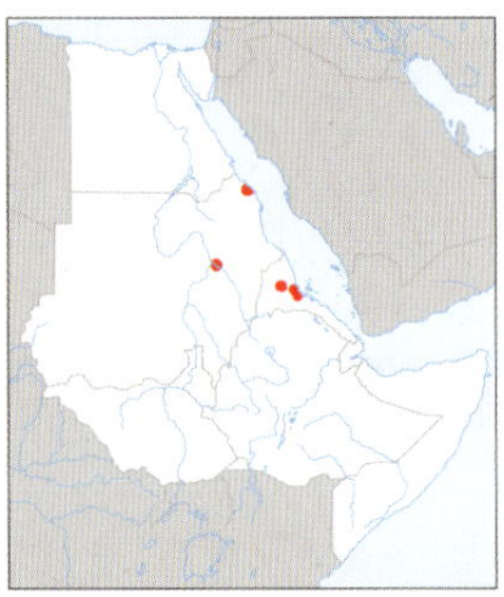

IDENTIFICATION: A small ground-dwelling gecko with coarse dorsal scalation, short distal joints on fingers and a distinct barred pattern, known from south-east Egypt and Eritrea. Head oval, eyes very large, prominent; ear-opening fairly small and round. Body slim, moderately depressed. Limbs short and slim; digits fairly long, moderately dilated, free distal joints of digits fairly short (an excellent aid to identification). Tail segmented, thin and tapering, cylindrical, not constricted at its base, about 52–56 per cent of total length. Dorsal scales heterogeneous; small, smooth, rugose or striated granular scales interspersed by 14–16 irregularly longitudinal rows of large, strongly keeled, sub-trihedral, slightly protruding, round or oval-shaped tubercles. Subcaudal scales transversely expanded. Anterior chin shields large and in broad contact. Males have 8–10 precloacal pores. Maximum total length about 10.7cm (maximum snout–vent length 4.7cm), average 7–9cm. Dorsum pinkish or yellowish-brown with darker brown mottling. There are 8–9 distinct transverse dorsal bands of dark brown tubercles, interspersed by wider bands of whitish tubercles. In some specimens the bands degenerate into small dark spots posteriorly. Tail has around 13 fine bands. A dark stripe extends from nostril across eye to above shoulder, bordered above by a pale stripe. Sometimes there are two additional short parallel dark stripes above shoulder; some have a dark arrow-shaped mark on the snout. Iris yellowish or greenish-grey. Pinkish-white below. Taxonomic Notes: Similar to *H. citernii* of Djibouti and Somalia, in a clade with sister species *H. afarensis* and *H. sinaitus*.

HABITAT AND DISTRIBUTION: Semi-desert to dry stony savanna, from near sea level to 1,350m. Originally described from Wadi Aideb at Gebel Elba in extreme south-east Egypt, recently recorded on Atbara River, about 15km south-east of Atbara town in the Sudanese Nile basin. Also Keren and coastal lowlands in Eritrea, between Gahtielay and Massawa. Probably more widespread in Nile River and Red Sea States of Sudan and lowlands of central and northern Eritrea, but unrecorded. Rare and local in Egypt but widespread and common on coast around Massawa in Eritrea.

CONSERVATION STATUS: Least Concern.

Hemidactylus foudaii, female, Keren.

Hemidactylus foudaii, male, Massawa.

NATURAL HISTORY: A small nocturnal and ground-dwelling species that favours stony habitats. The types were active at night in dry *Vachellia* woodland, one walking among low vegetation, the other perched on a small rock, another under a large rock on a steep densely vegetated slope by day. Specimens from Eritrea were found during the day under rocks on a hill slope in dry coastal semi-desert with a few low *Vachellia* trees and no herbaceous cover. The Keren specimen was under a rock in the shade of a large tree at the edge of field on the outskirts of the town. Two hard-shelled eggs, roughly 7mm diameter, are laid in any suitable refuge. Diet invertebrates.

RAHANUIN GECKO

Hemidactylus fragilis
Calabresi, 1915

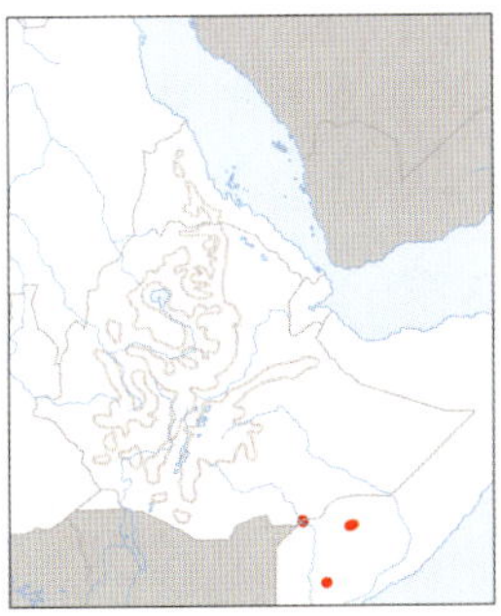

IDENTIFICATION: A small flat, banded gecko, confined to southern Somalia and extreme southern Ethiopia, known from just three specimens. Head narrow, depressed; eyes large, prominent; ear-opening small, fairly round. Body depressed. Limbs fairly short and stout. All three types had missing or regenerated tails, which were slim and conical, without a basal constriction. Dorsal scales almost homogeneous; small, smooth and slightly overlapping, but on the posterior flanks are a few enlarged, smooth, flat tubercles, in one or two rows. Anterior chin shields short, in narrow contact. The only male has 26 preano-femoral pores. Maximum size about 8cm (maximum snout–vent length 3.7cm). Dorsum pale brown, with five narrow wavy dark transverse bands; a narrow dark brown stripe runs from nostrils across eye to the temporal area above ear-opening, then continues above shoulder along flanks and ends at groin. Underside unmarked, pale. Taxonomic Notes: Lumped for a long time with the African invasive *H. frenatus*, but a recent revision confirms this is a valid species, although relationships unknown in absence of fresh genetic material.

Hemidactylus fragilis, preserved, Rahanweyn.

HABITAT AND DISTRIBUTION: Dry savanna at roughly 200m or slightly more. Endemic to southern Somalia and adjacent southernmost Ethiopia. Collected only twice. The type, a male, is from Bur Meldac (= Meel Daaq), a granitic outcrop in the Bay region, about 56km south of Dinsoor, Somalia. Two further specimens (both females) at Dolo on the border with Ethiopia and an imprecise locality in the Rahanweyn region (roughly corresponds to the present Bay region, between the Juba and Webi Shebelle Rivers). Probably more widespread, but overlooked, might be in north-east Kenya.

CONSERVATION STATUS: Data Deficient, but probably not threatened.

NATURAL HISTORY: Nothing known, but presumably similar to other *Hemidactylus*. It might be terrestrial or arboreal.

PACIFIC HOUSE GECKO

Hemidactylus frenatus
Duméril & Bibron, 1836

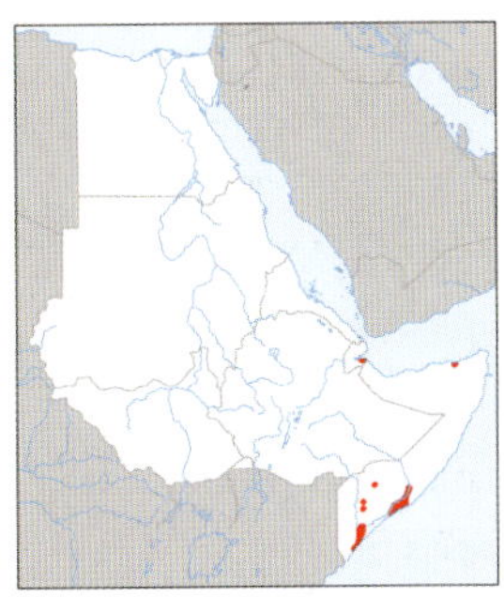

IDENTIFICATION: A small to medium-sized, moderately robust flat gecko with homogeneous-looking scalation, an invader into urban areas of southern Somalia. Head oval, long and narrow; eyes large, prominent; ear-opening moderate, round or slightly oval. Body stout and depressed. Limbs fairly short and slim. The innermost digit (particularly on front foot) is very short, and claw lies flat (an excellent aid to identification). Tail slightly segmented, the regenerated tail not segmented and covered by uniformly small scales. Subcaudal scales transversely enlarged. Dorsal scales look homogeneous, but close examination reveals small, granular, smooth scales, with a few enlarged, flat, tubercles in 2–8 rather irregular rows, these are enlarged posteriorly. Anterior chin shields in broad contact, posterior chin shields more or less equal in size. Males have 24–36 preano-femoral pores in a continuous series. Maximum known size about 12.5cm (maximum snout–vent length 6cm), average 8–11cm (males larger), hatchlings about 4.5cm. Dorsum yellowish, greyish or pinkish-brown, pattern-less or with three diffuse darker stripes, sometimes with visible light spots; a dark streak from nostril passes through eye to flank; iris brownish-grey. Underside pale to bright yellow. Taxonomic Notes: Its nearest relative in north-east Africa is *H. flaviviridis*, another non-native species in Africa, and both members of the tropical Asian clade of *Hemidactylus*.

HABITAT AND DISTRIBUTION: Confined to urban areas in north-east Africa, never further than 200km from coast, between sea level and 450m. An invasive from southern Asia (the type from Java), it was transported to Africa in cargoes. In our area recorded from Djibouti City, Bosaso in Puntland, and the Benadir coast of southern Somalia (localities include Giovai [Jofay] Island, Kismayo, Gelib and several records from vicinity of Mogadishu, along Webi Shebelle River to Eggi, with inland records at Dinsor and Baidoa in Bay region). Probably more widespread in towns along whole southern Somali coast. Very common in southern Somalia where it occurs. Its distribution needs monitoring; it is an aggressive coloniser that displaces indigenous geckos. Elsewhere, almost cosmopolitan, although the only other African record is at Lamu, Kenya.

CONSERVATION STATUS: Least Concern.

NATURAL HISTORY: Nocturnal, mostly arboreal, sometimes terrestrial. Somali specimens were common on walls, in buildings and in pole huts of local villages. Often near lights on walls, hunting insects. They eat small geckos, their own species included. Territorial; a dominant male establishes a territory and drives away smaller males. They have two calls, a series of chirps and a *chuck, chuck, chuck*. Three or four weeks after copulation two hard-shelled eggs, roughly 8×9.5mm, are laid in crevices, mostly in man-made structures. Communal egg-laying sites well known. Eggs hatch in about 50–90 days.

Hemidactylus frenatus, Djibouti.

ARCHER'S POST GECKO

Hemidactylus funaiolii
Lanza, 1978

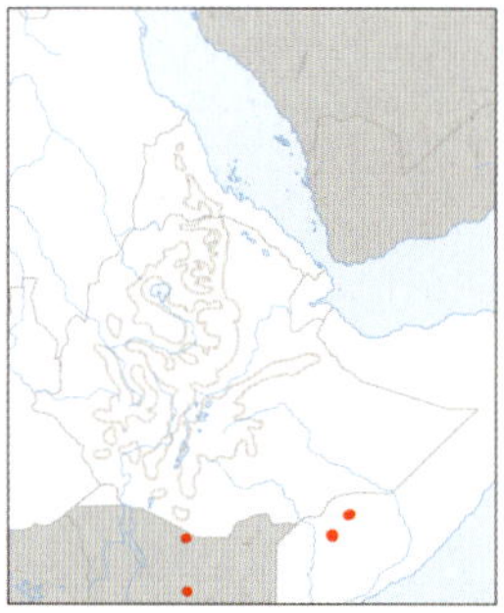

IDENTIFICATION: A small terrestrial stocky brownish gecko with homogeneous single-keeled scales on back, from northern parts of southern Somaliland and adjacent Ethiopia. Head oval, eyes large and prominent. Body sub-cylindrical, relatively long. Digits short, stout, equally wide along entire length. Tail not segmented, cylindrical, about 50 per cent of total length and tapering smoothly (although regenerated tails swollen and shorter), subcaudal scales slightly enlarged. The scalation is unusual; scattered smooth conical granular scales on neck, becoming overlapping

Hemidactylus funaiolii, Kenya.

and strongly keeled scales on body, no enlarged tubercles or scales on hind thighs. Anterior chin shields in broad contact. Males have 6–7 precloacal pores. Maximum size about 7.5cm (maximum snout–vent length 3.7cm), average length 6–7cm. Colour brownish, pinkish on head, dorsum marked by yellow or white spots, and brown streaks that may coalesce to form vague bars; there is a brown irregular streak with poorly defined edges from nostril to eye and on back of head, which forms a W-shaped bar on occiput. Top of snout pale. Underside uniform white. Taxonomic Notes: Molecular analysis groups this species in a clade with two very similar species, *H. isolepis* and *H. lanzai*.

HABITAT AND DISTRIBUTION: A rare Horn of Africa endemic; semi-desert and dry savanna, between 400–1150m. Records include Berdale and 'Rahanuin' country (probably from the Somalia–Ethiopia border and towards Baidoa). Also Huri Hills on Kenya's northern border, so probably in southern Ethiopia; in Kenya also from between Archer's Post and Laisamis. Widespread, but appears rare or secretive.

CONSERVATION STATUS: Least Concern.

NATURAL HISTORY: Poorly known. Kenyan specimens were under rocks. Biology presumably similar to other ground-dwelling *Hemidactylus* of dry country, i.e. nocturnal, shelters under ground cover or in holes, lays hard-shelled eggs and eats invertebrates.

GRANCHI'S GECKO

Hemidactylus granchii
Lanza, 1978

IDENTIFICATION: A rare Somali endemic, a stocky gecko, its back studded with huge keeled tubercles. Head triangular, short; eyes large, prominent; ear-opening large, elliptic. Body stout, slightly depressed. Limbs moderately long, fairly slim; digits moderately long, weakly dilated, adhesive pads well developed. Tail segmented, slim, moderately depressed, tapering, not constricted at its base; original tail length unknown. Subcaudal scales strongly enlarged. Dorsal scales heterogeneous, small granular scales interspersed by big, strongly keeled subcircular tubercles in 14–16 more or less regular longitudinal rows; enlarged tubercles on hind thighs. Anterior chin shields large, in broad contact. Males have five precloacal pores. Maximum total length about 12–13cm (maximum snout–vent length 6.2cm). Dorsum yellowish-brown with an inconspicuous pattern of a few randomly placed grey or brown spots that tend to form transverse bars on limbs and tail, and four somewhat indistinct X-shaped dark spots on neck and back; in some animals a vague dark streak passes from nostrils through eye to above ear-opening; iris golden-brown. Whitish below. Taxonomic Notes: Genetic analysis indicates close relationship to *H. pauciporosus* and *H. macropholis*.

HABITAT AND DISTRIBUTION: Dry, sandy plains with sparse vegetation between 50 and 1,110m. On fixed dunes with low bushes in coastal areas, in sandy steppe and grassland, and xerophilous open woodland inland. Known from just four localities in Somalia. The type was from Mudug region (about 80km north of Hobyo in central Somalia). Also recorded between Uargalo (or Wargalo) and Galkayo; in Mudug, the third specimen was from the coastal environs of

Hemidactylus granchii, north-east of Hargeisa.

Warshiikh, in the middle Shebelle in southern Somalia. Recently discovered 45km north-east of Hargeisa in the Woqooyi Galbeed region of Somaliland, the northernmost record. Presumably widespread between these four distant localities, but secretive and rarely encountered.

CONSERVATION STATUS: Least Concern; suitable habitat is extensive and thus it is not under threat.

NATURAL HISTORY: Poorly known. No field details for the first three specimens, the recently collected adult female was found in the day at an entrance to a sewer drain under the road, on a stony plain with sandy patches and scattered, stunted vegetation. The agility of this female on the drain wall and well-developed adhesive pads suggest this is an arboreal species. Biology presumably similar to other *Hemidactylus*, i.e. nocturnal, lays eggs and eats invertebrates.

HEYDEN'S GECKO

Hemidactylus granosus
Heyden, 1827

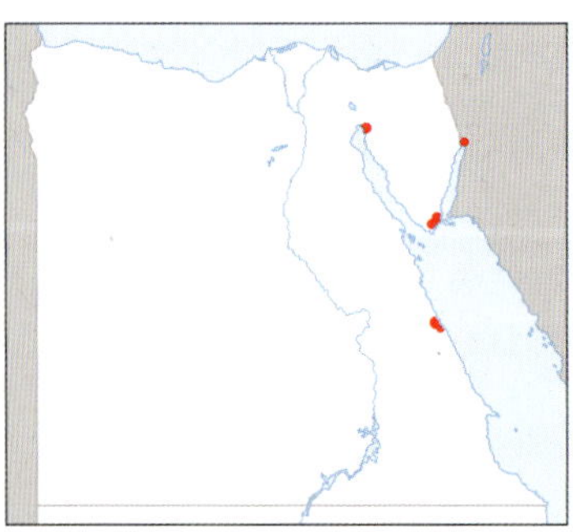

IDENTIFICATION: A medium-sized, rather slender gecko with clear rows of large dorsal tubercles, a high number of upper labials (9–11) and low number of precloacal pores, from the Egyptian Sinai and Red Sea coast. Head rather elongated, sub triangular, eyes large and prominent; ear moderately large, crescentic. Body moderately long, slim and depressed. Limbs moderately long and fairly slim; digits very long and rather narrow, moderately dilated, adhesive pads well developed and numerous, free distal joints of digits long. Tail strongly segmented, slender, depressed at base, then cylindrical, tapering smoothly, not constricted at its base, 52–57 per cent of total length. Subcaudal scales transversely enlarged. Dorsal scales heterogeneous, small granular scales interspersed by 14–15 fairly regular longitudinal rows of large, keeled, sub-trihedral, subcircular tubercles, flatter on back, conical on flanks. Anterior chin shields large, usually in broad contact. Males have 3–7 precloacal pores. Maximum total length about 12cm (maximum snout–vent length 5.4cm), average 10–11cm. Dorsum pale buff, with four patchy dark brown markings outlining an X-shape, or with narrow darker crossbands. Usually a narrow neck band, which may join temporal/eye-stripes to form a crescent on nape. Lateral tubercles usually whitish; tail with 12–14 dark bands, wider and more clearly delineated towards tip. Iris yellowish-brown. The regenerated tail is uniformly buff above. Pinkish-white below. Taxonomic Notes: Recently resurrected from synonymy of *H. turcicus*. Genetic analysis indicates it belongs with other Arabian species in *H. saba* species group, not the unrelated *H. turcicus* species group.

HABITAT AND DISTRIBUTION: Desert and semi-desert, from sea level to 1,600m. However, mostly found in dry anthropogenic habitats. Current African distribution poorly known, as it was long conflated with *H. turcicus*, but in our area it seems to be confined to Egypt. The only confirmed records with molecular data are from Sinai (Sharm El Sheikh and Rauabi village in Nabq, at the southern tip), Suez (Ayoun Musa, Gulf of Suez) and Red Sea coast (35km south of Quseir). Probably more widespread, particularly further south, but old records require verification. Elsewhere, to Saudi Arabia.

CONSERVATION STATUS: Data Deficient in Egypt. In Arabia widespread and fairly common.

NATURAL HISTORY: Biology similar to *H. robustus* and *H. turcicus*. Nocturnal, generally scansorial,

Hemidactylus granosus, south of Quseir.

sometimes terrestrial. Taken in rubbish dumps, road ditches, old walls and buildings, concrete culverts, abandoned construction sites, etc. Specimens from Sharm El Sheikh were found by day under empty barrels and inside buildings; Red Sea coastal specimens were on hotel walls at night in a large coastal resort. Saudi Arabian animals were in concrete culverts by day. In Egypt sympatric with *H. flaviviridis*, *H. robustus* and *H. turcicus*. Lays eggs and eats invertebrates.

UNIFORM-SCALED GECKO

Hemidactylus isolepis
Boulenger, 1895

IDENTIFICATION: A small stout terrestrial, brownish, banded gecko with homogeneous smooth dorsal scales, from the Ethiopian Ogaden and west-central Somaliland. Head bulky, relatively wide and high; eyes large, prominent. Body sub-cylindrical, fairly stout. Digits short, stout, equally wide along entire length. Tail not segmented, cylindrical, about 50 per cent of total length and slightly swollen behind cloaca. Subcaudal scales usually enlarged. The scalation is unusual; all dorsal scales are roughly the same size, smooth and overlapping, arranged in about 59–81 rows around mid-body. Head and neck scales smaller. Anterior chin shields in broad contact. Males have 6–8 precloacal pores. Maximum size about 10cm (maximum snout–vent length 4.9cm), average length 8–9cm. Colour pinkish or greyish-brown, with irregular dark brown and smaller white spots; both dark brown and white spots tend to coalesce into transverse bands. Dark and white dorsal spots present also on limbs and tail. A dark brown irregular temporal stripe, its edges poorly defined, runs from nostrils across eye to the temporal area. Scales around eyes vivid yellow or white. The regenerated tail is more swollen, shorter, with fewer, duller markings. Pinkish-white below. Taxonomic Notes: Recently re-examined, and now known only from northern Ogaden (including the type locality) and Somaliland. Specimens from the southern Ogaden and southern Somalia represent another (undescribed) form.

HABITAT AND DISTRIBUTION: Semi-desert savanna, dry grassland, shrubland and *Vachellia–Commiphora* bushland, at 145–2,000m. Occurs from the Gedo

Hemidactylus isolepis, Qool-Cadday.

Hemidactylus isolepis, Kebri-Beyah.

Hemidactylus isolepis, male, Borama.

Hemidactylus isolepis, female, Borama.

Region in Somalia across the Ethiopian Ogaden into Somaliland. Known localities include Lugh on the Juba River, Dolo, Gode, the Turfa River south of Fik, environs of Harar and Kebri Beyah. Occurs widely in Borama, across Ogo to Goolis Mountains (Ferdusa Ruins in Sheikh), Hargeisa and Qool-Cadday.

CONSERVATION STATUS: Least Concern. Widespread, locally common, although targeted to some extent in past by commercial animal collectors in Somaliland.

NATURAL HISTORY: Little-known. Terrestrial and nocturnal. They favour stony terrain at higher elevations (mostly 1,000–1,800m) where can be common. Tolerant of cold conditions, for example in Goolis Mountains in Somaliland where winter temperatures can fall to 5–10°C. They shelter under flat rocks by day. Twice, a true pair was found together under the same rock, another pair was observed outside their shelter at night, suggesting they live in pairs for at least part of the year. Diet small insects and other invertebrates.

ETHIOPIAN GECKO

Hemidactylus jubensis
Boulenger, 1895

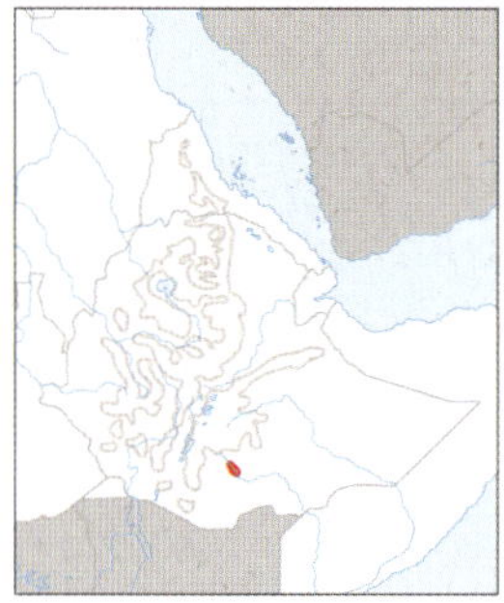

IDENTIFICATION: A medium-sized, surprisingly stocky rupicolous mottled gecko, endemic to southern Ethiopia. Head wide, eyes large, prominent; ear-opening large, oval. Body relatively short, stout and depressed. Limbs short and robust; digits long, moderately dilated, adhesive pads well developed. Tail segmented, tapering, the swollen basal portion marked off by a faint constriction, distinctly flattened, with slight median dorsal furrow. Subcaudal scales transversely enlarged. Dorsal scales heterogeneous, small granular scales interspersed by 14 irregular longitudinal rows of moderately enlarged, feebly keeled, flat subcircular tubercles. Anterior chin shields large, in broad contact. Males have eight precloacal pores. Maximum total length about 14cm (maximum snout–vent length 6.7cm). Dorsum greyish-brown or brown, mottled darker brown on head and limbs, and neck and back with five dark brown, transverse, wide X-shaped spots. A dashed line of narrow pale lozenges along spine is an excellent aid to identification. Tail barred; a diffuse dark stripe runs from nostril through eye to above ear-opening. Iris reddish-brown. Underside white. Taxonomic Notes: Similar and closely related to *H. barodanus*, but differs slightly in coloration and shape and size of dorsal and tail tubercles.

HABITAT AND DISTRIBUTION: Endemic to Ethiopia, on granite outcrops in evergreen thickets and shrub or savanna, at 1,100–1,400m. Long known from a single type, collected in 1893 on the upper Ganale River, probably somewhere east of Kebri Mangest in southern Ethiopia. Recently rediscovered between Negele and the Genale River to the north, and much suitable habitat with large granite boulders exists between Negele and Mena, so probably more widespread. A

Hemidactylus jubensis, Negele.

Hemidactylus jubensis, preserved, upper Genale River.

vague record (not shown on map) from Milmil east of Degeh Bur in northern Ogaden needs further study.

CONSERVATION STATUS: Data Deficient.

NATURAL HISTORY: Poorly known. Biology probably similar to *H. barodanus*. Rediscovered specimens were found at night, on trees or crawling on large grey-brown, extensively cracked granite boulders; presumably they hide in cracks by day. Named for the Juba River, which commences as the Genale (the type locality) in southern Ethiopia, becoming the Juba in Somalia.

KLAUBER'S GECKO

Hemidactylus klauberi
Scortecci, 1948

IDENTIFICATION: A small stout terrestrial gecko with homogeneous dorsal scales, known from a single (female) specimen collected at Gelib in southern Somalia. Head bulky and deep. Eyes large with a dark iris. Body cylindrical, stout and slightly elongate. Limbs short, stout. Digits short and thick, not equally wide along entire length; adhesive pads very small. The tail of the only specimen is regenerated and unsegmented, sub-cylindrical, slightly depressed, tapering smoothly. The scalation is very unusual; all dorsal and ventral scales on body are roughly the same size, strongly keeled (each scale with 3–6 keels, an excellent aid for determination), and overlapping, in 68 rows at mid-body. Head and neck scales smaller. The anterior chin shields are not in contact; separated by two small scales. Maximum size probably 7–8cm (snout–vent length 3.9cm in only known specimen with regenerated tail). As originally described, dorsum yellowish-grey with a black stripe extending from nape to tail, flanks greenish with yellow spots. A wide black stripe bordered by, and with internal, yellow spots runs from posterior edge of eye across ear-opening, fading on mid-flanks. Top of head and nape with several black spots; snout blackish with yellow spot between eyes; rostral, upper and some lower labials and front edge of eye yellow. Belly and underside of tail yellowish; throat, chest and part of chin speckled blackish-brown.

Hemidactylus klauberi, preserved type, Gelib.

HABITAT AND DISTRIBUTION: Endemic to Somalia, known only from Gelib in Middle Juba region, at an altitude of 25m in a semi-arid area with nearly 600mm annual rainfall. The exact collection locality is unknown, but at Gelib a narrow riparian forest along the Juba River is interspersed by cultivation and away from the river there are extensive areas of grassy xerophilous open woodland with dense low *Vachellia* trees and thornbush, all of which periodically flood.

CONSERVATION STATUS: Data Deficient.

NATURAL HISTORY: Unknown. Unusual among *Hemidactylus* in that almost all its scales have 3–6 strong keels (one is usual), for reasons unknown, but possibly connected with movement in dense grassy undergrowth of the river valley.

SOMALILAND MOUNTAIN GECKO

Hemidactylus laevis
Boulenger, 1901

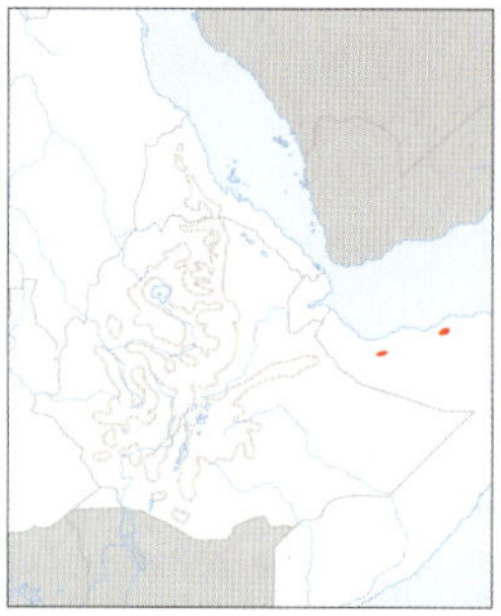

IDENTIFICATION: A small to medium-sized endemic of Somaliland mountains, body and head depressed, with a carrot-shaped swollen tail covered by small granules. Head wide, sub-triangular; eyes large but less prominent; ear-opening small, round. Body depressed, stocky. Limbs fairly short, slim; digits fairly short, strongly dilated, Tail segmented, strongly swollen, carrot-shaped (narrower in males), constricted at its base, covered by small granular juxtaposed scales, about 50–55 per cent of total length. Subcaudal scales enlarged. Dorsal scales heterogeneous; small juxtaposed granules with 10–14 rows of numerous small, round, flattened smooth tubercles. Anterior chin shields rather short, in narrow contact. Male has two pre-cloacal pores and inconspicuous hemipenial pockets. Maximum known size 11.5cm (maximum snout–vent length 5.7cm), average 9–11cm. Dorsum pale grey, pink or yellowish-brown, with four to five irregular transverse bands with sinuous margins; 4–5 dull spots on flanks, sometimes merged into a line; with about 9–11 transverse, high- or low-contrast bands on tail. There is a W-shaped dark marking on occiput and two dark spots just behind eyes, sometimes with additional black marbling; a pair of dark longitudinal markings over sacrum sometimes present; a dark stripe from nostril, through eye, becomes more diffuse posteriorly to level of ear-opening and is confluent with first crossbar on body. A dark spot is present on posterior part of snout. Regenerated tail not segmented, uniformly coloured with larger blackish spots. Underside, pinkish-yellow, tail marbled with darker pigment. Juveniles with similar pattern, but more greyish and a yellow tail. Taxonomic Notes: Long known solely from the type, but recently rediscovered, enabling study of this beautiful and rare species. Molecular analysis indicates it is a basal species within a large clade of all African terrestrial *Hemidactylus* (including *H. afarensis*, *H. albopunctatus*, *H. citernii*, *H. isolepis* and *H. somalicus*).

Hemidactylus laevis, Erigavo.

Hemidactylus laevis, juvenile, Erigavo.

HABITAT AND DISTRIBUTION: Endemic to Somaliland; associated with montane habitats, but also in adjacent drier savanna and semi-desert limestone and gypseous hills, at 1,600–1,850m. The type locality is montane evergreen thicket and scrub, dominated by *Buxus*, *Cadia*, *Dodonaea* and *Euclea shimperi*. Known from two widely separated localities; type from Gaan Libaah in Goolis range, recently rediscovered at Erigavo and Buq in Cal Madow Mountains in Sanaag region. Probably more widespread in Somaliland mountains, particularly the eastern Cal Madow and Wagar range. Recent research failed to find it on Gaan Libaah, in Erigavo it is fairly common.

CONSERVATION STATUS: Data Deficient.

NATURAL HISTORY: Although nested in a terrestrial clade of *Hemidactylus*, it is evidently arboreal and lives on rocks, although all the 30+ recently recorded animals were agile on building walls, hunting insects around lights. Shy and nocturnal. None seen in dry cold season between December and March, but active and abundant in warm wet season. The conspicuously swollen tail probably stores fat reserves from wet season. The specimens collected in Erigavo, oddly, were almost all females. One or two hard-shelled eggs are laid. In Buq village sympatric with the arboreal *H. pauciporosus* and ground-dwelling *H. ophiolepoides*.

LANZA'S GECKO

Hemidactylus lanzai
Šmíd, Mazuch, Nováková, Modrý, Malonza, Abdirahman-Elmi, Carranza & Moravec, 2020

IDENTIFICATION: A small slim terrestrial brownish gecko with homogeneous smooth dorsal scales, in south-west Ethiopia and adjacent South Sudan. Head oval; snout long, fairly pointed; eyes large, prominent. Body sub-cylindrical, elongate. Digits short, stout, fairly equally wide along entire length. Tail not segmented, cylindrical, slightly over 50 per cent of total length and tapering smoothly (although regenerated tails may become swollen). Subcaudal scales enlarged. The scalation is unusual; all dorsal scales roughly the same size, smooth and overlapping, those on head and neck smaller. Anterior chin shields in broad contact. Male has 4–8 precloacal pores. Maximum size about 8cm (maximum snout–vent length 4.2cm). The dorsal surface, upper limbs and tail are pink or pinkish-brown, with irregular dark brown spots and smaller white spots; dark brown spots tend to coalesce into transverse bands. A well-delineated narrow dark brown stripe runs from nostrils across eye to the temporal area, bordered above by a pale stripe. A conspicuous dark brown triangular spot on snout. The scales around eyes are vivid yellow or white. Underside pinkish-white. Juveniles are uniformly pinkish-brown with few scattered pale dots on occiput and anterior third of body. Taxonomic Notes: This species results from a recent re-examination of *H. isolepis*, which as originally defined was two species, the northern form (at Lugh in Somalia, north across the Ogaden to Somaliland) retains that name.

HABITAT AND DISTRIBUTION: Near-desert, semi-desert and dry savanna, below 1,300m. In very arid environments; rocky basalt terrain, sandy plains

Hemidactylus lanzai, Chew Bahir Lake.

Hemidactylus lanzai, Chew Bahir Lake.

Hemidactylus lanzai, Chew Bahir Lake.

with sparse vegetation or treeless lava deserts. Known from Lokomarinyang and Kokuro in the Ilemi Triangle in extreme south-east South Sudan, east to northern Lake Turkana and Lake Chew Bahr. Older records from further east (Malka Murri, Mandera) probably belong to an undescribed form with similar dorsal scalation. Elsewhere, south across northern Kenya to Baringo and Archer's Post. Locally common in our area, widespread and common in adjacent Kenya.

CONSERVATION STATUS: Least Concern.

NATURAL HISTORY: Poorly known. Nocturnal. The male holotype from Lake Chew Bahr was found c.1 hour after sunset at the edge of a dry riverbed, peering from a burrow, into which it retreated when disturbed. A gravid female nearby was under a stone by day in dry dense scrub in July. Usually under rocks in lava deserts of northern Kenya. One or two hard-shelled eggs are laid in soil under rocks. Diet invertebrates.

ANDERSSON'S GECKO

Hemidactylus laticaudatus
Andersson, 1910

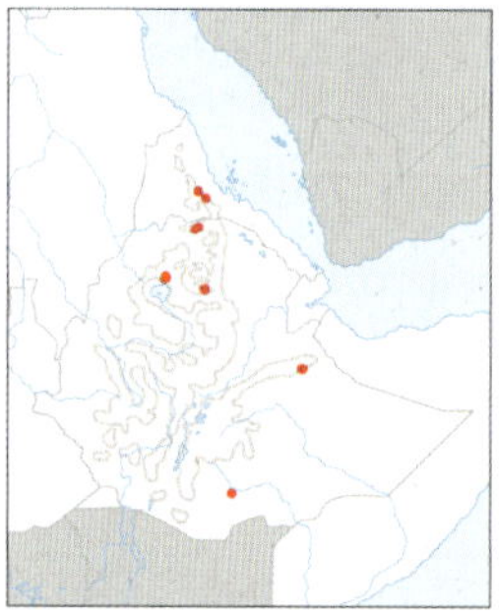

IDENTIFICATION: A smallish to medium-sized *Hemidactylus*, with a swollen carrot-shaped tail, endemic to highland Ethiopia and Eritrea, often seen on buildings. Head wide, pear-shaped; eyes fairly large, prominent; ear-opening moderate, oval. Body depressed and stocky. Limbs short, fairly slim. Tail strongly segmented, swollen, carrot-shaped, with a median dorsal furrow, constricted at its base, c.51–55 per cent of total length. Subcaudal scales usually enlarged. Dorsal scales heterogeneous; small granules interspersed by 14–18 longitudinal rows of larger, rounded, weakly keeled or flat tubercles. Anterior chin shields either narrowly in contact or not in contact. Males have 12–27 preano-femoral pores, interrupted in middle. Maximum known size 13.4cm (maximum snout–vent length 6.1cm), average 10–12cm. Colour variable, and changes from greyish-brown to greyish-yellow or white. Five irregular diffuse darker bands on body and 7–11 on tail. There is often a vague W-shaped mark on occiput, and a dark stripe of varying intensity extends from snout through eye to the temporal region. The regenerated tail is shorter, not segmented, with uniform small scales and less boldly marked. White or yellow-grey below. Juveniles similar but tail more yellow. Taxonomic Notes: Eritrean examples were regarded as a different species, *H. fossatii*. However, we have seen no characters that unequivocally distinguish them.

HABITAT AND DISTRIBUTION: On rocks in evergreen thicket and scrub to montane savanna, at higher elevations of 1,600–2,500m. Sporadically reported from several widely separated localities. In Eritrea at Asmara and Saganeiti. Northern Ethiopian localities include the environs of Lake Tana, Gondar, Azezo, Axum, Adua and Lalibela. Also at Harar, eastern Ethiopia, and Negele in the south. Probably more widespread but overlooked. Often locally common and tolerant of urbanisation.

Hemidactylus laticaudatus, Gondar.

Hemidactylus laticaudatus, Harar.

CONSERVATION STATUS: Least Concern.

NATURAL HISTORY: Poorly known. Nocturnal and arboreal. At Axum, specimens were active on rock outcrops at night, but most commonly seen on buildings, hunting insects near lights at night. In Asmara, observed at night on low walls on outskirts of city, together with *H. barodanus*; in Harar seen on high walls of the old city and on adjacent buildings (together with *H. robustus*). Lays two hard-shelled eggs. Eats invertebrates.

TROPICAL HOUSE GECKO

Hemidactylus mabouia
Moreau De Jonnès, 1818

IDENTIFICATION: A small, flat gecko, with heterogeneous dorsal scalation and a high number of preano-femoral pores, known only from extreme southern Somalia near the Kenya border. Head oval, eyes large, prominent; ear-opening small, oval. Body short, stocky, slightly depressed. Limbs moderately long and stout; digits long, moderately dilated. Tail segmented, slim and tapering, slightly over half total length, and original tail has six enlarged pointed tubercles on each whorl; the spiky tail is a good field character. Subcaudal scales transversely enlarged. Dorsal scales heterogeneous, the small granular or slightly overlapping scales are interspersed by 10–14 (elsewhere 8–16) irregular rows of larger flat or conical, weakly keeled, rounded tubercles. Anterior chin shields usually in wide contact. Males have 28–29 preano-femoral pores in a split or continuous series. Maximum total length about 11cm (maximum snout–vent length 5.2cm), although larger elsewhere, average 8–10cm, hatchling 4–4.5cm (based on the only three specimens from our area, from southern Somalia). They change colour, the dorsum may be pale brown, pinkish-brown, brownish-grey or almost white; neck and back with a series of 5–6 faint dark chevrons or sub-rectangular crossbars, and 7–9 additional dark bands on tail. Sometimes a diffuse dull darker stripe extends from nostril across eye to above ear-opening; top of head mottled black or grey, limbs barred; iris grey or red-brown. Underside whitish or pale yellow. Taxonomic Notes: A problematical species, described from the Caribbean, but originated in Africa, and has spread around the tropics. Several names have been used for East African populations, and molecular analysis indicates a complex picture. We retain *H. mabouia* until the status is clarified.

HABITAT AND DISTRIBUTION: Poorly known in our area; known only from two specimens at Ola Uager (or Hoja Wajeer) Zone, about 2km from the left bank of the Lak Bush-bush in extreme southern Somalia and one from Sar Uanle, about 18km south-west of Kismayu, Lower Juba region. All three were collected in dense coastal thicket and adjacent savanna, near sea level. Elsewhere, widespread throughout sub-Saharan Africa, although records sporadic in west; on islands in Indo-Pacific; and by natural or human agency it has dispersed along east coast of South America, and through the Caribbean into Florida.

CONSERVATION STATUS: Least Concern.

NATURAL HISTORY: Nocturnal and in our area ground-dwelling. The Ola Uager specimens were collected near or under the globular trunk of *Adenia globosa* in the underbrush of a secondary lowland evergreen thicket. The Sar Uanle specimen was close to the sea on a fixed dune with grassy vegetation and very low shrubs. Kenyan and Tanzanian animals often associated with habitation and palm groves, and shelter under fallen fronds. One or two small hard-shelled eggs, roughly 8–9mm in diameter, are buried separately under stones, in plant debris or in crevices. Diet small invertebrates.

Hemidactylus mabouia, Kenya.

BOULENGER'S GECKO

Hemidactylus macropholis
Boulenger, 1896

IDENTIFICATION: A large, usually rufous, stocky gecko with large dorsal tubercles, widespread in the Ethiopian Ogaden and Somalia. Head robust, oviform to pear-shaped in old males, eyes large, very prominent; ear-opening large, oval. Body stout and slightly depressed. Limbs short and stout; digits moderately long, strongly dilated. Tail segmented, nearly circular, with a slight median dorsal furrow, not constricted at its base tapering, about 53–58 per cent of total length, may be swollen, carrot-shaped and constricted at base in larger specimens. Subcaudal scales transversally enlarged The regenerated tail is similar, but covered by uniform small scales. Dorsal scales heterogeneous, small granular scales are interspersed by 12–16 rows of strongly keeled, striated, trihedral and slightly oval tubercles; giving the animal a striped appearance. Anterior chin shields large, in broad contact. Males have 6–13 precloacal pores. Maximum total length about 20cm (maximum snout–vent length 9.5cm), average 12–17cm, females smaller. Rufous, pink, pinkish or yellowish-brown in colour, dependent upon local soil colour. Some have five diffuse darker crossbars or X-shaped spots on neck and back, others rather irregularly spotted; tail has 9–11 dark bands or series of irregular spots; a vague dark stripe extends from nostril across eye to above ear-opening; top of head and limbs mottled by dark brown or red; iris yellowish to red-brown. Whitish below. Juveniles similar. Taxonomic Notes: The type was from Dolo Odo in Somalia. Ongoing molecular and morphological analysis indicates this taxon is actually a complex that includes undescribed cryptic species.

HABITAT AND DISTRIBUTION: Typical of Somali arid zone, in dry open *Vachellia–Commiphora* bushland or scrub and semi-desert with soft red sand, but also in adjacent rocky terrain, from near sea level to 1,610m (although usually at 400–1,200m). Found throughout much of Somalia and the Ethiopian Ogaden. Occurs widely south of Ogo and Goolis Mountains in west-central Somaliland, south and east of Hargeisa towards the Haud region and east through the Nogal Valley to Puntland, where mostly in western and northern regions along border with Somaliland and the north coast. Notable localities include: Hargeisa, Burao, Bohodle, Las Anod, Garowe, Gardo, Geriban and Galkayo. Not recorded from the Guban plain, north coast and mountain belt of northern Somaliland. Its absence from most of the east coast of Puntland and central Somalia probably due to under-collecting. Ogaden records include the environs of Degeh Bur, Ado, Warder, and a curious isolated record from Sof Omar caves, far to the west; probably has a broader distribution here. Widespread in

Hemidactylus macropholis, female, Burao.

Hemidactylus macropholis, juvenile, Hargeisa.

central and southern Somalia at many inland localities (Dusa Marreb, El Bur, Baidoa, Bardera, Belet Uen, Oddur, and Lugh), as well as coastal localities (Meregh, Mogadishu, Merca and Kismayo). Elsewhere, to northern and eastern Kenya. Old records from central Ethiopia and Eritrea are now assigned to other species. Tolerant of humans and urbanisation, although might be at risk from invasive species like *H. flaviviridis*.

CONSERVATION STATUS: Least Concern.

NATURAL HISTORY: Nocturnal, territorial and largely terrestrial. A shy species. In natural settings it shelters under ground cover, in rock cracks, under bark or in hollows and cracks of dead or old trees and in termitaria. At Degeh Bur in Ethiopia and Saraar plains south-west of Garadag in Somaliland, specimens were found in cavities of old *Vachellia* trees, at approximately 1m above ground. In remote flat sandy areas, without rocks, these geckos were active on the ground after dark. In urban areas it has been found at night on the walls of a concrete water reservoir at Toon south of Hargeisa, in holes in concrete bridges (along with its eggs) and ambushing insects on walls around lights; in Somaliland it often shares such locations with *H. robustus* and *H. flaviviridis*. In natural habitats in Somaliland it is sympatric with *H. smithi*. Two hard-shelled eggs, approximately 12–14mm in diameter, are laid in any suitable refuge (often tree cracks and under rocks). Juveniles have been found under rocks.

PARKER'S GECKO

Hemidactylus megalops
Parker, 1932

IDENTIFICATION: A very small ground-dwelling pinkish-brown gecko with homogeneous dorsal granular scales, known from just four males, collected in the Sol Haud region of Somaliland. Head oviform, eyes large, fairly prominent; ear-opening small, round. Body cylindrical, slightly depressed, fairly slim. Digits short, stout, equally wide along entire length. Tail not segmented, short and slim, cylindrical, with slight median furrow, about 44 per cent of total length. Subcaudal scales small. Back covered with small, smooth, granular, and juxtaposed scales; all roughly the same size. Anterior chin shields small, not in contact or just in point contact. The males have two precloacal pores. Maximum size about 7cm (maximum snout–vent length 3.6cm), average length 8–9cm. Dorsum pale pinkish-brown, back closely and uniformly dappled with small, circular, white dots; four broad, diffuse, dusky crossbars on body, another across tail base. The tail has six narrow, white crossbars. The diffuse irregular patches behind the eyes may extend back and meet to form a crescent on back of head. Underside whitish. Taxonomic Notes: Undoubtedly closely related to two species that also occur in northern and central Somalia, *H. somalicus* and *H. puccionii*.

HABITAT AND DISTRIBUTION: Endemic to Somaliland. Near-desert, in arid low hilly country with rocky limestone or gypseous soils, and sparse vegetation, at 570–1,000m. Collected in the 1930s at two widely separated localities in Somaliland on the Puntland border. Three (including the holotype) were from Buran in Sanaag region, the fourth from the northern slopes of Wadi Nogal, about 37km south-east of Halin (Sool region). Might occur in the intervening country, all suitable habitat, but unrecorded.

CONSERVATION STATUS: Data Deficient.

NATURAL HISTORY: Nothing known, but presumably similar to other *Hemidactylus* of dry country, i.e. nocturnal and terrestrial, shelters under ground cover or in holes, lays eggs and eats invertebrates.

Hemidactylus megalops, preserved type.

MOUNT SINAI GECKO

Hemidactylus mindiae
Baha El Din, 2005

IDENTIFICATION: A medium-sized, slim Egyptian gecko with notably depressed and long head and a strongly banded tail, in mountains of south Sinai. Head oval or sub-triangular, notably long and depressed; eyes large, prominent; ear moderately large, crescentic. Body slim, depressed. Limbs moderately long and slender; digits long. Tail segmented, slender, cylindrical, tapering smoothly, not constricted at its base, about 55 per cent of total length. Subcaudal scales slightly enlarged transversely. Dorsal scales heterogeneous, back covered with small granular scales interspersed by 14–16 longitudinal rows of rather small, weakly keeled, round tubercles. Anterior chin shields large and in broad contact. Males have 4–6 precloacal pores. Maximum total length about 12cm (maximum snout–vent length 5.7cm), average 10–11cm, males slightly smaller. Dorsum translucent greyish-pink, with a pattern composed of four dark brown X-shaped narrow crossbars on neck and back, the broad gaps studded with very distinct whitish tubercles. Tail has 8–9 broad alternating dark and light (distal half white) transverse bands. A clear dark band runs from nostril across eye to above ear-opening. Iris greyish-brown. Regenerated tail pinkish with scattered black spots, covered by small uniform scales. Underside white. Taxonomic Notes: Sister to *H. dawudazraqi* of the Middle East. It belongs (with *H. lavadeserticus*, *H. lemurinus* and *H. turcicus*) in the *H. turcicus* group, which lies within a so-called Arid Clade of mostly tuberculated, arboreal *Hemidactylus*; their distribution centred on Arabia and the Horn of Africa.

HABITAT AND DISTRIBUTION: Mountainous desert, largely in or near relatively mesic environments, between 600 and 2,000m. This is a highly restricted near-endemic species, known from a handful of records in the mountains of south Sinai in Egypt, within a radius of about 50km of the highest peak of Mt Katherine. Known localities are; Wadi Feiran, Sheikh Awad, El Tarfa, Wadi Gebal and 'Blue Valley' (south of Wadi Wazarah). Elsewhere, known from the Wadi Ramm region of south-west Jordan, and probably widespread over intervening country; all suitable habitat.

CONSERVATION STATUS: Least Concern, although generally uncommon, secretive and local.

NATURAL HISTORY: Poorly known. Nocturnal and rock-dwelling, on steep rocky mountains with boulders, cliffs and caves, and in surrounding shrubby or oasis-type vegetation. In Jordan found along walls of a canyon, with relatively high humidity and cool temperatures during daytime. Very agile on overhanging and smooth surfaces. When disturbed they retreat into narrow cracks and crevices. In St Katherine town several were observed inside abandoned and occupied traditional dwellings, and on some modern buildings. They lay hard-shelled eggs. Diet invertebrates.

Hemidactylus mindiae, Sinai.

TANA RIVER GECKO

Hemidactylus modestus
Günther, 1894

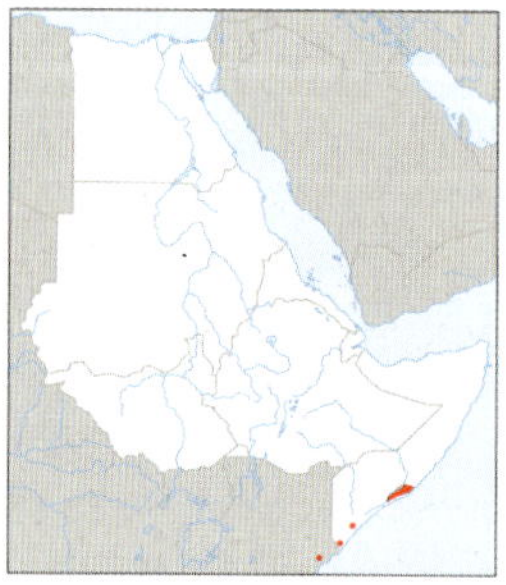

IDENTIFICATION: A small, brown, stout terrestrial gecko with homogeneous smooth dorsal scales, sometimes spotted light and dark, confined to a narrow coastal strip in southern Somalia. Head oval, fairly narrow; eyes moderate, less prominent, iris dark; ear-opening small. Body cylindrical; limbs stout and short. Digits short, stout, equally wide along entire length. Tail not segmented, cylindrical, short, about 45–50 per cent of total length, tapering smoothly, subcaudal scales slightly enlarged. Unusually, body and tail covered by small, overlapping scales of roughly equal size, arranged in about 80–100 rows around mid-body, head scales smaller with a few enlarged flat tubercles on occiput; enlarged tubercles on hind thighs, anterior chin shields in broad contact. Male has 10–18 precloacal pores. Maximum length about 9.5cm (maximum snout–vent length 4.8cm), average length 7–8cm. Coloration distinctive and fairly variable, with a uniform and spotted phase, intermediate examples also known. Uniform phase brown above, whiter below, with dark upper and lower labials; known from Ngatana (Kenya) and Ola Uager and Mareri, near Gelib in Somalia. The spotted phase (found further north) is light brown with darker longitudinal markings forming irregular dorsolateral lines; elongate dark markings from these lines extend to the middle of the back. Sometimes scattered white spots on neck, body and tail. These animals have a dark bar through each eye, which coalesce on back of neck to form a crescent, and extend as a thin line towards groin, fading beyond. Occasionally there are additional incomplete crescents behind eyes. Upper labials pale. Underside whitish. Regenerated tail swollen, shorter and less colourful.

HABITAT AND DISTRIBUTION: Coastal forest and thicket, and dry open woodland on consolidated coastal dunes, from just above sea level to 100m. Endemic to hot, humid coastal strip of southern Somalia and Kenya, not extending more than 50km inland. Southern Somali records include Ola Uager in Lower Juba region, around Mareri near Gelib, and on lower Juba. Widespread around Mogadishu (known from Genale, Afgoi, Balad and north along coast to Warshiikh). Elsewhere, coastal Kenya.

Hemidactylus modestus, Warshiikh.

CONSERVATION STATUS: Data Deficient. Probably more widespread as suitable habitat is extensive, so not under threat.

NATURAL HISTORY: Unknown. Kenya specimens were in piles of rubbish and rotting vegetation under mango trees. It seems that the darker, uniform-coloured phase occurs in well-vegetated habitats, the paler spotted form in more open country. Biology presumably similar to other *Hemidactylus*, i.e. nocturnal, lays eggs, eats invertebrates.

SNAKE-SCALED GECKO

Hemidactylus ophiolepis
Boulenger, 1903

IDENTIFICATION: A small terrestrial gecko with homogeneous smooth dorsal scales and pinkish-brown colour with transverse whitish lines, known only from low dry central Ethiopia. Head fairly long and wide, eyes large, prominent, iris dark. Body sub-cylindrical, slightly depressed. Digits relatively short, stout, equally wide along entire length; adhesive pads very small. Tail not segmented, cylindrical, about 50 per cent of total length and very slightly swollen behind cloaca, then tapering smoothly. Subcaudal scales slightly enlarged. The scalation is unusual; all dorsal scales on body, as well as on head and neck roughly the same size, fairly large, smooth, flat and overlapping, arranged in about 50 rows around mid-body. Anterior chin shields in broad contact. Males have eight precloacal pores.

Hemidactylus ophiolepis, female, Dire Dawa.

Hemidactylus ophiolepis, female, Dire Dawa.

Maximum size about 9.5cm (maximum snout–vent length 4.5cm), average length 7–9cm. Coloration distinctive, dorsum pale grey or pinkish-brown, with a number of narrow white crossbars, irregularly and discontinuously distributed; sometimes scattered dark brown or blackish scales present on back. A dark brown stripe extends from nostril through eye and sometimes reaches shoulder. Tip of snout, upper labials and scales around eyes usually yellowish or white, and typical yellow spot in front of eye is useful aid to identification. Regenerated tails more swollen, shorter, and lack transverse narrow white lines. Underside white. Juveniles resemble adults. Taxonomic Notes: Similar and closely related to *H. ophiolepoides*, and in same clade as *H. albopunctatus*, *H. modestus* and *H. ophiolepoides*.

HABITAT AND DISTRIBUTION: Endemic to Ethiopia, in dry grassland savanna with relatively few trees and rich herbaceous cover, on plains at 750–1,250m. Originally described from Amibarra in Danakil region of central Ethiopia. Several specimens from Ilala Sala plains of Awash National Park, recent records at Dire Dawa and a locality about 15km north-east of there, suggest it is more widespread.

CONSERVATION STATUS: Least Concern. Although secretive and uncommon, unlikely to be under threat.

NATURAL HISTORY: Nocturnal and ground-dwelling, as indicated by its short toes and poorly developed adhesive pads. On Ilala Sala plain, Awash National Park, several were found by day hiding beneath isolated boulders resting on soil, but never among rocks favoured by other species of *Hemidactylus*. A female north-east of Dire Dawa was concealed under a small stone in fairly flat, open, semi-desertic savanna with some grass cover and a few scattered stones. A juvenile had eaten a small grasshopper. Lays eggs, clutch details unknown.

SOMALILAND GROUND GECKO

Hemidactylus ophiolepoides
Lanza, 1978

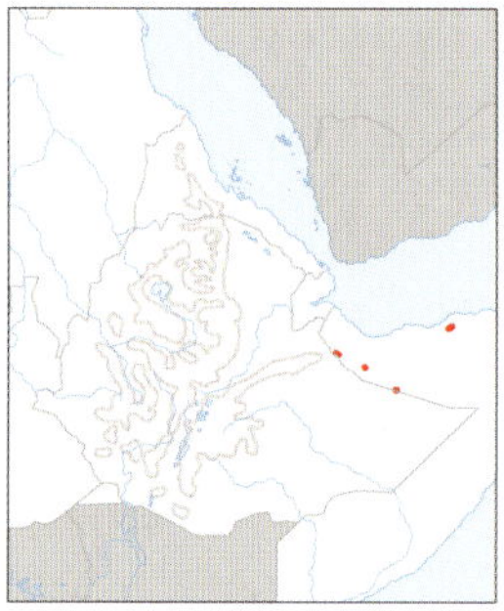

IDENTIFICATION: A small terrestrial gecko with homogeneous smooth dorsal scales all of similar size; endemic to Somaliland. Head fairly long and wide. Eyes large, prominent, iris dark. Body sub-cylindrical, slightly depressed. Digits short, stout, equally wide along entire length; adhesive pads very small. Tail not segmented, cylindrical, about 50 per cent of total length and tapering smoothly. Subcaudal scales slightly enlarged. The scalation is unusual; all the dorsal, head and neck scales are roughly the same size, fairly large, smooth, flat and overlapping, arranged in 53–59 rows around mid-body. Anterior chin shields very variable, may be in broad or short contact, or separated by a small scale. Male has 22–25 precloacal and femoral pores. Maximum size about 10.5cm (maximum snout–vent length 5.2cm), average 8–10cm. Coloration distinctive and fairly variable, seems to be dimorphic with a spotted (more common) and uniform (less common) phase. The uniform phase (from Erigavo region) is yellowish with a faint darker vertebral stripe and sometimes fine and irregular dark speckles. The spotted phase (Borama, Hargeisa and Haud regions) is violaceous-brown (sometimes whitish-yellow), with a thin vague vertebral stripe, bordered on both sides by a single row of scattered dark and white spots. In both phases a dark brown stripe extends from the nostril through eye and ends above shoulder (may continue as thin vague line up to groin). The upper labials and scales around eyes are usually yellow or white; yellow spot in front of and behind eye is a useful aid to identification. Regenerated tail more swollen, shorter and lacks white spots. Belly is pinkish-white, underside of tail yellowish. Juveniles of the spotted phase are pale violaceous-brown with numerous dark brown and whitish spots; the colour of juveniles of the uniform phase is unknown.

HABITAT AND DISTRIBUTION: Endemic to Somaliland and adjacent Ethiopia, in semi-desert, dry savanna and *Vachellia–Commiphora* bushland of the Somali arid zone, at 920–1,980m. In Borama and Erigavo regions it is a montane species, found at relatively high altitudes, 1,500–1,980m, but in the Haud it occurs in arid, low-altitude country. Somaliland records are sporadic. Known from Borama District between Borama town and Dilla village, eastwards from about 10km south of Hargeisa, also in the Haud region, about 24km west of Durruqsi village. Isolated records from Erigavo and Buq village in Sanaag region, Somaliland. Probably more widespread, but overlooked. Although data lacking, it is uncommon and secretive.

Hemidactylus ophiolepoides, female, Hargeisa.

Hemidactylus ophiolepoides, male, Borama.

CONSERVATION STATUS: Least Concern.

NATURAL HISTORY: Poorly known. Nocturnal and terrestrial. Favours rocky or stony habitats. Recently collected specimens were found under rocks. An adult female was crawling at night on the rocky edge of a dry wadi in Buq village. The Haud animals were in a termite nest. The specimen from between Borama and Dilla was under a stone at the edge of a field. None was found in the dry, cold season (December–March) so probably aestivates then; active in the same area between June and September, after rain, emerging from burrows and crevices Sympatric with other terrestrial *Hemidactylus*, e.g. *H. albopunctatus* and *H. isolepis*.

NORTH SOMALI HOUSE GECKO

Hemidactylus pauciporosus
Lanza, 1978

IDENTIFICATION: A medium-sized, stocky, yellowish-brown gecko with large dorsal tubercles and relatively few (4–8) precloacal pores, endemic to rocky areas of northern Somalia. Head fairly oval or sub-triangular, eyes large, prominent; ear-opening large, crescentic. Body stout, slightly depressed. Limbs moderately long and stout; digits fairly long and strongly dilated, adhesive pads well developed. Tail segmented, covered with small striated slightly overlapping scales, fairly slim, tapering, depressed, with a slight median dorsal furrow, not constricted at its base, about 50–55 per cent of total length. Subcaudal scales transversely enlarged. Dorsal scales heterogeneous, with small granular scales interspersed by 12–14 (occasionally 16) rows of strongly keeled, striated, sub-trihedral and rounded or slightly oval tubercles. Anterior

Hemidactylus pauciporosus, Gan Libah.

Hemidactylus pauciporosus, Borama.

chin shields large and in broad contact. Males have 4–8 precloacal pores. Maximum total length about 13cm (maximum snout–vent length 6.2cm), average 11–12cm, females slightly smaller. Dorsum yellowish-brown to hazel-grey, the impression is of a pale unicoloured gecko, but close inspection reveals pale X-shaped dark spots; one on neck, three on back and one at base of tail, with an additional two between hind limbs, and 7–10 distinct bands on tail. A vague dark stripe extends from nostril across eye to above ear-opening. The top of the head and limbs mottled darker brown. Iris yellowish-brown. Pinkish or yellowish-white below. Juveniles have similar coloration and pattern, but differ in that the posterior half of the tail is white with black bands. Taxonomic Notes: Originally considered a subspecies of Arabian *H. yerburii*, but recent genetic analysis indicates a closer relationship with *H. granchii* and then *H. macropholis*.

HABITAT AND DISTRIBUTION: A fairly wide range of habitats in rocky semi-desert and dry savanna, between sea level to about 1,800m, however most records are at 900–1,500m. Widely distributed endemic to northern Somalia, from south-east Puntland to western Somaliland and perhaps adjacent Ethiopia. The easternmost record is from Eyl in the Nogal Valley, thence north across the Sol plateau (including Garowe, Wadi Dudo, Darror Valley (Buran and Iskushuban), the Karin Gap to Galgalo Oasis and Candala in north coastal Puntland. In Somaliland mostly known from mountainous regions, including the western Cal Madow Mountains (Rugay, Mader Moge and Buq villages). Known from Illiad and west of the Goolis Mountains, to the environs of Sheikh. Also found on the top of Gan Libah and in Ogo Mountains north-east of Hargeisa. Fairly common in Borama, western Somaliland. In Somaliland not known from very dry and hot Guban plains and sandy regions of the Haud. Probably occurs in Ethiopia in the approximate triangle between Jijiga, Dire Dawa and Borama.

CONSERVATION STATUS: Not Assessed; however widespread and locally very common, one of the commonest *Hemidactylus* in northern Somalia.

NATURAL HISTORY: Nocturnal and arboreal, equally at home on rocks and in urban settings; often on walls at night near lights hunting invertebrates. Sometimes diurnal in dark places, like sewer channels or bridges, if prey is available. Rarely on ground. In natural settings usually rocks or large boulders, it shelters in crevices. In Somaliland sympatric at higher elevations with other rock-dwelling species such as *H. barodanus*, *H. granchii* and *H. laevis*. In hotter, sandier habitats it is replaced by the similar *H. macropholis*. Two hard-shelled eggs, roughly 10mm in diameter, are laid in any suitable refuge.

TREE GECKO

Hemidactylus platycephalus
Peters, 1854

IDENTIFICATION: A large, flat, arboreal gecko, known from southern Somalia and Ethiopia. Head oval; eyes large and prominent; ear-opening small, oval. Body slim and depressed. Limbs moderately long and stout; digits long and strongly dilated, and adhesive pads well developed. Tail segmented, slim, tapering smoothly, sub-circular, with a slight median dorsal furrow, not constricted at its base, moderately long, 51–54 per cent of total length. Subcaudal scales transversely enlarged. Dorsal scales heterogeneous, small granular scales interspersed by 8–12 longitudinal rows of larger flat or conical, smooth or weakly keeled, rounded tubercles. Anterior chin shields usually in broad contact. Males have 45–57 preano-femoral pores, females occasionally have four precloacal pores. Maximum total length about 19cm (maximum snout–vent length 9.4cm), average 12–17cm. Usually some shade of grey or brown, the neck and back with a series of 5–6 faint dark chevrons, and 10–11 additional dark bands on tail, flanks usually darker; a dull darker head stripe sometimes present, extending from nostril across eye to above ear-opening or shoulder. Top of head mottled black or grey, limbs barred; iris grey or red-brown. Underside whitish or pale yellow. Taxonomic Notes: Closely related to *H. smithi*, its sister species.

HABITAT AND DISTRIBUTION: In forest, woodland, moist and dry savanna, and semi-desert, from sea level to around 1,300m in our area. Widespread in southern Somalia between the Webi Shebelle and Juba Rivers, thence west towards the Kenyan border, north to Jesomma, El Mocoile (or Mukayle), Oddur and Dolo. Known localities include the Bajuni Islands (Chovaye, Ngumi and Coiama), Giohar, Afgoi, Mogadishu, Brava, Kismayo, Afmadu, Gelib, Bardera, Baidoa and Lugh; Kenya/Ethiopia border localities include Mandera, Ramu, Malka Murri and Moyale, thence into southern Ethiopia to Filtu, and between Hudet and Goba, south of the Daua River. In dry Somalia further north it is replaced by *H. smithi*, which has almost identical biology. Elsewhere, south to Zimbabwe and the Comoros. In southern Somalia widespread and common, in southern Ethiopia more local. It tolerates human activities.

CONSERVATION STATUS: Least Concern.

NATURAL HISTORY: Nocturnal. Often on large trees, also rocks; will enter buildings or on bridges. Rarely descends. In natural settings shelters under bark and in cracks in trees and rocks; in buildings in any suitable crevice, emerging at night to patrol the walls and ceiling in pursuit of insects attracted to lights. May congregate in places, a group of 11 was observed high on a trunk in southern Kenya, a similar number on a building in the centre of town in eastern Kenya. Lays two hard-shelled eggs, 10–13mm in diameter, in any suitable refuge. Eats a wide range of invertebrates. In vicinity of Mogadishu and along the Shebelle River as far north as Ghersale it is sympatric with similar, but much smaller *H. smithi*.

Hemidactylus platycephalus, Wachile.

Hemidactylus platycephalus, Wachile.

PUCCIONI'S GECKO

Hemidactylus puccionii
Calabresi, 1927

IDENTIFICATION: A small, slim, ground-dwelling, pinkish-brown gecko with heterogeneous dorsal granular scales, known only from central Somalia. Head oviform, eyes large, fairly prominent; ear-opening small, oval. Body sub-cylindrical, slightly depressed, slim. Digits short, stout, equally wide along entire length. Tail very slightly segmented, slim, cylindrical, with a slight median dorsal furrow, about 55 per cent of total length, subcaudal scales small. Dorsal scales heterogeneous, small, smooth, granules, interspersed by 10–12 rows of feebly keeled tubercles. Anterior chin shields small, almost circular and not in contact. Males have four precloacal pores. Maximum size about 11cm (maximum snout–vent length 5.3cm), average 8–9cm. Dorsum rosy-maroon, sprinkled with little black spots; five to six more or less defined dark red-brown crossbars on back and about 8–10 on tail; darker diffuse irregular patches behind eyes, iris bluish-grey in specimens. Underside pinkish-white. Taxonomic Notes: closely allied to *Hemidactylus somalicus*.

HABITAT AND DISTRIBUTION: In fixed sand dunes covered with mixed bushland on coast, xerophilous open woodland, subdesert scrub, steppe and grassland with scattered low shrubs further inland, between sea level and 200m. Endemic to central Somalia, mostly along coast but extends about 130km inland. Known from Geriban in the north, thence south along the coast (Afghedud [Afguduudle] and between Hobyo and Durgale). Widespread in Galguduud and Middle Shebelle, known from El Bur, Bud Bud, El Dere, the environs of Warshiikh and Mogadishu, and north on the Webi Shebelle to Burdere. On the Mudug plain, between Garowe and Geriban it meets the closely related, possibly sister species, *H. somalicus*. Widespread and locally common.

CONSERVATION STATUS: Least Concern.

NATURAL HISTORY: Poorly known. Nocturnal and terrestrial. Two specimens from near Mogadishu were in very open sandy country without shade plants, within 500m of sea; one in a mollusc shell, the other a rocky cairn. Biology unknown but presumably similar to other *Hemidactylus* of dry country, i.e. lays eggs and eats invertebrates.

Hemidactylus puccionii, preserved, El Bur.

RED SEA GECKO

Hemidactylus robustus
Heyden, 1827

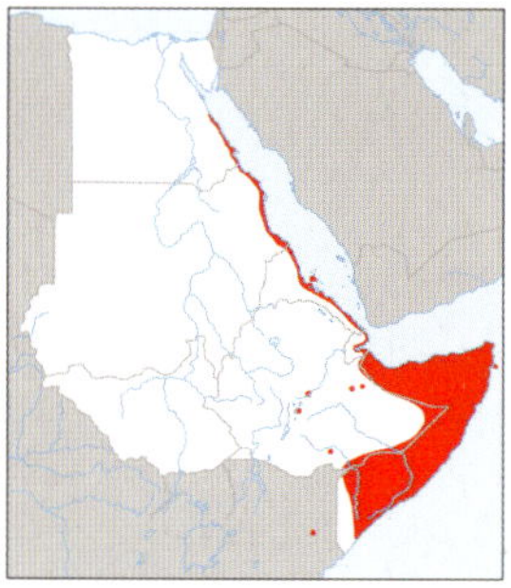

IDENTIFICATION: A smallish, fairly slim gecko, often striped, with heterogeneous dorsal scales, introduced along Red Sea coast from Egypt to Somalia. Head oval or sub-triangular in males, eyes large and prominent; ear moderately large, crescentic. Body slim, depressed. Limbs moderately long and stout; digits fairly long, moderately dilated. Tail weakly segmented, circular, tapers smoothly and appears smooth, with keeled tubercles that do not stand out, not constricted at its base, about 50–54 per cent of total length. Subcaudal scales slightly enlarged transversely. Dorsal scales heterogeneous; small granular scales interspersed by 13–18 rows of keeled and posteriorly pointed, slightly oval or almost round tubercles. Anterior chin shields large, in broad contact. Males have 5–8 precloacal pores. Maximum total length about 11cm (maximum snout–vent length 5.5cm), average 9–10cm, females smaller. The back is pale grey or yellowish to pinkish-brown, with darker spots, sometimes forming three longitudinal stripes, or blotches that may form irregular crossbars, or 4–5 indistinct cross-shaped marks on back, but tubercles are often white; tail has 15–18 irregular, narrow, dark bands or only spots. A clearly delineated dark band extends from the nostril across the eye to the ear-opening, continuing as an interrupted line to the shoulder. Lips whitish. Iris yellowish-brown. Pinkish-white below. Juveniles have similar coloration and pattern. Taxonomic Notes: Genetic analysis indicates that this species (with

Hemidactylus robustus, male, Burao.

Hemidactylus robustus, Marsa Alam.

several other species in south-west Arabia) belongs in the *H. robustus* clade, and is closely related to *H. awashensis*. The name *H. parkeri* (now relegated to synonymy of *H. robustus*) was formerly applied to some African populations of this species.

HABITAT AND DISTRIBUTION: Occupies a wide range of usually arid habitats, from semi-desert to dry savanna and xerophilous woodland, from sea level to about 1,850m (at Harar, Ethiopia), but mostly at lower elevations. Usually associated with man-made structures. Can be found under debris, among rocks or fallen logs, on walls inside or outside houses and ruins. It probably originated in Arabia and was transported to Africa in cargo. Extensively distributed along Red Sea coast from near Safaga in Egypt south across the coastal lowlands of Sudan, Eritrea (on several Dahlak Islands including Andeber, Dahlak Kebir, Dissei and Nokra), Ethiopia, Djibouti and most of the Somali peninsula. A limited penetration inland in Somalia; might occur in the eastern Ethiopian Ogaden, as it is known from Filtu. Reported from Lake Langano and Awash National Park in central Ethiopia. Present in cities across its wide range and often sympatric with other invasive species like *H. flaviviridis* or *H. frenatus*. Elsewhere, a couple of records from north-east Kenya, introduced on Zanzibar. Outside Africa occurs from western India to Arabia, introduced to Socotra. Widespread and locally very common.

CONSERVATION STATUS: Least Concern.

NATURAL HISTORY: Nocturnal and fast-moving, on ground or on rocks and buildings. Mostly synanthropic, often hunting insects around lights, in gardens, on walls, tree trunks, fallen timber and building debris. Active on ground at night at Geerissa in Awdal region, Somaliland, around piles of plant debris near houses, into which they retreated when disturbed. In natural settings in Egypt found under dense halophytic vegetation on low littoral dunes, away from rocks or buildings. In Candala in Puntland in rocky gorges. Two hard-shelled eggs, roughly 8mm in diameter, laid in a crevice or under bark. Communal clutches found in dead, hollow branches in a bush in a Somaliland garden. Call a series of short barks repeated at intervals of a few seconds. Often sympatric with *H. flaviviridis* (over entire range), *H. macropholis* (Burao, Somaliland), *H. laticaudatus* (Harar, Ethiopia) and *H. barodanus* (Las Geel, Somaliland); and probably eaten by those larger species.

PRINCE RUSPOLI'S GECKO

Hemidactylus ruspolii
Boulenger, 1896

IDENTIFICATION: A small chunky gecko, vivid black-and-yellow or grey, with a swollen, carrot-shaped tail, of the dry Somali peninsula. Head oval or sub-triangular in males, eyes moderately large, prominent; ear-opening fairly large, elliptic. Body stout, short and slightly depressed. Limbs short, fairly slim; digits fairly short, moderately dilated. Tail distinctive, segmented, carrot-shaped, wide and swollen, depressed, less distinctly constricted at its base, slightly less than 50 per

Hemidactylus ruspolii, Kenya.

cent of total length. Subcaudal scales not transversely enlarged. Regenerated tail shorter and wider, leaf-shaped, depressed and swollen. Dorsal scales heterogeneous, small granular scales interspersed by 14–18 rows of large, strongly keeled, trihedral and mostly rounded tubercles. Anterior chin shields large, in broad contact. Males have 28–36 precloacal pores in an uninterrupted series. Maximum total length about 10cm (maximum snout–vent length 5.7cm), average 6–8cm. Ground colour variable, may be pinkish, grey, brown or almost black. Usually there is a series of four or five dark sub-rectangular blotches along the spine, with two parallel rows of similar blotches either side. Many of the dorsal tubercles are white or yellow, and in striking contrast to the ground colour. A distinct dark stripe runs from the nostril across the eye to above the ear-opening; top of head and tail mottled black or dark brown. Underside white. Some juveniles are almost uniform black, with yellow or red tubercles, the tail appears black with fine yellow bands, and some adults retain this colour. Taxonomic Notes: Genetic data and morphological similarity indicate this species is closely related to *H. angulatus*, which is also known from north-east Africa.

HABITAT AND DISTRIBUTION: A typical inhabitant of the Somali arid zone, from semi-desert to dry savanna with sandy dry *Vachellia–Commiphora* bushland and thicket , between sea level and 1,000m. Occurs from the Haud region (environs of Balumbal, Bohodle, 35km west of Qoriley) in southern Somaliland, across the eastern Ogaden of Ethiopia (Ado and Galadi) to the Hiran region of south-central Somalia (Belet Uen), thence south to Mogadishu and Afgoi and along the coast to Kismayu, and Afmadu. Extends north along the Juba River to Bardera, Berdale and Lugh, as far as Dolo and Mandera regions, thence to northern Kenya. Probably more widespread in central Somalia and Ogaden than shown on map, but records lacking. Not in the pet trade or subject to specific pressures. Widespread and fairly common.

CONSERVATION STATUS: Least Concern.

NATURAL HISTORY: Nocturnal and apparently largely terrestrial. In Somalia and Ethiopia associated with fairly sandy habitats. By day, shelters under rocks, stones, plant debris or in logs on ground, often under tree bark or in tree cracks. A specimen from Galadi in the Ogaden was found in the wall of an abandoned building. In northern Kenya, geckos hidden under Doum palm leaves touching the ground emerged after sunset. Two hard-shelled eggs are laid. Eats a wide range of invertebrates.

SPECKLED GECKO

Hemidactylus sinaitus
Boulenger, 1885

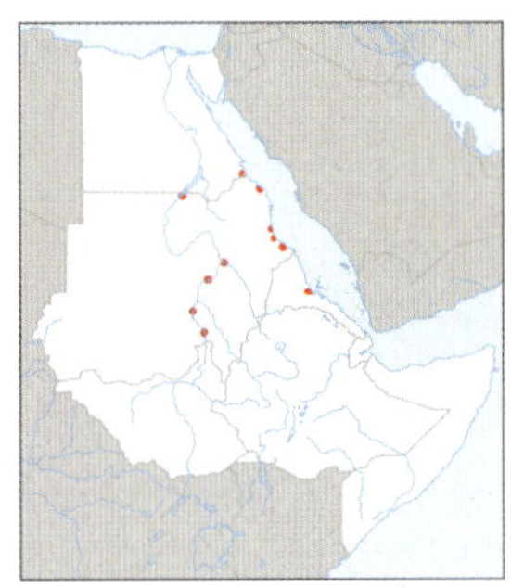

IDENTIFICATION: A small ground-dwelling gecko with heterogeneous dorsal scales and uniform subcaudal scales, from south-east Egypt to central Eritrea. Head oval, eyes large and prominent; ear-opening fairly small and oval. Neck distinct, rather short; body short, fairly stout, slightly depressed. Limbs somewhat short and thick; digits moderately long and slightly dilated. Tail weakly segmented, slightly depressed at base, with a slight median dorsal furrow, slim, tapering, not constricted at its base, about 50–54 per cent of total length. Subcaudal scales uniform (an excellent aid to identification). Dorsal scales heterogeneous, small, smooth or rugose granular scales interspersed by 14–16 rows of keeled, trihedral and more or less rounded moderately large tubercles. The anterior chin shields are fairly large, in short contact, or not in contact. Males have 4–7 precloacal pores. Maximum total length about 11cm (maximum snout–vent length 5.7cm), average 9–10cm. Dorsum brownish or yellowish-grey, with irregular and indistinct dark mottling, and an irregular zigzag paler vertebral stripe. The tubercles are dark on the back, white on flanks. Tail with numerous indistinct dark bands (10–15) interspersed by fine white rings, one tubercle wide. A fairly wide dark brown stripe extends from the nostril across the eye to above the ear-opening (may continue past the shoulder in extralimital populations). Limbs and top of head mottled dark brown and white or yellow. Iris yellowish-grey. The regenerated tail is greyish-yellow, mottled dark brown. The belly is white, the tail underside yellowish mottled dark brown. Taxonomic Notes: Closely related and similar to the recently described *H. afarensis*; both, along with *H. citernii* and *H. foudaii*, form a monophyletic group, typified by short distal joints on the digits.

HABITAT AND DISTRIBUTION: In mostly very dry, semi-desert sandy environments with scattered *Vachellia* trees, without or with sparse vegetation, along coasts or rivers, between sea level and 400m. Records somewhat sporadic. Known from Shalatein and Gebel Elba on the Egyptian Red Sea coast. An old record of a Wadi Halfa (Sudan) juvenile needs verification. Widespread along Sudanese Red Sea coast (Durrur, Suakin and Tokar), inland Sudanese records include Nile Valley near Atbara, Wad Ben Naga, Ab-Dareesh (or Abu Idres) about 40km south of Khartoum on the White Nile, and south from Asalaya. A handful of records from the eastern coastal lowlands of Eritrea include Ailet (just north of Gahtielay) and Green Island (Sheik-Said Island) near Massawa. Probably in intervening areas, all suitable habitats. In the Afar Triangle it is replaced by its sister species, *H. afarensis*. Elsewhere, coastal lowlands of extreme south-west Saudi Arabia (Farasan Kebir Island) and Yemen. Appears to be rare and local.

Hemidactylus sinaitus, Atbara.

CONSERVATION STATUS: Least Concern.

NATURAL HISTORY: Poorly known. Nocturnal and largely terrestrial on open ground or low rocks, although observed at night climbing *Vachellia* trees, also on a *Lycium* bush in *Vachellia* woodland. Presumably shelters by day in bushes, burrows under bushes or under rocks. An adult male from Ab-Dareesh in Sudan was found by day under the bark of a *Vachellia* tree on a dune beside the White Nile. They lay eggs and eat small invertebrates. At several localities they are sympatric with *H. foudaii* and *H. robustus*.

SMITH'S TREE GECKO

Hemidactylus smithi
Boulenger, 1895

IDENTIFICATION: A small to medium-sized attractive, flat, mottled ashy-grey arboreal gecko, known from northern and central Somalia and the Ethiopian Ogaden. Head oval, eyes large, prominent; ear-opening small, oval. Body fairly slim, depressed. Limbs moderately long, quite stout; digits long, strongly dilated. Tail segmented, slim, tapering smoothly, sub-circular, with a slight median dorsal furrow, not constricted at its base, fairly short, about 50 per cent of total length. Subcaudal scales transversely enlarged. Dorsal scales heterogeneous, small granular scales interspersed by 10–12 rows of larger, flat or conical, smooth, rounded tubercles. Anterior chin shields usually in broad contact. Males have 28–33 preano-femoral pores. Maximum total length about 12cm (maximum snout–vent length 5.7cm), average 10–11cm, hatchlings about 2.5cm. Dorsum grey or pale brownish-grey, neck and back having 5–6 darker, brown or blackish, wavy crossbars finely edged with black posteriorly, with 8–10 additional dark bands on tail; a dark stripe extends from nostril across eye and above ear-opening to shoulder, becoming a fairly thin streak to groin and onto tail. There is a distinct W-shaped dark marking on the occiput, the head mottled black or dark brown, limbs barred; iris sandy-grey. Underside cream, white or yellow. Juveniles are browner, but with an identical pattern. Taxonomic Notes: Genetic data and morphology indicate, interestingly, a close relationship to *H. platycephalus*, a larger arboreal form widespread further south.

HABITAT AND DISTRIBUTION: Associated with arid or semi-arid low-altitude country. Known from near-desert coastal regions, dry *Vachellia–Commiphora* bushland and thicket, and xerophilous inland woodland, from near sea level to about 1,500m. Endemic to the Somali peninsula. Widespread in northern Somalia, from Borama and Hargeisa in western Somaliland through the Ogo, Goolis and Wagar Mountains, thence south to the Haud plateau on the Ethiopian border; notable localities include Hudisso, Lafaruug, Burao, Shanshacade and Bohodle. Unknown from most of Sanaag region, perhaps due to under-collecting. Recorded from Las Korey, on coast of north-east Somaliland, thence souths along the Somaliland/Puntland border, from El Donfar and Buran in the Darror Valley to Sol Haud and the Nogal Valley. Records from central Somalia include Geriban, Galkayo, inland of the Hobyo coast (at Biomal dunes, between Wisil and Geedaley) and Bud Bud. In southern Somalia reported from Mogadishu and environs, thence north from Ghersale (Garsaale) to the upper Juba Valley south of Lugh. Two old and somewhat imprecise localities in Ethiopia, 'Shebeli' (Shebelle River valley) in the Ogaden, and between Imi and Fik on the 'upper Ganale' (River), which may be east of Kebri Mangest. Probably more widespread in the Ethiopian Ogaden, but this region is poorly explored. A vague record from Axum (not shown on map) in northern Ethiopia needs confirmation. Secretive, but widespread and locally common. Not in the pet trade, or subject to specific pressures.

CONSERVATION STATUS: Least Concern.

NATURAL HISTORY: Nocturnal, chiefly arboreal. Only on trunks and thicker branches of larger trees, it spends the day in shade of branches and leaves, usually at heights of 2–3m. It may feed by day in deep shade if insects are available. At night specimens have been observed moving onto thinner outer branches. Although solitary, several may live on a tree if it is the only suitably large one in the area. They rarely descend, and are not found in inhabited areas. A fairly shy and secretive species, easily found but difficult to catch. Two hard-shelled eggs, 10×8.9mm large, were found north of Burao in an abandoned termite mound, about 5m from a single *Vachellia* tree; three months later the eggs hatched in captivity. Diet small invertebrates.

Hemidactylus smithi, Lafaruug.

Hemidactylus smithi, young adult, Burao.

SOMALI GECKO

Hemidactylus somalicus
Parker, 1932

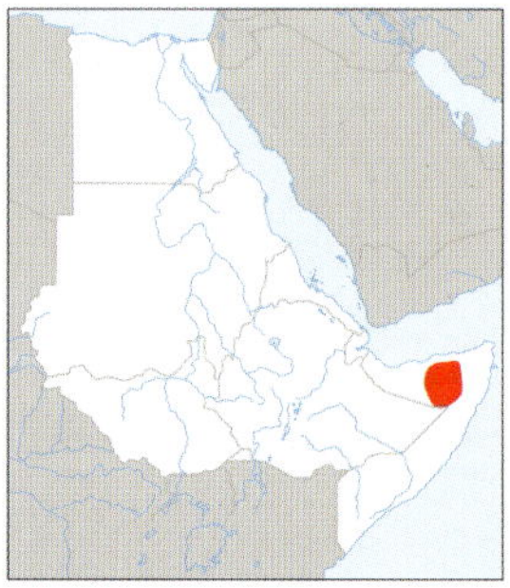

IDENTIFICATION: A small slim ground-dwelling pinkish-brown gecko with heterogeneous dorsal granular scales, known only from eastern Somaliland and adjacent Puntland. Head oviform, eyes large, fairly

Hemidactylus somalicus, female, Yuffleh.

Hemidactylus somalicus, male, Nogal Valley.

prominent; ear-opening small, oval. Body cylindrical, slightly depressed, slim. Digits short, stout, equally wide along entire length. Tail very slightly segmented, slim, cylindrical, with a slight median dorsal furrow, about 50 per cent of total length. Subcaudal scales small. Dorsal scales heterogeneous, small, smooth, granules (juxtaposed anteriorly, slightly overlapping posteriorly and on flanks), interspersed by a few irregularly scattered enlarged scales, especially posteriorly. Anterior chin shields small, almost circular, usually in point contact or not in contact. Males have two precloacal pores. Maximum size 8.6cm (maximum snout–vent length 4.3cm), average 7–8cm. Dorsum pinkish-white, dusted with darker golden-brown pigment that forms an indefinite pattern of four to five crossbars on the back and about 11 on tail, sometimes accompanied by numerous small white spots. There are darker diffuse and irregular patches behind the eyes, which extend backwards and meet to form a crescent on back of head, though often interrupted. Iris golden brown. Underside pinkish-white. Juveniles golden-yellow, dusted with brown. Taxonomic Notes: Recent genetic sequencing indicates it is most closely related to *H. citernii*; both are members of a large clade with *H. afarensis*, *H. foudaii* and *H. sinaitus*.

HABITAT AND DISTRIBUTION: Near-desert, semi-desert and dry savanna with rocky limestone or gypseous soils and sparse vegetation, especially low *Vachellia* trees, shrubs and succulent flora, particularly aloes, at elevations between 440 and 1,760m. Endemic to eastern Somaliland and adjacent Puntland. Its approximate range is from El Donfar in the Daror Valley in Puntland south across Sol Haud and the Carcar Mountains, along the border with Somaliland (near Xidda and Gardo) to the Nogal Valley, where reported at Garowe. Thence slightly west towards the Haud region; at Qoriley on the Somaliland/Ethiopia border and north to the environs of Las Anod and Hudun in the Nogal Valley. Recently found on the plains south of Yuffleh in Erigavo district. Probably more widespread in Sanaag (apart from the coast and mountains) and Sool regions of Somaliland. Eastern limits in Puntland unknown. Widespread and locally fairly common.

CONSERVATION STATUS: Least Concern.

NATURAL HISTORY: Poorly known. Nocturnal and terrestrial, it favours rocky or stony terrain between low hills in very arid, sparsely vegetated environments. A Haud specimen was found in a dead branch on the ground. Other specimens, including those recently collected south of Yuffleh, were under rocks. One or two hard-shelled eggs are laid. Diet small arthropods. A rare species, known only from about ten specimens.

NYIKA GECKO

Hemidactylus squamulatus
Tornier, 1896

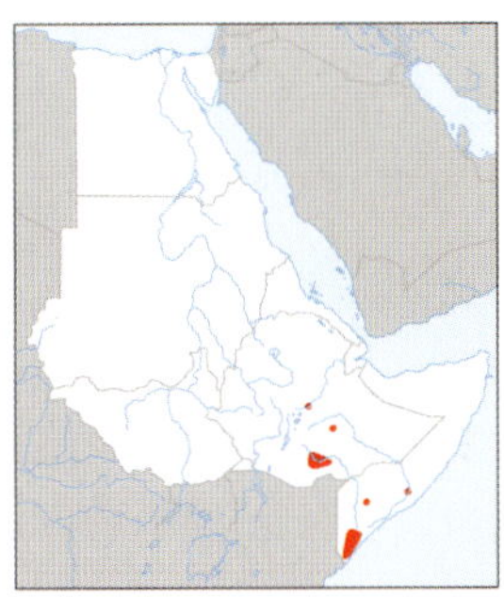

IDENTIFICATION: A small stout terrestrial gecko from southern Ethiopia and Somalia, with heterogeneous dorsal scales, the larger ones distinctly keeled. Head bulky, eyes large and prominent. Body sub-cylindrical, slightly depressed, fairly stout. Digits short,

Hemidactylus squamulatus, Sof Omar Caves.

Hemidactylus squamulatus, juvenile, Negele.

stout, equally wide along entire length; adhesive pads very small. Tail not segmented, cylindrical, about 50 per cent of total length, tapering smoothly or slightly swollen behind cloaca. Subcaudal scales usually enlarged. The scalation is unusual; most dorsal scales small, smooth and overlapping, but interspersed by 10–16 longitudinal rows of distinctly larger, more or less rhomboidal, keeled scales. Anterior chin shields in broad contact. Males have 6–20 precloacal pores. Maximum size about 10cm (maximum snout–vent length 5.3cm, females larger), average 6–8cm, hatchlings 2–2.5cm. Shades of brown, grey or rufous above, with a number of fine, vague darker crossbars or spots, some have dark patches behind the eyes, which may extend backwards and meet to form a crescent on back of head (an excellent aid to identification); some of the enlarged keeled dorsal scales are tipped posteriorly with cream or white. The regenerated tail is more swollen, shorter, less marked and colourful, with larger scales than the original. Underside whitish. Hatchlings and juveniles may be brown, pink or orange, finely speckled yellow or white. Taxonomic Notes: Very variable in both colour and appearance; genetic analyses indicate that as presently defined, there are several very distinct clades and at least five cryptic species concealed in this taxon, at least two in our area.

HABITAT AND DISTRIBUTION: Semi-desert, dry and moist savanna and grassland, from just above sea level to about 1,350m. Known from southern Ethiopia, apparently common in the vicinity of Negele and environs. Also near Sof Omar caves and an isolated record from the far side of the south-east Ethiopian mountains at Kereyu Lodge in Awash National Park. Sporadic records in southern Somalia include Ola Uager and north of Lak Bush-Bush River, Afmadu, near Mareri (close to Gelib), Dinsoor and north of Eggi. No records from the drier north Rahanuin region along the Somalia/Ethiopia border. Probably more widespread, but under-collected. Elsewhere, to dry north-east and east Kenya and north Tanzania; oddly absent from dry desertic northern Kenya. Widespread, locally common in southern Ethiopia.

CONSERVATION STATUS: Least Concern.

NATURAL HISTORY: Nocturnal and terrestrial. Emerges near dusk and hunts on surface at night; by day it shelters in holes and under ground cover. It lays two hard-shelled eggs, roughly 7–9mm in diameter, in any suitable refuge. Eats a wide range of invertebrates. Near Sof Omar caves in Ethiopia an adult pair was found by day under a flat rock in dry *Vachellia–Commiphora* bushland with red sandy soil.

TAYLOR'S GECKO

Hemidactylus taylori
Parker, 1932

IDENTIFICATION: A medium-sized stocky gecko with a unique, distinctive, short and carrot-shaped tail, known only from the Gardo region of Puntland. Head oviform to sub-triangular, eyes large and prominent; ear-opening fairly large, elliptic. Neck distinct, short; body stout, short and slightly depressed. Limbs short, quite muscular; digits fairly short, strongly dilated.

Original tail very distinctive, divided by annular grooves into narrow segments, covered by juxtaposed granular scales, about 5–6 transverse rows in a whorl, each with four additional slightly enlarged, pointed and keeled tubercles, strongly carrot-shaped and depressed, tapering, constricted at its base, original tail short, about 45 per cent of total length. Subcaudal scales large. Regenerated tail very short, depressed, leaf-shaped, swollen, covered with small, uniform scales. Dorsal scales heterogeneous, small granular scales interspersed by 13–15 longitudinal large, strongly keeled, trihedral and mostly rounded or slightly oval tubercles. Anterior chin shields large and in broad contact. Males have 8–9 precloacal pores. Maximum total length about 13cm (maximum snout–vent length 7.1cm). Dorsum pinkish-white, with an indistinct trace of four X-shaped dark spots, one on neck, three on back and one across base of tail; an indistinct dark stripe runs from eye above the ear-opening. Limbs and digits have dark transverse bars; the tail has numerous small regular black dots. Whitish below. Taxonomic Notes: Never sequenced but probably closely related to other species in the region with large keeled tubercles on dorsum, e.g. *H. granchii*, *H. macropholis* and *H. pauciporosus*.

HABITAT AND DISTRIBUTION: Endemic to border country between Puntland and adjacent Somaliland. Inhabits arid low hills, near-desert country with sandy and rocky limestone or gypseous soils and sparse steppe vegetation, stunted *Vachellia* trees and shrubs, at 800–900m. Known only from near Gardo in Puntland and about 35km south of there, 14km west of Yake village, both localities on the Sol Haud plateau. Gardo records are from 12km north-west of the town, and 'Heibogan', a cave or hole 3km north of town. Probably more widespread on Sol Haud but the area is under-explored.

CONSERVATION STATUS: Data Deficient.

NATURAL HISTORY: Poorly known, but probably nocturnal. This gecko is still known only from just six specimens captured in 1930 and 1931. The unique morphology of both the original and regenerated tail of this gecko distinguishes it from all other species of *Hemidactylus*, not just those of our region. Its range experiences year-round high temperatures and low annual rainfall, <150mm, and is one of the least hospitable places in the Horn of Africa. All specimens were found under rocks, suggesting it is terrestrial and possibly rupicolous. Presumably eats invertebrates and lays two hard-shelled eggs.

Hemidactylus taylori, preserved type.

OGADEN GECKO

Hemidactylus tropidolepis
Mocquard, 1888

IDENTIFICATION: A small terrestrial gecko from sandy areas of the Ethiopian Ogaden and Somalia, with heterogeneous dorsal scales, all of them keeled. Head oval, wide and deep, eyes large but less prominent. Body sub-cylindrical, slightly depressed, relatively long. Digits moderately long and thick, equally wide along entire length; adhesive pads small. Tail not segmented, cylindrical, slightly >50 per cent of total length, tapering smoothly (not swollen when regenerated). Subcaudal scales usually enlarged. The dorsal scalation is unusual; small, strongly keeled scales interspersed by 10–12 rows of large, strongly keeled scales. Anterior chin shields in broad contact. Males have 6–10 precloacal pores. Maximum size about 8.6cm (maximum snout–vent length 4.3cm, females larger), average 6–8cm, hatchlings 2–2.5cm. Two colour morphs exist, one patterned and the second, from red sandy soils, plainer. The former is olive-brown with 9–10 pairs of darker brown blotches on body, extending onto tail, where they gradually fade. The plain form is rufous; each enlarged keeled dorsal scale

Hemidactylus tropidolepis, Inaafmadow.

Hemidactylus tropidolepis, Shanshacade.

is brown or pale yellowish-orange. An irregular brown streak with diffuse edges runs from the nostril through the eye, it may not reach the shoulder or may continue to hind limbs. There is a distinct W-shaped dark spot on snout in some specimens. Underside pinkish-white. Hatchlings and juveniles may be brown or orange, finely speckled pale yellowish-orange.

HABITAT AND DISTRIBUTION: A typical species of the Somali arid zone, living in dry open *Vachellia–Commiphora* savanna and semi-desert with soft red sandy soil, but also found on fixed coastal sand dunes with mixed bushland and scattered trees, from near sea level to 860m. Known from two disjunct regions in northern Somaliland and adjacent parts of Ethiopia and Puntland, also southern Somalia into adjacent Kenya. Northernmost record north of Inaafmadow in Burao district, central Somaliland, thence south into Haud region (Shanshacade village), along Ethiopian border west of Bohodle. Two records in Nogal Valley (Bihen and about 30km west of Oisame). In the Ethiopian Ogaden it is known from Ado village and about 60km south-west of Bohodle. Southern Somali records at mesic localities between the Shebelle and Juba Rivers (Sguss village, 8km north of Balad), Dinsor, Bardera, Balli (perhaps Ballei Uen), Belet Amin, Afmadu and Mareri, near Gelib. A Dolo record needs confirmation, it might be *H. funaiolii*. Elsewhere, east and south-east Kenya. Widespread and locally very common.

CONSERVATION STATUS: Least Concern.

NATURAL HISTORY: Poorly known. Terrestrial and nocturnal. Favours sandy habitats in arid lowlands. The Haud specimens were under rocks and logs by day in red sand areas with scattered grass and thorn scrub. Occasionally sympatric with other geckos with similar habitat requirements, e.g. *H. albopunctatus*. Sympatric at Dinsoor (southern Somalia) with *H. squamulatus*. An adult from Sguss village was in a covered concrete silo. Dorsal colour depends on habitat; northern Somali animals are plain rufous, southern Somali animals spotted on a brown background. Two hard-shelled eggs laid in any suitable refuge; diet a wide range of invertebrates.

TURKISH GECKO

Hemidactylus turcicus
(Linnaeus, 1758)

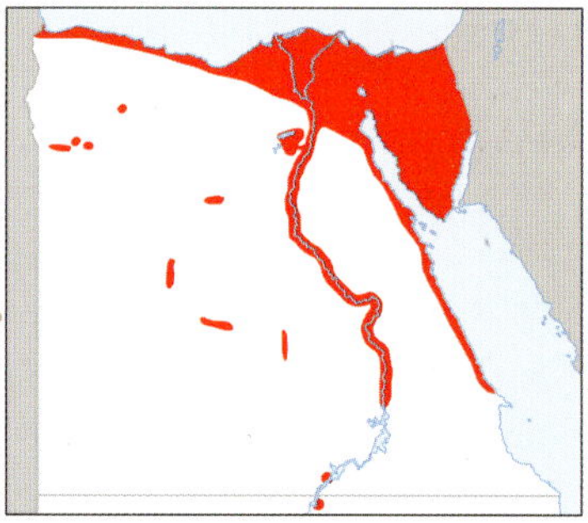

IDENTIFICATION: A common, medium-sized, fairly stout gecko with a deep head, small number of upper labials (7–10) and high number of preanal pores, widespread in Egypt except most of Red Sea Coast. Head sub-triangular, short and deep, eyes moderate but prominent; ear moderately large, oval or crescentic. Body stout and depressed. Limbs moderately long and fairly slim; digits long, moderately dilated. Tail strongly segmented, slender, depressed at base, cylindrical beyond, tapering smoothly, not constricted at its base, about 55 per cent of total length. Subcaudal scales transversely enlarged. Dorsal scales heterogeneous, small granular scales interspersed by 12–16 rows of large, strongly keeled, trihedral, round or slightly oval tubercles, flatter on the back, conical on flanks, and often whitish. Anterior chin shields usually in contact. Males have 6–10 precloacal pores. Maximum total length about 12cm (maximum snout–vent length 5.6cm), average 10–11cm, males slightly smaller. Shades of light brown to grey or yellowish above, with a series of 4–5 fairly distinct wide, dark and light transverse bands (often X-shaped) and scattered dark spots. Tail has 10–13 pairs of light and dark bands, which increase in contrast distally, being black and white over its terminal portion (especially in juveniles). A diffuse dark band, edged above by an indistinct pale stripe, runs from the nostril across the eye to above ear-opening. Iris yellow or grey brown. Yellowish or pinkish-white below, tail underside mottled brown. Taxonomic Notes: Long regarded as a very widespread, monotypic form, but recent molecular studies have revealed cryptic species within the taxon. Some have been described and synonymised (e.g. *H. granosus*, *H. robustus*) or are new (e.g., *H. dawudazraqi* and *H. lavadeserticus* from the Middle East). An additional cryptic form on the Red Sea Coast awaits description; the taxonomy of this fascinating species is not yet resolved.

HABITAT AND DISTRIBUTION: In our area usually in urban areas, where it favours buildings, ruins, walls and other man-made structures, also gardens and oases. In undisturbed areas it inhabits rocky cliffs, ravines and rocky debris in semi-desert, but usually in mesic microhabitats, from sea level to about 1,000m. Known with certainty from the Mediterranean coast in Egypt, Sinai, the Suez Canal and the Nile Valley south to Aswan. Records at Abu Simbel in Egypt and Wadi Halfa in northern Sudan need further evaluation. Occurs around most large oases in the Western Desert in Egypt, notably Siwa, Kharga, Dakhla and Bahariya. In light of the resurrection of *H. granosus* and the presence of the aforementioned undescribed form, original records of this species on the Red Sea coast south to the Hamata (Qulan Islands) need re-evaluation. Elsewhere: Mediterranean North Africa, the Middle East and Europe. An aggressively invasive species, it has been introduced to South and North America. Very common and widespread.

CONSERVATION STATUS: Least Concern.

NATURAL HISTORY: An excellent climber, but sometimes on the ground, sheltering in burrows by day. Observed in Ismailia City 30m up on walls of residential buildings. They emerge at sunset or later. Active all night between spring and autumn, with a shorter period of activity in winter and only on warm nights. Very active March to November in Egypt, but little winter activity; they may hibernate. Solitary and highly territorial, but often found in large concentrations in suitable places. Usually synanthropic, they hunt insects by a 'sit-and-wait' strategy around lights at night, in and outdoors. In northern Sinai, often on buildings and external brick walls; but also under rocks, on date palms in oases, sometimes under tree bark. On sea shores they shelter under beached artefacts and in numbers under dense halophytes. Diet insects and other invertebrates (mostly moths, beetles and flies). Two hard-shelled eggs, roughly 8.5×10mm, are laid; buried in soil, under stones or in rock crevices. In Egypt reproduction starts in March and continues to November; usually three or four clutches per season. Hatchlings are 15–20mm from snout to vent. The male's territorial call is a series of high-pitched chirps, *goh-goh-goh*; they also make a clicking call. Sympatric on Red Sea coast with *H. robustus* and *H. flaviviridis*; in Sinai with *H. granosus* (at Sharm El Sheikh) and *H. mindiae* (in Wadi Feiran).

Hemidactylus turcicus, male, Cairo.

Hemidactylus turcicus, gravid female, Cairo.

DWARF GECKOS *LYGODACTYLUS*

A genus of attractive, small (usually 6–9cm), diurnal geckos, mostly in sub-Saharan Africa, although a few occur on Madagascar and in South America. More than 70 species are known, divided currently into 13 species groups, six species in two groups occur in our area, one nearly endemic. Often highly visible and active, they are territorial, in colonies, usually under a single dominant male, but some species form colonies with several dozen members, and more than one adult male. Many are brown or grey but a number of species have brightly coloured heads and necks; they have a limited ability to change colour. Dwarf geckos have fairly large eyes with round pupils; a long snout; a short wide neck; depressed body. They have short strong limbs with clawed digits that are very unequal in length, a vestigial thumb and strongly dilated distal toe tips with paired lamellae. The tip of the unregenerated tail has a pair of scansors, which help the gecko grip; these are an excellent aid to identification of the genus. The dorsal scales are small, granular and smooth; males have 2–12 precloacal pores. Mostly arboreal (some will use rocks), they venture onto buildings, fences and ornamental trees; it has been shown that in savanna, trees felled by elephants improve available habitats for dwarf geckos. Any gecko active by day in towns or villages in southern parts of our area will probably be a dwarf gecko. They speciate rapidly, small climatic and vegetational changes isolate populations and their rapid rate of breeding mean that changes accumulate quickly to produce new forms; many species have very small ranges and there are many endemics. They are prone to accidental translocation, as eggs or living animals, in loads of firewood and timber, produce and building stone. Consequently, their taxonomy is not fully resolved; currently very few species are undergoing genetic analysis.

KEY TO NORTH-EAST AFRICAN MEMBERS OF THE GENUS *LYGODACTYLUS*

1a: Head some shade of yellow, cream or white, with dark markings; head and/or neck in clear contrast to body colour; nostril narrowly separated from rostral scale — 2

1b: Head not yellow, white or cream, and similar to dorsal coloration; nostril narrowly separated from or touching rostral scale — 4

2a: Throat of males white, with a black median streak which fuses with an inner dark chevron marking, uniting, or just failing to unite it with the outer chevron marking; female throat immaculate white; only in Sudan and extreme south-west Eritrea — *Lygodactylus sudanensis*, Sudan Dwarf Gecko

2b: Does not have this combination of characters — 3

3a: Head usually mustard-yellow, with sharp-edged dark lines; throat black in males, sometimes with a small pale chin spot and two additional spots in centre; throat in females pale with a distinctive black-and-white O-shaped chin spot and usually four irregular, more or less evident, dark stripes; coastal zone of southern Somalia — *Lygodactylus scorteccii*, Scortecci's Dwarf Gecko

3b: Head prominently marbled or more or less streaked; throat in both sexes white, usually with three longitudinal blackish streaks fused on chin, sometimes further back, nearly always a distinctive white O-shaped chin spot; southern Ethiopia and northern parts of south Somalia — *Lygodactylus keniensis*, Kenya Dwarf Gecko

4a: Three fine brown or black parallel lines on chin which don't meet, fused dark spots above shoulder; dark line through eye reaches shoulder or may continue further; extreme southern Ethiopia, between Moyale and Dolo Odo — *Lygodactylus grandisonae*, Bunty's Dwarf Gecko

4b: Does not have this combination of characters — 5

5a: Fairly stout; large (up to 9cm); nostril not in contact with rostral scale; darker brown, grey or almost black with three dark, often large ocellate flank spots; throat in male pale with three blackish chevrons; in mesic forest habitats at higher altitudes — *Lygodactylus gutturalis*, Chevron-throated Dwarf Gecko

5b: Fairly slim; small (up to 6.5cm); nostril in contact with rostral scale; light brown or grey-brown, usually with virtually no high contrast markings, except a dark thin stripe on side of head; throat plain, immaculate; in semi-deserts and dry savanna, at lower altitudes — *Lygodactylus somalicus*, Somali Dwarf Gecko

BUNTY'S DWARF GECKO

Lygodactylus grandisonae
Pasteur, 1962

IDENTIFICATION: A small pale greyish-coloured dwarf gecko with very distinct throat pattern, from dry savanna of extreme southern Ethiopia. A moderately stout animal with a wide neck, nostril in contact with rostral scale, large mental scale deeply fissured posteriorly. Tail about 50 per cent of total length, or slightly more. Male has 6–7 precloacal pores. Known only from single series of 29 type specimens, the largest 5.4cm in total length. Dorsum pale grey-brown with fine rufous spots, a dark brown or black line runs from snout through eye; on either side of the fore limb it widens into three dark blotches and then breaks up somewhere behind the shoulder. On the chin is a distinctive pattern of three fine parallel dark lines (an excellent aid to identification). Underside whitish, unmarked. Taxonomic Notes: Belongs to the *somalicus* species group.

HABITAT AND DISTRIBUTION: Described from dry savanna woodland at Malka Murri, on the Kenya/Ethiopia border at 900m altitude. Probably more widespread on adjacent sides of the border, but under-collected.

CONSERVATION STATUS: Data Deficient, but its habitat is protected by a national park on the Kenyan side.

Lygodactylus grandisonae, preserved paratype.

NATURAL HISTORY: Unknown but presumably similar to other dwarf geckos, i.e., diurnal and arboreal, lays two small spherical eggs, eats arthropods. Sympatric with *L. keniensis* and *L. somalicus battersbyi*. Not collected since its discovery and never photographed alive. The specific name honours Alice G. C. ('Bunty') Grandison, redoubtable former curator of herpetology at the Natural History Museum, London.

CHEVRON-THROATED DWARF GECKO

Lygodactylus gutturalis
Bocage, 1873

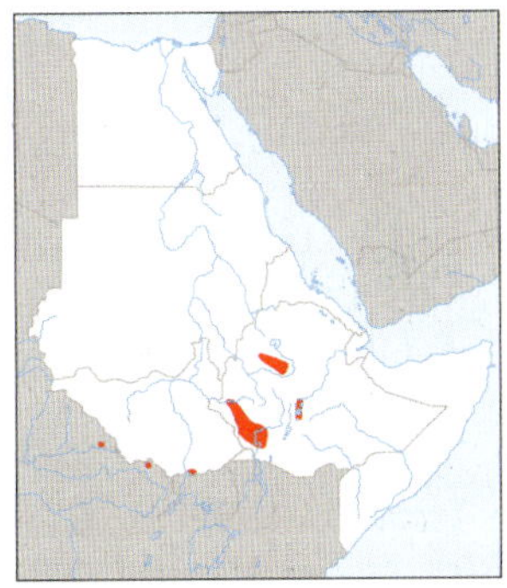

IDENTIFICATION: A medium-sized greyish-blotched dwarf gecko from medium to high-altitude forests in western Ethiopia and extreme southern South Sudan. A relatively stout species, with a short neck and head, nostril narrowly separated from rostral scale, mental scale undivided. Unregenerated tail about 50 per cent of total length. Males have 6–9 precloacal pores. Maximum size about 9cm. Dorsal colour quite variable and can change according to animal's temperament; it may be predominantly black, grey or brown, with little or no strong contrast between head and body, with lighter blotching, and these blotches may form a diffuse dorsolateral band. There is usually a dark line through the eye which ends above ear-opening or before reaching shoulder, above and on either side of shoulder there are at least three dark blotches or dark-edged pale ocelli. Tail has diffuse paler bars. Throat of male white (orange-yellow near neck) with three dark chevrons, or a solid arrowhead and spots; throat of female varies from similar to almost immaculate. Subcaudals with a grey-stippled median stripe. Ventrum vivid yellow or cream. Taxonomic Notes: The genetic variation across the huge range of this species suggests multiple species involved; at present

Lygodactylus gutturalis, Debre Marcos.

Lygodactylus gutturalis, Debre Marcos.

Lygodactylus gutturalis, male throat, Debre Marcos.

three subspecies are accepted. Belongs to the *picturatus* species group.

HABITAT AND DISTRIBUTION: Forest, woodland, gallery forest and wooded mesic savanna, at 450–2,500m in our area. Often common in suitable localities in towns and agricultural, deforested land. In Ethiopia, widespread in hill country of the south-west between Jinka and Gambela; also, Shashamene, Lake Langano and Nazret in Ethiopian Rift Valley. On north-west plateau between Debre Marcos and Chagni, south-west of Lake Tana. Unknown from highlands east of the Rift Valley. Probably more widespread in Oromia, Amhara and Benishangul-Gumuz regions, but under-explored. A single South Sudan record (Nimule), but probably more widespread, as known from north-west Uganda, north-east DR Congo and almost all of the Central African Republic. A museum specimen from Jebel Marra in Darfur, west Sudan (not on map) might be this species. Elsewhere, west to Senegal, south to Kenya and north-west Tanzania; it has the widest distribution of any *Lygodactylus* species.

CONSERVATION STATUS: Least Concern.

NATURAL HISTORY: Diurnal and arboreal; living on trees, particularly bases of large forest trees, although it will inhabit fences and buildings. Congo specimens were found in heaps of plant debris. The Debre Markos specimens were common on brick walls, wooden fences and wooden power poles along the main road in the town. At Jinka it lives in close association with *L. keniensis* and there seems to favour the trunks and larger branches of big trees in deep shade. One to two eggs, roughly 0.8cm diameter, are laid in a suitable refuge, often under bark or in a deep tree crack. It probably breeds throughout the year.

KENYA DWARF GECKO

Lygodactylus keniensis
Parker, 1936

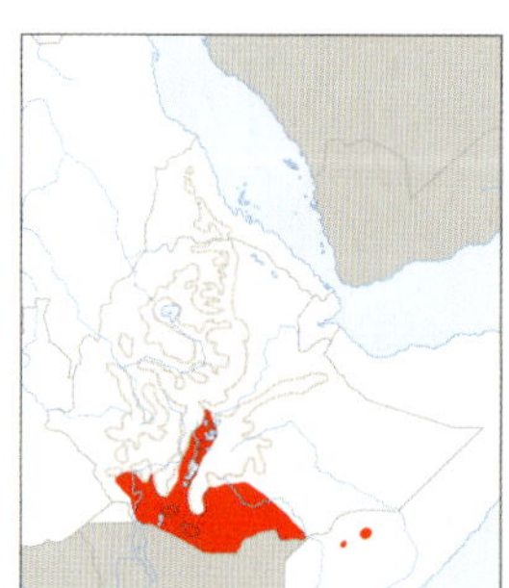

IDENTIFICATION: A medium-sized bicoloured dwarf gecko from southern Ethiopia and northern parts of south Somalia. Moderately stout, nostril narrowly separated from rostral scale, mental scale undivided. Tail roughly 50 per cent or slightly longer of total length. Males have 7–8 precloacal pores. Maximum length about 9cm, average 6–8cm, hatchlings 2.5cm. Variable in colour and pattern; head cream or yellow, with distinctive black markings, sometimes just large spots, or bold lines (these specimens resemble *L. scorteccii*). Usually there is a dark stripe from nostrils through eye to shoulder, two transverse bands or spots on snout and between the eyes link these. Head contrasts with dorsal colour, which is usually some shade of grey, but may vary from very pale to almost blue-grey, mottled to a greater or lesser extent with dark ocelli or blotches (some populations strongly resemble *L. gutturalis*). Tail can be uniform grey, or may have a series of dark chevrons and pale double spots. Underside may be immaculate white, sometimes yellow (particularly males) or have a yellowish median stripe. Throat and chin in both sexes white with three or four longitudinal blackish streaks which fuse on chin and sometimes touch each other. A distinctive O-shaped white chin spot is usually present; occasionally the throat and chin are completely black. Hatchlings pale grey with faint dorso-lateral paler head and body stripes, tail orangey. Taxonomic Notes: Belongs to the *picturatus* species group.

HABITAT AND DISTRIBUTION: Widespread in dry savanna and semi-desert of southern Ethiopia, sporadic in northern regions of south Somalia, at altitudes between 300 and 2,000m. In Ethiopia occurs from Debre Zeit in the Rift Valley south to Arba Minch. Widespread if sporadic (locally very common) throughout southern Ethiopia: localities include Jinka, Mui, Negele, Filtu, Moyale, Turmi, Bidre, Konso, Lakes Turkana and Chew Bahr. Thence extends through hilly dry savanna to Rahanuin country in southern Somalia, at Uegit and Oddur. Elsewhere, west to north-east Uganda and northern Kenya. Probably in the Ilemi Triangle in extreme south-east South Sudan, although unrecorded.

CONSERVATION STATUS: Least Concern.

NATURAL HISTORY: Diurnal and arboreal, active in morning and late afternoon at lower altitudes. It forages on all sizes of tree, but will also utilise buildings, walls and fences. A study in the lower Kerio Valley in Kenya found that it often shared habitat with the Somali Dwarf Gecko *L. somalicus*, but when both were on the same big tree, this species occupied the trunk, and was the only species on larger trees. One tree had up to 46 individuals on it, with 15 females, 21 males and ten juveniles. Two eggs are laid in bark cracks and fissures. Two eggs, roughly 6×8mm, found under a rock west of Yabello hatched in captivity after 73 days. At Lake Langano several eggs were found in the rosettes of *Sansevieria* leaves; >100 eggs were found under dry tree bark. The wide colour and pattern variation shown by this species suggests there are cryptic forms concealed within it: specimens from Konso, Key Afer and Turmi are reminiscent of *L. scorteccii*, those from further east (Bidre, Negele, Oddur and Uegit) resemble *L. gutturalis*, those from the Rift Valley (e.g., Shashamane, Debre Zeit) are intermediate. Sympatric at Jinka and Shashamane with *L. gutturalis*, and at Chew Bahr Lake and Filtu with *L. somalicus*.

Lygodactylus keniensis, male, Konso.

Lygodactylus keniensis, male, Konso.

SCORTECCI'S DWARF GECKO

Lygodactylus scorteccii
Pasteur, 1959

IDENTIFICATION: A medium-sized bicoloured dwarf gecko from coastal southern Somalia. A moderately stocky animal, nostril separated from rostral scale, mental scale undivided. Unregenerated tail about 50 per cent of total length or slightly longer. Males have 6–8 precloacal pores. Maximum size about 8cm, hatchlings about 2cm. Head bright yellow with brown and black striping, the stripes extend just past the shoulders, where the sharp-edged black stripes become diffuse and gradually fade towards the tail. The background of the back is bluish-grey. A series of diffuse darker ocellate markings present on flanks. Snout yellow with dark spot in centre, and a T-shaped dark marking between eyes. Female colour a subdued version of the male. Underside white; male has yellow lower labials and a black throat, usually with one pale chin spot (sometimes two, in centre), sometimes with a light U-shaped spot at the back of the throat. The female's throat is light with a distinctive black-and-white O-shaped chin spot and usually four irregular, more or less evident, dark stripes. Taxonomic Notes: Pasteur described Révoil's original type, from 'Somalia', as very dark dorsally. Recent specimens are light bluish-grey, indicating the type was either aberrant or darkened in preservative. A member of the *picturatus* species group.

HABITAT AND DISTRIBUTION: Usually in riverine forests and adjacent relatively mesic thicket and farmlands, within low arid savanna, between near sea level and about 150m. It will venture onto fences, walls and buildings, and is tolerant of urbanisation; occurs widely in Mogadishu. Most records in southern Somalia, along the lower courses of the Juba and Shebelle Rivers, notably at Kismayu, Gelib, between Lugh and Bardera, Afgoi, Balad, Giohar and Warshiikh. Elsewhere, in eastern Kenya on the Tana River south and north of Garissa. Populations referred to this species by Pasteur, from north-east Kenya and southern Ethiopia, at Lakes Turkana and Chew Bahir need verification.

CONSERVATION STATUS: Data Deficient.

NATURAL HISTORY: Poorly known but presumably similar to others of genus, i.e., diurnal and arboreal, lives in structured colonies, lays two eggs, eats small arthropods. Specimens from Garissa in Kenya and Balad in Somalia were common on branches of small isolated trees, in dense vegetation near river, where they were active on the trunk, at 1–2.5m above ground.

Lygodactylus scorteccii, Mogadishu.

SOMALI DWARF GECKO

Lygodactylus somalicus
Loveridge, 1935

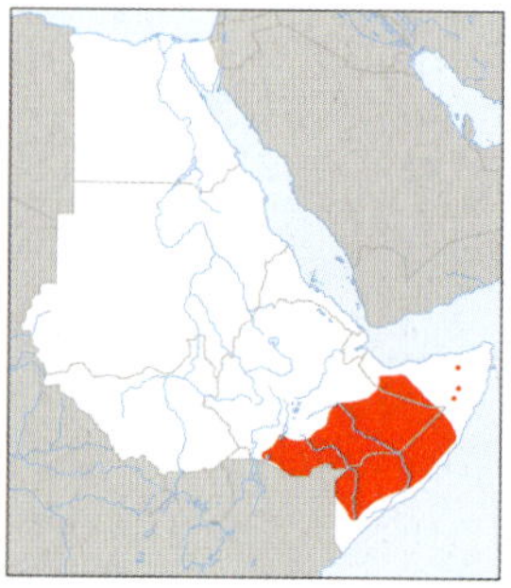

IDENTIFICATION: A small, pale, greyish-brown dwarf gecko with an immaculate throat, in semi-desert and dry savanna of the Horn of Africa. Fairly slender, nostril in contact with rostral scale, large mental scale deeply fissured posteriorly. Unregenerated tail about 50 per cent of total length. Males have 5–6 precloacal pores. Maximum size about 6.5cm, males seem to be larger than females. Head and body concolorous, pale brown to ashy-brown or grey; with irregular vague pale blotches along spine, and sometimes ocelli behind shoulder. Some individuals have a vague pale vertebral stripe. A diffuse darker stripe through eye ends before reaching shoulder. Flanks brownish, sometimes marbled pink, tail with fractured light and dark bands. Underside white, usually unmarked, rarely some spotting on chin and neck. Taxonomic Notes: Belongs to the *somalicus* species group. Two subspecies recognised: the nominate rare north-east form, known only from the Somaliland/Puntland border and south-western *L. s. battersbyi* everywhere else. The nominate differs from *battersbyi*, *inter alia*, in having a dark crossbar between eyes, fewer (178–203) dorsal scales (vs. 208–246, counted from rostral scale to first caudal whorl), no ocelli on flanks (usually present, more or less distinct), usually the tail is shorter than snout–vent length, and smaller size (snout–vent length 2.37–2.95cm vs. 2.63–3.28cm).

Lygodactylus somalicus, Hargeisa.

Lygodactylus somalicus, Lake Chew Bahir.

HABITAT AND DISTRIBUTION: Dry savanna and semi-desert of southern and eastern Ethiopia, including the Ogaden, and most of Somalia save the extreme north-east tip, from near sea level to around 1,500m. The *battersbyi* subspecies is found from Lake Chew Bahr and probably drier southern parts of Rift Valley (a vague early locality was 'Badditu to Dime') east through Moyale, Malka Murrii and Filtu to Dolo Odo in southern Ethiopia. The northernmost record in the Ogaden is at Odamuda (north-east of Ginir). Southern Somali records include Gelib, Bidi and Lugh in the Juba Valley, thence north to Giohar and Belet Uen in the Shebelle Valley, and Galmudug in central Somalia (El Dere, Haro near Hobyo, Beyra, Galkayo). In northern Somalia, widespread in the Haud region, north-west to Hargeisa and east to Burao in central Somaliland. Not known from the northern mountains and coastal zone. The much rarer nominate form, *L. s. somalicus* (including the synonym *L. s. annectens*) is confined to a small area of the hot arid Somaliland/Puntland border; recorded at Garowe in the Nogal Valley, Bar Madobe and the Daror Valley, Buran (the three dots in the top-right corner of map). Elsewhere, south to northern Kenya. Widespread and fairly common.

CONSERVATION STATUS: Least Concern.

NATURAL HISTORY: Arboreal and diurnal, a fairly shy animal. Active in morning and late afternoon, on trees and bushes in dry country. In Somaliland easily seen on ornamental trees and bushes in city gardens, even on main streets in Hargeisa and Burao. Often solitary in urban situations, but on larger trees, colonies with a dominant male, several females and juveniles can be observed. Males are highly territorial but dominance depends on individual size, not sex (in confrontations over food items); a female was seen to dominate a favourable position at a breach in a termite archway, she excluded the male within whose territory the breach occurred. They lay two eggs, often in same bark cracks as eggs of Kenya Dwarf Gecko, if occurring syntopically; a communal egg-nesting site with up to 40 eggs has been reported. Sympatric with *L. grandisonae* and widely with *L. keniensis*.

SUDANI DWARF GECKO

Lygodactylus sudanensis
Loveridge, 1935

Lygodactylus sudanensis, Gambela.

IDENTIFICATION: A medium-sized pale dwarf gecko from low altitude savannas of eastern Sudan, adjacent parts of Eritrea, and South Sudan. Moderately stout, nostril narrowly separated from the rostral scale, mental scale undivided. Unregenerated tail about 50 per cent of total length. Males have 7–9 precloacal pores. Maximum size about 7.5cm, most specimens 3.3–3.7cm long from snout to vent, hatchlings about 1.5cm long. Dorsal colour in males pale olive with little contrast between the head and the body, a series of indistinct diffuse darker and paler ocellate markings form two dorsolateral stripes; the female's dorsal pattern is more pronounced, on a yellowish brown background. The head is dull yellow or cream with a blackish line from the nostril through eye to the flank (though often interrupted on the neck), breaking up behind the forelimb into two or three spots; a second shorter line extends from the angle of the mouth to the ear opening. There is a black spot on the snout and an undulating transverse band between eyes. The male's throat is white, with two dark chevrons which may be linked by a black median streak; the female's throat is immaculate white. The limbs are pale yellowish brown; the tail pale yellowish brown to brownish grey, with several light bands or spots. Yellowish white below. Juveniles are yellowish grey-brown, with two darker dorsolateral bands that commence behind the eye, the lower shorter and thinner, otherwise speckled by whitish yellow; a dark stripe extends from the snout to eye and an undulating transverse band runs between the eyes. Taxonomic Notes: Belongs to '*picturatus*' species group.

HABITAT AND DISTRIBUTION: Wooded savanna at low elevations, roughly between 350–650m. A single Eritrean record (Cunama region east of Om Hajer). Sporadically recorded in southern Sudan and South Sudan, mostly associated with the Nile basin. South Sudan records include Chak Chak and Wau, in Bahr El Ghazal, Gondokoro and Mongalla (near Juba), Khor Attar and Melut in Upper Nile State. Sudanese records on both Nile valleys thus: El Misherat, Roseires, Dinder National Park, Sennar, Singa, and Abu Zor (the type locality) on the Blue Nile, Duem and Goz Abu Goma on the White Nile. Probably in south-west Ethiopia.

CONSERVATION STATUS: Not Evaluated.

NATURAL HISTORY: Poorly known. Arboreal, usually on trees (*Vachellia*, and others), sheltering in crevices or under bark. Observed on buildings at Mongalla. One found on a boat at Omdurman (near Khartoum). In Dinder National Park animals were active on trunks of trees and stout bushes. Mostly diurnal, but may be active at dusk. Probably lays two hard-shelled eggs. Diet small invertebrates.

PIGMY GECKOS *TROPIOCOLOTES*

A Saharo-Arabian genus of small, slender, ground-dwelling, mostly slow-moving geckos with pointed snouts, large eyes with vertical pupil, homogenous imbricate dorsal scales that may be smooth or keeled. Their toes are clawed, not dilated into pads. They are nocturnal, feed on small invertebrates and lay one hard-shelled egg several times a year. These are the smallest lizards of north-east Africa; with a snout–vent length no larger than 35mm. Pigmy geckos are typical of deserts and semi-deserts in North Africa and Arabia, with isolated populations in southern Iran. Fifteen species are known, seven in our area, one endemic. Their taxonomy is not yet totally clarified, new species are constantly being discovered. Their closest relatives are Clawed desert geckos *Stenodactylus*, which have a similar distribution, and the Gulf Short-fingered Gecko *Pseudoceramodactylus khobarensis*, along with the recently resurrected genus *Trigonodactylus*, Web-footed sand geckos, of Arabia and southern Iran.

KEY TO THE NORTH-EAST AFRICAN MEMBERS OF THE GENUS *TROPIOCOLOTES*

1a: Sub-digital scales unicarinate to almost smooth; dorsum usually with broad transverse dark and light bands; only in Sinai, Egypt 2

1b: Sub-digital scales tricarinate; dorsum spotted, sometimes forming irregular transverse bands or almost plain; Egypt to Somalia, not in Sinai 3

2a: Larger, maximum snout–vent length up to 3cm; longer head (up to 8.81mm); supralabials 8–11; infralabials 7–10; southern Sinai and along Gulf of Aqaba north *Tropiocolotes nattereri*, Natterer's Pigmy Gecko

2b: Smaller, maximum snout–vent length up to 2,87cm; shorter head (up to 7.57mm); supralabials 7–11; infralabials 7–9; north-east Sinai border with Israel/Palestine *Tropiocolotes yomtovi*, Yom-Tov's Pigmy Gecko

3a: Dorsals not keeled, relatively small 4

3b: Dorsals strongly keeled, relatively large 5

4a: Scales on dorsal side of head coarse, <14 between eyes; dorsal pattern consists of irregular often faint dark and light spots; more than 18 tail bands *Tropiocolotes nubicus*, Nubian Pigmy Gecko

4b: Scales on dorsal side of head fine, more than 14 between eyes; dorsal pattern consists of regular series of large dark and light blotches; fewer than 16 tail bands *Tropiocolotes steudneri*, Steudner's Pigmy Gecko

5a: Scales on dorsal side of head small, more than 13 between eyes, mostly smooth; ventral scales notably smaller then dorsals and less keeled *Tropiocolotes bisharicus*, Bishari Pigmy Gecko

5b: Scales on dorsal side of head large, fewer than 13 between eyes, mostly keeled and imbricate; dorsal and ventral scales strongly keeled and of roughly equal size 6

6a: Smaller, maximum size 6cm; larger dorsal scales, 35–40 rows at mid-body; anterior chin shields usually not in contact with second lower labials; tail without bands, orange-yellow; Eritrea to Somalia *Tropiocolotes somalicus*, Parker's Pigmy Gecko

6b: Larger, maximum size 8,5cm; smaller dorsal scales, 44–47 rows at mid-body; chin shields in broad contact with second lower labials; tail banded, yellowish with white tip; northern Egypt *Tropiocolotes tripolitanus*, Tripoli Pigmy Gecko

BISHARI PIGMY GECKO

Tropiocolotes bisharicus
Baha El Din, 2001

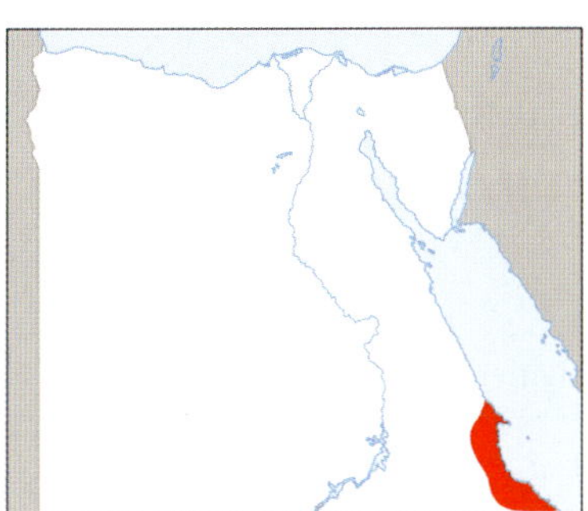

IDENTIFICATION: A small, slender gecko of the southern Egyptian coast, with keeled dorsals and an orange-yellow, short thick tail. Head large with 15–18 smooth scales between eyes; snout long, pointed; eyes notably large. Neck distinct. Limbs long and slender, with five thin clawed toes. Tail relatively short and thick, about 50 per cent of total length. Body scales of equal size, keeled and imbricate, 43–51 rows at mid-body. No precloacal pores. Maximum size about 6.5cm, hatchling about 3cm. Dorsum sandy-gray, with six ill-defined brownish transverse bands across back, each followed by several lighter yellowish spots. Iris bicoloured, upper third golden-brown, lower two-thirds dark brown. A well-marked, dark brown, wide line extends from rostral to well beyond shoulder,

Tropiocolotes bisharicus, Wadi El Rada.

Tropiocolotes bisharicus, Wadi El Rada.

upper labials white. Tail orange-yellow with 9–10 transverse bands. Regenerated tail plain orange-yellow with few dark specks, not barred. Underside white, except tail, which is yellowish. Taxonomic Notes: In a clade with the closely related *T. somalicus*, *T. steudneri* and *T. nubicus*.

HABITAT AND DISTRIBUTION: Desert or semi-desert, rocky or stony terrain in mountainous country, at 200–400m. Shelters among rocks at base of hillsides flanking sandy wadis with sparse vegetation, especially *Vachellia tortilis* trees and *Zilla spinosa* shrubs. Unlike the sympatric *T. steudneri*, they prefer elevated sites, not fine alluvium on wadi bottoms. Known only from south-east Egypt, localities are Wadi Sartout in Wadi El Gemal National Park, Wadi El Rada, Kulet El Dob, Bir Abraq, Bir Baneit and Gebel Elba. Probably present in adjacent north-east Sudan. Uncommon and local, not in the pet trade.

CONSERVATION STATUS: Least Concern.

NATURAL HISTORY: Known from a few specimens. All were active at night, lying on large rocks or walking slowly around. Terrestrial, but will climb rocks searching food. In other aspects probably similar to other pigmy geckos.

NUBIAN PIGMY GECKO

Tropiocolotes nubicus
Baha El Din, 1999

IDENTIFICATION: A small, slender gecko of the Egyptian and Sudanese Nile, with smooth dorsals and coarse dorsal head scales. Head narrow, notably depressed, 10–13 scales between eyes. Neck indistinct. Limbs rather short and slender, with five thin clawed toes. Tail relatively long and slender, about 60 per cent of total length. Body scales of equal size, smooth and imbricate, 49–59 rows at mid-body. Precloacal pores present in males and some females. Maximum size about 8cm, average 6.5–7.5cm, hatchling 3.5cm. Dorsum sandy-pink, translucent, with irregular brown blotches and a few scattered, indistinct, lighter specks. Iris golden-brown. A dark brown line extends from the nostril to a little beyond shoulder, upper labials white. Tail yellowish with 18–21 transverse bands. Regenerated tail plain yellow, not barred. Underside white, except tail, which is yellowish. Taxonomic Notes: This recently described species closely resembles *T. steudneri* and molecular analysis confirms the two as sister species.

HABITAT AND DISTRIBUTION: In our area seems to be associated with fluvial habitats and the transition zone between the Nile Valley and adjacent deserts, where recorded on low rocky hills, in shallow sandy wadis and near human settlements, at 70–400m (recently discovered, genetically confirmed populations from Niger and Algeria occur on isolated desert mountains at much higher altitudes). Originally described from Egypt, where known from Luxor in the north, south along the Nile Valley through Wadi Abbad east of Edfu to Abu Simbel, east and west of Lake Nasser. Also recorded at Asalaya in Sudan, about 250km south of Khartoum; recently found at Dongola and south-east of Atbara. Unconfirmed records from

Tropiocolotes nubicus, Dongola.

Wadi Halfa in northern Sudan. Probably much more widespread in central and northern Sudan in Nile Valley. Elsewhere, Tassili n'Ajjer and Hoggar Mountains in south-east Algeria and Termit Massif in east-central Niger.

CONSERVATION STATUS: Data Deficient.

NATURAL HISTORY: Terrestrial and nocturnal, encountered at night motionless under low vegetation or among rocks. One specimen was climbing on a low wall. By day hides under rocks and other ground cover, even paper rubbish. Makes a very audible call lasting five seconds, which can be heard at dusk, at up to 50m distance. A single egg, roughly 7×5.1mm, is buried in soil; up to five clutches were laid by a captive female in one season. At some localities it lives syntopically with *T. steudneri*. Very common at Abu Simbel in southern Egypt, in built-up areas of the city, even in houses.

PARKER'S PIGMY GECKO

Tropiocolotes somalicus
Parker, 1942

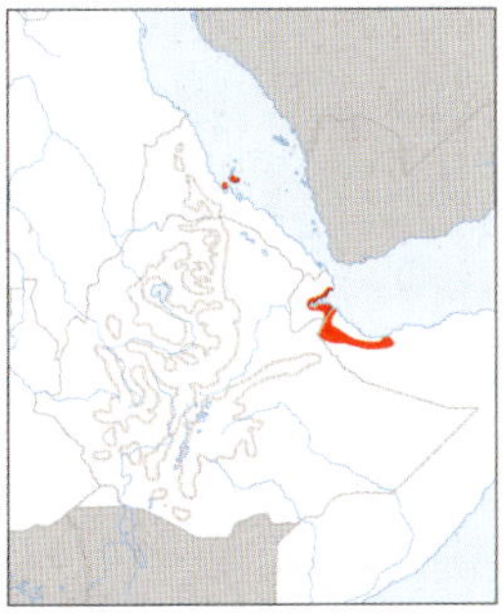

IDENTIFICATION: A small, rough-scaled gecko with plain brownish coloration and unbanded orange-yellow tail; the only member of the genus in Eritrea, Djibouti and Somalia. Head narrow, depressed, with 10–12 keeled scales between eyes; snout moderately long. Neck short, less distinct. Limbs rather short and slender, with five thin clawed toes. Tail relatively long and fairly thick, 54–56 per cent of total length. Body scales of equal size, large, keeled and imbricate, 35–40 rows at mid-body. No precloacal pores. Maximum size about 6.5cm, average about 6cm, hatchling about 3cm. Back grey-brown, with some darker flecks and several pale scales arranged in two or more longitudinal dorsolateral rows which extend from the eyes to tail. Iris golden or greyish-brown. A wide dark brown line extends from nostril to well beyond shoulder. Upper labials white. Tail orange-yellow, originally described as having alternating light and dark transverse bars (12–13), but tail of a recent series from Somaliland unmarked orange-yellow. Regenerated tail plain pinkish-yellow. Underside white, except tail, which is yellowish. Taxonomic Notes: Originally lumped with *T. tripolitanus*, another rough-scaled species, but recent molecular analysis indicates it is a separate species, and closely related to *T. nubicus* and *T. steudneri* (which have smooth dorsal scales).

HABITAT AND DISTRIBUTION: Semi-desert to dry savanna, primarily on sandy soils, between sea level and 970m. Inhabits sandy terrain with patches of stones and sparse vegetation, stony plains or hillsides with almost no vegetation, and wadi beds. Parker's Pigmy Gecko is the southernmost living representative of the genus, confined to the coastal zone. Two isolated Eritrea records (Dissei and Dahlak Kebir Islands in Dahlak archipelago), and a population extends from Obock, Tadjoura and southern environs of Djibouti City into north-west Somaliland and along border between Djibouti and Ethiopia. In Somaliland it occurs east through the sandy plains and low, dry hot coastal hills of the Guban; notable records include the plains east of Weeraar, Geerissa, Biji, Berbera and Lafaruug. The easternmost record is Wadi Faradeero, about 20km south of Berbera. Probably in southern Eritrea and adjacent Ethiopia. Widespread and locally common.

CONSERVATION STATUS: Least Concern.

NATURAL HISTORY: Like other pigmy geckos, ground-dwelling and nocturnal. Most Somaliland

Tropiocolotes somalicus, Lafaruug.

animals were found by day in very arid country, hidden under rocks in shade, which must provide cooling and higher humidity. In sandy habitats without rocks found at night around temporary fences of spiny *Vachellia* branches surrounding village houses. Fast-moving, alert and quick to escape on hot nights. When threatened it waves its tail rapidly. A gravid female from Geerissa contained one large egg in September.

STEUDNER'S PIGMY GECKO

Tropiocolotes steudneri
(Peters, 1869)

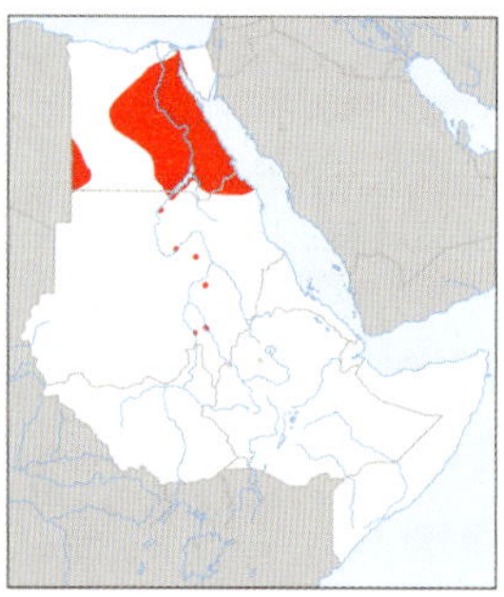

IDENTIFICATION: A small, slender gecko with smooth dorsals and fine dorsal head scales, widespread in desert in Egypt and Sudan. Head narrow, notably depressed; 14–18 scales between eyes; snout narrow and long. Neck usually distinct. Limbs rather long and slender, with five thin clawed toes. Tail relatively long and slender, about 55 per cent of total length. Body scales of equal size, smooth and imbricate, 45–60 rows at mid-body. Precloacal pores present in most males and some females. Maximum size about 7cm, hatchling about 3cm. Dorsum usually sandy or pale brown, semi-translucent, with diffuse brown blotches, interspersed by large white spots, forming vague light and dark crossbars. Iris golden-brown. A dark brown line extends from nostril almost to shoulder, upper labials white. Tail yellowish with 10–16 transverse bands. Regenerated tail plain yellow, not barred. Underside white, except tail, which is yellowish.

HABITAT AND DISTRIBUTION: Desert and semi-desert, gravel plains, wadis and oases in sandy habitats, in our area between sea level and about 600m. The most widespread pigmy gecko in Egypt, in the Western and Eastern Deserts south of Cairo. In the Western Desert more local, recorded at all major oases. In the Eastern Desert one of the most common geckos. Absent from Siwa and Qattara Depression in north-west Egypt. An isolated population occurs in south-west Egypt, on the Gilf El Kebir Plateau near the Sudanese and Libyan border. Recently confirmed in north-west Sinai, south of Ras Matarma. Sudanese records more sporadic, including Wadi Halfa, Kosha, Jebel Barkal near Merowe in the Nile Valley in the north, Bayuda Desert between Merowe and Atbara, and found as far south as Sennar on the Blue Nile and Asalaya on the White Nile. Probably more widespread in Sudan. Elsewhere, west to Mauritania across southern Libya, southern Algeria and Niger. The commonest member of the genus in Egypt and much harvested for the pet trade.

CONSERVATION STATUS: Least Concern.

NATURAL HISTORY: A ground-dwelling nocturnal species. By day they hide in holes, within thick vegetation or under rocks (and under rubbish, card and paper around habitation); sometimes two animals shelter together. Moves slowly and can make short jumps.

Tropiocolotes steudneri, Marsa Alam.

Tropiocolotes steudneri, Quseir.

Tropiocolotes steudneri, Quseir.

When walking it waves its tail in a slow serpentine motion. Relatively fast-moving. Diet includes small insects and other invertebrates. In Egypt, remains active and easy to find in winter, where larger species of reptile hibernate. Reproductive details unknown but probably lays a single egg like other species of pigmy geckos.

TRIPOLI PIGMY GECKO

Tropiocolotes tripolitanus
Peters, 1880

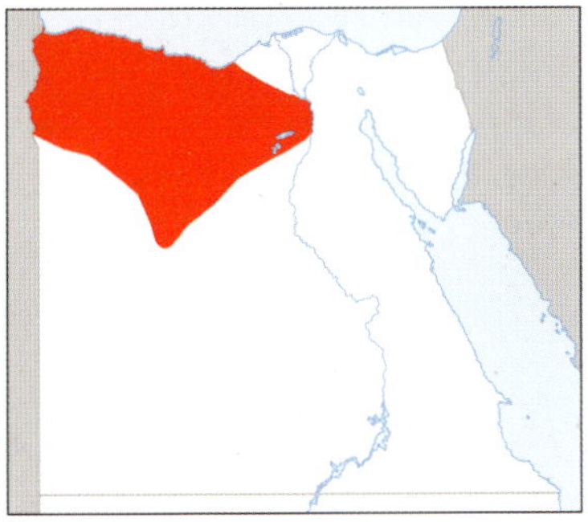

IDENTIFICATION: A small, slender gecko with keeled body scales, virtually without pattern or markings, in northern Egypt. Head narrow, notably depressed; 12 scales between eyes; snout narrow and long. Neck short, less distinct. Limbs rather short and slender, with five thin clawed toes. Body fairly elongate, slim. Tail relatively long and slender, about 55–60 per cent of total length. Body scales of equal size, strongly keeled and imbricate, 44–47 rows at mid-body. Precloacal pores not recorded. Maximum size about 8.5cm, average 6.5–7.5cm, hatchling 3.5cm. Dorsum sandy-brown or greyish, with small dark and whitish specks, appears plain. Iris golden or greyish-brown. A dark brown line extends from nostril to a little beyond shoulder. Upper labials white. Tail yellowish anteriorly, but often whitish or greyish posteriorly, with 11–14 bands. Underside white, except tail, which is yellowish. Taxonomic Notes: Three different subspecies currently recognised, but status questionable; the nominate, *T. t. tripolitanus*, occurs in our area, whereas the other two are known from further west.

HABITAT AND DISTRIBUTION: Sandy and stony deserts with generally sparse bushy or herbaceous vegetation, not usually in totally barren country. Common

Tropiocolotes tripolitanus, Tunisia.

Tropiocolotes tripolitanus, Tunisia.

in oases and their surroundings. In north-east Africa confined to north-west Egypt, west of Cairo and south to the Faiyum oasis, thence west to the Libyan border. Sporadic throughout the periphery of the Qattara Depression, from El Maghra to Siwa. Elsewhere, west in a broad band to Mauritania and Western Sahara. Common in the pet trade.

CONSERVATION STATUS: Least Concern.

NATURAL HISTORY: Similar to others in the genus; nocturnal, terrestrial, slow-moving. On the periphery of Qattara Depression found commonly in and under woody vegetation. It also shelters beneath rocks and in holes. Females lay one hard-shelled egg, approximately every four weeks during spring and summer. In winter they were found sheltering by day under the bark of dead trees in the direction of the sun, seeking warmth.

NATTERER'S PIGMY GECKO

Tropiocolotes nattereri
Steindachner, 1901

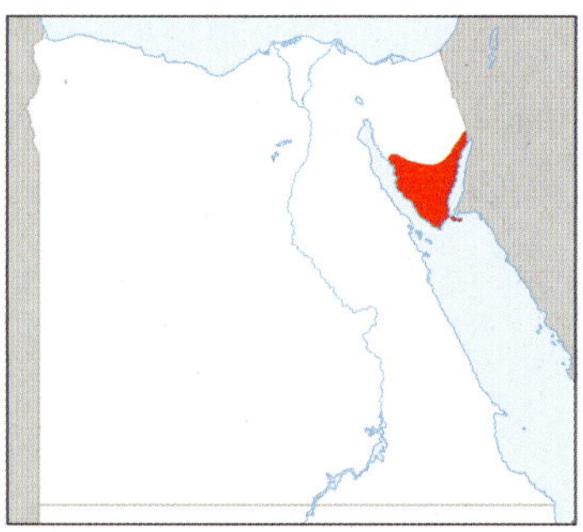

IDENTIFICATION: A tiny, slender gecko of southern Sinai, with smooth dorsals and usually broad transverse dark and light bands on back. Head elongate with narrow, long and pointed snout; eyes notably large. Neck distinct. Limbs long and slender, with five thin clawed toes and almost smooth unicarinate sub-digital scales. Tail relatively long and slender, about 55–60 per cent of total length. Body scales of equal size, smooth and imbricate, 42–59 rows at mid-body. No precloacal pores. Maximum size about 6.3cm, hatchling about 2.5cm. One of the smallest lizards in our area. Coloration distinctive, dorsum pale yellow, sandy-yellow or yellowish-grey with a dorsal pattern consisting of 4–5 transverse, W-shaped dark and light bands. Iris brown. A dark brown line extends from nostril almost to shoulder. Upper labials white. Tail sandy or yellow with 10–16 transverse bands, last third or quarter greyish-white. Regenerated tail dirty yellow, not barred. Underside white. Taxonomic Notes: Populations in northern Sinai, Arava Valley, Judean and Negev Deserts and adjacent Jordan and Saudi Arabia recently described as a new sister species, *T. yomtovi*.

HABITAT AND DISTRIBUTION: Desert or semi-desert, usually in rocky or stony terrain in hilly and mountainous country, often with sparse vegetation, from near sea level to 1,600m. Found on hillsides, wadi beds, rock-strewn plains with coarse alluvium, and in a palm oasis. Throughout much of southern Sinai, although it avoids sand dunes of the north and east. Recorded in coastal hills on Gulf of Aqaba from Eilat and Taba in north to Ras Mohamed in south, including Tiran and Sanafir Islands. Unevaluated records from Metla Pass and environs (not shown on map) may involve this species. Elsewhere, north and east to southern Arava Valley as far north as north-east of Aqaba in Sinai and Eilat Mountains in Israel/Palestine and adjacent Jordan. Widespread and fairly common in southern Sinai.

CONSERVATION STATUS: Least Concern.

NATURAL HISTORY: Strictly nocturnal. Terrestrial, though occasionally climbs rocks. Shelters by day under rocks and in rock crevices. Active and agile, moving with short dashes and long jumps. When threatened it waves its tail. Produces a little chirp at night, sometimes by day. A single, relatively large (about 6.7×5mm) white egg is laid. Feeds on a broad spectrum of small arthropods. These geckos can change tail colour; barred by day and uniform white at night.

Tropiocolotes nattereri, Eilat.

YOM-TOV'S PIGMY GECKO

Tropiocolotes yomtovi

Ribeiro-Júnior, Tamar, Maza, Flecks, Wagner, Shacham, Calvo, Geniez, Crochet, Koch & Meiri, 2022

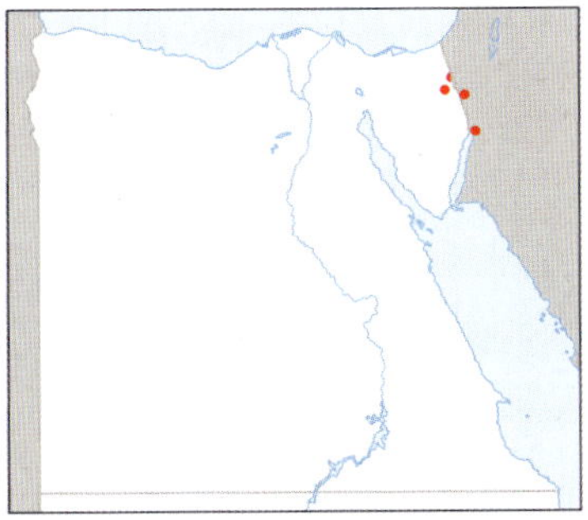

IDENTIFICATION: A tiny, slender gecko of north-east Sinai, with smooth dorsals and usually broad transverse dark and light bands on back. Head elongate with narrow, long and pointed snout; eyes notably large. Neck distinct. Limbs long and slender, with five thin clawed toes and almost smooth to unicarinate sub-digital scales. Tail relatively long and slender, about 55–60 per cent of total length. Body scales of equal size, smooth and imbricate, 48–60 rows at mid-body. No precloacal pores. Maximum size about 6.1cm, hatchling about 2.5cm. Probably the smallest lizard in our area. Coloration distinctive, dorsum pale to sandy-yellow, yellowish-grey, pink-brown or brown with a dorsal pattern consisting of 4–5 transverse, W-shaped dark and light bands, or less often pairs of smaller white or cream dots. Iris brown. A dark brown line extends from nostril almost to shoulder. Upper labials white. Tail sandy or yellow with 10–16 transverse bands, last third or quarter whitish. Regenerated tail dirty yellow to brownish, not barred. Underside white. Taxonomic Notes: Only recently described, split from its sister species, *T. nattereri*, from which it differs mostly by smaller size, shorter head and slightly lower counts of some head scales (e.g., supralabials, infralabials, gulars and lamellae under fingers and toes).

HABITAT AND DISTRIBUTION: Desert or semi-desert, usually of rocky or stony character in hilly and mountainous country, often with sparse vegetation, from near sea level to 830m in our area. Hillsides, wadi beds, rock-strewn plains with coarse alluvium, and in a palm oasis. Throughout much of southern Israel/Palestine, reaches its western boundary in our area (but probably more widespread than known) at two sites in Egypt: Gebel Urayf An Naqah and the flat hilly country east of El Quseima in north-east Sinai. Known from Eilat Mountains (where sympatric with *T. nattereri*), so its presence in Taba area cannot be excluded. An unevaluated record from Al Arish (not shown on map) probably involves this species. Elsewhere to south-west Jordan and extreme north-west Saudi Arabia.

CONSERVATION STATUS: Least Concern.

NATURAL HISTORY: Nocturnal, spends daytime in shelters, usually under stones, in rocky crevices, burrows or under debris. Ground-dwelling, may occasionally climb rocks. At night active and agile, moving with short dashes and long jumps. Said to have circadian changes of tail coloration, barred by day, mostly uniform by night. Diet tiny arthropods. Females lay one large egg (4.8–5.5×6.3–7mm) per clutch several times a year. When threatened it waves its tail and produces a typical distress call.

Tropiocolotes yomtovi, Eilat.

Tropiocolotes yomtovi, Eilat.

FAMILY SCINCIDAE SKINKS

A family of shiny-bodied lizards, often with small legs, varying in size from 5–40cm in north-east Africa. Their flat shiny scales are toughened by osteoderms, giving skinks a stout, fairly rigid but flexible coating which is resistant to wear, useful for burrowing or rock-dwelling animals. Most lose their tails very easily. They inhabit a variety of habitats; on rocks and trees, on ground or under it; any lizard found pushing its way with its nose through unconsolidated soil or sand, or leaf litter, will be a skink. Some are territorial and live in structured colonies. Worldwide there are >1,700 species of skink; they occur throughout the tropics and part of the temperate world. There is some debate about the taxonomic division of skinks and the status of families/subfamilies. We have followed the classification of the Reptile database, which splits them into seven subfamilies. In our area there are four subfamilies, 12 genera and about 50 species.

KEY TO THE NORTH-EAST AFRICAN GENERA AND SOME SPECIES OF SKINK

1a: Only has two legs, both at back	*Haackgreerius miopus*, Two-legged Skink
1b: Has four legs	2
2a: Eye cannot close, no movable eyelids	3
2b: Eyes can be closed, movable eyelids	5
3a: In our area, only on the Sinai Peninsula, Egypt	*Ablepharus rueppelli*, Rüppell's Snake-eyed Skink
3b: In our area, south of 20°N	4
4a: Active by day and only found in intertidal zone, on coral, in our area confined to southern Somali coast	*Cryptoblepharus africanus*, Coral Rag Skink
4b: Usually hidden in leaf litter, in savanna, woodland and forest	*Panaspis*, Snake-eyed skinks
5a: Relatively large, earhole wholly concealed by large scales, in desert and semi-desert, usually in sand, and usually strongly banded	6
5b: Earhole visible or only partially concealed; lacks above combination of characters	7
6a: Dorsal scales smooth, snout very elongate and wedge-shaped	*Scincus*, Sandfish
6b: Dorsal scales keeled, snout more rounded	*Scincopus fasciatus*, Tiger Skink
7a: Dorsal scales smooth	8
7b: Dorsal scales keeled	11
8a: Lower eyelid scaly, head deep, dorsal colour usually a mix of red, yellow and brown, in our area largely confined to Mediterranean coast of Egypt	*Eumeces schneideri*, Golden Skink
8b: Lower eyelid with transparent disc, does not have above combination of characters	9
9a: Small, <15cm maximum length, in our area confined to forest in south	*Leptosiaphos*, Leaf-litter skinks
9b: Large, >15cm total length, widespread in our area and not usually in forest	10
10a: Nostril pierced in or bordered by rostral	*Chalcides*, Bronze skinks
10b: Nostril pierced between 2–3 nasals, well separated from rostral	*Mochlus*, Writhing skinks
11a: Usually has postnasal scale, 17 species widespread in our area and often common	*Trachylepis*, Typical African skinks
11b: Usually no postnasal scale, in our area only on northern Egyptian coast (*Heremites vittatus*) or extremely rare and only in coastal Eritrea (*H. septemtaeniatus*)	*Heremites*, Golden and Bridled skinks

SUBFAMILY MABUYINAE

A widespread subfamily of more than 220 species of skink, in 25 genera. Two genera occur in north-east Africa; 17 species of Typical African skinks, genus *Trachylepis*, and two species of *Heremites*, the Golden and Bridled skinks.

TYPICAL AFRICAN SKINKS *TRACHYLEPIS*

One of Africa's most visible groups of lizards, *Trachylepis* is a genus of small to large diurnal skinks, living in a range of habitats. Originally in the genus *Mabuya*, a huge pantropical genus, typical skinks from the Afro-Malagasy region are now assigned to genus *Trachylepis*. They have large eyes and a visible external earhole. They have no preanal or femoral pores, but have cylindrical, shiny bodies with mostly keeled, overlapping cycloid scales; the number of keels on the scales appears to increase with age in some species. The long tail is easily shed. Most are shades of grey or brown, with stripes, but some are vividly coloured and in some species the male and female (and in some the juveniles) have different colour patterns. Most lay eggs, some give live birth, some do both. More than 85 species of *Trachylepis* are known, most in Africa. Seventeen species, in eight different species groups, occur in north-east Africa.

KEY TO THE NORTH-EAST AFRICAN MEMBERS OF THE GENUS *TRACHYLEPIS*

1a: Scales on soles of feet usually non-spinose, smooth or tubercular (except some juveniles of *T. quinquetaeniata*)	2
1b: Scales on soles of feet usually keeled and spinose	15
2a: Unregenerated tail extremely long, >70 per cent of total length, limbs relatively spindly	3
2b: Unregenerated tail <70 per cent of total length, limbs usually not spindly	4
3a: In our area only in south-west South Sudan	*Trachylepis buettneri*, Buettner's Skink
3b: In our area, confined to Ethiopia and south-east South Sudan	*Trachylepis megalura*, Grass Top Skink
4a: Dorsal scales without keels	5
4b: Dorsal scales with obvious keels	6
5a: Confined to a small area of highland Ethiopia, mid-body scales rows 26–29	*Trachylepis boehmei*, Ankober Skink
5b: Confined to northern Somalia, mid-body scales rows 32	*Trachylepis hemmingi*, Hemming's Skink
6a: Build robust, body thickset, always terrestrial	7
6b: Build moderate, body slim, may climb	9
7a: Mid-body scale rows 30–38, paravertebral scales 52–57, flank scales with two keels	*Trachylepis dichroma*, Bicoloured Skink
7b: Mid-body scale rows 30–34, paravertebral scales 40–55, flank scales with three keels	8
8a: Paravertebral scale rows 46–55, flanks often orange or red, range South Sudan	*Trachylepis perroteti*, Orange-flanked Skink
8b: paravertebral scales 40–49, flanks not usually orange or red, widespread	*Trachylepis brevicollis*, Short-necked Skink
9a: Dorsal scales usually more than five keels	10
9b: Dorsal scales 3–5 keels	11
10a: Flanks never speckled and lighter in colour than back; confined to Webi Shebelle River in Somalia north of 02°N	*Trachylepis ferrarai*, Webi Shebelle Skink

10b: Flanks usually speckled black and white, and darker than back, widespread *Trachylepis maculilabris*, Speckle-lipped Skink

11a: Mid-body scales 32–42, rock-dwelling, females and juveniles striped black and yellow, with a blue tail, adult males some shade of brown with several black patches behind earholes *Trachylepis quinquetaeniata*, Five-lined Skink

11b: Mid-body scales 25–34, not necessarily rock-dwelling, pattern not as above 12

12a: In our area, only in south-west South Sudan *Trachylepis affinis*, Senegal Skink

12b: Not known from south-west South Sudan 13

13a: In our area, known from Somalia and/or east of Ethiopian Rift Valley 14

13b: In our area in eastern South Sudan or west of Ethiopian Rift Valley *Trachylepis wingatii*, Wingate's Skink

14a: Maximum size about 23cm, a distinctive cluster of tooth-like scales partially cover front of ear-opening *Trachylepis hildebrandtii*, Hildebrandt's Skink

14b: Maximum size about 35cm, no cluster of tooth-like scales obscure ear-opening *Trachylepis planifrons*, Tree Skink

15a: Small, less than 18cm maximum length, with a conspicuous narrow sharp-edged white stripe low on flanks 16

15b: Larger, up to 25cm, without a white stripe low on flanks *Trachylepis striata*, Striped Skink

16a: Frontoparietal shields fused together, mid-body scale rows 30–34, confined to north-west highlands of Ethiopia and Eritrea *Trachylepis isselii*, Issel's Skink

16b: Frontoparietal shields paired, mid-body scale rows 27–36, largely absent from north-west highlands of Ethiopia and Eritrea but widespread elsewhere *Trachylepis varia*, Variable Skink

SENEGAL SKINK, BROWN-FLANKED SKINK

Trachylepis affinis
(Gray, 1839)

IDENTIFICATION: A slim striped skink just reaching our area in South Sudan. It has a short, rounded head and a large eye, rimmed yellow. Tail about ⅔ of total length. Dorsal scales usually have three keels (but up to seven). Scales in 26–34 rows at mid-body. Maximum size about 22cm or slightly more, average 15–18cm, hatchling size unknown. Colour quite variable: usually some shade of brown or olive, the head may be rufous or golden-brown, the back immaculate or with 4–6 punctuated lines of black. A lateral stripe, which may be black, light or dark brown, sometimes speckled, extends from behind eye to rear limb junction; below this on some specimens is a fine light (white or yellow) lateral stripe. The upper labials, throat and underside may be white or yellow.

HABITAT AND DISTRIBUTION: Forest, moist and dry savanna, and suburbia, although often associated with quiet water sources, between sea level and 1,300m. In our area, known only from Yei, in Central Equatoria, South Sudan, but probably widespread there, as recorded from just across DR Congo border in Garamba National Park, where it is abundant. Elsewhere, west to Senegal.

CONSERVATION STATUS: Least Concern.

Trachylepis affinis, Nigeria.

NATURAL HISTORY: Diurnal and terrestrial, active in leaf litter, sometimes climbs tree trunks, and sleeps out on thin branches. A study in Nigeria found they were most active in late morning and early afternoon. West African females contained 2–5 eggs, females in Nigeria were gravid in November and December and hatchlings observed in March and April; in Gabon eggs were laid in February.

ANKOBER SKINK

Trachylepis boehmei
Koppetsch, 2020

IDENTIFICATION: A small striped skink, described recently from specimens collected in 1975, at a single locality in highland Ethiopia, and closely resembling *T. megalura*. It has a short, rounded head. Tail about 56 per cent of total length. Dorsal scales smooth. Scales in 26–29 rows at mid-body. Maximum size about 12.5cm, average 9–12cm, hatchling size unknown; a small female was 6cm. Life colour unknown (it was described from previously misidentified preserved material); in preservative dorsal colour brown, with 6–8 dark brown stripes from nape to tip of tail, with a blue-grey vertebral stripe and rufous-brown flank stripe. Head and neck spotted brown, the ventral surface pale, in some specimens a series of 12 or so dark lines extend from chin to cloaca, the outer lines may extend along underside of the tail.

HABITAT AND DISTRIBUTION: Ankober, North Shoa, in high central Ethiopia, just west of the Rift Valley, at 2,450m. Originally the area was under *Juniperus/Podocarpus* forest; nowadays it is mostly a mixture of grassland and agriculture.

CONSERVATION STATUS: Not Assessed.

NATURAL HISTORY: Unknown. Never knowingly seen alive. Presumably diurnal and terrestrial, like other highland *Trachylepis*. Several females contained 3–4 foetuses, indicating that it gives live birth, as do some other highland lizards.

SHORT-NECKED SKINK

Trachylepis brevicollis
(Wiegmann, 1837)

IDENTIFICATION: A large robust skink with a cylindrical body. Despite common name, neck is not particularly short, but head is short, eyes large, pupil round. Tail stout at base and about half total length. Dorsal scales large with usually two keels; the flank scales have three keels, scales in 30–34 rows at

Trachylepis brevicollis, gravid female, Shanshacade.

Trachylepis brevicollis, juvenile, Shanshacade.

Trachylepis brevicollis, female, Dire Dawa.

Trachylepis brevicollis, juvenile, Kenya.

Trachylepis brevicollis, adult male, Kenya.

mid-body. Maximum size about 32cm (snout–vent length up to 15.8cm), average 18–26cm, hatchling about 7–9cm. Colour tremendously variable, with some sexual dichromatism and ontogenic changes. Colour phase of adults has no apparent effect on that of their offspring. Hatchlings black with bright yellow barring on anterior flanks, which change to clusters or crossbars of fine yellow dots as they grow, then become brown or grey, or dark brown with light crossbars. Some adults black with conspicuous rows of whitish spots covering entire dorsum. Adult females usually brown, with irregular dark specks or crossbars, light or heavy, sometimes with poorly defined light dorsolateral stripes, or uniform brown, or with a paler anterior half of body and head. Adult males usually have distinctive longitudinal black and brown stripes, sometimes vivid white speckling on front half of body. Uniformly paler below, except darker tail and throat in males. Specimens from rocky terrain in central and northern Ethiopia and Eritrea usually more spotted; those from sandy areas of Somalia and Ogaden typically uniform. Taxonomic Notes: Very variable; several species have been synonymised with this taxon, but recent molecular studies suggest multiple cryptic species are concealed within the group.

HABITAT AND DISTRIBUTION: Occupies a wide range of habitats usually in low dry country. Found in moist and dry savanna, woodland, semi-desert and coastal thicket, from sea level to about 1,500m. A single old record from Sudan, at 'Mogatta' (Muqatta) on the Atbara River, near the Eritrean border. Widespread in lowlands of Eritrea. Occurs from south-east South Sudan, at Boma in Jonglei State, and the Kapoeta and Lokomarinyang area, east across most of eastern Ethiopia and Somalia (sporadic in Puntland), thence through southern Djibouti right across the Afar Triangle, up the Awash River west as far as Sodere. Elsewhere, south to northern Tanzania, and to the Arabian Peninsula.

CONSERVATION STATUS: Least Concern.

NATURAL HISTORY: Terrestrial, often found amongst rocks and boulders and inclined to seek a prominent position to bask in the sun. Shelters beneath stones, logs, bark of trees, in termite hills or almost any other suitable hole in ground. Gravid females at Shanshacade in southern Somaliland were found under a dry camel carcass in July. Diet mostly invertebrates but will take larger items including small mammals. This skink is viviparous. A female from Awash National Park had three offspring in October, a Dageh Bur female had two, a Shanshacade female had five offspring in October; a Kenya female contained four embryos. Juveniles grow rapidly reaching adult size in a few months. Often in large groups, a big rotting *Vachellia* trunk in riverside bush at Garissa, eastern Kenya, housed 27 individuals. Although widespread, it is a shy species, not easy to see.

BICOLOURED SKINK, ZEBRA SKINK

Trachylepis dichroma
Günther, Whiting & Bauer, 2005

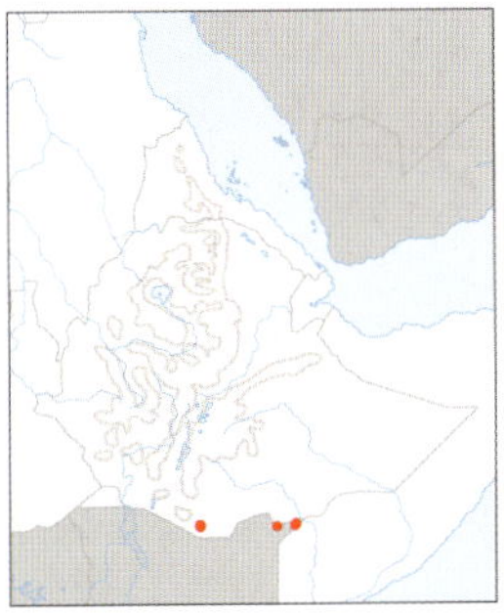

IDENTIFICATION: A large robust skink with a thick-set cylindrical body. The head is short, eyes large and prominent, the pupil round. Tail stout at base and about half total length. Dorsal scales usually have two keels, the flank scales two keels (some a weak third keel), scales in 30–38 (most commonly 34–38) rows at mid-body. Maximum size about 38cm (snout–vent length up to 18.6cm), average 20–34cm, hatchlings 7–9cm. Colour variable, with strong sexual dichromatism. Some adult males finely striped light and dark brown dorsally, with small white spots on flanks, others uniform brownish-bronze. In some (possibly breeding) males the lower flanks and side of tail are vivid orange-red. The throat may be black, blue or pale (possibly also connected to breeding). Belly cream shading outwards to yellow. Adult females brown, scales black-edged, body and tail have narrow crossbars consisting of black-and-white scales; juveniles resemble females but black crossbars finer, with fewer white scales. In some hatchlings, the upper dorsal surface is grey, with extensive white spotting, the flanks are barred black, grey and white, but others are finely barred black on brown or brick-brown. Taxonomic Notes: This species was described as a different member of the highly variable *T. brevicollis* complex; its validity has been confirmed molecularly. Its small range has been greatly expanded due to re-assignment of many former specimens of *T. brevicollis*.

HABITAT AND DISTRIBUTION: Due to earlier confusion with Short-necked Skink *T. brevicollis*, the range of Bicoloured Skink is not yet clarified. It inhabits low to high-altitude savanna and semi-desert, at altitudes of 200–1,800m outside our range. In north-east Africa known with certainly on the basis of one young specimen collected about 20km north of Mega in southern Ethiopia, at 1,529m; also, at Ramu and Mandera on the border between north-east Kenya and Ethiopia. Elsewhere, south to central and northern Tanzania. Several specimens that might be this species (typical coloration but no molecular confirmation) are known from Rahanuin region, southern Somalia (not shown on map).

CONSERVATION STATUS: Least Concern.

NATURAL HISTORY: Similar to *T. brevicollis*, and often found in same habitat. Mostly terrestrial but will climb rocks, fallen trees, etc. Diurnal, fond of basking in prominent position, but quite wary. It will utilise any suitable habitat, in holes on flat plains, squirrel

Trachylepis dichroma, juvenile, Mega.

Trachylepis dichroma, adult female, Tanzania.

warrens, rodent holes, termite hills, under logs, rocks, boulder clusters on hillsides, rock cracks, tree holes and so on. The Ethiopian specimen was found in daytime under a large flat rock in dense bushland along a road. Females in captivity gave birth to 5–7 offspring, which reached maturity after 18–24 months. A juvenile female showed thanatosis while being photographed, curling on her back and remaining motionless for several minutes.

Trachylepis dichroma, male, Tanzania.

BUETTNER'S SKINK

Trachylepis buettneri
(Matschie, 1893)

IDENTIFICATION: A slim striped skink with a massive tail; it might just reach our area in South Sudan. Short, rounded narrow head and a large eye. The unregenerated tail is huge, 75–80 per cent of total length. Dorsal scales have three keels. Scales in 26 (very occasionally 28 or 29) rows at mid-body. Maximum size about 36cm (females larger), average 24–32cm, hatchling unknown. Brown or rufous-brown above, West African specimens usually have around four longitudinal lines of elongate dark spots, not always present in Central African examples. A slightly contrasting broad brown flank stripe that darkens downwards is present; below this a narrow white stripe extends from lips to rear limb junction. Underside cream or white.

HABITAT AND DISTRIBUTION: Moist and dry savanna and gallery forest, between sea level and 1,200m. Known from Garamba National Park, just across border from South Sudan, in DR Congo, so almost certainly present in adjacent South Sudan. The 1893 type was from Togo, and was described in 1919 as *Mabuya sudanensis* by Karl Patterson Schmidt, who stated he had 'eleven specimens from localities in the Sudan', but were all from within DR Congo.

CONSERVATION STATUS: Least Concern.

NATURAL HISTORY: Diurnal and largely arboreal, hunting and basking in bushes and small trees, making use of their long tail for balance. Sometimes active on the ground, especially in grassland. Specimens in trees jumped down when pursued. At Lamto, Ivory Coast, entire populations are virtually exterminated by annual January bush fires, and the population is replenished when the buried eggs hatch.

Trachylepis buettneri, preserved, Ivory Coast.

WEBI SHEBELLE SKINK

Trachylepis ferrarai
(Lanza, 1978)

IDENTIFICATION: A medium-sized brown, fairly robust skink known only from vicinity of Webi Shebelle River near Mogadishu, Somalia. Body cylindrical, tail fairly long, about ⅔ of total length. Dorsal scales have 5–10 keels, scales in 30 rows at mid-body. Known from two male specimens, one 20cm, the other with a regenerated tail 18.4cm. Dorsal colour brown with small black sub-rectangular blotches forming two or four vague lines, with a trace of a paler vertebral stripe and a diffuse, narrow light dorsolateral stripe. A dark line extends from under the eye to ear-opening. One specimen has a narrow white lateral stripe, the other a broader blotched lateral band. Lips and underside ivory-white. Taxonomic Notes: Extremely similar in appearance to *T. maculilabris*.

HABITAT AND DISTRIBUTION: Habitat where types taken is dry open woodland, much degraded by agriculture, with patches of gallery forest along river; altitude just below 100m. One specimen was found on the ground at Afgoi university campus, the other in a hole near Balad, both within 50km of Mogadishu.

CONSERVATION STATUS: Data Deficient.

NATURAL HISTORY: Unknown.

HEMMING'S SKINK

Trachylepis hemmingi
(Laurent & Gans, 1965)

IDENTIFICATION: A medium-sized, brown, stocky skink with smooth dorsal scales, endemic to northern Somalia. Head rather small and moderately long, snout pointed, eyes large and prominent, the pupil round. Ear-opening large with one or two very short lobules on its anterior border, neck short and thick. Tail stout at base and about 53–60 per cent of total length. Dorsal scales rather large, smooth, with a very faint trace of three keels on those of posterior body, in 32 rows at mid-body. Maximum size about 20cm (snout–vent length up to 9.6cm), average 16–18cm, hatchling size unknown. Colour pattern distinctive: light brown above, with six longitudinal series of darker dots; a broad dark band originating in front of eye passes through the temporal region, above ear-opening and ends above fore limb, with the darkest area around eyes. Flanks paler than back, lower parts of head including labials white. Underside whitish. Taxonomic Notes: A relatively recently described species, formerly known from just one specimen; additional specimens were recently discovered, misidentified as *T. brevicollis*.

HABITAT AND DISTRIBUTION: Endemic to northern Somalia. Occurs in low semi-desert country and arid savanna with rocky and hilly terrain, at altitudes of 120–1,200m. The type was collected at Taleh on the slopes of Nogal Valley in generally gypseous country of the Sool region of eastern Somaliland. Additional specimens were found in same area and further north in Buran and on southern slopes of Cal Madow

Trachylepis hemmingi, Las Anod.

Trachylepis hemmingi, preserved, Berbera.

Mountains. Also recorded at Galgalo Oasis in Puntland, about 30km south-west of there (on Somaliland side of border), with an isolated record far to the west, on the coast near Berbera in the Sahil region. Probably widespread in the intervening country, which is under-collected.

CONSERVATION STATUS: Least Concern.

NATURAL HISTORY: Unknown, but probably similar to other similar-sized *Trachylepis*, i.e., diurnal, lays eggs, eats invertebrates, takes cover in any suitable refuge. Its arid, rock-strewn homeland suggests it is a terrestrial or rock-dwelling species. The specific name honours Christopher Francis Hemming, British entomologist, plant ecologist and explorer; an expert on locusts who worked for the Desert Locust Survey in Somalia, he collected the type and made other important collections of amphibians and reptiles there.

Trachylepis hildebrandtii, female, God Heeli.

Trachylepis hildebrandtii, male, Mogadishu.

HILDEBRANDT'S SKINK

Trachylepis hildebrandtii
(Peters, 1874)

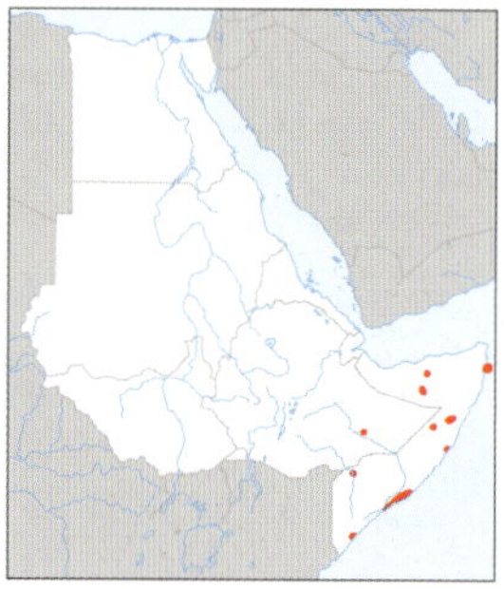

IDENTIFICATION: A fairly large, attractively marked, skink with several distinctive large lobules on anterior border of ear-opening, virtually confined to Somali peninsula. Head short, snout pointed. Ear-opening large, with 2–3 strikingly long and pointed scales projecting backwards from the front edge. The body is cylindrical; tail long, about 60–65 per cent of total length. Dorsal scales have three strong keels, scales form 29–34 rows at mid-body. Maximum size about 23cm (snout–vent length up to 8.2cm), average 16–20cm, hatchling size unknown. Colour pattern distinctive, with adult males very different to females and juveniles. Dorsum pale brown or grey-brown, uniform or with two longitudinal series of faded dots in adult males. Behind ear is a series of 3–4 large blackish blotches in a longitudinal row, interspersed with bright yellow, which extends onto anterior flanks. There is black marbling on the chin and side of neck. Females and juveniles usually have a more or less distinct pattern of dark blotches on back, often arranged in prominent transverse bars, accompanied by pale vertebral and dorsolateral stripes, sometimes with a similar but much weaker lateral stripe. Underside white.

HABITAT AND DISTRIBUTION: A typical species of arid coastal lowlands along Indian Ocean. Found in dry savanna, woodland, semi-desert and coastal thicket with consolidated dunes, even on beaches. Widespread, although records sporadic, in southern Somalia along Benadir coast from Kismayu to Mogadishu and Warsheikh, inland on Juba River at Lugh. One of the commonest reptile species in Mogadishu. In central Somalia known from the Hobyo coast, and inland at Geriban and Galkayo, in Puntland on the Hafun Peninsula, El Afweyn in Sanaag region and between Yagoori and God Heeli in Sool, Somaliland. A single old imprecise record (the specimen lost) from the Ogaden in Ethiopia, north of the Shebelle River, approximately between Imi and Gode, but it is not improbable considering the Somali distribution. Widespread and locally very common. Unlikely to be threatened by human activities.

CONSERVATION STATUS: Least Concern.

NATURAL HISTORY: A diurnal skink, mostly rock-dwelling but will move onto houses, huts and other man-made objects like walls and bridges. In Mogadishu it lives on rocks by the beach, and slightly further inland on cairns gathered for construction work. However, doesn't seem to move far from the coast in that area. One was regurgitated by a Brown House Snake *Boaedon fuliginosus*.

ISSEL'S SKINK

Trachylepis isselii
(Peters, 1871)

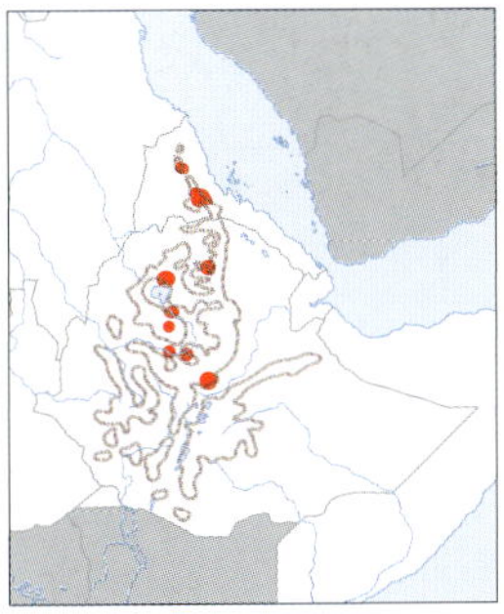

IDENTIFICATION: A small striped brown skink, known only from the north-west highlands of Ethiopia and Eritrea. Snout quite pointed. Body slightly depressed. Tail long, ⅔ of total length, and thin. Dorsal scales have three keels and are arranged in 30–34 rows at mid-body. Maximum size about 16cm, average 10–14cm, hatchling probably 5–6cm. Dorsal and flank colour variable, usually some shade of brown, bronze, olive or grey, speckled to a greater or lesser extent black and white. There is a distinct, narrow white dorsolateral stripe, and a similar lateral stripe low on the flanks. Underside usually white or cream. Juveniles have a blackish body and four pale stripes, a bronze head with white lips and a bluish-grey tail. Taxonomic Notes: Described from Eritrea, and regarded by some authorities simply as a variety of *T. varia*, from which it differs only in its fused frontoparietal scale.

HABITAT AND DISTRIBUTION: High-altitude grassland, moist savanna and agriculture, between 1,000 and 2,500m. In Ethiopia, occurs north from Ambo through the north-west highlands to the Simien Mountains; most records above 1,500m. A cluster of records from central Eritrea, some at lower altitudes (Ghinda), some high.

CONSERVATION STATUS: Not Assessed.

NATURAL HISTORY: Diurnal and terrestrial, sometimes climbs. Active on ground, often in rocky areas. Hides in holes, under rocks and under tree bark. Breeding details unknown, but probably similar to *T. varia*, i.e., gives live birth to a small litter.

Trachylepis isselii, Asmara.

Trachylepis isselii, Asmara.

Trachylepis isselii, juvenile, Asmara.

SPECKLE-LIPPED SKINK

Trachylepis maculilabris
(Gray, 1845)

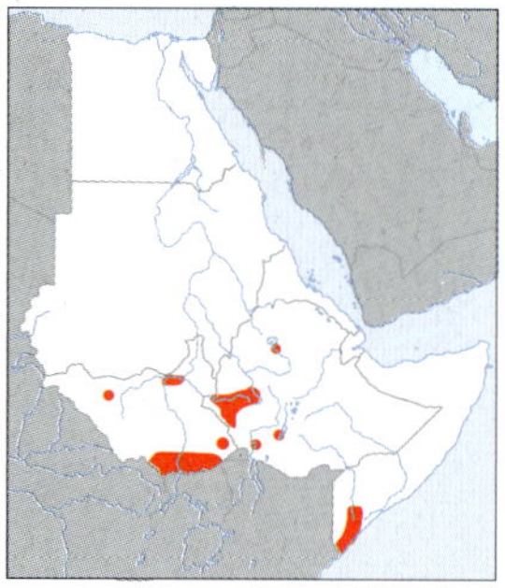

IDENTIFICATION: A fairly large, long-tailed arboreal skink, with often speckled lips. Snout pointed; eye fairly large with distinct yellow eyelids. Body cylindrical, toes long and thin, tail very long and slender, 60–70 per cent of total length, and used as a climbing aid. Dorsal scales keeled (3–9, usually 5–8 keels), arranged in 29–38 (usually 29–35) rows at mid-body. Maximum size about 30cm, average 15–25cm, hatchlings 5–8cm. Colour very variable, with location, although most have distinct white, black-speckled lips, hence the common name. Scales around the ear usually yellow, a useful field aid. Somali coast specimens are typically uniform brown above (sometimes with black speckling), and may have a broad orange flank stripe, the tail is brown, black speckled, the side of neck speckled white, underside yellow, and chin white with fine black dots. Those in Ethiopia are usually brown or olive above, often with black and/or white speckling on neck and flanks (decreasing towards tail) and a broad dull gold or brown flank stripe (not always present); some have vivid yellow flanks, chin and throat; sometimes the tail is bronzy. Occasional individuals have dark vertical flank bars. Taxonomic Notes: Molecular analysis indicates that. as presently defined, *T. maculilabris* is a species complex, and is in the process of being split.

HABITAT AND DISTRIBUTION: Forest clearings, woodland, coastal thicket, farmland and gardens, often near natural water sources. From sea level to about 2,000m, possibly higher. In Somalia, known only in south, from lower Juba River (south of 02°N) to Kismayu and thence south along coast to Kenya. In Ethiopia widespread in the south-west forests, and west from there to Gambela; also, Bahar Dar, the vicinity of Lakes Chamo and Abaya, and the lower Omo River. Probably quite widespread in South Sudan but records sporadic, several from Western, Central and Eastern Equatoria, Boma in Jonglei, a few records further north (Aweil, in northern Bahar El Ghazal, and Bahr El Zeraf). Elsewhere, west to Senegal, south to Angola and central Mozambique.

CONSERVATION STATUS: Least Concern.

NATURAL HISTORY: Diurnal and arboreal, active on trees and bushes. It climbs expertly, using its tail; will climb along even very thin twigs and on leaves. Also utilises buildings, walls, thatched roofs and rock outcrops. Often in waterside bushes and trees. In dense forest it is usually found around clearings. Sleeps at night in a leaf cluster, cracks and holes in trees, under palm fronds, etc. Clutches of 3–8 eggs, roughly 1 × 1.5cm, and in Central Africa it appears to have no definite breeding season, 5–6 clutches laid per year; juveniles found in October in south-west Ethiopia. Eats arthropods and snails, and observed to eat a juvenile Tropical House Gecko, *Hemidactylus mabouia*.

Trachylepis maculilabris, south of Gambela.

Trachylepis maculilabris, Kenya.

GRASS TOP SKINK, LONG-TAILED SKINK

Trachylepis megalura
(Peters, 1878)

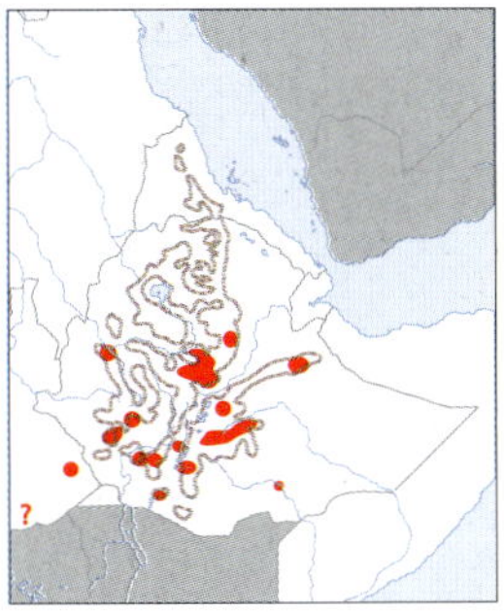

IDENTIFICATION: A fairly large, striped, slim skink with little limbs and a very long tail. Snout rounded, eye small, legs thin and spindly. Tail very long and slender, 60–75 per cent of total length and is used as a climbing aid. Dorsal scales have three weak keels; they may be smooth in large adults, in 24–27 rows at midbody. Maximum size about 25cm (7cm body, 18cm tail), average 15–22cm, hatchlings 5–7cm. Colour very variable, usually shades of brown, golden or rufous, a prominent white flank stripe runs from upper lip along entire body and may extend onto tail. Beneath this is a narrow black stripe, above it usually a broad black stripe. Back may be uniform brown, or have any number of fine or broad longitudinal black or white stripes; extralimital lowland specimens seem to have very pale stripes. Some females have broad red dorsal or dorsolateral stripes. Below white or yellow, sometimes up to nine fine black ventral stripes.

HABITAT AND DISTRIBUTION: Grassland and moist savanna, usually at fairly high altitude; reported from sea level elsewhere, in our area at 1,300–3,500m. In Ethiopia occurs largely in grassy highlands both east and west of the Rift Valley, but unrecorded north of 10°N. Occasional records from margins of the southwest forest in Ethiopia. A single South Sudan record, at Boma, in Jonglei, although a northern Uganda record suggests it might also occur in the Imatong Mountains. Elsewhere, the montane region of DR Congo, south to central Mozambique.

CONSERVATION STATUS: Least Concern.

NATURAL HISTORY: Diurnal, it lives in grassland and

Trachylepis megalura, Mulu.

Trachylepis megalura, Mulu.

ascends into clumps, using its long tail it slides swiftly through thick grass. In high-altitude areas it basks in the morning up in clumps before hunting. Shy and secretive, sliding rapidly away or hiding if approached. At night, it hides under rocks, in grass clumps or in holes. Gives live birth, up to 15 young recorded in a Kenya specimen, hatchlings recorded in November and December in highland Ethiopia. Its sliding movements mean it is liable to be mistaken for a snake if not seen clearly.

ORANGE-FLANKED SKINK

Trachylepis perroteti
(Duméril & Bibron, 1839)

IDENTIFICATION: A big robust skink with a cylindrical body, short head and small eye, the pupil round. Tail fairly long, about 55 per cent of total length. Dorsal scales usually have three keels, in 32–34 rows at mid-body. Maximum size about 45cm, average 21–35cm, hatchling size unknown. Ground colour dull green or bronze above, sometimes with fine longitudinal lines of brown dashes. Flanks dull orange (becoming vivid orange in males during breeding season) or brown, shading to orange or rufous-brown with extensive white spotting. Immaculate white or ivory below. Juveniles less vividly coloured. Taxonomic Notes: Two subspecies are recognised; the nominate in our area. Belongs to the *affinis* species group.

HABITAT AND DISTRIBUTION: Moist and dry savanna, and forest fringes, between sea level and 1,400m. In our area known only from Sudan (Western and Southern Darfur) and South Sudan, from Western Bahr El Ghazal to Lakes and all three Equatoria States. Elsewhere, west to Senegal and southern Mauritania, south and south-west to DR Congo and northern Uganda.

CONSERVATION STATUS: Least Concern.

NATURAL HISTORY.: Diurnal and terrestrial (although they will climb fallen logs and on termitaria), they live in holes, excavated under a bush, and mostly hunt by ambush, watching from cover until suitable prey passes by. Lays eggs but no further breeding details known.

Trachylepis perroteti, Ghana.

TREE SKINK

Trachylepis planifrons
(Peters, 1878)

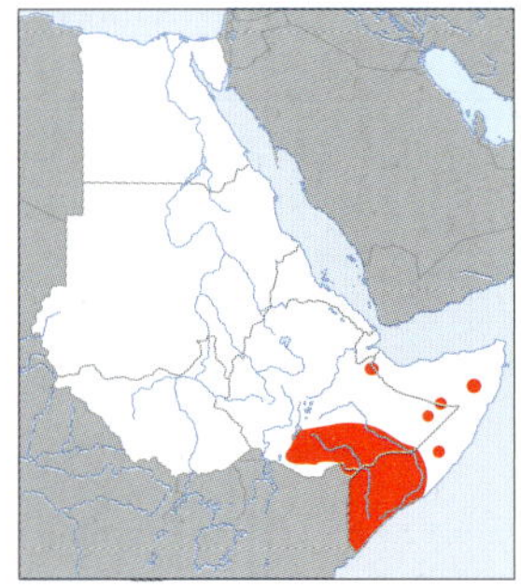

IDENTIFICATION: A large, long-tailed arboreal skink. Eye medium-sized, body slightly depressed. Tail very long and thin, ⅔ of total length. Dorsal scales mostly have three (sometimes five) weak keels, scales in 25–32 rows at mid-body. Maximum size about 35cm, average 22–30cm, hatchlings 6–7cm. Back grey or brown, usually with a distinct bronze, white or cream dorsolateral stripe and a black flank stripe extending from snout through eye to hind limbs, in some the dorsolateral stripe is poorly defined and the flank stripe fades to grey posteriorly. There are usually several rows of indistinct black speckles on centre of back, outside of these are often 1–2 rows of white dots. Belly grey or white, often with dark grey speckling; tail usually grey. Liable to be confused with a striped skink, but has a much longer tail.

HABITAT AND DISTRIBUTION: Coastal thicket, moist and dry savanna and semi-desert, provided there are reasonable-sized bushes present. From sea level to about 1,600m. Widespread in Somalia, but rare in north, with very few records from parts of the east and north-east (none at Mudug or Sanaag). Also widely distributed in lowlands of eastern Ethiopia, although the only records from the Ethiopian Rift Valley are around Lakes Chamo and Abaya, and there are few definite records from the Ogaden. Elsewhere, south-west to Malawi, northern Zambia and south-east DR Congo.

CONSERVATION STATUS: Least Concern.

NATURAL HISTORY: A diurnal, arboreal skink, living in both trees and bushes, on trunks and even quite thin branches. It climbs expertly, aided by its long tail.

Trachylepis planifrons, Kenya.

In dry areas it basks in the early morning sun, then hunts, retiring into shade when the day becomes hot. It is quick-moving, when approached it sidles around the trunk, keeping the wood between itself and danger, if pursued along a branch it will jump if threatened. Will descend to move to another tree, and sometimes hunts on ground. Hides in tree holes and cracks, will sleep on tips of branches, but will also live in burrows or cracks in rock, and on Kenya coast sometimes in coral rag. Males become reproductively active at a snout–vent length of 7cm, females at 9cm, clutches of 8–9 eggs reported for females from Kenya in June and July.

FIVE-LINED SKINK

Trachylepis quinquetaeniata
(Lichtenstein, 1823)

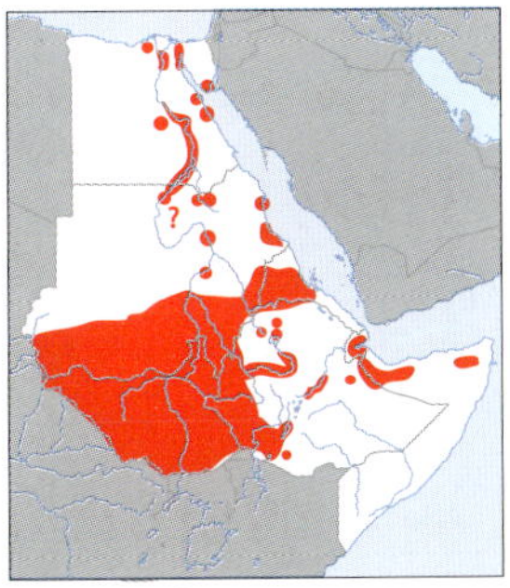

IDENTIFICATION: A large, conspicuous, robust rock-dwelling skink with a short head, females and juveniles have a bright blue tail, and males black dots behind earhole. Body cylindrical, tail fairly long, about 55–60 per cent of total length. Dorsal scales have three keels, scales in 32–42 rows at mid-body, usually 36–40. Maximum size about 25cm, average 18–22cm, hatchlings 6–7cm. Juveniles black, females brown, with five yellow or cream stripes on body, a blue tail, and the head may be reddish or bronze. Adult males are various shades of brown or black, with white speckling, they have several big black blotches, separated by white dots, on each side of neck. The lateral stripe persists anteriorly, where it may be yellow or orange on the neck and blue or blue-white on the chin. Breeding males develop a black-speckled chin.

HABITAT AND DISTRIBUTION: Low to medium-altitude semi-desert, dry and moist savanna and woodland, from sea level to about 2,200m, but almost always on rocks (or nearby substitutes such as buildings or large trees). Widespread in western Ethiopia and western Eritrea, especially in river gorges, but virtually no records east of Ethiopian Rift Valley. In Somalia known only from mountainous north-west Somaliland, thence into Djibouti. Appears to occur almost throughout South Sudan, although records sporadic in north. In Sudan, widespread in Western and Southern Darfur, Southern and Northern Kordofan, east through Khartoum to Kassala with a handful of records on the Red Sea coast. North of Atbara, it appears to be confined to the Nile Valley, and extends north to Cairo, east to Ismailia, west to Faiyum and just west of the Delta; isolated Egyptian records include south Sinai, Kharga Oasis, and Safaga, Ain Sukhna and around Hurghada on the southern Red Sea coast, where possibly introduced. Elsewhere, south to Tanzania, west to Senegal; introduced to northern Algeria, Florida and California.

Trachylepis quinquetaeniata, male, Heglig.

Trachylepis quinquetaeniata, male, Aykel.

CONSERVATION STATUS: Least Concern.

NATURAL HISTORY: A diurnal skink. Usually on rocky hills, rock outcrops, lava fields or in canyons. They will move onto buildings, huts, walls and bridges, and trunks of big trees; sometimes found in holes in ground, but rarely far from open rock. Tolerant of very hot conditions; in northern Kenya they live among lava boulders and tolerate diurnal temperatures of 42–44°C. Often visible in suitable habitat, males bask in a prominent position. They are territorial, big males dominate a space and do not tolerate other adult males. Quite wary, if approached they move away. Swim well, in Egypt they have been seen to jump off boulders into water and swim to escape. They lay 3–10 eggs roughly 1.7×1cm, in a suitable spot such as in leaf litter (especially leaf 'drifts' in rock cracks), under logs, deep in rock fissures, etc.) Communal egg clutches of up to 60 eggs have been found. In West Africa a clutch of seven eggs hatched after 43 days incubation. Diet arthropods, a Congo specimen had seeds, small fruit and an earthworm in its stomach, a Tanzanian animal had eaten the tail of a juvenile conspecific. A lizard with a huge range and is highly adaptable; although a sub-Saharan species, it has obviously rafted down the Nile and is now colonising Egypt.

Trachylepis quinquetaeniata, hatchling, Kenya.

Trachylepis quinquetaeniata, female, Las Geel.

STRIPED SKINK

Trachylepis striata
(Peters, 1844)

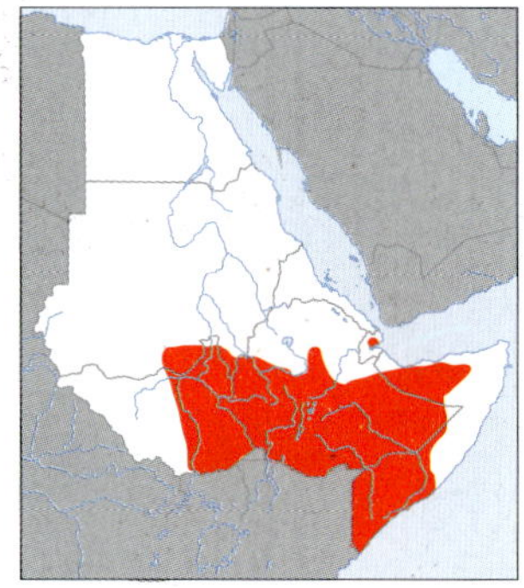

IDENTIFICATION: A medium-sized, robust, striped skink, very common in suitable habitat and often seen in towns. Head and body slightly depressed. Tail just over half of total length. Dorsal scales have 3–7 keels, scales arranged in 30–42 rows at mid-body. Maximum size about 25cm, average 18–22cm, hatchlings 6–7cm. Ground colour shades of brown, olive, rufous or dull green, with two cream, yellow or light greenish dorsolateral stripes, usually sharp-edged but may be diffuse, and fade posteriorly on some animals. Top of head and snout reddish, flanks usually speckled white or yellow, underside cream or white with some grey or black speckling. There may be white or black dorsal speckles. Some specimens are bronze-coloured. Similar in appearance to the Tree Skink, *Trachylepis planifrons*, but has a much shorter tail.

HABITAT AND DISTRIBUTION: Forest clearings, coastal thicket, moist and dry savanna, semi-desert and urban areas, common even in heart of cities like Addis Ababa and Hargeisa. Found from sea level to about 2,800m, or even higher. Throughout most of southern half of Ethiopia, south of 12°N, into Djibouti, and widespread in Somalia, although there are virtually no records from the centre of the country between 05 and 08°N. Absent from Eritrea. In South Sudan, known from Western, Central and Eastern Equatoria, particularly near the White Nile, also in Unity, Jonglei and Upper Nile State, but no records in the west. Fairly widespread in southern Sudan, in Southern Darfur, Southern Kordofan and Blue Nile States, records include Nertiti, Kadugli, El Obeid and Talodi, to about 12°N. Elsewhere, south to South Africa; a south-east African savanna form.

Trachylepis striata, Arba Minch.

CONSERVATION STATUS: Least Concern.

NATURAL HISTORY: Diurnal and essentially arboreal, but copes with urbanisation very well, living on walls and fences in gardens and plantations, and often seen in towns. Takes cover in any suitable refuge. Quickly becomes tame and confiding, in urban gardens rapidly learning to tolerate people. A very versatile species, specimens in Kenya colonised a new house within three months of completion, moving in from the surrounding bush where they were uncommon. In suitable habitat it may occur in large numbers, but sometimes rare, for example in undisturbed semi-desert. Gives live birth, clutches of up to nine recorded. Diet arthropods, but seen to eat a gecko; also recorded taking vegetable matter, fruit and carrion. Their main enemies are birds of prey and snakes. Several dicephalic (two-headed) examples of this lizard have been recorded.

VARIABLE SKINK

Trachylepis varia
(Peters, 1867)

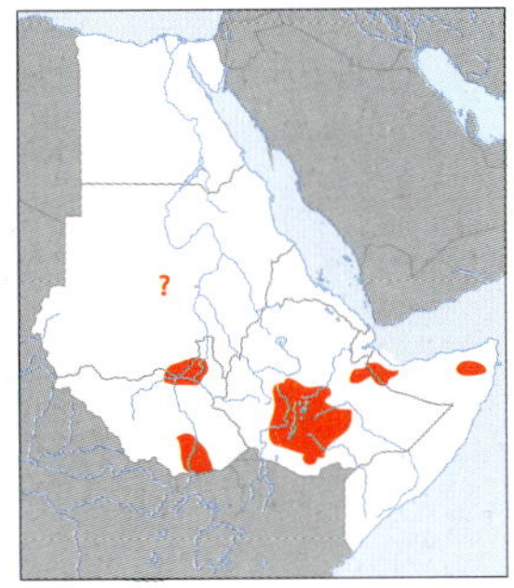

IDENTIFICATION: A small to medium-sized striped skink, variable in colour, hence the name. Snout quite pointed. Body slightly depressed. Tail long, ⅔ of total length, and thin. Dorsal scales have three keels and are arranged in 27–36 rows at mid-body. Maximum size about 18cm (6cm body, 12cm tail), possibly slightly larger, average 10–15cm, hatchlings 5–6cm. Back usually shades of brown or bronze, uniform or speckled to a greater or lesser extent with black, the black blotches sometimes forming broad vague crossbars. Nearly all have a white stripe low on flanks, a white dorsolateral stripe and usually a pale vertebral stripe, which are useful for identification. Underside normally white or cream. Taxonomic Notes: A species complex with at least four cryptic species; liable to be split in the near future.

HABITAT AND DISTRIBUTION: Coastal thicket, woodland, moist and dry savanna, and high-altitude grassland, from sea level to 3,600m. Not common in semi-desert. Occurs from the east side of Southern Kordofan in Sudan, into South Sudan (widespread east of 28°E), thence across southern Ethiopia (records from north-west usually *T. isselii*), but largely absent from the Ogaden. Sporadically recorded in northern Somalia (mostly Ogo and Goolis Mountains and Sol Haud Plateau in Somaliland, and Carcar Mountains in Puntland) but absent south of 08°N. Elsewhere, south to South Africa, west to the eastern DR Congo.

CONSERVATION STATUS: Least Concern.

NATURAL HISTORY: Diurnal and terrestrial, sometimes climbs. Active on ground, often in rocky areas, fond of broken country around base of small rocky hills, ridges, etc. Hides in holes, under rocks and under tree bark. Quite tolerant of urbanisation. Gives live birth, with 3–10 babies recorded.

Trachylepis varia, Chancho.

WINGATE'S SKINK

Trachylepis wingatii
(Werner, 1908)

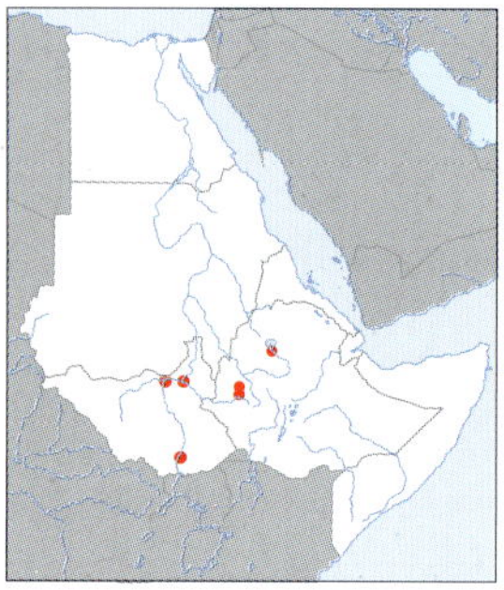

IDENTIFICATION: A fairly large, brown, rare skink, of western Ethiopia and South Sudan. The snout is rounded. Tail fairly long, 55–65 per cent of total length. Dorsal scales have three keels and are arranged in 30–32 rows at mid-body. Maximum size about 24cm, average 15–20cm, hatchling size unknown. Back bronzy or reddish-brown, sometimes with fine elongate blackish spots forming longitudinal lines. Dorsal colour darkens on flanks to form a dark lateral stripe; below this is a conspicuous narrow, sharp-edged white stripe from snout to tail; scales below this stripe may be rufous. Underside cream, yellow or greenish-yellow.

HABITAT AND DISTRIBUTION: Sporadically recorded at a disparate handful of habitats (high grassland, forest fringes, woodland, moist savanna) in western Ethiopia and South Sudan, between 400 and 2,000m. Localities include vicinity of Lake Tana, Ghimbi and Didessa River west of Nekemte in Ethiopia; Lake No, near Malakal (Unity and Upper Nile States) and near Juba in Central Equatoria, South Sudan. Presumably more widespread in the intervening country.

CONSERVATION STATUS: Least Concern.

Trachylepis wingatii, Bahar Dar.

NATURAL HISTORY: Poorly known. Diurnal and terrestrial. At Bahar Dar on Lake Tana it was active on the ground and earth banks, taking cover in vegetation, and also seen to jump into water when pursued and swim. On the east side, it was found in the woodland surrounding churches, and basking on rock walls. Reproductive details unknown. The author who discovered and described this species was Franz Werner (1867–1939), Austrian zoologist and explorer, at the University of Vienna. Among other topics, he studied the herpetofauna of Sudan, from where he described several new species of reptiles and published in 1919 the first major publication on the distribution of the reptiles of Sudan, still relevant today.

GOLDEN AND BRIDLED SKINKS *HEREMITES*

A genus recently resurrected for three species of skinks, formerly in *Trachylepis*; molecular work indicates they separated from *Trachylepis* 30 Mya, although anatomically they are very similar, differing only in certain minor scale differences. Two species occur in our area.

KEY TO NORTH-EAST AFRICAN MEMBERS OF THE GENUS *HEREMITES*

1a: Back with five stripes, slender, adults <20cm, in our area only in northern Egypt
Heremites vittatus, Bridled Skink

1b: Back with four incomplete stripes, robust, adults >20cm, in our area only in Eritrea
Heremites septemtaeniatus, Golden Grass Skink

BRIDLED SKINK

Heremites vittatus
(Olivier, 1804)

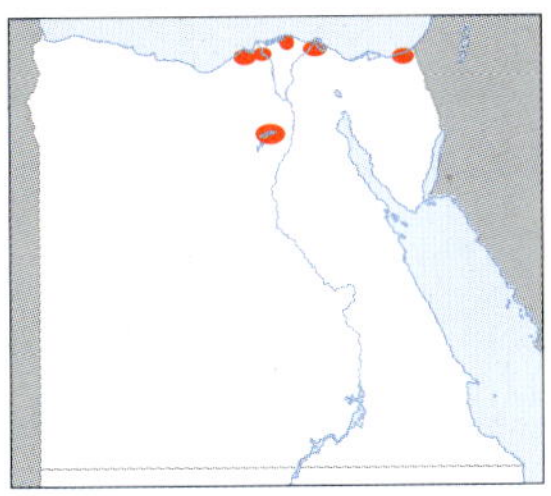

IDENTIFICATION: A little striped skink of northern Egypt. The snout is rounded. Tail fairly long, 60 per cent of total length. Dorsal scales have three keels and are arranged in 32–34 rows at mid-body. Maximum size about 18cm males, 20cm females (although larger elsewhere), average 12–16cm, hatchlings 3cm snout-vent length. Back brown, with five light, dark-edged longitudinal stripes, the vertebral stripe often most prominent and widest, the two dorsolateral stripes weakest. Some can be almost pattern-less. Underside white or greyish-white.

HABITAT AND DISTRIBUTION: In our area, largely occupies edges of cultivated land, orchards, palm groves, fields, and fringes of deserts, at low altitude. In Alexandria found in city parks and vegetated gardens, utilises hedges. In Egypt, associated with the north end of the Nile Delta and its margins, but also recorded from the Faiyum and Al Arish. Elsewhere, west to Algeria, east to Iran.

Heremites vittatus, Jordan.

CONSERVATION STATUS: Least Concern.

NATURAL HISTORY: Diurnal and terrestrial. Active on ground, darting around edge of vegetation, but sometimes climbs low shrubs or into hedges to bask; shelters under ground cover or in holes. They give live birth to 3–8 young in spring. Largely eats invertebrates; cannibalism observed.

GOLDEN GRASS SKINK

Heremites septemtaeniatus
(Reuss, 1834)

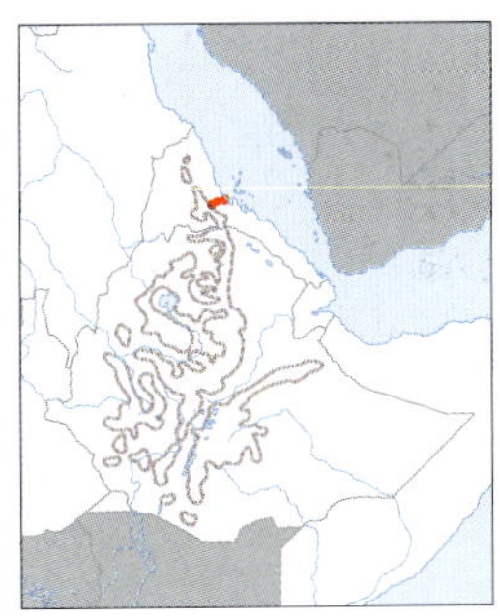

IDENTIFICATION: A robust, rare skink of Eritrea, possibly introduced and perhaps extinct there. Snout rounded. Tail fairly long, 55–60 per cent of total length. Dorsal scales have three keels and are arranged in 32–38 rows at mid-body. Maximum size about 24cm. Back brown or olive, sometimes with small white dots, and four dark stripes, or lines of dark blotches, that usually fade or break up into dots about halfway down back. A dark stripe from snout through eye becomes a series of shrinking dark spots. The white on lips may extend to a low white lateral stripe. Underside white or pinkish-white. Taxonomic Notes: Called *Mabuya aurata* in earlier literature.

HABITAT AND DISTRIBUTION: In our area, only in eastern Eritrea, from the type (at Massawa) and two specimens collected a long time ago at Ghinda and Saati, in semi-desert and dry savanna, from sea level to 1,000m. Elsewhere, across the Middle and Near East, from Turkey to Pakistan. The huge distance (>1,500km) to the nearest coherent population in the Middle East suggests that the Eritrean animals were perhaps introduced. No recent specimens have been collected.

Heremites septemtaeniatus, Oman.

CONSERVATION STATUS: Least Concern.

NATURAL HISTORY: Diurnal and terrestrial. Active on the ground in rocky and sandy areas, darting around the edge of vegetation. They give live birth.

SUBFAMILY LYGOSOMINAE

A subfamily of just over 55 species of skink, in six genera. Two genera occur in north-east Africa; the only known species of *Haackgreerius*, an unusual two-legged skink, endemic to Somalia, and ten species of writhing skinks, genus *Mochlus*. Writhing skinks have a chequered taxonomic history; various species having been placed in the genera *Riopa*, *Mochlus* and *Lygosoma*. All north-east African forms are now in *Mochlus*, the name *Lygosoma* is reserved for Asian species.

TWO-LEGGED SKINK *HAACKGREERIUS*

An African genus with a single species; a small slender burrowing skink with no front limbs and only two toes on hind limbs. Little is known of its lifestyle. It is endemic to central Somalia and restricted to sandy habitats. Probably closely related to the skinks of the genus *Mochlus*. The generic name honours Allen E. Greer, Australian herpetologist, who worked extensively on skinks, and Wulf D. Haacke, South African herpetologist, who worked for many years at the Ditsong National Museum (formerly the Transvaal Museum).

HAACKE-GREER'S TWO-LEGGED SKINK

Haackgreerius miopus
(Greer & Haacke, 1982)

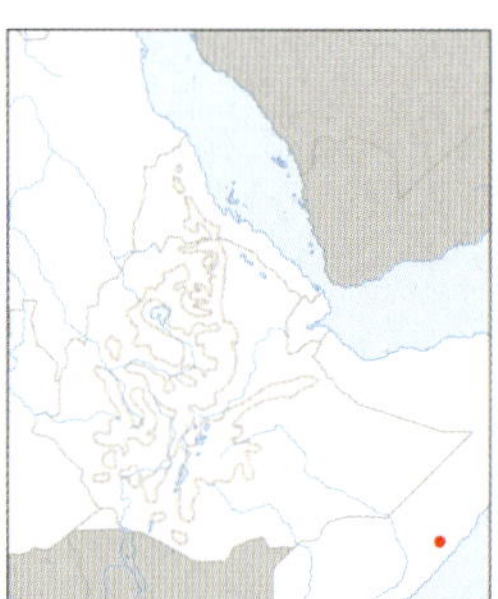

Haackgreerius miopus, preserved type.

IDENTIFICATION: A unique Somali endemic, a small short-tailed skink without front limbs; hind limbs reduced, with two toes and undifferentiated sub-digital lamellae. Body narrow, cylindrical, with a ventrolateral ridge. The unregenerated tail is short, thin and tapers rapidly, about 49 per cent of snout–vent length. Snout projecting, wedge-shaped. Vestigial eyes covered by an undifferentiated ocular scale. Nasal in broad contact, supranasal and postnasal fused to nasal, no prefrontals, frontoparietals and interparietal fused. Ear-opening moderate in size, obscured by free edges of two anterior head scales, situated just above level of mouth. There are 22 rows of smooth subequal scales at mid-body, 97–102 paravertebrals. Total size 9–11cm (snout–vent length 6.3–7.3cm), hatchling size unknown. Light brown above with a dark dorsolateral line extending from canthus posteriorly above ear, along body onto tail and a similar dark brown stripe along head and spine, prominent on frontoparietal-interparietal head scales and at base of tail. Uniform pale brown below with some dark brown spotting on chin and anterior throat.

HABITAT AND DISTRIBUTION: Flat mixed shrubby bushland with scattered trees, grassland and partially vegetated dunes. Restricted to sandy soils with comparatively open low thornbush and sparse grassy cover, at an altitude of 282m. Known from two specimens, both found 7km north-west of Rugno (Ruunyooc) Well, about 80km south-east of El Bur on road to El Dere, in Galguduud. Probably more widespread in surrounding areas of the Hobyo coast. Unlikely to be under threat, suitable local habitat is extensive.

CONSERVATION STATUS: Data Deficient.

NATURAL HISTORY: Unknown. A psammophilous (sand-loving) species, lives under surface, but is probably diurnal. The adult male and female were collected in the dry warm season in August–September.

WRITHING SKINKS *MOCHLUS*

A group of African burrowing skinks with small legs. As the name implies, when seized or handled they writhe and struggle fiercely, often jabbing the sharp snout into the restraining hand. They have scaly, mobile lower eyelids and small to medium-sized eyes, and most have sharp pointed snouts. Their smooth, shiny and elongated bodes are adapted for pushing through sand, soft soil or leaf litter, or for dwelling in holes. These skinks occupy a range of habitats, from semi-desert to woodland. Some species lay soft-shelled eggs, others give live birth. They feed on small insects and their larvae. About 18 species are known; ten occur in north-east Africa, six are endemic. The Horn of Africa, with the highest number of species, is the centre of evolution of this genus.

KEY TO NORTH-EAST AFRICAN MEMBERS OF THE GENUS *MOCHLUS*

1a: Unregenerated tail very long, >55 per cent of total length — 2

1b: Unregenerated tail short or moderately long, <55 per cent of total length — 3

2a: Limbs well developed in adults, length of hind limb contained almost three times in distance between axilla and groin; longer tail, 57–63 per cent of total length; 26–30 rows of scales at mid-body; head triangular — *Mochlus mabuiiformis*, Mabuya-like Writhing Skink

2b: Limbs small in adults, length of hind limb contained almost seven times in distance between axilla and groin; longer tail, 55 per cent of total length; 22–24 rows of scales at mid-body; head narrow — *Mochlus tanae*, Tana River Writhing Skink

3a: Unregenerated tail very short, 40–45 per cent of total length, nostril in a single nasal — 4

3b: Unregenerated tail moderately long, usually 47–52 per cent of total length, nostril in a single nasal or surrounded by two scales — 5

4a: Anterior part of ear-opening with one very small lobule concealed under scales; three upper labial scales in front of subocular, one preocular, a pair of more or less distinct dark longitudinal lines on back — *Mochlus vinciguerrae*, Vinciguerra's Writhing Skink

4b: Anterior part of ear-opening has an obvious, long and triangular lobule; four upper labial scales in front of subocular, two preoculars, no pairs of distinct dark longitudinal lines on back — *Mochlus grandisonianius*, Grandison's Writhing Skink

5a: Dorsally dark, with a spot on each scale; body elongate; southern Somalia — 6

5b: Dorsally pale, with a dark spot on each scale, or uniform; body stout; range Sudan, Ethiopia, Somalia — 7

6a: Frontonasal scale in contact with rostral, head and neck light speckled, longer tail (about 51–52 per cent of snout–vent length) — *Mochlus productus*, Boulenger's Writhing Skink

6b: Frontonasal scale not in contact with rostral, head uniform black with pale neck, shorter tail (about 47 per cent of snout–vent length) *Mochlus simonettae*, Simonetta's Writhing Skink

7a: Four upper labial scales in front of subocular, dorsal scales keeled in juveniles 8

7b: Three upper labial scales in front of subocular, dorsal scales smooth in juveniles 9

8a: Dorsal scales three-keeled in both adults and juveniles, in 28 rows at mid-body; southern Somalia *Mochlus laeviceps*, Common Writhing Skink

8b: Dorsal scales smooth in adults, but three-keeled over posterior trunk in juveniles, in 26 rows at mid-body, confined to Haud region and adjacent Ogaden of northern Somali peninsula *Mochlus paedocarinatus*, Lanza's Writhing Skink

9a: Larger, snout–vent length up 140mm; fifth (outer) toe clearly shorter than second and has only 5–6 sub-digital lamellae; chin, throat and ventral side of neck white or yellowish; widespread from Sudan to southern Somalia, but not known from most of Somaliland *Mochlus sundevalli*, Sundevall's Writhing Skink

9b: Smaller, snout–vent length up 95mm; fifth (outer) toe at least as long as second and has 7–9 sub-digital lamellae; chin, throat and ventral side of neck bright yellow; confirmed only in Somaliland *Mochlus somalicus*, Somali Writhing Skink

GRANDISON'S WRITHING SKINK

Mochlus grandisonianius
Lanza & Carfì, 1966

IDENTIFICATION: A very small, burrowing, short-tailed, writhing skink restricted to the north-east tip of the Somali peninsula. Head short, snout depressed and wedge-shaped. Nostril in a single nasal, frontonasal not in contact with rostral, prefrontals usually fused with posterior loreals. Ear-opening small, with an obvious long and triangular lobule. Eye small, with a scaly lower eyelid. Four upper labial scales in front of subocular, two preoculars. Body cylindrical, tail subcircular in cross-section, about 40–45 per cent of total length. Limbs short with five toes. The scales are smooth, in 24–27 rows at mid-body. Maximum size about 10cm (snout–vent length up to 5.9cm), average size unknown. Dorsum cinnamon-brown; a broad, dark brown, lateral band from nostril through eye, above ear, ends at fore limb. Some dorsal scales dark pigmented along their midline, more visible on posterior part of back and on tail, where tends to form indistinct stripes. Ventrum whitish. Taxonomic Notes: Closely allied to *M. vinciguerrae*, another short-tailed Somali writhing skink.

HABITAT AND DISTRIBUTION: Sandy areas in very arid steppe and grassland with scattered low shrubs, and in poorly vegetated coastal plains with extremely arid conditions, between sea level and 190m. Known from two localities; Geriban in Mudug region of central Somalia and Hafun Peninsula in Puntland.

CONSERVATION STATUS: Data Deficient.

NATURAL HISTORY: Poorly known. Presumably similar to other writhing skinks. Two specimens were collected in fine loose sand. Probably gives live birth. The female collected in Geriban in August contained two embryos 25mm long.

Mochlus grandisonianius, preserved.

COMMON WRITHING SKINK

Mochlus laeviceps
(Peters, 1894)

IDENTIFICATION: A medium-sized writhing skink with keeled dorsal scales, endemic to Benadir coast of southern Somalia. Head short and triangular; snout depressed and wedge-shaped. Nostril surrounded by two scales, frontonasal not in contact with rostral, four upper labials in front of subocular, two preoculars. Ear-opening small, with an obvious triangular lobule. Limbs well developed with five digits. Body stout and relatively short, cylindrical, tail subcircular in cross-section, about half total length. Dorsal body scales tricarinate in both adults and juveniles, in 28 rows at mid-body. Maximum size about 17cm (snout–vent length up to 8.5cm), average size unknown. Dorsum yellow-grey, with a dark spot on most scales; the overall appearance is of a speckled animal. A dull dark dorsolateral line may extend from the nostril, along the body and onto the tail. Ventrum yellowish-white.

HABITAT AND DISTRIBUTION: Consolidated coastal dunes and dry open savanna with thornbush, between sea level and 120m. Restricted to a small, sandy coastal area between Brava, Mogadishu and Afgoi in southern Somalia. Probably more widespread but records lacking, probably due to under-collecting.

CONSERVATION STATUS: Not Assessed.

NATURAL HISTORY: Poorly known. Specimens were found during the day, 10 and 40km south of Afgoi, together with *M. vinciguerrae*, but unlike that species they were closer to the surface and more often in holes; some animals were in stumps.

MABUYA-LIKE WRITHING SKINK

Mochlus mabuiiformis
(Loveridge, 1935)

IDENTIFICATION: A large, long-tailed, writhing skink from coastal southern Somalia. Head triangular, snout rounded, nostril in a single nasal, ear-opening large with two small rounded lobules at front. Four upper labial scales in front of subocular, two preoculars. Limbs well developed with five digits. Scales smooth, in 26–30 rows at mid-body. Males may reach a maximum size of 23.6cm (snout–vent length up 9.5cm), with tail 57–63 per cent of total length, average size unknown. Adults faintly striped or uniformly plumbeous-grey; each individual scale is paler at its base, the light area increasing in size towards tail so that scales on tail are light-centred with dark edges. Lips white with brown barring. Ventrum uniform white. Juveniles more vivid; the black crown of head mottled pale brown, the occipital scale conspicuously white with a dark centre; a vertebral line of pale brown, one scale in width, begins behind occipital scale and continues to base of tail where it disappears. Vertebral line flanked on either side by a dorsolateral stripe of same colour but two scales in width and commencing at last supraocular. Limbs uniformly black in some specimens; tip of tail transparent red,

Mochlus mabuiiformis, preserved.

each scale edged brown. Below, white, the internal organs visible through the scales; tail clear coral-pink.

HABITAT AND DISTRIBUTION: Coastal and dry savanna, in our area below 108m. In southern Somalia known from only two places, Giohar, and Sguss village (c.5km north of Balad), on the Shebelle River. Probably widespread along whole Benadir coast and further south towards Kenya. Elsewhere, in south-east Kenya, in Tana Delta.

CONSERVATION STATUS: Data Deficient.

NATURAL HISTORY: Little-known. Diurnal and surface0dwelling; one was seen in shade of thorn bush during day. The specific name refers to the fact that the author felt this species resembles typical skinks of the genus *Mabuya* (now *Trachylepis*).

LANZA'S WRITHING SKINK

Mochlus paedocarinatus
Lanza & Carfi, 1968

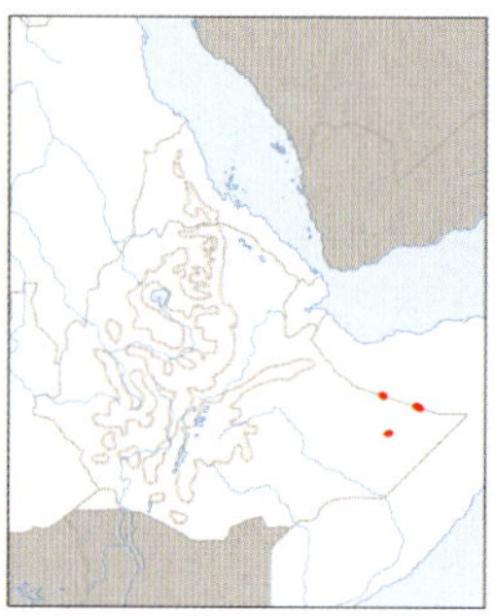

IDENTIFICATION: A poorly known medium-sized writhing skink with keeled dorsal scales in juveniles, restricted to the Haud region of Somalia, just entering Ethiopia. Head short and triangular, the snout is not depressed and wedge-shaped. Nostril surrounded by two scales, frontonasal not in contact with rostral, four upper labial scales in front of subocular, two preocular. Dorsal body scales keeled towards the rear limbs in juveniles (not adults), in 26 rows at mid-body. Maximum snout–vent length in males 8.3cm, females 8.4cm. Dorsum of preserved specimens cream to pale brown, sometimes with darker brown speckling on back and a suffusion of similar colour on upper head. Ventrum uniformly cream. Taxonomic Notes: Previously known as *M. laeviceps paedocarinatum*, now elevated to a full species.

HABITAT AND DISTRIBUTION: A species of the Somali arid zone, living in dry open *Vachellia–Commiphora* savanna and semi-desert with soft red sand, at altitudes of 670–990m. Known from the Haud region in southern Somaliland at several localities along the border with Ethiopia; between Bohodle and Balleh Dig, and Balumbal village (further west). Also known from Ado in the Ogaden in Ethiopia. Probably widespread in the intervening country (all suitable habitat) but unrecorded. Sympatric with *M. vinciguerrae*.

CONSERVATION STATUS: Data Deficient.

NATURAL HISTORY: Poorly known; never knowingly seen or photographed alive. Presumably similar to other writhing skinks, i.e., diurnal, burrowing, and eats invertebrates. All specimens were collected in 1932 by Colonel R. H. R. Taylor, found by day beneath logs, stones or in termite nests in an otherwise wholly sandy biotope. It probably gives live birth. Three females collected at end of March had two embryos in an advanced stage of development.

Mochlus paedocarinatus, preserved.

BOULENGER'S WRITHING SKINK

Mochlus productus
(Boulenger, 1909)

IDENTIFICATION: A large, short-legged and dark-coloured writhing skink with light speckling; confined to southern Somalia. Head short, the snout rather rounded. Nostril surrounded by two scales, frontonasal in contact with rostral. Ear-opening small with two small rounded lobules at front. Eye small, with a scaly lower eyelid. Four upper labial scales in front of subocular, two preoculars. Body elongate, cylindrical in cross-section. Tail moderately thick, subcircular in cross-section, and about 55 per cent of total length. Limbs short with five toes. Scales smooth, in 24–26 rows at mid-body. Maximum size about 22cm (snout–vent length up to 11.4cm), average size 18–21cm. Dorsal colour dark, reddish-brown, with numerous small round white spots, larger on tail. Top and sides of head white speckled. Ventrum brownish-white, tail below with scattered brown spots; chin and sides of neck spotted reddish-brown.

HABITAT AND DISTRIBUTION: Fixed sand dunes, with mixed bushes and scattered trees, and dry open woodland with dense vegetation, at elevations between 25 and 311m. Endemic to southern Somalia, on the Juba and Shebelle Rivers and the area between them. Known from Bardera and Dinsor in the north, at Gelib and Goscia (in a narrow band of riverine forest), and along the Juba River to the sea near Bardera. Also recorded further east at Giohar. An unconfirmed record at Lac Badana, between Kismayu and the Kenya border.

CONSERVATION STATUS: Least Concern.

NATURAL HISTORY: Unknown.

SIMONETTA'S WRITHING SKINK

Mochlus simonettai
(Lanza, 1979)

Mochlus simonettai, preserved.

IDENTIFICATION: A medium-sized, short-legged, black-headed and light-necked writhing skink restricted to southern Somalia. Head short; snout neither depressed nor wedge-shaped. Nostril in a single nasal, frontonasal not in contact with rostral. Ear-opening small with two small rounded lobules at front. The eye is small, with a scaly lower eyelid. Four upper labial scales in front of subocular, two preoculars. Body relatively long and narrow, cylindrical in cross-section. Tail subcircular in cross-section, about 47 per cent of total length. Limbs short with five toes. Scales smooth, in 24–26 rows at mid-body. Maximum size about 16cm (snout–vent length up to 8.8cm), average size unknown. A very colourful skink; dorsal colour very dark, brown or blackish, with a whitish spot in centre of each scale, the tail appears darker. Head and lower labials uniform black. Neck whitish to hazel. Underside white. Taxonomic Notes: Similar and perhaps most closely related to the sympatric *M. productus*.

HABITAT AND DISTRIBUTION: Dry open woodland on consolidated coastal dunes, at an elevation of 135m. Endemic to southern Somalia, collected only at one locality about 6km south-east of Afgoi on Benadir coast. Probably more widespread than it appears, suitable habitat is extensive and thus it is not under threat.

CONSERVATION STATUS: Not Assessed.

NATURAL HISTORY: Poorly known. The specimens were collected just after dark in the garden of the university campus, on red sandy ground. Generally cryptic in its habits, uses all four legs when moving slowly on surface; in a hurry it may use only the fore limbs, pressing the hind limbs along the tail, it can also move like a snake. Insect remains found in the stomach.

SOMALI WRITHING SKINK

Mochlus somalicus
(Parker, 1942)

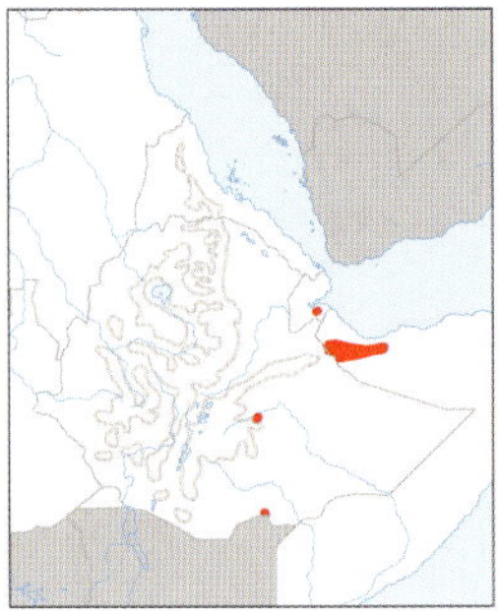

IDENTIFICATION: A small to medium-sized, finely speckled writhing skink, endemic to Somaliland, often in mountains. Head of moderate length, snout depressed and wedge-shaped. Nostril surrounded by two or three scales. Eye small, with a scaly lower eyelid. Three upper labial scales in front of subocular. Body cylindrical, tail subcircular in cross-section, about half of total length. Limbs short with five toes. Scales smooth, iridescent, in 26–28 rows at mid-body. Maximum size about 17cm (snout–vent length 5.5–9.5cm), average size 12–14cm. Dorsal colour usually shades of brown, finely speckled. Head strongly flecked black, the lips barred black and yellow. There is broad greyish lateral band, more or less dark. Tail dorsally speckled like body, underside white or speckled (especially in some males). Ventrum and flanks in females yellowish, the underside white, in mature males the ventrum and flanks are bright yellow. Juveniles pale brown, spotted darker brown above, the spots forming a longitudinal series, with a broad, white-spotted, dark, lateral band. Taxonomic Notes: Closely allied to *M. sundevallii*.

Mochlus somalicus, Awdal.

Mochlus somalicus, north-west of Borama.

HABITAT AND DISTRIBUTION: Probably endemic to Somaliland. Ranges from barren semi-desert country on coastal plains to dry *Vachellia* savanna, usually in rocky landscapes, at altitudes of 660–1,650m. All specimens assigned with certainty to this species (including the type) are from Somaliland. Known localities include Borama Hills in the west (at Dilla, north-west of Bawn, and Quljeet). Might occur further west in nearby Ethiopia. Ranges east across the Ogo Mountains to Goolis Mountains, and adjacent maritime hills inland of Berbera, observed in Hudisso and near Lafaruug. Occurs at Djalelo in Djibouti, in the Ogaden region in Ethiopia (a dubious record from Sheik Hussein), and along the Kenyan border at Murri; also doubtfully in southern Somalia near Afgoi. Fairly common in Somaliland. Strikingly different animals assigned to this species from central, south-east Kenya and north-east Tanzania are almost certainly another species.

CONSERVATION STATUS: Least Concern.

NATURAL HISTORY: Unknown. Most specimens from Somaliland were collected in soft sand under rocks in dry *Vachellia* savanna.

SUNDEVALL'S WRITHING SKINK

Mochlus sundevallii
(Smith, 1849)

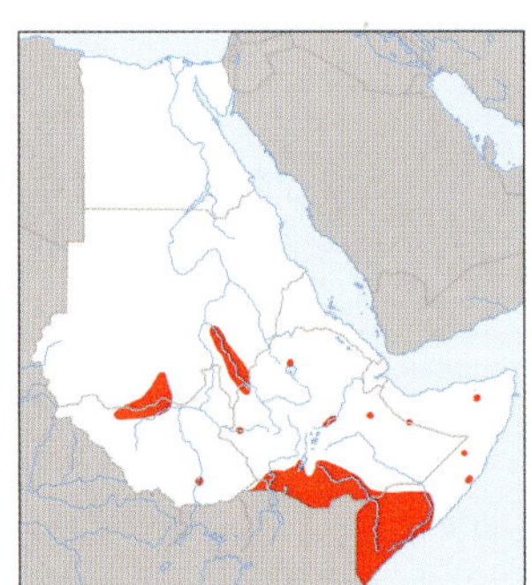

IDENTIFICATION: A medium to large, shiny-bodied, small-limbed writhing skink with a short, flat, wedge-shaped snout and small eyes, the lower eyelid is scaly. Three upper labial scales in front of subocular. Ear-opening tiny and deeply sunk. Body stout and cylindrical; Tail thick, about half or a little less the total length. Scales smooth, in 24–30 rows at mid-body. Maximum size 23cm (snout–vent length up to 9–14cm), average size 12–20cm, hatchling size unknown. Colour very variable; usually brown to grey, uniform or with a dark spot at base of each scale; the overall appearance is of a finely speckled animal. A dull dark dorsolateral line may extend from nostril, along body and onto the tail. The ventrum in females is cream or white, sometimes speckled, in mature males ventrum and flanks bright yellow. Juveniles often have a bright red tail, which fades with maturity. Taxonomic Notes: *M. afer*, long regarded as a separate species, has been shown by molecular analysis to be this species.

Mochlus sundevallii, Wachile.

HABITAT AND DISTRIBUTION: A wide range of habitats, including semi-desert savanna, *Vachellia–Commiphora* bushland and *Vachellia* woodland. In our area at elevations between sea level and 2,000m. Widespread and very common in southern Ethiopia, from Omo basin in the west to Dolo Odo in the east, thence south into the southern third of Somalia. Occurs along Sudanese Blue Nile River; in South Kordofan and on both sides of Sudan/South Sudan border along the Jur and Bahr El Ghazal Rivers. Also known from Juba. Sporadic records in Ethiopia include Gambela, Gorgora, the upper Awash River valley and Harar; in Somalia at Galkayo and near Hobyo, and Buran in eastern Somaliland. Elsewhere, south through Kenya and Uganda to northern South Africa and Namibia.

CONSERVATION STATUS: Least Concern.

NATURAL HISTORY: Burrowing, usually found under rocks and logs, in soft soil and sand; often taken in pitfall traps. They may be crepuscular, sometimes seen on the surface at dusk; often mistaken for a snake as they slide across the ground. They lay 4–7 eggs, roughly 1 × 1.5cm. Diet insects and other invertebrates.

TANA WRITHING SKINK

Mochlus tanae
(Loveridge, 1935)

IDENTIFICATION: A long, very slender writhing skink, from coastal southern Somalia. Head of moderate length, narrow, the snout slightly depressed, not wedge-shaped. Eye small; the lower eyelid has a transparent brille. Four upper labial scales in front of subocular, two preoculars. Body cylindrical, tail subcircular in cross-section, 55 per cent of total length. Legs small and very short, with five toes. Scales smooth, in 22–24 rows at mid-body. Maximum size 22cm (snout–vent length about 10cm), hatchling size unknown. Uniformly brown or plumbeous-grey above, the upper labials flecked lighter brown, similar light flecks on side of neck. Throat and lower labials white, heavily spotted dark brown.

HABITAT AND DISTRIBUTION: Coastal and dry savanna, at low altitude, in our area below 100m. Known from Belet Amin, on Juba River, and vicinity of Mogadishu, in southern Somalia. Probably widespread along whole Benadir coast and further south towards Kenya. Elsewhere, in south-east Kenya and north-east

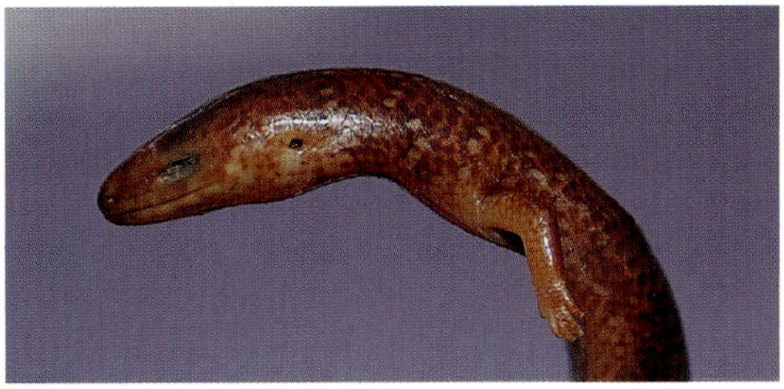

Mochlus tanae, preserved.

Tanzania. Probably more widespread than it appears, suitable habitat is extensive and thus it is not under threat.

CONSERVATION STATUS: Not Assessed.

NATURAL HISTORY: Unknown.

VINCIGUERRA'S WRITHING SKINK

Mochlus vinciguerrae
(Parker, 1932)

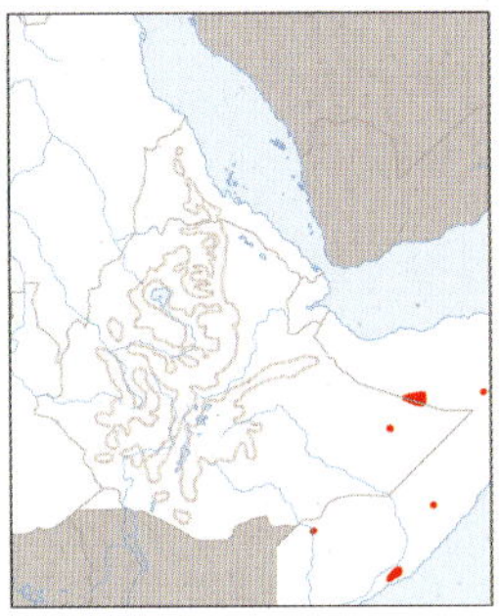

IDENTIFICATION: A small, short-tailed writhing skink restricted to the Somali arid zone of central Somalia and adjacent Ethiopia. Head short, the snout depressed and wedge-shaped. Nostril pierced in a single nasal, frontonasal not in contact with rostral, prefrontals usually fused with posterior loreals. Ear-opening very small and almost obscured by free edges of two anterior head scales, with one very small lobule concealed under that scale. Eye small, with a scaly lower eyelid. Three upper labial scales in front of subocular, one preocular. Body cylindrical, tail subcircular in cross-section, about 40–45 per cent of total length. Limbs short with five toes. Scales smooth, in 24–25 rows at mid-body. Maximum size about 11cm (snout–vent length up to 6.4cm), average size 9–10cm. Dorsum pale yellowish-brown; a broad, dark brown, lateral band extends from nostril through eye, above ear and limbs, with a pair of more or less distinct dark longitudinal lines on back, these markings are sometimes separated by a pale dorsolateral stripe. Uniform white below.

HABITAT AND DISTRIBUTION: A species of the Somali arid zone, living in dry open *Vachellia–Commiphora* savanna and semi-desert with soft red sand, at altitudes of 100–800m. Known from the Haud region in southern Somaliland, where recorded at several localities on border with Ethiopia between Bohodle and 60km west and 31km north of this town, also further east in the Nogal Valley in Puntland, about 9km north of Garowe, then south across the Haud and the Ethiopian Ogaden, at Ado. Recorded in central Somalia (at El Bur) and southern Somalia (Lugh; vicinity of Afgoi). It seems to be widespread and locally common. Suitable habitat extensive and thus it is not under threat.

CONSERVATION STATUS: Least Concern.

NATURAL HISTORY: A burrowing species, restricted to sandy habitats. In southern Somalia dug up from a depth of 30–80cm under thornbush debris. A female collected in July gave live birth. The specific name honours Decio Vinciguerra (1856–1934), Italian specialist on ichthyology and pisciculture, who also published on herpetology.

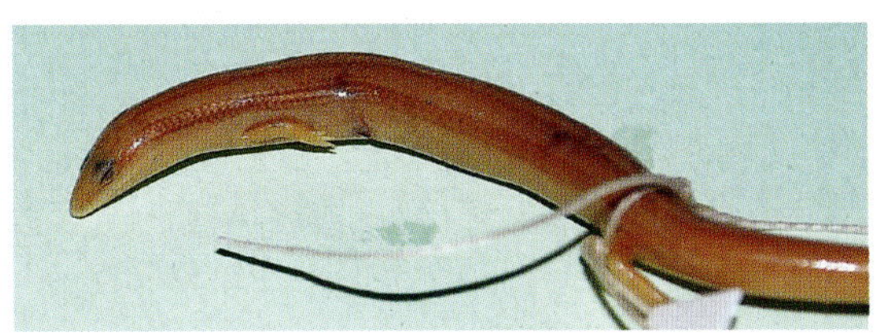

Mochlus vinciguerrae, preserved.

SUBFAMILY SCINCINAE

A subfamily of around 300 species of skink, in more than 30 genera, widespread in the Americas, Africa, Madagascar and Asia; some nine species in four genera occur in north-east Africa, most spend much of their lives buried.

BRONZE SKINKS *CHALCIDES*

An extremely variable group of shiny, elongate skinks with cylindrical bodies and smooth scales, they are terrestrial and most are burrowers (some excellent), living in holes or pushing through soft soil and sand. The snout is conical or wedge-shaped; eyelids functional, the lower with an undivided transparent disc; the limbs are small but functional and the nostril opening is located within the rostral scale, bordered to the rear by a small nasal. Several species with wedge-shaped snouts and projecting labial edges, very elongated bodies, and short, very reduced limbs were formerly included in the genus *Sphenops*, but recent molecular analysis indicates that members of the genus are all nested among other 'normal' members of the genus *Chalcides*. A Palearctic genus, *Chalcides* has radiated widely in northern Africa, southern Europe and the Middle East. About 33 species are known, six occur in north-east Africa. *Chalcides* taxa began to diversify around 10 Mya, probably in the western Maghreb region. The four species that were not in *Sphenops* belong to the *C. ocellatus* group.

KEY TO NORTH-EAST AFRICAN MEMBERS OF THE GENUS *CHALCIDES*

1a: Snout wedge-shaped, labial edge projecting, ventrum concave — 2

1b: Snout conical, pointed; labial edge not projecting; ventrum flat or convex — 3

2a: Fore limb with three fingers, hind limb with four toes; ear-opening without a fringe, more or less covered by scales; Sudan — *Chalcides delislei*, De l'Isle's Wedge-snouted or Sand Skink

2b: Limbs usually with five digits; ear-opening with fringe of pointed scales; Egypt and Sinai — *Chalcides sepsoides*, Audouin's Wedge-snouted or Sand Skink

3a: No vertebral body stripe, mid-body scale rows 26–34; northern Egypt, Sinai, introduced and rather isolated populations along Red Sea coast, Gulf of Aden and Indian Ocean from Sudan to Somalia — *Chalcides ocellatus*, Ocellated Bronze Skink

3b: Vertebral body stripe present, mid-body scale rows 22–28 — 4

4a: Vertebral scales extensively expanded (paravertebral width/vertebral width <0.9), high number of ventrals (81–92), north-west Ethiopia at higher elevations, lowlands of extreme south-east South Sudan and adjacent southern Ethiopia — *Chalcides bottegi*, Bottego's Bronze Skink

4b: Vertebral scales not expanded, or only weakly so (paravertebral width/vertebral width >0.9), low number of ventrals (68–80) — 5

5a: A very broad, vaguely defined brown, mid-dorsal band, typically covers entire width of four mid-dorsal scale rows (usually ½+4+½); head short, broad, triangular; mid-body scale rows 26–28, southern Egypt, Sudan, Eritrea and Afar region of Ethiopia — *Chalcides humilis*, Saharan Bronze Skink

5b: A moderately wide, sharply defined darker brown, mid-dorsal band, typically covers entire width of just two mid-dorsal scale rows (usually ½+2+½); head moderate length, narrow; mid-body scale rows 22–26, confined to extreme southern Eritrea, Djibouti, Somaliland and adjacent eastern Ethiopia — *Chalcides ragazzii*, Ragazzi's Bronze Skink

BOTTEGO'S BRONZE SKINK

Chalcides bottegi
Boulenger, 1898

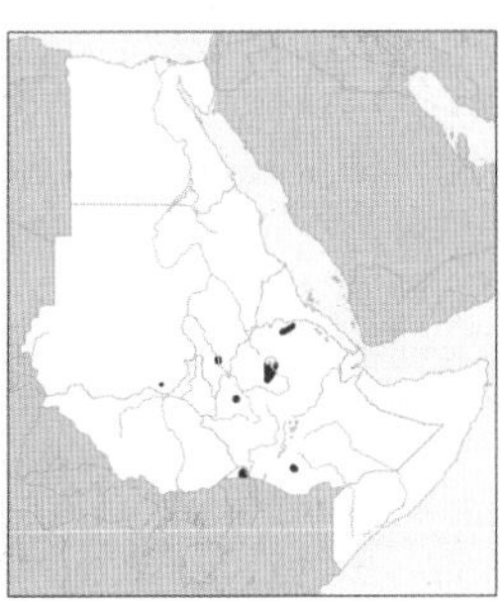

IDENTIFICATION: A large, stocky skink with very wide dorsal scales, mostly restricted to Ethiopian highlands. Head broad, with a slight neck and pointed nose; the body stout and cylindrical. Tail thick and tapers smoothly; roughly 50 per cent of total length. Scales smooth and dorsal scales broader than they are deep, in 22 (very occasionally 24) rows at mid-body. Vertebral scales extensively expanded. Large number of ventrals, 81–92 (counted from the mental, excluded, to precloacal scale, included). Maximum size about 27cm (snout–vent length up 13.7cm), average 18–25cm, hatchling unknown. Colour pattern distinctive: top of the head brown; lips barred black and white. A broad, sharply defined darker brown,

Chalcides bottegi, Bahar Dar.

mid-dorsal band typically covers entire width of just two mid-dorsal scale rows (usually ½+2+½), and extends from back of head to base of tail; bordered on either side by a light tan stripe and below this a fine darker stripe, which extends from nostril to hind limb. Arranged across these dorsal bands are more or less regular transverse rows of highly contrasting dark ocellate spots, with a white centre and dark edges. Underside uniform whitish to light tan.

HABITAT AND DISTRIBUTION: Montane grassland and savanna at higher elevations of 1,600–2,000m and dry savanna and semi-desert at lower elevations, usually 380–1,100m. Also cultivated land. Typically found in mesic highlands of north and north-west Ethiopia, in Tigray region from between Adua and Shire, in Amhara common around Lake Tana, thence south to the hills near Ghimbi in Oromia. Absent from moist tropical forest of western Ethiopia, but reappears at Lokomarinyang at Ilemi Triangle, south-eastern South Sudan.The type was collected at an imprecise locality between Amarr and Sancurar in southern Ethiopia, so presumably occurs in that area, but no recent records. Two Sudan records: from Bados in Blue Nile State and Tanguru (or Tungaru), south-east of Talodi, Southern Kordofan. Elsewhere, north-west Kenya. Probably more widespread but no records, probably due to under-collecting.

CONSERVATION STATUS: Least Concern.

NATURAL HISTORY: The Adua individual was under a small rock. All East African specimens were taken under logs, rocks and other debris in sandy areas during daylight hours; the species' habits are unknown as it has never been observed foraging, either at night or in the daytime. When uncovered, Bottego's Bronze Skink thrashes about in a serpent-like manner when attempting to escape; if held, a specimen will push its snout against its captor's hand as if attempting to burrow. These skinks give live birth, details unknown but similar species may have 2–20 offspring. Captives appear to be territorial, with males fighting in breeding season.

DE L'ISLE'S WEDGE-SNOUTED OR SAND SKINK

Chalcides delislei

(Lataste & Rochebrune in Lataste, 1876)

IDENTIFICATION: A small to medium-sized, thin but elongate burrowing skink, limbs with three fingers and four toes, from the Sudanese coast. Head relatively long, narrow and not distinct from neck; snout depressed, wedge-shaped, with a projecting labial edge, adapted for pushing through sand. Eye very small; ear-opening covered by small scales without a fringe. Body very elongate, slim, cylindrical. Limbs short and reduced, the front legs distinctly smaller than back legs, with three fingers and four toes. Tail long, conical and thick (not much thinner than body), about 45 per cent of total length. Scales smooth, narrow, usually in 24 (rarely 22 or 26) rows at mid-body. Maximum size about 16cm (snout–vent length up 9.4cm), average

Chalcides delislei, Senegal.

12–15cm, hatchling size unknown. Dorsum plain yellow or yellowish-brown, often with numerous (about 11) faint, dark longitudinal stripes, most obvious on tail; limbs pinkish-brown; tail pinkish-white or bluish; a dark stripe runs from nostril across eye to temporal region, sometimes towards shoulder; upper labials and underside white. Taxonomic Notes: Formerly included in genus *Sphenops*.

HABITAT AND DISTRIBUTION: Dry desert or semi-desert, in shifting or semi-shifting sand dunes or patches of loose sand in rocky areas, in our area near sea level. The only north-east African records are from the coastal plains along Red Sea coast in Sudan, notably at Suakin and Durrur, where it seems to be common. Elsewhere, widespread along southern edge of Sahara in the Sahel, west to Mauritania and Senegal.

CONSERVATION STATUS: Least Concern.

NATURAL HISTORY: Poorly known. Apparently nocturnal. Highly adapted for burrowing in loose sand, they move by 'sand-swimming', using a serpentine motion, but not the limbs, which are tucked in. Mali specimens were found under rocks or buried in sand below them. Sudanese specimens were found among loose sand on the maritime plain. When disturbed, they rapidly dived into the sand and disappeared. They give live birth to several juveniles.

SAHARAN BRONZE SKINK

Chalcides humilis
Boulenger, 1896

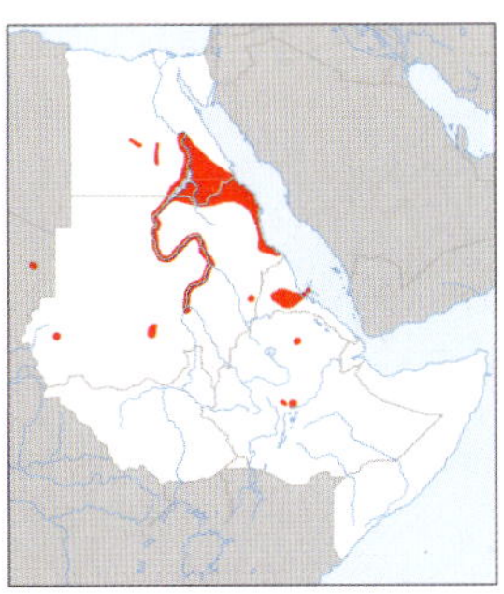

IDENTIFICATION: A medium to fairly large, stocky skink, with a wide longitudinal dorsal stripe, found in Ethiopia, Eritrea, Sudan and south-east Egypt. Head broad and triangular, with a slight neck and pointed

Chalcides humilis, Ghinda.

Chalcides humilis, Asmara.

nose; body stout and cylindrical. Tail thick and tapers uniformly; roughly 50 per cent of total length. Scales smooth, in 26–28 rows at mid-body. Vertebral scales are unexpanded. Ventrals 71–79. Maximum size about 25cm (snout–vent length up 12.4cm), average 15–22cm, hatchling size unknown. Colour pattern distinctive: top of head yellowish or greyish-brown; lips barred black and white. Dorsal coloration generally yellowish or greyish-brown. A very broad, vaguely defined brown, mid-dorsal band, typically covers the entire width of four mid-dorsal scale rows (usually ½+4+½); bordered on either side by a light tan stripe and below this a fine darker stripe, which extends from eye to behind fore limb. Arranged across these dorsal bands are more or less regular transverse rows of highly contrasting dark ocellate spots, with a white centre and dark edges. Underside uniform whitish to light tan. Occasional specimens without a dorsal stripe (but with a tail stripe) are not unusual. Taxonomic Notes: In the past treated in *C. ocellatus* or *C. ragazzii*, now elevated to a morphologically and genetically distinct species distributed between the natural ranges of both aforementioned forms.

HABITAT AND DISTRIBUTION: Favours a wide range of arid environments; dry grassland, open woodland, semi-desert scrub, coastal dunes, desert oases, also cultivated fields and gardens. In mountainous mesic habitats of Eritrea and northern Ethiopia it occurs at elevations up to 2,400m, although commonest between sea level and 1,200m. Originally described from the Dahlak Islands, it is widespread across northern Eritrea. In north-east Ethiopia, the species has been collected at Lake Ashangi and it is common in Awash National Park. Unconfirmed reports further west at Debre Zeit and Addis Ababa. In Sudan occurs from Khartoum north through Atbara and Dulgo (Delgo), towards Wadi Halfa near Egyptian border. Other Sudan records include Southern and Northern Kordofan, the vicinity of Kassala, far to the west at Jebel Marra in Darfur and hills along the Red Sea coast, where collected at Port Sudan, Tokar, and Durrur (Daror). Widespread in south-east Egypt, where it is known from between Gebel Elba and Wadi Halfa, north to Luxor along the Nile. Isolated records at Dakhla and Kharga oases in the Western Desert. Elsewhere, across the Sahel and central Sahara, to southern Algeria and Mali. Common and widespread.

CONSERVATION STATUS: Least Concern.

NATURAL HISTORY: Diurnal, though often most active at dawn and dusk, given to basking in sun, but always skulking close to some cover into which it withdraws rapidly in alarm, finding shelter in cavities among stones or plant roots, beneath logs, rocks and other debris. Recently collected in Awash, Ghinda and on Dissei Island, hidden beneath rocks by day. Females give live birth to a brood of 4–8 young.

Chalcides humilis, Awash National Park.

OCELLATED BRONZE SKINK

Chalcides ocellatus
(Forskål, 1775)

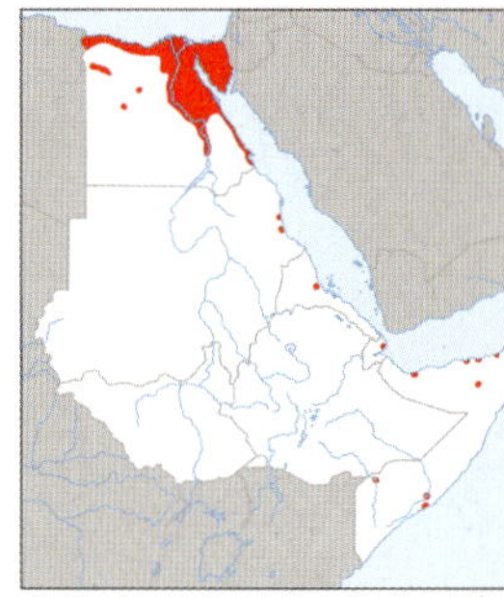

IDENTIFICATION: A medium to fairly large, stocky skink, without a longitudinal dorsal band, in our area largely in northern Egypt, sporadic elsewhere. Head short and broad, with a slight neck and pointed nose; body stout and cylindrical. Tail thick and tapers smoothly, roughly 50 per cent of total length. Scales smooth and dorsal scales broader than they are deep, in 26–34 (mostly 28–30) rows at mid-body. Vertebral scales are unexpanded. Ventrals 72–80. Maximum size about 23cm (snout–vent length up 10.8cm), average 18–20cm, hatchling 7–8cm. Dorsal coloration generally olive-grey or brown, with numerous small black-and-white ocelli, each occupying a single scale, scattered or arranged in loose diagonal transverse bands. Lips barred black and white. No broad dorsal stripe. Some specimens almost plain, except the tail. Underside uniform whitish to light tan. The blackish form *sachii* in southern Somalia is generally very dark with white speckles (ocelli) covering head, including lips, chin, body and tail. Throat, belly and tail white. Taxonomic Notes: Several subspecies known. The nominate occurs in Egypt and Sinai. Dark animals from Lugh and possibly also Mogadishu in southern Somalia were described as *C. o. sachii*, which might prove to be a valid taxon.

HABITAT AND DISTRIBUTION: A generalist, it inhabits various habitats, from rocky and sandy areas to cultivated land and margins of wetlands in oases. Locally common around human settlements and ruins. Naturally widespread in much of Egypt, save the hyper-arid interior of the Western Desert. Common in the Nile Valley and Delta, along the Mediterranean coast, and throughout Sinai, up to at least 1,700m. Found in Bahariya, Farafra and Siwa oases west of

Chalcides ocellatus sachii, preserved, Lugh.

Chalcides ocellatus, Cairo.

Nile River but also along Red Sea coast through Hurghada and Quseir. In the southern Egyptian Nile Valley it is sympatric with Saharan Bronze Skink. Several disjunct, possibly introduced, populations are known from the Red Sea coast in Sudan (Suakin), Eritrea (Saati), Djibouti (Obock) and the Gulf of Aden in Somaliland (Berbera, Las Khoreh) and Puntland (Bosaso, Gardo, Candala). In southern Somalia typically coloured skinks were also found in Balad and Mogadishu. All are probably due to human agency. The dark individuals (*C. o. sachii*) were originally described from Lugh, identical-looking specimens also found in Mogadishu, where normal and dark-coloured specimens must co-exist. Elsewhere, across much of North Africa, parts of southern Europe, east to Pakistan. Introduced to Sri Lanka, Florida and Arizona.

CONSERVATION STATUS: Least Concern.

NATURAL HISTORY: Partially fossorial and terrestrial; found under ground cover by day. A series from Mogadishu was dug from among plant root stocks in sandy soil. Active by day or night, depending upon climate. Very fast and agile, and burrows swiftly into loose sand or matted plant growth using serpentine locomotion. Diet arthropods and their larvae, occasionally plant matter. In North Africa it may hibernate. Mating observed in spring in Egypt. Females give live birth to 3–10 young about 6–12 weeks after mating.

RAGAZZI'S BRONZE SKINK

Chalcides ragazzii
Boulenger, 1890

IDENTIFICATION: A medium-sized skink with an elongate body, restricted to eastern Ethiopia and adjacent areas. Head of moderate length, narrow, with a slight neck and pointed nose; body relatively slim and cylindrical. Tail thick, about 50 per cent of total length and tapers smoothly. Scales smooth, in 22–26 rows at mid-body. Vertebral scales are unexpanded or slightly expanded. Ventrals 68–80. Maximum size about 25cm (snout–vent length up 12.7cm), average 15–22cm, hatchling unknown. Colour pattern distinctive: top of head dark brown with lighter edges; lips barred black and white. A moderately wide, sharply defined darker brown, mid-dorsal band, typically covers entire width of just two mid-dorsal scale rows (usually $\frac{1}{2}+2+\frac{1}{2}$), bordered on each side by a black stripe, bold on neck and thin on body, which extends from back of head to base of tail. Dorsal stripe usually black on neck, with a few scattered white ocelli; bordered on either side by a light tan stripe and below this a fine darker stripe, which runs from nostril to behind fore limb. Arranged across these dorsal bands are more or less regular transverse rows of highly contrasting dark ocellate spots with a white centre and dark edges. Underside uniform whitish to light tan. Occasional blackish or uniform specimens recorded.

HABITAT AND DISTRIBUTION: Occupies a wide range of habitats, including semi-desert on the coastal plains, open and dry *Vachellia* savanna and grassland in rocky landscapes, lava fields, and cultivated lands, between sea level and 2,300m, although commonest below 1,700m. Originally described from Assab in southern Eritrea. Widespread in Djibouti and very common in mountainous parts of western and central Somaliland, in the Ogo and Wagar ranges, east to

Chalcides ragazzii, Hargeisa.

Chalcides ragazzii, Daallo Forest.

Chalcides ragazzii, Huluul.

Las Korey in the Sanaag region. In Ethiopia known from the eastern edge of the Ahmar Mountains, in the vicinity of Harar then into the northern Ogaden (known from the Jijiga Plains and probably Sheik Hussein). Records from Awash not substantiated, despite recent collecting efforts, only *C. humilis* was found there. Common and widespread in Somaliland.

CONSERVATION STATUS: Least Concern.

NATURAL HISTORY: Poorly known. Presumably similar to other bronze skinks, i.e., diurnal or nocturnal, gives live birth, eats insects. Most Somali specimens were in soft sand under rocks in dry *Vachellia* savanna. Might be territorial, as always found singly under rocks (unless juveniles). A gravid female was seen in July. Captive specimens remained buried by day, and became active after dark.

AUDOUIN'S WEDGE-SNOUTED OR SAND SKINK

Chalcides sepsoides
(Audouin, 1829)

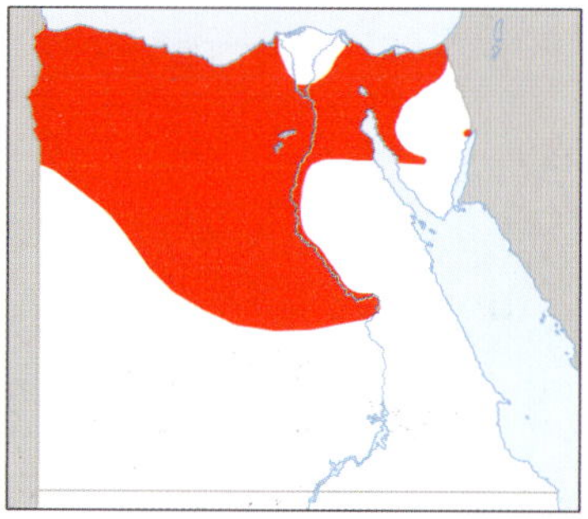

IDENTIFICATION: A small to medium-sized burrowing skink with a narrow and very elongate body, and limbs with five fingers and toes, in Egypt and Sinai. Head fairly long, narrow and not distinct from neck; snout wedge-shaped, with a projecting labial edge, adapted for penetrating into sand; eye very small; ear-opening obscured by three or four pointed scales. Body elongate, slim, cylindrical. Limbs small and reduced, with front legs distinctly smaller than back, with five digits (rarely four on fore limb). Tail elongate, conical and thick (not much thinner than body), usually

Chalcides sepsoides, Zaranik.

Chalcides sepsoides, north-west of Ismailia.

about 45–48 per cent of total length. Scales smooth, narrow, in 24–26 (rarely 22) rows at mid-body. Maximum size about 19cm (snout–vent length up 11.6cm), average 13–17cm, hatchling size unknown. Dorsum plain sandy, often with numerous (about ten) faint, dark longitudinal stripes, most obvious on tail; limbs pinkish- brown; a dark stripe from nostril across eye to temporal region, upper labials white and underside white. Taxonomic Notes: Formerly included in the genus *Sphenops*.

HABITAT AND DISTRIBUTION: Near-desert to semi-desert, restricted to sandy habitats with shifting or semi-shifting sand dunes, with some vegetation, at low altitudes. Also in rocky areas with loose sand. Not found in very dry desert. Widespread in northern third of Egypt (absent from Nile Delta) and in a broad swathe south-west from Libyan border to Nile Valley, as far south as Luxor. Also in low sandy regions of central and north Sinai. Elsewhere, west to central and eastern Libya, east to south-west Jordan. Fairly common and widespread.

CONSERVATION STATUS: Least Concern.

NATURAL HISTORY: Mostly nocturnal, from dusk to dawn, sometimes diurnal in winter. Ambushes prey from below the sand, where it freely moves by 'sand-swimming' (moving through sand by a serpentine motion, without using limbs, which are tucked in). Shelters by day in loose sand near bushes, in rodent burrows or under rocks and similar debris. Sometimes moves long distances on surface; if disturbed, rapidly dives into sand and disappears. Secretive, its presence is often revealed by its typical sinusoid tracks. Gives live birth to 2–5 juveniles.

LONG-LEGGED SKINKS *EUMECES*

A Saharo-Sindian genus of fairly large, heavily built skinks with a robust short head indistinct from neck, elongate stout body, well-developed limbs with five clawed digits adapted for burrowing and a relatively long tail. They have smooth dorsal scales, two broadened vertebral scale rows, a large ear-opening edged with several distinctly conical lobules. Most are colourful; dark above, paler below with dark stripes and often orange spots or streaks on body and tail. They are terrestrial and diurnal, inhabiting various dry habitats. *Eumeces* are active hunters and feed mainly on arthropods, snails and small vertebrates. About six or seven species are known, from north-west Africa to west India and Central Asia, two occur in Africa and one in our area. Genus *Eumeces* as currently understood is paraphyletic and has caused much taxonomic debate. *E. schneiderii* is sister to a clade comprising the genera *Scincopus*, *Scincus* and *E. algeriensis*, and the latter species is sister to *Scincopus* in all analyses.

SCHNEIDER'S SKINK

Eumeces schneiderii
(Daudin, 1902)

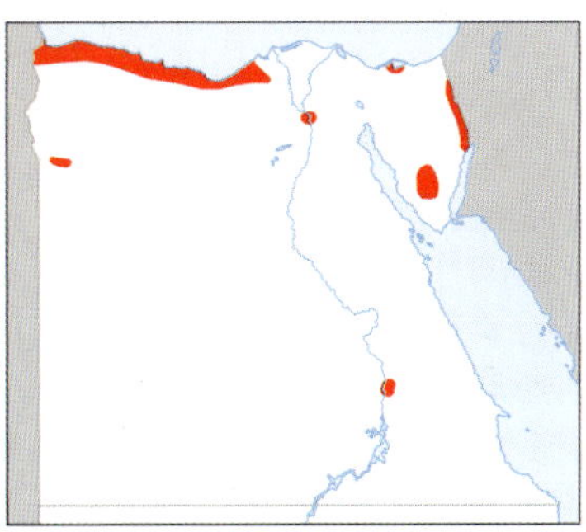

IDENTIFICATION: A large, colourful, stocky long-tailed skink from northern Egypt and Sinai. Head robust, triangular, not distinct from neck; snout short, obtuse; eye moderately large with round pupil; ear-opening large and open, with 3–5 long pointed lobules anteriorly. Body stout, elongated, cylindrical; limbs fairly long, strong, with five well-developed rounded digits without a fringe; tail relatively stout, tapering smoothly, about 60 per cent of total length. Dorsal scales smooth, slightly larger than ventrals, 26–28 rows at mid-body. Maximum size about 42cm (maximum snout–vent length 16.3cm), average 30–40cm. Very colourful, dorsum olive-grey interspersed by scattered orange scales, which may form vague crossbars, an orange lateral streak runs from posterior upper labials along flanks and onto tail, separating the greyish dorsum from yellowish venter; limbs brownish with orange spots. Juveniles distinctly coloured, yellowish-brown with yellowish-white ocellate pattern. Taxonomic Notes: This lizard has a wide distribution, several subspecies have been described, and their validity is debated. In our area the nominate occurs east of the Nile, *E. s. aldrovandii* west of the

Eumeces schneiderii, Cairo.

Nile. Future molecular analysis might split the group further.

HABITAT AND DISTRIBUTION: Usually in dry stony country with vegetation; in rocky wadis, coastal plains, dry steppe, cultivated lands, even rubbish dumps, from sea level to at least 1,600m. Known from northern Egypt, where confined to the Mediterranean coastal zone from the western margins of the Nile Delta to the Libyan border. Known also from Cairo and Siwa Oases. An isolated Kom Ombo record in the lower Nile Valley needs further evaluation, but might indicate it occurs along the Nile but has escaped attention. In Sinai, occurs in the high mountainous St. Katherine region and on coast between Al Arish and Romani, and along north-east border. Probably more widespread in Sinai, but poorly known and uncommon. Elsewhere, west to the Algeria/Tunisia border, east to north-west India, and north to Central Asia. In northern Egypt widespread and fairly uncommon but collected in large numbers for pet trade; in Sinai local and uncommon.

CONSERVATION STATUS: Least Concern.

NATURAL HISTORY: Diurnal and terrestrial, foraging for prey in bushed rocky areas and ruins. They shelter in burrows or under rocks. Animals resting have a strange posture, flat with their legs pointing backwards against the flanks. They have a broad diet, very fond of snails but take other invertebrates (especially beetles and spiders, an Egyptian animal contained a scorpion), small vertebrates like lizards and rodents, occasionally plants, especially fruit. A fast, agile but rather shy skink, not common. A clutch of 2–20 eggs, about 18×32mm are laid in deep burrows or under rocks and hatch after c.2 months, depending on temperature. Males are very aggressive during mating.

Eumeces schneiderii, Tunisia.

BANDED OR TIGER SKINKS *SCINCOPUS*

An African genus containing a single species; a large, rare, stout banded desert skink, banded yellow and black, it lives in burrows, and is widely distributed at the periphery of the Sahara. It is nocturnal and lays eggs. Interestingly, it is most closely related to another large-bodied skink, *Eumeces algeriensis*, of Algeria and Morocco, and then to the fairly similar, also sand-dwelling skinks of the genus *Scincus*. Banded Skink diverged from *E. algeriensis* about 9.6 Mya.

TIGER SKINK, PETER'S BANDED SKINK

Scincopus fasciatus
(Peters, 1864)

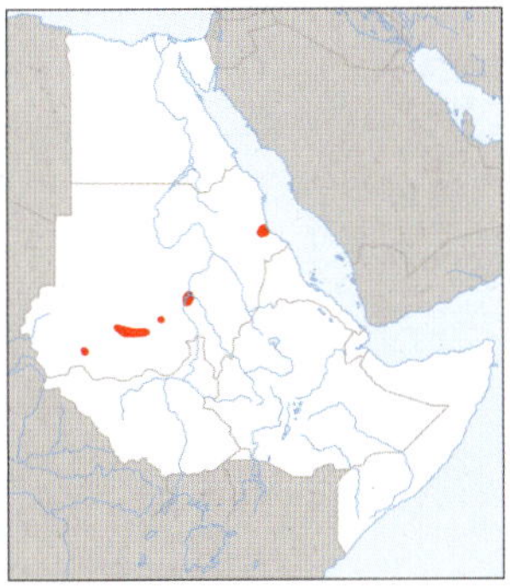

IDENTIFICATION: A big, robust, elegant banded skink with a short tail, from arid country in Sudan. Head short and pyramidal, slightly distinct from neck; snout obtusely truncate; eye very large and black, with no discernible pupil (in preserved animals vertically oval); ear-opening large, covered by two very big scales. Body thickset and cylindrical, limbs moderately long, of equal size; digits more or less rounded with short claws and no broad fringes. Tail short, thick and tapering rapidly, about 30–40 per cent of total length. Dorsal scales thick, broad and striated, transversally enlarged, the two vertebral series largest. Mid-body scales in 22–26 rows. Two anal scales cover the cloaca, in males these scales are more distinct, noticeably separated from each other and have round posterior edges; both scales in females are less distinct, almost united, smooth to touch and have straight posterior edges; males also have wider and more robust head. Maximum size about 30cm, average 20–28cm. Coloration distinct, yellowish or orange above, with typically seven (occasionally eight) equally spaced transverse black bold bands, tail tips (and regenerated tails) are black. Snout, upper and lower labials and venter whitish. Juveniles have identical coloration and pattern as adults. Taxonomic Notes: Two subspecies recognised, the nominate and *S. f. melanocephalus*, which has a black head, ear lobules and fore limbs, and may have no or vestigial bands. Both occur in our region.

HABITAT AND DISTRIBUTION: Dry open Sahelian savanna and semi-desert, from sea level to 1,080m. Outside of our area it inhabits desert and semi-desert, in vegetated areas with drifting sand. A rare skink, widespread but sporadic in the southern Sahara and north-west Africa. In north-east Africa it is limited to a fairly narrow band of very dry sandy country in Sudan, between the southern savannas and northern Nubian Desert. The black-headed subspecies *S. f. melanocephalus* occurs only at Suakin and Erkowit (0–1,080m) on the Red Sea coast. The nominate is reported at Khartoum, Bara, Wad Banda, En Nahud and between there and Umm Ruweba in northern Kordofan. Also at Ed Daein in South Darfur at 470–600m. Elsewhere, west to Mauritania, and along northern border of the Sahara.

CONSERVATION STATUS: Least Concern, but exploited for the pet trade, exported in large quantities from Togo (although it doesn't actually occur there).

NATURAL HISTORY: A terrestrial, nocturnal, rarely encountered secretive skink. Shelters by day in rodent or other animal burrow (Nigerian specimens at a depth of 60cm) or hidden in sand; Tunisian specimens were in *Opuntia* hedges or in tufts of *Ziziphus* shrubs with many rodent burrows. In West Africa they occur in farmland. Diet mostly insects, including termites, and other arthropods but known to take other lizards. Rather slow-moving, but may try to bite if grasped. Breeding details poorly known, but a captive female laid four soft-shelled oval eggs, which hatched after several weeks. A mating pair seen on a May night in Tunisia moved in a circle and bit their partner's neck. Often killed on roads at night in West Africa, and sometimes persecuted in the mistaken belief that it is venomous.

Scincopus fasciatus, Niger.

SANDFISH *SCINCUS*

A Saharo-Arabian genus of medium to large, fossorial, stout and unusual skinks highly adapted for desert life. They have a streamlined shape, wedge-shaped snouts, tiny eyes, a short tail and the ability to dive into and 'swim' through sand. Most are diurnal (one species is nocturnal) active foragers, they lay eggs and eat mainly invertebrates. The five species are widespread in the Sahara and Arabia, one occurs in north-east Africa. Sandfish skinks are most closely related to the genera *Eumeces* and *Scincopus*, but diverged >11.7 Mya.

COMMON SANDFISH

Scincus scincus
(Linnaeus, 1758)

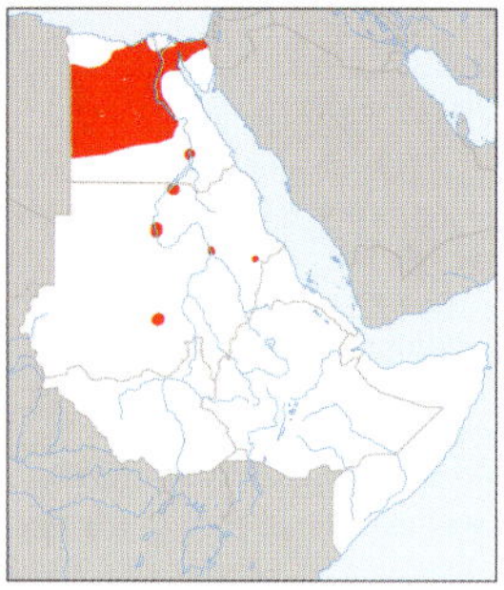

IDENTIFICATION: A medium-sized, robust skink, with long sharp-edged snout, smooth dorsal scales, an active sand-diver. Head moderately long, wedge-shaped; snout spatulate and overlapping the lower jaw; eye small; ear-opening small, covered by scales; neck thick, indistinct. Body robust, elongated, U-shaped in transversal section (venter flattened), with clearly defined ventrolateral edges. Limbs well developed, hind ones slightly larger, digits much depressed with broad lateral scale fringes; tail stout, shorter than body, about 35–40 per cent of total length. Dorsal scales large, smooth and of equal size, usually 26 (occasionally up to 30) rows at mid-body. Males longer, with a larger and thicker head. Maximum size about 22cm (maximum snout–vent length 14.7cm), average 13–16cm. Coloration distinct, dorsum sandy-yellow with 8–9 brownish (2–3 scales wide) transverse bands, sometimes interrupted in middle, the sandy and brown bands roughly equal in width; head brownish grey, upper labials whitish. Venter white. Juveniles plain sand-coloured without dorsal banding.

HABITAT AND DISTRIBUTION: Desert and semi-desert with wind-blown, loose aeolian sand, with or without bushes, also in dunes or patches of soft sand in areas of fairly hard compacted sand or gravel, from sea level to about 500m. Widespread in Egypt, in sandy country of north and central Sinai, the northern Eastern Desert and Western Desert, including most major oases, on periphery of Qattara Depression and at Aswan. Sporadically recorded on the Nile in Sudan, known from Wadi Halfa, Bayud near Dongola and Atbara, also in Hamashkoraib not far from the border with Eritrea. The southernmost record is at Bara in North Kordofan. Records from North Darfur and Red Sea coastal plains (not shown on map) need further evaluation. Elsewhere, west to Algeria, Mali and northern Nigeria, east across Arabia. In Egypt common and widespread but collected in large quantities for pet trade; in Sudan poorly known.

Scincus scincus, Egypt.

Scincus scincus, Egypt.

CONSERVATION STATUS: Least Concern.

NATURAL HISTORY: A highly sand-adapted diurnal (sometime crepuscular) lizard, actively forages on loose wind-blown sand dunes and feeds mainly on various invertebrates both on the surface or buried in sand. When it gets too hot, they hide in the sand. They have very sensitive hearing and can detect prey moving in sand at a distance of up 15cm. When disturbed, they quickly dive into the sand, plunging down 20–30cm, and can reach 40cm. Shy and fast-moving. Their diet includes invertebrates, smaller lizards and plants. Oviparous, a captive female laid six eggs with a mean size of 11.5×22.1mm. They mate in spring, copulation lasting 2–5 minutes, the male bites the female's neck and holds her tail with his legs while mating.

SUBFAMILY EUGONGYLINAE

A subfamily of about 460 species of skinks, in around 50 genera, mostly in Australasia and islands of the Pacific Ocean, and Asia, but some nine species in four genera occur in north-east Africa. Most are small, nondescript burrowers.

SNAKE-EYED SKINKS *PANASPIS*

An African genus of small, slim, cylindrical-bodied egg-laying skinks with well-developed five-toed limbs and cryptically coloured; some have immovable lower eyelids, in some the eyelids move. About 21 species are known, mostly in African forests, but five savanna species occur in sub-Saharan north-east Africa. Taxonomy not yet wholly clarified. The genus *Panaspis* is most closely related to a clade including the sister genera *Leptosiaphos* and *Lacertaspis* from tropical Central Africa, all three diverged from a common ancestor in the Paleocene, approximately 60 Mya.

KEY TO NORTH-EAST AFRICAN MEMBERS OF THE GENUS *PANASPIS*

1a: Frontoparietal scales fused	2
1b: Frontoparietal scales paired	3
2a: Strongly flecked back, mid-body scale rows 24, evergreen montane forest of western Ethiopia	*Panaspis annettesabinae*, Sabin's Snake-eyed Skink
2b: Uniform back, mid-body scale rows 22–28, mostly savanna in southern and eastern Ethiopia and all of Somalia	*Panaspis wahlbergii*, Wahlberg's Snake-eyed Skink
3a: Very long tail, approximately 2.5–3 times longer than snout–vent length; sub-digital lamellae keeled, bearing two spines; mid-body scale rows 20–22, in our area only southern Ethiopia	*Panaspis megalurus*, Blue-tailed Snake-eyed Skink
3b: Moderately long tail, approximately 1.5 times longer than snout–vent length; sub-digital lamellae obtusely keeled; mid-body scale rows 22; northern Ethiopia or south-central Sudan	4
4a: Two supraoculars; four supraciliaries; third and fourth toes equal; short fore and hind limbs, nowhere near touching each other when pressed to body; highlands of northern Ethiopia	*Panaspis tancredi*, Ethiopian Snake-eyed Skink
4b: 3 supraoculars; five supraciliaries; fourth toe longer than third; long fore and hind limbs, almost touching each other when pressed to body; lowlands of south-central Sudan	*Panaspis wilsoni*, Sudanian Snake-eyed Skink

SABIN'S SNAKE-EYED SKINK

Panaspis annettesabinae
Colston, Pyron & Bauer, 2020

IDENTIFICATION: A small skink with well-developed limbs and flecked back, confined to western highlands of Ethiopia. Head distinct. Immovable eyelids, each with a transparent spectacle. Scales smooth, close-fitting, in 24 mid-body scale rows. Tail cylindrical, tapering to a fine point, approximately 60 per cent of total length, although regenerated tail is shorter. Maximum known size about 11cm (snout–vent length 4.2cm), hatchlings unknown. Dorsum coppery-brown, flecked black, loosely arranged in 3–5 irregular lines which become less distinct posteriorly. There is usually a distinct dark side stripe lightly speckled, edged above by a whitish tan stripe. Light-coloured upper labials lack round black spots. Tail coppery-bronze. Regenerated part of tail coral-pink. Belly whitish. Juveniles have darker limbs and lateral coloration, and less evident dark flecking dorsally and light flecking laterally, with bright blue tail. Taxonomic Notes: Recently described, formerly within *P. wahlbergii* complex.

Panaspis annettesabinae, Bedelle.

Panaspis annettesabinae, Debre Zeit.

HABITAT AND DISTRIBUTION: Moist evergreen montane forest and associated savanna, at altitudes of 1,720–2,000m. The type was collected a short distance south-west of Bedelle in the Oromia region of western Ethiopia. Additional specimens were observed in Bedelle, Debre Zeit and Jimma. Probably widespread in the intervening country but no records. Suitable habitat is extensive and thus it is not under threat.

CONSERVATION STATUS: Not Assessed.

NATURAL HISTORY: Unknown, but presumably similar to other species. The type was found underneath a tarpaulin. The Jimma animal was under a rock in a forest clearing.

BLUE-TAILED SNAKE-EYED SKINK

Panaspis megalurus
(Nieden, 1913)

IDENTIFICATION: A small slim skink with a very long tail, in our area confined to southern Ethiopia. Scales smooth, in 20–22 mid-body scale rows. Tail cylindrical, tapering to a fine point, slightly over half the total length. Tail very long, 75 per cent of total length, although regenerated tail shorter. Legs short, with five digits and fourth toe longer than third. Maximum size about 17cm, average and hatchling sizes unknown. Iridescent brown or olive above, with a pale,

Panaspis megalurus, Kenya.

Panaspis megalurus, between Wachile and Mega.

black-edged dorsolateral stripe from neck to beyond hind limbs. A line of white blotches extends from behind eye along flanks to just behind fore limbs. The upper flanks are grey to orange, becoming paler below, the underside of the limb orange-pink, the belly white. The type specimen, a female, had an orange tail, a recently reported male had a blue tail. The presence of *P. megalurus* in our area is based upon a single recently observed specimen, tentatively assigned to this species, which corresponds to the species' description, but differs slightly in pattern and has a shorter original tail, 60 per cent of total length. The black-edged pale dorsolateral stripe is almost indistinct and there are no white blotches on neck, which is striped like the flanks. Underside of tail and posterior belly bluish-grey with dark stripes. Taxonomic Notes: We provisionally assign the Ethiopian individual to this species, but given its geographical isolation and slight but important morphological differences, it might prove to be an undescribed form.

HABITAT AND DISTRIBUTION: The only known subadult specimen collected in our area was in open savanna with thick grass cover, at 1,519m, between Wachile and Mega in southern Ethiopia. Elsewhere, on mid-altitude central plains of Tanzania and Maasai Mara in southern Kenya.

CONSERVATION STATUS: Least Concern.

NATURAL HISTORY: Unknown, but presumably similar to other small terrestrial skinks. The long tail and very elongate body suggest that it might live in long grass and use its tail to assist with rapid movement, like other long-tailed lizards. The Ethiopian specimen was found under a rock in grassland.

ETHIOPIAN SNAKE-EYED SKINK

Panaspis tancredi
(Boulenger, 1909)

IDENTIFICATION: A small skink with short limbs and paired frontoparietals, known from a single specimen, probably from the highlands of northern Ethiopia. Head short, neck distinct. Body elongate, cylindrical in cross-section, scales smooth, in 22 mid-body scale rows, the two vertebral rows are enlarged. Tail cylindrical in cross-section, slightly over half the total length. Legs short, with five digits obtusely keeled inferiorly. Snout–vent length of the only known specimen is 2.8cm, the regenerated tail is 24mm long. Dorsum brown, darker on flanks with blackish lines along scales, and a series of whitish spots on sides of head and neck. Underside greyish-white.

HABITAT AND DISTRIBUTION: Described in 1909 from 'Dabarif'. This might refer to Debarq, a small town in the foothills of the Simien Mountains north of Lake Tana, in Amhara Region, northern Ethiopia. The habitat there is montane grassland with broadleaf evergreen trees on steep hillsides, at 2,860m. A specimen recently observed but not collected some 22km north of Debarq, at an elevation of 1,640m probably belongs to this species. Might be more widespread than it appears, suitable habitat is extensive and thus it is not under threat.

CONSERVATION STATUS: Not Assessed.

NATURAL HISTORY: Unknown. The recently observed specimen was under a stone in high grass along the main road between Debarq and Axum.

WAHLBERG'S SNAKE-EYED SKINK

Panaspis wahlbergii
(Smith, 1849)

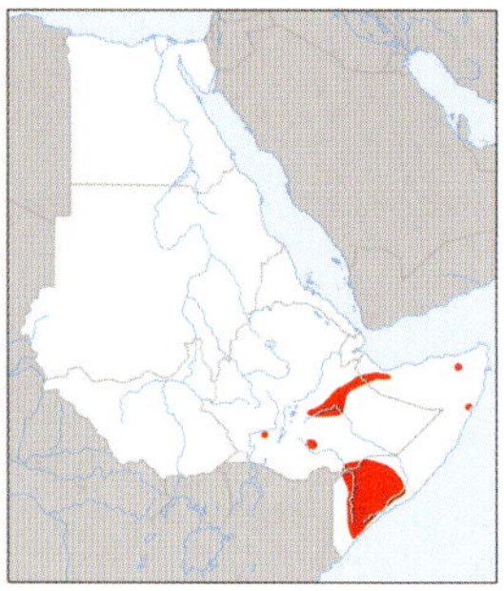

IDENTIFICATION: A small skink with well-developed limbs, widespread in southern and eastern Ethiopia and most of Somalia. Head distinct. Immovable eyelids, each with a transparent spectacle. Scales smooth, close fitting, in 22–28 (mostly 24–26) mid-body rows; scales on the soles rounded. Tail cylindrical, tapering to a fine point, slightly over half the total length. Maximum size about 14cm (snout–vent length up to 5.1cm), average 8–12cm, hatchlings about 3cm. Dorsum grey, brown or bronze, often with up to six fine dark lines. There is usually a distinct dark side stripe, edged above by a fine white stripe. Belly white or blue-white, but may be pinkish-orange in breeding males. Taxonomic Notes: Recent molecular studies suggest a high level of cryptic diversity within *P. wahlbergii*, with new species described, and more in progress of description. Populations around Negele in southern Ethiopia may belong to an as-yet undescribed form. Specimens from Mandera, in extreme north-east Kenya, and probably adjacent areas, are referred to the recently revalidated *P. massaiensis*. For now, we leave other populations in the Horn of Africa in *P. wahlbergii*, although this will probably change.

Panaspis wahlbergii, Negele.

HABITAT AND DISTRIBUTION: Occupies a wide range of habitats in Horn of Africa, including moist and dry savanna, high grassland, open *Juniperus* woodland and semi-desert and coastal bush, from sea level to 2,100m. A population occurs from Dolo Odo in south-east Ethiopia into southern Somalia, between the Juba and Webi Shebelle Rivers; another extends along the upper Webi Shebelle north of Ginir to Lake Alemaya, thence into Somaliland, recorded from Harirad in the Borama area and Gaan Libaah. Sporadic records in Ethiopia include Gadat (west of Lake Abaya), several from the Negele area and two widely separated records from Puntland, at El Donfar and Eyl. It is probably much more widespread but overlooked. Elsewhere, as presently defined a southern African species, see above.

CONSERVATION STATUS: Least Concern.

NATURAL HISTORY: Diurnal. Quick-moving and active, forages on ground, living in grass tufts, leaf litter and recesses in broken ground. They reach sexual maturity in 8–9 months. A clutch of 2–6 eggs, roughly 4×8mm, is laid. They are short-lived, with a lifespan of about two years. Diet termites and other small insects.

SUDANIAN SNAKE-EYED SKINK

Panaspis wilsoni
(Werner, 1914)

IDENTIFICATION: A small very slender skink with a long tail and legs, known from a single specimen from Talodi in Sudan. Head long, neck narrow. Body elongate, cylindrical in cross-section, scales smooth, in 22 mid-body rows. Tail cylindrical, tapering to a fine point, slightly over half the total length. Legs thin and long, with five toes. Snout–vent length of only known specimen 2.4cm, length of tail with missing tip 31mm. Dorsum bronzy-brown with four indistinct darker thin lines. Flanks with distinct dark side stripe, edged above by a fine white stripe. A line of white blotches extends from behind eye along flanks to just in front of fore limbs. Belly bluish-green.

HABITAT AND DISTRIBUTION: Sudanian savanna with *Combretum–Terminalia* woodland, in rocky terrain, at altitude 485m. The original and only known specimen was collected in Talodi in Southern Kordofan, Sudan, in 1914. The habitat where the type was collected is extensive in southern Sudan and northern South Sudan, therefore the species must be more widespread, but presumably overlooked.

CONSERVATION STATUS: Data Deficient.

NATURAL HISTORY: Unknown.

Panaspis wilsoni, preserved type.

LEAF LITTER SKINKS *LEPTOSIAPHOS*

A genus of 18 species of small (usually <15cm total length), slim, smooth-bodied burrowing skinks, with very long yet stout tails; a good field character. Most occur in tropical Central Africa, largely in forest. The scales are smooth and shiny. Two species just reach north-east Africa, in South Sudan.

KEY TO NORTH-EAST AFRICAN MEMBERS OF THE GENUS *LEPTOSIAPHOS*

1a: Five digits on forefoot	*Leptosiaphos kilimensis*, Kilimanjaro Five-toed Skink
1b: Four digits on forefoot	*Leptosiaphos aloysiisabaudiae*, Uganda Five-toed Skink

UGANDA FIVE-TOED SKINK

Leptosiaphos aloysiisabaudiae
(Peracca, 1907)

Leptosiaphos aloysiisabaudiae, Imatong Mountains.

IDENTIFICATION: A small, small-limbed skink with a short, rounded snout and medium-sized eyes, the lower eyelid has a large window. Ear-opening tiny and tympanum is deeply sunk. Body slim and cylindrical, tail thick and long, about 60 per cent of total length. Scales smooth, in 22–24 rows at mid-body. Limbs small and slender, four digits on fore limb, five on hind limb, lamellae beneath fourth finger 10–12, beneath fourth toe 13–15. Average size about 14cm, hatchling size unknown. Colour: brown, becoming reddish posteriorly, with darker median blotches on most scales, darker laterally, labials with light and dark vertical barring, chin and throat white, becoming yellow or reddish posteriorly. Within its limited range, only likely to be confused with other small Five-toed skinks; see key.

HABITAT AND DISTRIBUTION: In our area, only at Lotti forest, on west side of the Imatong Mountains, South Sudan, at just above 1,000m. Elsewhere, sporadically known from Uganda, eastern DR Congo, Cameroon and Nigeria.

CONSERVATION STATUS: Least Concern.

NATURAL HISTORY: Virtually nothing known, but presumably burrowing and diurnal, living beneath the surface of the soil or in leaf litter, lays eggs, eats small insects, insect larvae and other small arthropods that it finds beneath the leaf litter.

KILIMANJARO FIVE-TOED SKINK

Leptosiaphos kilimensis
(Stejneger, 1891)

IDENTIFICATION: A small, slim, tiny-limbed skink from South Sudan. A very short, deep head, rounded snout and fairly large eyes, iris golden-brown, lower eyelid scaly with a small window. Ear-opening small. Tail thick and long, but often regenerated, usually 50–70 per cent of total length, tapering gently to a blunt tip. Scales smooth, in 22–26 rows at mid-body. Maximum size about 18cm, average size 10–15cm, hatchling 56–58mm. Colour: iridescent golden-brown or bronze above, with fine black flecks that form broken longitudinal lines. There is an irregular dark line through eye, and lips clearly barred black and white, the head scales finely mottled black. Anterior flanks pink or pinkish-orange, sometimes finely speckled black and white. Tail darker, grey or blue-grey, the underside uniform pinkish-white.

Leptosiaphos kilimensis, Tanzania.

HABITAT AND DISTRIBUTION: Widespread in rainforest, moist savanna, high grassland and woodland, from near sea level up to about 2,000m, but in our area known only on Mt Korobe (03.58°N, 30.52°E, altitude 1,590m) south-east of Yei in Central Equatoria, South Sudan. Elsewhere, from Kenya and Uganda south-west to Angola.

CONSERVATION STATUS: Least Concern.

NATURAL HISTORY: Diurnal, sometimes found basking in forest in patches of sunlight, but often in leaf litter between buttress root of forest trees or under logs. Might also be crepuscular, often on surface at dusk and not long after dawn. Lays a clutch of 2–4 eggs, measuring about 1.4×0.8cm, many clutches may be laid together beneath a suitable log or in leaf litter between buttress roots. Eats small insects, insect larvae and other small arthropods that it finds beneath the leaf litter.

COASTAL SKINKS (SHINING SKINKS) *CRYPTOBLEPHARUS*

A genus of small, egg-laying, agile and fast-moving skinks with well-developed limbs and immovable lower eyelids. Largely terrestrial, but some are arboreal and some climb rocks. Many species occur around coasts. Originally only a single species was recognised, with many races/subspecies, but Australasian and Oceanic island forms have been elevated into dozens of species. One species occurs along the southern Somali coast.

CORAL RAG SKINK

Cryptoblepharus africanus
(Sternfeld, 1918)

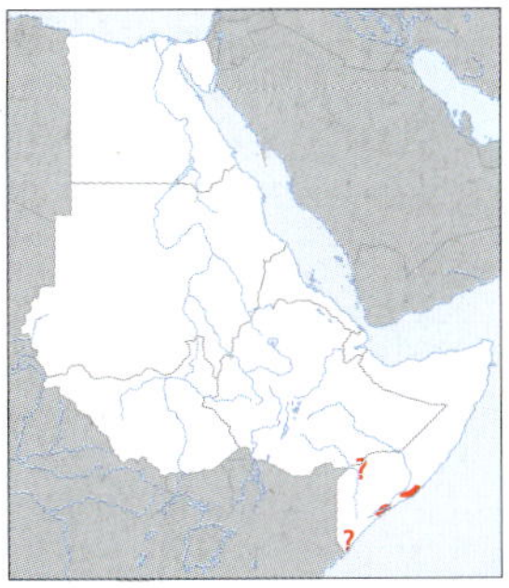

IDENTIFICATION: A small skink with well-developed limbs, five long-clawed toes. Head distinct. Immovable eyelids, each with a transparent spectacle. Scales smooth, close-fitting, in 21–24 mid-body rows. Tail cylindrical, tapering to a fine point. Maximum size about 12cm, 60 per cent of which is tail. Dorsum blackish-bronze, with two speckled gold dorsolateral stripes; paler below.

HABITAT AND DISTRIBUTION: On coral rag (old coral formations, on and above waterline) along coast and islands. Does leave immediate vicinity of the intertidal zone. In our area on the southern Somali coast, from a cluster of records around Mogadishu, and Baraawe, further south; probably occurs all the way along Somali coast south from Mogadishu to Kenya border, but not surveyed. A doubtful inland record at Lugh, 330km from the coast. Elsewhere, from Kenya south to southern Africa.

CONSERVATION STATUS: Least Concern.

NATURAL HISTORY: Diurnal. Quick-moving and active, forages on intertidal zone of coastal coral formations, where it may be seen darting up and down, following waves, catching insects and small crustaceans; may also take small fish. Swims well, and if threatened and unable to reach a hole will jump and swim to safety. Salt-tolerant, the ancestral species dispersed across the Pacific by rafting. One or two eggs are laid in a crevice above high-water mark in coral rag. Often in large densities, even on quite small offshore coral islets, with several hundred on a small (15m diameter) isolated coral outcrop in a bay at Watamu, Kenya. May come ashore a short distance to scavenge.

Cryptoblepharus africanus, Mogadishu.

EURASIAN SNAKE-EYED SKINKS *ABLEPHARUS*

A Palearctic genus of small, slim, striped, shiny brown or golden skinks with small but well-developed pentadactyl limbs and immovable lower eyelids. Largely terrestrial. Eleven species are known, from southern Europe, via the Middle East to India and the Himalayas, with isolated populations in southern Arabia. The genus commenced radiating in the Early Miocene, about 16–17 Mya. One species just reaches our area.

RÜPPELL'S SNAKE-EYED SKINK

Ablepharus rueppelli
(Gray, 1839)

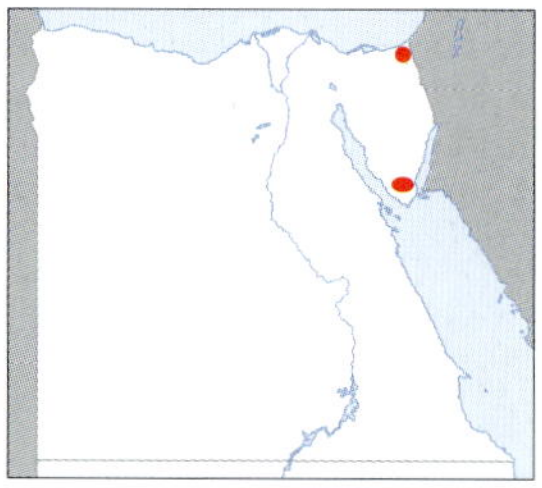

IDENTIFICATION: A small shiny skink without eyelids, in our area only in Sinai in Egypt. Head distinct. Immovable eyelids, each with a transparent spectacle. Scales smooth, 18 mid-body rows. Tail elongate, conical and thick (not much thinner than body), usually about 60–70 per cent of total length. Maximum size about 12cm (maximum snout–vent length 3.5cm in males, 4.3cm in females), average 7–11cm. Dorsum golden-brown, a dark dorsolateral stripe extends through the eye onto side of tail. Underside greyish. Juveniles often have a vivid coral-pink tail. Taxonomic Notes: Two subspecies known, the nominate form in our area.

HABITAT AND DISTRIBUTION: Usually in semi-desert and desert fringes, but with some vegetation. In our area, only in Sinai, between 500 and 1,600m, on mountains in the south, also at El Quseima on north-east border. Elsewhere, north to Syria, east to Jordan.

CONSERVATION STATUS: Least Concern.

NATURAL HISTORY: Diurnal and terrestrial. Quick-moving and active but secretive, forages on the ground in vegetated areas, quick to take cover. Up to three eggs are laid, probably in a single clutch per season. Diet small invertebrates.

Ablepharus rueppelli, Israel/Palestine.

AGAMAS

FAMILY AGAMIDAE

A family of Old World lizards, with nearly 60 genera and more than 550 species in six subfamilies, found in Australia, Asia, southern Europe and Africa; some species introduced in USA. Most African agamas belong in the subfamily Agaminae, except the big spiky Dabb or Spiny-tailed lizards, *Uromastyx*, which are placed in the Uromastycinae. Most agamas have large heads with prominent eyes and thin necks, depressed bodies and long tapering tails that are shed reluctantly. Some have a well-developed pineal eye, manifest as a small depression on top of the head. Some species are highly visible and tolerant of urbanisation. Several species are territorial, living in structured colonies under a dominant male, who drives away rivals. Others appear to live in pairs, or sometimes a male with 2–3 females. The males of several species are vividly coloured, and bask in prominent places; they can also change colour to a limited extent, and the colour displayed may be indicative of their position in the hierarchy. They feed largely on ants and termites. All of the African agamas lay eggs. Like crocodiles, the sex of the young is determined by the incubation temperature of the eggs, and evidence from some species indicates that females may be able to retain developing eggs in the body over a long time, until suitable weather conditions for laying arrive (usually a heavy storm). At present, the indigenous north-east African agamas are assigned to seven genera, *Acanthocercus* (six species), *Agama* (15 species), *Stellagama* (one species), *Pseudotrapelus* (three species), *Trapelus* (two species, possibly three), *Xenagama* (four species) and *Uromastyx* (six species). In some areas, they are incorrectly believed to be venomous. They diverged from the chameleons, with which they form a monophyletic clade, about 85 Mya in the Upper Cretaceous.

KEY TO THE NORTH-EAST AFRICAN GENERA OF *AGAMAS*

1a: Body extremely broad and flat, tail <50 per cent of total length and does not taper abruptly, clearly segmented into whorls of a single row of spiny scales	*Uromastyx*, Dabb lizards
1b: Body not extremely broad and flat, unregenerated tail not as above	2
2a: Tail very broad and spiky at base, tapering very abruptly to a little filament, or slightly broad and segmented but with only 1–2 scale rows in each whorl	*Xenagama*, Strange agamas
2b: Tail not very broad and spiky at base, not tapering abruptly	3
3a: Tail fairly broad at base, strongly segmented into whorls of scales	4
3b: Tail not particularly broad at base, not strongly segmented into whorls of scales	5
4a: Known only from northern Egypt	*Stellagama vulgaris*, Starred Agama
4b: In our area not in Egypt	*Acanthocercus*, Large agamas
5a: Ear-opening minute	*Trapelus*, Desert agamas
5b: Ear-opening large, tympanum visible	6
6a: Somewhat spindly limbed, spines absent or present in small quantities around earhole, on head or n eck	*Pseudotrapelus*, Uniform-scaled agamas
6b: Limbs not spindly, spines present in little clusters around ear, on head or neck	*Agama*, Agamas

SUBFAMILY AGAMINAE

A subfamily of nearly 140 species of agamas in ten genera, in Africa and Asia, with limited penetration into Europe. Six genera occur in north-east Africa.

AGAMAS *AGAMA*

An African genus of stocky, short-bodied, medium to large-sized lizards, often highly visible, with triangular heads, well-developed, strong, clawed limbs and thin toes. Males have precloacal pores. The scales are small, hard, overlapping and often spiny. They are diurnal, living on the ground, rocks and sometimes trees. Some live in structured colonies, with a dominant male, recognisable by its size and vivid colours; subordinate males are tolerated, provided they are smaller and have subdued colours. Gravid females also become very colourful. Others are solitary, these are small, plump and cryptically coloured, although males may show nuptial colours. Agamas eat mostly arthropods, especially ants, but may also take plants. Just fewer than 50 species are known; 15 species occur in north-east Africa (mostly in Ethiopia and Somalia), ten of these are endemic. Taxonomy is not yet fully clarified; see our notes on some species.

KEY TO THE NORTH-EAST AFRICAN MEMBERS OF THE GENUS *AGAMA*

1a: Dorsal scales heterogeneous	2
1b: Dorsal scales more or less homogeneous, none conspicuously larger than rest	3
2a: Very small, usually <13cm, in our area only in extreme south-west Sudan	*Agama gracilimembris*, Slender-limbed Agama
2b: Large, >25cm, in our area only in eastern Ethiopia and Somalia	*Agama robecchii*, Robecchi's Agama
3a: Ventral scales keeled	*Agama bottegi*, Bottegoi's Agama

3b: Ventral scales smooth 4

4a: No nuchal crest 5

4b: A distinct longitudinal crest of enlarged, lanceolate scales on nape 7

5a: No tufts of spines behind ear and on side of neck *Agama hartmanni*, Hartmann's Agama

5b: Tufts of spines behind ear and on side of neck 6

6a: Tympanum fully exposed, not encroached by surrounding clusters of elongate spines; snout–vent length of adult males <54mm and females <64mm *Agama persimilis*, Somali Painted Agama

6b: Tympanum partly occluded by encroaching clusters of elongate spines; snout–vent length of adult males at least 58mm, females at least 76mm *Agama rueppelli*, Rüppell's Agama

7a: Nasal shield rounded 8

7b: Nasal shield pear-shaped, prolonged into distinct point anteriorly 10

8a: Nasal scale not in contact with first canthal scale, 6–8 scales in nuchal crest, mid-body scale rows 55–59 *Agama lanzai*, Lanza's Rock Agama

8b: Nasal scale in contact with first canthal scale, 7–123 scales in nuchal crest, mid-body scale rows 58–79 9

9a: Medium-sized, to 28cm (snout–vent length up to 103mm), scales in nuchal crest low, mid-body scale rows 67–79 , precloacal pores in continuous row, range Somaliland *Agama somalica*, Somali Rock Agama

9b: Large, to 36cm (snout–vent length up to 126mm), scales in nuchal crest tall, mid-body scale rows 58–73, precloacal pores not in continuous row; Egypt to extreme north-west Somaliland *Agama spinosa*, Spiny Agama

10a: Small, snout–vent length of adults not more than 59mm, non-mucronate body scales, endemic to Ethiopia *Agama lucyae*, Lucy's Agama

10b: Medium to large-sized, snout–vent length >66mm, mucronate body scales 11

11a: Nostril clearly displaced beneath *canthus rostralis*; first canthal scale behind nasal separated from rostral by just two (rarely one or three) markedly elongate scutes; adult male has patch of intense blue-black pigment on posterior throat, underside pale grey or dirty white *Agama doriae*, Doria's Agama

11b: Nostril sited on *canthus rostralis*; adult male lacks prominent dark patch on posterior throat; underside dark, bluish or blackish 12

12a: Dorsal body scales very large, keeled, imbricate and strongly mucronate, in 55–58 rows at mid-body and 33–37 between limbs on vertebral line *Agama smithii*, Smith's Agama

12b: Dorsal body scales medium-sized, weakly keeled and mucronate, in 67–91 rows at mid-body and 56–80 between limbs on vertebral line 13

13a: Tail of male predominantly blue throughout, occasionally with white rings, but never red or yellow, female has two maroon or brick-red dorsolateral patches, widely separated, mid-body scales 67–91 *Agama lionotus*, Kenya Red-headed Rock Agama

13b: Tail of male red, orange or yellow at base, dark blue at tip, females with 1–2 elongate yellow or pale orange dorsolateral patches, not widely separated, mid-body scale rows 60–82 14

14a: Medium-sized, up to 28cm (snout–vent length up to 119mm), yellow to orange head, uniform pale reddish or reticulated throat, orange or body-like coloured fore limbs, bicoloured red (anteriorly) and black tail *Agama finchi*, Malaba Rock Agama

14b: Large, up to 40cm (snout–vent length up to 150mm), yellow to orange head, uniform to striated throat, body-like coloured fore limbs, a tricoloured (whitish to yellow, red and black) tail *Agama picticauda*, Rainbow Agama

BOTTEGOI'S AGAMA

Agama bottegi
Boulenger, 1898

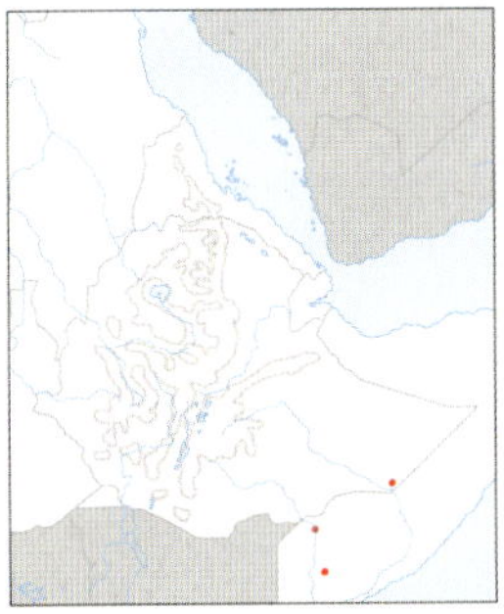

IDENTIFICATION: A large *Agama* with keeled ventral scales in southern Somalia. Head moderately convex and long, snout rounded or pointed, occipital scale enlarged; neck thin; body long, rather slender and depressed. Nasal scale pear-shaped, keeled and not in contact with first canthus scale, but separated by 1–2 smaller scales. Neck crest relatively short and consisting of few (6–7) but distinctly high crest scales. Earhole surrounded by three tufts of spiny scales, with two additional tufts on neck. Vertebral and lateral body scales strongly keeled, mucronate and shingled, ventral scales moderately keeled. Body scales in 52–60 rows around mid-body, 40–43 dorsal scales between limbs on vertebral line. Males have a continuous row of nine precloacal pores. Tail slightly compressed, about twice as long as snout–vent length. Dorsal and ventral tail scales strongly keeled, larger than body scales. Maximum size about 36cm (snout–vent length up to 12cm), hatchling size unknown. Originally described as yellowish-olive, throat greyish, gular fold orange; there is a round black blotch either side on neck in front of shoulder; in other specimens top of head is darker than sides, with a dark tapering stripe between eye and earhole. Females and juveniles have five dark crossbars and a vertebral stripe of irregular, sub-rectangular markings. Taxonomic Notes: Belongs to the *spinosa* species group.

HABITAT AND DISTRIBUTION: A species of the Somali arid zone, known only from Lugh and Dinsoor, near the Juba River in southern Somalia, in *Vachellia–Commiphora* dry, open, lowland savanna with red sandy soil, at 150–310m. An unconfirmed record from Galgalo, north-east of Mustahil, in extreme south-eastern Ethiopia near the Somalia border. Probably widespread in the intervening country, but no records.

Agama bottegi, preserved.

CONSERVATION STATUS: Data Deficient.

NATURAL HISTORY: Very poorly known; only seven specimens collected. Presumably, like other agamas, it is territorial, lays eggs, eat insects and other arthropods. The specific name honours Vittorio Bottego (1860–97), Italian army officer and explorer, during whose last expedition Ugo Ferrandi collected the type specimen of *A. bottegi*.

DORIA'S AGAMA

Agama doriae
Boulenger, 1885

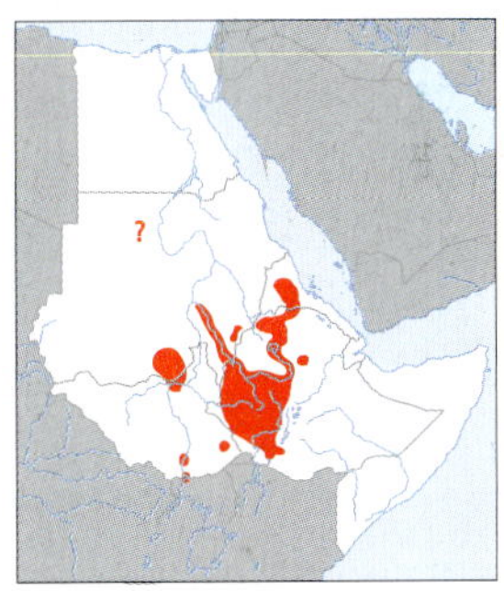

IDENTIFICATION: A fairly large colourful agama often associated with river gorges. Head broad, neck thin, snout short and rounded. Body squat and flattened. Tail thin and tapering, about 60 per cent of total length. Dorsal scales keeled, scales at mid-body in 74–98 rows, ventrals smooth. Maximum size about 30cm, average 18–25cm, hatchling size unknown. Males have a single row of 8–14 precloacal pores. Displaying males have a rufous head, the throat is intense blue with a bright orange or red patch on gular pouch; back blue-grey or purplish, with a broad diffuse

Agama doriae, male, Aykel.

Agama doriae, gravid female, Mulu.

Agama doriae, male throat, Mulu.

light vertebral line. Females are purplish-grey with a pale vertebral line that splits red and yellow crossbars. Two blackish stripes behind eye, the lower between eye and ear-opening. Underside pale or slightly darker reticulated, throat spotting pale with dark reticulations or network of dark lines, two connected dark spots at back. Juveniles grey with irregular narrow rufous crossbars, yellowish-grey head and two dark bars behind eye. Underside pale. When not displaying, general impression is of a dull grey-brown lizard with rufous blotches.

HABITAT AND DISTRIBUTION: A wide range of habitats, high grassland to medium and low-altitude savanna and semi-desert, between 300 and 2,700m. In Ethiopia strongly associated with rocky river gorges. In our area, in central highlands of Eritrea (where one of commonest reptiles), in Ethiopia from Lake Tana along the Blue Nile, also a handful of localities on western plateau, the Omo River and southern Rift Valley. A number of scattered South Sudan localities (Redjaf, Nimule, Boma); in Sudan along Blue Nile from Ethiopian border almost to Khartoum, with a cluster of records in Southern Kordofan, a curious record from Dongola. Elsewhere, Central African Republic west to Ghana (subspecies *A. d. benueensis*).

CONSERVATION STATUS: Least Concern.

NATURAL HISTORY: Poorly known. Diurnal and rupicolous; living on rock outcrops and in rocky valleys, but also seen on riverbanks and trees, even isolated ones in fields. They live in structured colonies under a dominant male who basks in a prominent position. No clutch details known, but juveniles were found in western Ethiopia in June, basking on rocks in the Omo Gorge. Diet presumably insects and other arthropods.

MALABA ROCK AGAMA, BRIAN FINCH'S AGAMA

Agama finchi
Böhme, Wagner, Malonza, Lotters & Kohler, 2005

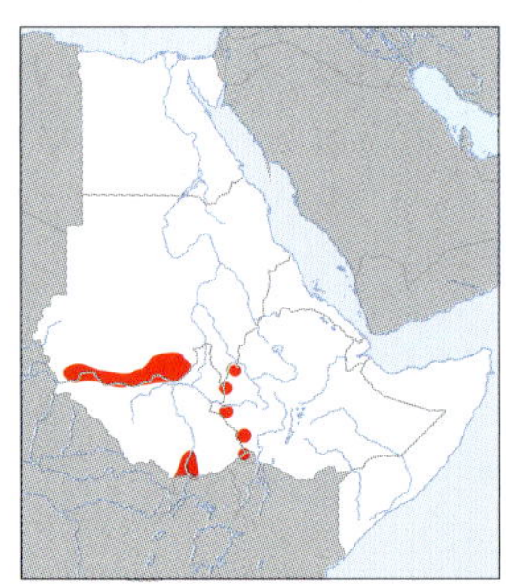

IDENTIFICATION: A fairly large, red-headed, rock-dwelling agama. Head broad, neck thin, snout short and rounded. Body squat and flattened. Tail thin and tapering, about 55–70 per cent of total length. Dorsal scales keeled, scales at mid-body in 70–79 rows. Maximum size about 28cm, average 18–25cm, hatchling size unknown. Displaying males have a red, orange or yellow head, often speckled black, and the orange extends on the spine, often to halfway between the pairs of legs. Body blue-black and tail starts blue-black, then orange, with a paler blue tip. Females pale or dark brown, with green head spots, and a broad, irregular and diffuse lateral stripe of orange, yellow and black. Juveniles resemble females, but flank stripe less prominent. Taxonomic Notes: In our area, the subspecies *A. f. leucerythrolaema* differs from the nominate in being slightly larger, with blue forelimbs as opposed to orange. See also taxonomic notes on *A. picticauda*. Belongs in the *agama* species group.

HABITAT AND DISTRIBUTION: Rocky hills in semi-desert, mid-altitude savanna and forest, between 300 and 1,600m. Ethiopian records include Gambela, Shimala River and Bench Maji in the south-west. In South Sudan widely recorded in Central Equatoria State, the Ilemi Triangle, and museum records of '*A. agama*' from Western and Eastern Equatoria will probably prove to be this species. Found in Sudan in Southern Kordofan, Southern and Western Darfur, and Fakoki (incorrectly called Fagogli), in nothern Kardofan. Elsewhere, west to the Central African Republic, south through Uganda and north-east DR Congo.

CONSERVATION STATUS: Least Concern.

Agama finchi, female, West Kordofan.

Agama finchi, West Kordofan.

NATURAL HISTORY: Poorly known. Diurnal and terrestrial; lives on rocky hills and outcrops, but will also utilise tree and buildings. Often relatively abundant and allows close approach. They live in structured colonies under a dominant male who basks in a prominent position; up to 14 females observed with one male. No clutch details known, but mating was observed in northern Uganda in late June; numerous juveniles also seen at this time. Diet probably insects and other arthropods.

SLENDER-LIMBED AGAMA

Agama gracilimembris
Chabanaud, 1918

IDENTIFICATION: A small, spindly limbed, solitary terrestrial agama, might just reach western Sudan. Head broad, neck thin. Body squat and flattened. Tail thin and tapering, about 62–67 per cent of total length. Dorsal scales keeled, occasionally spiny, scales at mid-body in 70–85 rows, ventrals keeled. No tufts of spines behind ear and on side of neck, dorsal head scales strongly keeled. Maximum size about 14cm (snout–vent length up to 5.7cm, males smaller), average 9–12cm, hatchling size unknown but juveniles in October–December averaged 26.7mm snout–vent length. Unusually, females larger than males. Males have a single row of 8–12 precloacal pores. Colour varies, but general impression is of a mottled grey, brown and rufous lizard, often with a diffuse, pale vertebral stripe and a tail that may be brown, or grey with dark crossbars, often two irregular darker crossbars on head. If stressed, they become very dark. Underside pale; throat has fine brown speckling or lines. Male's tail blue at sides, orange below, and there may be patches of deep blue and orange on limbs.

HABITAT AND DISTRIBUTION: Sudan and Guinea savanna, between 200 and 1,300m, although if it

Agama gracilimembris, Ghana.

occurs in western Sudan likely to be at 400–600m. No formal records in our area, but recorded in northern Central African Republic, including the village of Tizi-Fongoro, at border with Western Darfur, Sudan. Elsewhere, west to Mali.

CONSERVATION STATUS: Least Concern.

NATURAL HISTORY: Poorly known. Diurnal, terrestrial and solitary. Usually seen in flat open country, with areas of exposed soil, but also in fields, road margins, in woodland, on inselbergs and sheet rock. Basks during day, often on small termite mounds, rocks, up in leafless shrubs or at edge of bushes, waiting for insect prey to pass, shelters by night in a hole or under ground cover. When pursued may run down a hole, but also up into grass tussocks, a strategy observed in Australian agamids but rarely in Africa. Gravid females seen in April–May; it is suggested that adults all die after the rainy season and this is an annual species. Diet insects and other arthropods.

SUDANESE AGAMA

Agama hartmanni
Peters, 1869

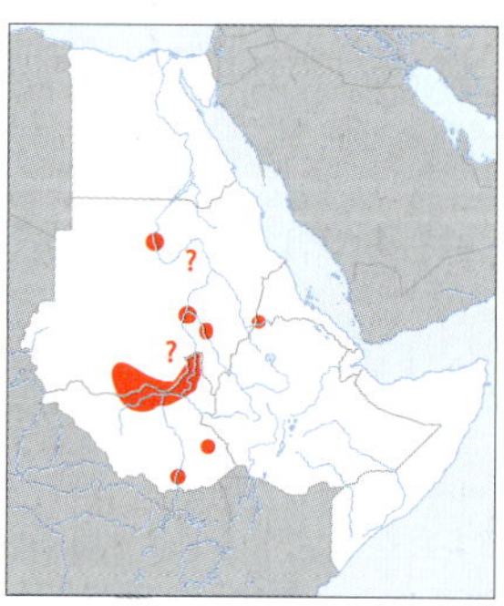

IDENTIFICATION: A large slim agama of Eritrea and Sudan. Head fairly slim, long and deep, the neck is thin. Body fairly slim and elongated, cylindrical, and limbs are long and not as muscular as in most agamas. Tail thin and tapering, about 60–63 per cent of total length. Dorsal scales keeled, scales at mid-body in 76–81 rows, ventrals smooth. No tufts of spines behind ear and on side of neck (a good character in the field). Maximum size about 23cm (snout–vent length up to 9.2cm), average 15–21cm, hatchling size unknown. Males have a single row of 9–12 precloacal pores. General impression is of slim agama, often with a vertebral stripe. Males blue-grey, with vague interrupted pale crossbars, and often have a series of poorly defined pale ocelli on flanks; orangey around eyes. Females more brown or rufous, with a series of paired dark sub-rectangular blotches on spine, a short white stripe behind eye and a pale cheek stripe that fades on flanks but is punctuated by a series of bright white spots; often with another lateral line of white spots. Both sexes have a pale distinct vertebral stripe from occiput to, at least, the middle of tail. Underside

Agama hartmanni, female, South Kordofan.

Agama hartmanni, male, South Kordofan.

cream, sometimes with fine brown lines on chest. Taxonomic Notes: Confused in past with *A. doriae*, so some old records are suspect.

HABITAT AND DISTRIBUTION: Dry savanna and semi-desert, between 250 and 600m. A single Eritrean record, at Omhajer in the low west, widespread but sporadic records from Sudan include Dongola (the type locality, in the north), Khartoum, Singa, Sennar and parts of Northern and Southern Kordofan (Baleela to Gole, east to Heglig and Abu Likri) and vicinity of the White Nile (Taufikia, Meschra El Rek [or Mashraar Raag] on Southern Kordofan border). In South Sudan, at Renk, Kodok, Rejaf, south of Juba, and the Boma Hills.

CONSERVATION STATUS: Least Concern.

NATURAL HISTORY: Poorly known. Diurnal, terrestrial but also climbs trees. Presumably lays eggs, diet insects and other arthropods. Recently observed specimens in Southern Kordofan were found on the ground in fairly high grassland at the edge of woodland, where sympatric with *A. doriae*. Individuals were basking on larger pieces of ploughed clay.

LANZA'S ROCK AGAMA

Agama lanzai
Wagner, Leaché, Mazuch & Böhme, 2013

IDENTIFICATION: A large rock-dwelling agama from hot arid coastal areas of northern Somalia. Head round, body scarcely depressed, hind limbs stout. Nasal scale round, not in contact with first canthus scale, but separated by two smaller scales. Neck crest well developed, consisting of 6–8 conical scales equal to diameter of ear-opening. Earhole surrounded by six tufts of spiny scales, with two additional tufts on neck. Vertebral and ventral scales smooth, lateral body scales keeled. Body scales in 55–59 rows around mid-body, 51–59 dorsal scales between limbs on vertebral line. Males have a single interrupted row of ten precloacal pores. Tail slightly compressed, more than twice as long as snout–vent length. Dorsal tail scales spiny and keeled, ventral tail scales smooth. Maximum size about 35cm (snout–vent length up to 10cm), hatchling size unknown. Displaying males have a vivid red-orange head; the throat is yellowish with an orange gular fold and pouch, and rest of underside blue-grey. Body blue with a tapering vertebral stripe. Two males in combat exhibited dark blue heads with creamy-white speckling, body blue with white and yellow speckles. Non-displaying males rufous, with 3–4 regular paired black marks on vertebrae. Juveniles resemble non-displaying males, but have greenish heads. Gravid females very colourful; body is entirely yellow with several red bands and blotches, head often blue, legs and tail brownish. Throat yellowish, vaguely reticulated. Taxonomic Notes: Recently described, previously lumped with *A. somalica*, both under *A. spinosa* in northern Somalia (Somaliland and Puntland). It is now recognised as a distinctive species belonging to the *A. spinosa* species group.

HABITAT AND DISTRIBUTION: A north Somali endemic in the environs of the Gulf of Aden. On stony coastal plains with rocky outcrops or sheet rock with sparse vegetation, at 300–600m. Occurs in the Guban strip, a hot dry landscape between the sea and mountain ranges of northern Somalia, from Aroweina and Santaalo (Somaliland) in the west, across Wadi

Agama lanzai, gravid female, Berbera.

Agama lanzai, non-displaying male, Berbera.

Agama lanzai, male, Wadi Faradeero.

Faradeero south of Berbera, and from the low hill country east of Berbera to Karin in Puntland. Also recorded at Wadi Goddalola, about 40km east of Bosaso. Probably widespread in the intervening country (all suitable habitat) but no records. In more upland and mesic habitats south of its range it is replaced by *A. somalica*.

CONSERVATION STATUS: Least Concern.

NATURAL HISTORY: A rupicolous lizard. Diurnal and lives in harems with a dominant male, several females and juveniles. Gravid females seen in December. Males fighting each other in nuptial coloration observed in July. At the same locality an adult male in non-displaying colours and several juveniles were observed in February. Dominant males bask in a prominent spot; females occupy lower regions of the rocks, and juveniles shuttle between rocks and the ground. Often sympatric with the Dabb lizards *Uromastyx macfadyeni* (in Somaliland) and *U. princeps* (in Puntland).

KENYA RED-HEADED ROCK AGAMA

Agama lionotus
Boulenger, 1896

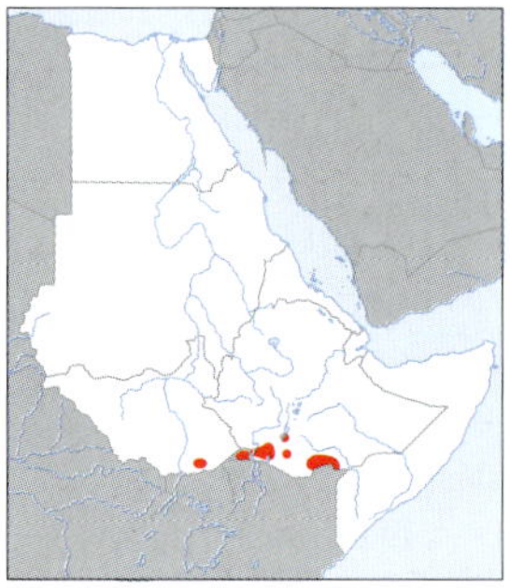

IDENTIFICATION: A big, rock-dwelling lizard. Head large and triangular, males have a small crest on nape. The eye is fairly large, ear-opening obvious. The body is depressed, legs strong and muscular, toes long and thin. Tail laterally compressed (in some males with a prominent ridge), tapering smoothly, roughly 60 per cent of total length. In some specimens there is a club-like bulge at tip of tail, used in fights. Body scales homogenous, in 67–91 rows at mid-body. Maximum size about 35cm (females are smaller), average 20–30cm, hatchling size 8–10cm. Displaying males have a vivid yellow or orange-yellow head, the throat is orange. Body blue, often with a pale tapering vertebral stripe and darker ocelli; tail often ringed with white or pale blue. Males with pale heads are usually non-dominant adults. Non-displaying males brown, or even whitish, with faint darker crossbars and green/yellow head speckles. Females and juveniles brown, with green head speckling, vague dark crossbars and a vertebral stripe of irregular, sub-rectangular markings; usually a maroon or rufous patch where upper limbs touch body, and these patches may coalesce to form a maroon stripe. Taxonomic Notes: Belongs to the *lionotus* species group.

HABITAT AND DISTRIBUTION: Woodland, moist and dry savanna and semi-desert, from sea level to about 1,500m. Largely absent from areas without rock outcrops or big trees. Widespread in southern Ethiopia, south-west and south-east from Arba Minch, to Mandera on Kenya border in south-east, and lower Omo River and around north end of Lake Turkana in the south-west; often common where there are rock outcrops. In South Sudan known from Talanga Forest and the Ilemi Triangle, but distribution there complicated by past confusion with *A. agama* and *A. finchi*. Elsewhere, south through Kenya to northern Tanzania.

CONSERVATION STATUS: Least Concern.

NATURAL HISTORY: A diurnal, rock- or tree-dwelling colonial lizard, but will use buildings and walls, where dense colonies may form due to abundance of suitable sites and refuges. The dominant male (and sometimes subordinate males) bask in a prominent spot; activity greatest in mid-morning. They hide in rock and tree cracks. If threatened, they run rapidly for cover, and can jump considerable distances (50cm or more). When basking, males bob the head up and down. Intruders in a dominant male's territory are attacked. The occupant bobs his head as he approaches; both sidle sideways and then rush in, biting savagely, and using the tail as a club. The loser flees. A study at Malindi found that if dominant male was removed from the colony, the subordinate male quickly assumed control and displayed bright colours. Females lay 5–10 eggs in a hole they excavate, usually at the start of the rainy season, but sometimes to coincide with unexpected storms. In gravid females the red patches on flanks become vivid, and may signal to a male that she is unavailable. In Ethiopia incubation was 50–60 days. Hatchlings and small juveniles are sometimes eaten by large adults, hence do not share refuges with them. Diet includes a wide range of insects, especially ants and beetles, plant material is sometimes taken; large adults may eat small lizards including their own species. Mongooses have been seen digging up eggs; adults are often taken by small birds of prey and snakes; predation by Red Spitting Cobras, *Naja pallida*, seen in Ethiopia, the snake carefully climbing a wall to catch sleeping agamas.

Agama lionotus, male, Yabello.

Agama lionotus, female, Yabello.

LUCY'S DWARF AGAMA

Agama lucyae
Wagner & Bauer, 2011

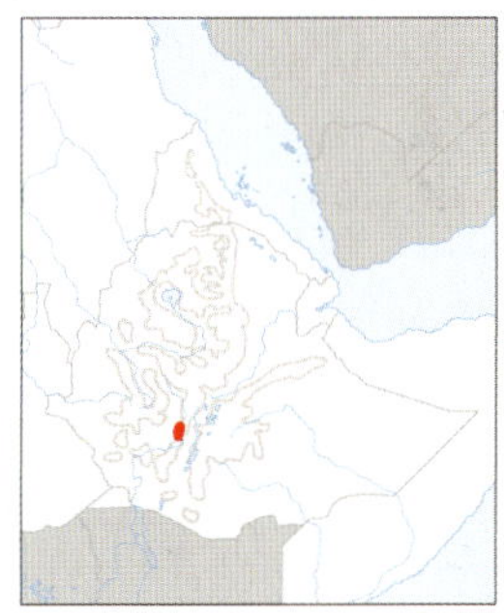

IDENTIFICATION: A small, lightly built agama, endemic to central Ethiopia. Head short and depressed, with smooth scalation and a large occipital scale. Earhole surrounded by single short conical scales. A few tufts of spinose scales behind ear and on lateral side of neck are present, tympanum large and shallow. Short neck, crest present in both sexes. Gular and ventral scales smooth. Body squat and flattened, with homogeneous scalation and strongly keeled but non- or only slightly mucronate dorsal scales, in 66–69 rows around mid-body and 47–52 between limbs on vertebral line. One row of eight precloacal pores in males. Tail long and thin, 62–80 per cent of total length, not arranged in distinct whorls. Maximum size about 13–14cm (snout–vent length up to 5.4cm in males, 5.9cm in females). Upper parts of body and head brownish. A pale vertebral stripe, sometimes edged by dark bands, extends from just behind occiput along body and onto tail. Two dark stripes behind eyes. Four faintly dark transverse bands between limbs in males. Throat, belly and hind limbs dark with extensive white spotting in males, uniformly whitish in

Agama lucyae, north of Soddo.

Agama lucyae, north of Soddo.

Agama lucyae, north of Soddo.

females (save some fine marbling at back of throat). Taxonomic Notes: Recently described, similar to *A. doriae*, their ranges overlap in central Ethiopia but can be identified by its very small size.

HABITAT AND DISTRIBUTION: Endemic to central Ethiopia; the types are from rocky, semi-arid, open bushland without dense vegetation, at about 1,300m, c.58km north-west of Sodo in the Omo Gorge in southern Ethiopia. A specimen from 40km south of the type locality, in similar habitat (open wooded savanna) at 1,075m, might be this species. Probably more widespread in south-west Ethiopia, especially along Omo River in suitable habitat.

CONSERVATION STATUS: Data Deficient.

NATURAL HISTORY: Poorly known. Diurnal, probably rock-dwelling. Freshly hatched juveniles (snout–vent length 2.5cm) and a gravid female with five eggs were collected in July. Diet presumably arthropods.

SOMALI PAINTED AGAMA

Agama persimilis
Parker, 1942

IDENTIFICATION: A beautiful little ground-dwelling agama. Head rounded, with a big cluster of prominent spiky scales on neck behind ear-opening; the longest of which is about half the width of the eye diameter. Eye about the same diameter as earhole; tympanum completely exposed. No neck crest. Body strongly depressed. Limbs long, hind toes very long. Tail broad at base, tapering sharply, 60–65 per cent of total length. Body scales homogeneous, keeled and overlapping, in 52–57 rows at mid-body. Ventral scales smooth. Maximum size about 16cm, average 10–14cm, hatchling probably 5–6cm. It seems that females (average snout–vent length 5.6–6.4cm) are larger than males (average snout–vent length 4.3–5.4cm), which is unusual in agamas. Males have one row of 10–12 precloacal pores. Overall rufous or rufous-brown, with a broad pale or brown bar between eyes, the snout is pink or brown above, a big pink or red patch on crown and a fine or broad grey, white, red or brown vertebral line. Either side of this are paired brown and black or orange blotches, the back is suffused grey or blue-grey towards the back legs. There is a big brown blotch on the anterior flanks. Limbs mottled brown. Underside white. Taxonomic Notes: Belongs to the *lionotus* species group, most similar to *A. rueppelli*.

HABITAT AND DISTRIBUTION: A species of the Somali arid zone, in dry open *Vachellia–Commiphora* savanna and semi-desert with soft red sand, at 80–970m (to 1,200m extralimitally). Occurs from the Haud region (Balumbal, Shanshacade and Bohodle) in southern Somaliland across the eastern Ogaden of Ethiopia to the Mudug of central Somalia (Geriban, Hàro or Xaaro), thence south to 04°N, to Jesomma and Bud Bud. Also at Mandera on the Kenya/Somali/Ethiopian border, probably in the intervening country but records lacking. Elsewhere, in east and south-east Kenya and extreme northern Tanzania.

CONSERVATION STATUS: Least Concern.

NATURAL HISTORY: Poorly known. Diurnal and ground-dwelling, in open country, living in holes sited under a shady, spreading bush. Hunts around the perimeter of the bush, rushing out to snap up prey. If threatened usually dashes back under the bush, and may dive into its hole, but occasionally remains motionless, hoping to avoid detection. Not colonial, usually solitary, but may live in true pairs or occasionally one male with two females. They sometimes ascend into bushes and small trees to bask, and may be seen hanging from a flimsy twig, looking dead. A gravid female was collected in late August in Somaliland (northern Somalia), but no other reproductive details known. Diet arthropods.

Agama persimilis, Shanshacade.

Agama persimilis, Shanshacade.

RAINBOW AGAMA

Agama picticauda
Peters, 1877

IDENTIFICATION: A big orange-headed, blue-bodied agama. Head large and triangular, males have a small crest on nape. Eye fairly large, ear-opening obvious. Body depressed, legs strong and muscular, toes long and thin. Tail laterally compressed; roughly 60 per cent of total length, sometimes with a club-like bulge at tip, used in fights. There are 8–10 precloacal pores in males. Body scales homogenous, in 59–82 rows at mid-body. Maximum size about 40cm, (females smaller), average 25–32cm, hatchling 9–12cm. Displaying males have a vivid red-orange or orange-yellow head, the orange tapering on the back, sometimes to a diffuse pale vertebral line. The male's deep gular pouch is orange. Body usually dark blue, often with a pale tapering vertebral stripe and darker ocelli; adult male's tail distinctive; blue-grey at base, to orange, with a deep blue tip. Non-displaying males brown. Females and juveniles brown, with green head speckling, vague dark crossbars and a vertebral stripe of irregular, sub-rectangular markings; mature females usually have orange flank patches. Underside paler, usually yellowy-grey. Taxonomic Notes: For many years, agamas of this species, widespread across West and Central Africa, were called *A. agama*. Recent research indicates that true *A. agama* is confined to an area centred on Cameroon, and the correct name for the widespread Rainbow Lizard is *A. picticauda*. Belongs to the *agama* species group.

HABITAT AND DISTRIBUTION: Woodland, moist and dry savanna and semi-desert, from sea level to about 1,500m. True distribution in our area complicated by difficulty of distinguishing this species from *A. finchi*, but as presently defined, it occurs west from Gambela in Ethiopia across most of South Sudan and south-west Sudan. A historical account (as *A. colonorum*) describes it as common in Sudan, in Kordofan and around Nuba Mountains, south of El Obeid, also from Kadugli and Talodi, south to Tungaru. Also known from Darfur. Reappears in South Sudan, in Bahar El Ghazal and Central Equatoria. Elsewhere, west to Senegal.

CONSERVATION STATUS: Least Concern.

NATURAL HISTORY: A diurnal, rock- or tree-dwelling colonial lizard; will use buildings and walls, where dense colonies may form due to abundance of suitable sites and refuges. Dominant males bask in a prominent spot; activity greatest mid-morning. They hide in rock and tree cracks. If threatened, they run rapidly for cover and can jump considerable distances (50cm or more). When basking, males bob the head up and down. Intruders in a dominant male's territory are not tolerated; the occupant bobs his head as he approaches; both sidle sideways and then rush in, biting savagely, and using the tail as a club. The loser flees. Females lay 3–10 eggs in a hole they excavate, usually at start of rainy season, but sometimes to coincide with unexpected storms. Eat a wide range of insects but a study in Nigeria found >95 per cent of diet was ants.

Agama picticauda, female, Nigeria.

Agama picticauda, male, Nigeria.

ROBECCHI'S AGAMA

Agama robecchii
Boulenger, 1891

IDENTIFICATION: A large, heavily built, spiky, ground-dwelling agama with a bird-like, greenish head and short tail, in northern and central Somalia. Head massive, with huge eye sockets sited high on its upper surface and a conspicuously short snout. Nasal scale round, not in contact with first canthus scale. Upper head scales small, no enlarged occipital scale. Ear about ⅔ of eye-opening diameter, superficial and exposed. No neck crest. Thin legs, hind limbs comparatively short. Body short, much depressed, covered by strongly heterogeneous dorsal scales and a scattering of thorn-like spinose scales, surrounded by a rosette of smaller spines. Gular and ventral scales smooth. Tail short, 44–54 per cent of total length (slightly longer in males), slender and tapering sharply. Males with a triple row of 23–26 precloacal pores. Maximum size about 29cm (snout–vent length up to 14cm), hatchling size unknown. Dorsum cream, coarsely marbled rufous-brown, many of the enlarged tubercles on back distinctly orange and head often suffused or veined greenish, whilst tail is conspicuously marked with pale and dark bands. Eyelids bluish to greenish. Blue pigment on throat forms short longitudinal lines in at least some specimens. Underside whitish. A female from a red sand area was red with a greenish head. Juveniles brown, head with white reticulations. Taxonomic Notes: Genetic analysis, coupled with

Agama robecchii, female, Qool-Cadday.

Agama robecchii, male, Sheikh.

Agama robecchii, juvenile, Qool-Cadday.

unusual morphology and behaviour suggest this species is not an *Agama*, and should be reallocated to a different genus.

HABITAT AND DISTRIBUTION: Semi-desert and arid savanna, from sea level to about 1,790m. Found on stony limestone plains, but also on sandy limestone plains and adjacent *ban* (treeless plains) without rocks. Widely distributed in Somaliland, from Hargeisa east to Sheikh and Erigavo, south of the mountains (an old record further north, from Las Korey in the Guban) into eastern Puntland and the Nogal Valley. Sporadic Somali records include Hobyo in the Mudug, Mogadishu area, and Aware in the eastern Ogaden in Ethiopia. Probably widespread in the eastern Ogaden but records lacking. It has a wide range, although usually uncommon.

CONSERVATION STATUS: Least Concern.

NATURAL HISTORY: Poorly known, yet its behaviour in many ways is unusual and unlike other agamas. Diurnal and ground-dwelling. In Aware in Ethiopia, it was found living in holes on open, sandy plains. When agitated curls its tail defensively over its back. In similar habitat in the Haud of southern Somaliland a female and a juvenile were found under roadside rocks, 2m apart. Adult males have a most remarkable defence: when approached they press their bodies to the ground and freeze, becoming almost invisible, moving only if almost trodden on. No other agama is known to do this. They are also quite unusually camouflaged, the adults creep about with their brown bodies close to the ground and their heads raised, imitating small bushes. Not a colonial species, it is usually solitary and shy. Behaviour closely resembles agamas of the genus *Xenagama*, with which it shares a similar distribution and habitat preferences. Diet insects and other arthropods. Faeces of fresh captives were large, black and contained only remnants of wing cases (elytra) and legs of beetles (probably Tenebrionidae). In captivity they favour large cockroaches or mealworm beetles (but not their larvae); their large heads and powerful jaws suggest they are adapted to feed on big beetles. Unusually, too, captive females are aggressive towards males, unlike most agamas.

RÜPPELL'S AGAMA

Agama rueppelli
Vaillant, 1882

Agama rueppelli, displaying male, Somaliland.

Agama rueppelli, gravid female, Fadhi-Gaab.

Agama rueppelli, Dire Dawa.

IDENTIFICATION: A medium-sized, attractive, dry-country agama. Head triangular, with a prominent cluster of big spiky scales around the ear-opening, partially concealing it. Spiky scales as large as diameter of eye-opening. No nuchal crest. Body squat and depressed, limbs slender and fairly long. Tail 60–65 per cent of total length, very broad at its base, tapering sharply. Body scales keeled, occasionally spiky, in 54–64 rows at mid-body. Ventral scales smooth. There are 9–13 precloacal pores (fewer in females). Maximum size about 28cm (snout–vent length up to 8.8cm), average 18–25cm, hatchling size unknown. Ground colour brown or grey, sometimes uniform and pattern-less, other have a faint or distinct pattern. Usually a pale brown bar between eyes with a chocolate-brown bar in front of and behind this. Back of head grey-brown. Two broad sub-rectangular blotches either side of vertebral scales on neck; a fine pale vertebral stripe runs length of body, with three pairs of big irregular warm brown blotches either side. On the flanks there is either a brown bar or series of brown blotches, the limbs are mottled, brown and yellow, the tail barred light and dark brown. Yellow or cream below. Juveniles have similar coloration. Displaying males have top of head and snout orange, sides bright yellow. Some gravid females retain usual adult pattern, others develop a green or turquoise head and vivid orange flank patches. Taxonomic Notes: Three different subspecies currently recognised, but status questionable. Two of these, *occidentalis* and *septentrionalis*, occur in East Africa (Kenya, south-west Ethiopia and South Sudan), whereas the nominate is endemic to the Horn of Africa. Belongs to the *lionotus* species group.

HABITAT AND DISTRIBUTION: Dry savanna and semi-desert in open country, mostly at low elevation (but up to 1,920m in northern Somalia). Prefers stony ground. In our region known from Ilemi Triangle in extreme south-east South Sudan, east to southern Ethiopia. Sporadic records from Dolo, the Ogaden and Awash River. Occurs more or less continuously from Dire Dawa east across Somaliland to easternmost tip of Puntland (Bender-Meraya), thence south to Mudug, Bay and the middle Shebelle regions of central and southern Somalia. Seemingly absent from Djibouti; replaced further south on red sandy soils of *Vachellia–Commiphora* savanna by similar *A. persimilis*.

Elsewhere, south to Kenya and north-east Uganda. Fairly common and widespread.

CONSERVATION STATUS: Least Concern.

NATURAL HISTORY: Poorly known. Diurnal and ground-dwelling, although it will climb into shrubs and bushes to bask. Lives in open country, in holes under a shady, spreading bush or in thickets, hunting around the thicket and out in the open, dashing back to cover if threatened. Will also live on sheet rock, provided there are crevices in which it can hide, but not usually on vertical rocks, or anywhere where larger, rock-dwelling agamas live. When threatened presses its body to the ground, blending with the surroundings. Not a colonial species, usually solitary, sometimes in pairs. No breeding details known, presumably lays eggs in a moist hole. Pregnant females seen in Somaliland in July–August. Males in display colours seen in the hot rainy season (May–September). Diet insects and other arthropods. The specific name honours Eduard Rüppell (1794–1884), German naturalist and explorer, who made several expeditions to north-east Africa and collected the types of a number of species including *Sclerophrys regularis*, *Hemorrhois nummifer*, *Acanthocercus cyanogaster*, *Pseudotrapelus sinaitus*, *Cyrtopodion scabrum*, *Hemidactylus granosus* and *H. robustus*.

SMITH'S AGAMA

Agama smithii
Boulenger, 1896

IDENTIFICATION: A large, rare agama with strongly mucronate dorsal scales in dry savanna of the Somali arid zone. Head convex and triangular. Nasal scale pear-shaped and in contact with first canthus scale. Neck crest short, consisting of a few long scales, ear-hole surrounded by three tufts of spiny scales, with two additional tufts on neck. Body squat and flattened, limbs long and muscular. Dorsal body scales large, keeled, imbricate and strongly mucronate, in 55–58 rows at mid-body and 33–37 between limbs along vertebral line. Ventral scales small and smooth. Tail rounded, long and thin, 64 per cent of total length. Dorsal tail scales keeled and as large as dorsal scales. Maximum size about 28cm (snout–vent length 11–11.5cm), hatchling size unknown. Females pale olive-brown above, with traces of dark crossbars; a broad, irregular and lateral stripe of maroon or rufous. Underside light grey, throat with dusky longitudinal streaks and a diffuse dark maroon stripe behind eye. Head of a recently collected female had indistinct green speckling in occipital region. Coloration of males and hatchlings unknown. Taxonomic Notes: In past regarded as synonymous with *A. spinosa* or *A. agama*, now recognised as a distinctive species within the *A. spinosa* complex.

HABITAT AND DISTRIBUTION: A species of the Somali arid zone, in dry *Vachellia–Commiphora* lowland savanna on red sandy soils, at 130–220m. The type is from 15km west of Mandera, near the Dawa River on the Ethiopia/Kenya border; a recent specimen was collected at Garissa in Kenya, about 540km south-west from the type locality. Another specimen, probably of the same species, was collected in similar habitat 13km west of Afmadu in southern Somalia. Probably more widespread in the intervening country (all suitable habitat) but no records.

CONSERVATION STATUS: Not Assessed.

NATURAL HISTORY: Poorly known. Diurnal and probably ground-dwelling. The adult female was found behind a branch of a small *Vachellia* tree, about 1.5m

Agama smithii, Kenya.

up. The specific name honours Arthur Donaldson Smith (1866–1939), American physician, hunter, African explorer and collector of the type.

SOMALI ROCK AGAMA

Agama somalica
Wagner, Leaché, Mazuch & Böhme, 2013

IDENTIFICATION: A medium-sized rock-dwelling species from semi-arid upland areas of northern Somalia. Head broad, moderately convex; body slightly depressed; hind limbs muscular. Nasal scale round, in contact with first canthus scale. Neck crest long, consisting of 9–12 conical scales distinctly shorter than half of the ear-opening diameter. Earhole surrounded by six tufts of spiny scales, with three additional tufts on neck. Vertebral scales smooth to feebly keeled, lateral body scales keeled, ventrals smooth. Body scales in 67–79 rows around mid-body, 59–75 dorsal scales between limbs along vertebral line. Males have one continuous row of 10-12 precloacal pores. Tail slightly compressed, more than twice as long as snout–vent length. Dorsal tail scales spiny and strongly keeled, ventral tail scales smooth and smaller. Maximum size about 28cm (snout–vent length up to 10.3cm), hatchling size unknown. Displaying males have a vivid yellow head, the throat is yellow or slightly orangey with an orange gular fold and pouch, and rest of underside blue-grey. The body is blue, usually with a paler vertebral stripe. Non-displaying males resemble females; dorsum grey-brown with a pattern of darker brown or reddish blotches that may form transverse bars, flanks with white-centred brown ocelli. Juveniles are similar, but head and shoulders have indistinct green speckling, or has a greenish, white-speckled head. Gravid females colourful; body yellowish to orange with 5–6 red bands with a whitish central spot, a broad irregular lateral stripe of red; head turquoise-green to blue, legs and tail brownish.

Taxonomic Notes: Recently described, previously subsumed, with *A. lanzai* in northern Somalia (Somaliland and Puntland), in *A. spinosa*; now recognised as a distinctive species in the *A. spinosa* complex.

HABITAT AND DISTRIBUTION: A north Somali endemic. Lives on rocky hills and outcrops in a semi-arid landscape, in medium to high savanna, at 770–1,670m.

Agama somalica, juvenile, Sheikh Pass.

Agama somalica, Amoud University.

Agama somalica, Amoud University.

Agama somalica, gravid female, Las Geel.

In low hot coastal areas of Guban it is replaced by *A. lanzai* and in extreme north-west Somaliland by *A. spinosa*. Confirmed records are from western and central Somaliland, from Borama in the west and Sheikh in the east. Might occur further east (Cal Madow Mountains), but not documented. Abundant and easily observed at Las Geel, Naaso Hablood in Hargeisa and Ferdusa Ruins in Sheikh. Its proximity to the Ethiopian border suggests it may also occur there. One of the most common and easily seen reptiles in Somaliland in suitable habitats.

CONSERVATION STATUS: Least Concern.

NATURAL HISTORY: Diurnal and rupicolous; lives in harem groups with a dominant male and several females and juveniles. Often on quartz rocks or boulders, with the syntopic *Acanthocercus phillipsii*. Single males sometimes bask on *Vachellia* trees, around 2m above ground in areas without rocks. Juveniles partially terrestrial, not strictly rupicolous like adults. Pregnant females seen July–August. Males in display colours observed during hot rainy season (May–September), and in non-display colours mostly in winter (January–February). Presumably, like other agamas, territorial, lays eggs, and eats insects and other arthropods.

SPINY AGAMA

Agama spinosa
Gray, 1831

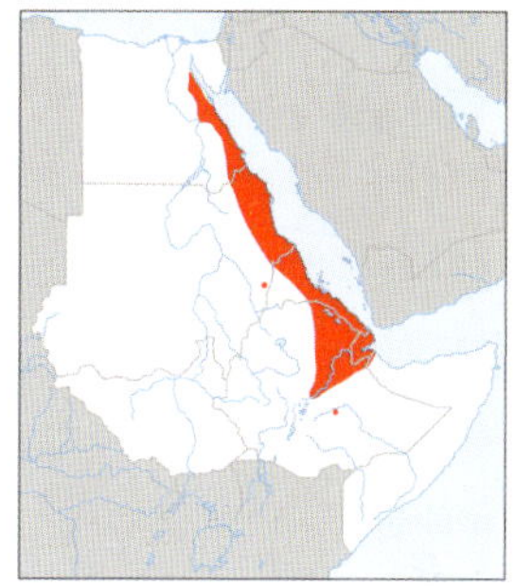

IDENTIFICATION: A comparatively large rock-dwelling species found in mountains overlooking the Red Sea coastal plains and hinterland. Head moderately convex, body slightly depressed and hind limbs muscular. Nasal scale round, in contact with first canthus scale. Neck crest relatively short, consisting of 7–11 conical scales smaller than half of the ear-opening diameter. Earhole surrounded by six tufts of spiny scales, with three additional tufts on neck. Vertebral and lateral scales keeled, imbricate and mucronate (especially in females); ventrals smooth. Body scales in 58–73 rows around mid-body, 48–61 dorsal scales between limbs along vertebral line. Males have one discontinuous row of 7–12 precloacal pores. Tail slightly compressed, about twice as long as snout–vent length. Dorsal tail scales spiny and strongly keeled, ventral tail scales smooth and smaller. Maximum size about 36cm (snout–vent length up to 12.6cm), females smaller than males, hatchling size unknown. Colour and pattern vary greatly according to age, sex and breeding condition. Displaying males have a vivid yellow (populations from Eritrea, Ethiopia, Djibouti and extreme north-west Somaliland) to purple-red head (Sudan, Egypt), the throat is yellow or slightly orangey with an orange gular fold and pouch, and rest of underside bluish. Body cobalt-blue with a paler vertebral stripe of varying intensity. Non-displaying males resemble females; dorsum grey-brown with a pattern of darker brown or reddish blotches that may form transverse bars, flanks with white-centred brown ocelli. Juveniles similar, but head and neck have indistinct green speckling, or head is green with whitish speckling. Gravid females colourful; body yellow with five to six red bands with a whitish central spot, a broad irregular lateral stripe of red; head turquoise-blue, legs and tail brownish to grey.

HABITAT AND DISTRIBUTION: A wide-ranging rupicolous species. Inhabits dry mountainous or hilly areas in desert, semi-desert savanna and arid *Vachellia* scrub. In Egypt occurs south of 29°N in granitic and limestone mountain regions of the Eastern Desert at relatively high elevations (up to 2,187m), where it lives on boulders and rocky slopes overlooking wadis, or on large stones on wadi beds. In Sudan, only known from a corridor along the Red Sea coast. Records from around the Merowe Dam need verification. In Eritrea, *A. spinosa* occurs in warmer mountainous areas (e.g., Keren, Ghinda, Af Abed, Gahtielay). In Ethiopia widespread in Danakil region, where often seen on lava flows; known from Awash, Gewane and Asayita. Extends east to Djibouti, where found close to sea, and extreme north-west Somaliland, where replaced further inland by *A. somalica*.

CONSERVATION STATUS: Least Concern.

NATURAL HISTORY: A diurnal, colonial species, very tolerant of hot arid regions. Usually found on rocks, although specimens on buildings at Asayita allowed close approach. The dominant male basks in a prominent position, often right on top of a boulder. Gravid females seen in July–August in the Awash region and in extreme north-west Somaliland. Males in displaying colours were seen in the same area during the hot rainy season (May–September). Further north (mostly in Egypt) they hibernate in winter. Like other agamas, territorial and lays eggs (6–7). It eats insects (beetles, ants) and other arthropods, and in smaller quantities also plants and their fruits, for example *Capparis* sp.

Agama spinosa, gravid female, Gewane.

Agama spinosa, Erkowit.

Agama spinosa, male, Gewane.

Agama spinosa, Sinkat.

LARGE OR TREE AND ROCK AGAMAS *ACANTHOCERCUS*

Big agamas lacking an occipital scale. Males have large, heavy heads. Genetically close to Strange agamas (*Xenagama*), but in appearance they are most similar to typical agamas (*Agama*). They differ from the latter in that they have heterogeneous dorsal scalation, dorsolateral skin folds and tail scales arranged into clearly visible whorls, so the tail looks segmented. They are territorial and diurnal, most are rock-dwellers, but some are excellent tree climbers too. Males are usually larger and more vividly coloured than females. There are about 13 species, all in Africa and the Middle East; six occur in our area. The lineage including *Acanthocercus* and *Xenagama* diverged from other agamas approximately 21 Mya in Africa. Taxonomy not yet wholly clarified. Recent molecular studies suggest that, as presently defined, at least three closely related genera are concealed within *Acanthocercus*.

KEY TO MEMBERS OF THE GENUS *ACANTHOCERCUS*

1a: Flanks covered with uniformly small, smooth scales *Acanthocercus annectans*, Eritrean Rock Agama

1b: Flanks with some conspicuously enlarged, keeled and spinose scales in addition to smaller ones 2

2a: Region between dorsolateral folds has at least some clearly enlarged, keeled and spinose scales in addition to numerous smaller ones; no conspicuous pale vertebral stripe (sometimes a diffuse pale band, lacking sharp lateral margins) 3

2b: Region between dorsolateral folds has only small, more or less uniform scales; always a prominent pale vertebral stripe, clearly defined by sharp lateral margins 5

3a: Scales on anterior throat more or less distinctly and sharply keeled; vertebral scales rather homogenous, smooth or weakly keeled; mid-dorsal scales on back not sharply differentiated from those of tail; nostril directed posteriorly; head without swollen cheeks *Acanthocercus cyanogaster*, Blue-bellied Rock Agama

3b: Scales on anterior throat not distinctly and sharply keeled; vertebral scales heterogeneous, with strongly keeled enlarged scales; at base of tail, a more or less abrupt transition between smaller scales of back and larger caudal scutes; nostril directed laterally; head with swollen cheeks 4

4a: Large, males reaching 36cm (snout–vent length up to 13.8cm); enlarged dorsal body scales arranged in distinct rows along vertebral line, on flanks sometimes form up to five transverse rows; ventral scales in 77–92 longitudinal rows (from shoulders to cloaca) *Acanthocercus gregorii*, Gregory's Blue-headed Tree Agama

4b: Small, usually <29cm (snout–vent length 9.3–11.7cm); enlarged dorsal body scales not arranged in rows along vertebral line, more scattered; ventral scales in 71–75 longitudinal rows *Acanthocercus minutus*, Ethiopian Tree Agama

5a: Enlarged, keeled scales on flanks in distinct transverse rows, most closely compacted into a single restricted field; scales on postero-dorsal face of thigh essentially homogeneous, all of moderate size, keeled, spinose and regularly arranged; caudal scale rows (distance behind vent equivalent to maximum breadth of tail) 16–21 (mostly 18–19) *Acanthocercus phillipsii*, Phillips' Rock Agama

5b: Enlarged, keeled scales on flanks widely dispersed or in weak and isolated transverse rows; scales on postero-dorsal face of thigh clearly heterogeneous, with enlarged, keeled and spinose scales irregularly arranged and intermingled with numerous smaller ones; caudal scale rows 20–29 (mostly 22–26) *Acanthocercus guentherpetersi*, Peters' Rock Agama

ERITREAN ROCK AGAMA

Acanthocercus annectans
(Blanford, 1870)

IDENTIFICATION: A huge, elegant agama, with a very long thin tail, vividly coloured. Head triangular and large, neck narrow. Eye fairly large and there is a cluster of small spiky scales above and below ear-opening. Body slightly flattened, legs strong and muscular. Tail broad at base (especially in males), then thin, tapering smoothly, long, roughly 65–70 per cent of total length. Mid-dorsal scales weakly keeled, but lateral scales uniformly small and smooth, an excellent aid to identification. Tail scales not usually arranged in visible whorls, if so, these comprise four rings of scales. Males have a double row of 18–28 precloacal pores, females a single row of 8–10. Maximum size about 50cm (snout–vent length up to 15.2cm), average 25–40cm, hatchling but probably around 6–7cm. Displaying males have a blue-and-yellow head, the back is blue with yellow ocellate-type spotting, the tail base is yellow, shading to blue or turquoise, the limbs are blue. Throat yellowish, with a blue reticulated pattern, which occasionally extends to entire belly. Some larger males have a broad pale or bluish vertebral stripe, which tapers towards tail. Non-displaying males can be brown, pink or rufous, with a black neck. Females greenish-grey, with black speckling and pale dorsal crossbars; juveniles have diffuse brown crossbars on a grey background, and a black neck band (an excellent aid to identification).

HABITAT AND DISTRIBUTION: On hills and rock outcrops in dry rocky savanna and semi-desert from near sea level to about 2,100m, although in Daallo forest in Cal Madow Mountains of northern Somaliland it inhabits relatively mesic habitats of relict

Acanthocercus annectans, female, Daallo Forest.

Acanthocercus annectans, male, Cal Madow Mountains.

Juniperus forest at about 2,000m. Specimens found at *c.*1,300m along road from Mekelle to Dallol in Danakil (northern Ethiopia) live on very dry and hot rocky slopes. Occurs in several widely separated populations. Described from Eritrea in 1870, where it ranges south into northernmost Ethiopia. Another population occurs from Djibouti and far western Somaliland into Ethiopia, along the eastern escarpment of the western Rift Valley; where it is plentiful in the Dakhata Valley east of Harar (and sympatric with *A. guentherpetersi*), thence into dry hilly savanna between the Bale Mountains and Negele highlands, and south to extreme north-east Kenya. Another population occurs in the Sanaag, Sool and Nugaal regions of north-east Somalia; it is common in the Cal Madow Mountains, and was recently found at several localities in the low hilly country between there and Berbera, e.g., Laas Dhuure, Huluul and Daarbuduq. An isolated populations occurs at Jesomma in the Hiran of south-central Somalia. Absent from central Somalia. Elsewhere, in northern Kenya.

CONSERVATION STATUS: Least Concern.

NATURAL HISTORY: Poorly known. In northern Ethiopia diurnal, living on rocky hills, in loosely structured colonies. In Ethiopia and Somaliland often seen on old stone buildings, ruins or bridges. Fast and wary, running away at high speed if threatened. Dominant males bask on prominent rocks. Males in display colours observed in Somaliland between July and September, and in cryptic coloration in February. Females lay 8–12 eggs, 2 x 1.3cm. Diet mostly insects and other arthropods. At Buq, in Sanaag region of Somaliland, adult agamas preyed on Fruit Chafers *Pachnoda iskuulka* and cockroaches of the genus *Periplaneta*.

Acanthocercus annectans, male, Fadhi Gaab.

BLUE-BELLIED ROCK AGAMA

Acanthocercus cyanogaster
(Rüppell, 1835)

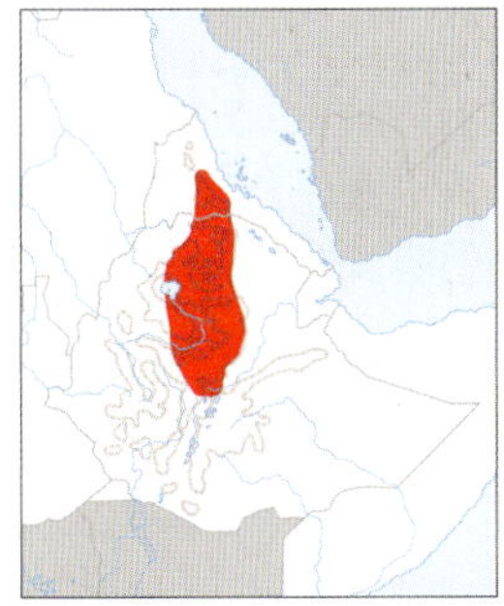

IDENTIFICATION: A fairly large, vividly coloured yellowish-blue agama with a flat head, found on rocks. Head triangular, wide and rather depressed; forehead concave in profile; snout rather long, obtusely pointed; cheeks not swollen, with several clusters of spines below and behind the earhole; neck narrow, short. Ear-opening clearly visible, large and equal to diameter of the large eye. Few enlarged scales around ear but a row of distinctly enlarged scales lie between it and posterior edge of eye. Body broad and flattened; legs strong and muscular. Tail broad at base, round, tapering smoothly, moderately long, roughly 60–65 per cent of total length. Mid-dorsal scales weakly keeled, more or less homogeneous in size, flat and enlarged, scales on flanks small but interspersed with some large ones; usually no sharp line of transition between scales on back and those on upper surface of tail. Tail scales arranged in obvious whorls, each comprising four (sometimes three) rings of scales. Males with 4–6 rows of up to ten precloacal pores. Maximum size about 30cm (snout–vent length up to 11.8cm), average 25–30cm, hatchling size unknown. Adult males vividly coloured, back dark grey with extensive yellow or cream marbling, limbs and tail deep turquoise-blue, head yellowish, snout totally or partially blue. Underside blue, throat blue or yellowish-cream with dark blue longitudinal lines. Females brown or brownish-grey, with slightly paler head and tail, latter with darker bands or spots, the enlarged flank scales may be paler yellowish, the underside pale yellowish; gravid females also have two conspicuous red spots on upper part of tail base.

HABITAT AND DISTRIBUTION: A widespread highland rock agama, known only from north-central

Acanthocercus cyanogaster, Asmara.

Acanthocercus cyanogaster, Kohaito Plateau.

Ethiopia and Eritrea, where it favours mountainous habitats in moist and dry savannas, montane woodland, grassland or near human settlements, at elevations of 1,400–2,700m. Extends from the Eritrean highlands, in the vicinity of Keren south through Asmara, Nefasit, Dekemhare, Mendefera, Kohaito Plateau and Senafe, to northern Ethiopia. Although described from Massawa on Eritrean coast, it does not seem to occur there; the types were probably collected at higher elevations inland. Widespread in Tigray and Amhara regions of north-central Ethiopia, as far south as upper Rift Valley, at Debre Zeit and Nazret; other well-known localities include lower slopes of Simien Mountains (Debarq), Bahar Dar, Gondar, Axum, Lalibela and vicinity of Addis Ababa. Unconfirmed records from Ahmar Mountains, east of Rift Valley, need confirmation. Historic records from southern Ethiopia (*A. annectans*, *A. gregorii*, *A. minutus*) are misidentifications, it does not occur there. Widespread and fairly common.

CONSERVATION STATUS: Least Concern.

NATURAL HISTORY: Usually on rocky outcrops, hillsides and large boulders but also stone buildings (it lives on the ancient castles of Gondar), walls, bridges and dry stone walls in farmland. A fairly shy diurnal species. Lives in loosely structured colonies with one male and several females and juveniles. Males bask on prominent rocks; females are active lower down the rocks. Feeds on arthropods. Females lay up to 15 eggs in damp soil in the ground. In Nazret this species was syntopic on the same rocks with the similar *A. minutus*.

GREGORY'S BLUE-HEADED TREE AGAMA

Acanthocercus gregorii
(Günther, 1894)

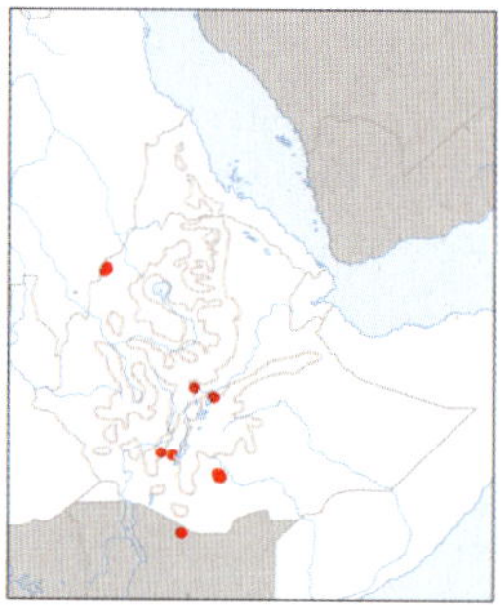

IDENTIFICATION: A big, stocky tree agama, males blue with a broad yellow vertebral stripe and tail base, and distinctly enlarged vertebral scales, mostly in central and southern Ethiopia. Head of adult male huge, triangular, broad and deep; forehead flat or convex in profile; snout short, rounded; distinctly swollen cheeks with several scattered spinose scales below and behind earhole; neck wide and short. Ear-opening sometimes partly concealed, and smaller than diameter of eye. Body squat, not flattened; legs strong and muscular. Tail broad at base, round, tapers smoothly, relatively thick and short, roughly 57–60 per cent of total length, or ⅓ to ½ longer than snout–vent length. Mid-dorsal scales heterogeneous in size, small and large, interspersed with strongly keeled or spiky scales; flank scales much smaller, with intermixed enlarged keeled or spiky scales, which may be arranged in transverse scale rows. Pale-coloured enlarged dorsal body scales are mostly arranged in distinct rows along vertebral line. Usually there is a sharp size transition between scales on back and those on upper surface of tail. Tail scales arranged in somewhat indistinct whorls comprising four (sometimes three) rings of scales. Males have 1–3 rows of precloacal pores, with a total number of 7–37; females lack pores. Maximum size up to 37cm (snout–vent length up to 13.8cm), females smaller, average 22–30cm, hatchling 8cm. Displaying males almost wholly turquoise-blue, except yellow basal portion of tail; back sometimes finely speckled greenish-yellow, with narrow rufous crossbars. Non-displaying males greyish-brown with pale greenish-yellow speckling on enlarged scales of back; throat pale, reticulated with blue. Females less bright, back grey-brown or grey, with dark crossbars (with pale centres) and pale speckling; gravid females have bright rufous dorsal crossbars. Taxonomic Notes: The formerly widespread *A. atricollis* was recently split into several species, two of which occur in Ethiopia, *A. gregorii* and *A. minutus*; they are very similar and difficult to distinguish.

HABITAT AND DISTRIBUTION: Moist and dry savanna and, at higher elevations, woodland or agricultural land with patches of trees, at elevations of 700–2,100m (mostly 1,200–1,600m). Recent work indicates a widespread but sporadic distribution; from Metemma on Sudan border, Akaki (near Addis Ababa), Wonji, south of Nazret, Arba Minch, Kamba and Negele. Probably more widespread, especially at lower elevations in southern Ethiopia. Elsewhere, south through Kenya to northern Mozambique. Widespread and common in southern Ethiopia.

CONSERVATION STATUS: Least Concern.

NATURAL HISTORY: Diurnal and arboreal, living on trunks of big trees, sometimes on rocky hills and termitaria. Around Negele, in southern Ethiopia, terrestrial, in grassland with scattered trees, and sheltered under rocks beside a dirt road. Strongly territorial, they live in structured colonies, controlled by a big, dominant male who challenges and fights intruders. In suburbia they will live on ornamental trees, walls, bridges, etc. They quickly become used to humans, males will continue to display even if approached, but in the bush they are more wary and will move surreptitiously to the far side of the trunk if approached, using one eye to peer over the barrier. If restrained or threatened, they display the orange interior of their mouth, and they bite with enough force to draw blood. Four to 15 eggs, roughly 1.5×2.5cm are laid. The mother may dig a hole in soft soil or utilise an

Acanthocercus gregorii, Kenya.

Acanthocercus gregorii, Kenya.

Acanthocercus gregorii, Tanzania.

existing one; in Negele a clutch was found buried in the soil under large flat stone. Diet a wide range of invertebrates and occasionally small vertebrates (rodents, birds, other reptiles). In the Ethiopian Rift Valley sympatric with *A. minutus*. Widely (but incorrectly) believed to be venomous and dangerous.

PETERS' ROCK AGAMA

Acanthocercus guentherpetersi
Largen & Spawls, 2006

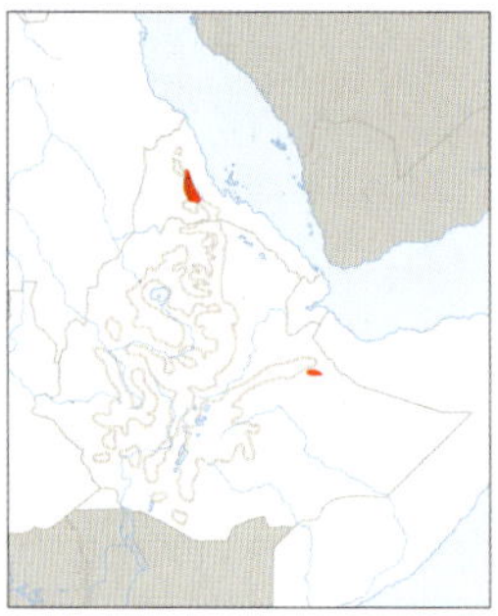

IDENTIFICATION: A medium-sized, flat agama, vividly coloured, on rocks in eastern Ethiopia and Eritrea. Head triangular, broad and flat; snout pointed; neck narrow. Eye large, with a cluster of small spiky scales around big ear-opening. Nostril displaced below *canthus rostralis*. Body strongly flattened, legs strong and muscular. Tail broad at base, tapering smoothly, roughly 60–63 per cent of total length. Mid-dorsal scales enlarged, smooth or very weakly keeled, but lateral scales smaller with widely scattered, enlarged, keeled scales. Scales on back of thighs large; keeled and spinose scales arranged irregularly and intermixed with numerous smaller ones, while base of tail has scales in 20–29 (mostly 22–26) rows. Tail scales arranged in whorls comprising three rings of scales. Males have a double row of 21–25 precloacal pores. Maximum size about 30cm (snout–vent length 7.7–10.2cm in males, 7.2–8.1cm in females), average 23–27cm, hatchling size unknown. Two striking colour phases in displaying males. In one, head, limbs and tail are orange; there is a stripe between the nostril and ear-opening, which has a light orange margin, the body is velvet-black, with a golden-yellowish vertebral stripe and somewhat darker flanks. The enlarged lateral scales are golden-yellowish. In the second, head yellowish-brown, sometimes with pale speckling; body and thighs brownish, with a pale vertebral stripe that is characteristically narrow and sharp-edged; calves, fore limbs and posterior tail bluish. Non-displaying males brown above, with a yellow vertebral stripe, paler on flanks, limbs and tail ochre brown. Underside pale; throat has a somewhat striated pattern. Females harlequin patterned. Head dark above with white speckling, and a poorly defined vertebral stripe. Body, limbs and tail have bold but messy whitish-grey and black (occasionally brown) markings. Pale flank spots may coalesce to form a bold stripe between limbs. Juveniles have similar coloration to females.

HABITAT AND DISTRIBUTION: Has a curiously disjunct distribution, but in similar habitats, in eastern Ethiopia and Eritrea, at 1,500–2,300m. Occupies granitic boulders and rocks in areas of dry *Vachellia* scrub in hill country. Widespread in Eritrea; localities

Acanthocercus guentherpetersi, Adi Teklezan.

Acanthocercus guentherpetersi, Dekheta Valley.

Acanthocercus guentherpetersi, female, Dekemhare.

include Afabet in the north, south along Anseba Valley, around Adi Teklezan and Dekemhare, and Saganeiti in the south. In eastern Ethiopia currently known from the mountains east of Harar towards Babile, and the Dakhata Valley ('Valley of Marvels'). No other records; presumably the huge gap between localities is due to inadequate fieldwork. At drier and lower elevations further east in Somaliland it is replaced by the very similar but distinct *A. phillipsii*.

CONSERVATION STATUS: Least Concern.

NATURAL HISTORY: Poorly known. Diurnal and strongly rupicolous, favouring large granitic or other boulders; not usually on ground. When threatened runs into crevices between boulders. Often abundant, but territorial, they live in structured colonies under a dominant male who basks in a prominent position; up to ten females observed with one male. No clutch details known. Gravid females and males in display colours were observed in November in Eritrea. In Dakhata Valley they live in sympatry with *A. annectans*; in Eritrea with *A. doriae*. Diet insects and other arthropods.

ETHIOPIAN TREE AGAMA

Acanthocercus minutus
(Klausewitz, 1957)

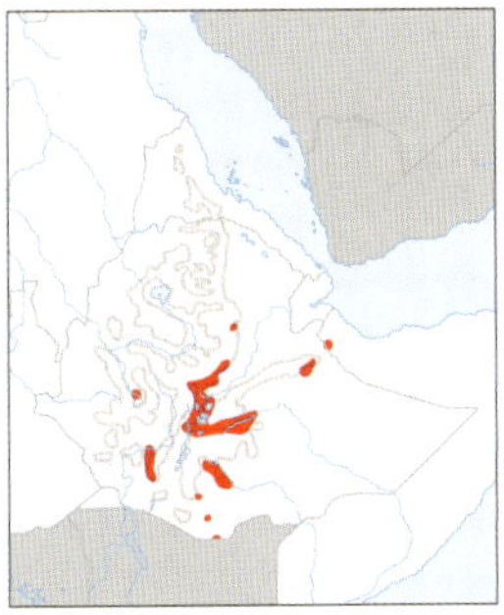

IDENTIFICATION: A big, stocky tree agama, males blue with a yellow tail base and a poorly developed vertebral band of enlarged scales, mostly in central and southern Ethiopia. Head of adult males huge, broad, triangular and relatively deep, the forehead flat or convex in profile; snout short, almost rounded; cheeks distinctly swollen with several scattered spinose scales; neck wide and short. Ear-opening sometimes partly concealed, smaller than diameter of eye. Body squat and slightly depressed; legs strong and muscular. Tail broad at base, round, tapers smoothly, it is relatively thick and short, roughly 55 per cent of total length. Mid-dorsal scales heterogeneous in size, composed of big and small, strongly keeled or spiky scales; flank scales much smaller, intermixed with enlarged keeled or spiky scales. Two rows of scales along spine slightly larger than flank scales, and bordered by a series of enlarged keeled scales; enlarged dorsal scales more evenly scattered than in the similar *A. gregorii*. There is usually a sharp line of transition between scales on back and those on upper surface of tail. Tail scales arranged in indistinct whorls of four rings of scales (three at base of tail). Males usually have two rows of 14–17 precloacal pores; females lack pores. Maximum size up to 29cm (snout–vent length up to 11.7cm), female smaller. Average 20–25cm, hatchling about 7cm. Displaying males wholly turquoise-blue, save a yellow tail base; back sometimes finely speckled greenish-yellow. Non-displaying males greyish-brown above with pale greenish-yellow speckling, throat pale, reticulated with blue. Females less vivid, back grey-brown or grey, with dark crossbars (pale centres) and pale speckling; gravid females have bright rufous dorsal crossbars. Juveniles brick-red or brown with diffuse darker crossbars on back and tail. Taxonomic Notes: Recently elevated from a subspecies of *A. atricollis* to species; in older literature often confused with the similar species *A. cyanogaster* and *A. annectans*.

Acanthocercus minutus, male, Lake Langano.

HABITAT AND DISTRIBUTION: Moist and dry savanna, woodland, forest clearings, agricultural land with patches of trees and grassland, at elevations of 1,500–3,000m. Near-endemic to Ethiopia, a single western

Somaliland record near Ethiopian border, in hills north of Quljeet (and a curious record from central Kenya). In Ethiopia widespread in and around Rift Valley, occurs along road from Addis Ababa to Nazret and south via the lakes, thence east to Bale; at Dinsho, Dodolla and Sheikh Hussein. Sporadically recorded north along eastern slope of the western plateau, through Ciaffedenza (type locality) and Let Marefia north to Batie, east of Dessie. Apparently isolated populations around Negele, Yabello, Mega, El Sod and Moyale, and west of the Rift at Key Afer, Jinka, Felenguai and Omo Valley as far north as Didessa. Elsewhere; sporadic records from Kenya. Widespread and common, tolerant of human activities.

CONSERVATION STATUS: Least Concern.

NATURAL HISTORY: Unlike members of *A. atricollis* complex, this species appears to be partly terrestrial, particularly in savanna and grassland at lower elevations in southern Ethiopia. Seen on houses and surrounding walls further north, between Lake Awassa and Modjo; at Nazret seen on rock outcrops in fields, together with *A. cyanogaster*. Shelters in holes, beneath rocks or in their crevices. Several were found on large old trees (mostly *Vachellia*) or termite hills. Strongly territorial, they live in structured colonies, controlled by a big, dominant male, who perches prominently (rocks, high walls, termitaria) and observes his territory, usually with one or more females in close attendance. Eats a wide range of invertebrates, occasionally small vertebrates. Females lay 10–25 (usually fewer than 18) soft-shelled eggs, approximately 1.5×2cm in size.

Acanthocercus minutus, female, Lake Langano.

PHILLIPS' ROCK AGAMA

Acanthocercus phillipsii
(Boulenger, 1895)

IDENTIFICATION: A flat, rupicolous agama with a pale vertebral stripe, from Somaliland and eastern Ethiopia. Head triangular, broad and depressed; snout pointed; neck narrow. Eye large, with a cluster of small spiky scales around the big ear-opening. Nostril displaced on the *canthus rostralis*. Body strongly flattened, legs strong and muscular. Tail broad at base, tapering smoothly, roughly 60–64 per cent of total length. Mid-dorsal scales enlarged, smooth or very weakly keeled, while enlarged and keeled scales on flanks are usually in distinct transverse rows or mostly packed into a single small area (an excellent aid to identification). Scales on back of thighs all approximately the same size and arranged regularly, whilst base of tail has scales in 16–21 (mostly 18–19) rows. Tail scales arranged in whorls comprising three rings of scales. Males have a double row of 18 precloacal pores. Maximum size about 25cm (snout–vent length 6.2–8.5cm in males, 5.5–7.0cm in females), average 18–24cm, hatchling size unknown. Rather similar to *A. guentherpetersi* but less vividly coloured. Head, limbs and tail olive above. A white stripe connects the nostril and ear-opening, which is yellow-edged. Eyelids greyish. Body blackish or brown, with small greenish-white or yellow spots which sometimes coalescence to form a sharp-edged dorsolateral and vertebral stripe, which reaches the tail base. White below, either unmarked or has a network of dark pigments, and an orange throat patch. Interestingly, displaying males, non-displaying males and females are all similarly coloured; but juveniles have a more striped appearance and orange tail base.

HABITAT AND DISTRIBUTION: A north Somali near-endemic, in dry *Vachellia* scrub in hill country, at

Acanthocercus phillipsii, Ferdusa Ruins.

Acanthocercus phillipsii, Arabsiyo.

altitudes of 1,200–1,800m. Unlike *A. guentherpetersi*, inhabits more arid and warm habitats at lower elevations. Most records from central and western Somaliland, ranging from Sheikh (where abundant at Ferdusa Ruins) west across the Goolis Mountains, Gan Libah and Ogo Mountains, north and west of Hargeisa, to Borama in western Somaliland. An isolated but dense population occurs in eastern Ethiopia at Dire Dawa. There is a single historic but imprecise Eritrean record (the holotype of *Agama trachypleura*, not shown on map), and it probably does not occur there. One of the commonest agamas of Somaliland.

NATURAL HISTORY: Poorly known. Diurnal and strongly rupicolous. Favours granite boulders or other large, usually limestone, rocks. Unlike *A. guentherpetersi* often found on lower rocky outcrops. When threatened they retreat into rock crevices. Often abundant, but territorial, living in structured colonies under a dominant male who basks in a prominent position with a raised tail. No clutch details known. Gravid females observed in June in Somaliland. Sympatric virtually throughout range with *Agama somalica*, with males living side by side on the same boulder. Diet insects and other arthropods.

CONSERVATION STATUS: Least Concern.

STARRED OR ROUGH-TAILED AGAMA *STELLAGAMA*

This formerly monotypic genus was recently split into three; all large rock-dwelling agamas of the eastern Mediterranean. They have been placed in several genera and the debate continues, the present move into *Stellagama*, from the genus *Laudakia*, is not accepted by some taxonomists.

STARRED AGAMA, ROUGH-TAILED AGAMA

Stellagama vulgaris
(Sonnini & Latreile, 1803)

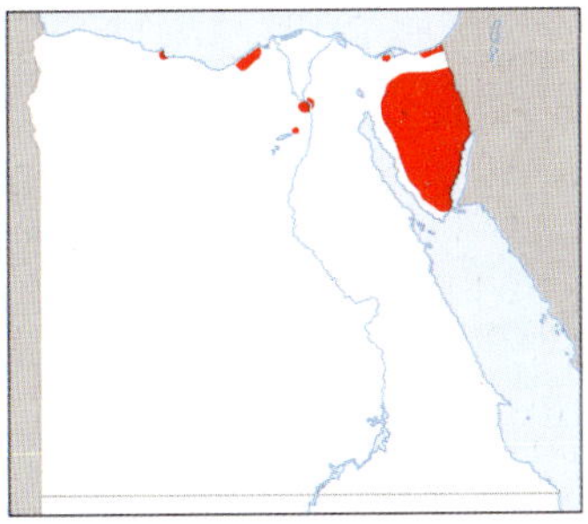

IDENTIFICATION: Head triangular, wide and moderately depressed; nasal scale round or slightly pear-shaped, nostril displaced below *canthus rostralis*, occipital scale not enlarged; ear-opening large, equal to eye, exposed, with small spinose rosettes behind and on sides of neck. Neck crest absent. Body relatively short, wide and fairly depressed, with a distinct dorsolateral fold. Limbs strong and fairly long. Tail round, depressed at base, tapering smoothly, about 55–60 per cent of total length. Mid-dorsal scales heterogeneous, irregular and keeled; flanks have much smaller scales interspersed by regular bands of spiny rosettes; ventral scales smooth. Tail scales arranged into distinct circular segments, each consisting of two whorls of enlarged mucronate scales. Males with precloacal patch of actively secreting scales (up to 60) and callous glands, consisting of 3–5 rows of modified scales along central line on belly. Maximum size about 30cm (maximum snout–vent length 14cm in males, 14.6cm in females), average 20–25cm, hatchling about 6–7cm. Two subspecies in our area, with greatly different colouration. In *S. v. brachydactyla* (Sinai) dorsal colour is olive-grey, sometimes even bluish, back with about 4–6 yellow or orangey transverse bands. Tail conspicuously banded yellow and black. Venter greyish; throat usually dark. Displaying males show greater contrast, being blackish with orange bands. The less colourful *S. v. vulgaris* (Egyptian coast) is dull pale olive-grey, with four whitish dorsal patches, tail bars less conspicuous, throat spotted. Taxonomic Notes: Several subspecies but their status is debated. Some authors make the distinction between more widespread subspecies *brachydactyla* and the recently described *salehi* from southern Sinai.

HABITAT AND DISTRIBUTION: Dry, near-desertic, rocky, well-vegetated areas. Usually inhabits hilly or mountainous country, from near sea level to 2,640m. Mostly found on vertical rock outcrops, cliffs, rocky hills, group of large stones and banks of dry wadis, buildings and walls, sometimes on old trees. The better-known subspecies *S. v. brachydactyla* occurs in southern and central Sinai, where found at higher elevations around St. Katherine, Wadi Feiran and as far north as El Quseima, Gebel Magara and adjacent mountains; elsewhere north and east to Jordan and Saudi Arabia. *S. v. vulgaris* is known from the coastal zone of northern Egypt, from Alexandria west to Burg El Arab, with an isolated population at Ras El Hekma. Introduced to Cairo and Kom Oshim but probably now extinct. Reported between Al Arish and Rafah region, extending about 15km inland, also at Bir Al-Abd, 35km west of Al Arish.

CONSERVATION STATUS: Least Concern, but collected for the pet trade, and in places its habitat is under threat.

NATURAL HISTORY: A typical rock-dwelling, rather shy, agamid with a diurnal, heliophilous lifestyle. The Sinai subspecies inhabits mountainous or hilly country at higher elevations (above 500m) and at lower altitudes is mostly found in or near oases. The Mediterranean coastal zone form inhabits coastal ridges and cliffs, as well as ruins and buildings; often in gardens and orchards with large trees, on garden walls and on stony hillsides at low elevation. It is excellent climber, often basking in hottest part of day and males usually overlook their territory from a prominent place on a rock. Both sexes maintain territories. Feeds mostly on invertebrates (especially beetles), but small vertebrates (other lizards, rodents, birds) and occasionally flowers and other plant material are also taken. Several clutches of usually 7–9 (range 4–20) eggs, roughly 1.4×2.2cm, laid per year, hatch 1–2 months after oviposition. Newly hatched juveniles observed in August. In winter they hibernate in holes and crevices.

Stellagama vulgaris, Sinai.

UNIFORM-SCALED AGAMAS *PSEUDOTRAPELUS*

A largely Arabian genus of medium-sized rock-loving agamas, with large triangular heads and long thin limbs, exposed ear-openings, no enlarged occipital scale; and a narrow whip-like unsegmented tail. Body scales feebly keeled, small and uniform. No dorsolateral fold. Males have a single row of 4–8 large, rounded precloacal pores. Displaying males are noticeably vivid turquoise-blue, other animals cryptically coloured with 3–4 distinct bands on dorsum. Usually in dry rocky hill country, from Eritrea and Sudan to Sinai, southern Syria and Iraq, south to Arabia. They feed on arthropods and occasionally on plants. Six species are known, four of these described in the last ten years, but debate on their taxonomy continues; three occur in our area. Closely related to the paraphyletic *Acanthocercus* and short-tailed *Xenagama*, but split from nearest Arabian *Acanthocercus* around 15.9 Mya.

KEY TO NORTH-EAST AFRICAN MEMBERS OF THE GENUS *PSEUDOTRAPELUS*

1a: Third toe equal in length to fourth; Sinai	*Pseudotrapelus sinaitus*, Sinai Agama
1b: Third toe much longer than fourth; Eritrea and Sudan to Sinai	2
2a: Males with four well-separated precloacal pores, 13 lamellae under fourth toe, non-displaying specimens with brownish-grey body and fairly indistinct speckles; Sinai	*Pseudotrapelus aqabensis*, Aqaba Agama
2b: Males have six continuous precloacal pores, 16–17 lamellae under fourth toe, non-displaying specimens have distinct pinkish-red body and fairly large whitish ocelli, widespread in Sudan, Egypt and Eritrea, also south Sinai	*Pseudotrapelus chlodnickii*, Chłodnicki's Agama

AQABA AGAMA

Pseudotrapelus aqabensis
Melnikov, Nazarov, Ananjeva & Disi, 2012

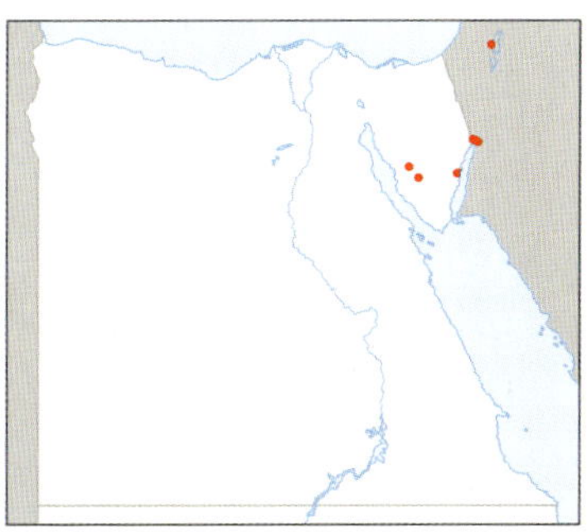

IDENTIFICATION: A medium-sized, slender agama with big head, fine, almost uniform scalation, in Sinai. Snout short, pointed; nostrils on *canthus rostralis* or slightly above; occipital scale not enlarged; ear-opening large, slightly smaller than eye, exposed. No nuchal crest. Body slender, moderately long and slightly depressed. Third toe much longer than fourth; 13 lamellae under fourth toe. Tail almost round or slightly laterally compressed at base, then more laterally compressed towards tip, tapering smoothly, long, about 65–70 per cent of total length. Vertebral dorsal scales slightly larger than scales on flanks, all relatively small, slightly overlapping and keeled; ventral scales smooth; tail scales equal or about two times larger than dorsal body scales, keeled and not arranged in distinct circular segments. Males have four well-separated precloacal scales. Maximum size about 26cm (maximum snout–vent length 10.4cm), average 20–25cm, hatchling size unknown. Head, throat, neck and fore limbs of displaying males bright turquoise-blue; back brownish to yellowish-grey without ocelli or only finely speckled; a pale yellowish stripe extends down middle of back. Females greyish-brown with a bluish-grey head and several brick-red bands on back. Males and females in non-displaying colours or stressed animals may be brownish-grey with faint whitish spots and a yellowish or brownish head, neck and shoulders. Three dorsal body transverse bands are present; tail has about 20 dark bands. Gravid females more vividly coloured, with distinctly reddish body bands. Taxonomic Notes: Recently described, elevated from within *P. sinaitus*.

HABITAT AND DISTRIBUTION: Rocky desert and mountainous or hilly country, from near sea level to about 2,200m. On flat coastal plains often found on rocky outcrops or scattered boulders. An Arabian species, just reaching central and southern Sinai, confirmed at Wadi Balah near Barakat, about 11km north-east of Feiran and Nuweiba, provisionally known from coastal hills between Nuweiba and Taba and slightly north of Dahab (not mapped), and Aqaba. *Pseudotrapelus* in Sinai are widespread and common, but split of *P. sinaitus* means old records need reassessing. Elsewhere, north to extreme south-west Jordan and western Saudi Arabia.

CONSERVATION STATUS: Not Assessed.

NATURAL HISTORY: Very similar to others in genus, i.e. strictly diurnal, heliophilous, rupicolous, active, and diet consists of arthropods and occasionally plants. Both sexes are largely solitary but hold territory perched on tops of boulders of rocks, even in full sunlight and hottest hours of day. Five to six relatively large soft-shelled eggs are laid. Widespread and sympatric or narrowly parapatric with *P. chlodnickii* and *P. sinaitus*.

Pseudotrapelus aqabensis, Gulf of Aqaba.

CHŁODNICKI'S AGAMA

Pseudotrapelus chlodnickii
Melnikov, Śmiełowski, Melnikova, Nazarov & Ananjeva, 2015

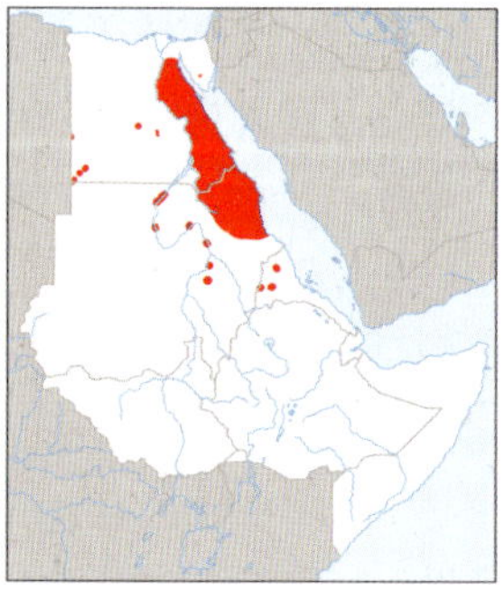

IDENTIFICATION: The only *Pseudotrapelus* in mainland Africa (excluding Sinai), with a very distinct pinkish-red coloration. Body moderately long and depressed, relatively slender. Third toe longer than fourth; 16–17 lamellae under fourth toe. Tail round at base, then laterally compressed, tapering smoothly, long, about 62–70 per cent of total length. Dorsal scales uniformly sized (several rows of vertebrals slightly enlarged), relatively small, slightly overlapping and keeled; ventral scales smooth or feebly keeled; tail scales about three times larger than dorsal body scales, keeled and not arranged into distinct circular segments. Males have six equally large, sometimes four larger and two outermost smaller, continuous precloacal scales. Maximum size about 35cm (maximum snout–vent length 11.5–11.7cm in males, 9cm in females), average 20–30cm, hatchling size unknown. Head, throat, neck and fore limbs bright turquoise-blue in displaying males. Back brownish-grey with large white ocelli, up to eight scales wide. A whitish to yellow-brownish vertebral stripe is present. Females grey with a bluish head and several brick-red bands on back. Young males, females in non-displaying colours or stressed animals may be pale chocolate-red with large white ocelli and 3–4 dark bands on back, or pale greyish-pink with ocelli and no dorsal body bands. Tail has about 25 dark bands. Juveniles grey with brown dorsal bands. Taxonomic Notes: Recently described, formerly known in Egypt, Sudan and Eritrea as *P. sinaitus*.

HABITAT AND DISTRIBUTION: Desert and semi-desert, rocky areas in mountainous and hilly regions, also sandy or stony plains with rocky outcrops, from near sea level to about 1,200m. Often found on boulders in wadis and hillsides. Occurs along Red Sea coast and inland from Ain Sukhna to south of Port Sudan. Sporadic records elsewhere: in Eritrea (Sabderat, Zara Gold and Bisha Mine areas), northern Sudan (near Wadi Halfa, Dongola, Merowe and Abu Hamad, between Atbara and Shendi, El Muswarat), Egypt (Kharga and Dakhla Oases, Jabal Awaynat, Wadi El Firaq and Wadi Abd El Malek on Gilf El Kebir plateau) and Sinai (Wadi Feiran). Elsewhere, to south-east Libya. Widespread but uncommon, harvested in Egypt for the pet trade.

CONSERVATION STATUS: Least Concern.

NATURAL HISTORY: Diurnal and heliothermic. Those

Pseudotrapelus chlodnickii, male, Meroe.

Pseudotrapelus chlodnickii, Akasha.

from Nubian and Bayuda deserts of northern Sudan occur in more sandy habitats than animals further north. Often basks during hottest part of day. A Sudanese animal was found sleeping on open sand at night, at Meroe, south of Atbara. Males are territorial but only during breeding season, perching on highest point of their territory. Diet arthropods, sometimes plants, ants were found in faeces of Sudan animals. Eggs laid in deep burrows in damp soil between rocks; a Dongola female had nine eggs, about 1.5cm in diameter, in her oviducts. As usual in agamid lizards, gravid females with distended abdomens are very colourful. Probably hibernates in colder northern deserts.

SINAI AGAMA

Pseudotrapelus sinaitus
(Heyden, 1827)

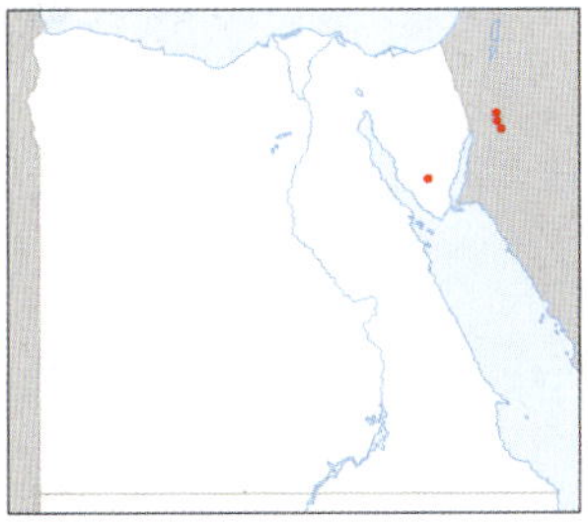

IDENTIFICATION: A medium-sized, slender agama with a large head, only in southern Sinai. Body scales small and roughly uniform; has four well-separated precloacal pores. Head large and heart-shaped, slightly depressed; snout short, pointed; nostrils on *canthus rostralis*; occipital scale not enlarged; ear-opening large, equal or slightly smaller than eye, exposed. Third toe equal in length to fourth, and 15–17 lamellae under fourth toe. Tail almost round at base, then sub-cylindrical towards tip, tapering smoothly, long, about 65–75 per cent of total length. Vertebral dorsal scales slightly larger than scales on flanks, all relatively small, slightly overlapping and feebly keeled; ventral scales smooth; tail scales about two times larger than dorsal body scales, keeled and not arranged into distinct circular segments. Males have six continuous precloacal pores (sometimes eight, the extra two small). Maximum size about 30cm (maximum snout–vent length 92mm in males, 79mm in females), average 20–25cm, hatchling size unknown. Displaying males may be completely turquoise-blue (distal tail sometimes yellow or has a diffuse yellow vertebral stripe). Females brown, head greenish, yellow or brown with several brick-red bands on back, the tail is faintly banded. Males and females in non-displaying colours

Pseudotrapelus sinaitus, male, Jordan.

Pseudotrapelus sinaitus, female, Jordan.

or stressed animals are less colourful, usually in shades of grey-brown. Taxonomic Notes: Following revisionary splits, true *P. sinaitus* now has a much smaller range. Two subspecies; in our area nominate *P. s. sinaitus* occurs, with *P. s. werneri* in the flat basalt deserts of northern Jordan and southern Syria.

HABITAT AND DISTRIBUTION: Open rocky desert areas, preferring places with herbaceous and other low vegetation, in our area at about 1,640m. A single genetically confirmed population in Sinai, at St. Katherine's Monastery at base of Mt Sinai; (but might be more widespread); old records from Sinai are other species.

CONSERVATION STATUS: Not Assessed in Egypt.

NATURAL HISTORY: Similar to both preceding species. A diurnal, active, rupicolous lizard, they bask in prominent positions on rocks or boulders, and are territorial. When disturbed, they flee to a refuge; sleep in stony crevices or under rocks. Five to six eggs are laid in a single clutch. Diet invertebrates and plant material.

GROUND OR DESERT AGAMAS *TRAPELUS*

A mainly Palearctic genus of small, slim, pale-coloured, solitary agamas, with thin tails. They live on the ground, occasionally rocks. Some 13 species are known at present, across North Africa, the Middle East, and east to western China. Two species in our area, but there is debate about the true status of subspecies *T. mutabilis pallidus*, see relevant taxonomic notes.

KEY TO NORTH-EAST AFRICAN MEMBERS OF THE GENUS *TRAPELUS*

1a: Ventral scales keeled, dorsal scales more or less uniform — *Trapelus savignii* Sinai Agama

1b: Ventrals smooth, dorsal scales heterogeneous, with many enlarged spiky scales — *Trapelus mutabilis*, Changeable Agama

SAVIGNY'S AGAMA

Trapelus savignii
(Duméril & Bibron, 1837)

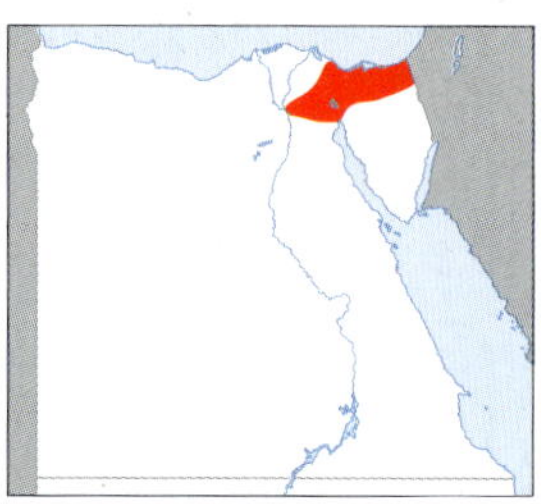

Trapelus savignii, male, Israel/Palestine.

IDENTIFICATION: A pallid, thin-limbed ground-dwelling agama, which reaches eastern Egypt. Head fairly broad, neck thin. Body slightly flattened. Tail thin and tapering, about 55–60 per cent of total length. Dorsal scales fairly uniform (occasional larger scales present), keeled, scales at mid-body in 77–95 rows. Maximum size about 25cm, average 15–21cm, hatchling size unknown. Precloacal pores absent. General impression is of flattish agama. Male brown or grey, sometimes slightly rufous with poorly defined light crossbars, and in breeding season has a distinctive blue throat. Female shades of brown, with a series of paired rufous-brown sub-rectangular blotches on spine; when gravid these are brick-red. Underside white.

HABITAT AND DISTRIBUTION: Desert and near-desert, on sandy and gravel plains, largely below 200m. In our area, confined to north-east Egypt, east of the

Trapelus savignii, female, Israel/Palestine.

Nile, from the Khanka Dunes east to northern Sinai. Elsewhere, to Israel/Palestine.

CONSERVATION STATUS: Vulnerable.

NATURAL HISTORY: Diurnal and terrestrial, usually in areas with some vegetation, not on rocky ground, and they often climb onto bushes to bask, watch for prey and monitor their territory. A clutch of 1–5 eggs is laid in a hole dug by female, and hatches a couple of months later; hatchlings reported to appear in August and October. Diet insects, other arthropods and plant material.

CHANGEABLE AGAMA

Trapelus mutabilis
(Merrem, 1820)

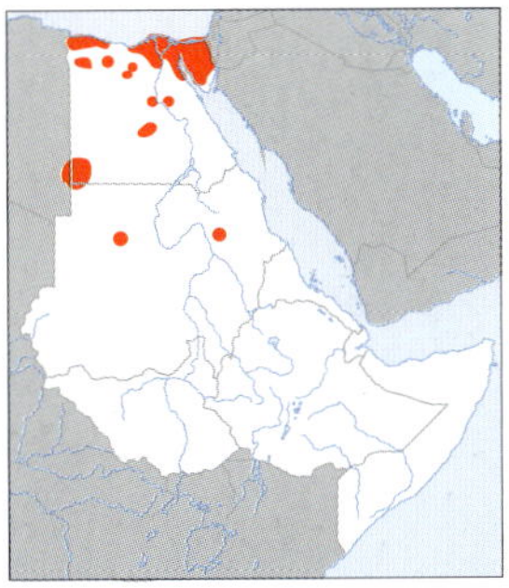

IDENTIFICATION: A pallid, thin-limbed, ground-dwelling agama with a spiky back, and usually distinct crossbars. Head fairly broad, neck thin. Body slightly flattened. Tail thin and tapering, about 50–55 per cent of total length. Dorsal scales heterogeneous, smooth but intermixed with larger spiky scales. Scales at mid-body in 72–118 rows (usually 91–118 in Egypt). Maximum size about 19cm, (snout–vent length to 9cm), average 15–21cm, hatchling size unknown. Two rows of 21 precloacal pores. Colour various shades of brown, grey or yellowish, sometimes slightly rufous, usually with irregular dark crossbars; these may be absent but tail almost always has clear dark narrow bands. In breeding season male has a vermiculated blue throat, the female may develop a rufous head. Underside white. Taxonomic Notes: For a long time, this species and *T. pallidus* were treated as separate species, in different habitats, in Egypt *T. pallidus* was found east of the Nile, *T. mutabilis* to the west. Recent work indicates that *T. pallidus* is a subspecies; *T. m. pallidus*, but this is not accepted by all. *T. m. pallidus* has 105–118 mid-body scale rows, and scalation of the hindleg and tail base is heterogeneous, but dorsal scalation largely heterogeneous; *T. m. mutabilis* has 91–97 mid-body

Trapelus mutabilis, male throat colours, east of Cairo.

Trapelus mutabilis, Libya.

scale rows, scalation of hindleg and tail base is homogenous, dorsal scalation largely homogeneous.

HABITAT AND DISTRIBUTION: Desert and near-desert, on sandy and gravel plains, wadis, rock pavement, rocky hills and plateaux, from sea level to above 1,000m. Usually in areas with some vegetation. Widespread across most of northern Egypt, north of 27°N, a couple of records from south-west Egypt (Gilf El Kebir, Jabal Awaynat), a handful of records in northern Sudan (Rahib Wells, Darfur, just on Sudan side of border at Jabal Awaynat, Bayuda, north-west of Atbara), and a curious old record from Djibouti. Elsewhere, west across the Sahara to Western Sahara, east (as *T. m. pallidus*) to Iraq and Arabia.

CONSERVATION STATUS: Least Concern.

NATURAL HISTORY: Diurnal and terrestrial, often perches on vegetation, rocks or a piece of elevated ground to bask, watch for prey and monitor its territory. Heat-tolerant and active even on very hot days. When approached they may initially remain motionless, or run, tail elevated, and then freeze, flat on the ground, hoping to avoid detection. A clutch of one to twelve eggs is laid in an excavated hole, more than one clutch may be laid per year. Hatchlings seen from July onwards. Diet arthropods, particularly ants.

Trapelus mutabilis, desert, east of Cairo.

STRANGE AGAMAS *XENAGAMA*

Four medium-sized species are included in this spectacular but rather poorly known genus, three of them strictly ground-dwelling and confined to a limited arid region of northern Somalia (Somaliland and Puntland) and immediately adjacent Ethiopia. The fourth species is endemic to the Ethiopian highlands. They are readily distinguished from all other agamids in our area by the extraordinary structure of their tails. Their closest relatives are in the genus *Acanthocercus*, and these two genera diverged in 6–7 Mya.

KEY TO NORTH-EAST AFRICAN MEMBERS OF THE GENUS *XENAGAMA*

1a: Tail much shorter than head and body	2
1b: Tail longer than head and body; only in highlands of Ethiopia	*Xenagama zonura*, Ethiopian Strange Agama
2a: Base of tail discoidal, longer than broad, merging more or less gradually into terminal filament; marginal spines not larger than those on upper tail surface, all upper scales on broadened part of tail have strongly developed spines	3
2b: Base of tail discoidal, usually broader than long, abruptly differentiated from terminal filament with large marginal spines; about half of all upper scales on broadened part of tail obtusely keeled, without spines; confined to soft sandy areas of west-central Somaliland	*Xenagama taylori*, Taylor's Strange Agama
3a: Tail relatively long, basal disc usually feebly developed, merging gradually into filament; spines on dorsum of tail base oblique; on underside four scales between last two large marginal spines before filament is attached; dry grasslands of Ethiopia/Somaliland border	*Xenagama wilmsi*, Wilm's Strange Agama
3b: Tail distinctly divided into a basal disc and distal filament, disc merging more or less gradually into filament; tail spines on dorsum of broadened tail base directed dorsally; on underside only two scales between last two large marginal spines before filament is attached; only on semi-desert plains of east-central Somaliland and Puntland	*Xenagama batillifera*, Vaillant's Strange Agama

VAILLANT'S STRANGE AGAMA

Xenagama batillifera
(Vaillant, 1882)

IDENTIFICATION: A relatively large ground-dwelling agama with a broad, spiky tail, endemic to gypseous regions of northern Somalia. Head fairly flat and wide, with a distinctly depressed body, short but powerful limbs and remarkable tail which is expanded at its base into a flattened disc (not as strongly developed as *X. taylori*), as wide as long or little longer, more or less gradually merging into terminal filament. Scales at the back of the disc just about equal in size and possess spines directed dorsally (unlike *X. taylori*). Terminal filament noticeably thick, with apparently enlarged marginal spines at its start. Body scalation heterogeneous to nearly homogeneous, with smooth matrix scales, intermixed with slightly larger, smooth or very feebly keeled scales. The largest of the genus, up to 14.5cm total length (snout–vent length 7.8–10.1cm). Dorsum cream to brown with rather obscure darker mottling, younger animals have four distinct transverse bands with paler centres. Underside whitish. In adult males the chin, throat, shoulders and chest become brilliant blue. Mature females have irregular blue lines on chin and throat. Young animals have a dark stripe between eye and occiput. Taxonomic Notes: Prior to 2013, this species was conflated with the recently described *X. wilmsi*. Also likely to be confused with *X. taylori*.

HABITAT AND DISTRIBUTION: Widespread and common in central and eastern Somaliland, from the Sogsoda Plains south of the Wagar Range across the Sanaag region as far east as Bender-Merayo in Puntland; at 800–1,800m. Not known from lower altitudes in Sool region south of Sanaag. Replaced by *X. taylori* further south and west. Found in flat semi-desert and savanna, on sandy soils in the west, gypseous or clay soils further east. Common south of Erigavo in Sanaag (one of coldest places in Somalia). Of all the *Xenagama*, this lizard (with *X. taylori*) favours the driest habitats.

CONSERVATION STATUS: Least Concern.

NATURAL HISTORY: Diurnal and solitary outside breeding season. On sandy soil they live in self-dug burrows about 80–120cm in length, 50–80cm deep and with a mouth diameter of 3.5–5cm. Burrows of adults longer and deeper than those of juveniles or subadults and always end in humid soil. Usually, burrows spaced about 1.5m apart (on stony plains,

Xenagama batillifera, adult male, Erigavo.

Xenagama batillifera, juvenile, Yuffleh.

Xenagama batillifera, female, Erigavo.

however, they shelter in short holes under rocks). They use their spiky tails to block their burrows against predators (as do *X. taylori* and *X. wilmsi*). A shy animal, their observed flight distance is about 10m. In June, gravid females were found near Sheikh, and the estimated population density there was around 20 animals in a 50×50m square.

TAYLOR'S STRANGE AGAMA

Xenagama taylori
(Parker, 1935)

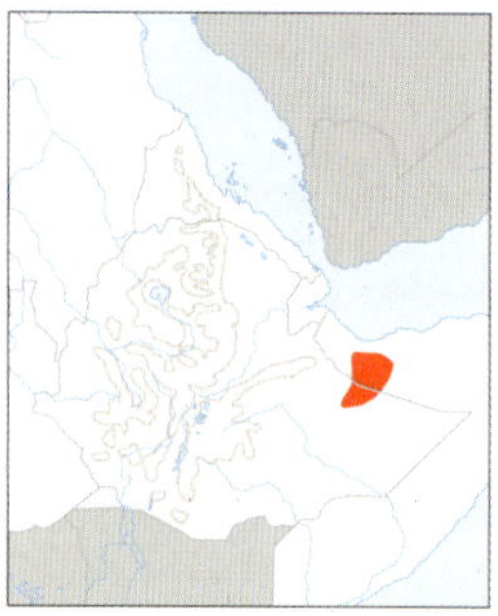

IDENTIFICATION: The smallest *Xenagama*, with a discoidal tail, of sandy areas in north-west Somaliland and adjacent Ethiopia. Head fairly flat and wide, body distinctly depressed, short but powerful limbs and a remarkable tail expanded at its base into a flattened disc that is usually broader than long, has enormously enlarged spines at its margins and is abruptly differentiated from a slender terminal filament. Scales at back of disc all more or less equal in size, smooth or feebly keeled (unlike *X. batillifera*). Terminal filament whip-like and lacks enlarged marginal spines at its base. Body scalation heterogeneous to nearly homogeneous, with smooth dorsal scales intermixed with slightly larger, smooth or very feebly keeled scales. Maximum total length about 11.5cm (snout–vent length 6.8–8.2cm) in males, 10.5cm (snout–vent length 6.1–7.8cm) in females. Newly hatched juveniles measured 3.9–4.3cm in total (snout–vent length 2.5–2.7cm). Dorsum reddish with rather obscure darker mottling and underside whitish, but in adult males the chin, throat, shoulders and chest are brilliant blue. In mature females the blue is confined to some irregular lines on chin and throat. In juveniles, the limbs and marginal enlarged spines on the broadened tail are whitish.

Xenagama taylori, juvenile, Dhubaato.

Xenagama taylori, male, Dhubaato.

Xenagama taylori, Hargeisa.

HABITAT AND DISTRIBUTION: Semi-desert, dry savanna and woodland, with sparse vegetation, at 1,050–1,500m. Described in 1930s from border country between Ethiopia and northern Somalia (Somaliland), more recently recorded in flat semi-desert around Degeh Bur and Aware in Ethiopia. Common and widespread in Somaliland on open sandy plains around Hargeisa, north to Gaan Libaah and in the east to Burao, south to the Haud. In Somaliland usually found on sandy (not clay) soils without rocks or grass, in semi-arid savanna with trees (*Vachellia*, *Commiphora*), small bushes (about 20–30cm high) and some shrubs (typically *Aloe megalacantha*, *A. rigens*).

CONSERVATION STATUS: Least Concern, but is harvested for the international pet trade and thus needs reviewing.

NATURAL HISTORY: Strictly terrestrial and diurnal, shelters in self-dug burrows in sandy soil, into which it retreats head-first if disturbed. The broad and extraordinary spiny tail serves as an effective deterrent to any predator that might be tempted to follow this lizard into its lair. Burrows are 80–120cm long, descend at an angle of c.30° and are about 30–75cm deep. The mouth of these burrows is 3–5cm in diameter and usually sited under a small bush or shrub. A very shy species, with a minimal flight distance of about 10m, and is quick to hide before an observer gets near. Most active in the hot season, between April and September. Captive females laid 8–9 eggs. Diet invertebrates.

WILMS' STRANGE AGAMA

Xenagama wilmsi
Wagner, Mazuch & Bauer, 2013

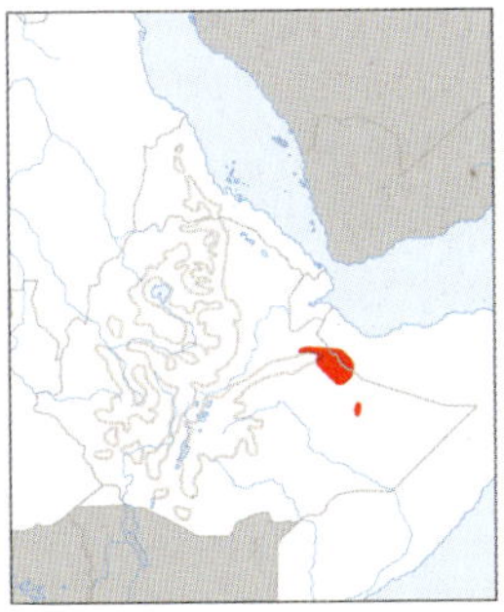

IDENTIFICATION: A large *Xenagama* of the grassy plains of the northern Ethiopia/Somalia border. Head robust and comparatively deep, body scalation heterogeneous, with small dorsal scales interspersed by keeled to spinose enlarged scales, those on flanks usually very spinose. Gular and ventral scales smooth. Tail shorter than body and head, discoidal part not well developed, narrower than in *X. batillifera* or *X. taylori*, but broader than *X. zonura*, gradually merging into terminal filament. All dorsal and lateral tail scales strongly keeled and spinose. Three rows of very well-developed preanal pores in males. Maximum length about 15cm (snout–vent length up to 9.7cm). Newly hatched juveniles measured 4.2cm total length (snout–vent length 2.9cm). Upper parts of body and head shades of brown. An indistinct pattern of four brownish transverse bands with a few dark-margined pale ocelli between limbs. Enlarged dorsal scales pale, and looks speckled. A more or less distinct pale vertebral stripe extends onto tail. Belly and lower parts of tail dirty white or yellowish. Tail usually has three dark bands on upper discoidal part. Chin with some irregular blue lines, throat and chest blue. There is a vivid pale stripe between nostril and tympanum and two or three often indistinct dark stripes behind eyes. Taxonomic Notes: Recently described, previously subsumed within *X. batillifera*.

Xenagama wilmsi, Kebri Beyah.

HABITAT AND DISTRIBUTION: Dry grassland with clayey soil, low bushes or plants and stunted *Vachellia* trees, at 880–1,800m. Known from eastern Ethiopia and western Somaliland (northern Somalia). Occurs east from Gildessa in eastern Ahmar Mountains in Ethiopia, in border region on Borama and Wajaale plains, even in heavily farmed areas, and south into the Ogaden as far south as Sassabane. Replaced by *X. taylori* further east. Prefers more mesic habitats than *X. batillifera* and *X. taylori*. Records from central

Xenagama wilmsi, Dilla.

Xenagama wilmsi, Kebri Beyah.

Ethiopia almost certainly result from confusion with *X. zonura*.

CONSERVATION STATUS: Least Concern.

NATURAL HISTORY: Diurnal, like its two foregoing close relatives. Although they seem to prefer sheltering under rocks in body-length holes, they may dig burrows in clayey soil. Burrow entrance largely oval but flat on the basal side, roughly 5–6×3–4cm in diameter, they are 40–70cm long and reach 30–50cm deep. The spiny tail serves as a very effective deterrent to any predator that might be tempted to follow this lizard into its lair. Between late November and February (winter), only adult males found, while in the hot rainy season, in June, males, gravid females and fresh juveniles were seen. In captivity females laid 5–8 eggs, roughly 1.2×1.7cm.

ETHIOPIAN STRANGE AGAMA

Xenagama zonura
(Boulenger, 1895)

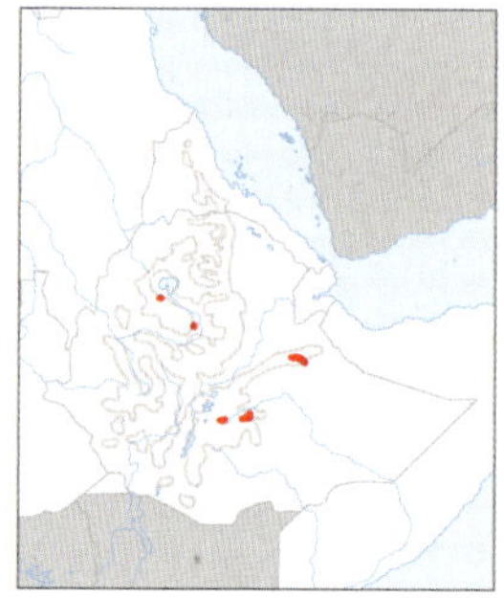

IDENTIFICATION: An unusual *Xenagama*, like one of the big tree agamas (genus *Acanthocercus*). Head large, triangular and depressed. Some enlarged, keeled and spinose scales on both back and flanks, those between dorsolateral folds tend to have weak or no keels. There is usually a rather sharp transition between scales on back and those on upper surface of tail; which is divided into segments, each composed of just two rings of scales. Tail slightly longer than body and head, not discoidal but base distinctly broad. Three rows of very well-developed preanal pores in males. At centre of throat scales are smooth and nostril is directed sideways. Maximum total length of males 19cm (snout–vent 8.5cm), although few specimens recorded. Females smaller, snout–vent length 6.5–7.7cm. Dorsum dark grey to black, olive or brownish, with a pale but rather diffuse vertebral stripe (especially in males) and back is often patterned with scattered yellowish spots, sometimes accompanied by dark marbling that may form irregular transverse bars. In adult males upper surface of the limbs and snout becomes turquoise-blue, the sides of the head may be metallic green, and the throat, chest, abdomen and lower tail are a vivid deep blue. Chin

Xenagama zonura, Robe.

Xenagama zonura, Dodolla.

and throat with reticulated pattern. An indistinct pattern of two rows of blackish transverse bands or lines of spots on upper surface of tail. Two often indistinct dark stripes behind eyes. Gravid females are uniformly dark, except vivid blue throats with a reticulated pattern and orange-red body blotches on back.

HABITAT AND DISTRIBUTION: Endemic to highlands of Ethiopia, in both humid and dry grasslands at 2,000–2500m. Sporadically known from both sides of Rift Valley; in the north recorded at Piccolo Abai bridge, Gojjam and north of Biccena, in the south at Sheikh Mahomed near Delo and around Lake Alemaya near Harar, and probably occurs throughout Ahmar Mountains. Recently observed along main roads in farmlands of Bale region, around Dodolla and Robe; where single pairs were basking on roadside rocks. Near Harar they live on large boulders or stone walls in farmland on slopes of Ahmar Mountains, in areas with *Euphorbia abyssinica*, *Dodonaea viscosa*, *Vachellia* and *Juniperus* trees. Widespread and common in both areas. Doesn't burrow like congenerics.

CONSERVATION STATUS: Least Concern.

NATURAL HISTORY: Poorly known. Diurnal, partially ground-dwelling and rupicolous. Heavily gravid females were found in July. In captivity females laid 4–6 eggs. East of Robe observed to be sympatric with the much bigger, vividly blue *Acanthocercus minutus*.

SUBFAMILY UROMASTYCINAE

A subfamily of 18 species in two genera, one genus in our area.

SPINY-TAILED OR DABB LIZARDS, MASTIGURES *UROMASTYX*

An African and Middle Eastern genus of large, strongly built, herbivorous lizards, in desert and semi-desert. They have bulbous heads, a distinct tympanum, an obvious neck without a crest, a depressed body covered by small uniform scales, well-developed, strong, clawed limbs, and a short flattened tail covered with whorls of strongly spinous scales. Pre-anal and femoral pores are present in most species. The tail does not autotomise (break off) and is used both in defence and to exclude potential predators from the burrow or crevice into which the animal retreats when threatened. Diurnal, terrestrial and solitary, most shelter in burrows that they dig themselves. Some climb well on rocks. Males are slightly larger than females and more brightly coloured during the breeding season. They lay eggs. Currently some 15 species are recognised, six species occur in north-east Africa, two of them endemic with very small ranges. The genus began diverging during the Middle Miocene, approximately 16 Mya, in south-west Asia with a later radiation in North Africa.

KEY TO NORTH-EAST AFRICAN MEMBERS OF THE GENUS *UROMASTYX*

1a: Without preano-femoral pores, confined mostly to Puntland	*Uromastyx princeps*, Somali Spiny-tailed Lizard
1b: With preano-femoral pores	2
2a: Last 12–21 tail whorls formed of continuous scale rows, scales on successive whorls in line, not juxtaposed	3
2b: Only last 2–5 tail whorls formed of continuous scale rows, above that scales in successive whorls are juxtaposed	5
3a: Anterior margin of ear-opening lacks enlarged scales	*Uromastyx ocellata*, Ocellated Spiny-tailed Lizard

3b: Anterior margin of ear-opening with enlarged scales 4

4a: 17–29 (very rarely 31) gular scales (counted from mental scale to a line between anterior margins of ear-openings); tail width between fourth and fifth whorls equivalent to 63–79 per cent of maximum tail width at fifth whorl; temporal region and labials of head with several dark blue or black and yellow streaks; southern Sinai Peninsula in Egypt *Uromastyx ornata*, Ornate Spiny-tailed Lizard

4b: 29–32 gular scales; tail width between fourth and fifth whorls equivalent to 56–62 per cent of maximum tail width at fifth whorl; temporal region and labials of head lack dark blue or black and yellow streaks; endemic to Somaliland *Uromastyx macfadyeni*, MacFadyen's Spiny-tailed Lizard

5a: Very large, reaching 75cm; 247–322 scales at mid-body, 112–193 ventrals between gular and inguinal fold; north-east Egypt and Sinai Peninsula *Uromastyx aegyptia*, Egyptian Spiny-tailed Lizard

5b: Relatively small, reaching 40cm; 187–227 scales at mid-body, 68–112 ventrals between gular and inguinal fold; Sudan *Uromastyx dispar*, Southern Saharan Spiny-tailed Lizard

EGYPTIAN SPINY-TAILED LIZARD

Uromastyx aegyptia
Forskål, 1775

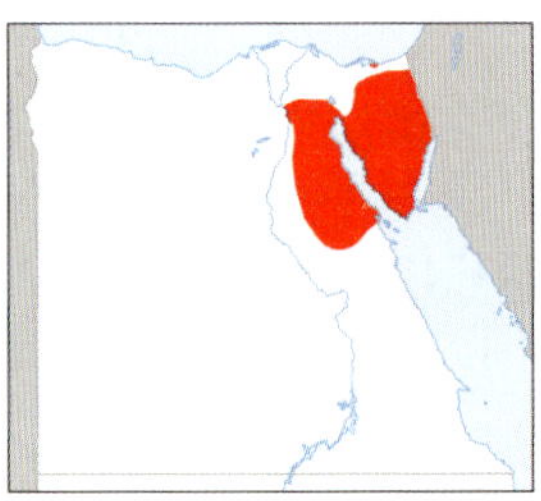

IDENTIFICATION: A huge, dun-coloured robust terrestrial lizard of north-east Egypt and Sinai. Head looks relatively small. Front edges of ear-openings have enlarged scales; 4–7 scales between sub-oculars and supralabials, and 33–59 gulars. Body broad and very depressed. Dorsal scales conspicuously very small and smooth, 247–322 around mid-body; irregularly interspersed, enlarged spinose tubercular scales present on flanks. Fore limbs covered dorsally with small granular scales, hind limbs with enlarged spinose tubercles. Tail flattened and long, about 60–103 per cent of snout–vent length, and consists of 20–23 primary whorls, with 29–46 scales around fifth whorl. Dorsal vertebral tail scales elongated and rectangular. There are 14–20 preano-femoral pores on each side. Maximum total length about 75cm (snout–vent length up to 59cm, sexes similar size), hatchling usually 9–11cm. Coloration uniform and slightly varies with temperature. At low temperatures dark grey, at higher temperatures they become light yellowish-grey, sometimes with faint yellowish vermiculations. Underside is uniform pale yellowish-grey, with blackish throat and neck. Young dark brown or brownish-grey with 5–6 rows of yellow markings (often ocelli) arranged in crossbars. Taxonomic Notes: Three subspecies recognised. Nominate *U. a. aegyptia* occurs in our area. Recent phylogenetic work indicates this species is well separated from other African species, forming a discrete clade, most closely related to *U. ocellatus*.

HABITAT AND DISTRIBUTION: Flat desert and semi-desert, with open sandy habitats and a hard surface layer covered with gravel and stone, between sea level and 1,500m. Typically in broad wadis with a coarse substrate. Occurs in north-east Egypt east of Nile Valley, approximately from the Delta in the north, to Wadi Qena (north of Qena in Eastern Desert) in the south. Widespread in Sinai, except most of sandy north with dunes (although found at southern margin

Uromastyx aegyptia, Sinai.

of Zaranik Protected Area); in south more scattered, mostly in lower, coastal areas. Elsewhere, to north-west Saudi Arabia (same subspecies, other subspecies occur across the Arabian Peninsula, north to Syria, east to Iran). Uncommon. Under threat from pet trade, development activities, killing by traffic, and harvested for food and traditional medicine outside our area.

CONSERVATION STATUS: Vulnerable.

NATURAL HISTORY: Very tolerant of high temperatures and aridity. Diurnal and ground-dwelling, digs its own deep burrows, which may be 10m long, to a depth of 1.8m, and usually occupied by single animal. Young observed climbing small *Vachellia* trees. Usually occurs in loose colonies of up to four animals. Forages in morning and afternoon, but seldom far from their burrows. In colder months, they are fairly active on warm afternoons. When disturbed, they flee surprisingly quickly to their burrows. Largely herbivorous, like other mastigures, feeding on plants, blooms, fruits and seeds; juveniles also take insects. They mate post-hibernation (November–February), a few weeks later females lay a single large clutch of 13–41 (mean c.20) eggs, with an average size 2.4×4.2cm, in deep damp burrows. Incubation about 60 days. They live at least 20 years in the wild.

SOUTHERN SAHARAN SPINY-TAILED LIZARD

Uromastyx dispar
Heyden, 1827

IDENTIFICATION: A large impressive ground-dwelling spiny-tailed lizard of deserts west of the Red Sea coastal hills in Sudan. Head small, the front edges of the ear-openings have enlarged scales; 4–8 scales between sub-oculars and supralabials, 30–43 gulars.

Uromastyx dispar, Chad.

Body depressed and broad. Dorsal scales homogenous, small and granular, except conical scales of variable size on sides of neck, scales at mid-body in 187–227 rows. Fore limbs covered dorsally with small granular scales, hind limbs with enlarged spinose tubercles. Long flat tail (about 63 per cent of snout–vent length) consists of 16–21 primary whorls with 30–36 scales around fifth whorl. Dorsal vertebral tail scales elongate and rectangular. There are 10–18 preano-femoral pores on each side. Maximum total length about 40cm (snout–vent length up to 25cm), hatchling around 8cm. Dorsal coloration sandy-yellow or yellowish-green with dark spots or vermiculations, tail more strongly and irregularly speckled; males have black head, limbs and tail. Underside uniform whitish in females. In males throat blackish, other parts white with scattered black bars and blotches. Young darker, yellowish-brown, with vermiculations and several transverse bold bands interrupted in vertebral area. Taxonomic Notes: Four subspecies. Nominate *U. d. dispar* occurs in our area, the others in West Africa. According to recent phylogenetic studies, *U. dispar* groups with other similar but extralimital species (*U. geyri*, *U. alfredschmidti*, *U. acanthinura* and *U. nigriventris*) in the *U. acanthinura* group. Most closely related to *U. acanthinura* and *U. nigriventris* from north-west Africa.

HABITAT AND DISTRIBUTION: Rocky areas, oases, stony or gravel plains but also flat sandy habitats without rocks but with a fine sand substrate, even in sand dunes, at 180–370m in Sudan. In our area, recorded at Wadi Halfa near Egyptian border (so might occur in Egypt), south through the eastern Libyan and western Nubian desert along the Nile Valley to the Bayuda desert north of Khartoum, localities include Ambukol, Dongola, Abu Hamad, about 150km from Debba towards Khartoum and El Matamma west of Shendi. Recorded also at Al Fashir in North Darfur. Elsewhere; widespread across the southern Sahara and Sahel to southern Morocco and Mauritania.

CONSERVATION STATUS: Least Concern. Collecting for pet trade is a major possible threat in Sudan.

NATURAL HISTORY: Poorly known. Diurnal and terrestrial, never far from its deep burrow to which it retreats. Diet herbs and other plants, juveniles take insects. Clutches of 12–16 eggs, c.21×36–38mm in size, hatching after 95–98 days, in captivity. Probably in loose colonies.

Uromastyx macfadyeni, juvenile, Berbera.

MACFADYEN'S SPINY-TAILED LIZARD

Uromastyx macfadyeni
Parker, 1932

Uromastyx macfadyeni, Berbera.

Uromastyx macfadyeni, Berbera.

IDENTIFICATION: A small bluish-yellow spiny-tailed lizard, endemic to coastal Somaliland. Head small, front edges to ear-openings with enlarged scales, 4–6 scales between sub-oculars and supralabials, 29–32 gulars (from mental scale to a line between anterior margins of ear-openings). Body depressed. Dorsal scales small, smooth, sub-imbricate, 157–182 around mid-body. All limbs covered dorsally with small granular scales, except upper thighs and tibia which have some spinose tubercles. Tail consists of 22–23 primary whorls with 23–26 scales around fifth whorl; flattened and long, about 74–84 per cent of snout–vent length. Dorsal vertebral tail scales slightly elongate to square in shape, with very reduced spikes. There are 13–15 preano-femoral pores on each side. Maximum total length 25cm (snout–vent length up to 14cm), hatchling size unknown. One of the smallest members of genus. Males have a series of diffuse green or blue and yellow crossbars, overlain by a net of dark-edged, yellow-centred ocelli. Head yellow above, mottled blue on sides. Tail usually yellow. Underside yellowish-white, with a few scattered dark bold spots. Females less vivid, pale brownish or yellow dorsally; juveniles express adult pattern with pale brownish tones and clear bluish transverse bands. Taxonomic Notes: Sister species to *U. princeps*. Although confused in the past with *U. ocellata*, it is not related.

HABITAT AND DISTRIBUTION: A northern Somali endemic of dry and hot arid areas along Gulf of Aden. Usually on stony coastal plains with sparse vegetation, at 53–500m. Occurs in so-called Guban strip between the sea and mountain ranges of Somaliland, from Santaalo (south-west of Berbera) in west, across Wadi Faradeero and Dagah Shabell south of Berbera, thence east to Heis village and Maydh. Probably widespread in the intervening country (all suitable habitat) but no records. In west of its range it is replaced in similar habitats by *U. ocellata*, and in the east by *U. princeps*. Locally fairly common, but might be exploited for the pet trade.

CONSERVATION STATUS: Least Concern.

NATURAL HISTORY: Diurnal and terrestrial. Recent observations indicate that when disturbed on open plains these lizards run to shelter under rocks; in rocky country they hide in crevices. Largely active during warmer months (May–September); rarely seen in November–March, presumably due to low temperatures and absence of fresh plant food. Small juveniles often found in same rock cracks as *Hemidactylus citernii*. A dead adult was found near Maydh in the branches of a bush at a height of 1.5m, probably cached by a predatory bird. Adult diet plant material; feeding on leaves, fruits and seeds, juveniles take some insects. Females lay 5–11 eggs.

OCELLATED SPINY-TAILED LIZARD

Uromastyx ocellata
Lichtenstein 1823

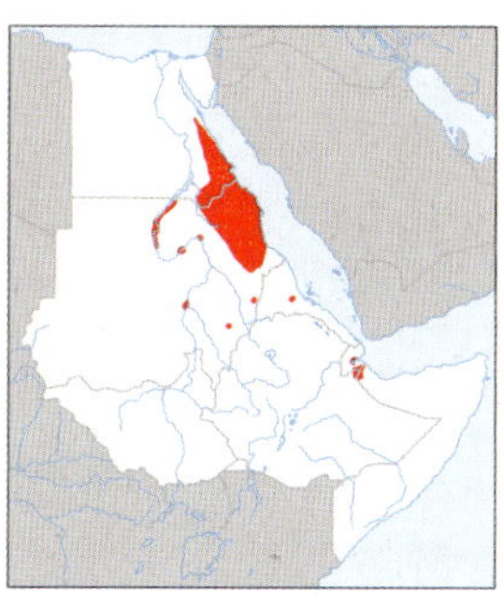

IDENTIFICATION: A small to medium spiny-tailed lizard, the back covered by black-fringed white ocelli, from Egypt to north-west Somaliland. Head small, front edges of ear-openings lack enlarged scales, 4–6 scales between sub-oculars and supralabials, 29–42 gulars. Body depressed. Dorsal scales small, granular, 189–256 around mid-body. All limbs have small granular scales, except lower hind limb with some spinose tubercles. Tail flattened and long, about 91 per cent of snout–vent length, and consists of 22–29 primary whorls with 24–29 scales around fifth whorl. Dorsal vertebral tail scales significantly elongate. 12–17 preano-femoral pores on each side. Maximum total length 32cm (snout–vent length up to 17.6cm), hatchling size 7–9.1cm. One of smallest members of genus. Coloration most vivid in breeding males, where back is bright green to reddish, usually with 7–8 transverse rows of large, white or yellow spots (ocelli), each with a dark margin and head is olive-green or dark red, sometimes with contrasting spots of red or green. Underside yellow or white. Females less vivid, with an either pale brownish, green, reddish or grey ground colour. Juveniles show adult pattern but are browner. Taxonomic Notes: According to recent phylogenetic studies, *U. ocellata* is part of a clade that includes several extralimital but similar species (*U. benti*, *U. shobraki*, *U. yemenensis* and *U. ornata*) within the *U. ocellata* group. Most closely related to *U. ornata*.

Uromastyx ocellata, Jidhi.

Uromastyx ocellata, Jidhi.

HABITAT AND DISTRIBUTION: A wide-ranging species of arid country. In lowland mountains, hills, rocky desert and lava fields, generally below 1,000m. Favours wadis with steep rocky sides, almost always in areas with *Vachellia* trees. A continuous population occurs south from 27°30'N in Egypt along Red Sea coast almost to Suakin, thence inland to around 17°N. Sporadic records from Nile Valley south of Wadi Halfa, as far as Omdurman, Kassala and Gedaref. Also extreme north-west Somaliland in lowlands of Awdal region, Djibouti (58km south-west of Tadjoura, Djalelo, near Lake Assal and Qued Wada Weyn south-east of Ali Sabieh) and immediately adjacent north-east Ethiopia. An old and somewhat dubious record exists from Asmara in Eritrea. Heavily exploited for the pet trade in Egypt and Sudan.

CONSERVATION STATUS: Near Threatened.

NATURAL HISTORY: A diurnal, rock-dwelling lizard. Males usually have prominent lookouts on small hills and rocky outcrops, to survey their territory. If threatened, they retreat into rock crevices or under rocks. Usually descend from rocks to wadi beds to feed; they climb on *Vachellia* trees to eat vegetation. Observed sleeping on branches of *Vachellia* in Egypt. Usually live in pairs. Females lay 11–16 eggs, roughly 3.5cm long. They hibernate in winter. Largely herbivorous, feeding on leaves, fruits and seeds, but young take some insects.

ORNATE SPINY-TAILED LIZARD

Uromastyx ornata
Heyden, 1827

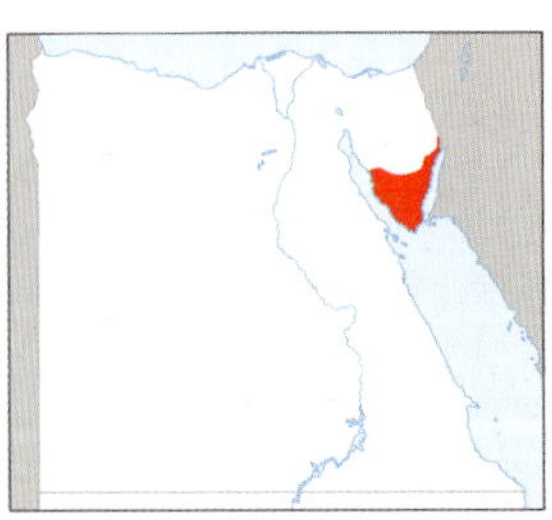

IDENTIFICATION: A medium rock-dwelling spiny-tailed lizard, vividly coloured, on Sinai Peninsula in Egypt. Head small, front edges of ear-openings with enlarged scales, 3–5 scales between sub-oculars and supralabials, 22–31 gulars. Body depressed. Dorsal scales homogenous, granular, 149–185 around

Uromastyx ornata, Sinai.

Uromastyx ornata, Sinai.

mid-body. All limbs covered dorsally with small granular scales, except lower hind limb which has some spinose tubercles. Tail slightly flattened and long, about 70 per cent of snout–vent length, and consists of 20–23 primary whorls with 19–25 scales around fifth whorl. Dorsal vertebral tail scales elongate; 7–14 preano-femoral pores on each side. Maximum total length 38.5cm (snout–vent length up to 21cm), hatchling 7–10cm. Colour very variable; males green, blue or red, with an irregularly reddish-brown net-like pattern and yellow spots on back. Sometimes yellow crossbars are present. Head blue with some black spotting, temporal region and labials with several dark blue or black and yellow streaks, an excellent aid to identification. Underside has dark pattern. Females and juveniles less colourful; pale brown with dark brown spots and sometimes pale yellow or pale red spots. Belly without pattern, yellowish or white. Taxonomic Notes: Two subspecies; nominate occurs in our area.

HABITAT AND DISTRIBUTION: Almost identical to *U. ocellata*, but seems less tolerant of extremely arid conditions. Rocky desert and volcanic rubble, in areas with less than 50mm annual rainfall. In our area, only in southern Sinai, south of 29o40'N, from Ras Mohamed in south, north to north-east border near Eilat in east and Abu Rudeis in west. Elsewhere, to adjacent north-west Saudi Arabia. Heavily exploited for the pet trade.

CONSERVATION STATUS: Near Threatened.

NATURAL HISTORY: Diurnal and heliophilic. Usually on and around rocks, and shelters in holes and crevices, but may excavate a burrow, up to 1m deep. Most time is spent in observation close to burrow, due to risk of predation. Strongly associated with *Vachellia* trees, rarely found above 1,000m. Largely inactive in winter, when hibernates. Eat a variety of mostly succulent plants, but seasonally may be dependent upon a single species; climb trees and shrubs to feed on their foliage and fruits. Usually solitary, but may live in a small group. Females lay 7–17 eggs. Sexual maturity at two years of age.

SOMALI SPINY-TAILED LIZARD

Uromastyx princeps
O'Shaughnessy, 1880

IDENTIFICATION: A small, pale brown short-tailed mastigure, endemic to north-east and east Somalia. Head short and broad, front edges of ear-openings lack enlarged scales, with 2–5 scales between sub-oculars and supralabials, and 28–43 gulars. Body squat. Dorsal scales small, smooth, interspersed with additional smaller scales. Mid-body scale rows 150–226. All limbs covered dorsally with small granular scales, except upper thighs and tibia with some enlarged smooth scales and spinose tubercles. Tail flattened, short and usually wide, about 35–53 per cent of snout–vent length, and consists of 9–14 whorls, made up dorsally of 4–8 enlarged, spiky scales; with 20–27 scales around fifth whorl. Dorsal vertebral tail scales slightly elongate to square, with weakly reduced spikes. No preano-femoral pores. Maximum total length around 27cm (snout–vent length up to 18cm), hatchling around 7cm. Pronounced sexual colour dimorphism. Dorsal ground colour pinkish-brown to green with small brown markings. Males have a yellowish-red to green dorsum with small, scattered dark spots. Top of head brownish, sides whitish to yellow, eyebrow and snout pale brown to orange. Venter yellowish with blue-grey marbling on chest and throat. Tail yellow-green or brown. Females grey-brown dorsally with small dark spots; venter immaculate white. Juveniles usually have several pairs of dark brown spots scattered on back, with similar pale spots on neck and flanks. Taxonomic Notes: Currently no subspecies are recognised, although *U. p. scorteccii*, with a comparatively wider tail, was described from Eyl in Puntland; ranging along the coast to Meregh in central Somalia.

HABITAT AND DISTRIBUTION: Rocky and stony biotopes with sparse vegetation, in hot arid country between sea level and 1,300m, but mostly below

Uromastyx princeps, juvenile, Somaliland.

Uromastyx princeps, adult, Somaliland.

Uromastyx princeps, Somaliland.

800m. In lowland mountains and hills, in wadis with rocky sides, and on inland limestone plateaux. In Somaliland known only from a few localities, most west of Waqadariya, on the coast, about 75km east of Maydh, and from several localities in the Daror Valley and Buran, along border with Puntland. Most widespread in Puntland, where it extends from coastal plains around Bosaso east to Candala and Bargal. Known from foothills of the Bari and Cal Miskaat Mountains (very common around Karin village) thence south from the Hafun Peninsula and Daror Valley to the Nogal Valley. Isolated records at Eyl, Wadi Run, Hobyo and Meregh, and along coast to Hobyo in Mudug and Meregh in Galguduud of central Somalia. Recently reported from a new locality in north-central Somaliland, (not shown on map), far from main range. Might reach Ethiopia. Erroneously reported in past from Zanzibar and Yemen. Locally very common and widespread in suitable habitats.

NATURAL HISTORY: Diurnal, terrestrial and rock-dwelling. Usually has a prominent lookout on small hills and rocky outcrops, from where they survey their territory. When disturbed they shelter in cracks and volcanic blowholes, often resting within 30cm of surface, with spiny tail plugging the aperture. They also climb and bask on top of old *Vachellia* trunks, dodging into cracks and holes when disturbed. On the coarsely pebbled bottom of a wide wadi near Karin in Puntland, solitary individuals basked on large stones (30–50cm in diameter), and when disturbed they quickly hid under them. On the inland limestone plateau they were observed perching on large water-eroded rocks. Sometimes very close to human settlements, in Bosaso between buildings, living on artificial stone and concrete bollards. Usually solitary or in pairs. Females lay 7–12 eggs. During winter (December–February) significantly less active. Largely herbivorous.

CONSERVATION STATUS: Least Concern.

LACERTIDS, OR 'TYPICAL' LIZARDS

FAMILY LACERTIDAE

A Eurasian-African family of fairly small, slender and diurnal lizards, usually with long tails and well-developed limbs, many are striped. About 360 species, in just over 40 genera are known. The north-east African forms are nearly all striped, terrestrial, fast-moving, arid-adapted species (one is a rare forest-dweller) and some are modified for this lifestyle by having fringes or keels on their toes, and often show leg-waving or leg-tucking behaviour, presumably to ensure rapid muscular response. All African species lay eggs, and some are very short-lived, hatching, breeding and dying in a single year, so that the entire population exists only as eggs for part of the year. Just over 40 species in eight genera are known from north-east Africa; all in the subfamilies Lacertinae and Eremiadinae, with Afro-Arabian distributions; they began diverging about 45 Mya.

KEY TO THE NORTH-EAST AFRICAN GENERA AND SOME SPECIES OF LACERTIDS

1a: Eyes cannot close; eye covered by a spectacle	*Ophisops*, Snake-eyed lizards
1b: Eyes with movable lids that close	2
2a: Largely arboreal, no narrow dorsal stripes, lime-green underside, confined to forest and woodland in South Sudan	*Adolfus africanus*, Multi-scaled Forest Lizard
2b: Terrestrial, does not have the above combination of characters	3
3a: Fewer than eight longitudinal rows of scales across the belly	4
3b: More than eight longitudinal rows across the belly	6
4a: Several clearly expanded rows scales along vertebral line	*Philochortus*, Shield-backed ground lizards
4b: Dorsal scales homogeneous	5
5a: Nostril bordered by 3–5 nasals and first labial, or narrowly separated from the latter	*Latastia*, Long-tailed sand lizards
5b: Nostril bordered by 2–4 nasals, well separated from first labial	*Heliobolus*, True sand lizards
6a: Toes usually with fringes, nostril in contact with first supralabial	*Acanthodactylus*, Fringe-toed lizards
6b: Toes without fringes, nostril not in contact with first supralabial	7
7a: Supraoculars isolated from other head shields by row of small granules	*Pseuderemias*, False sand lizards
7b: Supraoculars in direct contact with other head shields	*Mesalina*, Desert racers

FRINGE-TOED LIZARDS *ACANTHODACTYLUS*

A Palearctic genus of more than 45 species of fast-moving, slim, largely striped, semi-desert and desert lizards, widespread across the northern half of Africa, east through the Middle East to western India. Most species have elongated fringes on their toes (and sometimes on the fingers), enabling them to sprint on loose sand. Five, possibly six, species in our area. Their taxonomy is not fully resolved, and zoogeographically complicated, due to the fluctuation in size and extent of the sandy regions of the Western Palearctic; see our taxonomic notes.

KEY TO THE NORTH-EAST AFRICAN SPECIES OF *ACANTHODACTYLUS*

1a: Fringes on toes largely rudimentary, six longitudinal stripes, the lower two on each side unbroken and highly contrasting, in our area confined to extreme south-west	*Acanthodactylus guineensis*, Guinea Fringe-toed Lizard
1b: Does not have above combination of characters	2
2a: Pattern usually striped, usually less than 13 rows of ventrals	3
2b: Pattern usually spotted, usually more than 13 rows of ventrals	4
3a: Dorsal scales granular, smooth and uniform	*Acanthodactylus pardalis*, Leopard Fringe-toed Lizard

3b: Dorsal scales imbricate, keeled, increasing in size posteriorly *Acanthodactylus boskianus*, Bosc's Fringe-toed Lizard

4a: Dorsal scales on tibia large and strongly keeled, femoral pores usually <20, habitus small and slender, snout looks short *Acanthodactylus aegyptius*, Egyptian Fringe-toed Lizard

4b: Dorsal scales on tibia small and largely smooth, femoral pores usually >20, habitus larger and stocky, snout looks long 5

5a: Ventrals 12–15, femoral pores 18–24, larger, snout–vent length to 7.5cm, dorsal pattern usually consistent across back and flanks *Acanthodactylus scutellatus*, Nidua Fringe-toed Lizard

5b: Ventrals 14–18, femoral pores 19–26, smaller, snout–vent length to 6.5cm, dorsal pattern often low contrast along spine, darker on flanks *Acanthodactylus longipes*, Long-footed Fringe-toed Lizard

EGYPTIAN FRINGE-TOED LIZARD

Acanthodactylus aegyptius
Baha El Din, 2007

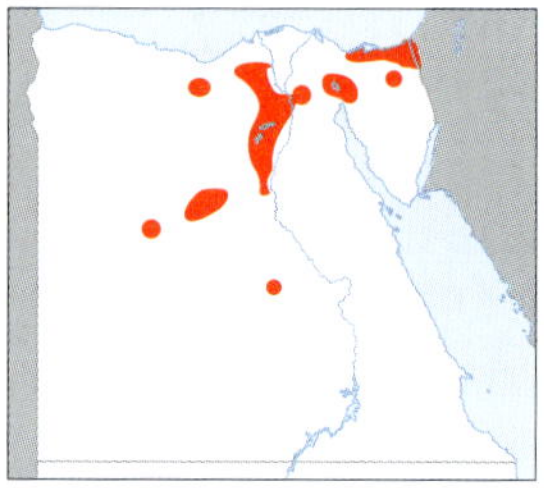

IDENTIFICATION: A rather delicate-looking short-snouted lizard of northern Egypt. Snout pointed, body slim, tail relatively long, two-thirds of total length. Body scales small and slightly keeled, arranged in 68–88 transverse rows at mid-body, enlarged ventrals in 12–15 rows. The dorsal scales have a slightly shiny aspect. Digits strongly fringed, although this isn't always obvious. There are 15–22 femoral pores on each side. Maximum size 16cm, average 12–14cm, hatchling size unknown. Colour: light brown or biscuit-coloured, covered with small yellow spots, sometimes darker on flanks. The lateral fold gives the impression of a side-stripe; the segmented tail looks barred. Underside cream. Juveniles have a yellow tail. Taxonomic Notes: Recently described, previously 'hidden' within *A. longipes*.

Acanthodactylus aegyptius, Lake Maghra.

Acanthodactylus aegyptius, Lake Maghra.

HABITAT AND DISTRIBUTION: Desert and near-desert at low altitude, so long as there is a small amount of vegetation, usually on sand, often in dunes. Almost endemic to Egypt; it occurs across north-east Sinai, the northernmost part of the Eastern Desert, and most of the northern Western Desert (west to 28°E). Known from several oases (Bahariya, Kharga, Wadi El Natrun, Farafra), the easternmost edge of Qattara Depression, and along western edge of Nile Valley. Elsewhere, just reaches western Israel/Palestine.

CONSERVATION STATUS: Not Assessed.

NATURAL HISTORY: Diurnal and terrestrial, fast-moving and wary, largely active in morning, even when very hot, and early evening. Shelters in burrows overnight. Two or three clutches of 2–3 eggs are laid between May and July each year, and juveniles appear from June onwards. It is suggested that it only lives for a year or so. Diet small arthropods.

BOSC'S FRINGE-TOED LIZARD

Acanthodactylus boskianus
(Daudin, 1802)

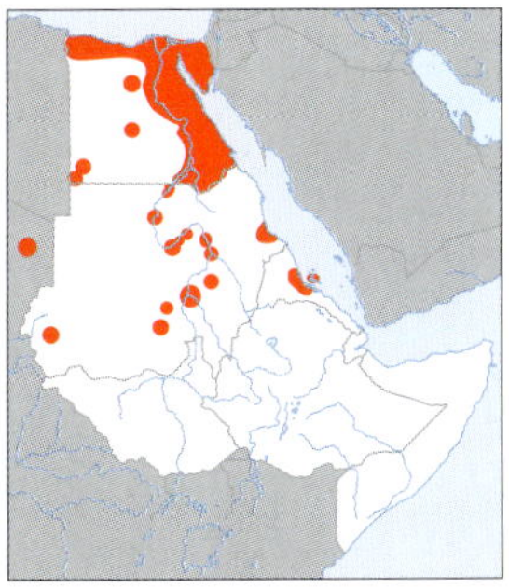

IDENTIFICATION: A big, alert, short-snouted striped desert lizard with a huge range. Snout pointed, body slim, tail relatively long, two-thirds of total length. Body scales small, imbricate and strongly keeled, arranged in 29–55 transverse rows at mid-body, enlarged ventrals in 10–12 rows. Digits strongly fringed. Femoral pores 15–26 on each side. Maximum size 27cm, average 18–25cm, hatchling size about 9cm. Colour: quite variable, but usually has three pairs of yellow or yellow-brown stripes on dark grey, rufous or brown background, sometimes the stripes interrupt narrow darker crossbars, giving a spotted appearance. Underside white or cream, in juveniles and females the tail is often orange below. Striping often very vivid in juveniles, and may fade in adults. Taxonomic Notes: Two subspecies recognised in Egypt; *A. b. boskianus* (Nile Delta and adjacent areas including northern Sinai) and *A. b. asper* (Eastern and Western Deserts). Recent molecular analysis indicates a more complex picture, with nine lineages, and two as-yet-unnamed cryptic species in Egypt.

Acanthodactylus boskianus, Edfu.

Acanthodactylus boskianus, Ambukol.

HABITAT AND DISTRIBUTION: Desert, near-desert, semi-desert and dry savanna, usually below 1,000m in our area, although to 1,500m in the Sinai and 2,400m elsewhere. Occurs virtually throughout north and east Egypt, records patchy in south and west, in Sudan sporadically recorded from Nile Valley south to Khartoum, records include parts of Kordofan, Darfur (Gebel Marra, Gebel Rahib), White Nile (Ab-Dareesh), Nile State (Atbara) and the Red Sea coast (Erkowit, Suakin), with a population in Eritrea, on the northern coastal strip and Dahlak Islands. Elsewhere, west to Mauritania, north and east to Iraq.

CONSERVATION STATUS: Least Concern.

NATURAL HISTORY: Diurnal and terrestrial, very fast-moving and wary. Often common and visible in lightly vegetated country, dashing bush to bush. In some areas active throughout the hottest hours of the day, but also sometimes at twilight. Where there is vegetation, they shelter in holes dug under bushes, or ground cover, but in desert they may dive into sand or dig their own burrows. Females excavate a hole and lay 2–8 eggs in spring, which hatch after 2–3 months. Diet largely arthropods, but they take small vertebrates, including geckos and juveniles of their own species.

GUINEA FRINGE-TOED LIZARD

Acanthodactylus guineensis
(Boulenger, 1887)

IDENTIFICATION: An alert, tiger-striped lizard of West African savanna, might just reach the south-west of our area. Snout pointed, body fairly slim, tail relatively long, 60–70 per cent of total length. Body scales small, sub-imbricate, the extent of keeling increasing posteriorly, arranged in 46–66 transverse rows at mid-body, enlarged ventrals in ten rows. Digits weakly fringed. Femoral pores 17–22 on each side. Maximum size 18cm, average 13–16cm, hatchlings 5–6cm. Colour: usually has three pairs of yellow or cream stripes (sometimes the pair bordering midline of back faded and/or narrow) on a rufous or brown background, the stripes interrupting to a greater or lesser extent blotchy black crossbars. Underside white or cream. Juveniles have a powder-blue tail.

HABITAT AND DISTRIBUTION: Dry and also moist savanna (unlike other fringe-toed lizards), just extending north into southern Sahel, between sea level and 1,300m, although usually below 800m. Although unrecorded from our area, its presence at Ouanda-Djalle in Central African Republic, close to the border with Southern Darfur, suggests it probably occurs in south-west Sudan. Elsewhere, west to Burkina Faso.

Acanthodactylus guineensis, Nigeria.

Acanthodactylus guineensis, Nigeria.

CONSERVATION STATUS: Least Concern.

NATURAL HISTORY: Poorly known. Diurnal and terrestrial on patches of open ground, prefers sandy habitats but also on soil and rock, very fast-moving and wary. Presumably lays eggs, diet largely arthropods.

LONG-FOOTED FRINGE-TOED LIZARD

Acanthodactylus longipes
Boulenger, 1921

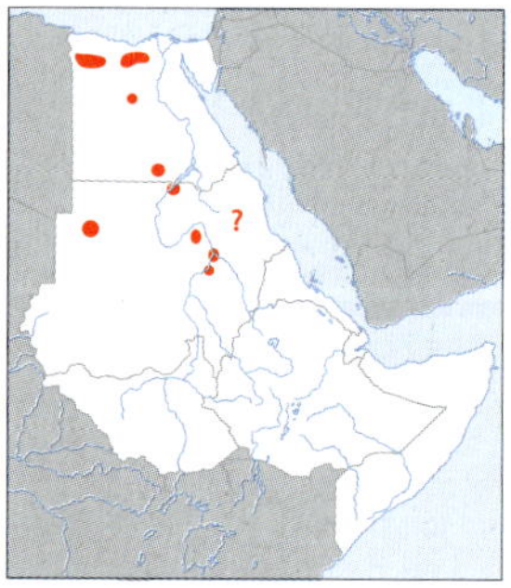

IDENTIFICATION: An alert, spotted desert lizard, in our area confined to Egypt west of the Nile. Snout

Acanthodactylus longipes, Niger.

pointed, body fairly slim, tail relatively long, 60–70 per cent of total length. Body scales small, granular and smooth, arranged in 68–94 transverse rows at mid-body, enlarged ventrals in 14–18 rows. Digits strongly fringed. Femoral pores 19–26 on each side. Maximum size 19cm, average 13–16cm, hatchling size unknown but an Egyptian juvenile was 10cm. Colour: usually shades of brown or biscuit-coloured, back marked with poorly defined light-centred ocelli, interspersed by dark speckling, from a distance looks vaguely striped. Often the central back markings are less obvious than on flanks. Tail light above, darker on sides. Underside white or cream.

HABITAT AND DISTRIBUTION: Desert, near-desert and semi-desert at low altitude, usually below 300m in our area, but higher elsewhere. In Egypt, sporadically recorded from Faiyum west through the Qattara Depression to Libya, also known from most oases in Egyptian Western Desert. Few records from southern Egypt, save Bir Kurayim, close to Sudan's northern border. Sudanese records include Merga, Nukheila, Wadi Halfa, Bayuda Desert, Merowe and Karima on northern Nile; probably extends further into desert in that area. Elsewhere, west across the Sahara to Mauritania.

CONSERVATION STATUS: Least Concern.

NATURAL HISTORY: Diurnal and terrestrial, associated with soft sand in dunes and undulating sand habitats. Very fast-moving and wary. When threatened, it often escapes by 'sand diving', sprinting and then plunging into sand; and it may shelter, buried within sand, as well as in burrows. A small clutch of eggs is laid. Diet largely arthropods, especially ants.

LEOPARD FRINGE-TOED LIZARD

Acanthodactylus pardalis
(Lichtenstein, 1823)

IDENTIFICATION: An alert, spotted and striped desert lizard, a rare inhabitant of north Egyptian coast. Snout relatively short and somewhat blunt for an *Acanthodactylus*, body fairly slim, tail 60–65 per cent of total length. Body scales small, granular and smooth, arranged in 50–71 transverse rows at mid-body, enlarged ventrals in 10–14 rows. Digits strongly fringed. Femoral pores 15–26 on each side. Maximum size 19–20cm, average 14–18cm, hatchling size unknown. Colour: usually shades of brown or biscuit-coloured, back marked with six or eight rows of white or cream ocelli; fine black bars link ocelli (sometimes lengthwise as well as crosswise); from a distance it looks spotted, hence name. Occasional very dark individuals are known. Juveniles more vivid, and may have blue tails. Underside white or cream. Taxonomic Notes: Recent research indicates that this species is confined to Egypt and coastal north-east Libya; historical records from elsewhere are other species.

HABITAT AND DISTRIBUTION: Near-desert, semi-desert and Mediterranean scrub, usually below 300m. In Egypt, occurs from the Faiyum and west side of the Nile Delta, north and west to Alexandria,

Acanthodactylus pardalis, Egypt.

thence west along coast to Libyan border. Elsewhere, in coastal north-west Libya.

CONSERVATION STATUS: Vulnerable, but endangered within Egypt. Its habitat is disappearing under development, agriculture, overgrazing and it is harvested for the pet trade.

NATURAL HISTORY: Diurnal and terrestrial, in flat open semi-desert with a Mediterranean rainfall of 50–150mm annually. It favours areas of hard clay substrates, with scattered stones and some vegetation. Fast-moving and wary. In summer, active in morning and later afternoon, in spring and autumn during middle of day; with little winter activity. A small clutch of eggs is laid. Diet largely arthropods, especially ants.

NIDUA FRINGE-TOED LIZARD

Acanthodactylus scutellatus
(Audouin, 1829)

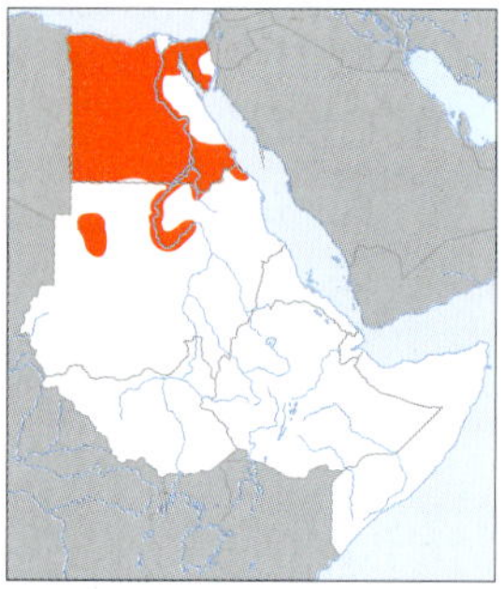

IDENTIFICATION: An alert, big, robust spotted desert lizard, widespread in Egypt and north Sudan. Snout long, body fairly slim, tail 60–70 per cent of total length. Body scales small, granular and smooth or slightly keeled, arranged in 56–78 transverse rows at mid-body, enlarged ventrals in 12–15 rows. Digits strongly fringed. Femoral pores 18–24 on each side. Maximum size about 25cm, average 18–22cm, hatchling size 7–8cm. Colour: usually shades of brown or biscuit-coloured, typically with fine dark reticulation but some specimens appear almost uniform, and a population in northern Nile Delta is dark grey. Pattern tends to disappear on tail. Underside of tail of adult females turns salmon-red in breeding season, and juveniles have a blue tail. Underside white or cream. Taxonomic Notes: Two subspecies in our area: *A. s. scutellatus* in Egypt, Sinai and Israel/Palestine, and *A. s. audouini* in west of species' range, east to north-east Sudan and southern Egypt.

HABITAT AND DISTRIBUTION: Desert, near-desert, semi-desert and dry savanna, at low altitude, usually below 800m, although occasionally higher. Occurs almost throughout Egypt, but absent from the Eastern Desert and parts of Sinai, and records sporadic in the south-west. A handful of records from northern Sudan, many associated with Nile in Northern Province, or Red Sea coast. Elsewhere, west across Sahara to Mauritania, east to Arabia.

CONSERVATION STATUS: Least Concern.

NATURAL HISTORY: Diurnal and terrestrial, in range of habitats; sand and gravel plains, dune areas, wadis, often in areas with some vegetation but also in quite barren areas without plants. Fast-moving and wary. If cornered, it can bury itself at quite remarkable speed, using its legs. Two to four clutches of eggs are laid in spring, 1–4 eggs per clutch, and first juveniles appear in June. It is suggested that adults live no more than a year. Diet largely arthropods, especially ants.

Acanthodactylus scutellatus, Aswan.

Acanthodactylus scutellatus, Meroe.

FOREST AND MEADOW LIZARDS *ADOLFUS*

A genus of six species of cold-adapted central African montane and forest lizards, of which a single species just reaches our area.

MULTI-SCALED FOREST LIZARD

Adolfus africanus
(Boulenger, 1906)

IDENTIFICATION: A slender forest lizard, with a rather knobbly appearance due to the strongly keeled back scales. Snout pointed; eye fairly large. Slim, with a long tail, approximately twice body length. Body scales rhombic with strong diagonal keels which converge towards mid-line of back; mid-dorsal scale rows distinctly larger than on flanks, in 18–24 transverse rows at mid-body. Ventral scales in six longitudinal rows. There are 12–17 femoral pores under each thigh. Maximum length about 16cm, average 10–14cm, hatchling size unknown. The entire top of the head is metallic, copper-bronze in colour. A continuous mid-dorsal band, bronze or brown, extends the length of body and tail, within which is a series of random black spots; a dark lateral band is bordered below by yellow spots and lines. Tail brown, with faint lighter bands. Underside immaculate lime-green.

HABITAT AND DISTRIBUTION: A true Guineo-Congolian primary forest dweller, usually associated with clearings where sunlight penetrates forest, at middle elevations (from 580m to about 1,200m). In our area, known only from Imatong Mountains in South Sudan. Elsewhere, west to Cameroon, south to Zambia, although few records in lowland Central African Forest.

Adolfus africanus, Kenya.

Adolfus africanus, Rwanda.

CONSERVATION STATUS: Least Concern. However, it appears to be a true inhabitant of primary forest, thus vulnerable to forest loss.

NATURAL HISTORY: Diurnal and arboreal, basks in dappled sunlight on fallen tree limbs, trunks and exposed roots within a few metres of ground in clearings within forest. Rarely on vertical tree trunks more than 3m above ground, suggesting that this species is primarily an inhabitant of undergrowth. It appears to prefer lower rather than higher montane forest. Presumably lays eggs, but no breeding details known.

TRUE SAND LIZARDS *HELIOBOLUS*

An African genus of six small terrestrial lacertids, living largely in dry savanna and semi-desert; three species occur in southern Africa, the other three are found in our area. There is debate about their taxonomy, and it is suggested that the genus as presently defined is polyphyletic; see our taxonomic notes. Nearest relatives *Pseuderemias* (from which it split 14–15 Mya) and then *Philochortus*.

KEY TO THE NORTH-EAST AFRICAN SPECIES OF *HELIOBOLUS*

N.B. The colour aspects of this key will not work with juveniles, all of which are striped black and yellow, but locality may give clues.

1a: Longitudinally striped, with broken black crossbars, frontal not in contact with supraoculars, head shields flat and strongly but finely striated (save north-east Somalia where subspecies *H. s. scorteccii* has smooth head scales), widespread — *Heliobolus spekii*, Speke's Sand Lizard

1b: Longitudinally striped, without broken black crossbars, frontal in contact with supraoculars, head shields smooth, not striated — 2

2a: Flanks of adult with distinctive large green or yellow ocelli, mid-body scale rows 52–64, in South Sudan — *Heliobolus nitidus*, Sahel Sand Lizard

2b: Flanks of adult without ocelli, mid-body scale rows 40–42, rare, not in South Sudan — *Heliobolus neumanni*, Neumann's Sand Lizard

NEUMANN'S SAND LIZARD

Heliobolus neumanni
(Tornier, 1905)

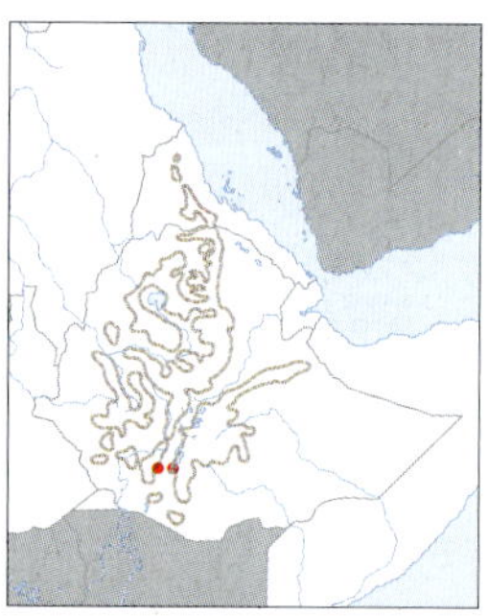

IDENTIFICATION: A small, rare lacertid of doubtful status. Snout pointed, body slim, tail relatively long, 60–70 per cent of total length, all tail scales keeled. Body scales small, strongly keeled but not pointed, not overlapping and arranged in 40–42 transverse rows at mid-body. Ventral scales in eight rows. Maximum size about 15cm, average and hatchling sizes unknown. Colour: a longitudinal white line runs over upper lip shields, through underside of ear and from armpit onwards becomes a number of white dots in a longitudinal row. A longitudinal white line from eye, touching the ear at the top, also dissolves into a longitudinal row of narrow white lines on side of body. In the black band between these two lines are eight blue spots, two of which are placed near the base of the tail. A white line, starting from the outer edge of the parietal, goes without significant interruption to the root of the tail, where it merges with the corresponding line on the other side of the body. The dorsal area between these two lines is not brown, but brown-grey with black spots. Taxonomic Notes: Our colour notes above are a translation of Tornier's original description. The type specimen of this species, from the Barssa Valley in south-west Ethiopia, is apparently lost, and a morphological analysis of more recent specimens places them in different genera; it is suggested they are misidentified, and that this species may not exist.

HABITAT AND DISTRIBUTION: In our area, known only from two localities in Ethiopia, Nechisar National Park in the southern Rift Valley at 1,200m altitude, and the type, from Barssa Valley, about 80km west of there, at 900m. Elsewhere, sporadic records in Kenya and Tanzania.

CONSERVATION STATUS: Data Deficient.

NATURAL HISTORY: Virtually nothing is known of this oddly distributed species. If valid, presumably similar to other members of the genus, i.e., diurnal and terrestrial. A coastal Tanzanian specimen was taken on white kaolin sand.

GLITTERING SAND LIZARD

Heliobolus nitidus
(Günther, 1872)

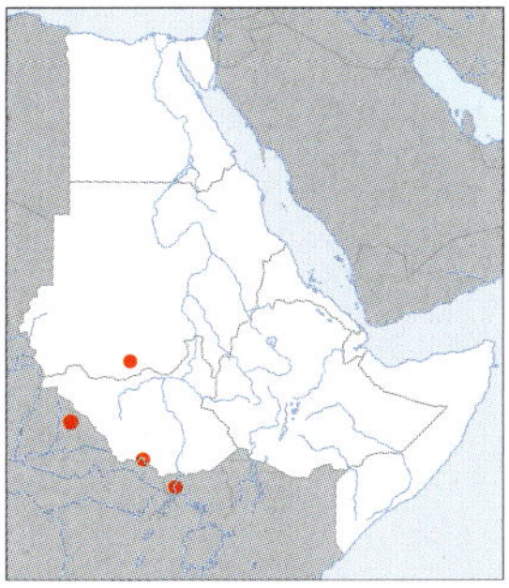

IDENTIFICATION: A beautiful lacertid with flank ocelli. Snout pointed, body slim, tail relatively long, 65–75 per cent of total length. Body scales small, strongly keeled but not pointed, not overlapping and arranged in 52–64 transverse rows at mid-body. Maximum size 23cm, average 14–22cm, hatchling size unknown. Colour: adults have four yellow dorsal stripes on a dark background, the central two formed by two coalescing stripes on neck, the flanks distinctly marked with vivid green and yellow ocelli, interspersed by black-and-white speckling; tail rufous, with a diffuse dark lateral line. Yellowish-white below. Juveniles have a red or orange tail, ocelli small and poorly defined, with a broad brown mid-dorsal stripe. Taxonomic Notes: Morphological and molecular analysis suggest this species is wrongly placed in *Heliobolus*, and skull shape and general appearance indicate a closer relationship with Long-tailed sand lizards, *Latastia*.

HABITAT AND DISTRIBUTION: Moist and dry savanna. Only a single definite record from our area, at the

Heliobolus nitidus, juvenile, Ghana.

Heliobolus nitidus, Neem.

Neem oilfield area in the Muglad Basin, South Kordofan, in Sudan. However, records from Murchison Falls National Park in Uganda, Garamba National Park, and northern Central African Republic, indicate it is probably widespread across the western side of South Sudan, albeit secretive. Elsewhere, west to Mali.

CONSERVATION STATUS: Least Concern.

NATURAL HISTORY: Diurnal and terrestrial, fast-moving and wary, it is a difficult species to observe. A two-year herpetofaunal survey in northern Ghana obtained only a single specimen. Shelters in holes; lays eggs, and eats small arthropods.

SPEKE'S SAND LIZARD

Heliobolus spekii
(Günther, 1872)

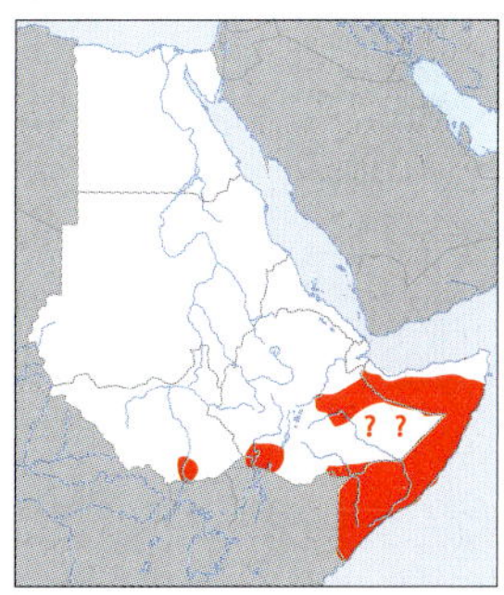

IDENTIFICATION: A small, relatively common, dry-country lacertid with a pointed snout. It is slender; tail relatively long, 65–70 per cent of total length. The scales at mid body are small, rhombic and strongly keeled, usually arranged in 63–71 transverse rows; ventrals in six straight rows. Maximum size about 18cm, average 10–16cm, hatchlings about 6–7cm. The back and flanks are usually rufous or dark brown with diffuse fine black crossbars, interrupted by five or six fine light yellow longitudinal stripes, the lowest on

Heliobolus spekii, male, Turmi.

Heliobolus spekii, juvenile, Weyt'o.

each flank breaking up into yellow or white spots. White below. Juveniles have no barring; they are black, with fine yellow stripes and a red or orange tail. Specimens from Degeh Bur in eastern Ethiopia are vividly rufous. Taxonomic Notes: Three subspecies recognised, two in our area. *H. s. scorteccii* is known from three localities in Somalia; Hobyo in the centre and Gardo and El Mur, in Puntland. It has smooth head scales and the pattern and coloration are darker and more pronounced. All other records from our area are *H. s. sextaeniata*.

HABITAT AND DISTRIBUTION: Coastal thicket and light woodland, moist and dry savanna, and semi-desert, from sea level to about 2,000m, although most common in dry savanna below 1,000m. Occurs virtually throughout Somalia, (see comments above about *H. s. scorteccii*), rare in the north but seen in the Ogo and Goolis Mountains, between the Ethiopian border and Sheikh. Widespread in Ethiopia east of the Rift Valley, very common at Degeh Bur. A handful of records from Eastern Equatoria in South Sudan (Nimule, Torit, Mongalla, Lokomarinyang in the Ilemi Triangle). Elsewhere, south to central Tanzania.

Heliobolus spekii scorteccii, preserved.

CONSERVATION STATUS: Least Concern.

NATURAL HISTORY: Terrestrial and diurnal, shuttling between shade and sun in order to avoid lethal high temperatures; often common in suitable habitat, where it can be seen waiting in ambush in shade of a bush. Fast-moving, and hard to catch. Shelters in holes and occasionally under ground cover. *H. spekii* is an egg-layer, probably laying 2–6 eggs per clutch. Diet invertebrates; and preyed upon heavily by snakes and small raptors.

LONG-TAILED SAND LIZARDS *LATASTIA*

An African genus of long-tailed, fast-moving lizards, mostly striped, inhabiting dry savanna and semi-desert. They are diurnal, feed on invertebrates and lay eggs. Most species are medium to large-sized, but two are small. Ten species are known, most in north-east Africa, although two are West African, one East African, and one also extends to south-west Arabia. Seven species occur in our area. Three of them have three distinct subspecies each. taxonomy is not yet wholly clarified. Their closest relatives are the African genera *Philochortus*, *Pseuderemias* and *Heliobolus*. These lizards are among the most frequently encountered lacertids in the Horn of Africa.

KEY TO NORTH-EAST AFRICAN SPECIES OF *LATASTIA*

1a: Small, snout-vent length up to 5.7cm; centre of pectoral region covered by regularly arranged scales similar to those on belly — 2

1b: Medium to large, snout–vent length 7.8–11cm; centre of pectoral region usually with a group of small, irregular scales that interrupt the linear arrangement of ventral plates — 3

2a: Frontal scale with a median groove; dorsal scales keeled or tectiform; back usually with longitudinal stripe — *Latastia boscai*, Bosca's Long-tailed Lizard

2b: Frontal scale without median groove; dorsal scales smooth and flat; back without longitudinal stripes — *Latastia taylori*, Taylor's Long-tailed Lizard

3a: Ground colour of back often distinctly reddish or golden, with darker markings frequently feeble or absent; 2–3 rows of obvious blue spots on each flank and similar spots often present also on back; dorsal scales strongly keeled — 4

3b: Ground colour of back not conspicuously reddish or golden, and dark markings usually more or less distinct; blue spots absent or confined to single row on each flank; dorsal scales smooth to strongly keeled — 5

4a: Dorsal scales not larger than laterals, in 50–70 rows at mid-body; does not occur on Benadir coast of southern Somalia — *Latastia caeruleopunctata*, Blue-spotted Long-tailed Lizard

4b: Dorsal scales larger than laterals, in 45–51 rows at mid-body; confined to Benadir coast of southern Somalia — *Latastia petersiana*, Peters' Long-tailed Lizard

5a: Colour and pattern very distinctive; dorsal colour greyish-beige, back and flanks covered by dense large blackish-brown dark vertical bars resembling a zebra pattern; mid-body scale rows 45–57 — *Latastia cherchii*, Cherchi's Long-tailed Lizard

5b: Dorsal pattern with longitudinal stripes or spots, not a zebra pattern; mid-body scale rows 47–70 (usually above 52) — 6

6a: Shorter head; smaller, reaching 8.7cm snout–vent length; 4–8 femoral pores on each thigh; dark spots in paravertebral lines numerous and usually pronounced — *Latastia doriai*, Doria's Long-tailed Lizard

6b: Longer head; larger, reaching 11cm snout–vent length; 6–14 (>10 where sympatric with *L. doriai*) femoral pores on each thigh; dark spots in paravertebral lines less numerous and usually reduced or absent (especially in nominotypical subspecies) — *Latastia longicaudata*, Common Long-tailed Lizard

BOSCA'S LONG-TAILED LIZARD

Latastia boscai
Bedriaga, 1884

IDENTIFICATION: A small, slim, long-headed and long-tailed lizard with obtusely keeled dorsal scales, very variable in colour. The head is long, narrow and flat, snout rounded, body slim and depressed, limbs and toes long and thin. Tail long, 65–75 per cent of total length. Dorsal scales obtusely keeled or tectiform (roof-like), in 36–48 rows at mid-body. Ventral scales rectangular, in six rows, sometimes with evidence of an additional row of smaller and more rounded scales on either side, but without an irregular scale cluster (characteristic of some species). Femoral pores 8–14 on each side. Maximum size about 20cm (snout–vent length up to 5.7cm), average 15–17cm, hatchling size unknown. Three subspecies differ mostly in coloration. The nominate, strongly

Latastia boscai burii, near Oog.

Latastia boscai boscai, Keren.

striped form, *L. b. boscai*, is pale brown above, with five dark brown or black longitudinal stripes, of which the outermost are broadest, and these and the median dorsal almost always extend onto tail, the flanks are black with one to three series of round yellowish spots. Upper parts of hind limbs blackish with numerous distinct round paler spots. Tail greyish-blue or reddish pale brown. Top of head uniformly greyish-brown. *L. b. burii* is bluish-grey or whitish above, becoming pale reddish-brown posteriorly, with three black dorsal longitudinal stripes, which disappear towards tail, and two additional short dark parietal lines; one black bold stripe on the flanks, with an additional ventrolateral dull brown streak. Upper parts of hind limbs and tail pale orange-brown. Top of head greyish-brown marbled red. The least striped form, *L. b. arenicola* uniformly reddish-brown above, or with merest trace of longitudinal stripes; there is a darker fuzzy stripe on flanks sometimes marked with regular narrow vertical bars of black and white, and an additional ventrolateral pale whitish stripe among which several blue spots are scattered in males. Upper parts of hind limbs and tail pale orange-brown. Top of head brown. Taxonomic Notes: Apparently an undescribed striped form occurs in southern Somalia on the coast from Afmadu to Giohar (not shown on map).

HABITAT AND DISTRIBUTION: Both sandy and rocky environments, extending from sub-desert country on coastal plain to arid savanna and dense scrub further inland. The nominate subspecies is known from two disjunct areas; in central Eritrea and eastern Ethiopia, Djibouti and adjacent parts of Somaliland. In Eritrea it is common in hills around Keren and Elabaret and near Gahtielay, at 300–1,500m. In Ethiopia it is reported from the vicinity of Dire Dawa and then west towards its type locality in Rugdeia Sogheira, it is also common in rocky terrain around Borama in western Somaliland and at the boundary towards Djibouti, where it has been recorded from Forêt du Day, at 90–1,600m. *L. b. burii* occurs further east, in rocky environments around Berbera, the Cal Madow Mountains (common west and north of Erigavo), east to Madashon district in Puntland; also the Nogal Valley, south-east Somaliland,

Latastia boscai arenicola, preserved, Haud.

Latastia boscai burii, Buq.

at 100–2,000m. *L. b. arenicola* occurs further south in sandy habitats in the southern Nogal Valley in the Haud region, between Burtinle in Puntland and Bohodle in Somaliland, thence south into the Ogaden. Isolated records from Ado and Dolo in Rahanuin country, southern Somalia, at 180–680m.

CONSERVATION STATUS: Least Concern.

NATURAL HISTORY: Diurnal, terrestrial and very fast-moving. The nominate and *burii* subspecies favour rocky habitats in hill country and can be found at high altitudes with relatively low winter temperatures. Both were observed moving around and sheltering beneath large rocks. Gravid females were observed at the end of March, when the rainy season begins and temperatures increase. A pair of *L. b. arenicola* was observed mating in May. Presumably lays eggs but no clutch details known.

BLUE-SPOTTED LONG-TAILED LIZARD

Latastia caeruleopunctata
Parker, 1935

IDENTIFICATION: An elegant, vividly coloured, relatively large, long-tailed lizard from the Haud and Ogaden regions of the Horn of Africa. Head and snout moderately long, the body slightly depressed, limbs short, toes long and thin. Tail long, 65–75 per cent of total length. The dorsal scales are strongly keeled, in 50–70 rows at mid-body. Ventral scales rectangular, in six rows, sometimes with an additional row of smaller and more rounded scales on either side. There are 6–11 femoral pores. Maximum size about 33cm (snout–vent length up to 8.3cm), average 20–30cm, hatchling size unknown. Dorsal ground colour usually brick-red, sometimes yellowish-brown, the back and flanks covered with discrete green, turquoise or blue spots, which are usually denser on the flanks. Often, between the flank spots are fine black vertical bars, among which several yellow spots are scattered. The black vertical bars sometimes merge into a single dark longitudinal stripe that continues to the end of the tail. Adult tail greenish-beige, or red above and white below; largely greenish in juveniles. Uniform white below.

Latastia caeruleopunctata, juvenile, Shanshacade.

Latastia caeruleopunctata, adult, Shanshacade.

HABITAT AND DISTRIBUTION: A typical species of the Somali arid zone, in dry open *Vachellia–Commiphora* savanna and semi-desert scrub with soft red sand, at 150–1,300m. It avoids stony terrain in hilly country. Widespread in the Haud region, from Degeh Bur east along the Somaliland/Ethiopia boundary, from Davegoriale through Bohodle to the easternmost tip of Ethiopia and east to Geriban in southern Puntland. Widespread in the Ogaden (known from Ado and around Warder), thence south to south-central Somalia (El Bur, Jesomma and Buulobarde). Isolated records from Lugh, Mandera to Malka Murri and about 15km east of Moyale on the Ethiopia/Kenya border. A specimen from Gildessa, north-east of Dire Dawa in eastern Ethiopia, might not be correctly identified. Widespread and locally very common. Unlike to be threatened by human activity.

CONSERVATION STATUS: Least Concern.

NATURAL HISTORY: Poorly known. Diurnal and terrestrial, reported to inhabit both open country and dense *Vachellia* scrub; in alarm it seeks cover in holes among tree roots. In Somaliland observed in burrows under piles of dry branches or shrubs. An adult in the Haud region of southern Somaliland lived around a large low dense shrub, with fallen leaves where it sheltered when alarmed. One was observed foraging in fallen leaves under a *Vachellia* shrub, a juvenile there was seen to catch and eat a spider. North of Degeh Bur it is sympatric with two common lacertids, *Heliobolus spekii* and *Pseuderemias brenneri*. Presumably lays eggs. Diet invertebrates.

CHERCHI'S LONG-TAILED LIZARD

Latastia cherchii
Arillo, Balletto & Spano, 1967

IDENTIFICATION: A relatively small, long-tailed lacertid with zebra stripes, confined to the Hobyo area in central coastal Somalia. Head and snout fairly short, the body slightly depressed, limbs short, toes long and thin. Tail long, 65–75 per cent of total length. Dorsal scales strongly keeled, in 45–57 rows at mid-body. Ventral scales rectangular, in six rows, sometimes with an additional row of smaller and more rounded scales on either side, and curiously broken up between the fore limbs to form an irregular cluster of smaller scales. There are 12–14 femoral pores. Maximum size about 27cm (snout–vent length up to 7.8cm), average 20–25cm, hatchling size unknown. Colour and pattern very distinctive; dorsal ground colour greyish-beige, the back and flanks covered by dense large blackish-brown vertical bars resembling a zebra pattern, sometimes broken up on the back to delimit two light longitudinal stripes, which begin on the neck behind the parietal scales and continue to the start of the tail. Neck and back covered by discrete light round spots in some specimens. Upper hind limbs brown with distinct round lighter spots. Uniform white below.

HABITAT AND DISTRIBUTION: In coastal plains with sand dunes and sparse vegetation (small bushes) in extremely arid conditions, and adjacent arid shrubland and grassland, between sea level and about 50m or higher. A poorly known species; apparently restricted to coastal region of central Somalia. Known from a few localities around town of Hobyo (between Hobyo and Sissib Wells and between Hobyo and 'Magangib' [Magajebe]). The types were collected by L. Bricchetti Robecchi from 'the route from Obbia to Berbera', probably near Hobyo as later records suggest. Probably more widespread along the coast north and south of Hobyo, but not collected. Unlike to be threatened by human activities due to the low population in this region.

CONSERVATION STATUS: Least Concern.

NATURAL HISTORY: Unknown. Probably similar to others of the genus, i.e., diurnal, terrestrial, diet composed of invertebrates. Presumably lays eggs but no clutch details known. The specific name honours Maria Adelaide Cherchi (1927–85), an Italian herpetologist from Genova University, whose speciality was the herpetofauna of arid lands; she published widely on rare Somali species.

Latastia cherchii, preserved, Hobyo to Sissib.

DORIA'S LONG-TAILED LIZARD

Latastia doriai
Bedriaga, 1884

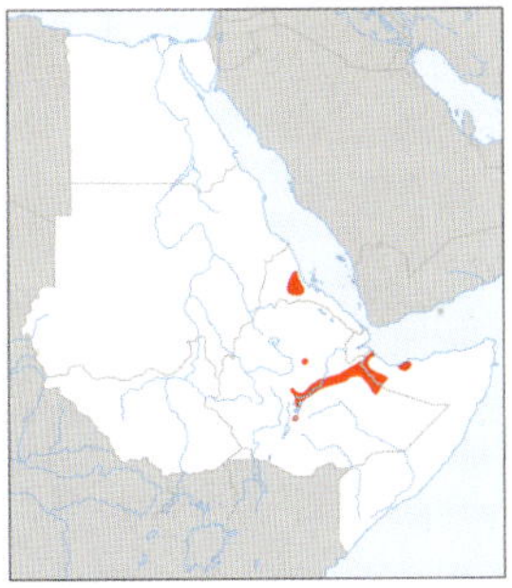

IDENTIFICATION: A relatively big, short-headed brown lizard with pattern of several longitudinal lines consisting of dark bold spots. Head fairly short, robust, snout pointed, body slim but muscular, with stocky rear limbs. Tail long, 65–75 per cent of total length. Dorsal scales tectiform or obtusely to strongly keeled, in 47–70 rows at mid-body. Ventral scales rectangular, in six rows, often with an additional row of smaller and more rounded scales on either side and curiously broken up between fore limbs to form an irregular cluster. There are 4–8 (usually 5–7) femoral pores on each side. Maximum size *c.*33cm (snout–vent length up to 8.7cm), average 25–30cm, hatchling size unknown. Three subspecies are recognised, which differ mostly in coloration, scale counts and distribution. The nominate form, *L. d. doriai*, is pale grey-brown and commonly has a dark brown or black vertebral line extending the length of the back, on either side of which are two or three rows of regularly spaced dark spots or blotches, and the innermost sometimes fuse to form a pair of longitudinal stripes, while those nearest the flanks show a more pronounced tendency to fuse into transverse stripes, creating a distinctive zebra pattern that may incorporate a single row of vivid yellow-green spots. Three longitudinal stripes cross the temporal region between the eyes and ear-opening, the upper and lower are dark, the middle one light, these are vertically overlapped by two or three blackish bars. Underside white and unmarked. Juveniles spotted like adults. Body scales in 54–70 (usually 66–70) rows at mid-body, femoral pores 4–7 (usually 5–6). The northern form, *L. d. martensi*, is similar to the previous form, but the dorsal lines are better developed and dorsum more intensively mottled with comparatively smaller spots; dark vertical bars in temporal region absent or reduced to one or very rarely two. Body scales in 47–61 (usually 56–60) rows at mid-body, femoral pores 5–8 (usually 7). The eastern form, *L. d. scorteccii*, is very distinctively patterned. The dorsal ground colour is light brown with up to eight very light lines formed by fusion of whitish spots or ocelli; the dorsum, flanks and upper parts of the tail are covered by indistinct dark bars, more evident on flanks, or spots, more evident on dorsum; the dark longitudinal lateral band incorporates a series of vivid blue or bluish-yellow spots. Body scales strongly keeled in 54–62 rows at mid-body, femoral pores 5–7. Taxonomic Notes: The very different biology and morphology of the last subspecies suggest that molecular analysis will probably reveal it to be a separate species.

Latastia doriai scorteccii, Berbera.

HABITAT AND DISTRIBUTION: Ranges from sub-desert habitats on coastal plains to arid *Vachellia* savanna and bushland further inland; often common around lava fields, as well as in agricultural landscapes. The nominate subspecies occurs from Sodere along the Awash River to Awash National Park, thence east past the type locality in Rugdeia Sogheira to Dire Dawa, and east into Somaliland to the Ogo Mountains

Latastia doriai martensi, Dekemhare.

Latastia doriai doriai, Metahara.

and Hargeisa, and south to Obele in the northern Ogaden. An isolated record from Batie, on the northern Ethiopian plateau, and also known from the vicinity of Addis Ababa, Lake Langano and Awassa in Ethiopia, might extend south in the Rift Valley, to Lake Chamo. Occurs at altitudes of 820–2,300m. The subspecies *L. d. martensi* occurs in the central highlands of Eritrea, at elevations between 270 and 2,100m; localities include Keren (the type locality), Sciotel, Arbascio, Mendefera, Dekemhare, south of Saganeiti and low hills along Red Sea coast west of Massawa. *L. d. scortecii*, a markedly different animal, is known only from the hot dry coastal Guban, north-west Somaliland; the type was from Mandah, also recorded at Zeila, near Berbera, in Berbera town and Bihin Gaha about 65km east of Berbera, between sea level and 250m. *L. d. doriae* probably occurs in nearby Djibouti.

CONSERVATION STATUS: Least Concern.

NATURAL HISTORY: Poorly known, but probably similar to the better-known *L. longicaudata*. Diurnal and terrestrial. Specimens in Awash National Park were active on open stony ground, and sheltered at night under lava rocks, both large and small. Juveniles were found here in November. In Somaliland a gravid female was observed in July. In Ethiopia and Somaliland, it occurs side by side with Common Long-tailed Lizard *Latastia l. longicaudata*, although the latter usually prefers more arid habitats and lower elevations. Diet invertebrates.

COMMON LONG-TAILED LIZARD

Latastia longicaudata
(Reuss, 1834)

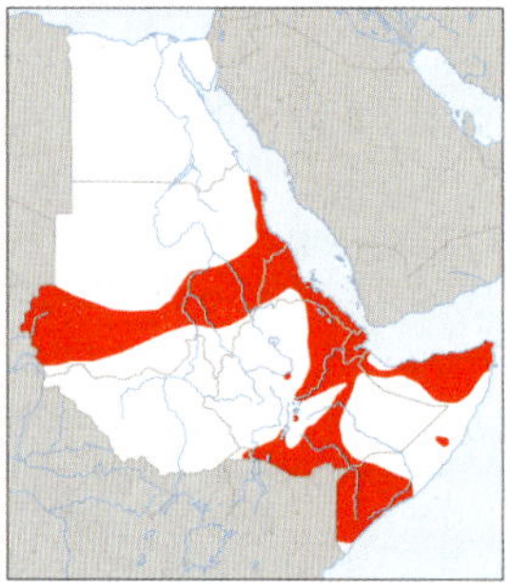

IDENTIFICATION: A big, long-headed, usually orange-and-blue lacertid, often common. The head is long, robust, snout pointed, body slim but muscular, with stocky rear limbs. Tail long, 65–75 per cent of total length. Body scales small, smooth or feebly keeled, in 51–70 (rarely up to 80) rows at mid-body. Ventral scales rectangular, in six or eight rows, often with an additional row of smaller and more rounded scales on either side and curiously broken up between the fore limbs to form an irregular cluster. There are 6–14 femoral pores on each side. Maximum size over 40cm (snout–vent length up to 11cm), average 25–35cm,

Latastia longicaudata longicaudata, Awash National Park.

hatchling size unknown. Three subspecies that differ in coloration, scale counts and distribution are recognised in our area. Nominate *L. l. longicaudata* is pale grey-brown with five or seven longitudinal rows of regularly spaced darker brown or reddish-brown spots or blotches which, particularly on sides of body, tend to fuse into transverse stripes and create a distinctive zebra pattern, particularly on sides of neck and head (an excellent aid to identification). Dark markings in midline unite longitudinally to form a continuous vertebral stripe, occupying most of the back, while on flanks a single row of vivid yellowish-green spots may be present or absent. Underside white and unmarked. Juveniles have similar markings but lateral transverse stripes more pronounced; males more vivid than females. Body scales tectiform or feebly keeled in 51–67 rows at mid-body, ventrals usually in six rows, femoral pores 10–14. Eastern *L. l. revoili* differs in more vivid coloration, is pale grey or bluish above with seven brick-red longitudinal stripes, the one in midline often occupies only anterior region of back. The red or black transverse stripes are well developed, giving appearance of a fine, almost reticulated pattern; on flanks interspaced by yellow or blue spots. Body scales smooth or tectiform in 54–70 rows at mid-body, ventrals usually in eight, less often six rows, femoral pores 6–11. Southern *L. l. lanzai* is most similar to the preceding form, but is more reddish, the midline dorsal stripe is usually well developed and extends from nape to tail, ventral scales in third transverse row wider than in second one. Body scales keeled in 53–69 rows at mid-body, ventrals usually in six rows, femoral pores 6–11.

HABITAT AND DISTRIBUTION: Wide-ranging, known from semi-desert, low moist and dry savanna on coastal plains, in thickets and large wadis. The nominate subspecies is known in our area from the Sahel in Sudan (most records from Kordofan, Darfur and along the Blue Nile), east to Gadarif and Kassala, north-east to the Red Sea coast hills, and north to Gebel Elba in Egypt. It extends south over most of the Eritrea lowlands (known from Dissei Island) into the Danakil Depression in Ethiopia, between sea level up to 1,400m. Also, Djibouti and the narrow Guban strip east to the coastal plains of Puntland, around Bosaso. Elsewhere, west to Senegal. *L. l. revoili* occurs in the Somaliland and Puntland mountains and south of there; from the Goolis Mountains west to Las Geel, and Sheikh, east from the Cal Madow Mountains (common around Erigavo) south across the Garadag plains to Las Anod in the Nogal Valley. Widespread and common on the Somaliland/Puntland border as far north as Karin, then east and south of Cal Miskaat Mountains via the Daror Valley to Iskushuban and Hordio. Isolated records further south inland of Hobyo at Uarandi and between Durgale and

Latastia longicaudata lanzai, Negele.

Latastia longicaudata revoili, betwen Garadag and El Afweyn.

Latastia longicaudata revoili, Las Geel.

Dolobscio (Lulubsho). Found at elevations between sea level and 1,900m. *L. l. lanzai* is known from southern Somalia, in the Hiran region and Mogadishu west to the Kenya border, notable localities include Giohar, Afmadu, Dinsor, Uegit and Dolo. Widespread in southern Ethiopia in red sand lowlands as far north as the Ghelbi Mountains (at Goulf), thence south-wests on the eastern highland plateau to Yabello, Lakes Chew Bahr and Turkana, to the Ilemi Triangle in South Sudan, at elevations between sea level and 1,500m. Elsewhere, to northern Tanzania.

CONSERVATION STATUS: Least Concern, but *L. l. longicaudata* is Vulnerable in Egypt.

NATURAL HISTORY: Diurnal, ground-foraging and very fast-moving, probably the fastest north-east African lizard, and is alert, wary and hard to approach. Active during hottest hours; it waits under bushes, dashing out onto open ground to seize prey, but will venture into thickets and cover, even foraging in leaf litter. Shelters in holes. Clutches of 3–12 eggs laid by captive specimens take from 8–12 weeks to hatch at temperatures of 24–28°C. The eggs hatch to coincide with the start of the rainy season. Diet invertebrates, including scorpions.

PETERS' LONG-TAILED LIZARD

Latastia petersiana
(Mertens, 1938)

IDENTIFICATION: An elegant, vividly coloured, large, long-tailed lizard from the Benadir coast of southern Somalia. Head and snout moderately long, the body slightly depressed, limbs short, toes long and thin. Tail long, 65–75 per cent of total length. Dorsal scales large, rhombic, imbricate and strongly keeled, in 45–51 rows at mid-body. Ventral scales rectangular, in six rows, sometimes with an additional row of smaller and more rounded scales on either side. There are 7–11 femoral pores on each side. Maximum size about 35cm (snout–vent length up to 9.5cm), average 25–32cm, hatchling size unknown. Dorsal ground colour usually greyish-yellow, sometimes golden, the back and flanks covered by discrete blue spots in six or eight series, and blackish vertical bars between the outer series of blue spots. In younger animals dorsal ground colour pale brown with blue spots merged into three longitudinal lines, the flanks completely blue with blackish vertical bars. Upper parts of hind limbs brown with distinct small round lighter spots. Uniform white below. Taxonomic Notes: Originally described as *Lacerta* (currently *Latastia*) *carinata* by W. Peters in 1874. At that time, another species with the name of *Lacerta carinata* Schinz, 1833 (currently a junior synonym perhaps of *Psammodromus algirus*) already existed. Therefore, it was necessary to establish a new name, *petersiana*, for this lacertid from southern Somalia.

Latastia petersiana, preserved, Mogadishu.

HABITAT AND DISTRIBUTION: A poorly known species, inhabiting a narrow coastal line of fixed sand dunes a little distance from the sea, with mixed bushland and scattered trees, below 100m. A Somali endemic, confined to Benadir coast of southern Somalia (records from northern Somalia and eastern Ethiopia are erroneous). Known from a handful of specimens, originally described from Brava, also recorded at Kismayu, around Mogadishu and further north towards Balad. Not known inland and elsewhere.

CONSERVATION STATUS: Data Deficient.

NATURAL HISTORY: Unknown, never knowingly observed alive. Closely related to *L. caeruleopunctata*, which occurs in similar habitats, so its behaviour may be similar too.

TAYLOR'S LONG-TAILED LIZARD

Latastia taylori
Parker, 1942

IDENTIFICATION: A very small, long-headed and long tailed lizard with smooth dorsal scales, from Puntland and adjacent Somaliland territories. Head flat, snout rounded, body slim and depressed, limbs and toes long and thin. Tail long, 70–75 per cent of total length. Dorsal scales smooth, oval, in 33–41 rows at mid-body. Ventral scales rectangular, in six rows; without an irregular cluster of smaller scales between fore limbs. There are 9–13 femoral pores on each side. Maximum size about 20cm (snout–vent length up to 4.7cm), average 14–17cm, hatchling size unknown. Pale reddish-brown or olive above and on tail, almost uniform, faintly marbled with grey anteriorly and on head; anterior flanks and side of neck marked with very irregular brown and greyish-white vertical marbling. Two, more or less distinct, dark stripes extend from behind eye along flanks to just before hind limbs. Snout orange-yellowish. Uniform white below.

HABITAT AND DISTRIBUTION: Rocky environments, extending from subdesert country on coastal plains to arid savanna further inland, up to a maximum elevation of about 1,900m, but mostly at 500–1,500m. At lower elevations it inhabits hilly country or rocky limestone plateaux with sparse vegetation, at higher elevations in mountainous regions it is found in mesic habitats with a dense growth of succulent plants, *Dracaena* and *Olea* or even *Juniperus* trees. Known mostly from Puntland but also adjacent parts of Somaliland. Southernmost locality Wadi Run between Garowe and Gardo, thence extends along border to Buran region in Somaliland and north to Galgalo Oasis and Qoow (Banda Ziada District) near coast. Widespread in Cal Miskaat Mountains, notably localities include Sukorre, Habeno, Abal and the Gid Daa Wells, north of Dadar in the east.

CONSERVATION STATUS: Least Concern.

Latastia taylori, preserved type.

NATURAL HISTORY: Poorly known, probably similar to others of the genus, although it is much more of a high-altitude species than any other *Latastia* in our area. A Qoow female collected in December was gravid. Its mountainous habitat is characterised by high daytime temperatures (37°C) and stagnant air, with cool winds and low temperatures at night. The specific name honours Captain R. H. R. Taylor, British officer in the Royal Artillery, and administrator in British Somaliland. A member of the Anglo-Ethiopian Boundary Commission, Taylor was responsible for the most important collection of reptiles and amphibians ever made in Somaliland; now housed in the Natural History Museum, London.

DESERT RACERS *MESALINA*

A Saharo-Sindian genus of fairly small slender active terrestrial lacertids, that favour desert and semi-desert. They have a pointed snout and prominent nostrils between three nasals. Their coloration is cryptic; mostly sandy grey-brown with an ocellated or more or less evident striped dorsal pattern. They occur in northern Africa and south-west Asia, from Senegal and Morocco to Pakistan and western India, Somalia to Turkmenistan. About nineteen species are known, eight in our area, two are endemic. The estimated divergence times place the origin of the genus in the early Miocene (c.22 Mya). They are closely related to lacertids of the genera *Acanthodactylus* and *Ophisops*.

KEY TO NORTH-EAST AFRICAN MEMBERS OF THE GENUS *MESALINA*

1a: Occipital scale absent or very small, extreme south Sinai — *Mesalina bernoullii*, Bernoulli's Short-nosed Desert Racer

1b: Occipital scale present — 2

2a: Rostral scale in contact with frontonasal, snout very prominent and nasal scales very swollen — *Mesalina rubropunctata*, Red-spotted Desert Racer

2b: Rostral scale separated from frontonasal by upper nasals, snout moderately prominent with nasal scales not or only slightly swollen — 3

3a: Dorsal scales large, imbricate, quite pointed and keeled; endemic to south-central Somalia — *Mesalina ercolini*, Somali Desert Racer

3b: Dorsal scales small to medium-sized, granular or sub-imbricate, smooth; not in south-central Somalia — 4

4a: Subcaudals smooth; ventral side of lower arm covered with small granules; lower eyelid with a window mainly of two large transparent scales of equal size, edged black — 5

4b: Subcaudals more or less keeled; ventral side of lower arm covered with large imbricate scales; lower eyelid with a window of several transparent or semi-transparent scales, not edged black — 6

5a: Dorsal pattern of scattered light and dark spots; mid-body scale rows 37–51; Egypt, west of Suez Canal — *Mesalina guttulata*, Small-spotted Desert Racer

5b: Dorsal pattern usually more or less striated, with obvious ocelli in our area; mid-body scale rows 44–55; only in Sinai — *Mesalina bahaeldini*, Mount Sinai Desert Racer

6a: Lower eyelid window of two large transparent scales; dorsal pattern striated with no obvious ocelli; only Siwa region of north-west Egypt — *Mesalina pasteuri*, Pasteur's Desert Racer

6b: Lower eyelid window of 4–5 or more sub-equal semi-transparent scales; not known from Siwa region of north-west Egypt — 7

7a: Mid-body scale rows 40–50, northern Egypt and Sinai — *Mesalina olivieri*, Olivier's Desert Racer

7b: Mid-body scale rows 37–40, south-east Egypt along Red Sea coast, south to Somaliland — *Mesalina martini*, Martin's Desert Racer

MOUNT SINAI DESERT RACER

Mesalina bahaeldini
Segoli, Cohen & Werner, 2002

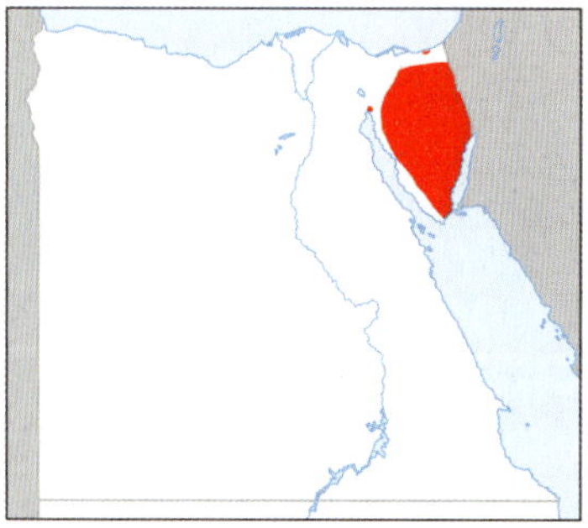

IDENTIFICATION: A small, slender, grey-brown lacertid, usually with a striated dorsal pattern and scattered white spots; Sinai. Head and snout fairly long, slightly flattened; snout obtusely pointed, with moderately prominent nostrils; usually four upper labials anterior to subocular; lower eyelid with a window of two large transparent scales of equal size, edged black; temporal scales small and smooth. Body almost cylindrical, slightly depressed, slim; tail long, about 65–75 per cent of total length. Body scales medium-sized, granular, smooth, in 44–55 rows at mid-body; ventral scales in ten longitudinal rows, and 28–32 lengthwise transverse rows. Ventral side of lower arm covered with small granules. There are 10–14 (usually 13) femoral pores on each side. Subcaudals smooth. Maximum length about 16cm (snout–vent length 5.8cm), hatchlings about 6cm. Specimens from montane Sinai pale grey-brown with several darker and paler stripes, interspersed by scattered whitish spots, tail irregularly banded. Extralimitally (e.g., Jordan) may lack stripes but retain scattered ocelli. A distinct dark spot on subocular scale. Underside white. Juveniles have a bluish tail, and may or not be strongly striped. Taxonomic

Notes: Belongs in the *M. guttulata* complex, along with that North African species and three species (one still undescribed) from Arabia. Two subspecies are recognised, the nominate and *M. b. curatorum* from Suez in Egypt, which has larger body scales and is slightly different in colour; and to date has not been analysed molecularly.

HABITAT AND DISTRIBUTION: Desert and semi-desert, usually in stony or rocky areas with sparse vegetation in mountainous or hill country, usually at medium or higher altitudes to 2,070m). Regarded initially as a montane Sinai endemic, subsequent work indicates that populations of *M. guttulata* from Israel/ Palestine, Jordan and northern Saudi Arabia belong to this species as well. Widespread in mountainous central and southern Sinai, but also known from flatter, mostly non-rocky areas in Suez and Al Arish (elevation 27m) in northern Sinai. Very common around St. Catherine's Monastery from where it was originally described. Elsewhere, east to northern Saudi Arabia. Fairly common and widespread.

CONSERVATION STATUS: Least Concern.

NATURAL HISTORY: A diurnal and terrestrial species that favours bare stony or rocky habitats with sparse vegetation, mostly active in morning and afternoon. When inactive, it shelters under stones or in burrows. Diet arthropods; observed to eat scorpions and desert ground-dwelling mantids. Females lay 1–7 (average four) eggs, roughly 5×10mm in size, probably several times in season, juveniles hatch one or two months later. The specific name honours Sherif Baha El Din (b. 1960), prominent Egyptian herpetologist, ornithologist, environmental consultant, and author of the first (and to date, only) comprehensive, accessible guide to the Egyptian herpetofauna.

Mesalina bahaeldini, Sinai.

BERNOULLI'S SHORT-NOSED DESERT RACER

Mesalina bernoullii
(Schenkel, 1901)

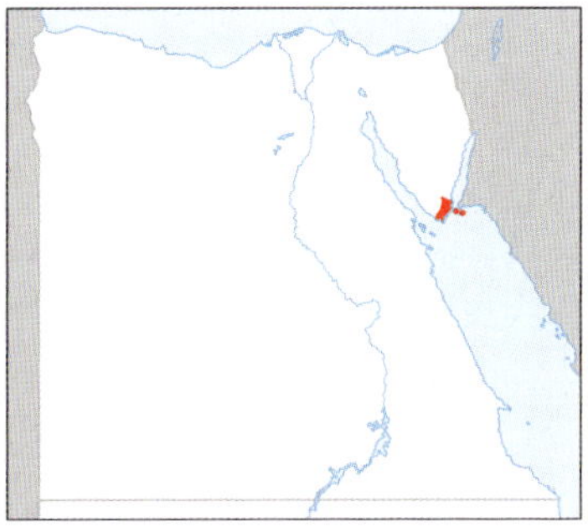

IDENTIFICATION: A small to medium-sized lizard with depressed body, short head and snout, and no or very small occipital scale; only in South Sinai. Head short, broad and relatively high; forehead distinctly concave; snout short, obtusely pointed, with more prominent nostrils; usually four upper labials anterior to subocular; lower eyelid with two large semi-transparent scales of equal size; temporal scales small, smooth; occipital scale absent or very small. Body rather strongly, depressed, slim; tail long, about 70 per cent of total length. Body scales fine, granular and smooth, in 46–55 (mostly 52 in our area) rows across mid-body; ventral scales in 12 longitudinal rows and 29–36 transverse series. Femoral pores 11–17 (mostly 15) on each thigh, subcaudals smooth. Maximum length 16–17cm (snout–vent length 5.7cm), hatchlings 6–7cm long. Dorsum pinkish-grey with numerous whitish and pale brownish or reddish dots, resembling ocelli, larger on flanks and sides of tail; some specimens may be almost uniformly plain. Underside whitish. Juveniles usually brownish and densely spotted, or uniformly plain with a blue tail. Taxonomic Notes: Recently re-elevated to species level following analysis and a four-way split of the *M. brevirostris* species complex.

HABITAT AND DISTRIBUTION: Desert and semi-desert, with sandy soil or gravel and a moderate amount of vegetation; on coast found in broken country on uplifted fossil reefs. Restricted to southernmost tip of south Sinai; recorded from Ras Mohamed National Park and nearby wadis as far north as Wadi Kid north of Sharm El Sheikh and Nabq. Also, on adjacent Tiran (Egypt) and nearby Sanafir Islands (Saudi Arabia).

Elsewhere, east and north to Syria, Iran, Kuwait and northern Saudi Arabia. Local and uncommon in Egypt.

CONSERVATION STATUS: Not Assessed.

NATURAL HISTORY: Diurnal, even in hottest parts of day. Active after initially basking, shuttling between foraging and sheltering under rocks or in shade of bushes. Ground-dwelling, but may occasionally climb low branches of vegetation in its territory. Diet small insects; small scorpions, spiders and solpugids also recorded. Reproduction poorly known; after spring mating, the female lays 1–6 eggs, roughly 10–7mm. Sympatric with *M. olivieri*.

Mesalina bernoullii, Syria.

SOMALI DESERT RACER

Mesalina ercolini
(Lanza & Poggesi, 1975)

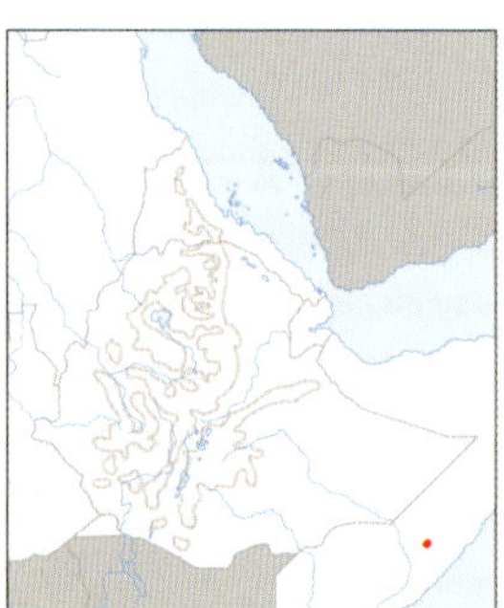

IDENTIFICATION: A Somali endemic known from a single specimen; a medium-sized, short-legged lacertid with two rows of dark bold spots on dorsum and keeled dorsal scales. The head is moderately long and depressed; snout slightly concave and pointed, nostrils not prominent. Four upper labials anterior to subocular which is excluded from lip; lower eyelid semi-transparent with about 12 enlarged scales; upper head scales nearly smooth; temporal scales rather large, smooth. Body slightly depressed. Tail is absent in type. Body scales large, imbricate, keeled at mid-body and smooth on flanks, in 45 rows at mid-body. Ventral scales in eight longitudinal rows and 50 transverse rows. Limbs short, the ventral side of lower arm covered with moderately large imbricate scales. Femoral pores 16 on each side. Maximum length probably about 22cm; snout–vent length 6.6cm. Dorsum hazel-brown with two paravertebral rows of 9–10 darker brown, light-edged sub-rectangular blotches; a narrow vague brownish dorsolateral stripe. A light supraciliary stripe, and a white, brown-edged subocular stripe extends onto neck. Limbs hazel-brown above; the hind limb has some small whitish, blackish-edged spots which are barely distinct on fore limb; underside white.

HABITAT AND DISTRIBUTION: Known from a single female specimen from one of Lanza's field collectors, taken at Bud Bud village, 70km south of El Bur in Galguduud region of south-central Somalia. The habitat there is xerophilous open woodland and subdesert scrub on red sand, elevation 138m. Subsequent detailed fieldwork has failed to find any more.

CONSERVATION STATUS: Data Deficient.

NATURAL HISTORY: Unknown. Presumably similar to other *Mesalina* (terrestrial, diurnal, lays eggs, eats arthropods). Its relatively short legs and protective, large, keeled overlapping dorsal scales suggest that it spends substantial time in dense low ground cover.

Mesalina ercolini, preserved, Bud Bud.

SMALL-SPOTTED DESERT RACER

Mesalina guttulata
(Lichtenstein, 1823)

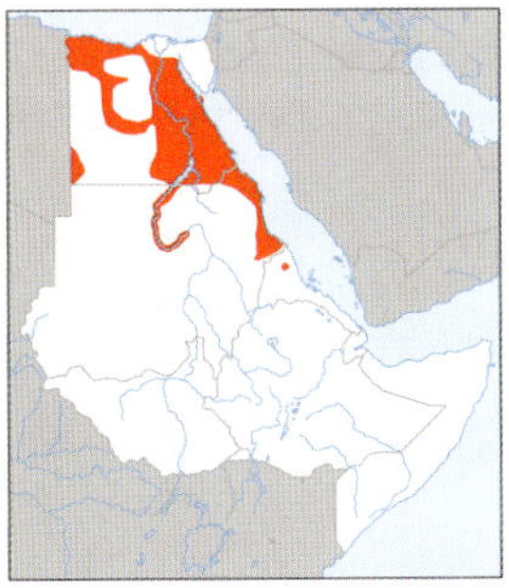

IDENTIFICATION: A small, slender, grey-brown lacertid with long depressed head, dorsum with numerous paler and darker spots; known from Egypt west of the Suez Canal, Sudan and probably Eritrea. Head narrow, snout long, pointed with moderately prominent nostrils; usually four (exceptionally five) upper labials anterior to subocular; lower eyelid with a window of two large transparent scales of equal size, edged with black; temporal scales small and smooth. Body depressed, slim; tail long, about 70 per cent of total length. Body scales medium-sized, granular, sub-imbricate, in 37–51 rows across mid-body; ventral scales in ten longitudinal rows and 27–33 transverse rows. Ventral side of lower arm covered with small granules. There are 10–15 (usually 12) femoral pores on each side. Subcaudals smooth. Maximum length about 16cm (snout–vent length 5.2cm), hatchlings about 6cm. Dorsum light grey-brown, with scattered white and dark brown or blackish spots (ocelli), sometimes ordered in longitudinal rows and forming a weak striated pattern; tail irregularly banded or spotted; a distinct dark spot on subocular scale below eye. Underside whitish. Almost pattern-less or rather strongly striated specimens are known. Juveniles have a blue tail and vivid ocelli. Taxonomic Notes: A member of the *M. guttulata* species complex, together with *M. bahaeldini*; a recent revision of the complex restricted this species to North Africa.

HABITAT AND DISTRIBUTION: Rocky areas in desert and semi-desert; including mountains, hillsides, gravel plains, rocky wadis, and other similar biotopes with a hard substrate, at low and medium altitudes. The most widespread *Mesalina* in our area; occurs

Mesalina guttulata, Wadi El Gemal.

Mesalina guttulata, Barramiya.

widely in Egypt west of the Suez Canal (but absent from cultivated areas of Delta and sandy parts of Western Desert). Occurs on most Red Sea islands, including the Hurghada archipelago. Extends along Red Sea coast and hills to Sudan, where reported from Dongonab Bay and Um Elsheikh Island, Durrur, Port Sudan, Suakin, Erkowit, Akik and Hamashkoraib. A vague old record from Eritrea (the dot on map shows the approximate position). Extends into Sudan along the Nile (Wadi Halfa, Ambukol, Kerma, Dongola to Merowe Water Reservoir, where reported near Es Suweigat). Tentative sight records (not mapped) from Kassala; El Mattama near Shendi; El Fula, Baleela and El Muglad in north-western South Kordofan and some localities in Darfur in the west. Elsewhere, west to Mauritania across the Sahara. Widespread and common.

CONSERVATION STATUS: Least Concern.

NATURAL HISTORY: A secretive, diurnal, ground-dwelling lizard, sometimes climbs low bushes. Favours rocky or stony habitats with sparse vegetation, avoids large sandy areas. Shelters in burrows in loose substrate or under rocks. Reproduction starts early in spring; males bite the female's flanks and hold on during mating. Clutches of 2–7 eggs laid several times per year. Diet small invertebrates. At various places of Egypt and Sudan sympatric with *M. olivieri*, *M. pasteuri* and *M. rubropunctata*.

MARTIN'S DESERT RACER

Mesalina martini
(Boulenger, 1897)

IDENTIFICATION: A small, slender, sandy grey-brown lacertid with a striated dorsal pattern, associated with the Red Sea coast. Head long; snout moderately long, obtusely pointed, with less prominent nostrils; usually four or five upper labials anterior to subocular; lower eyelid with a window of 4–8 semi-transparent scales of sub-equal size; temporal scales small, smooth. Body strongly depressed and slim, tail long, about 70 per cent of total length. Body scales medium-sized, granular or sub-imbricate, obtusely tectiform, in 37–40 rows across mid-body; ventral scales in eight or ten longitudinal rows and 26–34 transverse rows. Ventral side of lower arm covered with large imbricate scales. Femoral pores 9–16 on each side. Subcaudals smooth. Maximum length about 16cm (snout–vent length up to 4.7cm), hatchling size unknown. Dorsum pale grey or yellowish to dark brown, with a longitudinal pattern consisting of darker stripes, lines or darker spots, or rows of small blackish blotches each associated with a prominent white spot; a thin pale dorsolateral stripe is present. Occasionally pattern-less and uniform. Cream or dirty white below. Males more brightly coloured, especially during mating season.

HABITAT AND DISTRIBUTION: Deserts and semi-deserts, lava fields and wadis, and other suitable littoral habitats, along Red Sea coast and Gulf of Aden, between sea level and 1,080m, but usually at lower altitudes. Widespread and abundant in much of Djibouti (type from Obock) and nearby Somaliland and Ethiopia (where extends to Biye-Gurgur and Aroweina, the most inland records, 120km from sea). Occurs in eastern lowlands of central Eritrea, also on Green Island near Massawa and the Dahlak archipelago, on Nocra and Dahlak Islands, and found along entire Red Sea coast of Sudan, north to Marsa Alam in south-east Egypt. Distribution probably continuous along Eritrean coast and also the Guban in Somaliland, but records lacking. Elsewhere, south-west Yemen.

CONSERVATION STATUS: Least Concern, but Vulnerable in Egypt; its habitat is being lost and degraded by tourism development on the Red Sea coast.

NATURAL HISTORY: Poorly known. Terrestrial, solitary and diurnal, basks in morning on ground near vegetation, active during hottest times of day. Lays eggs. Diet insects and other arthropods. Found on sandy and stony soil. A dark brown Djibouti specimen was observed foraging on ground among stones in lava desert with sparse vegetation; Egyptian specimens on a sandy substrate among halophytic vegetation beside the sea.

Mesalina martini, Eritrea.

OLIVIER'S DESERT RACER

Mesalina olivieri
(Audouin, 1829)

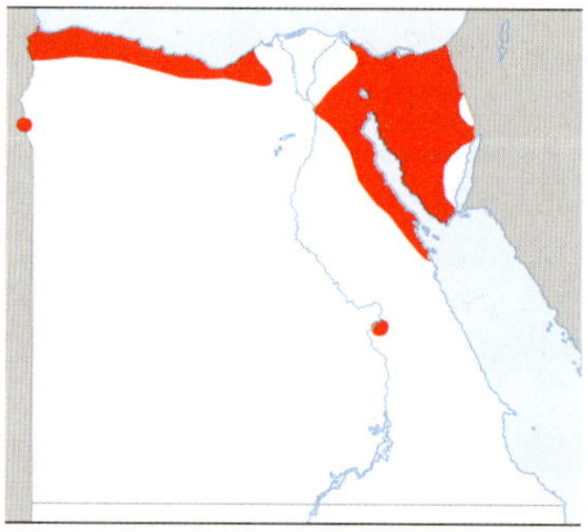

IDENTIFICATION: A small, slender, sandy grey-brown lacertid with striated dorsal pattern and obvious ocelli, in northern Egypt. Head small; snout relatively elongated, obtusely pointed, with moderately prominent nostrils; four or five upper labials anterior to subocular; lower eyelid with a window of 4–5 semi-transparent scales of sub-equal size, not edged with black; temporal scales small and smooth. Body almost cylindrical, slightly depressed, slim; tail long, about 65–70 per cent of total length. Body scales medium-sized, granular, smooth, in 40–50 (mostly 44–45 in our area) rows across mid-body; ventral scales in ten longitudinal rows and 27–36 transverse rows. Ventral side of lower arm covered with large imbricate scales. Femoral pores 9–15 (mostly 12) on each thigh, subcaudals more or less keeled. Maximum length about 16.5cm (snout–vent length 5.3cm), hatchlings about 6cm. Coloration and pattern fairly variable; dorsum light grey-brown, with four darker brownish stripes and rows of small ocelli (some animals look distinctly spotted). The vertebral zone usually differs in colour (grey, blackish, dark brown) from rest of body. Underside whitish. Juveniles always intensively striped. Taxonomic Notes: Belongs to *M. olivieri* species complex, with two subspecies but molecular analysis indicates a more complex situation of cryptic, undescribed species.

HABITAT AND DISTRIBUTION: Open sandy desert, semi-desert and gravel plains, usually with rocks and some vegetation, sometimes in dunes, and saltmarshes, usually at low altitudes. Restricted in Egypt to the north; along Mediterranean coast from Libyan border east to Nile Delta, reappears east of there, occurs north-east and south-east of Cairo, across most of Sinai (absent from mountainous areas) and along eastern coast to Hurghada. Isolated records west of Siwa and at Luxor. Elsewhere, east to southern Syria, western Iraq and northern Saudi Arabia, and west to Morocco and Senegal. Widespread and locally common, but possibly at risk from construction on Mediterranean coast.

CONSERVATION STATUS: Least Concern.

NATURAL HISTORY: Diurnal, mostly active in morning, starting the day by basking, when warm enough they forage actively; resting occasionally in shade. In midsummer may be active also during twilight. Ground-dwelling, but may climb low vegetation and even sleep there. Shelters in burrows and under rocks. Feeds mostly on small insects, including ants. After spring mating (April), female lays usually 2–4 (but up to eight) eggs, roughly 10–6–8mm, several times per season, at intervals of 3–4 weeks. Sexual maturity reached when 7–9 months old. Sympatric with *M. guttulata*.

Mesalina olivieri, Zaranik Protected Area.

PASTEUR'S DESERT RACER

Mesalina pasteuri
(Bons, 1960)

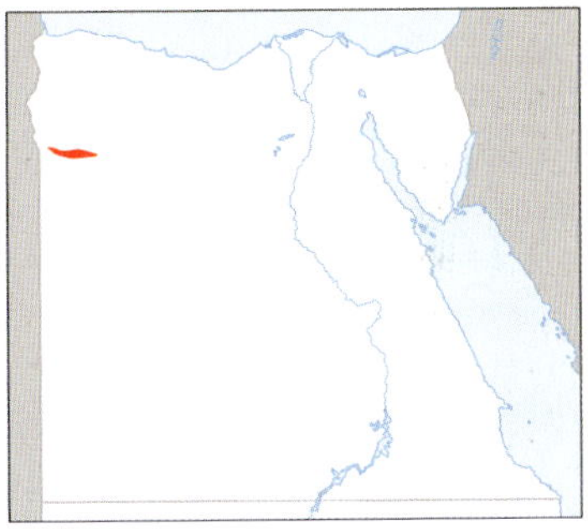

IDENTIFICATION: A small, slender, sandy grey-brown lacertid with striated dorsal pattern and no obvious ocelli, only in north-west Egypt. Head small, fairly short; snout moderately long, obtusely pointed, nostrils not prominent; five (rarely four) upper labials anterior to subocular; lower eyelid with a window of two (sometimes three plus smaller) transparent scales, and 2–3 smaller ones on low anterior margins; temporal scales rather large, slightly carinate. Body strongly depressed, elongate and slim; tail long, about 70 per cent of total length. Body scales medium-sized, granular or sub-imbricate, smooth or slightly keeled, in 32–41 (mostly 36 in our area) rows across mid-body; ventral scales in ten longitudinal rows and 28–33 transverse rows. Ventral side of lower arm covered with large imbricate scales. Femoral pores 10–14 pores each side, subcaudals strongly keeled. Maximum length about 16cm (snout–vent length up to 4.6cm), hatchlings 6–7cm. Dorsum light, sandy-grey, with four fairly low-contrast brownish stripes; vertebral zone between the two inner stripes is pale brownish-grey, usually with white spots (not ocelli). Underside white. Males more brightly coloured. Juveniles darker than adults and more brightly coloured with stronger striation; tail yellowish. Taxonomic Notes: Belongs to *M. olivieri* species complex, sister to *M. olivieri* (which largely occurs north of the Sahara).

Mesalina pasteuri, Siwa Oasis.

HABITAT AND DISTRIBUTION: Desert with mobile, soft sand and some vegetation. Known only from Siwa region of western Egypt, the easternmost limit of its range, where it occurs at and around several oases, from Shayata in the west to Tabaghbagh in the east; in a depression about 15m below sea level, bounded to the north by an escarpment and to the south by dunes. Elsewhere, west, right across the Sahara to Mauritania, although interestingly not recorded from Libya or northern Chad.

CONSERVATION STATUS: Data Deficient; very local in Egypt, should be monitored for threats due to habitat loss.

NATURAL HISTORY: Poorly known. Generally diurnal, on extremely hot days becomes active after sunset. Terrestrial. In Egypt said to have a particular preference for small undulating sand dunes with bushes of *Alhagi graecorum* and the grass *Stipagrostis*, spending much of time in the vegetation. Sympatric here with two species of *Acanthodactylus* and *M. guttulata*. Oviparous, in captivity females laid three eggs, roughly 5–6–10mm in size. Diet small invertebrates. The specific name honours Georges Pasteur (1930–2015), French herpetologist, geneticist and evolutionary biologist, generally considered to be the father of *Lygodactylus* systematics, and collector of the type specimens of *M. pasteuri*.

RED SPOTTED DESERT RACER

Mesalina rubropunctata
(Lichtenstein, 1823)

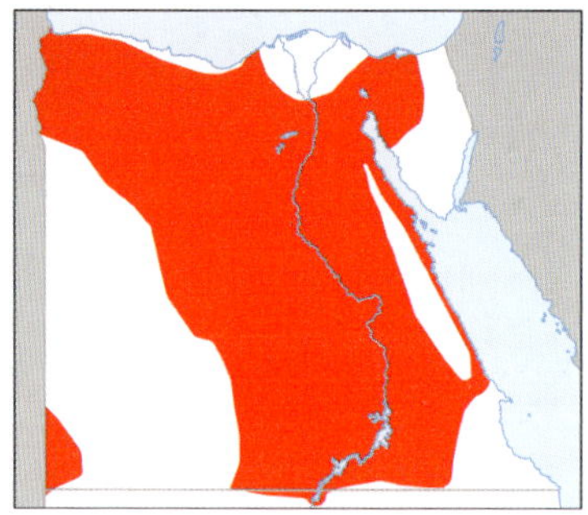

IDENTIFICATION: A small to medium-sized, strongly depressed, distinctly patterned lizard with swollen nasal scales. Head fairly short and wide, depressed; snout rather short, obtusely pointed with very prominent nostrils (an excellent aid to identification); usually four (sometimes five) upper labials anterior to subocular; lower eyelid with a window of 5–8 semi-transparent scales of roughly equal size; temporal scales small, smooth. Body rather strongly depressed, wide; limbs long, slim; tail about 60 – 65 per cent of total length. Body scales small, granular, in 53–67 (average 60 in our area) rows across mid-body; ventral scales in 12 longitudinal rows and 32–37 transverse rows. The ventral side of the lower arm is covered with small granules. Femoral pores 13–22 (average 16) on each thigh, subcaudals smooth. Maximum length about 18cm (snout–vent length 6.7cm), hatchlings about 7cm. Coloration and pattern distinct; dorsum pale pinkish or olive-grey with distinct white spots bordered dark red or brown, usually arranged in more or less regular longitudinal rows; tail plain above, strongly spotted on sides. Underside whitish. Juveniles plain pinkish-grey with no pattern and a bluish tail. Pattern-less specimens are not uncommon.

HABITAT AND DISTRIBUTION: Desert or semi-desert with little or no vegetation, usually in sandy or rocky hyper-arid areas, at low and medium elevations; harsh country where no other reptiles occur. In Western and Eastern Deserts of Egypt and north-east Sinai, (west and south of Al Arish, southernmost records from Wadi Gharandal and Nekhel). Absent from Nile Delta, the north coast, the great sand sea of the south-west and eastern maritime mountains. Just enters Sudan around Wadi Halfa. Elsewhere, west throughout the Sahara to Mauritania. Widespread, an uncommon and fairly secretive species.

CONSERVATION STATUS: Least Concern.

NATURAL HISTORY: Similar to other desert racers, i.e., ground-dwelling, diurnal, lays 3–7 eggs in two or three clutches, eats invertebrates. During night shelters in a self-dug burrow under a rock, during hottest hours of day hides under rocks or in shade of plants. Of all the *Mesalina* in our area, it is the most tolerant of harsh hyper-arid desert conditions.

Mesalina rubropunctata, Morocco.

SNAKE-EYED LACERTIDS *OPHISOPS*

A Palearctic genus of around 11 species of fast-moving, slim, largely striped, semi-desert and desert lizards, widespread across the Near and Middle East, east to India. They lack closable eyelids. Three species in our area.

KEY TO NORTH-EAST AFRICAN SPECIES OF *OPHISOPS*

1a: In our area, only south of 27°N; supraciliary scales in contact with supraoculars, dorsal scales in 17–19 rows — *Ophisops elbaensis*, Mount Elba Snake-eyed Lacertid

1b: In our area, only north of 27°N; a row of tiny granular scales separate supraciliary scales and supraoculars, dorsal scales in 19 or more rows — 2

2a: In our area, only from north-east Sinai; usually nine or more femoral pores — *Ophisops elegans*, Elegant Snake-eyed Lacertid

2b: In our area, only from Egyptian coast west of Nile Delta, usually <9 femoral pores — *Ophisops occidentalis*, Western Snake-eyed Lacertid

MOUNT ELBA SNAKE-EYED LACERTID

Ophisops elbaensis
Schmidt & Marx, 1957

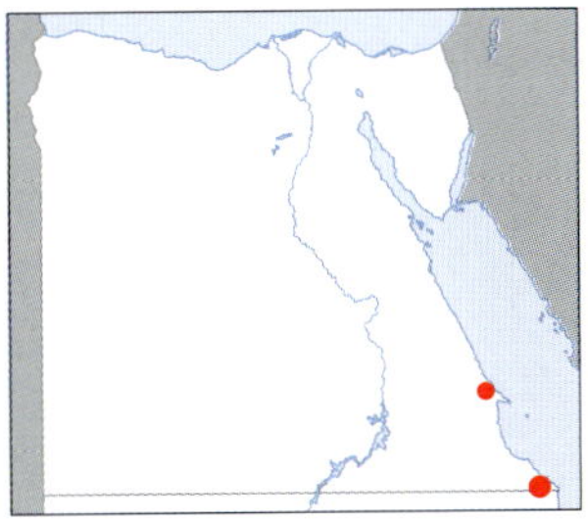

IDENTIFICATION: A small long-tailed lacertid of the southern Egyptian Red Sea coast. Very slender; tail relatively long, 72–75 per cent of total length. Scales at mid-body large, imbricate and strongly keeled, usually arranged in 17–19 transverse rows; ventrals in six straight rows. Femoral pores 9–10 on each side. Maximum size about 13cm. Back and flanks strikingly marked with black and yellow stripes that fade on the tail, which is olive-brown, as is the head. Greyish-white below.

HABITAT AND DISTRIBUTION: Usually on rocky hill slopes with some vegetation, above 300m. Known only from Egypt's southern Red Sea coast, at Wadi El

Ophisops elbaensis, Gebel Elba.

Gemal National Park and Gebel Elba, but probably in the intervening country on suitable hills, and almost certainly extends into Sudan from Gebel Elba. Elsewhere, in Saudi Arabia and Yemen.

CONSERVATION STATUS: Data Deficient.

NATURAL HISTORY: Terrestrial and diurnal, readily climbing on rocks and vegetation. Active but secretive. Presumably lays eggs and preys on invertebrates.

ELEGANT SNAKE-EYED LACERTID

Ophisops elegans
Ménétriés, 1832

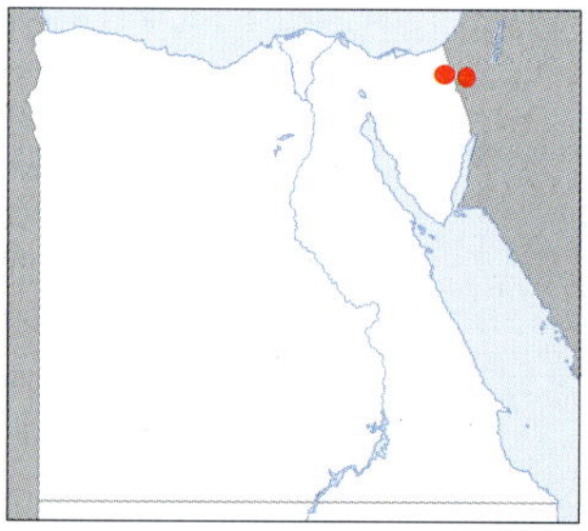

IDENTIFICATION: A small long-tailed lacertid of extreme north-east Sinai. Very slender; tail long, 65–72 per cent of total length. Scales at mid-body fairly large, imbricate and strongly keeled, arranged in 22–40 (usually <32 in Egypt) transverse rows; ventrals in six straight rows. Femoral pores 7–16 on each side (usually 9–11). Maximum size about 20cm, average 15–17cm, hatchlings about 6cm. Four narrow yellow stripes divide rufous or rich brown back; the brown overlain by irregular black markings, sometimes just a few dots, sometimes almost forming crossbars. Underside white. Juveniles vividly striped yellow and black. Taxonomic Notes: Several subspecies described, *O. e. ehrenbergii* was recognised in Egypt, but molecular taxonomy indicates a different picture, so revisions are on the cards.

HABITAT AND DISTRIBUTION: Egyptian specimens were found in vegetated small runnels with dense spiny vegetation, elsewhere in lightly to fairly well-vegetated rocky hillsides and wadis, open woodland, even lava fields. In our area, recorded from El Quseima, north-east Sinai, at 300m. Elsewhere, north to Turkey.

CONSERVATION STATUS: Least Concern, although Vulnerable in Egypt on account of its tiny range there.

NATURAL HISTORY: Terrestrial and diurnal, active into twilight during summer. Shelters in burrows, often at base of bushes. Territorial, the males fight in the breeding season. Up to three clutches of 3–6 eggs are laid; hatchlings from June onwards. Diet small arthropods.

Ophisops elegans, Israel/Palestine.

WESTERN SNAKE-EYED LACERTID

Ophisops occidentalis
Boulenger 1887

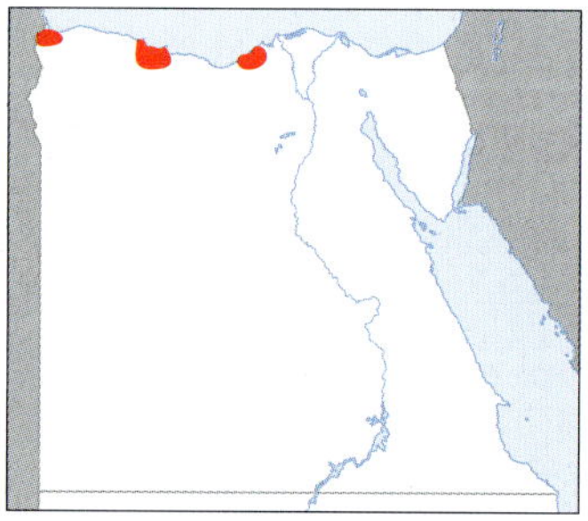

Ophisops occidentalis, Tunisia.

Ophisops occidentalis, Tunisia.

IDENTIFICATION: A small long-tailed lacertid of the north-west Egyptian coast. Very slender; tail relatively long, about ⅔ of total length. Dorsal scales large, imbricate and strongly keeled, arranged in 19–24 transverse rows; ventrals in six straight rows. Femoral pores 7–9 on each side. Maximum size about 13cm, average 10–12cm, hatchling size unknown. Shades of brown or grey-brown above, with four somewhat diffuse dorsal and dorsolateral stripes; ground colour overlain by irregular dark blotches and speckles. Scale keeling gives the tail a strongly ridged appearance. Underside greyish-white or white; sometimes dark spots on chin.

HABITAT AND DISTRIBUTION: In Egypt occurs on lightly vegetated calcareous ridges parallel to sea, particularly where there is surface rock, at altitudes below 200m, although elsewhere in better-vegetated landscapes, at altitudes up to 1,500m. Found along north Mediterranean coastline, never more than 10km from the sea, from Alexandria to Libyan border. Elsewhere, west to Morocco.

CONSERVATION STATUS: Least Concern, although under threat in Egypt on account of quarrying and coastal development in its small range.

NATURAL HISTORY: Terrestrial and diurnal, active on rocky slopes with surface rock and bushes. Agile and wary, rushing to hide in vegetation or burrows. Clutches of 3–4 eggs are laid. Diet small arthropods.

SHIELD-BACKED GROUND LIZARDS *PHILOCHORTUS*

A predominantly African genus of medium-sized, slim, fast-moving lacertids; most are striped, with very long tails. In the middle of the back, they have 2–6 longitudinal rows of enlarged scales. Their distribution centres on north-east Africa; one form reaches Egypt and other countries in the Sahara, and another occurs in Arabia. Six of the seven known species occur in our area. Their closest relatives are in the sister genera *Heliobolus* and *Pseuderemias*. The divergence between these genera occurred 17–18 Mya. Our understanding of the entire genus is poor. Shield-backed ground lizards are widespread, but are often secretive or rare.

KEY TO NORTH-EAST AFRICAN MEMBERS OF THE GENUS *PHILOCHORTUS*

1a: Frontal shield usually separated from supraoculars by a more or less complete row of small granular scales; dorsal scales keeled, in 22–28 rows at mid-body; 6–8 enlarged scales between hind limbs; dorsum with five longitudinal pale lines, the median bifurcating on nape *Philochortus hardeggeri*, Hardegger's Shield-backed Lizard

1b: Frontal shield usually in full contact with supraoculars and not separated by small granular scales; dorsal scales smooth or keeled, in 28–46 rows at mid-body; 10–16 enlarged scales between hind limbs 2

2a: Dorsum with five longitudinal pale lines, the median bifurcating on nape; dorsal scales smooth or weakly keeled; parietal shields most commonly completely separated through contact between interparietal and occipital *Philochortus phillipsii*, Phillip's Shield-backed Lizard

2b: Dorsum with six longitudinal pale lines, the median two bifurcating on nape	3
3a: Parietal shields in contact, the interparietal either absent or too small to reach occipital; dorsal scales usually smooth; Eritrea and central Ethiopia	*Philochortus spinalis*, Peters' Shield-backed Lizard
3b: Parietal shields completely separated through contact between interparietal and occipital; not in Eritrea and central Ethiopia	4
4a: Dorsal scales distinctly keeled, in 35–40 rows at mid-body	*Philochortus intermedius*, Boulenger's Shield-backed Lizard
4b: Dorsal scales smooth or very weakly keeled, in 30–37 rows at mid-body	5
5a: Mid-body scale rows 30–32; 10–14 femoral pores on each thigh; lower side of hind limbs, upper side of fore limbs and digits deep yellow; confined to southern Ethiopia	*Philochortus rudolfensis*, Turkana Shield-backed Lizard
5b: Mid-body scale rows 31–37; 11–18 femoral pores on each thigh; lower side of hind limbs, upper side of fore limbs and digits pinkish-brown or whitish; confined to northern Egypt	*Philochortus zolii*, Grass-loving Shield-backed Lizard

HARDEGGER'S SHIELD-BACKED LIZARD

Philochortus hardeggeri
(Steindachner, 1891)

IDENTIFICATION: A medium-sized slender lacertid with scaly granules between supraoculars and frontal, found in sandy habitats of northern and south-central Somalia. Head rather small, flat in profile, snout short and pointed. Unlike other species, frontal shield usually separated from supraoculars by a more or less complete row of small granular scales. Body elongate, feebly depressed, tail long, about 70–75 per cent of total length. Body scales large and distinctly keeled, in 22–28 (occasionally 30) rows; 2–4 mid-dorsal rows of scales noticeably enlarged. There are 6–8 enlarged scales between bases of hind limbs, 11–13 femoral pores on each side. Maximum size about 25cm (snout–vent length up to 7cm), average 20–23cm. Back pale brown, red-brown to blackish with fine darker speckling and five more or less distinct longitudinal pale lines, of which the outermost tend to be rather irregular and ill-defined or formed by a series of large blackish spots or bars, whilst the median one is divided anteriorly to create a V-shaped mark on the neck. The enlarged dorsal scales each have a small blackish spot. Underside pale and unmarked. Juveniles and young more brightly coloured, almost white and black, with a pinkish-red tail; this striping doesn't usually persist in adults.

HABITAT AND DISTRIBUTION: Seems to frequent sandy, semi-desert country, from the coastal plain inland to a maximum altitude of about 1,350m. Originally described from the sandy Guban plain south of Zeila, near the Djibouti border; it extends east to the vicinity of Berbera in Somaliland and further south across the dry low corridor between the Ogo and Goolis Mountains, into the sandy country around Hargeisa and Burao. Thence to the Haud region along the neighbouring border with Ethiopia, where it is fairly widespread in open dry *Vachellia–Commiphora* bushland. Known also from the lower and middle Shebelle regions (several localities around Mogadishu), and Galguduud (at Bud Bud) in south-central Somalia.

CONSERVATION STATUS: Least Concern.

NATURAL HISTORY: Poorly known; presumably similar to others of the genus. Active during day, ground-dwelling, solitary; specimens from Shanshacade (Haud region) in Somaliland ran

Philochortus hardeggeri, juvenile, near Berbera.

between low shrubs on red sandy soil, from time to time they climbed shrubs and rested on the lower branches, their tails hanging down. Probably lays eggs and eats insects and other arthropods. The specific name honours Dr Dominik Kammel von Hardegger (1844–1915), Austrian physician who explored Somalia and the Harar area, and collected the types of *P. hardeggeri*.

BOULENGER'S SHIELD-BACKED LIZARD

Philochortus intermedius
Boulenger, 1917

IDENTIFICATION: A medium to large-sized lacertid with six pale dorsal lines and keeled scales, mostly found in stony habitats of northern Somalia. Head rather small, relatively long, flat in profile, snout short and rounded. Neck as broad as head or slightly narrower. Body stout, feebly depressed, tail long, about 70–75 per cent of total length. Body scales small and distinctly keeled, in 35–40 rows; two to four mid-dorsal rows of scales noticeably enlarged. There are 12–15 enlarged scales between bases of hind limbs, 14–18 femoral pores on each side. Maximum size about 29cm (snout–vent length 5.5 – 8.5cm), average 23–26cm. Back dark brown to black, sometimes with a network or spots of still darker pigment, and six more or less distinct pale yellow lines, each of the median ones being divided anteriorly to create a pair of V-shaped marks on neck. Inner half of two enlarged rows of dorsal scales dark, outer half pale yellowish. Two outer dark lines extensively mottled ochre. Top of head greyish-brown with black spotting. Fore limbs yellow with black marbling, hind limbs with numerous small round light spots, tail greyish-brown or reddish towards tip. Underside pale and unmarked. Juveniles and younger animals more brightly coloured, almost white and black, with a red tail; this vivid striping fades in adults.

HABITAT AND DISTRIBUTION: Arid stony environments mostly in mountainous dry country, sea level to approximately 1,600m. The type specimens were collected in the vicinity of Berbera and further inland in the Goolis Mountains of central Somaliland. Also known from similar habitats in the hills of Borama and between Quljeet and Bawn in the west of the country. It may also enter the eastern Haud, at Dharkayn Geenyo. Range extends south from the Nogal Valley (at Eyl) to central Somalia, at Galguduud (Bud Bud) and further south to Bakool (Uegit) and Bay (Dinsor), Somalia, with an old record from Mogadishu, although there is some doubt that these specimens are correctly identified. Some populations in the Haud probably represent a new, undescribed species.

CONSERVATION STATUS: Least Concern.

NATURAL HISTORY: Poorly known. Diurnal, terrestrial, solitary. Probably lays eggs, eats insects. An adult male from Sheikh in Somaliland was observed stalking on the ground on a small hill in a rocky area, among

Philochortus intermedius, Sheikh.

dense bushes of *Dodonaea viscosa* and *Buxus hildebrandtii* at midday. It is sympatric with *P. spinalis* in Borama.

PHILLIPS' SHIELD-BACKED LIZARD

Philochortus phillipsii
(Boulenger, 1898)

IDENTIFICATION: A medium-sized, long-tailed lacertid with five pale dorsal lines and smooth scales, in Somaliland and east-central Ethiopia. Head rather small, flat in profile, snout short and pointed. Body elongate, feebly depressed, tail long, about 70–75 per cent of total length. Body scales small, smooth or only weakly keeled, in 28–38 rows; the two mid-dorsal rows of scales noticeably enlarged. There are 12–16 enlarged scales between hind limbs, 11–17 femoral pores on each side. Maximum size about 23cm (snout–vent length up to 6cm), average 18–21cm. Back pale red-brown to blackish with five more or less distinct longitudinal pale lines, the median one being divided anteriorly to create a V-shaped mark on neck. Dark interspaces uniform or with darker vertical spots. Juveniles and younger animals more brightly coloured, almost white and black, with a vermilion tail. Top of head light brown. Fore limbs yellow-brown with black marbling, the hindlimbs with round light spots, orange in juveniles, the tail orange-red almost the entire length in young, pinkish-beige in adults. Underside immaculate white. Striping of young individuals is retained in adults.

HABITAT AND DISTRIBUTION: Lowland semi-deserts and arid savanna with fairly stony terrain and grassland among lava fields, between sea level and 1,000m, usually below 750m. Described from Berbera in Somaliland. Common in the hot and dry Guban strip between the sea and inland Goolis and Wagar Mountains. Known from several localities on Somaliland/Puntland boundary in the east, from the coast across the Daror and Nogal Valleys, south to the Ethiopian border, in and around Qoriley, from there extending marginally to the Mudug region of central Somalia. In central Ethiopia widespread and rather common in Awash National Park.

CONSERVATION STATUS: Least Concern.

NATURAL HISTORY: Poorly known. Presumably similar to other shield-backed ground lizards, i.e., diurnal, terrestrial, solitary and very agile. Probably lays eggs, eats insects. In Awash, several younger specimens were found hidden under small stones by day. Some adults are relatively dark in colour, especially those associated with local lava fields. A juvenile was observed running on a stony plain at midday near Lafaruug in Somaliland; another juvenile at Dagah Shabell in Somaliland was seen resting with eyes closed in the shade of a small bush.

Philochortus phillipsii, juvenile, Awash National Park.

Philochortus phillipsii, Lafaruug.

TURKANA SHIELD-BACKED LIZARD

Philochortus rudolfensis
Parker, 1932

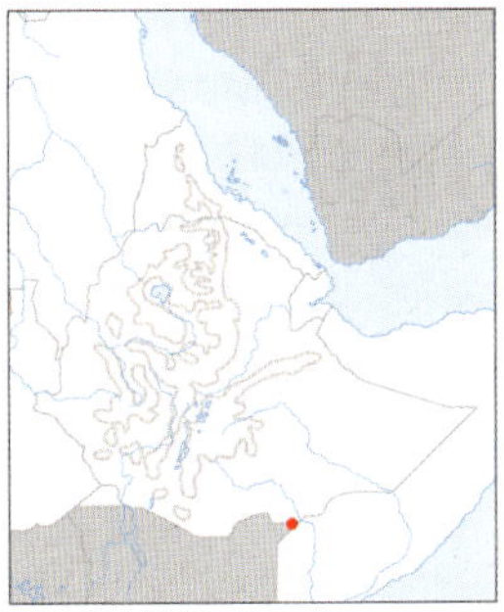

IDENTIFICATION: A medium-sized, finely striated, slender lacertid, confined to southern Ethiopia near border with Kenya and Somalia. Head short but snout looks sharp from above. Body thin, tail very long, 65–75 per cent of total length. Body scales smooth and granular, or feebly keeled posteriorly, in 30–32 rows; two mid-dorsal rows of scales noticeably enlarged. There are 14–16 enlarged scales between hind limbs, 11–18 femoral pores. Maximum size 18–20cm (snout–vent length 4.6–5cm), average 15–18cm. Top of head uniform yellow-grey, the body distinctly striped with five brown and six yellow stripes, which coalesce beyond tail base; the final 75 per cent of tail is vivid orange in youngsters, beige in adults. Fore limbs pinkish-brown with black marbling, hind limbs with small round light spots. Underside immaculate white. Striping of young individuals is usually retained in adults.

HABITAT AND DISTRIBUTION: Favours open, rocky and sparsely vegetated semi-desert or *Vachellia–Commiphora* bushland. Known from a handful of widely scattered localities in north and north-east Kenya, usually below 800m. In our area known only in the vicinity of Mandera on border with Ethiopia, but seems probable that it will eventually be found to occupy similar environments in southern Ethiopia. Obviously very widespread but secretive or rare. Probably occurs in South Sudan as well as in southern Somalia, in areas adjacent to the Kenya border.

Philochortus rudolfensis, Kenya.

CONSERVATION STATUS: Data Deficient.

NATURAL HISTORY: Poorly known. Diurnal, agile and terrestrial. Diet arthropods. Presumably lays eggs.

PETERS' SHIELD-BACKED LIZARD

Philochortus spinalis
(Peters, 1874)

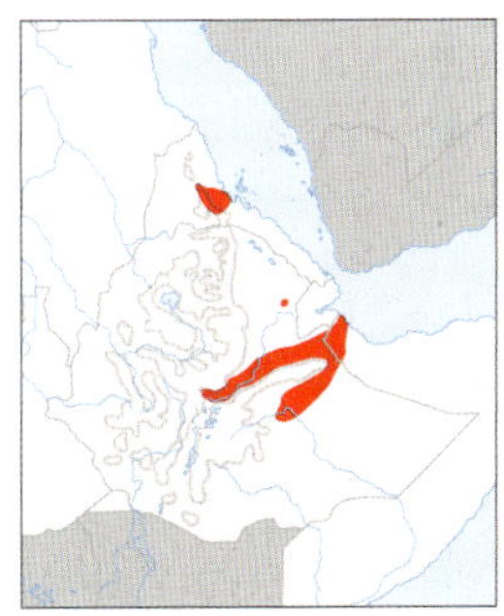

IDENTIFICATION: A medium-sized lacertid with six pale dorsal lines and smooth scales, in Ethiopia and Eritrea. Head short, relatively deep and concave in profile, with rather thick neck. Body elongate, feebly depressed, tail long, about 70 per cent of total length. Body scales small and smooth but become slightly keeled towards mid-back, in 30–46 rows; two mid-dorsal rows of scales noticeably enlarged. There are 10–16 enlarged scales between hind limbs, 12–15 femoral pores on each side. Maximum size about 20cm (snout–vent length 4.5–5.8cm), average 15–18cm. Colour and pattern distinctive. Back brown to blackish with pigmentation usually heaviest along vertebral midline, but dark areas may also have extensive pale mottling and there are six more (in younger animals and females) or less (males) distinctive longitudinal pale lines, each of the median ones being divided anteriorly to create a pair of V-shaped marks on neck. Top of head finely speckled light brown or black. Tail red in young, beige-brown in adults. Underside white. The vivid juvenile striping fades in adults.

HABITAT AND DISTRIBUTION: A wide-ranging species of semi-desert, dry savanna or grasslands among lava fields in low country, and mosaics of montane grassland/agricultural land on high plateaux; sea level to approximately 2,350m, with most records around 1,000m. The only species of *Philochortus* in Eritrea, where recorded at Keren, Ghinda, Massawa and the Kohaito Plateau. Also in eastern lowlands of Ethiopia, extending to the Awdal region in Somaliland, where reported from the plains near Zeila and Borama. Two records south of Ahmar Mountains in Ethiopia, Moggio River and Odamuda, north-east of Sheikh Hussein, but distribution between them probably continuous. Common in Awash National Park, around Mount Fantalle. Not recorded in Djibouti, but probably occurs.

CONSERVATION STATUS: Least Concern.

NATURAL HISTORY: Virtually unknown; presumably similar to other shield-backed ground lizards, i.e., diurnal, terrestrial, solitary. Probably lays eggs and eats insects. An adult male in Awash National Park was observed running between bushes in grassland at midday.

Philochortus spinalis, Fantalle.

Philochortus spinalis, Fantalle.

GRASS-LOVING SHIELD-BACKED LIZARD

Philochortus zolii
Scortecci, 1934

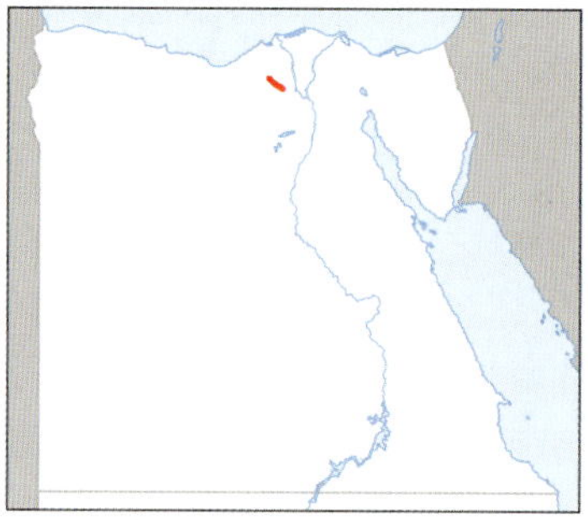

IDENTIFICATION: A medium to large, long-tailed, slender lacertid, confined to northern Egypt. Head short with an obviously long neck. Body thin, tail very long, 65–75 per cent of total length. Body scales smooth but become slightly keeled towards mid-back, in 31–37 (average 33) rows; two mid-dorsal rows of scales noticeably enlarged, slightly keeled. There are 10–14 femoral pores. Maximum size about 28cm (snout–vent length 5.2–7.3cm), average 20–24cm. Top of head uniform yellow-grey, the body distinctly striped with five brown bold and six white to yellow stripes, the two lateral dark streaks enclose a series of large black blotches, more evident in adults. There is an additional very narrow ventrolateral dark stripe on each side. Limbs yellow with black marbling, the tail red in young, reddish in adults or beige in old animals. Juveniles and hatchlings are more brightly coloured, almost white and black, with a vermilion tail. Underside white. Striping of young is usually retained in adults. Taxonomic Notes: The most widespread of the genus. *P. lhotei*, described from Niger, was synonymised with this species.

HABITAT AND DISTRIBUTION: In Egypt inhabits semi-desert near oases with grassy vegetation and sparse bushes like *Tamarix* or manna tree (*Alhagi graecorum*), and desert where barley is cultivated. Strongly associated with microhabitats where Halfa grass (*Desmostachya bipinnata*) grows. In our area, only at Wadi El Natrun in northern Egypt. Elsewhere, sporadic records from north and south-west Libya, north-west Niger, south-east Algeria, central Mali and south-west Mauritania.

Philochortus zolii, juvenile, Wadi El Natrun.

Philochortus zolii, Wadi El Natrun.

CONSERVATION STATUS: Endangered.

NATURAL HISTORY: Diurnal, solitary and terrestrial. An avid climber, the long tail aiding balance on vegetation. Digs a burrow in sandy soil below grass clumps. Eats insects or other arthropods; residues of small beetles, ants and spiders found in faecal samples. Most active early in morning, it basks and then forages actively in grass tussocks. In North Africa it hibernates during winter, from November to February. Breeding details unknown, but probably lays several eggs, like other *Philochortus*.

FALSE SAND LIZARDS *PSEUDEREMIAS*

A genus of small active terrestrial lacertids with a very narrow snout and four nasal scales. They favour dry lowland environments. Seven species are known; all occur in north-east Africa, three are endemic to Somalia. Together with genus *Latastia* they are one of the most common lacertids, even reptiles, in Somalia. Their closest relatives are members of the genus *Heliobolus* from sub-Saharan Africa. These genera split 14–15 Mya.

KEY TO MEMBERS OF THE GENUS *PSEUDEREMIAS*

1a: Upper head shields smooth or rugose, not striated; scales between last upper labial and ear-opening round; dorsal scales smooth or keeled 2

1b: Upper head shields striated; scales between last upper labial and ear-opening elongate, rhombic; dorsal scales keeled 6

2a: Dorsal pattern of irregular spots in adults, dorsal scales keeled; confined to coastal central Somalia and Puntland *Pseuderemias erythrosticta*, Boulenger's Sand Lizard

2b: Dorsal pattern of 5–7 dark stripes in adults, dorsal scales smooth; not in coastal central Somalia and Puntland 3

3a: Ventral scales in eight longitudinal series, with an additional row of smaller and more rounded scales either side *Pseuderemias savagei*, Savage's Sand Lizard

3b: Ventral scales in six longitudinal series, with an additional row of smaller and more rounded scales either side 4

4a: Seven dorsal stripes; restricted to eastern Somaliland and adjacent Puntland, central Somalia *Pseuderemias septemstriata*, Parker's Sand Lizard

4b: Five dorsal stripes; not in eastern Somaliland and adjacent Puntland, central Somalia 5

5a: Ventral scales in six longitudinal series, with an additional row of smaller and more rounded scales either side; subocular usually entering lips; posterior subcaudals smooth *Pseuderemias mucronata*, Blanford's Sand Lizard

5b: Ventral scales in eight longitudinal series, with an additional row of smaller and more rounded scales either side; subocular usually excluding lips; posterior subcaudals keeled *Pseuderemias smithi*, Smith's Sand Lizard

6a: Head relatively short, slightly concave in profile; upper head scales striated or rugose; subocular usually entering lips; five dark longitudinal stripes without any light spots; 13–19 femoral pores each side *Pseuderemias striata*, Peters' Sand Lizard

6b: Head long, flat in profile; upper head scales striated; subocular excluding lips; five dark longitudinal stripes of which at least outermost encloses row of white spots; 17–25 femoral pores each side *Pseuderemias brenneri*, Brenner's Sand Lizard

BRENNER'S SAND LIZARD

Pseuderemias brenneri
(Peters, 1869)

IDENTIFICATION: A small lacertid with very acutely pointed snout and keeled dorsal scales, confined to sandy habitats of southern Somaliland, the Ogaden and Somalia. Head long, flat in profile. Snout long, wedge-shaped, with very sharp canthus rostralis, without bulbous nostrils. Scales on top of head finely or strongly striated, subocular scale excluded from lip, temporal scales between upper labials and ear-opening elongated and rhombic. Body strongly depressed, tail long, about 70 per cent of total length. Body scales keeled (often tricarinate), rhombic and juxtaposed, in 53–72 rows across the mid-body. Ventral scales in six straight rows (with an additional row of smaller and more rounded scales either side) and in 25–27 transverse rows. Femoral pores 17–25 on each side, posterior subcaudals keeled. Maximum length about 18cm (snout–vent length up to 5.3cm). Pattern similar to the often sympatric *P. smithi*; the back has five more or less well-defined longitudinal bands of pale grey-brown, brick-red or blackish pigment, each of which usually incorporates whitish spots and is separated from its neighbours by complete or broken pale strips, sometimes uniform red with only pale spots present. Upper parts of hind limbs brown or brick-red with distinct, rather large, round paler spots. Underside whitish or cream. Juvenile orange above, finely striped with incorporated pale spots, round light spots on hind limbs, beige-orange tail.

HABITAT AND DISTRIBUTION: A wide-ranging species, similar in distribution and habitat to *P. smithi* and *P. striata*. A typical species of Somali arid zone, living in dry open *Vachellia–Commiphora* savanna and semi-desert with soft red sand but also on coastal sand dunes and grasslands, between sea level and 1,200m. From

Pseuderemias brenneri, Degeh Bur.

Pseuderemias brenneri, Shanshacade.

Pseuderemias brenneri, Shanshacade.

Burao in central Somaliland it spreads south-east along Nogal Valley, south through the Haud region and across the eastern Ogaden in Ethiopia (Degeh Bur, Milmil and Ado). Widespread in most of central Somalia (including Hobyo and Galkayo), south to the Benadir coast. An isolated record from Rahanuin country at Lugh. Historic records from northern Somaliland erroneous. Widespread and locally very common. Unlikely to be threatened by human activities.

CONSERVATION STATUS: Least Concern.

NATURAL HISTORY: Poorly known. Diurnal, terrestrial, favours sandy habitats, diet insects and other invertebrates. Digs a burrow in sandy soil below small bushes, presumably lays eggs but no clutch details known. A newly hatched juvenile was north of Degeh Bur in Ethiopia in July. The specific name honours Richard Brenner (1833–1874), German explorer and forester, who was on von der Decken's second and last African expedition to Somalia, and collected the type specimen of *P. brenneri*.

BOULENGER'S SAND LIZARD

Pseuderemias erythrosticta
(Boulenger, 1891)

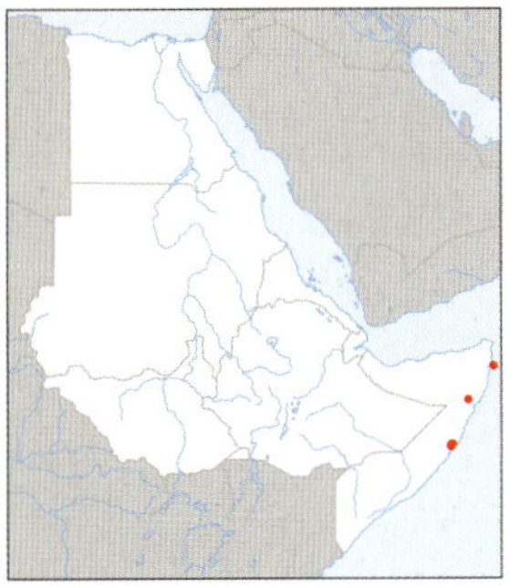

IDENTIFICATION: A small, long-headed lacertid with dorsal pattern of irregular spots, confined to coastal Somalia. Head strongly depressed, flat in profile. Snout long and narrow, wedge-shaped, without bulbous nostrils. Upper head scales smooth, subocular scale excluded from lip, temporal scales between upper labials and ear-opening round. Body strongly depressed, tail long, 70–75 per cent of total length. Body scales keeled, granular and juxtaposed in 53–60 rows across mid-body; ventral scales in six straight rows (with an additional row of smaller and more rounded scales either side) and in 24–27 transverse rows. Femoral pores 15–20 on each side, posterior subcaudals keeled. Maximum length about 19cm (snout–vent length up to 5.2cm). Adults sandy-grey or buff above, with brick-red or black-and-red dots; tail uniform or spotted bluish-grey or black, one spot per scale; juveniles have five grey longitudinal streaks, which may be broken up by round whitish spots; traces of dark streaks may persist in female. Upper parts of hind limbs grey and spotted, without distinct round paler spots. Underside uniform white.

Pseuderemias erythrosticta, preserved, Hobyo to Berbera.

HABITAT AND DISTRIBUTION: Sparsely vegetated dry coastal plains and sand dunes, arid steppe and grassland with scattered low shrubs, between sea level and about 200m. A poorly known species, recorded at a handful of localities in coastal central Somalia and Puntland (Hobyo, between Hobyo and Sissib, an imprecise locality in the Nogal Valley, the Hafun Peninsula). The type was collected from 'the route from Obbia to Berbera', probably near Hobyo as suggested by subsequent specimens.

CONSERVATION STATUS: Least Concern.

NATURAL HISTORY: Unknown, presumably similar to other *Pseuderemias*. The types of this rare lizard were collected by Luigi Robecchi-Bricchetti

(1855–1926), an Italian explorer, geographer, cartographer and naturalist, and the first European to extensively visit the Horn of Africa. Other new taxa based on his material were *Agama robecchii* and *Rieppeleon kerstenii robecchii*.

BLANFORD'S SAND LIZARD

Pseuderemias mucronata
(Blanford, 1870)

IDENTIFICATION: A small, long-headed lacertid with a dark vertebral band, reticulated flanks and smooth dorsal scales, confined to coastal Egypt to Somaliland. Head flat in profile, snout long, almost wedge-shaped, without bulbous nostrils, upper head scales smooth or roughly textured, subocular scale enters lip, temporal scales between upper labials and ear-opening round. Body slim, tail long, 70–75 per cent of total length. Body scales smooth, granular and juxtaposed in 61–78 rows across mid-body; ventral scales usually in six straight rows (often with an additional row of smaller and more rounded scales either side) and in 24–29 transverse rows. Femoral pores 15–23 on each side, posterior subcaudals smooth. Maximum length about 19cm (snout–vent length up to 5.2cm). Back has five dark brown longitudinal bands

Pseuderemias mucronata, juvenile, Berbera.

Pseuderemias mucronata, Daarbuduq.

separated by whitish to pale brown stripes. Two outermost dark bands are widest and usually incorporate pale marks, often as a row of small white spots, but sometimes in form of blotches which can be so large and numerous that they partly or completely disrupt the heavier pigment. In Somaliland a common pattern involves one dark vertebral band with two pale adjoining bands, flank markings often irregular and ill-defined, sometimes a series of large blackish or rufous-brown spots or bars. Some males have yellowish-green blotches in line on flanks. Juveniles finely striped, with a bluish tail. Underside uniform white.

HABITAT AND DISTRIBUTION: The northernmost of the genus. Dry savanna, semi-desert and coastal dunes with sandy soil further north, in south more often on consolidated stony soil; from sea level to about 1,500m. Largely confined to maritime plains. Distribution disjunct; from the Gebel Elba region in south-east Egypt, Sudan's Red Sea coast (Durrur and Suakin), the Eritrean coast (Massawa, Beylul), and Assab into Djibouti, eastern Ethiopia and the Guban region of Somaliland; the furthest inland record is near Dire Dawa in eastern Ethiopia. In Somaliland mostly known from maritime plain north of the Ogo, Goolis and Wagar Mountains, as far east as Xagal, east of Berbera. Probably occurs even further east, but not recorded. Southernmost confirmed record at Naaso Hablood near Hargeisa. Unconfirmed old records from the Ogaden (not shown on map).

CONSERVATION STATUS: Least Concern.

NATURAL HISTORY: Poorly known, but presumably similar to other *Pseuderemias*. Two specimens collected in Gebel Elba were on a plain with sandy hummocks stabilised by *Lycium* sp. bushes, with about 40 per cent vegetation cover, also found in coastal dunes. In coastal Somaliland one of the commonest reptiles and, together with *Pristurus crucifer*, the only species tolerant of and active during the intense midday heat. They dig burrows in sandy soil below small bushes. Mating observed between August and September in Somaliland, and freshly hatched juveniles seen in November.

SAVAGE'S SAND LIZARD

Pseuderemias savagei
(Laurent & Gans, 1965)

IDENTIFICATION: A small, relatively long-headed lacertid with smooth dorsal scales and large rectangular ventral scales in eight rows, confined to coastal eastern Somaliland and Puntland. Head flat in profile, snout long, almost wedge-shaped, without bulbous nostrils, upper head scales smooth, subocular scale enters lip, temporal scales between upper labials and ear-opening round. Body slim, tail long, 70–75 per cent of total length. Body scales smooth, granular and juxtaposed in 55–72 rows across mid-body; ventral scales in eight straight rows (with an additional row of smaller and more rounded scales either side) and in 23–28 transverse rows. Femoral pores 15–22 on each side, posterior subcaudals smooth. Maximum length about 19cm (snout–vent length up to 5cm). Colour pattern variable. Back greyish-beige with five dark brown longitudinal bands separated from one another by whitish to pale brown stripes. Main vertebral stripe thin, bordered on each side by a much wider uniformly pale stripe. Two additional dark flank stripes usually have roundish paler spots. Some specimens, generally the largest, almost uniform, smaller individuals have dark longitudinal zones more sharply outlined. Upper parts of hind limbs grey with distinct round lighter spots. Underside uniform white. Taxonomic Notes: Relatively recently described, most similar to and probably related to *P. mucronata*, and considered by some to be a subspecies of (or conspecific with) latter.

Pseuderemias savagei, preserved, Habo.

HABITAT AND DISTRIBUTION: A north Somali endemic with a small range. Inhabits very hot and arid coastal plains, and sand dunes with sparse vegetation, between sea level and 350m. Most records from a very narrow area between Gulf of Aden and nearby escarpments of Cal Madow and Cal Miskaat Mountains; localities include Las Korey in Somaliland, the Bosaso and Karin area in Puntland, and Candala and Habo at the north-eastern tip of Somali peninsula.

CONSERVATION STATUS: Least Concern.

NATURAL HISTORY: Unknown. The types from Candala were living in sand dunes dotted with small bushes about 30cm in diameter, together with *Pristurus crucifer*.

PARKER'S SAND LIZARD

Pseuderemias septemstriata
(Parker, 1942)

Pseuderemias septemstriata, El Afweyn.

Pseuderemias septemstriata, juvenile, Las Anod.

IDENTIFICATION: A small lacertid with seven dark brown longitudinal stripes, confined to eastern Somaliland and Puntland. Head rather short, depressed and slightly concave in profile. Snout long, pointed, without bulbous nostrils, upper head scales smooth or very feebly rugose, subocular scale enters lip, temporal scales between upper labials and ear-opening round. Body slim, tail long, 65–73 per cent of total length. Body scales smooth, granular and juxtaposed in 50–68 rows across mid-body; ventral scales in six straight rows (with an additional row of smaller and more rounded scales either side), and in 26–29 transverse rows. Femoral pores 14–20 on each side, posterior subcaudals smooth. Maximum length about 17.5cm (snout–vent length up to 5cm). Colour pattern distinct. Back yellowish-beige with seven dark brown longitudinal bands, median three usually very thin, but two outer are wider. Outermost dark wide band has small white spots, the dorsum sometimes has dark irregular markings between the median thin and adjoining dark stripes. Upper parts of hind limbs brown with distinct, large, round light spots. Underside uniform white (males have a yellowish chin and neck). Juveniles have bluish tails.

HABITAT AND DISTRIBUTION: Semi-desert, arid open savanna with scattered trees and small bushes, limestone or gypseous stony plains, at 50–1,850m (mostly above 400m). Widely distributed in eastern Somaliland, south of Cal Madow Mountains in western Puntland. Known from the southern Daror Valley at the boundary with Somaliland; notable localities include the gypseous plains around Erigavo and El Afweyn, around Buran, Gardo in Sol Haud, God Heeli, Halin and Las Anod in Nogal Valley. Isolated records further south in Mudug region in central Somalia between Hobyo and Galkayo. Probably more widespread, but under-collected. Widespread and locally very common, unlikely to be threatened by human activities.

CONSERVATION STATUS: Least Concern.

NATURAL HISTORY: Poorly known. Presumably similar to others of genus, i.e., diurnal, terrestrial, diet invertebrates. Dig burrows in soil below small bushes. Probably lays eggs, no clutch details known. Known from around Erigavo, one of the coldest places in Somalia, where in winter (November–February) temperatures can fall to a little above 0°C.

SMITH'S SAND LIZARD

Pseuderemias smithi
(Boulenger, 1895)

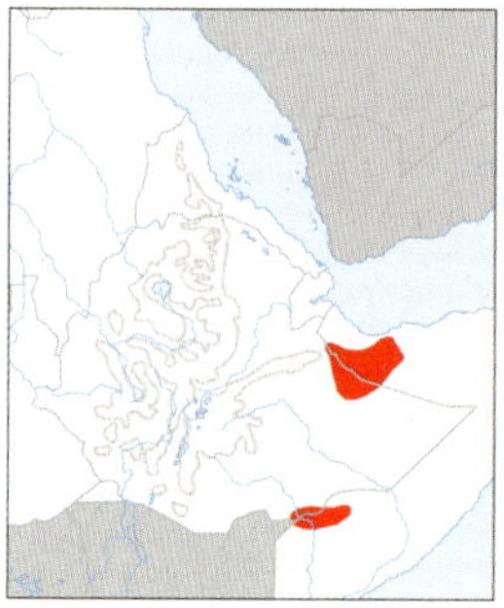

IDENTIFICATION: A small, relatively short-headed lacertid with white spots on dorsal lines and smooth dorsal scales; two disjunct populations. Head slightly concave in profile, snout short, pointed with slightly bulbous nostrils, upper head scales rugose, subocular scale excluded from lip, temporal scales between upper labials and ear-opening round. Body slim, tail long, 65–75 per cent of total length. Body scales smooth, granular and juxtaposed in 68–82 rows across mid-body; ventral scales usually in eight straight rows (often with an additional row of smaller and more rounded scales either side), and in 26–30 transverse rows. Femoral pores 17–23 on each side, posterior subcaudals keeled. Maximum length about 17cm (snout–vent length up to 4.7cm). Has six light longitudinal stripes, the two in centre of back are close. Ground colour rich rufous-brown to dark brown or grey. Back usually unmarked or has a single line of spots between central dorsal lines, longitudinal lines of heavy pale spots between other stripes. White below. Juvenile said to be darker than the adult.

HABITAT AND DISTRIBUTION: A wide-ranging species of dry savanna and semi-desert with sandy or sandy-loamy soil, between sea level and 1,600m. Occurs south of the Ogo, Goolis and Wagar Mountains in central Somaliland. Relatively widespread between Borama and Hargeisa, east to near Burao, thence south across western Haud to the northern Ogaden at Milmil (the type locality). A doubtful record from Berbera (not shown on map). Apparently absent in most of the Ogaden, reappears in southern Somalia, from Mandera on Kenya border east to Lugh in Rahanuin country. Elsewhere, in north-east Kenya.

Pseuderemias smithi, Borama.

Pseuderemias smithi, Sheikh.

CONSERVATION STATUS: Least Concern.

NATURAL HISTORY: Little known. Diurnal and terrestrial, very fast-moving and alert. Presumably lays eggs, no clutch details known. Gravid females observed in Kenya in June. Often seen in Somaliland by day, running and foraging without a break among bushes, or stands of *Aloe megalacantha*.

PETERS' SAND LIZARD

Pseuderemias striata
(Peters 1874)

IDENTIFICATION: A small, rather short-headed lacertid with keeled dorsal scales and five to seven dark dorsal bands completely devoid of pale spots. Head slightly concave in profile, snout fairly short, pointed, without bulbous nostrils, upper head scales striated or less frequently rugose, subocular scale usually enters lip, temporal scales between upper labials and ear-opening elongated, rhombic. Body depressed, tail long, about 70–75 per cent of total length. Body scales keeled (often tricarinate), rhombic and juxtaposed in 53–67 rows across mid-body; ventral scales in six straight rows (with an additional row of smaller and more rounded scales either side), and in 25–28 transverse rows. Femoral pores 13–19 on each side, posterior subcaudals keeled. Maximum length about 16cm (snout–vent length up to 4.7cm). Back has five or seven more or less well-defined longitudinal bands of pale brown to black pigment, which are almost or completely devoid of pale spots and separated by whitish to pale buff stripes. Upper parts of hind limbs brown or brick-red with distinct round lighter spots. Underside whitish or cream. Juveniles conspicuously striped black and yellow, top of head brown with bold pale markings, white below, posterior tail greenish-beige. Taxonomic Notes: Two subspecies recognised, the southern nominate form, south of the

Pseuderemias striata, Qoryale.

Pseuderemias striata, juvenile, Inaafmadow.

Shebelle River, and *P. s. gardoensis* north of it. The latter differs from nominate subspecies in having two additional short dark parietal lines and a longer head, twice as long as wide.

HABITAT AND DISTRIBUTION: A wide-ranging species similar in distribution and habitat requirements to *P. brenneri*. A typical species of the Somali arid zone, living in dry open *Vachellia–Commiphora* savanna and semi-desert with soft red sand but also on coastal sand dunes and grassland, or stony, poorly vegetated coastal and inland plains, between sea level and 1,150m. From vicinity of Burao and Kiridh in central Somaliland, thence south across the Haud, where recorded on the boundary with Ethiopia. Extends east to Puntland, known from Garowe, Gardo, Wadi Dudo south of Bender Beyla and Hafun, and as far north as El Gaal. Reported in central Somalia at Galkayo, Hobyo, Belet Uen, along the Benadir coast to Mogadishu and Brava, thence north and west to Rahanuin country at Lugh. Just enters the southern Ogaden along the Gestro River. Elsewhere, a vague record on the Tana River, Kenya. Widespread and locally very common. Unlikely to be threatened by human activities.

Pseuderemias striata, Kiridh.

CONSERVATION STATUS: Least Concern.

NATURAL HISTORY: Poorly known. Diurnal and terrestrial, very fast-moving. Shelters in holes. In Somaliland often seen hunting around perimeter of large flat bushes, rushing out to snap up prey. If threatened usually dashes back under bush and freezes, hoping to avoid detection.

GIRDLED LIZARDS AND THEIR RELATIVES

FAMILY CORDYLIDAE

The only exclusively African family of lizards, with ten genera and about 70 species. A single genus and species in our area; Ethiopian Girdled Lizard *Cordylus rivae*.

GIRDLED LIZARDS *CORDYLUS*

An African genus of diurnal, secretive, largely small, spiny, squat brown lizards, with regular, girdle-like scale rows. Their external ears are sometimes hard to see, concealed by spines. The body scales have osteoderms and are heavily keeled, the tail is spiky, but it can be shed and regenerated. Males have femoral pores. In defence they jam themselves in crevices by inflating their body and using an unusual hinge structure to shorten and thicken the skull. Most live on rocks, some on trees. They eat invertebrates, some eat plant material. They give live birth to a few large young. Some species are territorial, in colonies under a large dominant male. They may use chemical cues in their social behaviour, hence the glandular pores. Captive specimens have lived 30 years and they make charming and confiding pets. They are at risk from collection for the pet trade. Girdled lizards are secretive and hard to find. They have radiated in the rocky southern African landscape, with >40 species in the group. However, only one species of *Cordylus* occurs in our area, an Ethiopian endemic and the northernmost representative of this genus.

ETHIOPIAN GIRDLED LIZARD

Cordylus rivae
(Boulenger, 1896)

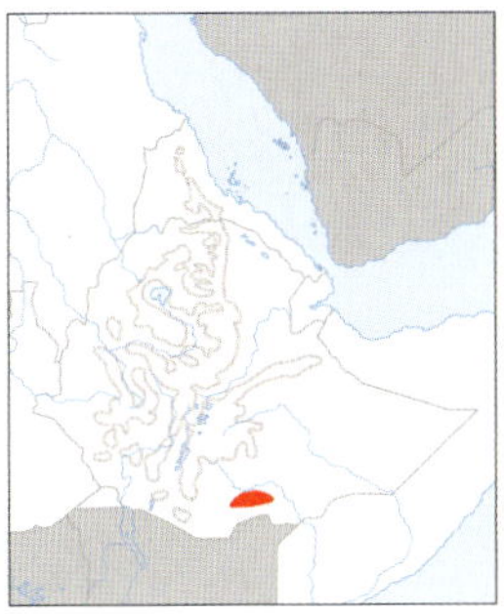

IDENTIFICATION: A medium-sized, stout, reddish-brown spiky lizard. Head triangular, slightly longer than broad, flattened, the snout is sharp, nasals in contact. Gulars small, imbricate, keeled and in 16–18 rows between angle of jaws. It has seven upper labials and six lower labials. Head scales smooth to moderately rugose, interparietal large and separates the anterior parietals. The eye is medium-sized, pupil round, with brownish-yellow iris (hard to see). A distinctive feature is a lateral and ventral 'ruff' of spinose scales immediately behind the ear-opening. Body robust, slightly depressed; limbs strong. Tail short and slender, about 50 per cent of total length, tapering sharply, and has 27–29 regular whorls of very spiny scales. Dorsal scales large and strongly keeled, in 22–25 rows at mid-body, ventral scales 22–23 rows, lateral scales rectangular, mucronate, juxtaposed, with horizontal keels. There are 2–5 femoral pores, which are well developed in both sexes. Maximum size about 21cm (snout–vent length up to 10.5cm), average 14–17cm, hatchling 5–7cm. Colour varies greatly, in freshly sloughed animals, the dorsum, flanks and tail are black or blackish-brown, sometimes marbled by paler flecks. A paler dorsolateral stripe is often present on each side. Upper labials, chin, belly and lower parts of tail whitish. Throat yellowish. Over time, specimens became discoloured by soil, to ochre-brown with blackish scale keels. Hatchlings and juveniles express the adult pattern with browner tones, more pronounced pale dorsolateral stripes with a lighter marbled tail.

Cordylus rivae, hatchling, Filtu.

Cordylus rivae, Filtu.

Cordylus rivae, Filtu.

Underside cream. Taxonomic Notes: According to recent phylogenetic studies, most closely related to *C. beraducii* of southern Kenya and northern Tanzania, from which it diverged c.7 Mya.

HABITAT AND DISTRIBUTION: Restricted to *Vachellia–Commiphora* open woodland and bushland or evergreen scrub with *Barbeya oleoides* trees, in hill country, at 900–1,250m. Prime habitat seems to be places without surface rock, although sometimes found in stony areas. An Ethiopian endemic, known from a few localities south and east of Negele town. The type locality is 'Giacorsa, in Gallaland', which is roughly 22km south-east of Salole village or 80km south of Negele. Rediscovered in 1975 about 100km east of Negele along the road to Filtu, and more recently found both there and further east, about 42km south-east of Filtu on the road to Dolo Odo. Probably widespread in the intervening region (all suitable habitat) but no records. An unconfirmed report from Mago National Park (not shown on map).

CONSERVATION STATUS: Least Concern. At present its habitat is thinly populated; and it is not collected for the pet trade.

NATURAL HISTORY: Poorly known. A diurnal, ground-dwelling lizard but also climbs trees and hides under exfoliating bark. Collected beneath roadside boulders and found hidden under the bark of old fallen *Vachellia* trees. Their colour depends on local soil colour; those from south-east of Filtu are reddish. Shy, secretive and activity is seasonal; in the dry season (July) no specimens were observed, in the wet season (October–November) they were abundant. They give live birth, clutches of one or two per year have been observed in captivity; hatchlings reached maturity in a year. Diet mostly insects and other arthropods.

PLATED LIZARDS AND THEIR RELATIVES

FAMILY GERRHOSAURIDAE

These relatively large, tough-scaled lizards were originally regarded as a subfamily of the Cordylidae. They are ancient and their presence on both mainland Africa and Madagascar indicates that they evolved before the two separated, in the Cretaceous, 80–100 Mya. There are seven genera in the family, five African and two Madagascan. Two genera occur in north-east Africa: *Broadleysaurus* and *Gerrhosaurus*. The head has large, symmetrical scales with osteoderms. Body scales hard and rectangular, also with osteoderms, overlapping, sometimes keeled. Along the flanks there is a curious lateral fold of soft, granular-scaled skin. They have well-developed eyes and obvious ear-openings. The femoral glands have clearly visible pores in most species, especially in males.

KEY TO THE NORTH-EAST AFRICAN GENERA AND SPECIES OF THE FAMILY GERRHOSAURIDAE

1a: Belly scales in ten (rarely nine) longitudinal rows, body thickset, stripes if present not continuous
Broadleysaurus major, Great Plated Lizard

1b: Belly scales in eight longitudinal rows, relatively slim, body stripes usually continuous, fine, black and yellow
Gerrhosaurus flavigularis, Yellow-throated Plated Lizard

GREAT PLATED LIZARD *BROADLEYSAURUS*

A genus of sub-Saharan Africa containing a single species, Great Plated Lizard *B. major*. Originally placed in the genus *Gerrhosaurus*, molecular analysis now indicates that this species and another huge southern African form, *Matobosaurus validus*, are well separated from the smaller plated lizards, and are placed in new genera. A ten-million-year-old fossil of this species is known from Mfangano Island in Lake Victoria.

GREAT PLATED LIZARD, ROUGH-SCALED PLATED LIZARD

Broadleysaurus major
(Duméril, 1851)

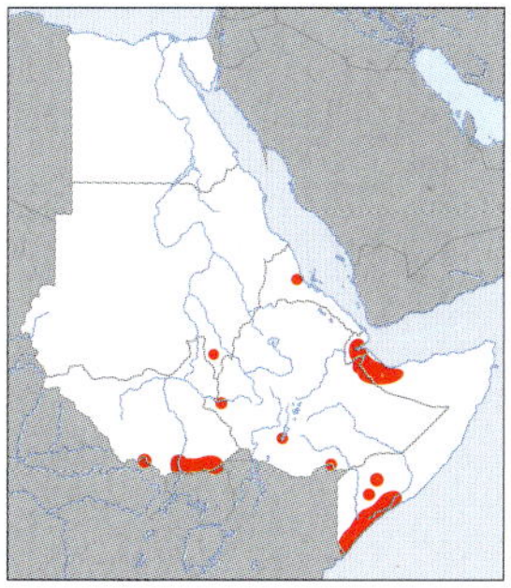

IDENTIFICATION: A large, stout, armoured lizard, with a short head and rounded snout; the eye and earhole are large and obvious. The body is thickset and compressed, limbs short and powerful. Tail about 60 per cent of total length, thick, tapers smoothly and not easily shed. Scales hard, rectangular, in 14–18 rows at mid-body, with ten rows of ventral plates. Males and females have 9–17 femoral pores; these are more obvious in males. Maximum length about 55cm, average 30–45cm, hatchlings 14–16cm. The two subspecies are coloured differently; *B. m. major* is brown, the throat red, pink or blue. *B. m. bottegoi* is black above, dark brown on the flanks, with a yellow dorsolateral stripe and the dorsal scales are yellow-edged. Some juveniles of the nominate resemble the uniformly coloured adults, others of both subspecies are black with yellow speckling, as they become adult the black may fade from the head backwards.

HABITAT AND DISTRIBUTION: Coastal woodland, thicket, moist and dry savanna, often around hills, from sea level to about 1,500m. Records very sporadic in our area. The nominate subspecies occurs in southern Somalia, from about 04°N, south to the Kenya border, and at a few localities in southern Ethiopia (Bisan River south of Lake Chamo, Murri). *B. m. bottegoi*, the black-and-yellow form, occurs in Djibouti, and directly east of there in western Somaliland (localities include Borama and the Wagar and Goolis Mountains). A single Eritrean record, at Ghinda; known from Gambela in western Ethiopia; the Tabi Hills in Blue Nile State, Sudan; and a handful of South Sudan records (Torit, Juba, Ngangala, Didinga Hills), and just across the DR Congo border in Garamba National Park. Elsewhere south to South Africa, west to Ghana.

CONSERVATION STATUS: Least Concern.

NATURAL HISTORY: A terrestrial, diurnal lizard, sheltering in holes, termitaria, rock piles and rock crevices. Sometimes found around the base of rocky hills, in coastal areas it may live in coral rag, hiding in crevices. Occasionally climbs, one found in a tree hole 3m above ground level. Quite tolerant of urbanisation. Males are territorial; territory-holding males develop vivid pink throats and will fight encroaching males, biting each other's legs and tail, trying to flip their opponents over. Females lay a small clutch of 2–7 eggs, 4–5cm long, usually beneath a rock, in a deep, damp hole or a termitarium; a clutch of seven eggs was found in southern Tanzania in June. Omnivorous, eating arthropods, fruits, flowers and other lizards. They make excellent, confiding pets. Known predators include birds of prey, large snakes and mongooses.

Broadleysaurus major, Kenya.

Broadleysaurus major, Djibouti.

PLATED LIZARDS *GERRHOSAURUS*

An African genus of seven species of diurnal, largely striped, medium-sized lizards, with thick armoured scales and a curious skin fold on the flanks. A recent revision elevated several forms out of synonymy or subspecific status to full species. One species in our area.

YELLOW-THROATED PLATED LIZARD

Gerrhosaurus flavigularis
Wiegmann, 1828

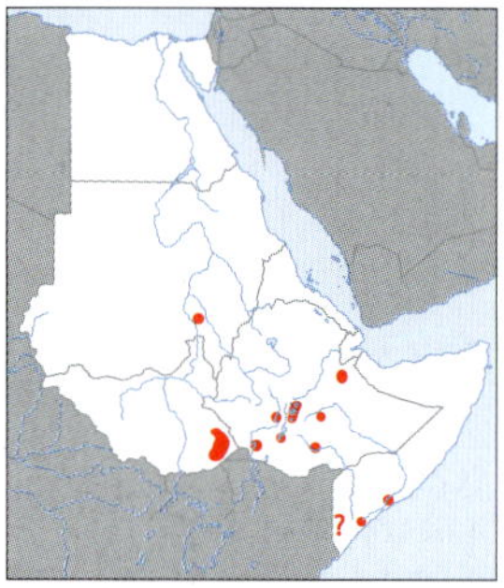

IDENTIFICATION: A medium-sized striped slender lizard. Head short, snout pointed, neck long. The eye is large, pupil round. The ear opening is triangular and obvious. The body is flattened dorsally, toes thin, some of the rear toes are very elongate. Tail long, two-thirds of total length, tapering smoothly, and easily shed. Scales strongly keeled, in 22–24 rows at mid-body. Males have 11–17 femoral pores, females none. Maximum length about 40cm, average 25–35cm, hatchlings 10cm. Colour quite variable, usually brown; there is a pair of fine yellow and black dorso-lateral stripes, sometimes a fine yellow, black-bordered vertebral stripe, the back may be speckled yellow. The flanks can be vertically barred black and cream, black and yellow, uniform or blotched red. The body stripes extend onto the tail if it is not regenerated. In breeding males, the chin and throat may turn green, pink, red, orange or yellow. Belly may be cream, white, yellow or blue-grey.

HABITAT AND DISTRIBUTION: Woodland, high grassland, moist and dry savanna and coastal bush, from sea level to about 2,000m. A species of south-east African savanna, it penetrates Ethiopia along the southern Rift Valley, known localities include Arba Minch, Omo River, Lakes Awassa, Langano and Zwai; but also recorded near Negele, Sheikh Hussein and Harar. Known from the southern Somali coast, along the middle and lower Shebelle River, but probably extends to the Kenya border, as it occurs all the way along the Kenya coast. A single Sudan record (Sennar) and a couple from South Sudan (Kapoeta, Boma Hills); presumably more widespread but under-collected. Elsewhere, south to South Africa.

Gerrhosaurus flavigularis, Kenya.

CONSERVATION STATUS: Least Concern.

NATURAL HISTORY: A terrestrial, diurnal lizard, often living in holes, under big rocks, leaf piles and in hollow logs. Secretive and fast-moving, often mistaken for a striped snake as it rushes through the vegetation. In the breeding season, the male's throat becomes suffused with bright colours. Females lay 3–8 eggs, roughly 1.5×2cm, in a moist hole, or in leaf litter under a thick bush. No clutch details for our area, hatchlings were collected in north-west Kenya in July. Diet: a wide range of arthropods, including grasshoppers, termites and millipedes.

CHAMELEONS

FAMILY CHAMAELEONIDAE

Chameleons are a family of unusual, diurnal arboreal lizards, famous for their ability to change colour. Tail prehensile in larger species and cannot be shed. They have binocular vision and their eyes move independently. They have no external ears and poor hearing. Their feet are uniquely adapted for climbing, with opposed bundled toes and sharp claws. Chameleons have a telescopic tongue, longer than their bodies, which can be shot at prey, an adaptation to silent predation on a flimsy branch. They have small, non-overlapping scales without bony plates. The sex organs of the male are located in sheaths at the base of the tail, which is thus broad in adults. Most can change colour rapidly, their colour being connected to their emotional/hormonal state, and sensitive to light, shade and temperature. Slow-moving; their active defences are limited to hissing, biting, squirming, jumping from the perch and surreptitiously moving to keep the branch between them and danger (some pygmy chameleons can also vibrate). So, their camouflage, by colour and body shape, has to be good. However, if sexually aroused or facing a rival (some are territorial and asocial), vivid, stunning colour patterns may appear, some become black with rage. In cold areas, they will also turn black in the early morning, flatten themselves to increase surface area and align themselves perpendicular to the light, in order to efficiently absorb heat and light from the sun (black is the best absorber of light and heat).

Some have horns, rival males use these in fights, they may also bite and claw. Chameleons occupy a wide range of north-east African habitats, even desert provided there is some vegetation. Low-altitude chameleons lay eggs; most high-altitude ones give live birth. They are useful animals, eating insect pests. Preyed on by many animals, including snakes and birds, particularly hornbills, shrikes and starlings. Predation can cause chameleon populations to peak and crash. Some species are social, especially horned species, where males are intolerant of conspecifics, and only come together to mate; but the small side-striped chameleons tolerate each other, and will co-exist in close proximity.

There are more than 220 known species of chameleon, in 12 genera, found in Africa, Madagascar and some Indian Ocean islands, a few reach the Middle East, one occurs in India and Sri Lanka. Greatest diversity is in eastern tropical Africa, Tanzania has more than 40 species. Seventeen species in three genera occur in our area. In some areas, they are greatly feared, for irrational reasons. They are also at risk from harvesting for the pet trade, as they are easily collected at night, showing up in torchlight. Although the pet trade does provide a living to some, there is a need to monitor chameleon populations and protect vulnerable habitats.

The key below should aid identification, but chameleons can be often fairly easily identified by a combination of locality, size, colour and pattern, head shape, presence or absence of horns, ear-flaps and ventral and gular crests. If you see a chameleon, photograph it.

KEY TO THE NORTH-EAST AFRICAN GENERA OF CHAMELEONS

1a: Tail 40 per cent or less of total length, often short and stumpy and rarely or weakly prehensile, rarely concentrically coiled; never larger than 11cm, overall colour usually predominantly brown or grey, with fine horizontal or oblique stripes ***Rieppeleon*, Rieppel's pygmy chameleons**

1b: Tail 50 per cent or more of total length, long and tapering smoothly, strongly prehensile and can be coiled up, maximum size always >12cm, overall colour usually predominantly green, without fine body stripes **2**

2a: No horns or rostral projection, body scales largely homogenous and not varying greatly in size, total length at least 25cm or more, usually in low to mid-altitude semi-desert, savanna or woodland ***Chamaeleo*, savanna chameleons**

2b: May have horns, or rostral projections, body scales heterogeneous, total length nearly always less than 25cm, in savanna, woodland or forest ***Trioceros*, African side-striped chameleons**

RIEPPEL'S PYGMY CHAMELEONS *RIEPPELEON*

A genus of three species of little striped chameleons in eastern Africa, one occurs in our area. All African pygmy chameleons were originally in the genus *Rhampholeon*, which was recently split into *Rhampholeon* (with three subgenera) and *Rieppeleon*.

KENYA PYGMY CHAMELEON

Rieppeleon kerstenii
(Peters, 1868)

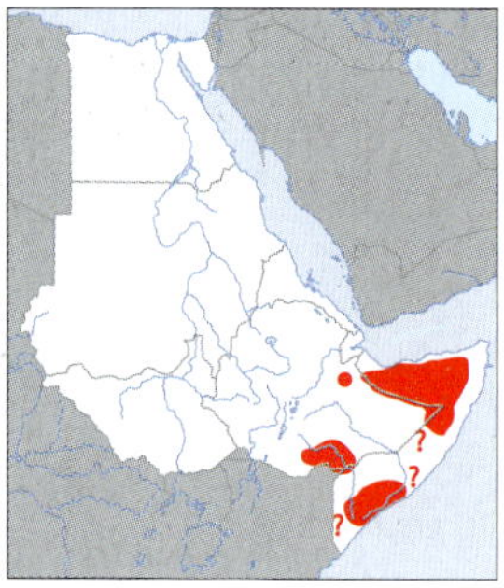

IDENTIFICATION: A small brown or grey chameleon, often with longitudinal stripes, but no rostral process. Casque flattened. Tail short, 25–42 per cent of total length. No true gular, ventral or dorsal crest, but has little spiky scales and scale clusters under chin, on tail, and on prominent raised eyebrows and face. Maximum size about 10cm, average 6–8cm, hatchling 3cm, females often larger than males. Colour quite variable, may be grey, brown or yellow-brown, sometimes uniform, often there is a series of light and dark horizontal body stripes or lines. A fairly broad dark stripe usually extends through eye, even on mono-coloured specimens. Taxonomic Notes: The subspecies in our area is *R. k. robecchii*.

HABITAT AND DISTRIBUTION: An unusually adaptable pygmy chameleon; occupies coastal woodland and thicket, moist and dry savanna and semi-desert, from sea level to 1,900m. Widespread in eastern Ethiopia and Somalia, but records very sporadic. In Ethiopia, there is a cluster of records on Kenya border (Mandera, Malka Murri, Moyale), also Negele and Dire Dawa, so presumably in the Ogaden. Probably occurs throughout Somalia; localities include Mareri Sugar Project at Jilib, Bardera and Mogadishu, and it goes right up the Kenya coast to the Somali border. One or two sporadic records from central Somalia, and a cluster of records along the north-east boundary of the Ogaden, in the Haud, and a few from northern Somalia, including Berbera, Borama, the vicinity of Erigavo and Gardo. Elsewhere, to Kenya and north-east Tanzania. With such a huge range, in often dry country, in conservation terms it is under no threat.

CONSERVATION STATUS: Least Concern.

NATURAL HISTORY: Lives in bush and grass, hides in thickets. Seems crepuscular, active at dawn and dusk in some areas, but on coast it moves around during day. Hides in holes and termite hills in dry savanna and semi-desert, in open country fond of thickets and thornbush. Camouflage superb, even in tiny leafless thorn bushes it is virtually impossible to spot. Lays eggs, a relatively large clutch of 9–12 eggs recorded, roughly 1×0.6cm in size. Six gravid females were collected in eastern Kenya in October–December, one clutch was laid in early November and took just over 50 days to hatch. Eats invertebrates. Can vibrate violently and does so when held or poked; this presumably is an anti-predator device but it is not known exactly how the vibration is produced; possibly by exhaling minute pulses of air. A Kenyan naturalist who picked one up, then dropped it, and said he had received an electric shock!

Rieppeleon kerstenii, Dire Dawa.

TYPICAL OR SAVANNA CHAMELEONS *CHAMAELEO*

At one time, all African 'non-pygmy' chameleons were placed in this genus, but after recent taxonomic changes, only 14 species remain there, including those that occur around the Mediterranean, in the Middle East and India. Seven species occur in our area; they are all relatively large, hornless, egg-laying savanna dwellers, with prehensile tails, mostly have large ranges, and are somewhat similar in appearance. There is debate about their status, see the relevant taxonomic notes. Head shape and size/shape and presence/absence of occipital lobes (ear-flaps) can vary, and some individuals may be hard to identify with certainty (particularly *C. gracilis*) but locality, ear-flap details and tail length should give useful clues.

KEY TO THE NORTH-EAST AFRICAN MEMBERS OF THE GENUS *CHAMAELEO*

1a: Occipital lobes (ear-flaps) absent — 2

1b: Occipital lobes (ear-flaps) present — 5

2a: Tail significantly shorter than snout–vent length, averaging 40–45 per cent of total length, dorsal crest virtually flat, in our area always west of Ethiopian Rift Valley — *Chamaeleo laevigatus*, Smooth Chameleon

2b: Tail just shorter, equal to or longer than snout–vent length, dorsal crest not flat, not necessarily west of the Ethiopian Rift Valley — 3

3a: Casque low, almost flat, with a low or indistinct parietal crest — *Chamaeleo gracilis*, Slender Chameleon (some specimens; see couplets 5 and 6)

3b: Casque strongly elevated with a prominent parietal crest — 4

4a: Males with prominent tarsal spurs — *Chamaeleo africanus*, Basilisk Chameleon

4b: Males with absent or rudimentary tarsal spurs — *Chamaeleo calcaricarens*, Spurless Basilisk Chameleon

5a: Casque tall, elevated above lateral crest by a height greater than or equal to diameter of the eyeball — *Chamaeleo chameleon*, Mediterranean Chameleon

5b Casque only moderately elevated or low, less than diameter of the eyeball — 6

6a: Casque hardly elevated above nape, parietal crest low, occipital lobes thin/rudimentary — *Chamaeleo gracilis*, Slender Chameleon (some specimens)

6b: Casque somewhat elevated posteriorly, parietal crest distinct, occipital lobes usually well developed — 7

7a: Occipital lobes and temporal region covered with large, plate-like scales, known only from a small area of the Ogaden on Ethiopia/Somalia boundary — *Chamaeleo ruspolii*, Ogaden Flap-necked Chameleon

7b: Occipital lobes and temporal region covered with small to medium-sized scales; in our area known from south-east Ethiopia and southern Somalia — *Chamaeleo dilepis*, Flap-necked Chameleon

BASILISK CHAMELEON

Chamaeleo africanus
Laurenti, 1768

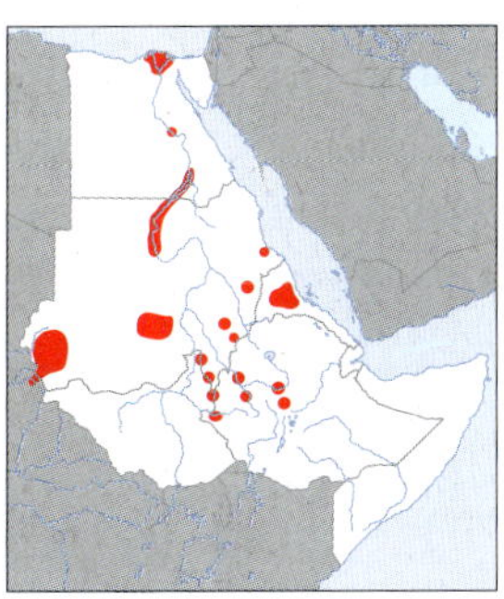

Chamaeleo. africanus, Mulu.

IDENTIFICATION: A big savanna chameleon, without ear-flaps. Casque tall and sharply raised to a point posteriorly. Tail prehensile, about 50–55 per cent of total length. Scales homogenous. A prominent gular crest of white spiny scales becomes a prominent ventral crest. Dorsal crest quite low. Males have prominent tarsal spurs. Maximum size up to 45cm (northern Egypt specimens), average 25–35cm, hatchlings about 6cm. Colour very variable, uniform or blotched, all shades of green from light to very dark, yellow, blue-green, blue-grey, brown and rufous. Often shows two flank stripes, solid or broken. Often a series of deeper green V-shapes along spine. When stressed may appear dark with black spots.

HABITAT AND DISTRIBUTION: Moist and dry savanna, semi-desert and vegetated areas in desert, between sea level and 2,400m. In Egypt and Sudan occurs sporadically along the Nile and environs. Known from Erkowit, north-east Sudan, scattered records at Kassala, Darfur, parts of Northern and Southern Kordofan and the Blue Nile. A handful of records in south-west Ethiopia (Blue Nile Gorge, Mulu), widespread in north-central Eritrea (Keren, near Saganeiti, Mendefera). Elsewhere, west to Mali across the Sahel.

Chamaeleo africanus, Neem.

Chamaeleo africanus, Amle.

Chamaeleo africanus, Mendefera.

CONSERVATION STATUS: Least Concern.

NATURAL HISTORY: Arboreal but will descend to ground to move from tree to tree, lives in trees, shrubs, even quite small bushes and thick grass clumps. When angry, hisses furiously, bites and inflates gular pouch; it can vibrate, and will jump from branches if threatened, turn dark and squirm down into the vegetation to hide. Major enemies include big rear-fanged savanna snakes like boomslang. In Ethiopia usually only found in wet season, so probably aestivates. Lays eggs in a tunnel excavated by female, clutch sizes of 49 recorded but 10–20 more usual. Captive eggs incubated at around 28°C took just under six months to hatch. Diet invertebrates.

SPURLESS BASILISK CHAMELEON

Chamaeleo calcaricarens
Böhme, 1985

IDENTIFICATION: A medium-sized chameleon, often vividly spotted, it resembles *C. africanus* but without tarsal spurs in males. Has a conspicuously raised helmet-like casque with a strong parietal crest. No earflaps. Lateral crest on head fades before occiput. Gular/ventral crest feebly developed with tubercles barely larger than surrounding scales. Body slender; scales homogeneous. Dorsal keel may be almost smooth or have low crest, most apparent anteriorly. Tail prehensile, roughly as long as body. Maximum size about 29cm (snout–vent length up to 15cm), but mostly 20–25cm, hatchling 5cm. Colour variable, but usually pale to dark green, brownish or grey, often with 1–3 variable white or yellow lateral stripes and a tendency to develop (when stressed) a dramatic pattern of broad dark stripes on side of head and sometimes large dark and oval blotches or spots on body. Eyeball with radiating stripes. Tubercles of gular/

ventral crest white. Taxonomic Notes: Previously synonymised with *C. africanus*; recent phylogenetic analyses have confirmed genetic differences to species level; some even indicate this form is more closely related to *C. chameleon* than *C. africanus*.

HABITAT AND DISTRIBUTION: Widespread in a considerable variety of habitats. Associated mostly with semi-arid scrub in dry savanna at lower elevations (sea level to approximately 1,000m), but extends well into montane grassland at higher altitudes (up to 2,400m) in central and northern Ethiopia. Precise distribution unclear, resulting from its recent description and close resemblance to *C. africanus* (even in the hand females of the two cannot be distinguished). Known with certainty from the Anseba Valley and Ghinda in Eritrea, thence south across northern Ethiopia, through Gondar to central Ethiopia (Addis Ababa, near Welenchiti), and east across Awash region. An isolated record from Forêt du Day in Djibouti. Widespread in Somaliland, well known from Tog Wajaale Plains, around Borama, and east across the Ogo Mountains (Hargeisa) to the Goolis Range inland of Berbera and Dagah Shabell. Also, the Dadar Valley in Cal Miskaat Mountains of Bari region in Puntland. The precise line of geographic separation between this species and *C. africanus* is unclear, especially in north-west Ethiopia and Eritrea. It has a relatively large area of distribution, tolerates high levels of human activity, and it is not currently under commercial collection pressure.

CONSERVATION STATUS: Least Concern.

NATURAL HISTORY: Diurnal and arboreal, but will descend to move from tree to tree in lightly vegetated country. In Mendefera in Eritrea and on Tog Wajaale Plains in Somaliland, observed in open savanna with grassy undergrowth on relatively small shrubs 0.7–1.5m high. Adult females slept at night in bushes at heights of about 0.7m. A subadult was asleep on an ornamental shrub in a hotel garden and two adult males slept on a *Vachellia* tree about 2.5m high and about 1m apart. Found in tall trees in villages, between buildings. In Awash National Park, large adults usually occupied fairly large, often isolated *Vachellia* trees on the plain. In Somaliland and Ethiopia active and abundant during warmer wetter months (May–September). Rarely seen in dry cold months, they probably aestivate. Lay eggs. A hatchling was found in May. Diet invertebrates.

Chamaeleo calcaricarens, Quljeet.

MEDITERRANEAN CHAMELEON

Chamaeleo chameleon
Linnaeus, 1758

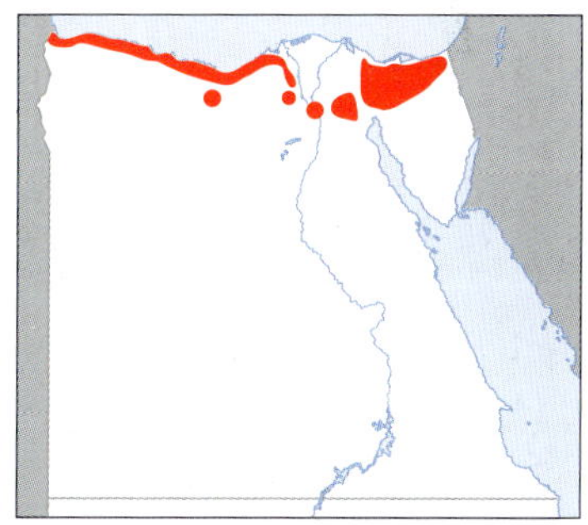

IDENTIFICATION: A big high-casqued chameleon, in our area confined to northern Egypt. Casque tall, ear-flaps present (sometimes fairly small). Tail prehensile, about half or slightly less than half total length. Scales homogenous. A low gular crest of white spiny scales is present. Ventral crest vestigial or absent. Dorsal crest absent or low. Tarsal spurs small or absent. Maximum size up to 30cm, average 20–25cm, hatchlings 5.5–7cm. Colour variable, usually shades of green or brown, often shows two flank stripes, both may be solid or broken, often a series of deeper green U-shapes along spine. When stressed head striping becomes obvious, and body may appear dark with black spots. Taxonomic Notes: Subspecies *C. c. musae* occurs east of Cairo, through Suez and into northern Sinai; it has larger ear-flaps and the casque is lower and more elongate.

HABITAT AND DISTRIBUTION: Dry savanna, semi-desert and vegetated areas in desert, farmland

and Mediterranean scrub, between sea level and 1,700m, although generally at low altitudes in our area. Occurs in Egypt, along western Mediterranean coastal plain, within 30km of coast, from Libyan border to northern Sinai. A curious record from Wadi Halfa in Sudan is probably misidentified. Elsewhere; on south-west coast of Arabian Peninsula, and west to Morocco, with a limited distribution around northern Mediterranean.

CONSERVATION STATUS: Least Concern.

NATURAL HISTORY: Diurnal and partially arboreal, often in small trees, bushes and grass clumps, but will descend to move from bush to bush and sometimes on the ground in cover; waiting quietly for food to pass. When angry, hisses, bites if handled and inflates gular pouch. Mates in summer, July–September, eggs laid between September and November, in a tunnel excavated by female, and hatch the following summer. Clutch sizes of 4–40 eggs recorded, in general the older the female, the more eggs. Diet invertebrates.

Chamaeleo chameleon, Morocco.

Chamaeleo chameleon, Morocco.

FLAP-NECKED CHAMELEON

Chamaeleo dilepis
Leach, 1819

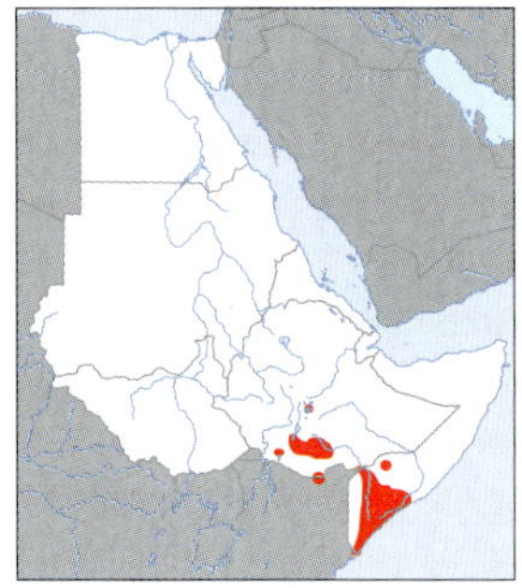

IDENTIFICATION: A big savanna chameleon, without horns or a rostral process, but with small to large ear-flaps. Casque is slightly raised at back. Tail prehensile, roughly as long as the body. Body scales heterogeneous, scales on ear-flaps often large. A prominent gular crest, consisting of a ridge of pale (often white) spiky scales becomes a prominent ventral crest, a dorsal crest of smaller spiky scales shrinks posteriorly and vanishes on tail. The interstitial skin in the grooves on chin may be vivid orange or yellow. Maximum size about 43cm, but most specimens 15–25cm, hatchlings 4–5cm. Colour very variable, mostly shades of green but can be brown, yellowish or grey; spotted black when angry, often a white lateral stripe is present. Taxonomic Notes: Various forms have been described as subspecies or even species, mostly based on flap size, many are just variants of a single polymorphic species, but we have chosen to recognise *C. ruspolii* as a valid species, for reasons detailed there.

HABITAT AND DISTRIBUTION: Coastal forest, woodland, thicket and both moist and dry savanna, from sea level to about 1,500m. In our area, from coastal Kenya to southern Somalia, to about 04°N, and a handful of records in south-east Ethiopia, including Negele, Lake Chamo, Yabello, Moyale and the Bisan River. Elsewhere, south to South Africa, west to Cameroon.

CONSERVATION STATUS: Least Concern; its huge range, relative abundance and willingness to utilise suburbia means it is under no threat.

NATURAL HISTORY: Lives in bushes, shrubs and trees, but will descend to ground. Often seen crossing roads. Sleeps on outer branches of trees at night. If threatened and unable to escape – it can move quite rapidly for a chameleon – it will inflate its body, distend its throat, raise its ear-flaps, hiss and bite furiously; it can also vibrate. In dry season it may aestivate, in a hole or some other refuge, such as a tree hole or weaver's nest. In the mating season, the male's gular groove orange skin becomes pale, and the female will then permit him to approach for mating; they mate for several days, with the female becoming increasingly intolerant of the male. Females lay on average 20–40 eggs (but up to 73 recorded), 10–15mm long, in a tunnel dug by the female in damp soil. Eggs hatch after several months.

Chamaeleo dilepis, Kenya.

OGADEN FLAP-NECKED CHAMELEON

Chamaeleo ruspolii
Boettger, 1893

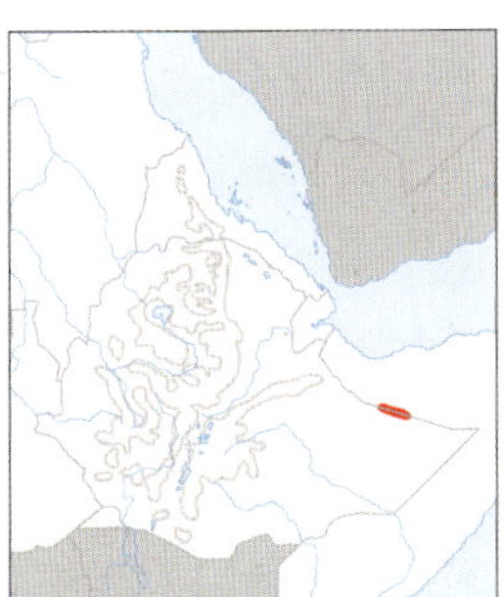

IDENTIFICATION: A medium-sized savanna chameleon, closely related to *C. dilepis*, but most of head covered by large, plate-like scales; confined to Haud region of Ethiopia and Somaliland. Casque slightly raised at back. Ear-flaps (occipital lobes) in contact mesially, well developed, and the flap and temporal scales are large, almost plate-like tubercles (an excellent aid to identification). The head's lateral crest may appear shortened, parietal crest very distinct with a single row of laterally compressed tubercles. Gular/ventral crest feebly developed with tubercles barely larger than surrounding scales. Tail prehensile, a little longer than body. Body scales homogeneous. Males possess prominent tarsal spurs. Maximum size about 27cm (snout–vent length 10.6–12.3cm), but most specimens 12–25cm, hatchling unknown. Colour of live specimens essentially unknown, no living specimen has been photographed, but presumably similar to *C. dilepis*. The types are grey in preservative, with a pale lateral stripe separating belly from flanks. Posteriorly, a white elongated spot points upwards from this white stripe. Taxonomic Notes: Usually regarded as a variant of *C. dilepis*, but recent molecular analysis indicates this is a species complex. Based on these findings and significant morphological differences we tentatively regard *C. ruspolii* as a valid species.

Chamaeleo ruspolii, preserved, Haud.

HABITAT AND DISTRIBUTION: A species of the Somali arid zone, in flat, open and dry *Vachellia–Commiphora* savanna with soft red sand soil, at altitudes of 875–1,090m. Collected only at nine precise and one imprecise locality in the Haud of southern Somaliland and eastern Ethiopia in the northern Ogaden. The types were collected by Prof. Carlo Keller during Eugenio Ruspoli's first expedition at unspecified locality in Ogaden. An additional 12 specimens were collected by Colonel R. H. R. Taylor on the Ethiopia/Somaliland boundary, between the villages of Candhodhexe in the east and Davegoriale in the west, *c.*70–185km west of Bohodle in the Togdheer region of Somaliland or immediately adjacent Ethiopia. Not known elsewhere.

CONSERVATION STATUS: Not Assessed.

NATURAL HISTORY: Very poorly known. Probably lives in trees and, like other species of the genus, lays eggs. The specific name honours Prince Eugenio Ruspoli (1866–1893), Italian aristocrat, explorer and naturalist. During his second expedition to the Horn of Africa he was killed by an elephant.

SLENDER CHAMELEON, GRACEFUL CHAMELEON

Chamaeleo gracilis
Hallowell, 1841

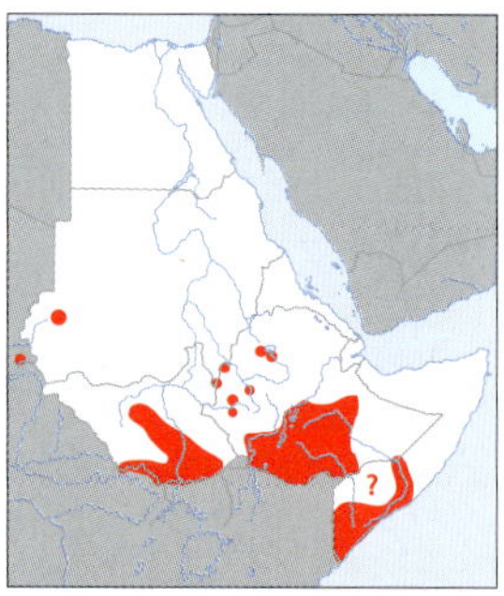

IDENTIFICATION: A big savanna chameleon, without horns, some have tiny ear-flaps. Casque slightly raised to a point posteriorly. Tail prehensile, about half total length. Scales homogenous. A prominent gular crest of white spiny scales becomes a prominent ventral crest, a dorsal crest is also present, shrinking in size towards the tail. Males have tarsal spurs. Maximum size up to 35cm, average 18–30cm, hatchlings 5–6cm. Colour very variable, all shades of green from light to very dark, blue-green, blue-grey, brown and rufous, usually with a prominent light brown or white flank stripe, a white or pale patch on shoulder and another pale spot in middle of flanks. Angry animals show black spotting. Skin of gular pouch often vivid orange, sometimes duller orangey-yellow. Taxonomic Notes: Two subspecies recognised, in our area the nominate form occurs.

HABITAT AND DISTRIBUTION: Moist and dry savanna, fond of acacia; in arid regions favours riverine thicket. From sea level up to 1,800m. In our area, quite widespread in southern Somalia, south of 04°N, and found over much of the land below 1,800m in southern half of Ethiopia. A dubious record from Puntland in northern Somalia (10.15°N, 49°E) needs to be verified. Widespread but sporadic in southern South Sudan (records include Torit, Yei, Bagangai, Imatong Mountains and surroundings, Yambio, Li Rangu). A single Sudanese record, Jebel Marra in Darfur. Elsewhere, south to Angola, west to Senegal.

CONSERVATION STATUS: Least Concern.

NATURAL HISTORY: Arboreal but will descend to ground to move from tree to tree, lives in trees, shrubs, even quite small bushes and thick grass clumps. When angry, hisses furiously, bites, and inflates the gular pouch; it can vibrate, and will jump from branches if threatened. Major enemies include big rear-fanged savanna snakes like boomslang. Lays eggs, clutches of 44 recorded but 10–25 more usual. In West Africa, eggs are recorded as taking 3–4 months to develop in the female's body and up to seven months to hatch, but breeding details unknown in north-east Africa. Captive males will fight each other, biting throats. Diet invertebrates.

Chamaeleo gracilis, Lake Langano.

Chamaeleo gracilis, Lake Langano.

SMOOTH CHAMELEON

Chamaeleo laevigatus
Gray, 1863

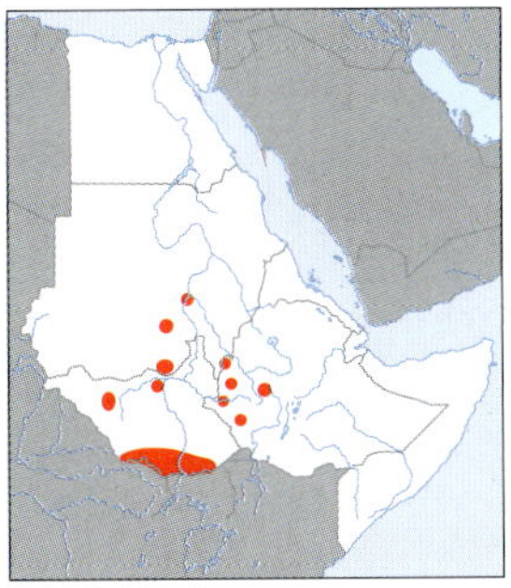

Chamaeleo laevigatus, Uganda.

IDENTIFICATION: A fairly large savanna chameleon, without horns, rostral processes or ear-flaps. Casque very slightly raised at the back. Tail prehensile, but about two-fifths of the total length, shorter than in most green chameleons. Body scales homogeneous. A white or pale gular crest of cone-like soft spiny scales and extends onto belly, to vent and sometimes beyond, below the tail. Dorsal crest is low, almost non-existent. Maximum size about 25cm, average 15–22cm, hatchling 4–5cm. Colour quite variable, but usually shades of green, greenish-brown or yellow-brown, with irregular darker spots and blotches; when animal is irritated these spots become larger and very dark. In some specimens, on either side of the vertebral ridge is a series of 4–6 large oval saddles, which may be darker green or brown. A white, cream or light grey colour form is also known, with scattered fawn-coloured spots.

HABITAT AND DISTRIBUTION: Moist savanna, at 300–1,300m in our area, but higher elsewhere. Distribution poorly understood; confusion with congenerics has created problems in mapping its range; for example, Eritrean records have now been reassigned to *C. africanus*. Two records in western Ethiopia (Didessa River Bride, Gambela area), sporadic in Sudan (Khartoum, El Obeid, Southern Kordofan, eastern Blue Nile) and South Sudan (Juba, Torit, Yei, Imatong Mountains, Sobat River, Northern Bahr El Ghazal, and just across the border in Garamba National Park in DR Congo). Elsewhere, south to Zambia, west to Chad.

CONSERVATION STATUS: Least Concern.

NATURAL HISTORY: Arboreal, living in bushes, small trees and thickets, sometimes in grassland, sedge or reedbeds. Will readily descend to ground to cross treeless areas, thus often seen crossing roads in savanna, especially in morning in rainy season. Lays eggs, clutch sizes of up to 60 eggs known but 15–30 more usual; taking 4–5 months to hatch. A series of specimens from DR Congo was captured in dry savanna as they apparently do not aestivate and their colour (yellow or grey) rendered them conspicuous. Diet invertebrates.

SIDE-STRIPED CHAMELEONS *TRIOCEROS*

A genus of about 40 species of mostly small (but one huge), tree- and bush-dwelling chameleons, widely distributed throughout tropical Africa. Nine species occur in our area, most are montane with small ranges. Males of some species have annular horns; others lack horns or have small rostral projections. Many are small with lateral stripes, the 'side-striped' group. They give live birth. Often present in huge numbers and largely tolerant of other members of same genus; their populations appear to peak and crash. Climatic change and the rising of the East African dome have to some extent fragmented north-east African populations and led to the evolution of the various forms, many in small highland refugia, with attached conservation concerns.

KEY TO THE NORTH-EAST AFRICAN MEMBERS OF THE GENUS *TRIOCEROS*

Couplet	Go to / Species
1a: Has one or two annular (ringed) horns	2
1b: Lacks annular horns	3
2a: A single tiny annular horn, in our area only in Imatong Mountains	*Trioceros conirostratus*, Sudanese Unicorn Chameleon
2b: Has two annular horns, confined to Bale Mountains, Ethiopia	*Trioceros balebicornutus*, Bale Two-horned Chameleon
3a: Midline gular crest (beard) present	4
3b: Midline gular crest absent	8
4a: Flank scalation homogenous, or finely heterogeneous (scales much same size)	5
4b: Background scalation strongly heterogeneous (large scales among the small)	7
5a: Only in Bale Mountains, Ethiopia, gular crest long and spiky	*Trioceros harennae*, Harenna Hornless Chameleon
5b: Occurs in South Sudan, gular crest fairly short	6
6a: Ventral crest of distinct white tubercles, usually some blue on body in life, black or white gular grooves	*Trioceros ellioti*, Montane Side-striped Chameleon
6b: Ventral crest of low white tubercles, no blue on body, gular grooves not black	*Trioceros kinetensis*, Mount Kineti Chameleon
7a: Never shows any green, nearly always has two light flank stripes, gular crest very small	*Trioceros bitaeniatus*, Side-striped Chameleon
7b: Often shows some green or yellow, usually only one obvious flank stripe, gular crest varies in size, often some tubercles large	*Trioceros rudis*, Rwenzori Side-striped Chameleon
8a: Dorsal crest extends well onto tail, casque slightly raised, many large flank scales among smaller ones, confined to northern slopes of Bale Mountains, Ethiopia	*Trioceros wolfgangboehmei*, Wolfgang Boehme's Chameleon
8b: Dorsal crest fades near base of tail, casque fairly flat, a few large flank scales among smaller ones, widespread in highland Ethiopia	*Trioceros affinis*, Ethiopian Highland Chameleon

ETHIOPIAN HIGHLAND CHAMELEON

Trioceros affinis
(Rüppell, 1845)

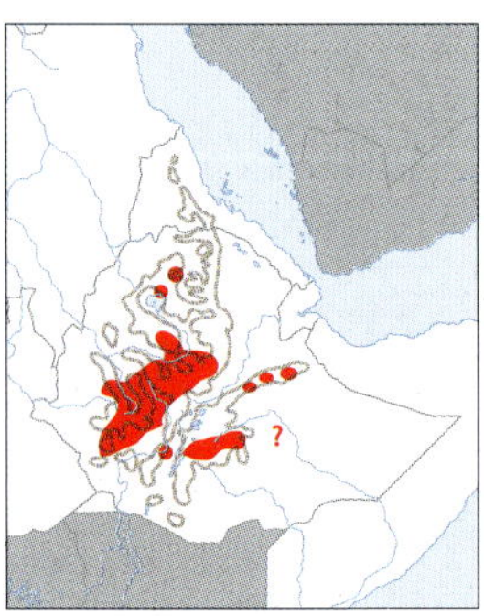

IDENTIFICATION: A small, fat green chameleon of highland Ethiopia. Casque very slightly raised posteriorly. Tail prehensile, 53–60 per cent of total length. Body scales heterogeneous, with a scattering of slightly larger scales on flanks. There is no true gular crest, although some individuals have a whitish row of scales in middle of chin. A smallish dorsal crest of spiny scales tends to disappear on the tail. Maximum size about 16cm, average 10–15cm, hatchlings 4–5cm. Colour predominantly shades of green, often deep green, males may be brown. There is usually an enlarged pale blotch behind eye, and a broad light stripe often shows on flanks, both may be white or pink. Some animals have big pale X- or Y-shapes on the flanks. When stressed shows a light diamond pattern on deep green, high on flanks.

Trioceros affinis, Addis Ababa.

Trioceros affinis, mating, Addis Ababa.

HABITAT AND DISTRIBUTION: Mid- to high-altitude forest, woodland, grassland and savanna, between 1,900 and 3,100m. An Ethiopian endemic, occurs throughout highlands of the country, west of the Rift Valley from the foothills of the Simien Mountains south to Bench Maji; and in high country east of the Rift Valley Lakes, easternmost record Harar.

CONSERVATION STATUS: Least Concern.

NATURAL HISTORY: Diurnal and arboreal, usually in bushes and small trees, in forest tends not to ascend very high in trees, usually less than 4m. Quite tolerant of agriculture and urbanisation, still present in some parks and wooded suburbs of Addis Ababa, and will tolerate exotic conifer plantations. Often in churchyards where religious custom prevents the felling of trees. Can be present in numbers, five were found in a small tree in Addis Ababa. Becomes active just after dawn, despite very low temperatures. They give live birth, after a gestation period of 3–5 months, to up to 20 young. Diet invertebrates.

WOLFGANG BOEHME'S CHAMELEON

Trioceros wolfgangboehmei
Koppetsch, Nečas & Wipfler, 2021

IDENTIFICATION: A small, fat green chameleon recently described from Bale Mountains, Ethiopia. Casque very slightly raised posteriorly. Tail prehensile, 53–60 per cent of total length. Body scales heterogeneous, with a number of slightly larger scales on flanks. No true gular crest. A smallish dorsal crest of spiny scales extends onto tail. Maximum size just under 16cm, average 12–14cm, hatchlings 4–5cm. Colour predominantly shades of green, sometimes almost turquoise, occasionally brown. There is usually an enlarged pale blotch behind eye, and a broad light stripe often shows on flanks. Occasional animals have big pale X- or Y-shapes on flanks. When stressed shows a mix of deep green and black on flanks. Taxonomic Notes: Elevated from within *T. affinis*; molecular analysis indicates separation between this species and Addis Ababa examples of *T. affinis*.

HABITAT AND DISTRIBUTION: High grassy savanna with stands of indigenous coniferous forest. Endemic to Ethiopia and described from two small towns; Dinsho, at 3,130m on north side of Bale Mountains, and Goba, north-east of Bale Mountains, at 2,740m.

Trioceros wolfgangboehmei, Goba.

Trioceros wolfgangboehmei, Goba.

The local area is extensively farmed, but some protected areas remain.

CONSERVATION STATUS: Not Assessed.

NATURAL HISTORY: Diurnal and arboreal, usually in bushes and small trees, often in clearings within coniferous forest, and quite common around Dinsho and Goba, found in hedges and sleeping on wire fences. Becomes active just after dawn, at low temperatures. Fifteen individuals were found on a small hedge at a Goba hotel. Hatchlings collected in December and January. Habits presumably similar to *T. affinis*.

BALE TWO-HORNED CHAMELEON

Trioceros balebicornutus
Tilbury, 1998

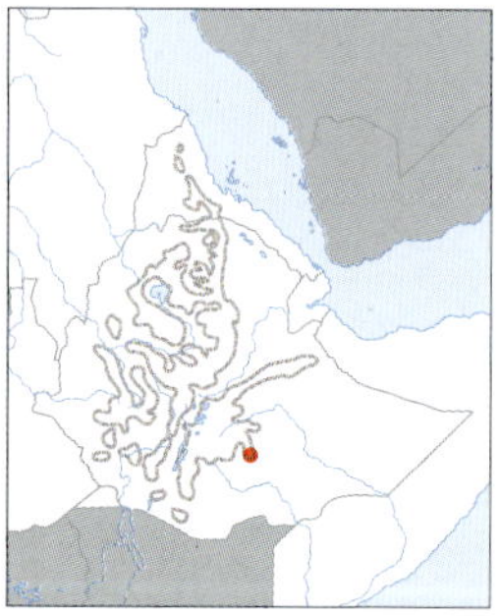

IDENTIFICATION: A unique two-horned chameleon, known only from forest south of the Bale Mountains in Ethiopia. Casque flat with no parietal crest. On the snout are two tightly opposed annular horns, short in females (up to 1.5mm), longer in males (up to 4mm). Tail prehensile, slightly more than half total length. Body scales heterogeneous, with several vague rows of enlarged plate-like tubercular scales on sides, two or three times the size of the surrounding scales. A dorsal crest of spiky triangular scales extends a short way on back, then shrinks, with no crest on tail. A gular crest or beard of fairly large scales becomes a ventral midline of white tubercles. Maximum size about 16cm, usually 10–14cm, hatchling unknown. Colour variable, shades of brown, olive, often with some yellow speckling and irregular darker flank blotches. Taxonomic Notes: DNA indicates this is a sister species to *T. harennae*.

HABITAT AND DISTRIBUTION: Known only from Harenna Forest on south slopes of Bale Mountains, south-east Ethiopia, between 1,500 and 2,400m, in a mix of coniferous and broadleaf forest. At present, only recorded near the national park road that descends from the Sanetti Plateau to the town of Dolo Mena, but probably extends east and west of the road, at similar altitudes.

CONSERVATION STATUS: Near Threatened. Although its entire habitat lies within a nominally

Trioceros balebicornutus, Harenna Forest.

Trioceros balebicornutus, Harenna Forest.

Trioceros balebicornutus, Harenna Forest.

protected national park, increasing numbers of people live in the park, along the road, and damage the forest while grazing cattle and gathering firewood.

NATURAL HISTORY: Little-known. Diurnal and arboreal, usually in lower shrub layer of the forest, at perch heights of 1–5m above ground, also in isolated bushes in clearings. The types were collected at night, sleeping on outer branches of shrubs. A museum specimen contained 16 eggs, but they probably give live birth in their relatively cold habitat. A juvenile was found in October.

SIDE-STRIPED CHAMELEON

Trioceros bitaeniatus
(Fischer, 1884)

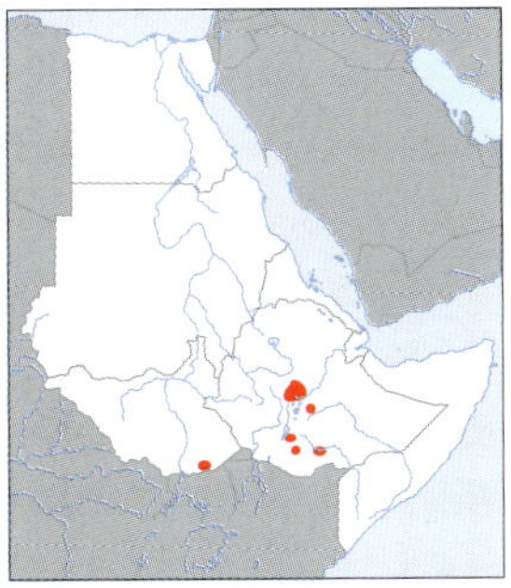

IDENTIFICATION: A fairly small brown or grey chameleon, usually with two side stripes, but no horns or rostral appendages. Casque is slightly raised posteriorly, Head narrow. Tail prehensile, roughly equal to or slightly less than half the total length (females usually 42–48 per cent, males 44–51 per cent). Body scales heterogeneous, with usually two rows (sometimes only one) of enlarged tubercular scales along sides. A weak gular crest becomes a weak ventral crest; prominent dorsal crest of spiky triangular scales. Maximum size about 17cm, usually 11–14cm, hatchling 3–4cm. Colour variable, often brown in Ethiopia, sometimes grey. Usually has two prominent pale flank stripes. When mono-coloured, head sometimes conspicuously darker than body. May also be blotched black, white and grey, with three or four large pale blotches horizontally in middle of flanks.

HABITAT AND DISTRIBUTION: Mid- to high-altitude open savanna and grassland, woodland and isolated hills, from 1,300m to 2,500m. East African, just reaching our area at a handful of localities. A cluster of Ethiopian records in vicinity of Addis Ababa (Akaki, Debre Zeit, Sodere, Asela), also known from eastern shore of Lake Chamo, the high country south-east of there, and around Yabello and Negele. In South Sudan known only from Imatong Mountains (Mt Kinyeti, Gilo, Talanga Forest). Elsewhere, south to northern Tanzania.

CONSERVATION STATUS: Least Concern.

NATURAL HISTORY: Diurnal, rarely found in large trees, usually in small trees, bushes, tall grass and sedges. Also hedges and ornamental plants, sometimes in large numbers. Capable of withstanding extreme cold. May be abundant in some areas, but populations appear to peak and crash, perhaps connected with the fact that this species gives live birth, and possibly also due to predation by birds, especially starlings and

Trioceros bitaeniatus, Debre Zeit.

shrikes. Usually 6–15 young are born, although one had 25 babies. Two broods of differing sizes recorded in female oviducts. Females have been known to eat their own young. Diet invertebrates.

SOUTH SUDANESE UNICORN CHAMELEON

Trioceros conirostratus
(Tilbury, 1998)

IDENTIFICATION: A small slender chameleon, rather similar to Side-striped Chameleon. Males have a tiny, long cone-shaped horn on tip of snout. A short gular crest. Tail equal to or slightly less than half total length. Body scales heterogeneous, with enlarged scales on flanks, particularly those within the flank stripes. Maximum size about 18cm, usually 12–14cm, hatchling around 3cm. Usually a shade of brown or green, with pale side stripes and orange tints. In middle of each flank, between the upper and lower stripe, a series of three distinctive oval-shaped darker markings often present. A member of the *T. bitaeniatus* complex.

HABITAT AND DISTRIBUTION: Hill forest, woodland, wooded grassland and farmland, between 1,000

Trioceros conirostratus, male, Kenya.

Trioceros conirostratus, female, Kenya.

and 2,600m. Described in 1998 from a single specimen at Lomoriti in the Imatong Mountains in South Sudan, at 1,200m. Although not seen in South Sudan since, it has now been recorded from a chain of hills along the Kenya/Uganda border and Uganda/South Sudan border, including Agoro-Agu Forest reserve, immediately south of the Imatong Mountains.

CONSERVATION STATUS: Least Concern.

NATURAL HISTORY: Little-known. Diurnal and arboreal. East African examples found in shrubs bordering fields in farmland on a plateau that was originally covered with *Vachellia–Commiphora* woodland. Presumably similar to others of genus, i.e., gives live birth, populations peak and crash, eats invertebrates. When handled, Ugandan specimens opened their mouths, exposed the pouches at the jaw angle and thrashed their heads, causing a foul-smelling exudate to fly out; the smell persisted for more than 24 hours.

MONTANE SIDE-STRIPED CHAMELEON

Trioceros ellioti
(Günther, 1895)

Trioceros ellioti, Uganda.

IDENTIFICATION: A medium-sized chameleon, without horns or a rostral process. No ear-flaps. Casque very slightly raised posteriorly, with one to four distinct longitudinal grooves on gular pouch, and skin between scales distinctively black. Tail prehensile, slightly greater than body length in males, slightly shorter in females. Scales weakly heterogeneous, with enlarged scales lying within lighter flank stripes. Very prominent gular crest of long, pale spiky scales becomes a prominent ventral crest, and a fairly prominent dorsal crest is present, the spines getting shorter towards tail. Maximum size about 19cm, average 10–15cm, hatchlings 3–4cm. Colour quite variable, but usually shades of vivid blue, green or brown, often with a broad, prominent, light-coloured (white, orange or yellow) flank stripe.

HABITAT AND DISTRIBUTION: High, moist savanna and grassland, between 1,000 and 2,800m in East Africa. In our area known solely from Imatong Mountains in South Sudan; one of four members of the *bitaeniatus* group found there.

CONSERVATION STATUS: Least Concern.

NATURAL HISTORY: Behaviour unstudied in Imatong Mountains, but elsewhere in East Africa, usually in low trees, bushes, shrubs and tall grass, fond of reedbeds and sedge in swampy areas. Often abundant and tolerant of urbanisation and agriculture; can also tolerate cold. Gives live birth, 4–12 babies, and the females can retain viable sperm for up to 18 months if not mated. Eats invertebrates.

HARENNA FOREST CHAMELEON

Trioceros harennae
(Largen, 1995)

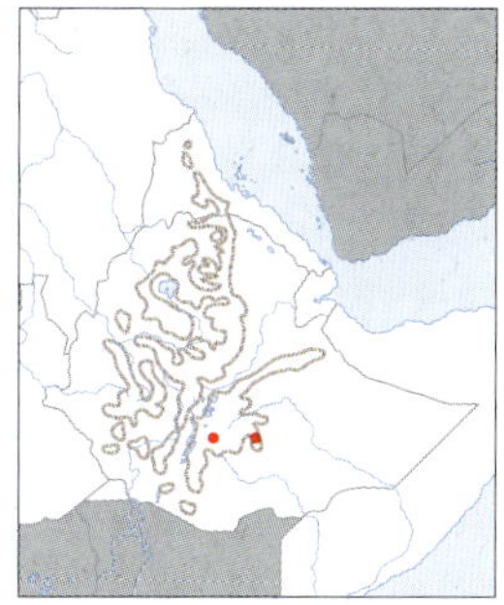

IDENTIFICATION: An Ethiopian endemic; a squat, bearded chameleon, known only from environs of Harenna Forest in the Bale Mountains and Arbegona village west of there. The casque is slightly elevated posteriorly; no horns or rostral projections. Tail prehensile, between 50–60 per cent of total length. Body scales heterogeneous, with a scattering of scales on flanks (some specimens have very few). A dorsal crest of spiky triangular scales extends onto the tail, decreasing in size posteriorly. A prominent gular crest or beard of fairly large scales becomes a ventral midline of white tubercles. Maximum size about 14cm, usually 10–12cm, hatchling unknown. Colour very variable, usually shades of green, but can be orange, brown or bluish, the dorsal crest is sometimes purple. When stressed, irregular darker or blue flank blotches appear and may coalesce to form a vague broad band on flanks. In early morning can be almost black when warming up in sun. Inside of mouth vivid orange or yellow, exposed in threat. Taxonomic Notes: Specimens from Arbegona are subspecies *T. h. fitchi*, with more heterogeneous scalation.

Trioceros harennae, Arbegona.

Trioceros harennae, Arbegona.

Trioceros harennae, Harenna Forest.

HABITAT AND DISTRIBUTION: Known only from vicinity of road through Harenna Forest on south slopes of Bale Mountains, south-east Ethiopia, and around village of Arbegona, roughly 80km west of there. It probably occurs between the two localities. In Harenna, it was found from 3,375 down to 2,400m. Arbegona is at 2,600m.

CONSERVATION STATUS: Least Concern. However, its exact distribution and tolerance of agriculture needs quantifying; the Bale Mountains National Park is under threat, with increasing numbers living people there.

NATURAL HISTORY: Little-known. Diurnal and arboreal, specimens from stands of giant heather at top of forest were observed thermoregulating not long after dawn, having turned black, sideways to the sun, at perch heights of 1–2m. They occur from the giant heather down through coniferous forest into broadleaf forest. Arbegona specimens were in hedges, bushes and crops in and around the village. Gravid females observed in April, and gave live birth to 8–12 young.

MOUNT KINETI CHAMELEON

Trioceros kinetensis
(Schmidt, 1943)

IDENTIFICATION: A small squat chameleon, from Mt Kinyeti in Imatong Mountains, never knowingly seen alive. Has a rounded nose. Casque slightly elevated posteriorly. A short low gular crest. Tail of male very slightly less than half total length, female's tail very slightly more, unlike most chameleons where male has a longer tail than female. Body scales slightly heterogeneous, with enlarged scales near spine. Maximum size about 13.5cm. Colour of living animal unknown. A member of the *T. bitaeniatus* complex, molecular analysis indicates it is the sister species to *T. bitaeniatus*.

DISTRIBUTION: Known from four specimens collected in 1939 on Mt Kinyeti, Imatong Mountains of South Sudan, above 3,000m; specimens were found in both *Podocarpus* forest and grassland above treeline.

CONSERVATION STATUS: Data Deficient.

NATURAL HISTORY: Unknown. Eight eggs present in oviducts of type, but probably gives live birth.

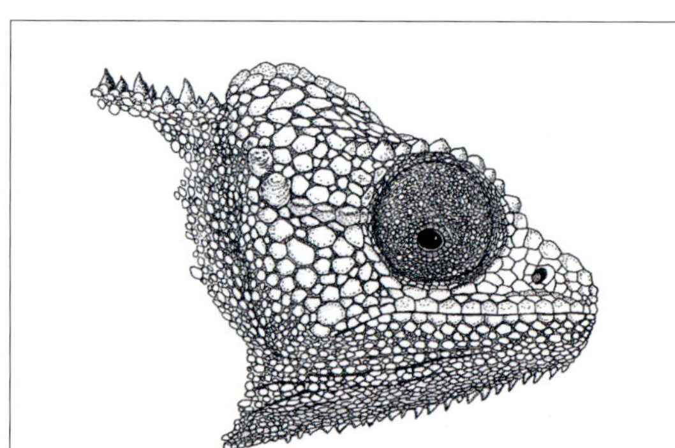

Trioceros kinetensis.

RUWENZORI SIDE-STRIPED CHAMELEON

Trioceros rudis
(Boulenger, 1906)

IDENTIFICATION: A small, fat chameleon, without horns, rostral process or ear-flaps. Casque slightly raised posteriorly. Tail prehensile, about half total length or slightly more than half. Body scales heterogeneous, with two distinctive flank lines of larger granular scales. A medium to small gular crest is present, becoming a short ventral crest, which rarely reaches the vent. A medium-sized dorsal crest of spiny scales extends the length of back and onto tail. Maximum size about 17cm, average 10–15cm, hatchlings 4–5cm. Colour usually blotchy green or yellow-green, with 1–2 distinctive light flank stripes following line of the big flank scales; these can be narrow or broad, diffuse or with sharp edges, or like a chain of interconnected elongate diamonds. Sometimes broad vertical bars of dark green and greenish-yellow appear on flanks and tail, or prominent rufous blotches along vertebral ridge. Occasional specimens are grey or blue-grey. DNA indicates its closest relatives are *T. kinetensis* and two montane forms from northern Tanzania.

HABITAT AND DISTRIBUTION: Medium to high-altitude grassland and savanna, usually above 2,000m, up to 4,000m in the Rwenzori. In our area, known only at Gilo, Imatong Mountains, elsewhere associated with Albertine Rift Valley.

CONSERVATION STATUS: Least Concern.

NATURAL HISTORY: Arboreal, fond of bushes and thicket, will also live in tall and tussock grass, reedbeds and sedge. Like other members of side-striped chameleon complex, often present in huge numbers in suitable habitat, especially open grassland with medium-sized, thick bushes. Tolerant of low temperatures. Adapts well to cultivation and urbanisation, living in hedges, shrubbery, tall crops and roadside vegetation. Gives live birth, a clutch of eight recorded. Eats invertebrates.

Trioceros rudis, Rwanda.

MONITOR LIZARDS

FAMILY VARANIDAE

A lizard family and genus found in Australia, Asia and Africa, many of which are very large. The world's heaviest lizard is a monitor, the South-East Asian Komodo Dragon *V. komodoensis*, it reaches 160 kg and 3.1m length; Asian Water Monitor Lizard *V. salvator* is the longest lizard in the world and grows to 3.5m long; north-east Africa's biggest lizard is Nile Monitor *V. niloticus*, reaching 2.5m or more. A fossil monitor nearly 6m long is known from the Pleistocene of southern Asia and Australia.

Monitors are carnivores, eating a wide range of prey, the Komodo Dragon is said to have killed and eaten humans. Monitors have powerful jaws and long retractile forked tongues, an obvious ear-opening and a long flexible neck. Their limbs are well developed, with large claws. The tail cannot be shed but is long and powerful; used as a climbing aid, in swimming and as a defensive weapon. The tough and leathery skin, covered with bead-like scales is exploited in some countries, leading to their persecution; in some they are eaten; in rural north-east Africa monitors may be killed by smallholders as they raid chicken runs. In Egypt long-distance lorry drivers may have a stuffed monitor on the radiator grille, to bring luck. All monitors lay eggs. Male combat has been observed in some species. Just over 80 species of monitor are known, split into 11 subgenera. Australia has the greatest diversity of monitor lizards with more than 30 species. At present only four species are known from Africa, all four occur in our area.

KEY TO NORTH-EAST AFRICAN MEMBERS OF THE GENUS *VARANUS*

1a: Relatively slim, snout quite pointed, head not very deep — 2

1b: Relatively stocky, snout rounded, head deep and somewhat bulbous — 3

2a: Mostly yellow and black, tail with broad dark and narrow yellow bands, tongue blue, usually near water — *Varanus niloticus*, Nile Monitor

2b: Mostly shades of brown, grey and black, tail with broad light and narrow black bands, tongue pink, usually in dry country — *Varanus griseus*, Desert Monitor

3a: Mid-body scale rows 110–167, scales small and granular on neck, in our area largely east of Rift Valley — *Varanus albigularis*, White-throated Savanna Monitor

3b: Mid-body scale rows 75–100, scales large and cobblestone-like on neck, in our area largely west of Rift Valley — *Varanus exanthematicus*, Western Savanna Monitor

DESERT MONITOR

Varanus griseus
(Daudin, 1803)

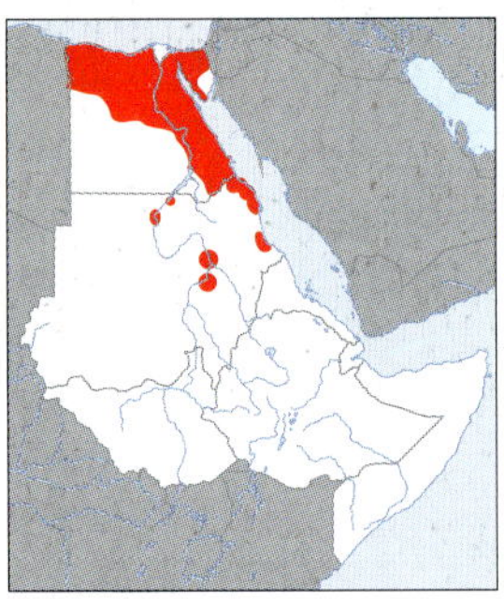

IDENTIFICATION: A brown, dark-banded monitor lizard of desert and near-desert, with a huge range. Head pointed, snout long and sharp, eye fairly large, pupil round, iris orange, earhole a slit. Tongue long and forked, pink with a grey tip. Neck long, body fairly thick, with well-developed stout and muscular limbs, and strong claws. Tail tough and whip-like, laterally compressed, cannot be shed, about 60 per cent of total length. Scales coarse, non-overlapping and bead-like, in 134–169 rows at mid-body. Maximum size about 1.4m, average 80cm to 1.2m, hatchlings 22–25cm. Both sexes have up to five precloacal pores on each side. Ground colour and pattern very variable; usually shades of brown with dark broad or narrow crossbars (tend to fade with age), between which are scattered pale circular spots, sometimes speckled black to greater or lesser extent. Tail strongly banded yellow and black, particularly in juveniles. A series of black lines usually extends backwards on head. Underside white or cream, sometimes with dark flank bars. General appearance changes with how dirty it is, and how near sloughing. Taxonomic Notes: Three subspecies recognised; nominate occurs in our area.

HABITAT AND DISTRIBUTION: Desert, near- and semi-desert; usually in areas with some vegetation. Will venture into saltmarshes, cultivated areas and rocky hills. Usually at low altitudes, but up to 1,200m in Morocco. In our area, widespread in Egypt, although absent from heart of Nile Delta, extensive mountainous areas and sand seas of the south-west. Extends into Sudan along Red Sea coast (Erkowit and Suakin), and the Nile (Dongola, Shendi), almost to Khartoum. Elsewhere west to Mauritania, east to India.

CONSERVATION STATUS: Least Concern, although exploited for its skin and for pet trade; and there is a small trade in stuffed specimens in North Africa.

NATURAL HISTORY: Diurnal and terrestrial. Shelters in burrows dug by other animals but also dig their own; these may be several metres long and more than 1m deep. They emerge a couple of hours after sunrise, bask and then hunt. In Egypt active between March and November, and during the hottest time of year active twice a day, sheltering in the midday. Early and late in the year active once at warmest times. Monitors have home ranges of several hundred ha, and forage actively, often walking 500–600m or more. In the breeding season, males may walk several kilometres in search of females, which they then chase, with ritual neck waving, bite the female and hold on. If cornered, they stand up, stiff-legged, lash with their tails, inflate the throat, and bite powerfully. Five to 20 eggs are laid in late summer and hatch the following spring. They eat any small animal they can overpower; faecal examination of Asian animals found they favoured rodents, agamas, beetles and snakes. Also ingest small stones, possibly for digestive purposes.

Varanus griseus, Tunisia.

NILE MONITOR, WATER MONITOR

Varanus niloticus
(Linnaeus, 1766)

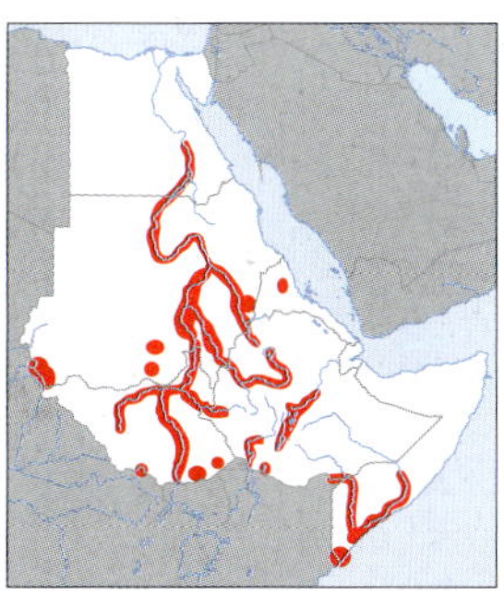

IDENTIFICATION: A huge, dark green and yellow lizard, usually near water sources. It has a pointed snout, the eye is large, pupil round, iris yellow and black, eyelids yellow. The tongue is long and forked. Neck very long, body long, relatively slim and cylindrical, with well-developed muscular limbs, and strong claws. Tail cannot be shed, tough and whip-like, laterally depressed, triangular in section, with a hard vertebral ridge, about 60 per cent of total length. Skin tough and leathery, the scales small and button-like, in 128–183 rows at mid-body (usually 137–165). Maximum size about 2.5m, possibly larger, average 1.5–2.2m, hatchlings 25–33cm. Ground colour black, brown, dark green or grey-green, spotted to a greater or lesser extent with yellow. There are 4–11 dorsal, yellow crossbars, which may be solid or ocelli. Limbs spotted yellow on black, flanks vertically barred or blotched green/black and yellow, the tail has 9–18 vertical yellow bars on a dark background. Belly dirty yellow or cream, with black or dark blue crossbars, vermiculations or blotches. The juveniles are vividly coloured. Big adults can become very dull, especially if foraging away from water and in process of sloughing.

HABITAT AND DISTRIBUTION: Usually near freshwater sources in all sorts of habitat, from sea level to about 2,000m in our area. Southern Somalia (lower Juba and Shebelle Rivers, and near Kenya border), rivers and lakes in Ethiopia (lower Omo, Daua River at Mandera, most Rift Valley lakes, upper Awash River,

Varanus niloticus, Awash National Park.

Blue Nile from Lake Tana downstream, the Baro and associated wetlands in Gambela area). A single Eritrean record, from the river network east of Arbascio, at surprisingly high altitude, 2,000m. Widespread in South Sudan, in White Nile and its tributaries, and in Sudan, Darfur, Kassala, Gadaref, the Blue Nile and White Nile, downstream to Egypt, as far north as Qena. Historically recorded further north on Nile but has now disappeared. Elsewhere, virtually throughout sub-Saharan Africa beside permanent water.

CONSERVATION STATUS: Least Concern. Eaten in places, and harvested for leather and the pet trade, but has benefitted from dams and associated reservoirs.

NATURAL HISTORY: A diurnal, versatile lizard. Runs fast, climbs trees and rocks, swims superbly, using its tail, can stay submerged for 20 minutes or more. Basks on riverbanks, rocks and trees. Often wary, if approached may jump into water and swim away, but can become habituated to humans. They lay eggs, often in an active termite nest. The female claws her way in and deposits 20–60 eggs, the hole is then resealed by the termites and the eggs are kept warm and moist while they develop. They forage on land and in water, eating any small animal, vertebrate or invertebrate, that they can overpower. Fond of mussels and freshwater crabs. Observed 'herding' fish in shallow water; they raid crocodile and sea turtle nests. They have an excellent sense of smell and will eat carrion; they raid chicken runs. Like other monitors, if cornered they inflate the throat and hiss loudly, raise themselves up high, stiff-legged and lash with their tails, and if seized they bite savagely and scratch with their claws.

Varanus niloticus, juvenile, Debre Zeit.

WHITE-THROATED SAVANNA MONITOR

Varanus albigularis
Daudin, 1802

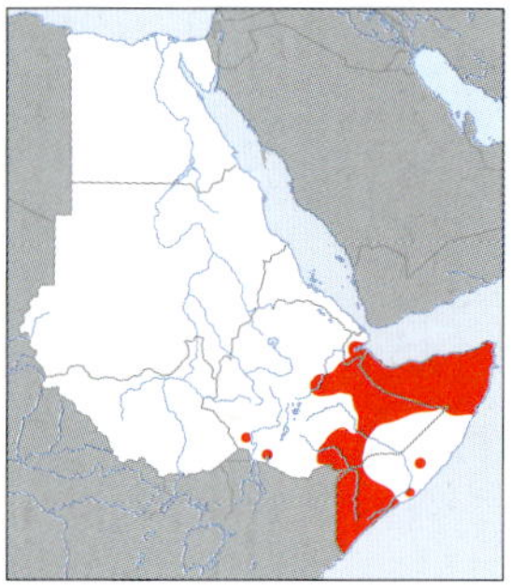

IDENTIFICATION: A big, heavily built, grey or brown monitor lizard. Head large and square, snout bulbous, with an obvious big nostril, eye is small, pupil round, earhole large. Tongue forked, long and blue. Body broad, somewhat triangular in section, with well-developed stout and muscular limbs, and strong claws. Tail slightly longer than body and cannot be shed, it is tough and whip-like, laterally depressed (although in big adults may be flattened horizontally across base), triangular in section, with a hard ridged top. Scales coarse, small, non-overlapping and bead-like, in 110–167 rows at mid-body. Two precloacal pores are present. Maximum size about 1.6 m, average 1–1.4 m, hatchlings 23–26cm. Colour and pattern often obscured by dirt and unsloughed skin patches, but clean specimens look grey, brown or black above, with 3–8 white crossbars, rows of white spots or ocelli. A broad dark bar on neck spreads out onto back; the side of head, neck and flanks are lighter. Tail barred grey or brown on yellow or cream. Underside paler with fine black vermiculations. Juveniles have a conspicuous black chin. Taxonomic Notes: In northern Uganda this species meets *V. exanthematicus*; the two were long regarded as subspecies, and intermediates exist. For now, we regard specimens from South Sudan and Sudan as *V. exanthematicus*. Three subspecies are recognised, *V. a. microstictus* in our area, which molecular studies indicate is more closely related to Arabian *V. yemenensis* than to the nominate form.

HABITAT AND DISTRIBUTION: Dry and moist savanna, coastal thicket and woodland, and semi-desert, from sea level to about 1,500m, although usually below 1,000m. Occurs virtually throughout Somalia,

Varanus albigularis, Huluul.

Varanus albigularis, Hargeisa.

although few records from Mudug and Galgaduud (apart from Bud Bud) in central Somalia. Widespread but sporadic in low country of eastern Ethiopia, east of Rift Valley, with a surprising high-altitude record from Harar, although no specimen preserved. Known in Djibouti from Obock and Forêt du Day. Status of records in Sudan and South Sudan problematic, see above. Elsewhere, south to South Africa.

CONSERVATION STATUS: Least Concern.

NATURAL HISTORY: Diurnal, mostly terrestrial, but climbs well if clumsily, ascending big trees and rock outcrops, often to considerable height. When inactive, it will hide in thickets, in holes (especially Aardvark, *Orycteropus afer*, and porcupine, *Hystrix* species, burrows), in rock fissures, in hanging beehives, tree holes and cracks, and abandoned termite hills. In the dry season it may aestivate. Tolerant of humans in some areas, more nervous in others. If cornered, they hiss very loudly and ominously and stand up high, stiff-legged. They can lash powerfully and accurately with their tails, and have a very hard bite (although the teeth are peg-like and blunt), they will hang on with bulldog-like tenacity to anything they seize. Males have home ranges of up to 25km2 (females smaller) and may walk 3–6km per day, foraging. They lay 8–51 eggs, roughly 3.5×6cm, in a hole dug by female, sometimes a termite hill hole, hollow tree or rock crevice. The diet includes a wide range of vertebrate and invertebrates (basically, any animal they can overpower), including small mammals, birds, snakes, other lizards, eggs, tortoises, insects, millipedes, other arthropods and even carrion. They catch and kill snakes by seizing them and violently lashing the snake side to side while clawing at the body and crushing it in their jaws. They seem to have an excellent sense of smell, and will excavate buried nests and holes containing live prey, especially mammals. Enemies include large birds of prey (especially Martial Eagles, *Polemaetus bellicosus*), ratels, *Mellivora capensis*, and big snakes like the Brown Spitting Cobra, *Naja ashei*; mongooses take a heavy toll of the eggs.

WESTERN SAVANNA MONITOR

Varanus exanthematicus
(Bosc, 1792)

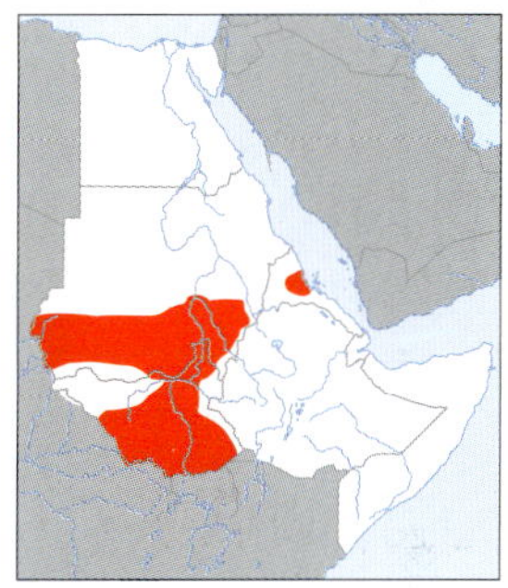

IDENTIFICATION: A relatively small, grey monitor lizard. Head thick; snout short and bulbous, with an obvious big nostril, eye fairly large, pupil round, iris orange, earhole large. Tongue long and forked, pink with a grey tip. Body broad and short, somewhat triangular in section, with well-developed stout and muscular limbs, and strong claws. Tail slightly less than half total length, tough and whip-like, laterally depressed (but in big adults may be flattened horizontally across base), triangular in section, with a hard ridged top. Dorsal scales coarse, non-overlapping and bead-like, they are big and cobblestone-like on neck, in 75–100 rows at mid-body. Two preanal pores are present. Maximum size about 1m, average 50–80cm, hatchlings 18–20cm. Ground colour light or dark grey, sometimes brown, with several light white spots or ocelli across back, the tail has light and dark vertical bars, the belly is dirty white with dark crossbars; there are often fine vertical bars on the lips. Like the White-throated Savanna

Varanus exanthematicus, captive.

Varanus exanthematicus, Chad.

Monitor its general appearance changes with how dirty it is, and how near sloughing.

HABITAT AND DISTRIBUTION: Dry and moist savanna from sea level to about 1,400m. A handful of records from central Eritrea. Widespread but sporadic across South Sudan, records include Paloich, Lafon, Torit, Juba, Yambio, Wau; a similar small number from Sudan, mostly in the south; known from Jebel Marra, En Nahud, Sungikai, Khartoum, Kosti, Sennar and Er Roseires; the type of *V. ocellatus* (a junior synonym of this species) was from 'Kordofan'. Elsewhere, west to Senegal.

CONSERVATION STATUS: Least Concern, although heavily exploited for its skin, the pet trade, traditional medicine and for food; Franz Werner reported that the only Kordofan specimen he obtained was a tail left over from a family meal!

NATURAL HISTORY: Diurnal, terrestrial, climbs well on trees and rocks, shelters in any suitable hole, termitaria, rock and tree cracks. Abundant in some areas, rare in others. Like White-throated Savanna Monitor it is confident in defence, if cornered, it will hiss very loudly and ominously, lash powerfully and accurately with its tail and bite hard, hanging on with bulldog-like tenacity to anything it seizes. Females lay from 6–20 eggs, roughly 3×5cm, usually in a hole they dig themselves. Hatchlings appear near start of rainy season. Diet includes the usual wide range of vertebrate and invertebrates; they will catch and eat any smaller animal they can overpower.

AMPHISBAENIANS OR WORM LIZARDS

CLADE AMPHISBAENIA

Probably the least known of north-east Africa's reptiles, the worm-like Amphisbaenians or Worm lizards are a group of bizarre squamate reptiles, specialised for burrowing life. Their relationships have been much debated, but they are now recognised as lizards within their own monophyletic clade. They are often pallid, pale yellow, purple or pink. The north-east African species have no ears or limbs (some species have front legs). Most African species are small, less than 30cm. Their smooth, rectangular scales are arranged in external rings or annuli, which superficially resemble those of earthworms; their heads are modified into wedges or keels, to assist with burrowing. The skin of amphisbaenids is only loosely attached to the muscles below; this permits the head and body to move in such a way that the head can be used as a ram when burrowing. Hence their skulls are relatively short and strong, often with a hardened blade-like edge to the snout, which can be used like a shovel and a ram, scraping soil from the front of the burrow and compacting it on the sides. They have large strong teeth used to seize their prey, mostly invertebrates, which they detect by scent and vibration, the victim is ripped to shreds before being dragged into the burrow. Some lay eggs, some give live birth. They live in savanna and woodland, and are rarely seen above ground, unless forced up by flooding. There are around 200 species; in our area there are two genera and four species; all endemic. They are very hard to identify to species level, the key below (and the specific keys) may require the use of a binocular microscope. Anyone finding a Worm lizard should take it to a local museum, little is known of their distribution or variation and museum specimens are needed.

KEY TO THE GENERA OF NORTH-EAST AFRICAN WORM LIZARDS

1a: Head modified into a vertical keel	*Ancylocranium somalicum*, Somali Sharp-snouted Worm Lizard
1b: Head modified into a horizontal keel	*Agamodon*, Blunt-faced worm lizards

SHARP-SNOUTED WORM LIZARDS *ANCYLOCRANIUM*

An African genus of small to moderate-sized Worm lizards. Three species are known, one in Somalia, the other two are endemic to savanna and woodland of south-eastern Tanzania. As the name suggest, some of the scales on the head are modified into a distinctive sharp-edged snout, visible from above. They live underground, but little is known of their lifestyle.

SOMALI SHARP-SNOUTED WORM LIZARD

Ancylocranium somalicum
(Scortecci, 1931)

IDENTIFICATION: A sharp-snouted worm lizard endemic to Somalia. Looks like a worm. Has a stout, elongate body and short head; viewed from above the head comes to a distinctive sharp keel. Nasal scales fused, rostral with straight posterior edges. Two to four pre-cloacal pores. Two subspecies. *A. s. somalicum* has 194–202 body annuli, with 26–33 dorsal and 19–25 ventral segments to a mid-body annulus and 5–7 caudal annuli. *A. s. parkeri*, Parker's Sharp-snouted Worm lizard has 187–189 body annuli with 31–33 dorsal and 20–23 ventral segments to a mid-body annulus and 7–8 caudal annuli. A series of eight nominotypic specimens were 13.8–23.4cm long; body diameter 0.5–1cm. Only three specimens of the Parker's subspecies are known; they measured 13.5–16.7cm; body diameter 0.65–0.7cm. Colour of the living animal not known, but probably uniformly pinkish-brown, unspotted.

Ancylocranium somalicum, underside, preserved type.

HABITAT AND DISTRIBUTION: Known from a number of records over a wide area. *A. s. somalicum* is recorded at several localities from the lowland sandy Benadir coast of southern Somalia; Kismayu and between Balad, and Eggi (or Ejim), at altitudes between sea level and 121m. *A. s. parkeri* was collected at three localities in the Haud area of northern Somaliland, on Ethiopian border, roughly between Odenleh and Candhodhexe villages in Togdheer region, at 710–787m. This is the typical Somali arid zone, with open and dry *Vachellia–Commiphora* savanna on soft red sandy soil. The two populations may be connected but the intervening region is under-collected.

CONSERVATION STATUS: Data Deficient.

NATURAL HISTORY: Unknown. Probably similar to other worm lizards, i.e., burrowing, lays a single elongate egg, eats invertebrates, fond of termites. The two specimens taken near Afgoi were found on the surface, apparently flooded out of their tunnels and drowned during protracted and unseasonal rains. They probably live very secretive lives deep in the soil.

BLUNT-FACED WORM LIZARDS *AGAMODON*

One of four genera in the Trogonophidae, distributed in North Africa, the Horn of Africa and eastern Arabia, *Agamadon* is an Afro-Arabian genus of small to moderate-sized worm lizards. Four species known, three in Somalia, one endemic to Yemen. As the name suggests, their teeth are acrodont, like those of agamid lizards. They live underground, in sandy soils, often in coastal dunes, and are characterised by two large, concavely curved or flat rostral and parietal scales on the front of the head, with which they ram through sand.

KEY TO MEMBERS OF THE GENUS *AGAMODON*

1a: Rostral and parietal scales strongly upturned at edges, body markedly compressed, dorsally flat; three upper labials; 22–24 caudal annuli; no pattern, pinkish-red in life; southern Somalia
Agamodon compressus, Flat Blunt-faced Worm Lizard

1b: Rostral and parietal scales flat, body compressed or not, dorsally rounded; four upper labials; 10–17 caudal annuli; pattern present or not; southern or central Somalia 2

2a: Body not noticeably compressed; dorsal pattern of brownish blotches almost always present; 124–138 body annuli; medium size (up to 21cm); southern Somalia *Agamodon anguliceps*, Angled Blunt-faced Worm Lizard

2b: Body slightly compressed; no dorsal pattern; 142–144 body annuli; small size (up to 10.2cm); central Somalia
Agamodon immaculatus, Calabresi's Angled Worm Lizard

ANGLED BLUNT-FACED WORM LIZARD

Agamodon anguliceps
Peters, 1882

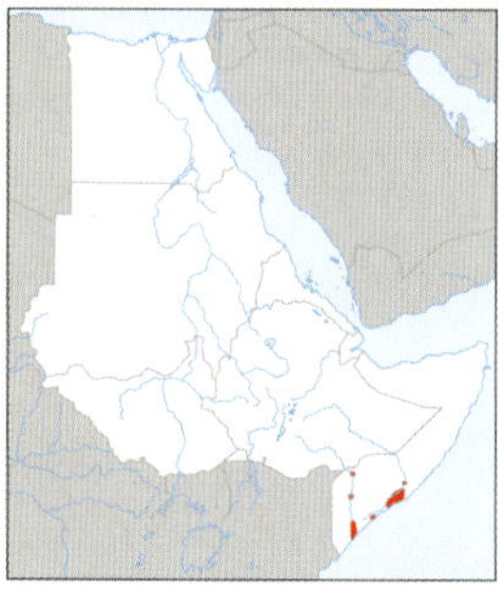

IDENTIFICATION: A patterned, medium-sized blunt-faced worm lizard with a plane scale above the eye, endemic to southern Somalia. Looks like a fat yellow elongate grub. A stout and not markedly compressed body, sub-triangular in cross-section. Head short, horizontally depressed; the edges of the very large frontal and rostral scales are not upturned, but plane or slightly concave. Four upper and three lower labials. Up to six pre-cloacal pores when present. Tail markedly reduced in depth from that of trunk, relatively short (less than 11 per cent of total length). It has 124–138 body annuli, 10–16 caudal annuli, and 42–55 segments to a mid-body annulus. Maximum size about 22.5cm (snout–vent length up to 20.5cm), average 13–18cm. Colour of living animal generally bright yellow dorsally, blotched with brown saddles and dots of varying sizes and shapes, which may coalesce to form an interrupted longitudinal line; bright pink ventrally. A very few specimens have a reduced pattern or are unspotted.

HABITAT AND DISTRIBUTION: On the coast in fixed sand dunes, with mixed bushland and scattered trees, between sea level and 100m; further inland in dry open woodland with denser vegetation, at 100–150m. Endemic to inter-riverine area and along the Juba and Shebelle Rivers in southern Somalia. Localities on Juba River include Lugh, Bardera and Alessandra Island near Gelib, Kismayu and Mofi. Coastal records include Giumbo, Brava, Merka, Afgoi and Mogadishu (where known from within the city). Also, Mahadday Weyne on the Shebelle River. Locally very common; and the most widespread of all species in our area, unlikely to be threatened by human activities.

CONSERVATION STATUS: Least Concern.

NATURAL HISTORY: Poorly known. Burrowing. Sympatric with *A. compressus*, examples of both species found near each other on the same dune, excavated in two hours after sunrise in Mogadishu. Most were hoed up from roots of shrubs and grasses, others dug up during building construction. Usually found in soft sand or sheltering under debris, e.g., logs, wooden boards, even found under shipping containers when they were moved.

Agamodon anguliceps, Kismayu.

CALABRESI'S ANGLED WORM LIZARD

Agamodon immaculatus
Calabresi, 1927

IDENTIFICATION: A uniform, small, blunt-faced worm lizard with a plane scale above eye, endemic to central Somalia. Looks like a worm. Has a stout and not markedly compressed body, sub-triangular in cross section. Head short, horizontally depressed; edges of very large frontal and rostral scales not upturned, but plane or slightly concave. Four upper and three lower labials. Two pre-cloacal pores when present. Tail markedly reduced in depth from that of trunk, relatively long (usually >11 per cent of total length). Has 142–144 body annuli, 15–17 caudal annuli, and 38–45 segments to a mid-body annulus. Maximum size about 10.2cm (snout–vent length 9cm), body diameter 0.5cm. Colour of living animal not known, perhaps yellow-ochre. Preserved specimens uniform white and show no trace of pattern or countershading. Taxonomic Notes: Originally described as a subspecies of *A. anguliceps*, then synonymised with *A. compressus*, but regarded as a separate species by amphisbaenid expert Carl Gans.

HABITAT AND DISTRIBUTION: Known only from two specimens on the Hobyo coastal plains and dunes. The type was from Afghedud (or Afguduudle) in Mudug, about 84km north-west of Hobyo town, 80km inland, altitude 282m; the second 7km north-west of Rugno (Ruunyooc) Well, about 80km south-east of El Bur on the road to El Dere, in Galguduud, 40km inland, altitude 110m. where there is low grassland along the fixed dune areas close to the sea and rich mixed shrubby bushland with scattered trees inland to the west.

CONSERVATION STATUS: Not assessed.

NATURAL HISTORY: Unknown. Probably similar to other worm lizards.

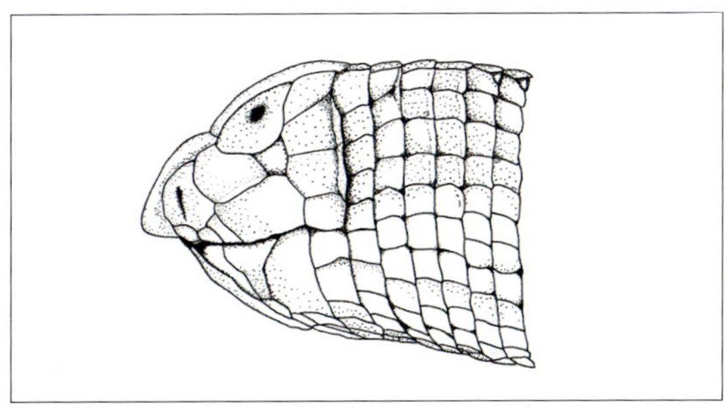

Agamodon immaculatus.

FLAT BLUNT-FACED WORM LIZARD

Agamodon compressus
Mocquard, 1888

Agamodon compressus, Mogadishu.

IDENTIFICATION: A Somali, Benadir coast endemic; a uniform pink, small-sized blunt-faced worm lizard, the scales above the eye and the rostral project. Looks like a fat shiny worm. A stout and strongly compressed body. Head short, horizontally depressed; edges of very large frontal and rostral scales upturned. Three upper and lower labials. Two pre-cloacal pores when present. Tail slightly or not reduced in depth from that of trunk. It has 147–157 body annuli, 22–24 caudal annuli, and 44–52 segments to a mid-body annulus. Total length around 9–11cm (snout–vent length 8–10cm). Colour of living animal pinkish-red and unspotted in life, whitish in preservative.

HABITAT AND DISTRIBUTION: Known from coastal fixed sand dunes a little distance from the sea, with mixed bushland and scattered trees, below 100m. Recorded at Brava, Mogadishu (including within the city) and the vicinity of Afgoi on the Benadir coast of southern Somalia.

CONSERVATION STATUS: Data Deficient, but locally common. Eight specimens were recently collected within 25m^2 on outskirts of Mogadishu. Unlikely to be threatened by human activities.

NATURAL HISTORY: Very poorly known. Sympatric with *A. anguliceps*, both found within meters of each other on same dune, excavated during the two-hour period just after sunrise in Mogadishu. Most were hoed up from roots of shrubs and grasses; others were dug up during building construction.

ORDER CROCODYLIA; CROCODILES

Familiar to all, crocodiles are big reptiles with four limbs, a long and powerful tail, and long jaws with sharp teeth. Unlike other reptiles, they have a four-chambered heart which enables efficient blood oxygenation, and a third eyelid, the nictitating membrane, which sweeps dirt from the eyeball. Some are dangerous to humans, including the Nile Crocodile, found in our area. There are 27 modern species of crocodilians (including alligators, caimans and gavials), in three families, mostly in the tropics. Six species occur in Africa, two in our area.

FAMILY CROCODYLIDAE

There are 18 species in this family, six in Africa, two in our area, Nile Crocodile *Crocodylus niloticus* and West African Crocodile *C. suchus*. It is possible that Central African Slender-snouted Crocodile *Mecistops leptorhynchus* (formerly *Crocodylus cataphractus*) might occur in south-west South Sudan. It is known from north-east DR Congo, <100km from the border, and there are a couple of museum specimens collected by sportsmen in 'Lado', in the early 1900s; Lado Enclave was an elephant-hunting area centred around Juba in present-day South Sudan, south of Jebel Lado. However, the term covered a wide area, and no subsequent specimens have emerged. Likewise, it is possible that Osborn's Dwarf Crocodile *Osteolaemus osborni* might also occur in South Sudan, as it is known from north-east DR Congo and northern Uganda, but as yet there are no records.

TRUE CROCODILES *CROCODYLUS*

A tropical genus, of some 13 species, two are found in Africa, both in our area. These two look very similar in general appearance, colour, body and tail shape, but molecular analysis indicates that they are separate species.

KEY TO NORTH-EAST AFRICAN MEMBERS OF THE GENUS *CROCODYLUS*

1a: Large, reaching more than 5m, ventral scales small, head narrower, widespread in large rivers and lakes in our area *Crocodylus niloticus*, Nile Crocodile

1b: Smaller, rarely more than 4 m, ventral scales larger, head wider and flatter, confined to Awash River and parts of Sudanese White Nile in our area *Crocodylus suchus*, West African Nile Crocodile

NILE CROCODILE

Crocodylus niloticus
Laurenti, 1768

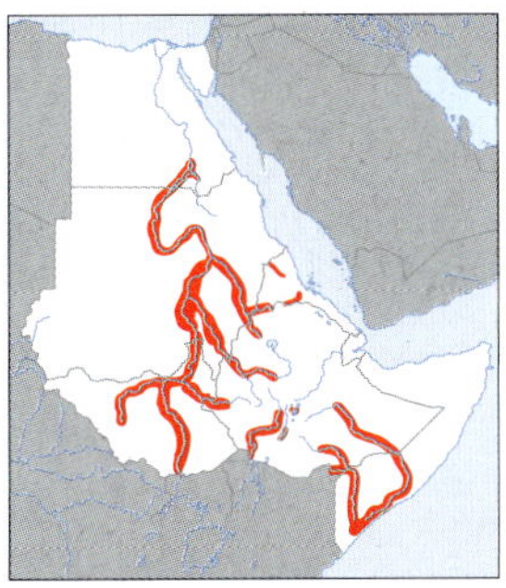

IDENTIFICATION: A big, thickset crocodile with a fairly broad snout. Males are larger than females. Maximum size about 5.5m (anecdotal reports of larger specimens exist), average 2–3.5m, hatchlings 28–32cm. Big adults may weigh as much as one tonne. The teeth are prominent, sticking out from both the upper and lower jaws, a big incisor on each side near the front of the lower jaw fits into a groove on the outside of the upper jaw, and acts as a locking mechanism. Crocodile teeth do not interlock, but they can cut; humans have had their arms bitten off. The skin is covered with regular, thick, horny scales, those on back are like armour plate, keeled and bony. Limbs short and powerful, with thick claws, the rear toes have partial webbing. Tail about 40 per cent of total length, with two raised keels posteriorly. Adults usually grey, brown or tan, sometimes greenish. Juveniles vivid yellow or tan with prominent black blotching and speckling. Some adults are bright yellow with black blotches.

HABITAT AND DISTRIBUTION: Widespread in suitable waterbodies in our area. Known localities include most suitable rivers and lakes, from sea level to about 1,500m, possibly higher. In Eritrea it occurs in the Anseba, Mareb and Setit Rivers. In Ethiopia, in the Blue Nile downstream of Shafartak Bridge, in the Omo River downstream of Abelti Bridge, in the Baro River and tributaries in the low west, and Webi Shebelle from about 800m downstream, likewise the Genale and Daua Rivers, and thence downstream across southern Somalia to the coast (the Juba and Shebelle are the only rivers in Somalia where Nile Crocodile is found). Lakes Chamo, Abaya and Langano, but not the Awash River, where *C. suchus* occurs. Widespread in South Sudan, in the White Nile and all its tributaries, in Sudan in the Blue Nile, White Nile and tributaries, downstream into Egypt, to Lake Nasser. It has disappeared further north in the last 100 years, although it did extend to the sea. Elsewhere, south to eastern South Africa. Building of large dams benefits Nile Crocodiles, in Sudan large numbers can be seen on the six major dams.

CONSERVATION STATUS: Least Concern, but under threat for its perceived risk to humans and domestic stock.

NATURAL HISTORY: Active by day and night; basking, sleeping and hunting. Big adults utilise open water and prime sandbars, juveniles tend to hide in shallow vegetated backwaters, and may dig burrows for shelter. When basking, they finely regulate their temperature by opening and closing their mouths. At night, they

Crocodylus niloticus, Arba Minch.

Crocodylus niloticus, Sai Island.

may gather in size classes. Their eyes reflect light and they can be thus seen from a distance with a spotlight or torch. Nile Crocodiles swim powerfully using the tail, limbs tucked in. They can submerge for 45 minutes, possibly longer, their heart rates slowing. Large males are territorial, defending a suitable beach or stretch of water and a group of females (usually 6–10). Nile Crocodiles have excellent sight, hearing and taste. They are also vocal, young ones can yelp loudly and they have a number of communication calls. Nile Crocodiles have elaborate courtship rituals, the male rubs the underside of his jaw and throat on the female's neck, this releases a flow from the glands under the lower jaw, the smell of which stimulates the female. Mating occurs in the water. The eggs are usually buried at night on a well-drained sandbank, open to the sky and above flood level. Between 20 and 95 eggs are laid. The female tends the nest, and excavates it when she hears the hatching youngsters chirping; she carries the babies to a quiet backwater and may guard them. The sex of the hatchling depends on the incubation temperature, lower temperatures produce mostly females. Nile Crocodiles become sexually mature at 6–10 years. Small crocodiles have many enemies but large ones are only at risk from humans and hippos. Crocodiles have a varied diet; hatchlings eat mostly insects, tadpoles and frogs, but they soon begin to take fish, which form the major part of the diet of all age classes. They also take any swimming creature, including waterbirds and smaller crocodiles. Adults snatch animals at the water's edge, waiting in ambush to drag prey in, or rush out to seize prey. In some areas in Africa, river crocodiles are more dangerous than lake crocodiles; possibly because lakes are often clear and fish can be seen, in muddy rivers they cannot, so animals at the waterside (including humans) are taken. This is the case in Lakes Chamo and Abaya in Ethiopia, where fishermen in very low freeboard boats tolerate the huge Nile Crocodiles that occur there, and vice versa. In Somalia, the lower Juba and Shebelle Rivers were notorious for crocodile attacks on drinking cattle, and occasionally humans, although poorly documented in recent years. In Sudan there seem to be relatively few documented cases of human/crocodile conflict, although occasionally swimming children have been attacked. In South Sudan fishermen who enter the water at night regard crocodiles as a risk! Cattle sometimes taken. Large prey is torn to bits, either by shaking or jerking the carcass, in the water they may seize part of the prey and then rotate on their long axis, twisting the bit off. They do not, as legend has it, store prey in underwater caverns until it decomposes. Crocodiles may also rotate quickly when they have seized part of a large living target, including humans. Crocodiles are attracted to splashing and jerky movements in the water, and in murky rivers may attack anything causing such a disturbance. In the wild, Nile Crocodiles are estimated to reach 70 years of age. Note well; Nile Crocodiles are dangerous to humans; you should never swim or enter the water where they occur, or sit or stand within 10m of the water's edge.

WEST AFRICAN NILE CROCODILE

Crocodylus suchus
É. Geoffroy Saint-Hilaire, 1807

IDENTIFICATION: A big, thickset crocodile with a fairly broad snout. Essentially similar to Nile Crocodile, but does not reach such a large size, maximum about 4m or slightly more, average 1.5–2.5m in our area. Although not yet fully quantified, ongoing research also indicates that the skull of this species has a more gnarled appearance, so it looks older than Nile Crocodiles of a similar age, at the same age this species has a wider and flatter head than Nile Crocodile, larger ventral scales and thus fewer scale rows, and lacks the occipital ridges at the back of the skull that Nile Crocodile has. Awash River animals look slimmer than Nile Crocodiles in other parts of Ethiopia. Taxonomic Notes: Described in 1807 by French naturalist Étienne Geoffroy Saint-Hilaire from the Niger and Nile Rivers. However, the name was long regarded as a junior synonym of *C. niloticus*, until 2003, when analysis indicted that West African 'Nile Crocodiles' clustered in a separate clade from other African Nile Crocodiles, and are worthy of specific status. Further research indicates that crocodiles in the Awash River in Ethiopia, and some (mummified) examples from Egypt and Sudan may belong to this species.

Crocodylus suchus, Awash National Park.

Crocodylus suchus, Awash National Park.

HABITAT AND DISTRIBUTION: In our area, known from the Awash River, in the vicinity of Sodere downstream to Lake Gemeri, virtually on the Djibouti border, and also (at least from mummified animals) on the White Nile in Kordofan in Sudan (and originally around Kom Ombo in Egypt, but now extinct there). Elsewhere, sporadic across parts of Central and West Africa.

CONSERVATION STATUS: Not Assessed.

NATURAL HISTORY: Essentially similar to *C. niloticus*. Largely associated with permanent water (although Mauritanian animals aestivated in holes and deep rock cracks if the water supply dwindled). However, most African crocodile research has been on Nile Crocodiles on the eastern side of the continent, so this needs reassessing in terms of the split. Seems to be less ready to attack drinking animals, humans included; West African specimens lack the fearsome reputation that Nile Crocodiles have in the rivers of East Africa. In places they are both respected and venerated.

SNAKES

SUBORDER SERPENTES

Snakes are highly specialised, legless squamate reptiles. They are found worldwide, absent only from some oceanic islands, Antarctica and very high latitudes. They have a very long spine and many ribs (>400 in one species) that support the rounded body and are used in locomotion. Snakes have no legs, although a few have vestiges of a pelvic girdle and little spurs, which represent vestigial legs. All species lack eyelids, they have a forked tongue that retracts into a sheath, no external ears, sharp but non-interlocking teeth, and a greatly enlarged internal cavity, with no left lung (retained but very reduced in a few species). All are carnivores, swallowing their prey whole; they can't bite bits off. They are ectotherms, getting their heat from outside; consequently, their energy budgets are low and they eat sparingly, some exist on ten or more good meals a year. The bones of the lower jaw are not fused at the front; they also have a flexible trachea, an extensible and flexible junction between the lower and upper jaw, and an enormously elastic skin, all adaptations allowing them to swallow animals much wider than their heads.

Some have evolved specialised teeth, fangs, designed to transfer venom. The venom is designed to immobilise prey rapidly and to aid digestion, not to keep humanity on its toes. A few snakes, for example spitting cobras, use venom in defence, particularly against predatory birds and humans. Non-venomous snakes kill their prey by constriction, the tight squeeze means the prey cannot breathe and blood cannot circulate, thus death is rapid.

Other snakes simply swallow their prey alive. Most snakes lay eggs in a warm, damp place. A few species give live birth, in north-east Africa notably sand boas, the Slug-eater, *Duberria lutrix*, and some vipers. There is no parental care. In north-east Africa snakes occupy a range of habitats, from sea level to high-altitude grasslands, and desert to rainforest. In general, the number of species decreases with increasing altitude, increasing aridity and increasing distance from the equator.

There are around 4,000 species of snake. In north-east Africa, there are just in excess of 200 species, in 12 families. Forty-six species in our area are dangerous to humans, 45 of these are venomous, the other is a large python. Of these 46, about half are known to have killed people. Thus, most people are afraid of snakes, a natural reaction. A dangerous snake near one's home must be killed or removed. However, the danger of snakebite depends on circumstances. The rural poor in remote regions are at significant risk, the better-off visitor to remote areas of north-east Africa is at much less risk, and, should they get bitten, at much less risk of dying. However, those whose work or pleasure takes them deep into remote country should get to know which snakes are dangerous and which are not. Snakes have an important part in food webs and wholesale destruction of them will upset the ecological balance. In 1985, Thailand exported 1.3 million snakes, mostly to other South-East Asian countries whose people were under the totally erroneous impression that eating snake parts is good for the health and affords long life, in much the same way that in the same region they believe that the meat of an owl is good for the eyes. Thailand's rat population exploded, freed from snake predators, destroying an estimated 400,000 ha of rice fields. House snakes, *Boaedon*, are widespread in the south of our area, and large adults feed mainly on rodents; these are snakes that local farmers should learn to identify and tolerate.

KEY TO THE NORTH-EAST AFRICAN SNAKE FAMILIES

Taxonomy of snakes is in a state of flux. A major revision in 2019, by Hussam Zaher and his team, published in the *PLoS One* journal, changed the structure of several superfamilies. Further changes are on the cards. This is fascinating but not directly relevant to a guide like this. We have chosen to use the useful category 'family' for the time being (all family names end in '-idae'), although users should note that, on account of their huge front fangs, the burrowing asps (*Atractaspis*) appear in this key in a different place to the rest of the family Atractaspididae (centipede-eaters, etc.). We have also followed what might be termed a 'traditional' running order, with blind and worm snakes first, then pythons and boas, followed by 'ordinary' snakes, and finally elapids and vipers.

1a: Body worm-like, head very short, round and blunt, body only tapers right at tail tip, tail blunt, eyes invisible or only visible as minute dark dots under head skin, body scales all same size	2
1b: Body not worm-like, head not very short or round, body tapers towards tail, tail not blunt, eyes well developed, enlarged belly scales present	3
2a: Small, usually <15cm, <0.5cm diameter, fewer than 18 mid-body scale rows, teeth only in lower jaw	Leptotyphlopidae, Worm snakes
2a: Larger, up to 70cm, body usually >0.5cm diameter, >20 mid-body scale rows, teeth only in upper jaw	Typhlopidae, Blind snakes
3a: Ventral plates broader than body scales but much narrower than body, vestiges of hind limbs present as short claws either side of vent	4
3b: Ventral plates as broad or almost as broad as body, no vestigial limbs	5
4a: Small, adults less than 1.2m, subcaudal scales single, no obvious neck	Boidae, Boas
4b: Large, adults >1.2m, subcaudal scales paired, obvious neck	Pythonidae, Pythons
5a: One of more pairs of enlarged venom-transporting fangs at front of upper jaw	6
5b: No pairs of enlarged fangs at front of upper jaw	8
6a: Fangs relatively short, immobile	Elapidae, Cobras, mambas and allies
6b: Fangs relatively long, and mobile, fold flat	7

7a: Uniformly dark; black or grey (sometimes with white head markings, but never with patterns or any other body markings), head same width as neck, live largely in holes or underground
Atractaspis, Burrowing asps (N.B. see couplet 12)

7b: Never uniformly dark or grey, nearly always patterned, head usually wider than neck, do not live underground Viperidae, Vipers

8a: Small, usually less than 35cm total length and some shade of grey, snout with an angular horizontal or upturned edge for burrowing, a single band-like prefrontal scale Prosymnidae, Shovel-snouts

8b: Does not have above combination of characters 9

9a: Small (<40cm), tail short (14–22 per cent of total length), stocky snakes with large eyes and short bullet-shaped heads, usually with a broad brown dorsal stripe and fine black vertebral line, give live birth, eat slugs
Duberria lutrix, Slug-eater (only representative in our area of family Pseudoxyrhophiidae)

9b: Does not have above combination of characters 10

10a: Nocturnal, terrestrial, relatively slow-moving harmless snakes, without fangs, head rarely conspicuously broad, colour usually grey, brown or black, usually lack patterns or stripes (some house snakes, *Boaedon*, may have a light dorsolateral stripe, and *Pseudoboodon*, Ethiopian Mountain snakes, may be striped but have a labial pit) Lamprophiidae, African house snakes and allies

10b: Do not have above combination of characters 11

11a: Rear-fanged, diurnal, fast-moving, very often striped, virtually always 17 rows of scales at mid-body (except some smaller sand snakes), eye pupil circular; mostly large, >1m long Psammophiidae, Sand snakes and allies

11b: Does not have above combination of characters 12

12a: Small, often dark, burrowing snakes, sometimes with bars, stripes or neck collars, rear-fanged, no loreal scale Atractaspididae, African burrowing snakes
(N.B. the *Atractaspis*, Burrowing asps, also belong in this family but are not rear-fanged)

12b: Not necessarily small, rarely burrow, loreal scales, some with rear fangs, others fangless
Colubridae, Colubrid or 'ordinary' snakes

FAMILY TYPHLOPIDAE BLIND SNAKES

A largely tropical family of over 270 species, in 18 genera. All have a unique appearance; they have a very short, blunt, usually rounded head, with a big rostral (snout) scale and their tails are also very short, only 1–3 per cent of total length. Their bodies do not taper like those of 'ordinary' snakes, but resemble metal rods with rounded ends. They have primitive, sometimes invisible or near-invisible eyes, and their body scales are all the same size; they lack the enlarged ventral scales of typical snakes. Largely live below the surface, in holes, soft soils or leaf litter, where they feed on arthropods. They are fond of termites and ants' eggs and may enter or live in termitaria, where they hunt for brood chambers with eggs. Seventeen species occur in north-east Africa. Their taxonomy is changing and some distribution records and patterns are zoogeographically inexplicable, suggesting some species are not properly defined. At present the north-east African forms are in five genera.

KEY TO THE GENERA OF TYPHLOPIDAE IN NORTH-EAST AFRICA

1a: Usually pink in life, or with yellow head, relatively slender, eye not visible or visible only as small pigmented spot 2

1b: Lacking above combination of characters 3

2a: In our area, known only in countries south of Egypt *Letheobia*, Gracile pink blind snakes

2b: In our area, known only from northern Egypt *Xerotyphlops*, Desert blind snakes

3a: Snout strongly overlapping (hooked) in profile, mid-body scale rows 24, usually a broad yellow or orange stripe on centre of rostral, with fine black side stripes (not this colour in *R. leucocephalus*) *Rhinotyphlops*, African beaked blind snakes

3b: Lacks above combination of characters 4

4a. Third supralabial overlaps ocular shield; postoculars 1–2; inferior nasal suture contacts second supralabial or preocular, mid-body scale rows 20 *Indotyphlops braminus*, Flower-pot Blind Snake

4b. Third supralabial not overlapping ocular shield; postoculars 3–7; inferior nasal suture contacts first supralabial or rostral, mid-body scales rows almost always >20 *Afrotyphlops*, African blind snakes

AFRICAN BLIND SNAKES *AFROTYPHLOPS*

A sub-Saharan African genus of c.30 species of largely thick, relatively big blind snakes, often striped, spotted or blotched. Six species in north-east Africa. Snouts rounded and eyes visible as light-sensitive, black spots beneath ocular scale. Many species eat termites, a large South African species was observed using its head to bore into an active termitarium, and at same time dribbling fluid, possibly to soften the soil and assist entrance. Within the descriptions, you will find a ratio of length/diameter – how long the snake is in comparison to its width. This ratio varies (and 1 is unstated), so for example in the description of Blanford's Blind Snake below, it says 'length/diameter ratio 29–41'. In this case, the fattest Blanford's Blind Snake has a length/diameter ratio 29:1, the thinnest 41:1.

KEY TO THE NORTH-EAST AFRICAN SPECIES OF *AFROTYPHLOPS*

Note: final identification of some of species here may require this key, a dead specimen, a binocular microscope and our figures showing head scalation, although we have tried to supply some easy-to-use characters. Note that *A. punctatus* appears twice, on account of variation.

1a: Third supralabial overlaps ocular shield, dorsal rostral narrow to moderate (less than half interocular head width); nostril pierced laterally; postoculars 1–2; inferior nasal suture contacts second supralabial or preocular 2

1b: Third supralabial not overlapping ocular shield, dorsal rostral broad (greater than half interocular head width); nostril pierced ventrally; postoculars 3–7; inferior nasal suture contacts first supralabial or rostral 3

2a: Mid-dorsal scales 257 or fewer, relatively short snout (snout/head width ratio <0.8), in our area only in southern Somalia *Afrotyphlops cuneirostris*, Somali Wedge-snouted Blind Snake

2b: Mid-dorsal scales 257 or more, relatively longer snout (snout/head width ratio >0.85), in our area in eastern Ethiopia and northern Somalia *Afrotyphlops calabresii*, Calabresi's Wedge-snouted Blind Snake

3a. Snout sharply angular in profile, with a keratinised horizontal edge in adults; ventral rostral broad (greater than half internarial snout width *Afrotyphlops brevis*, Angle-snouted Blind Snake

3b. Snout more or less rounded in profile, lacking a keratinised edge; ventral rostral narrow to moderate (less than half internarial snout width) 4

4a: Supraocular with lateral apex between ocular and preocular 5

4b: Supraocular with lateral apex between preocular and nasal 6

5a: Preocular usually overlapped by second supralabial *Afrotyphlops blanfordii*, Blanford's Blind Snake

5b: Preocular not overlapped by second supralabial *Afrotyphlops punctatus*, Spotted Blind Snake

6a: Preocular overlapped by second supralabial *Afrotyphlops punctatus*, Spotted Blind Snake

6b: Preocular not overlapped by second supralabial *Afrotyphlops lineolatus*, Lineolate Blind Snake

BLANFORD'S BLIND SNAKE

Afrotyphlops blanfordii
(Boulenger, 1889)

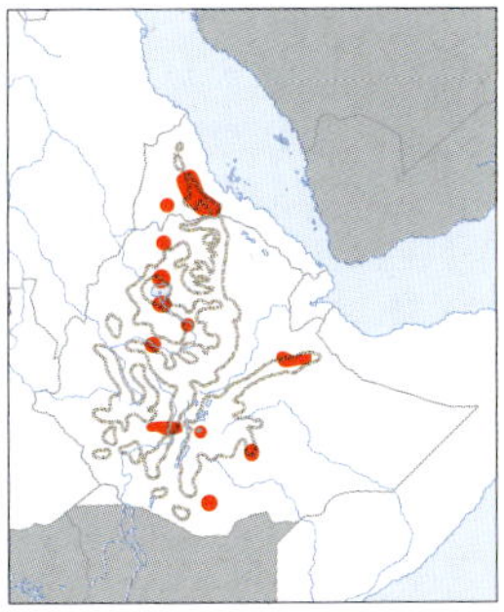

IDENTIFICATION: A big thick grey blind snake, endemic to Eritrea and Ethiopia. Head blunt, snout rounded. The eye is visible. Tail short and ends in a spike. Mid-body scale rows 28–32; number of dorsal scale rows 343–424; length/diameter ratio 29–41. Maximum size about 35cm, averages 25–30cm. Appears grey or almost black, close up each scale has a pale centre, so looks finely striped. Taxonomic Notes: Long regarded as a northern variety of *A. lineolatus*, and differs only in minor scalation details.

Afrotyphlops blanfordii, Wachile.

HABITAT AND DISTRIBUTION: Woodland, moist and dry savanna, grassland and farmland, at altitudes between 1,000 and 2,450m. Known from a cluster of records in highland Eritrea, widespread on western plateau of Ethiopia, south to the vicinity of Soddo; also, Lake Awassa. A handful of records east of Ethiopian Rift Valley includes Wachile in the south, Harenna Forest, Harar, and between Harar and Dire Dawa further north. Probably much more widespread but under-collected.

CONSERVATION STATUS: Data Deficient.

NATURAL HISTORY: Poorly known, but lives in holes and soft soil and sand; sometimes under ground cover (logs, rock, vegetation heaps etc.) but rarely seen except in rainy season, when flooded out of burrows. Sometimes dug up during ploughing. No breeding details known. On certain nights in rainy season males emerge and move around on ground. A number were found at Bahar Dar in close proximity, under logs and rocks in vicinity of flooded fields.

Afrotyphlops blanfordii, Wachile.

ANGLE-SNOUTED BLIND SNAKE

Afrotyphlops brevis
(Scortecci, 1929)

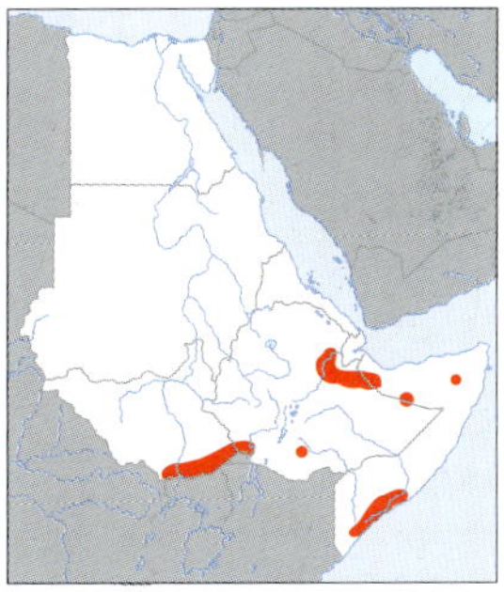

IDENTIFICATION: A big thick brown blind snake of savanna and semi-desert. Head blunt, snout pointed and quite sharp. Eye visible. Tail short and ends in a spike. Mid-body scale rows 29–40; number of dorsal scale rows 288–557; length/diameter ratio 17–38. Maximum size about 70cm, average 40–60cm. Juveniles often striped grey and white or brown and white, but big adults usually (not always) lose the striping, becoming uniform brown.

HABITAT AND DISTRIBUTION: Moist and dry savanna and semi-desert, from sea level to above 2,000m, although more common at low altitudes. Sporadically recorded in South Sudan (Idali, Kapoeta, Torit, Lado Enclave – a vague area around Juba); most Ethiopian records in the south and east (Omo National Park, Negele, Harar, Tendaho, Moullou River north-east of Mieso, and on Ethiopian/Somali border in the Haud at 8.24°N 45.48°E). In northern Somalia at Hargeisa and Gardo, records from southern Somalia include Kismayu, Mogadishu, Merka, Afmadu, Giohar and the Juba Sugar Project at Gelib. Elsewhere, south to Kenya.

CONSERVATION STATUS: Least Concern.

NATURAL HISTORY: Poorly known, but lives in holes and soft soil and sand; they have been collected in pitfall traps and dug out of termitaria, a pair was found poorly concealed at the entrance to a rodent burrow in Kenya. On rainy season nights the males emerge and move around on the ground.

Afrotyphlops brevis, Kenya.

CALABRESI'S WEDGE-SNOUTED BLIND SNAKE

Afrotyphlops calabresii
(Gans & Laurent, 1965)

IDENTIFICATION: A small wedge-snouted blind snake from dry country of Somalia and eastern Ethiopia. Snout prominent and distinctly wedge-shaped. Eye relatively large and visible. Tail short and ends in a spike. Mid-body scale rows 22; number of dorsal scale rows 257–302; length/diameter ratio 18–37. Maximum size about 19cm, average 14–16cm. In life, often looks quite silvery, but is rufous-brown, the scales pigmented at the edges, so the snake looks speckled. Taxonomic Notes: Southern Somali specimens previously known as *Typhlops cuneirostris*.

HABITAT AND DISTRIBUTION: Dry savanna and semi-desert, between 200 and about 1,100m. Known from a cluster of records around the border where Kenya, Ethiopia and Somalia meet (Mandera, Dolo, Lugh) and Somaliland, between the Haud and Goolis Mountains; known localities include 90km north-east of Hargeisa, Burao, Shanshacade, and Las Anod (easternmost record).

CONSERVATION STATUS: Least Concern.

NATURAL HISTORY: Burrows in sand, although may emerge at night; most specimens collected under ground cover, such as logs, rocks, brush piles, etc. One was dug out 30cm below the surface in sandy soil; in northern Kenya three were found under a large rock that sheltered an ant colony. Subadults from Shanshacade were active on the sand surface about an hour after sunset; when disturbed they rapidly buried themselves in a side-oscillating motion. A Kenyan example was regurgitated by a North-east African Carpet Viper *Echis pyramidum*.

Afrotyphlops calabresii, Somaliland.

SOMALI WEDGE-SNOUTED BLIND SNAKE

Afrotyphlops cuneirostris
(Peters, 1879)

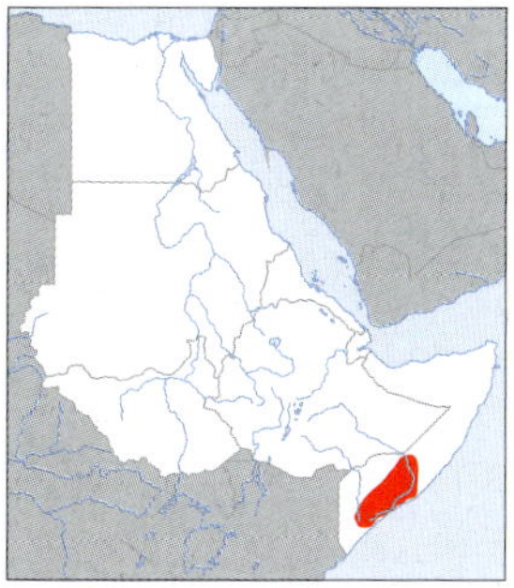

IDENTIFICATION: A small wedge-snouted blind snake, endemic to southern Somalia. The snout is prominent and distinctly wedge-shaped. Eye relatively large and visible. Tail short and ends in a spike. Mid-body scale rows 20–24 (usually 22), number of dorsal scale rows 216–257, length/diameter ratio 21–40. Maximum size about 16cm. In life, often looks quite silvery, but is rufous-brown, the scales pigmented at the bases and lateral edges of the scales, so looks striped.

HABITAT AND DISTRIBUTION: Dry savanna and semi-desert of southern Somalia, between 05°N and just on equator, between sea level and 200m. Extends from Belet Uen south to Afgoi and the coast at Mogadishu, thence through Brava (the type locality) to Kismayu; inland records include Afmadu and Dinsor.

CONSERVATION STATUS: Data Deficient.

NATURAL HISTORY: Poorly known. Burrows in sand, although it may emerge at night; most specimens collected by digging, at depths of up to 70cm, underneath decaying bushes and in root tangles. Habits presumably similar to other blind snakes; lays eggs, feeds on invertebrates, etc.

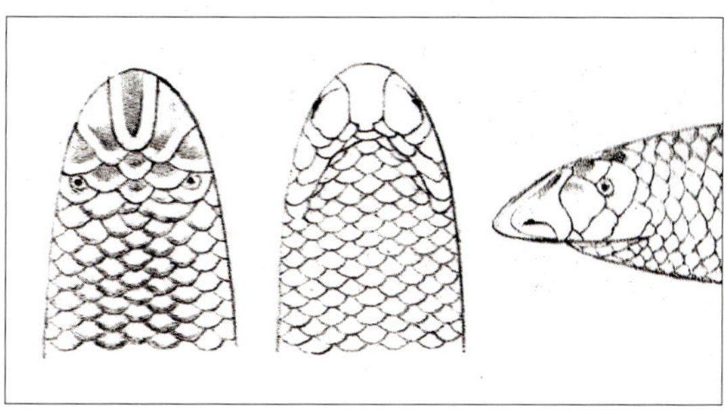

Afrotyphlops cuneirostris.

LINEOLATE BLIND SNAKE

Afrotyphlops lineolatus
(Jan, 1864)

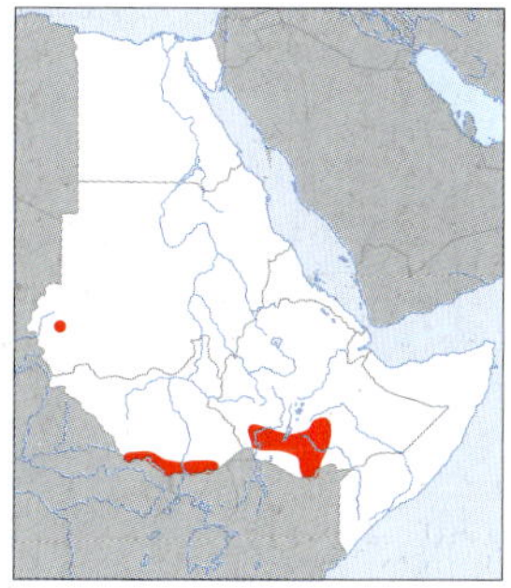

IDENTIFICATION: A big brown or grey, usually spotted, blind snake. Snout prominent and rounded. Rostral shield broad, ¾ width of head, truncated at rear. Mid-body scale rows 26–30; number of dorsal scale rows 323–503, length/diameter ratio 21–47. Average size 30–50cm, maximum 64cm. Colour: appears brown, red-brown or grey, on close examination the centre of each scale is paler, so appears finely striped with lines of tiny dots; often the pigmentation decreases so the lower flanks look lighter, and the underside may be colourless. In some specimens (especially in forest or woodland) the light spotting may darken or be absent on all or parts of the body, and the snake just looks uniformly dark above; occasional specimens are dark above and blotched below.

HABITAT AND DISTRIBUTION: Tolerant of a wide range of habitats, fairly widespread in forest, woodland and moist savanna, from sea level to 2,100m. Occurs largely in south of our area: records somewhat fragmented; known from Suni, in Darfur in Sudan, South Sudanese records mostly in Equatoria include Kagelu, Kajiko North, Katire, Talanga Forest and Yambio. Ethiopian records all in south; including Mt Bibita, Omo National Park, Lakes Chamo and Abaya, Genale River at the Siddam-Bale bridge and Moyale. Older records within our area are now assigned to other taxa. Elsewhere, south and west to Angola, west to Senegal.

CONSERVATION STATUS: Least Concern.

NATURAL HISTORY: Nocturnal and fossorial, living in holes, soft soil and leaf litter. Usually found in the rainy season, under ground cover. May emerge on wet nights and move around on the ground. Diet invertebrates, often termites. In forest of central Kenya this species has been observed moving along a safari ant (*Dorylus molestus*) migration trail towards the new nest site, has also been found hidden in the ants' nest when it was opened in search of the queen, and following an old trail to find a colony.

Afrotyphlops lineolatus, Tanzania.

SPOTTED BLIND SNAKE

Afrotyphlops punctatus
(Leach, 1819)

IDENTIFICATION: A thick blind snake in moist savanna of South Sudan. The snout is rounded and prominent, rostral broad above. Mid-body scale rows 28–32, number of dorsal scale rows 374–465, length/diameter ratio 26–33. Average size 35–55cm, largest specimen 66cm. Two main colour phases. One appears brown, or grey, on close examination the centre of each scale is paler, so the snake appears finely striped with lines of tiny dots, this breaks up and fades low on the flanks, the underside is reddish or light whitish-grey. The other form is white, pale pink or grey-white, with many irregular black blotches, the underside may be immaculate or have a few blotches.

HABITAT AND DISTRIBUTION: Forest and savanna, at low altitudes, between about 500 and 700m in our area. Known only from south-west Western Equatoria, South Sudan, at Yambio, Li Rangu and Bengangai Game Reserve. Elsewhere, south to Uganda, west to Senegal.

CONSERVATION STATUS: Least Concern.

NATURAL HISTORY: Poorly known. Like other members of its genus, lives in holes, leaf litter or loose soil, shelters under ground cover, feeds on invertebrates, and lays eggs. Nigerian specimens preferred slightly acidic, loamy soils, and fed on ants and termites. Specimens collected in Ghanaian savanna were found on the surface during the rainy season.

Afrotyphlops punctatus, Nigeria.

Afrotyphlops punctatus, Nigeria.

ASIAN BLIND SNAKES *INDOTYPHLOPS*

A genus of around 23 species of thin blind snakes, mostly in South-East Asia. A single introduced species is known in north-east Africa, mostly in coastal towns, where it arrived in cargoes, in soil or ballast.

FLOWER POT BLIND SNAKE, BRAHMINY BLIND SNAKE

Indotyphlops braminus
(Daudin, 1803)

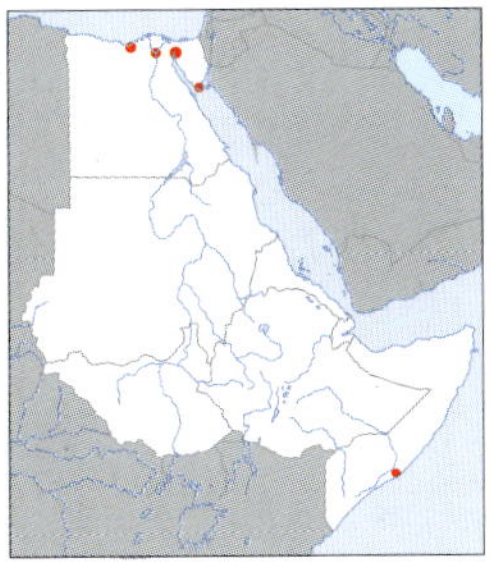

IDENTIFICATION: A small blind snake, introduced on north-east African coast. Snout distinct, rounded with a very narrow rostral that is less than ⅓ width of head, and does not reach level of the eye. It has an anterior constriction that is visible dorsally. Mid-body scales in 20 rows, mid-dorsal scales 261–368, length/diameter ratio 30–57. Average size 10–16cm, maximum 21cm. Can be grey, dark or light brown or black above, sometimes with a purplish sheen, paler below. Some specimens have a darker, sharp-edged vertebral stripe. The lower snout, chin, cloacal region and tip of tail may be pale cream or white. Some that are lighter in colour on dorsum have a dark apical spot on anterior part of dorsal scales. Taxonomic Notes: Previously in genus *Ramphotyphlops*.

HABITS AND DISTRIBUTION: An introduced species, in our area known from a handful of Egyptian localities, largely on the coastline (Suez Canal, Ain Sukhna, 60km west of Alexandria, Sharm El Sheikh), one or two inland records (Cairo, New Cairo City). Recorded also at Mogadishu. Elsewhere; originally described from India, and now occurs worldwide in the tropics.

CONSERVATION STATUS: Least Concern.

NATURAL HISTORY: A parthenogenic species, all individuals are female, hence it can spread easily (usually in flower pots or bags of soil, hence the common name), as only a single individual is needed to start a population. Often found in sand or sandy soil near sea, sheltering under ground cover. Feeds on the larvae, eggs and pupae of ants and invertebrates. Two to six tiny eggs are laid and are self-fertilising.

Indotyphlops braminus, Seychelles.

GRACILE PINK BLIND SNAKES *LETHEOBIA*

A sub-Saharan African genus, recently revived, of about 35 species of thin, relatively long blind snakes. Essentially a tropical genus, only five species occur in north-east Africa, and three are confined to a single locality each. A sixth species, *Letheobia sudanensis*, is known from Garamba National Park, so might just reach our area in south-west South Sudan. Most are pink in life, and thus sometimes called worm-like blind snakes. Their eyes are primitive, invisible, or visible only as indistinct darker dots beneath the scales. In preservative they turn colourless or tan. There are relatively few museum specimens; when they become better known a clearer taxonomic/zoogeographic picture may emerge.

KEY TO THE NORTH-EAST AFRICAN MEMBERS OF THE GENUS *LETHEOBIA*

1a: Mid-body scale count 20; only central Eritrea	*Letheobia erythraea*, Eritrean Gracile Blind Snake
1b: Mid-body scale count >20, not known from Eritrea	2
2a: Body brown, not pink, head yellow, only in southern coastal Somalia	*Letheobia jubana*, Juba Gracile Blind Snake
2b: Body pink, head not yellow, not known from southern coastal Somalia	3
3a: Mid-dorsal scale rows >554; highland Ethiopia	*Letheobia somalica*, Ethiopian Gracile Blind Snake
3b: Mid-dorsal scale rows <490; not in highland Ethiopia	4
4a: Mid-dorsal scale rows 432, relatively stocky; low western Ethiopia	*Letheobia largeni*, Largen's Gracile Blind Snake
4b: Mid-dorsal scale rows 427–487, very slim; southern South Sudan	*Letheobia toritensis*, Torit Gracile Blind Snake

ERITREAN GRACILE BLIND SNAKE

Letheobia erythraea
(Scortecci, 1928)

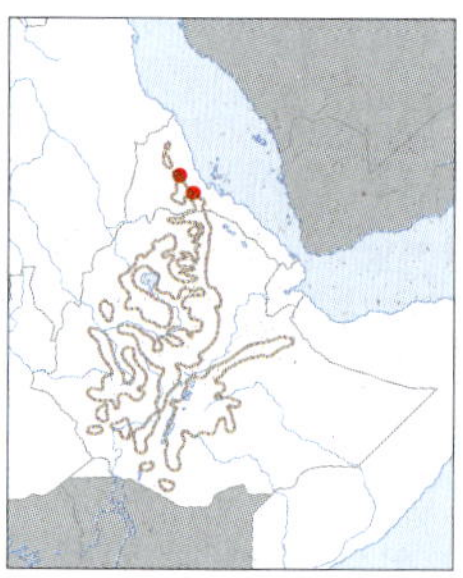

IDENTIFICATION: A pink blind snake from Eritrea; known from just two specimens. The snout is pointed, both from the side and above. The rostral scale is very broad and cuneiform. The eye is invisible. Mid-body scale count 20; 443–462 mid-dorsal scales; length/diameter ratio 68–72. Largest specimen 24.5cm. Life colour unknown, but probably pink, in preservative it is brown above, gold below.

HABITAT AND DISTRIBUTION: The two specimens are from Saganeiti (2,200m) and Gheleb (1,800m), in high-altitude savanna of central Eritrea.

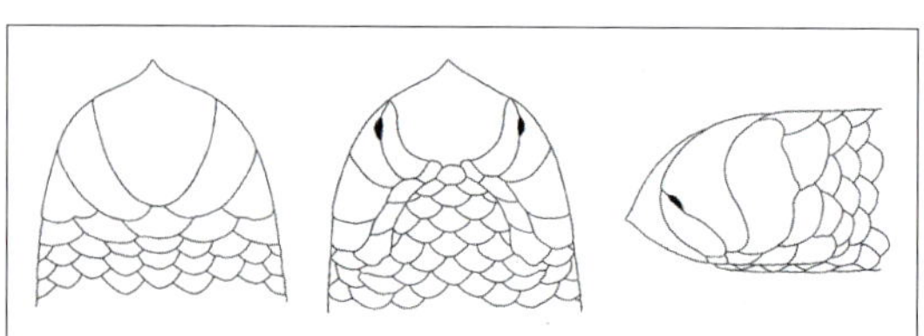

Letheobia erythraea.

CONSERVATION STATUS: Data Deficient.

NATURAL HISTORY: Unknown.

JUBA GRACILE BLIND SNAKE

Letheobia jubana
Broadley & Wallach, 2007

IDENTIFICATION: A yellow-headed blind snake known only from the lower Juba River in Somalia. Has a prominent, obtusely angled snout, and the very broad rostral shield tapers posteriorly to a point. Eye faintly visible below the preocular. Mid-body scale count 24, with 391–430 mid-dorsal scales; length/diameter ratio 44–57. Maximum length 51cm.

HABITAT AND DISTRIBUTION: Described from the now-defunct Juba Sugar Project, on the banks of the Juba River, a few kilometres north-west of Gelib town, at an altitude of 20m. So far unknown elsewhere.

CONSERVATION STATUS: Data Deficient.

NATURAL HISTORY: Poorly known, presumably similar to other members of the genus, i.e., burrowing, eats invertebrates, lays a few eggs. The type series was taken from bulldozed termitaria in what was riverine forest.

Letheobia jubana.

ETHIOPIAN GRACILE BLIND SNAKE

Letheobia somalica
(Boulenger, 1895)

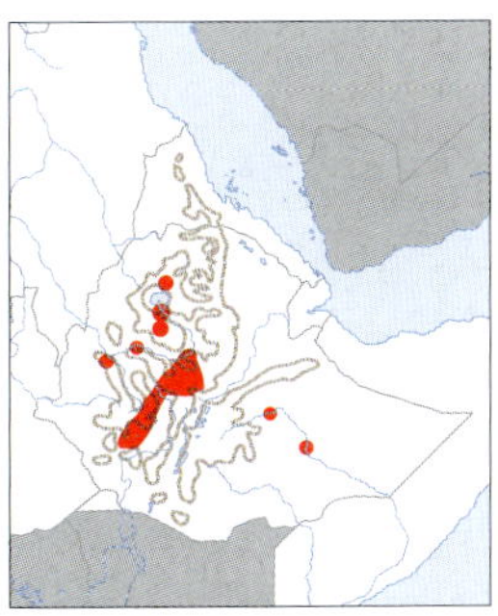

IDENTIFICATION: Despite the specific name, a pink gracile blind snake, endemic to Ethiopia. Snout pointed. The rostral scale is very broad above and below. Eye invisible. Mid-body scale count 24–30, with 510–696 mid-dorsal scales. Length/diameter ratio 54–66. Maximum length 67cm. In life, unpigmented and thus usually appears various shades of pink, due to the blood and organs being visible through the skin; occasional animals are pinkish-brown. Taxonomic Notes: the type was received by Boulenger from Arthur Donaldson Smith who in 1895 had been travelling in 'Somali country' (the Ogaden), hence the somewhat misleading specific name.

HABITAT AND DISTRIBUTION: Endemic to Ethiopia, with almost all records from highlands of western plateau, from Gondar south to Bonga, in moist savanna, grassland and forest, between 1,800 and 2,500m. Only two records east of the Rift Valley; the type locality at 06.53°N, 42.18°E, 600m, and Sheikh Hussein at 1,500m.

Letheobia somalica, Bonga.

CONSERVATION STATUS: Least Concern.

NATURAL HISTORY: Poorly known. Presumably similar to others of genus, living underground and feeding on small invertebrates, laying eggs. A specimen from near Bonga was dug out of leaf litter in forest; two Addis Ababa specimens were under cardboard in a city park.

LARGEN'S GRACILE BLIND SNAKE

Letheobia largeni
Broadley & Wallach, 2007

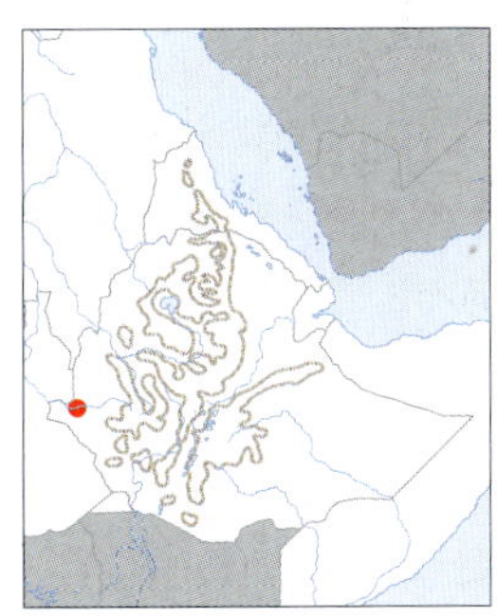

IDENTIFICATION: A pink blind snake from Gambela, in western Ethiopia, known from a single specimen. Snout rounded. Rostral scale broad but tapers towards mouth. The supraocular shield is separated from the rostral. Eye invisible. Mid-body scale count 22; 433 mid-dorsal scales. Length/diameter ratio 65. The specimen is 27.4cm long. Probably pink in life; the preserved specimen was creamy-white above and below.

HABITAT AND DISTRIBUTION: The single specimen was taken at Gambela, at 515m, in the low-altitude wet savanna of western Ethiopia.

CONSERVATION STATUS: Data Deficient.

NATURAL HISTORY: Unknown.

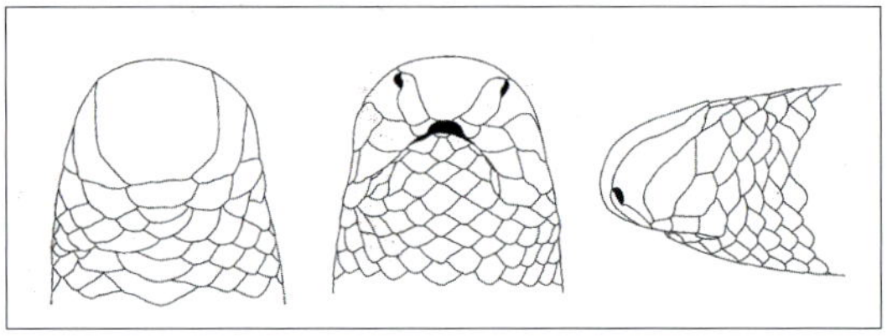

Letheobia largeni.

TORIT GRACILE BLIND SNAKE

Letheobia toritensis
Broadley & Wallach, 2007

IDENTIFICATION: A very slim pink blind snake from Eastern Equatoria, South Sudan. The snout is rounded but slightly elongate. Rostral scale very broad and sub-rectangular from above. Eye invisible. Mid-body scale count 22; 427–487 mid-dorsal scales. Length/diameter ratio 62–82. The length of known specimens varied from 11 to 27.7cm. Pink in life.

HABITAT AND DISTRIBUTION: Known from a series of specimens at Torit (altitude 625m) and a single from Lokwi (possibly Logwi), 70km south-west of Torit, 860m altitude, in Eastern Equatoria, South Sudan.

CONSERVATION STATUS: Least Concern.

NATURAL HISTORY: Poorly known. Presumably similar to others of genus, i.e., burrowing, eats invertebrates, lays a few eggs. The Torit animals were dug from humus among the buttress roots of large trees beside a river, the Lokwi snake was crawling on a road by day.

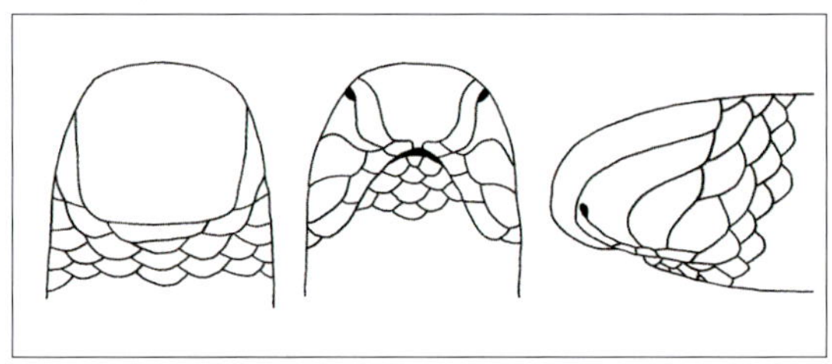

Letheobia toritensis.

AFRICAN BEAKED OR SHARP-SNOUTED BLIND SNAKES

RHINOTYPHLOPS

An African genus of long-bodied blind snakes, with a beaked snout and broad rostral scale. A number of species have been moved in and out of the genus since it was named in 1974; at present seven species remain, four found in our area. If *Rhinotyphlops* is a natural group, it may represent an 'arid corridor' genus, one of a group of organisms found in the dry lands of north-east and south-west Africa, but not in between; like the dik-diks and oryxes.

KEY TO NORTH-EAST AFRICAN MEMBERS OF THE GENUS *RHINOTYPHLOPS*

1a: A distinctive yellow vertebral stripe present	2
1b: No distinctive yellow vertebral stripe	3
2a: Yellow vertebral stripe 44–82 per cent of total length, number of vertebral scales 352–390, confined to small area around Mogadishu	*Rhinotyphlops scorteccii*, Scortecci's Blind Snake
2b: Yellow vertebral stripe 95–97 per cent of total length, number of vertebral scales 315–334, widespread in southern Somalia	*Rhinotyphlops unitaeniatus*, Yellow-striped Blind Snake

3a: Head with short yellow or orange median stripe, rostral from above at level of eyes around 75 per cent width of head; from the Haud south through Somalia and eastern Ethiopia
Rhinotyphlops ataeniatus, Somali Blind Snake

3b: Head white, rostral from above at level of eyes c.50 per cent width of head; only at Las Anod, Puntland
Rhinotyphlops leucocephalus, White-headed Blind Snake

SOMALI BLIND SNAKE

Rhinotyphlops ataeniatus
(Boulenger, 1912)

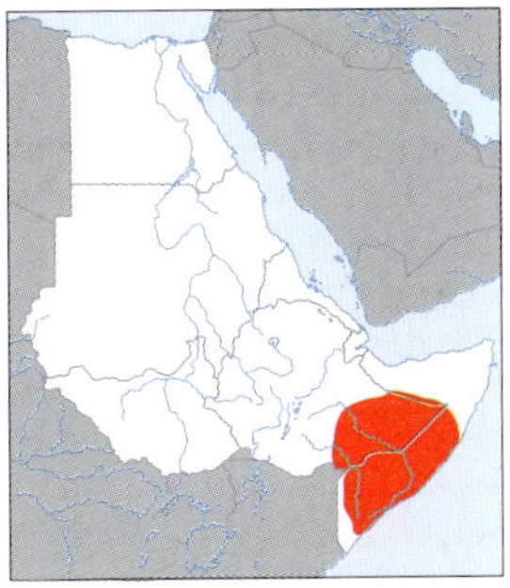

IDENTIFICATION: A dry-country blind snake with a striped snout, in Somalia and eastern Ethiopia. The snout is prominent; rostral is broad, around 66 per cent of head width at level of eyes, which are visible. Mid-body scale count 24, dorsal scales number 443–531, length/diameter ratio 34–72. Back uniform yellow-brown to dark brown or black; top of head bears a short, median orange or yellow stripe. Average size 25–40cm; largest specimen recorded 45.5cm.

HABITAT AND DISTRIBUTION: Dry savanna and semi-desert, from sea level to around 1,100m. A probable sight record from Hargeisa may be the northernmost record; occurs from the Haud (where several specimens collected by Colonel R. H. R. Taylor) south through the Ogaden and most of Somalia to Malka Murri in northern Kenya and Kismayu in Somalia. However, records are widely spaced and sporadic; known localities (in addition to the above) include Kelafo and west of Daror in the Ethiopian Ogaden, and Afgoi, Juba Sugar Project at Gelib, Mogadishu and near Hobyo in Somalia.

CONSERVATION STATUS: Least Concern.

NATURAL HISTORY: Not well known but presumably similar to others of genus. A specimen was found in the stomach of a Lizard Buzzard *Kaupifalco monogrammicus*. A captive Ethiopian specimen found in sand of a dry watercourse and kept in a sand-filled tank emerged during rainstorms and crawled around on the surface.

Rhinotyphlops ataeniatus, Daror.

YELLOW-STRIPED BLIND SNAKE

Rhinotyphlops unitaeniatus
(Peters, 1878)

IDENTIFICATION: Unmistakable on account of its colour. Snout very prominent. Rostral large, longer than is broad and extends beyond level of eyes. The eye is visible beneath the anterior edge of the nasal scale. Mid-body scale count 24, dorsal scales 467–586. Length/diameter ratio 38–77. Average size 25–40cm, largest specimen 43.5cm. Colour pattern distinctive; dorsum dark brown to black, with a bright yellow vertebral stripe 3–5 scales wide that runs from back of head to about 1cm before tip of tail. Ventrum slightly lighter than the dorsum.

HABITAT AND DISTRIBUTION: Dry and moist savanna and coastal thicket, from sea level to around 1,200m (but usually below 200m). In our area, known only from southern third of Somalia, from around 04°N south to Kenya; localities include Lugh, Dinsor, Gelib, Merca and Balad. Elsewhere, to north-east Tanzania.

CONSERVATION STATUS: Least Concern.

NATURAL HISTORY: This species presumably has the same habits as other blind snakes, it lives in burrows or loose soil, sand or leaf litter, where it feeds on small invertebrates, and lays eggs. Occasionally active by day, one was found crossing a path on the Kenya coast during an afternoon rainstorm, others found inside houses.

Rhinotyphlops unitaeniatus, Kenya.

SCORTECCI'S BLIND SNAKE

Rhinotyphlops scorteccii
(Gans & Laurent, 1965)

IDENTIFICATION: A Somali endemic and, like the previous species, unmistakable on account of the yellow stripe. Snout very prominent. Rostral large and notably longer than is broad. The eye is visible beneath the anterior edge of the nasal scale. Mid-body scale count 23–25, dorsal scales 352–390. Length/diameter ratio 47–73. Known specimens vary from 12.6 to 27.5cm. Colour pattern distinctive; dorsum is dark brown to black, with a bright yellow vertebral stripe that extends back from the nape, and is between 44 and 82 per cent of length of back. Ventrum is slightly paler than the dorsum.

HABITAT AND DISTRIBUTION: Known only from a few localities near Mogadishu, on the Somali coast (Mogadishu, Afgoi, Merca, lower Shebelle River), close to sea level.

CONSERVATION STATUS: Least Concern.

NATURAL HISTORY: Poorly known, presumably has same habits as other blind snakes. The Afgoi specimens were buried in sand, 30–60cm under a mat of vegetable debris; Mogadishu animals were dug up under grass tufts, one was crawling on a road after a storm.

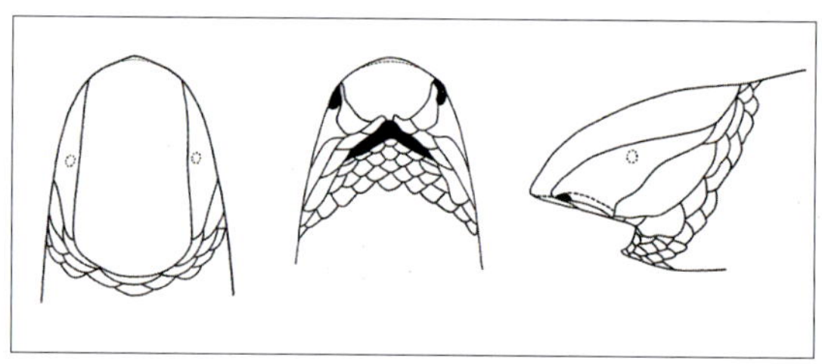

Rhinotyphlops scorteccii.

WHITE-HEADED BLIND SNAKE

Rhinotyphlops leucocephalus
(Parker, 1930)

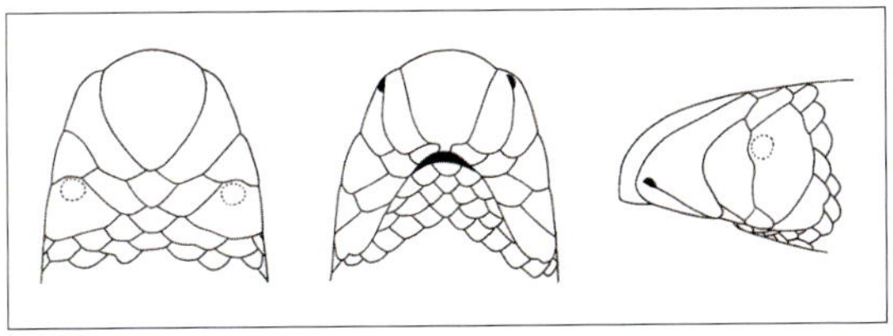

Rhinotyphlops leucocephalus.

IDENTIFICATION: A Somaliland endemic; a blind snake with a white head, known from a single specimen from the Haud. Snout very prominent and hooked. Rostral fairly large; from above it curves smoothly to a rounded point just in front of level of eyes. The eye is visible. Mid-body scale count 24, dorsal scales 369. Length/diameter ratio 43. Length 21.2cm. Parker described the specimen (presumably preserved) as plumbeous above, the lateral borders of the scales black, creating a striped effect; head and underside white with a sharp but irregular demarcation between dorsal and ventral colour.

HABITAT AND DISTRIBUTION: Known only from the semi-desert of Las Anod, Sool, southern Somaliland.

CONSERVATION STATUS: Date Deficient.

NATURAL HISTORY: Unknown.

DESERT BLIND SNAKES *XEROTYPHLOPS*

A recently named genus, erected to describe a monophyletic clade of six blind snakes associated with dry land in Europe, the Middle East and western Asia. One species just reaches our area.

EURASIAN BLIND SNAKE

Xerotyphlops vermicularis
(Merrem, 1820)

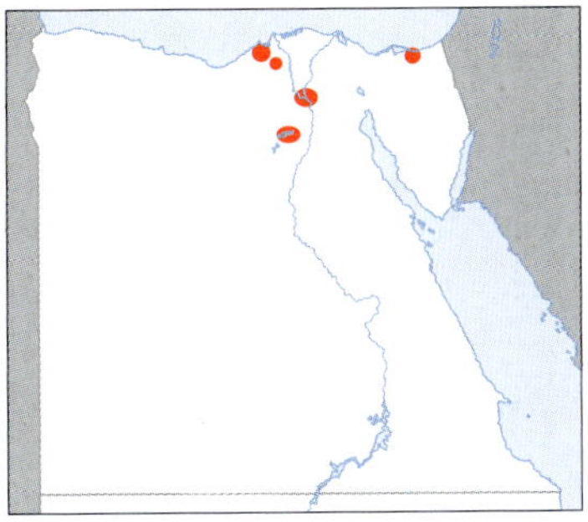

IDENTIFICATION: A slim pinkish blind snake, in our area confined to north-east Egypt. The body is cylindrical, snout rounded. Tail short, ending in a small conical spine-like scale. Mid-body scale rows 21–24, dorsal scale rows 200–2,400, length/diameter ratio 55–72. Maximum size around 40cm, average 18–30cm. Colour pinkish-grey to pinkish-brown, underside slightly paler.

HABITAT AND DISTRIBUTION: Semi-desert, near-desert and cultivated land in our area, at low altitude. Known from a handful of records in northern Egypt (Faiyum, near Cairo, west side of Nile Delta and Misfaq in north Sinai). Elsewhere, east to Afghanistan, north-west to Balkans in Europe.

CONSERVATION STATUS: Least Concern.

NATURAL HISTORY: Nocturnal and burrowing, spending most of its life in holes, soil cracks or buried in soft sand or soil, but sometimes on surface at twilight, or after heavy storms. In winter and spring can be found under logs and rocks, sometimes in numbers, but moves deeper in other seasons and cannot be found. Mate in spring and lay 2–8 elongate eggs. Diet mostly ants and their pupae but will take other invertebrates. They become agitated if handled and may secrete a foul-smelling discharge from the cloaca.

Xerotyphlops vermicularis, Israel/Palestine.

FAMILY LEPTOTYPHLOPIDAE WORM OR THREAD SNAKES

Found in the Americas, Africa and western Asia, indicating a West Gondwanan origin, the worm snakes are a family of just over 140 species. Sometimes called thread snakes, or bootlaces (which they resemble), the 11 north-east African species are split among four genera; *Epacrophis* (one species), *Leptotyphlops* (two), *Myriopholis* (seven) and *Tricheilostoma* (one). The family includes the world's smallest snake, *Tetracheilostoma carlae* of Barbados, reaching only 10cm; the second smallest is found in our area, Tana Delta Worm Snake *Myriopholis tanae*, at 10.2cm. Worm snakes are tiny, thin with non-tapering bodies, blunt heads and tails. They have no teeth in the upper jaw, a single lung and oviduct, and their eyes are reduced to a black light-sensitive cell beneath the head shields (a few species have a well-developed eye). They live in burrows and cracks underground and largely feed on the eggs, larvae and workers of social insects (although some have been found, strangely, up trees in bird nests, feeding on avian fleas). They avoid ants' attacks by chemical camouflage, coiling up and producing a pheromone. One to seven eggs are laid. Savanna and forest forms tend to be dark brown or black, those in semi-desert areas usually pink; their organs visible as dark regions under the skin. They are poorly known.

KEY TO NORTH-EAST AFRICAN MEMBERS OF THE FAMILY LEPTOTYPHLOPIDAE

Our key does not split the 11 species into genera (although they are morphologically different), but into individual species. As with blind snakes, you will need a dead specimen, a binocular microscope and our figures showing head scalation to identify an animal to a species, but the locality may also provide useful clues.

1a: Mid-body scale rows 16	*Myriopholis parkeri*, Parker's Worm Snake
1b: Mid-body scale rows 14	2
2a: Ocular shield excluded from the lip by supralabials	*Tricheilostoma dissimilis*, Sudan Worm Snake
2b: Ocular shield bordering lip	3
3a. Posterior supralabial reaching level of eye; precloacal shield heart-shaped or subtriangular; dark brown to black above and below	4
3b: Posterior supralabial not reaching level of eye; precloacal shield crescent-shaped; light brown above, lighter below	5
4a: Dorsal scale rows fewer than 245; South Sudan	*Leptotyphlops emini*, Emin Pasha's Worm Snake
4b: Dorsal scale rows more than 245; Ethiopia	*Leptotyphlops aethiopicus*, Ethiopian Worm Snake
5a: Anterior supralabial much larger than infranasal; a thick thorn-like apical spine on tail	*Epacrophis reticulatus*, Reticulate Worm Snake
5b: Anterior supralabial subequal to infranasal; small needle-like apical spine on tail	6
6a. Ten scale rows on tail	7
6b. Twelve scale rows on tail	10
7a: Snout strongly hooked in profile	*Myriopholis macrorhyncha*, Hook-snouted Worm Snake
7b: Snout not hooked, or only feebly hooked in profile	8
8a: Dorsal scale rows more than 300, adults exceed 15cm long	*Myriopholis cairi*, Cairo Worm Snake
8b: Dorsal scale rows fewer than 300, adults less than 15cm long	9
9a: Dorsal scale rows 278–300, subcaudals 30–32	*Myriopholis braccianii*, Scortecci's Worm Snake

9b: Dorsal scale rows 227–260, subcaudals 25–30 *Myriopholis tanae*, Tana Delta Worm Snake

10a. Prominent tubercles on head shields, tail short, subcaudals 28–34, weak beak extends below level of lip in lateral profile, anterior supralabial width along lip equal to that of infranasal *Myriopholis erythraeus*, Eritrean Worm Snake

10b. Tubercles absent on head shields, tail longer, subcaudals 34–48, weak beak not extending below level of lip in lateral profile, anterior supralabial width along lip twice that of infranasal *Myriopholis nursii*, Nurse's Worm Snake

RETICULATE WORM SNAKE

Epacrophis reticulatus
(Boulenger, 1906)

IDENTIFICATION: A stocky little worm snake, with a reticulate pattern, known only from Somaliland. The body is cylindrical, snout rounded. Tail short, ending in a spine that is recurved ventrally. Mid-body scale rows 14, number of dorsal scale rows 227–238, length/diameter ratio 41–57. Maximum size around 20cm. Colour dark brown above, lighter below, the dorsal scales are distinctly white-edged.

Epacrophis reticulatus, preserved.

HABITAT AND DISTRIBUTION: Dry savanna, semi-desert and dry mountains, between 900 and 1,250m. Endemic to Somaliland. The type was from the Goolis Mountains and Wagar Range, but also known from Hargeisa.

CONSERVATION STATUS: Data Deficient.

NATURAL HISTORY: Unknown, but presumably similar to other worm snakes, as described in the family introduction above.

ETHIOPIAN WORM SNAKE

Leptotyphlops aethiopicus
Broadley & Wallach, 2007

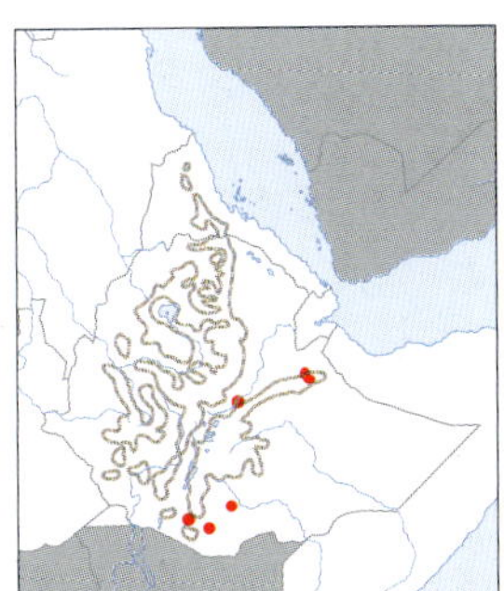

Leptotyphlops aethiopicus, Awash National Park.

Leptotyphlops aethiopicus, Awash National Park.

IDENTIFICATION: A thin, long, red-brown worm snake of eastern and southern Ethiopia. The body is cylindrical, snout rounded. Tail short, ending in a spine that is recurved ventrally. Mid-body scale rows 14, number of dorsal scale rows 239–261, length/diameter ratio 61–77. Maximum size around 16cm, although average 10–14cm. Colour red-brown to brown, some specimens have a distinctive white patch under the tail tip.

HABITAT AND DISTRIBUTION: In our area, known in the eastern Ahmar Mountains (near Dire Dawa, Harar), Awash National Park, and savannas in the south (Yabello, Arero and Negele), at altitudes between 1,000 and 2,100m. Elsewhere, a single record from Mt Marsabit, northern Kenya.

CONSERVATION STATUS: Least Concern.

NATURAL HISTORY: Presumably similar to other worm snakes. The Awash specimen was under a large stone in the shade of *Vachellia* trees in dense scrub, on the north slope of Fantalle Volcano.

EMIN PASHA'S WORM SNAKE

Leptotyphlops emini
(Boulenger, 1890)

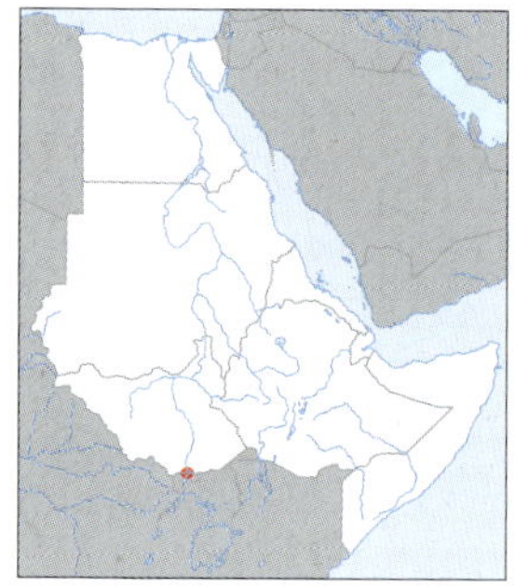

IDENTIFICATION: A thin black worm snake, in our area only in southern South Sudan. The body is cylindrical, snout rounded. Tail short, ending in a spine that is recurved ventrally. Mid-body scale rows 14, number of dorsal scale rows 198–244, length/diameter ratio 38–74. Colour uniform dark brown to black (red-brown in preservative) with white patches on lips and chin and a white cloacal shield. Reaches a maximum of about 16cm, although most specimens average 8–14cm.

HABITAT AND DISTRIBUTION: Savanna and forest-savanna mosaic, between sea level and 2,200m, although in our area known only from Loa, at 700m, north of Nimule in Central Equatoria, South Sudan. Elsewhere, south to northern Zambia, south-west to eastern DR Congo.

CONSERVATION STATUS: Least Concern.

NATURAL HISTORY: Lives in holes underground, sometimes emerging at night, or flooded out of its hole during the rainy season. Eats invertebrates, lays eggs.

Leptotyphlops emini, Tanzania.

SCORTECCI'S WORM SNAKE

Myriopholis braccianii
Scortecci, 1929

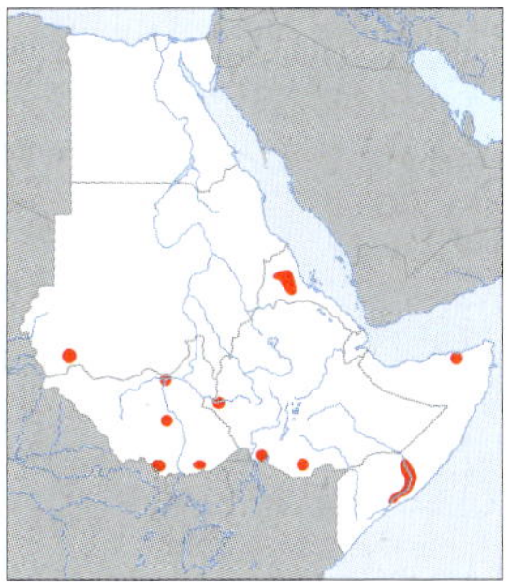

IDENTIFICATION: A distinctive, tiny, pink worm snake with a huge distribution. The body is cylindrical. Tail tapers slightly to a thick spine. Mid-body scale rows 14, number of dorsal scale rows 277–305. Length/diameter ratio 63–111. Largest specimen measured was 14.4cm. In life it is pink, preserved specimens have turned light brown to tan, pale yellow below.

HABITAT AND DISTRIBUTION: Moist and dry savanna, between sea level and 1,900m. Sporadically recorded over a huge range; Sudan (Nyala), South Sudan (Rumbek, Torit, Yei), Ethiopia (Gambela, Wachile and Turmi), Eritrea (Agordat, Mendefera, Keren) and Somalia (Afgoi, Belet Uen, Bosaso, Giohar, Mogadishu). Elsewhere, south to the Tana River in Kenya.

CONSERVATION STATUS: Least Concern.

NATURAL HISTORY: Poorly known, presumably similar to other worm snakes.

Myriopholis braccianii, Turmi.

CAIRO WORM SNAKE

Myriopholis cairi
(Duméril & Bibron, 1844)

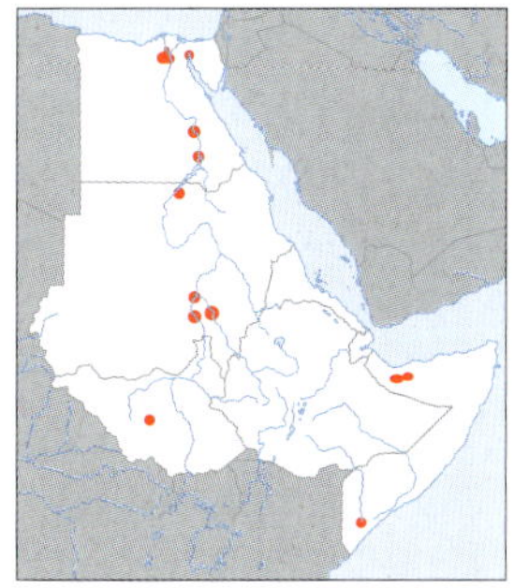

IDENTIFICATION: A distinctive, tiny, pink worm snake with a wide but disjunct distribution. Body cylindrical. Tail tapers slightly to a small downturned spine. Mid-body scale rows 14, number of dorsal scale rows 267–370. Length/diameter ratio 46–117. The largest specimen measured 25.4cm but usually smaller south of the Sahara. Colour pink, pinkish-brown or light brown, slightly paler below.

HABITAT AND DISTRIBUTION: Dry savanna, semi-desert, near-desert and farmland, between sea level and 400m. Sporadically recorded over a massive range. In Egypt, known from Suez, the southern Nile Delta and environs (Cairo, Giza, Faiyum), Luxor and Aswan. Sudan records include Omdurman and Wadi Halfa, in South Sudan at Rumbek in Lakes State, known from the Goolis Mountains and Sheikh in Somaliland, and the Juba Sugar Project at Gelib, southern Somalia. Records from elsewhere in our area now assigned to other species. Elsewhere, the subspecies *M. c. bilmaensis* is known from Niger.

Myriopholis cairi, Sheikh.

CONSERVATION STATUS: Least Concern.

NATURAL HISTORY: Poorly known but presumably similar to other worm snakes, i.e., nocturnal and burrowing, lays eggs, etc. Found under mud clumps, dung heaps, other ground cover. Reported to swim in Egypt. Sometimes several have been found in the same place.

ERITREAN WORM SNAKE

Myriopholis erythraeus
(Scortecci, 1929)

IDENTIFICATION: A thin red-brown or pinkish-beige worm snake, known from few specimens in Eritrea, and one Ethiopian record. The body is cylindrical, snout rounded and very slightly hooked. Tail short, but ends in a cone, not a spine. Mid-body scale rows 14, number of dorsal scale rows 311–335, length/diameter ratio 59–99. Maximum size around 17cm. Colour red-brown to pinkish-beige; an Awash National Park specimen has the seven dorsal scale rows brown, but chin and anterior underside are white, becoming pale brown.

HABITAT AND DISTRIBUTION: A Horn of Africa endemic, in dry savanna between sea level and 1,100m. In Eritrea, recorded at Assab, Massawa, the Dahlak Islands (Entedebir), and from Awash National Park in Ethiopia.

Myriopholis erythraeus.

CONSERVATION STATUS: Least Concern.

NATURAL HISTORY: Unknown, presumably similar to other worm snakes.

HOOK-SNOUTED WORM SNAKE

Myriopholis macrorhyncha
(Jan, 1860)

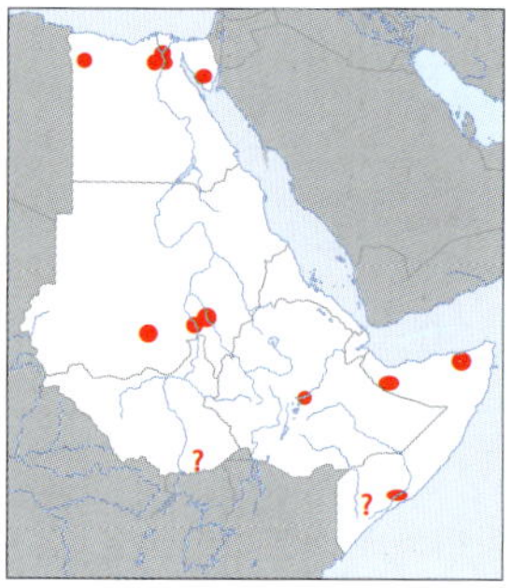

IDENTIFICATION: A long thin pink worm snake with a distinctive hooked snout (visible in hand), with a huge range. Body cylindrical. Tail tapers to a conical terminal spine. The eye is a clearly visible black dot. Mid-body scale rows 14, number of dorsal scale rows 315–492. Length/diameter ratio 49–133. Maximum size about 22cm, average 10–18cm. Usually pink, pinkish-white, purplish-pink or red-brown in life, slightly paler below in some animals, the scales sometimes light-edged to give a reticulate appearance.

Myriopholis macrorhyncha, Sodere.

HABITAT AND DISTRIBUTION: Dry savanna, semi-desert and desert, from sea level to around 1,300m. Known from a handful of widely scattered localities. Egyptian records largely in north (St Katherine area, Sinai, Siwa Oasis, the Faiyum and southern Cairo). Sudan records include near El Obeid,

Kosti and Sennar. A single Ethiopian record, Sodere, on the Awash River. Known from just east of Hargeisa (about ten kilometres south of Bandar Wanaag) in Somaliland, Bosaso in Puntland, and a vague record from 'southern Somalia', although this isn't unexpected as it occurs in nearby north-east Kenya. Probably in South Sudan as known from far north-west Kenya. Elsewhere, north to the Middle East, Turkey and Pakistan.

CONSERVATION STATUS: Least Concern.

NATURAL HISTORY: A nocturnal fossorial species, although it may emerge in twilight, living in cracks and holes and also soft sand, can be found under rocks in daytime. Feeds on ant larvae and other small invertebrates.

ANDERSON'S WORM SNAKE, NURSE'S WORM SNAKE

Myriopholis nursii
(Anderson in Boulenger, 1896)

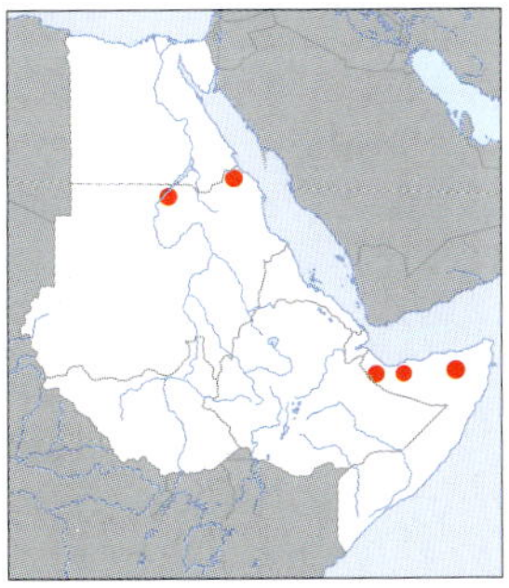

IDENTIFICATION: A long thin pink worm snake with a rounded snout, near the coast of the Red Sea and Gulf of Aden. Body cylindrical. Tail tapers to a small terminal spine. The eye is a clearly visible black dot. Mid-body scale rows 14, number of dorsal scale rows 281–378. Length/diameter ratio 48–105. Maximum size about 24cm, average 12–20cm. Usually pink or sandy-coloured, slightly lighter below in some, the scales very delicately light-edged.

HABITAT AND DISTRIBUTION: Semi-desert between sea level and 1,400m. Essentially a species of the fringes of the Arabian Peninsula, with a few African records: north of Gebel Elba, Egypt; Wadi Halfa, Sudan; the Ethiopian side of the Ethiopian/Somali border at 10.17°N, 42.83°E, near Borama and Wagar in the Goolis Mountains, Somaliland, and south-west of El Donfar, 10.55°N 48.97°E, Puntland.

CONSERVATION STATUS: Least Concern.

NATURAL HISTORY: Poorly known. Specimens found on loose sand and on ground. Presumably similar to other worm snakes.

Myriopholis nursii, Borama.

PARKER'S WORM SNAKE

Myriopholis parkeri
(Broadley, 1999)

IDENTIFICATION: A rare pink worm snake, of eastern Ethiopia, distinguished from all other worm snakes in our area by having 16 mid-body scale rows. Body cylindrical. Tail tapers slightly to a small blunt terminal cone. Mid-body scale rows 16. Length/diameter ratio 47–67. The two measured specimens were 11.3cm and 16cm. In life appears transparent pink, the snout slightly brownish, and the tiny eyes are obvious; preserved specimens are yellow-brown, lighter below.

HABITAT AND DISTRIBUTION: This distinctive little snake is known from only three specimens, in dry savanna. The type was from Degeh Bur, eastern Ethiopia, altitude 1,044m, and Sheikh Hussein, south-east

Ethiopia, altitude 1,386m. A specimen is also known from Tsavo National Park, Kenya.

CONSERVATION STATUS: Data Deficient.

NATURAL HISTORY: Unknown, presumably similar to other worm snakes. The Kenya example was active on the ground at night in December, the rainy season.

Myriopholis parkeri, Sheikh Hussein.

TANA DELTA WORM SNAKE

Myriopholis tanae
(Broadley & Wallach, 2007)

IDENTIFICATION: A distinctive, tiny, pink worm snake. The body is cylindrical. Tail tapers slightly to a thick spine. Mid-body scale rows 14, number of dorsal scale rows 227–260. Length/diameter ratio 46–87. The largest specimen measured was 10.2cm, the smallest 4.8cm. In life appears pink, preserved specimens have turned light brown to tan, paler below.

HABITAT AND DISTRIBUTION: Dry savanna and semi desert, Known from a handful of localities between 80 and 700m; Murle on the lower Omo River in Ethiopia, Ceel Cilan, near Barder on the Juba River, and near and at Afgoi near Mogadishu, in Somalia. Elsewhere, on the lower Tana River in Kenya.

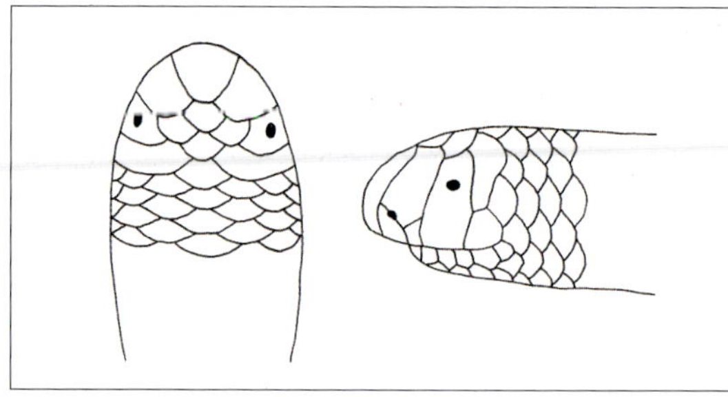

Myriopholis tanae.

CONSERVATION STATUS: Data Deficient.

NATURAL HISTORY: Poorly known, presumably similar to other worm snakes. Kenyan specimens were dug from black cotton soil, less than 15cm below surface, and also found in termitaria.

SUDAN WORM SNAKE

Tricheilostoma dissimilis
(Bocage, 1886)

IDENTIFICATION: Probably the most enigmatic species of our area, a little worm snake described in 1886 from a single specimen, now lost, from the 'White Nile'. Presumably similar in appearance to other worm snakes, it differed in having three upper labials and the second excludes the ocular scale from the lip (in most, the ocular scale touches the mouth). The sketchy type description indicates it was 10.4cm long, had 30 subcaudals and the length/diameter ratio was 42.

HABITAT AND DISTRIBUTION: Said to be from the White Nile ('Nil-blanc'), although whether this refers to the area now in White Nile State, Sudan, or somewhere further south, is unknown.

CONSERVATION STATUS: Data Deficient.

NATURAL HISTORY: Unknown.

BOAS

FAMILY BOIDAE BOAS

A diverse family of about 66 species in 14 genera, found in the Americas, Africa, Europe, Asia and parts of Australasia, but not Australia itself. Some are small, but some huge; anacondas are boas. Unlike pythons, they give live birth. Two genera occur in Africa, one genus, the sand boas, *Eryx*, occurs in our area.

SAND BOAS *ERYX*

A genus of 13 small, harmless arid-adapted boas that live underground. They have blunt heads and tails, narrow ventral scales and small claws either side of the vent, like pythons. They occur in northern Africa, south-west Europe and the Middle East, extending east to India. Most have a dorsal pattern of irregular darker blotches on a light background. Feared in some areas, possibly due to confusion with similarly patterned desert vipers. Five species occur in Africa and all are found in north-east Africa, two are endemic.

KEY TO NORTH-EAST AFRICAN MEMBERS OF THE GENUS ERYX

1a: Dorsal scales smooth	2
1b: Dorsal scales smooth anteriorly, becoming keeled posteriorly	3
2a: Tail terminating with a blunt rounded scale; ground colour sandy yellow-grey or light brown with irregularly shaped, darker brown-grey transverse bands	*Eryx jaculus*, Javelin Sand Boa
2b: Tail with a sharp strongly curved claw-like terminal scute; ground colour usually orange, with scattered large rounded or oval dark brown blotches	*Eryx muelleri*, Müller's Sand Boa
3a: Large, usually growing to more than 30cm; number of scales across head between eyes 9–15	*Eryx colubrinus*, Kenya Sand Boa
3b: Small, usually growing to less than 30cm; number of scales across head between eyes 4–6	4
4a: 193 ventrals, occurs north of Somaliland mountain chain	*Eryx borrii*, Guban Sand Boa
4b: 156–163 ventrals, occurs south of Somaliland mountain chain, to southern Somalia	*Eryx somalicus*, Somali Sand Boa

GUBAN SAND BOA

Eryx borrii
Lanza & Nistri, 2005

IDENTIFICATION: A small burrowing snake with a slender body and large scales on the top of head between the eyes, confined to coastal Somaliland. Eyes relatively large, on the angle between the top and side of the head. Head long, narrow, top covered with relatively large scales, with five between eyes. Enlarged rostral scale broad with an angular edge, for shoving through sand. Neck slightly narrower than head. Body slender, tail short (about 8 per cent of total length) and conical, pointed, slightly curved downward at tip. Dorsal scales smooth anteriorly, becoming keeled posteriorly, arranged in 39 rows at mid-body, ventrals wider than long, 193, subcaudals single, 26, anal scale entire. Maximum size about 39cm. Dorsal ground colour brown, lighter on head and sides of body, with longitudinal or more or less oblique off-white fragmented stripes, sometimes fused to form undulating narrow bands. Lower flanks off-white with dark, irregular, small spots and

Eryx borrii, Laas Dhuure.

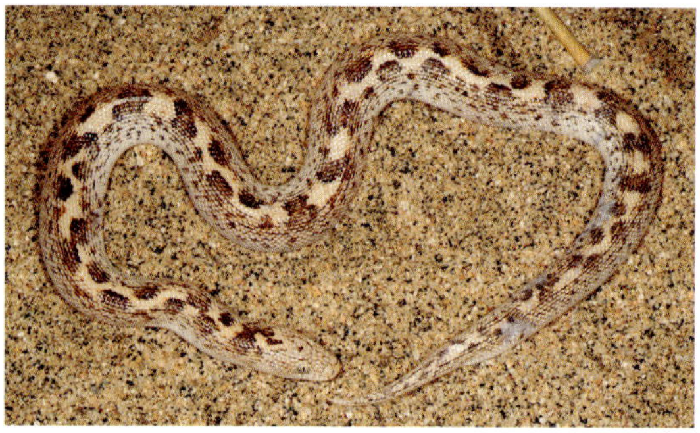

Eryx borrii, Laas Dhuure.

sometimes a series of brownish-black, larger roundish spots. Taxonomic Notes: Recently described, originally the type and an old specimen were identified as *E. somalicus*, which has higher ventral counts.

HABITAT AND DISTRIBUTION: In 'shrub-steppe', dry savanna, at an altitude of 400m, according to the type description. Endemic, only reported from the hot and arid Guban coastal plain, at Biji, on border of Awdal and Woqooyi Galbeed Regions in north-west Somaliland, where the only perennial water is in wadis. The second known specimen (see photograph) was recently found at Laas Dhuure in Sahil Region, 200km east of the type locality.

CONSERVATION STATUS: Least Concern; its habitat is sparsely populated and the species not in demand for pet trade.

NATURAL HISTORY: Unknown, but probably very similar to congeners. The Laas Dhure male was found after sunset on the surface of fine sand near a bush. The type, a female, was collected by A. Donaldson Smith. Sympatric with the widespread *E. colubrinus*.

KENYA SAND BOA, THEBAN SAND BOA

Eryx colubrinus
(Linnaeus, 1758)

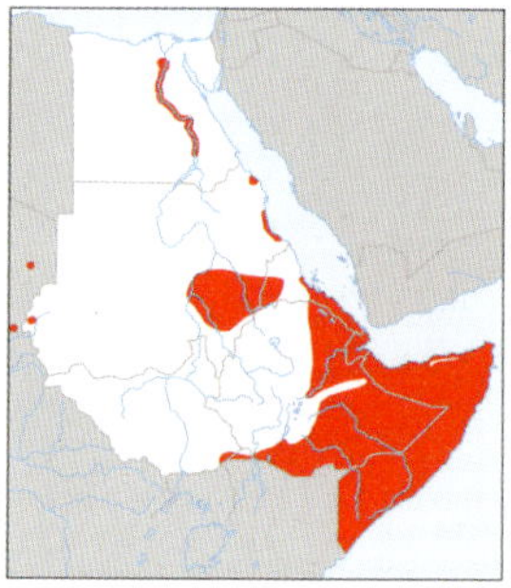

IDENTIFICATION: A small to medium-sized boa with keeled dorsal scales on tail. The tiny eyes, with a yellow iris and slim vertical pupil are set well forward, on the angle between the top and side of head. The top of head is covered with small scales, unusual in a harmless snake; 9–15 scales across head between eyes. Enlarged rostral scale broad with an angular edge, for shoving through sand. No obvious neck. Body short and stout, tail short (less than 10 per cent of total length) and rounded, but with a pointed tip. Scales on tail and final fifth of body are strongly keeled, unlike other body scales; this may aid locomotion or have a defensive purpose. Mid-body scales in 44–59 rows, ventrals narrow, 165–205, subcaudals single, 19–28, anal scale entire. Maximum size about 1m, average 30–60cm, hatchlings 15–19cm. Males are more slender and considerably smaller than females, with a longer tail. Ground colour usually grey, yellow or orange (depending on local soil colour), with sub-circular brown, black or grey blotches, belly immaculate white, cream or pinkish. A uniform brown, rufous or yellow-brown form exists, originally given the specific (then subspecific) name *rufescens*, but it appears to be simply a colour morph.

HABITAT AND DISTRIBUTION: Near-desert, semi-desert and dry savanna, from sea level to about 1,500m. Widely distributed over most of Somalia, Djibouti and Eritrea. In Ethiopia found in dry southern and eastern regions and Danakil depression, but absent from the highlands. In Sudan widely distributed in dry Sahelian *Vachellia* savanna, where recorded at Khartoum, Sennar, Wad Medani, Kassala and Darfur Region, but does not occur in the northern Sahara

Desert. Reported at several localities (Suakin, Port Sudan, Tokar, Durrur) along Red Sea coast. Absent from most of South Sudan, save the dry south in Torit District. In Egypt known from Gebel Elba in extreme south-east and along Nile River approximately between Aswan and Wadi Faiyum in the north. Elsewhere, west to Niger and Chad; possibly south-west Arabia.

CONSERVATION STATUS: Not Assessed; collected for the pet trade in Egypt, but widespread in arid areas and probably under no threat.

Eryx colubrinus, Wad Bena Naga.

Eryx colubrinus, *rufescens* phase, Wachile.

Eryx colubrinus, Awash National Park.

NATURAL HISTORY: A burrowing snake, probably nocturnal but might hunt by day in burrows. Occasionally basks. Lives in holes, under ground cover or buried in sand, usually in lightly vegetated areas. Hunts by ambush, striking from below the sand with remarkable accuracy. Rarely bites if handled but may emit a foul-smelling fluid from glands near the cloaca and coil tightly around the restraining hand. About 4–6 months after copulation it gives live birth to 4–20 young, each about 17–20cm long. Juveniles eat mostly lizards (especially *Heliobolus* and *Latastia*), large adults take rodents and Naked Mole-rats *Heterocephalus glaber* and birds. Captive snakes can reach sexual maturity in 14 months. They often live 8–10 years. Sand boas are widely feared in the Horn of Africa and Sudan on account of their 'dangerous' appearance; their Somali name is Apris, legend has it that if bitten, you take seven steps and die. This may be due to confusion with carpet vipers or burrowing asps.

JAVELIN SAND BOA

Eryx jaculus
(Linnaeus, 1758)

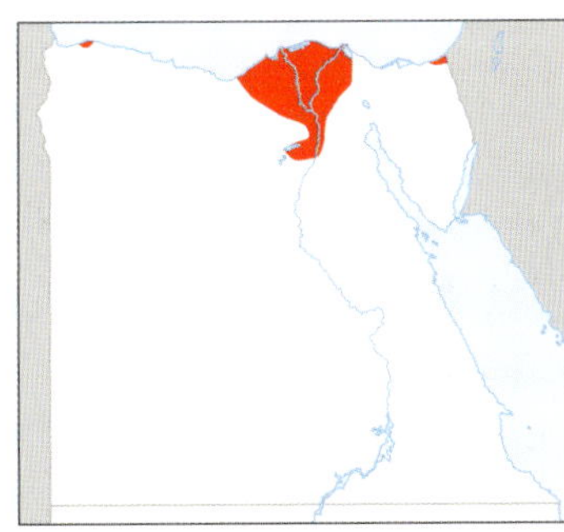

IDENTIFICATION: A small to medium-sized burrowing snake with a stout body, confined to the Mediterranean coast of Egypt in our area. Small eyes with vertical pupils, set high on the side of the head. Head short, with relatively large scales on top, and 5–9 scales between the eyes. The enlarged rostral scale is broad with an angular edge, for shoving through sand. Neck is as broad as head. Body short and thick, tail short (about 7 per cent of total length in females, 9.5 per cent in males) and terminating with a blunt rounded scale. Dorsal scales smooth and arranged in 40–51 rows at mid-body, ventrals very narrow, 165–200, subcaudals single, 15–34, anal scale entire. Maximum size about 90cm, average 30–50cm, hatchlings 9–14cm. Dorsum sandy yellow-grey or light brown with irregularly shaped, darker brown-grey transverse bands; belly immaculate, whitish. Occasional striped specimens recorded. Taxonomic Notes: A polytypic species; the nominate subspecies *E. j. jaculus* occurs in our area.

HABITAT AND DISTRIBUTION: A species of loose sandy and alluvial soils, in semi-desert, coastal scrub and agricultural land under Mediterranean influence, below 200m (extralimitally to 1,600m or more). In our area confined to the western Mediterranean coastal plain, the edges of the Nile Delta and lower Nile Valley, and parts of north Sinai in Egypt. In mesic areas of the Delta and lower Nile Valley it is found in marginal lands and on desert outskirts, as well as in isolated drier sandy pockets. In Sinai it is reported from Rafah and Gaza. Elsewhere, west to Libya, to south-east Europe, and east to Iran.

CONSERVATION STATUS: Least Concern, but in Egypt it is Critically Endangered, being rare and local, and under pressure from the pet trade and habitat destruction.

NATURAL HISTORY: A burrowing snake, mostly in holes or under stones. Active in evening or at night, when it may be encountered on surface. In colder periods of the year, it is diurnal. It forages for prey both actively and by ambush below the soil surface, waiting concealed with only top of head above ground. The prey is constricted in coils. Diet: lizards, small rodents, sometimes birds and large insects. It is ovoviviparous, mating occurs in April–May; 3–12 live young are born around September. Placid and very rarely bites.

Eryx jaculus, Jordan.

MÜLLER'S SAND BOA

Eryx muelleri
Boulenger, 1892

IDENTIFICATION: A fairly small blotched snake with a bullet-shaped head and smooth dorsal scales, only in Sudan in our area. The tiny eyes, with a yellow to grey iris and slim vertical pupil are set well forward, on the angle between the top and side of head. Top of head covered by larger scales, with five scales between eyes. Enlarged rostral scale broad with an angular edge, for shoving through sand. Neck a little narrower than the head. Body short and rather slender, tail relatively long (about 11–20 per cent of total length, longest in males) with a sharp, strongly curved, claw-like terminal scute. Dorsal scales smooth and arranged in 37–48 rows at mid-body, ventrals narrow, 172–188, subcaudals single, 14–24, anal scale entire. Maximum size about 70cm, average 40–50cm, hatchling size unknown. Ground colour usually orange, with scattered large rounded or oval dark brown blotches, belly immaculate cream or pinkish. Hatchlings are pinkish-white with brown blotches.

HABITAT AND DISTRIBUTION: A savanna species, found in generally less dry habitats than *E. colubrinus*. Occurs in a broad belt of Sudanian savanna and partially in drier Sahel, associated with well-vegetated spots and sandy soils, in our area below 500m. Occurs in Sudan in a rather narrow strip from Sennar (the type and easternmost locality) and Khartoum west through Bara and Umm Ruweba in northern Kordofan to the borders with Central African Republic and Chad, as far west as Ivory Coast. At several localities in Sudan as well as in Chad it is sympatric with *E. colubrinus*. Conservation Status. Not Assessed. Collected for the pet trade in Togo, but widespread in arid areas and probably under no threat.

NATURAL HISTORY: Poorly known. A terrestrial, burrowing and probably nocturnal species with generally similar biology to *E. colubrinus*. Lizards (Five-lined Skink *Trachylepis quinquetaeniata*; *Agama* sp.) were found in stomachs of juveniles and males, and rodents in other adults. Unlike other sand boas, lays near full-term eggs which hatch 10–22 days after laying; interestingly they lack the egg-tooth that most egg-laying species possess.

Eryx muelleri, Nigeria.

Eryx muelleri, Nigeria.

SOMALI SAND BOA

Eryx somalicus
Scortecci, 1939

IDENTIFICATION: A small burrowing snake with a stout body, large scales on the top of head between eyes and narrow whitish crossbars on body. Eyes relatively large, on side of head. Head short, narrow, snout fairly long, top covered with relatively large scales, 4–6 between the eyes. The enlarged rostral scale is broad with an angular edge, for shoving through sand. Neck as broad as head. Body short, stout, tail short (about 8–11 per cent of total length) and conical, pointed, more or less curved downward at tip. Dorsal scales smooth anteriorly, becoming keeled posteriorly, arranged in 34–40 rows at midbody, ventrals wider than long, 156–163, subcaudals single, 21–25, anal scale entire. Maximum size 39cm, average 25–30cm, hatchling size unknown. Dorsum pale to dark brown or red-brown with a series of narrow whitish or yellowish crossbars, sometimes dark-edged, which tend to be more or less evenly arranged along length of body, though sometimes fused in such a way as to reduce the ground colour to randomly scattered dark blotches. In most specimens all dorsal scales are darker at the edges and lighter in the middle, which creates a reticulate basic pattern. Lower flanks pale and irregularly patterned with dark spots or blotches, and underside is whitish.

HABITAT AND DISTRIBUTION: A typical species of the Somali arid zone, in desert, semi-desert, coastal dunes, shrub steppes and xerophilous open woodland, particularly in sandy areas, between sea level and 1,150m. Nearly endemic, known from sandy regions of northern Somaliland south of the mountainous belt, where it has been recorded from the vicinity of Hargeisa, Davegoriale in Haud, Garbaqabat in the Nogal Valley in Somaliland, and Gardo in Puntland, south across the Ogaden in Ethiopia to Bud Bud in central Somalia, then to Mahaddei Uen and Mogadishu in the south. Widespread and mostly in areas of low

Eryx somalicus, preserved, Bud Bud.

human population, and not in pet trade thus unlikely to be threatened by human activities.

CONSERVATION STATUS: Least Concern.

NATURAL HISTORY: Poorly known. Like others of genus, it is terrestrial, burrowing and probably a largely nocturnal species. A juvenile had eaten a *Hemidactylus*. A specimen from Bud Bud was found under a rock in broken xerophilous open woodland; a Mogadishu specimen was found on the beach, about 10m from the breaker zone, coiled in a small space in sand between two rocks. A Haud animal was found during daylight half-in and half-out of a hole in the ground, in sandy country with patches of grass and scattered large thorn trees. In Hargeisa area these snakes are feared and confused with *Abeeso* (vipers), notably Puff adders *Bitis arietans* and Carpet Vipers *Echis pyramidum*.

PYTHONS

FAMILY PYTHONIDAE OLD WORLD PYTHONS

A family of just under 40 species, found in Australasia, Asia and Africa, it includes some of the world's largest snakes.

PYTHONS *PYTHON*

A genus of ten relatively large constricting Old World snakes, six occur in Asia, four in Africa, two in our area. In 1999, the African Rock Python, long regarded as a single species, was split into two species, *P. sebae* and *P. natalensis*, on the basis of some physical differences (including head scales and head patterns). This has been generally accepted, but intermediates exist; molecular evidence may clarify the situation; in our area *P. sebae* occurs.

KEY TO THE NORTH-EAST AFRICAN MEMBERS OF THE GENUS *PYTHON*

1a: Maximum size <2m, sensory pits in first to fourth upper labial scales, head pear-shaped — *Python regius*, Royal Python

1b: Maximum size up to 5m or more, sensory pits only in first and second upper labial scales, head not pear-shaped — *Python sebae*, Central African Rock Python

CENTRAL AFRICAN ROCK PYTHON

Python sebae
(J. F. Gmelin 1789)

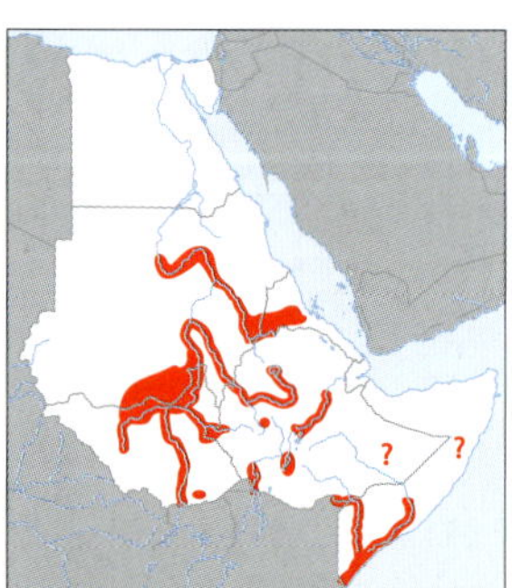

IDENTIFICATION: A huge, thickset snake with a subtriangular head. Snout rounded, eyes fairly large, pupil vertical (although hard to see). Top of head covered with medium to large scales. There are heat-sensitive pits at the front of the head on the lip scales. Body cylindrical, tail short, 9–10 per cent of total length. Scales smooth, in 76–99 rows at mid-body, 265–283 narrow ventrals, 67–76 subcaudal scales. There are small claw-like spurs either side of the anal scale (vestigial legs), which are largest in males. Maximum size about 6m (anecdotal reports of larger specimens). Colour usually black, yellow and warm brown, a pattern of big, light brown, yellow-edged blotches on back and flanks. There is a dark arrowhead on the crown, a yellow stripe runs through the eye and in front of and behind the eye is a broad dark patch.

Underside of tail white, and this is sometimes curled up in the air as a distraction technique, attracting predators such as leopards away from the head.

HABITAT AND DISTRIBUTION: Coastal thicket, grassland, moist savanna, woodland, forest, rocky hills and along rivers and around waterbodies in semi-desert. Occurs from sea level to 2,000m, although commonest below 1,500m. Published records lacking, due to problems of preserving such large snakes. In Somalia known from the Juba and lower Webi Shebelle Rivers and along the coast between Kismayu and the Kenya border. In Ethiopia, from the following rivers: Daua, lower Omo, Baro, Blue Nile (probably the entire length in country, although the only record is from Bahar Dar) and Awash. Also, Lakes Zwai, Abaya and Chamo. A few records from forests of the west. A handful of Eritrean records largely in the catchment of the Gash, Mareb, Barca and Anseba Rivers. Probably occurs virtually throughout South Sudan in vicinity of water, but few published records. In Sudan, reported from the following states: Kassala, Khartoum, White Nile, Northern State, Gezira, Gadarif, Blue Nile, Sennar (where known from Dinder National Park), Kordofan and Darfur. Elsewhere, south to Tanzania, west to Senegal and Mauritania.

CONSERVATION STATUS: Near Threatened. Doesn't co-exist well with humanity, and is harvested for leather and meat; python skin artefacts are often sold in tourist areas; elegant python skin shoes (*markoob*) are bought by the affluent in Sudan. Often killed for its perceived danger to stock.

NATURAL HISTORY: Nocturnal, but basks and hunts opportunistically by day. Mostly terrestrial, but juveniles will climb trees. Often aquatic, and adults may spend a lot of time in water, hunting and feeding there (they are excellent swimmers), emerging to bask. When inactive, shelters in holes, thickets, reedbeds, up a tree, in a rock fissure or underwater. Often curls up in a heap of coils, with head resting on top. Juveniles actively prowl, adults hunt by ambush. They eat a range of prey, adults mostly mammals, and juveniles consume mammals and birds, but other vertebrates taken. They lay 16–100 tennis-ball sized eggs, usually in a hole. Female guards eggs, and elevates her body temperature by basking and an unexplained mechanism. They are bad-tempered snakes and do not co-exist well with domestic stock. Avoid confrontation, but will bite if cornered. Often blamed for killing humans; there are a few documented cases of adults being killed but not eaten, and small children are at risk.

Python sebae, Lake Zwai.

Python sebae, Kenya.

ROYAL PYTHON, BALL PYTHON

Python regius
(Shaw 1802)

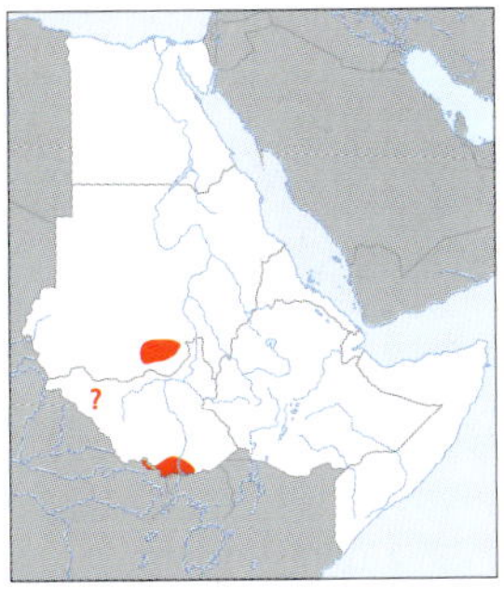

IDENTIFICATION: A small python with a pear-shaped head, thin neck and stout body. The iris appears dark

Python regius, Sierra Leone.

Python regius, Ghana.

but is yellow, the pupil is vertical. Small to medium fragmented scales on the crown. On the front upper labial scales are 3–4 big, obvious, heat-sensitive pits. Body fairly stout and subtriangular, with a prominent vertebral ridge. Tail short, 7–10 per cent of total length. Scales smooth, 53–63 rows at mid-body, 191–207 narrow ventrals, 28–47 paired subcaudals, anal scale usually entire. Maximum size about 1.5m, possibly larger, average 80cm to 1.2m, hatchlings 35–42cm. Colour black or blackish-brown, covered with brown or fawn (rarely orange or yellow) sub-circular, light-edged blotches, sometimes black spots within these blotches. Head dark at front, two light lines run through top of eye to back of head. Ventrum white in the centre, with black speckling and blotches.

HABITAT AND DISTRIBUTION: Grassland, dry and moist savanna, not in forest (but will colonise clearings) or very dry country, usually absent from regions with <600mm annual rainfall. Surprisingly few records in our area for such an easily identified snake. In South Sudan, at Juba, Nimule and Torit, and reported from Bahr El Ghazal; records in Sudan include Dilling, Kadugli and Nuba Mountains. Known from north-west Uganda and Garamba National Park in DR Congo, so almost certainly in Western Equatoria. Probably much more widespread but unreported. Elsewhere, west to Senegal through Guinea and Sudan savanna.

CONSERVATION STATUS: Near Threatened. Heavily exploited for pet trade further west.

NATURAL HISTORY: A slow-moving, nocturnal snake, occasionally basks. Hunts on ground and in small trees, looking for prey in tree holes and down burrows. When inactive, it hides in these burrows, tree cracks, hollow tree trunks or rock fissures. Mostly active in rainy season. A clutch of around 4–10 eggs (range 1–18) laid in February–April in a deep moist hole. The eggs are relatively large, roughly 5–6cm wide and 8–9cm long; incubation period 2–3 months. Diet rodents and birds, especially hole-nesters.

AFRICAN SNAKES (HOUSE SNAKES, FILE SNAKES, WOLF SNAKES AND ALLIES)

FAMILY LAMPROPHIIDAE

A family of African snakes with 16 genera and about 90 species. All are harmless, slow-moving, nocturnal animals. About 20 species in eight genera occur in our area (uncertainty due to the unresolved status of some species, under review).

KEY TO THE GENERA OF LAMPROPHIIDAE IN NORTH-EAST AFRICA

1a: Scales of vertebral row conspicuously enlarged	2
1b: Scales of vertebral rows same size as other dorsal scales	4
2a: Head elongate and depressed, weak to smoothly developed secondary keels on dorsal scales	*Mehelya*, Small and Forest file snakes
2b: Head broad or slightly elongate, weak to strongly developed secondary keels on dorsal scales	3
3a: Head shields rugose, head broad, as wide as long, strong keels on dorsal scales, gaps between dorsal scales, 193–268 ventrals	*Limaformosa*, Savanna file snakes
3b: Head shields mostly smooth, head slightly longer than wide, weakly developed keels on dorsal scales, no gaps between dorsal scales, 164–183 ventrals	*Gracilimima nyassae*, Black File Snake
4a: Distinctive pit in upper labial scales, pupil of eye small and round; highlands of Ethiopia, Eritrea or Somaliland	*Pseudoboodon*, Ethiopian mountain snakes
4b: No pit in labial scales, pupil of eye vertically elliptic, not necessarily in highlands	5
5a: Mid-body scales 15–17	6
5b: Mid-body scales 21 or more	7
6a: Head distinctly pear-shaped, mid-body scale count 15	*Hormonotus modestus*, Yellow Forest Snake
6b: Head not pear-shaped, mid-body scale count 17	*Lycophidion*, Wolf snakes
7a: Mid-body scale counts 21–23, has a vertebral stripe	*Lamprophis*, Montane house snakes
7b: Mid-body scale counts usually 25 or more, no vertebral stripe	*Boaedon*, African house snakes

AFRICAN HOUSE SNAKES *BOAEDON*

A genus of medium-sized (60cm to 1.3m), harmless, aglyphous, mostly African colubrid snakes (one species reaches Arabia) with smooth, shiny dorsal scales and vertical pupils. They are nocturnal, terrestrial and found in a variety of habitats, mostly savanna and forest, but are tolerant of urbanisation and often enter dwellings; hence their name. They lay eggs and eat small vertebrates. Taxonomy is complicated and species have been moved back and forth between the genera *Boaedon* and *Lamprophis*. At least three, possibly more, species of *Boaedon* occur in our area (see *B. fuliginosus* account).

KEY TO NORTH-EAST AFRICAN MEMBERS OF THE GENUS *BOAEDON*

1a: Loreal scale completely separated from upper labials by an additional scale or more often row of scales; dorsum pale, almost obscured by 4–5 longitudinal rows of large, dark brown or reddish blotches, each roughly circular in shape but showing a marked tendency to fuse with its neighbours	*Boaedon maculatus*, Somali House Snake
1b: Loreal scale always in contact with upper labials; coloration not as above	2
2a: Brown with bold yellow or yellowish-orange stripe, two scales deep and edged dark brown, that starts from snout, continuing through upper half of eye and extending uninterrupted along anterior half of body on both sides; only in extreme western Sudan (West Darfur)	*Boaedon longilineatus*, Long-lined House Snake
2b: Body uniformly coloured, with or without 1–2 pale head stripes; pale stripe on flank (if present) narrow, less than one scale deep, and usually separated from head stripes	*Boaedon fuliginosus*, Brown House Snake

BROWN HOUSE SNAKE

Boaedon fuliginosus
(Boie, 1827)

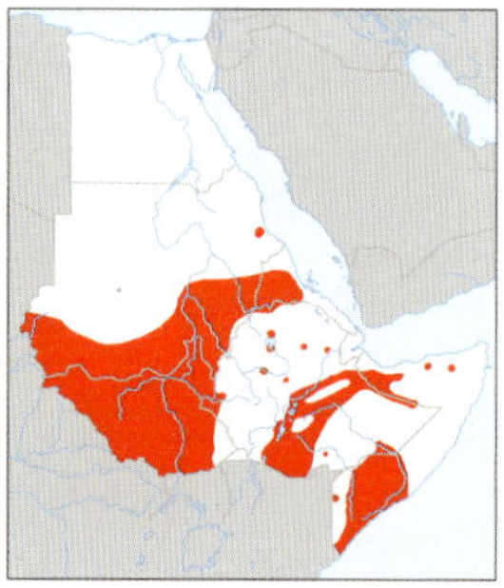

IDENTIFICATION: A medium-sized, fairly slim, dark-coloured, harmless colubrid. Head sub-triangular, like head of a rock python. Eye medium-sized (prominent in some arid-country specimens, leading to suggestions they are a separate species), pupil vertical, iris brown or yellow. Body cylindrical, muscular, tail 10–15 per cent of total length in females, 15–18 per cent in males. Scales smooth, in 23–35 mid-body rows, ventrals 181–250, subcaudals paired, 42–74, cloacal scale undivided. Maximum size about 1.2m, average 40–70cm, hatchlings 17–25cm. Colour very variable, usually brown, black, olive or rufous, often darkening with age. Usually has a pair of pale lines each side of head, one through eye, the other along cheek, and these may extend along body to greater or lesser extent. Several colour forms may occur in same area. A juvenile from Sanaag region in Somaliland was uniformly blackish-brown above. Belly white or cream. Taxonomic Notes: Taxonomy of the *B. fuliginosus* complex is unresolved; there are probably at least four taxa in our area, including an undescribed Ethiopian highland form. Some authorities believe that *B. fuliginosus* should be restricted to western Africa but we retain the name for now. Recent work indicates three additional species probably occur in Sudan/South Sudan, as follows. Central African Striped House Snake *B. paralineatus*; dorsal scales in 31–35 rows at mid-body; ventrals 225–243 in males, 239–250 in females; subcaudals 64–72 in males, 54–62 in females; brown in colour, two well-marked head stripes on each side, one additional on flanks, not connected to head stripes; preocular in contact with frontal. Forest Fringe House Snake *B. perisilvestris*; dorsal scales in 29–31 rows at mid-body; ventrals 199–213 in males, 217–232 in females; subcaudals 59–69 in males, 48–55 in females; brown or dark grey-green coloration, head stripes largely absent in adults, if present, thick but low contrast, no light stripe on flanks; often three upper labials entering eye. Yellow House Snake *B. subflavus*; dorsal scales in 29–33 rows at mid-body; ventrals 213–229 in males, 234–247 in females; subcaudals 63–73 in males, 48–57 in females; yellowish or brownish, head stripes absent or short and narrow, no light stripe on flanks; only two upper labials entering eye. Pending re-examination of museum specimens, we leave them under *B. fuliginosus*.

Boaedon fuliginosus, Lake Langano.

Boaedon fuliginosus, Jimma.

HABITAT AND DISTRIBUTION: In a range of habitats; semi-desert, moist and arid savanna, woodland and forest margins, between sea level to about 2,400m. In our area, virtually everywhere except Egypt and the northern Sudan desert, the Danakil region of Ethiopia and Eritrea, Djibouti, Puntland, the Ogaden of eastern Ethiopia and central Somalia. Largely absent also from south-west Ethiopia. Until a detailed review is carried out, it is not possible to describe precise distributions in Sudan and South Sudan of the three additional species listed above. However, Yellow House Snake appears to occur in dry Sahelian savanna, in southern Sudan (Darfur, Kordofan, and towards Ethiopia) and Upper Nile State of South Sudan. Central African Striped House Snake probably occurs in extreme southern South Sudan in forested areas along the border, and Forest Fringe House Snake seems to occur in the woodland-savanna mosaic of most of South Sudan, with specimens from Torit and Lohutok fitting its description. Elsewhere, the complex occurs west to Senegal, south to South Africa, with an isolated population in Morocco.

Boaedon fuliginosus, Lake Langano.

Boaedon fuliginosus, Erigavo.

CONSERVATION STATUS: Least Concern.

NATURAL HISTORY: Nocturnal and terrestrial, emerging at dusk (sometimes late afternoon) to hunt. By day shelters in holes or under ground cover (log, rocks, debris), in termitaria, in unoccupied huts or houses. Quick to bite with needle-sharp teeth if picked up or molested, but tames readily. Clutches of 2–16 eggs, roughly 2×3cm, are laid, incubation 2–3 months, depending on altitude, and they may lay more than one clutch a year; hatchlings usually emerge near start of rainy season. Young snakes eat mostly lizards (as do adults in dry areas) but adults eat many rodents, and do tremendous good in controlling rat population. Other prey includes birds, frogs and snakes. Although often uncommon in undisturbed habitat, this species, due to its wide choice of prey and secretive nocturnal habits, is efficient at thriving in peri-urban habitats where other snakes cannot survive.

LONG-LINED HOUSE SNAKE

Boaedon longilineatus
Trape & Mediannikov, 2016

IDENTIFICATION: A medium-sized, fairly slim but muscular, brownish colubrid snake with distinctly long, rather thick, yellowish-orange, and uninterrupted lines extending from snout to posterior body. Head sub-triangular, elongated, with a blunt, moderately long snout. Eye medium-sized, pupil vertical, iris hazel-brown. Loreal scale not separated from upper labials. Body cylindrical, muscular, tail 11–13.5 per cent of total length in females, 16–18 per cent in males. Scales smooth, in 25–29 mid-body rows, ventrals 211–233 (211–214 in males, 226–233 in females), subcaudals paired, 42–63 (55–63 in males, 42–49 in females), cloacal scale undivided. Maximum size slightly exceeds 112cm, average 60–100cm (males smaller). Colour distinctive: brown with a bold yellow or yellowish-orange stripe, two scales deep and edged dark brown, that starts from the snout, continuing through upper half of the eye and extending uninterrupted along anterior half of body on both sides, then fading. Lower and upper labials pale, whitish. Underside uniformly whitish. Taxonomic Notes: Described in a recent revision of Central African *Boaedon*, previously lumped with wide-ranging *B. lineatus*.

HABITAT AND DISTRIBUTION: Largely in transition zone between the Sahel and Sudanian low-rainfall savanna, at low to medium elevations, *c.*300–850m. In our area known from dry south-western Sudan, a confirmed record from Geneina in West Darfur, not far from border with Chad. Probably more widespread in this part of Sudan, but under-collected.

Elsewhere, from northern Cameroon to southern Chad.

CONSERVATION STATUS: Least Concern.

NATURAL HISTORY: Poorly known. Geneina specimens were inside buildings or in rubbish heaps in their vicinity; one was found by day in open grassland. One contained the tail of an unidentified snake. Behaviour presumably similar to other dry-country *Boaedon*.

Boaedon longilineatus, Chad.

SOMALI HOUSE SNAKE

Boaedon maculatus
Parker, 1932

IDENTIFICATION: A medium-sized, fairly slim, harmless colubrid with a distinct reticulate pattern, confined to arid lowlands of the Horn of Africa. Head sub-triangular and fairly narrow, with a blunt snout. Eye medium-sized, pupil vertical, iris brown or yellow. The loreal scale is completely separated from upper labials by an additional scale or, more often, row of scales, a unique character in genus. Body cylindrical, muscular, tail 13–15 per cent of total length in females, 17–18 per cent in males. Scales smooth, in 27–29 mid-body rows, ventrals 200–232, subcaudals paired, 54–66, cloacal scale entire. Maximum known size slightly exceeds 78cm, average 50–70cm, hatchling unknown, but probably around 17–20cm. Coloration distinctive, the pale grey or pinkish dorsum of young animals almost obscured by 4–5 longitudinal rows of large, dark brown or reddish to salmon blotches, each roughly circular and showing a marked tendency to fuse with its neighbours; adults have a reduced or faded version of the same pattern. Underside very pale brownish-grey.

Boaedon maculatus, Djibouti.

HABITAT AND DISTRIBUTION: Semi-desert and arid savanna environments, at 480–900m. Associated with other species of *Vachellia–Commiphora* bushland of the Somali arid zone. Rare and secretive, known only from several widely separated records. Djibouti records from Randa and Arta on Gulf of Tadjoura. Known from Hadaden, about 20km west of Bohodle in the Haud region on border of Ethiopia and Somaliland, and further east at Las Anod and north at Badhan. In Puntland recorded at Bihen (west of Garowe) and Galgalo Oasis in the north. A distant record from Oddur in Rahanuin country, in Bakool region of southern Somalia. Almost certainly occurs in Ethiopian Ogaden.

CONSERVATION STATUS: Least Concern.

NATURAL HISTORY: Terrestrial, nocturnal, slow-moving and secretive. By day shelters in holes or under ground cover (log, rocks, debris, etc.). A Djibouti animal was in a cave. Diet unknown, but probably takes lizards and rodents. Feared in Somaliland, probably due to confusion with carpet vipers. In the Haud it is sympatric with *B. fuliginosus*.

MONTANE HOUSE SNAKES *LAMPROPHIS*

A genus of seven smallish harmless snakes with smooth, shiny dorsal scales and vertical pupils. As presently defined, they are a disparate (and possibly not natural) group; four species occur in South Africa, one in the Seychelles and two in Ethiopia. The group name 'Montane house snakes' applies only to the Ethiopian species, both highland endemics.

KEY TO NORTH-EAST AFRICAN MEMBERS OF THE GENUS *LAMPROPHIS*

1a: Ventral scales fewer than 190, body brown, with several stripes — *Lamprophis abyssinicus*, Ethiopian House Snake

1b: Ventrals more than 210, body black, a single yellowish vertebral stripe or no stripe — *Lamprophis erlangeri*, Erlanger's House Snake

ETHIOPIAN HOUSE SNAKE

Lamprophis abyssinicus
Mocquard, 1906

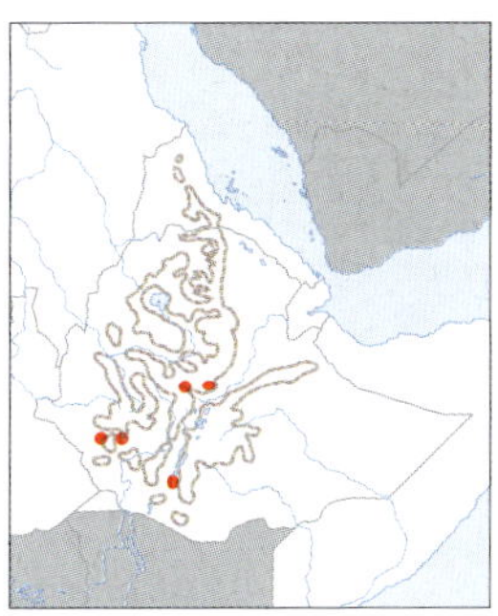

IDENTIFICATION: A small striped house snake of central and south-west Ethiopia. Eye fairly obvious. Head slightly distinct from neck. Body cylindrical, tail fairly short, 12–17 per cent of total length (longer in males). Scales smooth, in 21–23 rows at mid-body, ventrals 169–187, subcaudals paired, 34–51. Maximum size 64cm, average and hatchling size unknown. Colour brown or rufous-brown, with darker brown vertebral and dorsolateral stripes; the broad vertebral stripe may enclose a fine lighter stripe that commences on nape. Underside cream or orange, with dark pigmentation on outer margins of vertebral scales.

HABITAT AND DISTRIBUTION: An Ethiopian endemic, in savanna, woodland and farmland, between 1,200 and 2,300m. Known only from a few localities; Akaki, Ambo, Tepi, Wush-Wush and the Adoshebai Valley (west of Lake Chamo).

CONSERVATION STATUS: Least Concern.

NATURAL HISTORY: Poorly known, presumably similar to other house snakes, i.e., nocturnal and terrestrial, emerging at dusk to hunt, when inactive hides in holes or under ground cover (log, rocks, debris, etc). Diet probably lizards and rodents.

Lamprophis abyssinicus, Tepi.

ERLANGER'S HOUSE SNAKE

Lamprophis erlangeri
(Sternfeld, 1908)

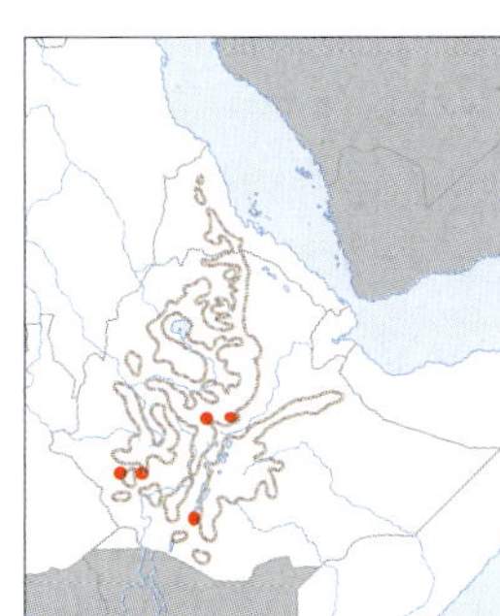

IDENTIFICATION: A large dark house snake, usually with a distinctive dull yellow dorsal stripe, of central and south-west Ethiopia. Eye fairly obvious. Head slightly distinct from neck. Body cylindrical, tail fairly short, 12–17 per cent of total length (longer in males). Scales smooth, in 21–23 rows at mid-body, ventrals 217–240, subcaudals paired, 38–64. Maximum size about 1.2m (large specimens usually females), average 60–90cm, hatchling size unknown but a Jimma juvenile was 22cm. Colour glossy black or very dark grey; a broad dull yellow, olive, tan or greenish vertebral stripe, 6–18 scales wide, extends from snout or occiput to tail tip. In large adults this may fade, leaving a uniformly glossy black snake. Underside black or grey, the ventrals often pale-edged. Taxonomic Notes: A recent revision places this snake in a monotypic genus, *Bofa*.

HABITAT AND DISTRIBUTION: An Ethiopian endemic, in forest, woodland, high savanna, grassland and farmland, between 800 and 2 800m. Records somewhat sporadic; known from Mt Choke, Ghimbi, Godare Forest, Bedelle, Jimma, Bonga and Metu west of the Rift Valley, and Shashamene, Kebri Mangest, Dodolla and Wondo Genet to the east. Probably fairly widespread in forest or deforested areas, and seemingly tolerant of light urbanisation, several specimens were found in and around Shashamene town.

CONSERVATION STATUS: Least Concern.

NATURAL HISTORY: Poorly known. Nocturnal and terrestrial, most active in wet season, shelters by day under ground cover, in holes, etc. Presumably lays eggs. Diet small vertebrates; rodents and lizards, possibly frogs. A road casualty at Jimma had eaten a lizard, and captive specimens readily took rodents.

Lamprophis erlangeri, Bedelle.

Lamprophis erlangeri, dark adult, south-east of Bonga.

ETHIOPIAN MOUNTAIN SNAKES *PSEUDOBOODON*

An African genus of medium-sized, aglyphous, harmless, colubrid snakes, endemic to high country of the Horn of Africa, all have a curious deep triangular pit of unknown function between the fifth and sixth upper labials. Closely related to *Lamprophis* and *Boaedon* and resembles them in having a python-shaped head, moderately distinct from the neck, snout rounded in profile, and nostrils situated between two nasal scales. The pupil is, unusually small and sub-circular and closes sideways in a vertically elliptic fashion. The dorsal scales are smooth, cloacal scale entire, subcaudals paired. They are nocturnal, terrestrial and slow-moving; feed mostly on frogs and rodents, but lizards and birds are also taken. They lay eggs. Four species are known, all confined to north-east Africa, two are endemic to Ethiopia.

KEY TO MEMBERS OF THE GENUS *PSEUDOBOODON*

1a: Ventral count >215, head distinctly pear-shaped, two anterior temporals, pattern low-contrast brown blotches on grey	*Pseudoboodon sandfordorum*, Sandford's Ethiopian Mountain Snake
1b: Ventral count <215, head not pear-shaped, single anterior temporal, pattern not as above	2
2a: Body slender, stripes always low contrast	*Pseudoboodon boehmei*, Böhme's Ethiopian Mountain Snake

2b: Body stout, high-contrast stripes or blotches nearly always present 3

3a: Vertebral stripe continuous and straight-edged *Pseudoboodon lemniscatus*, Striped Ethiopian Mountain Snake

3b: Vertebral markings form a zigzag stripe or longitudinal series of spots, or combination of both *Pseudoboodon gascae*, Gasca's Ethiopian Mountain Snake

BÖHME'S ETHIOPIAN MOUNTAIN SNAKE

Pseudoboodon boehmei
Rasmussen & Largen, 1992

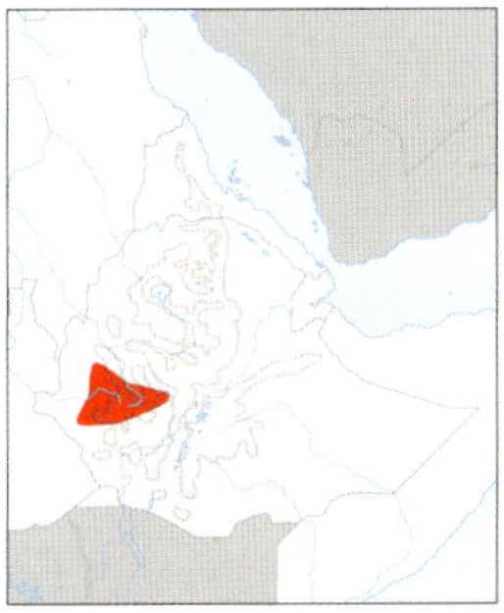

IDENTIFICATION: A medium-sized, slender colubrid, golden-brown in colour with a very feeble pattern, endemic to south-western Ethiopia. Head long, narrow, with a rounded snout, the neck is thin. Eye medium-sized, pupil round (but vertical in very bright light), iris yellow. There is a single anterior temporal. Body slender, cylindrical; tail long, about 20 per cent of total length. Scales smooth, without apical pits, in 21 mid-body rows, ventrals 184–195, subcaudals paired, 45–59 (57–59 in males, 44–45 in females), cloacal scale entire. Maximum known size slightly exceeds 90cm (males smaller), average 50–80cm, hatchling size unknown. Coloration distinctive; dorsum golden-brown, with a very feeble pattern consisting of three slightly darker longitudinal stripes extending length of body. Underside cream to bright yellow, only occasionally with faint dark streaks arranged in two longitudinal rows. Juveniles are orange to red with whitish ventrals, ornamented with two well-developed longitudinal stripes fusing on tail.

HABITAT AND DISTRIBUTION: Tropical forest or nearby moist grassland, at 800–2,150m. Endemic to Ethiopia, where recorded in south-west, known localities include Godare Forest, Bedelle and Metu in Illubabor Province, Ghimbi in Wollega Province, and Abelti, Asendabo and Jimma in Kaffa Province. Seems to be more closely associated with forest than other species of genus. Probably not threatened as it is fairly widespread and locally abundant, able to tolerate human activity and lives partially in protected or inaccessible areas with low human population.

Pseudoboodon boehmei, Mizan Teferi.

Pseudoboodon boehmei, Jimma.

CONSERVATION STATUS: Least Concern.

NATURAL HISTORY: Nocturnal, slow-moving and terrestrial (although the slim body suggests it might climb), emerging at dusk to hunt. Found crossing roads in forested areas or in grassland associated with forest. Several adults were observed in villages among houses at night. Other animals were found in short grass. Bedelle juveniles found under rocks on shore of a small lake, and at night feeding on the abundant local frogs (*Afrixalus clarkei*, *Hyperolius papyri*, *Hemisus microscaphus* and *Ptychadena* sp.). In captivity, juveniles continued to eat frogs, but adults preferred small rodents which they killed by powerful constriction. Not an aggressive snake, they do not try to bite when handled. Presumably lays eggs but no breeding details known.

GASCA'S ETHIOPIAN MOUNTAIN SNAKE

Pseudoboodon gascae
Peracca, 1897

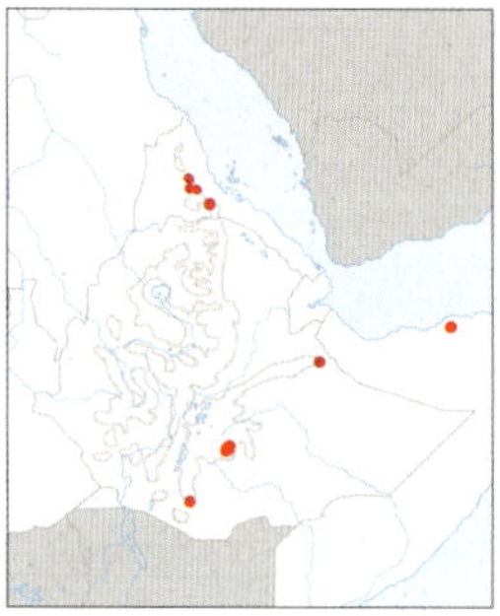

IDENTIFICATION: A small stout, rare snake with a distinctive dorsal mix of brown blotches and zigzags. Head short and sub-triangular, distinct from neck, and resembles that of a rock python. Snout rounded. Eye small, pupil usually vertical, iris yellow to brown. There is a single anterior temporal. Body stout, cylindrical; tail fairly short, about 11–20 per cent of total length. Scales smooth, without apical pits, in 19–25 (35 in one animal) mid-body rows, ventrals 168–214, subcaudals paired, 33–59 (45–59 in males, 33–41 in females), cloacal scale entire. Maximum size 66cm (males not significantly smaller, up to 62cm), average 40–60cm, hatchling size unknown. Colour pattern distinctive: an arrow-shaped rich brown mark on top of head, dorsum cream, tan or pale brown with a series of large, rounded or oval dark brown blotches along midline of back, completely or partially fused into a zigzag vertebral stripe, accompanied on flanks by separate row of distinct dark spots on a greyish-white background. Underside cream, with numerous small dark spots and streaks usually arranged in ill-defined longitudinal rows; underside of tail usually has a median stripe. Taxonomic Notes: Several disjunct populations vary considerably, so cryptic forms may be concealed within the taxon.

HABITAT AND DISTRIBUTION: A range of habitats; dry *Vachellia* savanna or scrubland at lower elevations, montane grassland, dry farmland without tree cover (at 2,240m), margins or clearings in submontane forest dominated by large trees, such as *Podocarpus*, *Celtis* or *Croton*, and montane forest with dense bushes of *Buxus hildebrandtii*, *Cadia purpurea*, *Dodonaea viscosa* and large *Juniperus* trees, at 1,500–2,450m. First recorded at Maldi in Eritrea, known also from Asmara, between Adi Caie and Mendefera, and near Tsazega. In Ethiopia at scattered localities along southern edge of eastern plateau and mountains, notably at Chinahasen, about 3.5km north of Manyate village in Harenna Forest, Bale Mountains, and at Yabello. A recently collected specimen at Daallo Forest north of Erigavo in Somaliland extends the known range of this species and genus to a third country, and some 535km east of the nearest known locality.

CONSERVATION STATUS: Least Concern.

Pseudoboodon gascae, Daallo Forest.

Pseudoboodon gascae, Chinahasen.

NATURAL HISTORY: Little-known. Nocturnal and slow-moving. Diet unknown, but probably eats frogs, lizards and rodents. The specimen from Harenna Forest was found at end of rainy season, in October, in late afternoon, under a piece of wood on grass near the forest. The Somaliland animal was active on the ground at night following a brief rainstorm, at the edge of montane forest. Probably lays eggs. A passive snake, they do not bite when handled.

STRIPED ETHIOPIAN MOUNTAIN SNAKE

Pseudoboodon lemniscatus
(Duméril, Bibron & Duméril, 1854)

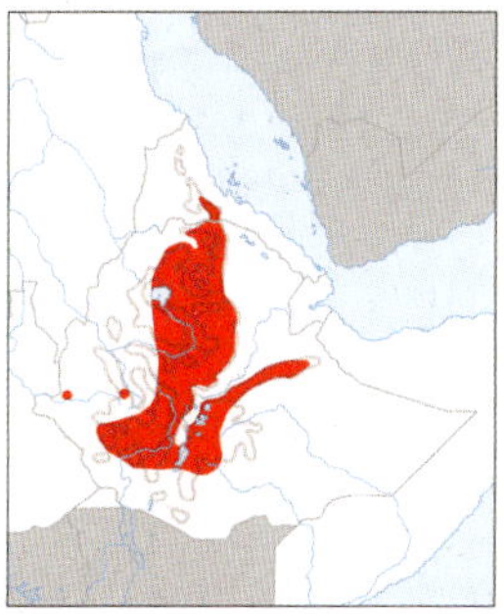

IDENTIFICATION: A medium-sized, fairly robust colubrid with pattern of high-contrast stripes, endemic to highlands of Ethiopia and Eritrea. Head sub-triangular and distinct from neck, the snout rounded. Eye small, pupil oval (vertical in bright light), iris yellow or brown. There is a single anterior temporal. Body stout, cylindrical; tail short, about 11–12 per cent of total length. Scales smooth, without apical pits, in 19–25 (usually 21) rows at mid-body, ventrals 182–208, subcaudals paired, 32–64 (47–64 in males, 32–46 in females), cloacal scale entire. Maximum size 97cm (males smaller, up to 60cm), average 40–70cm, hatchling size 16–20cm. Dorsum typically some shade of yellow-brown or grey-brown, rarely bright golden-yellow, usually with a well-defined pattern of three straight-edged brown bands, one vertebral and two lateral, extending along length of body. The median dark band originates as an arrow-shaped mark on top of head that sometimes encloses a thin pale line, while on lower flanks there is frequently a whitish band with a dark stripe below, which runs from upper lip to tail tip. Underside white or yellowish, with dark spots or streaks often arranged in distinct longitudinal rows; underside of tail usually has a median stripe. Some specimens are less boldly marked, with low-contrast stripes of light and dark brown. Juveniles resemble adults.

Pseudoboodon lemniscatus, Addis Ababa.

Pseudoboodon lemniscatus, Mulu.

HABITAT AND DISTRIBUTION: An Ethiopian and Eritrean endemic, associated with relatively cool and humid environments, particularly montane grassland and light woodland, but extends into tropical forest and edges of Afro-alpine moorlands. Altitudinal range 1,600–3,300m. In Eritrea at Saganeiti and Senafe, probably more widespread but under-collected. In Ethiopia confined to plateaux and mountains either side of Rift Valley, including Simien Mountains, Gondar, Lake Tana, Debre Marcos, Addis Ababa, Jimma, Bedelle, Dodolla, Bale Mountains and Harar. One of commonest highland snakes above 2,500m.

CONSERVATION STATUS: Least Concern.

NATURAL HISTORY: Largely nocturnal and terrestrial, but occasionally seen both basking and prowling by day; one found in a tree. Shelters by day under ground cover, in holes, vegetation heaps, etc. A slow-moving snake, usually in open country, favours grassland. Tolerant of urbanisation, found in wooded suburbs of Addis Ababa. Females lay clutches of up to 21 eggs. A clutch incubated at 17–20°C hatched after 111–119 days in Addis Ababa, where soil temperatures at 30cm are 10–17°C; in central Shoa eggs usually laid December–January and hatchlings appear May or June. Young eat mostly lizards and frogs, adults take rodents, frogs and lizards, probably birds. Two animals were observed feeding on toads *Sclerophrys kerinyagae*, one contained a rodent. Captive accepted freshly killed small birds.

SANDFORD'S ETHIOPIAN MOUNTAIN SNAKE

Pseudoboodon sandfordorum
Spawls, 2004

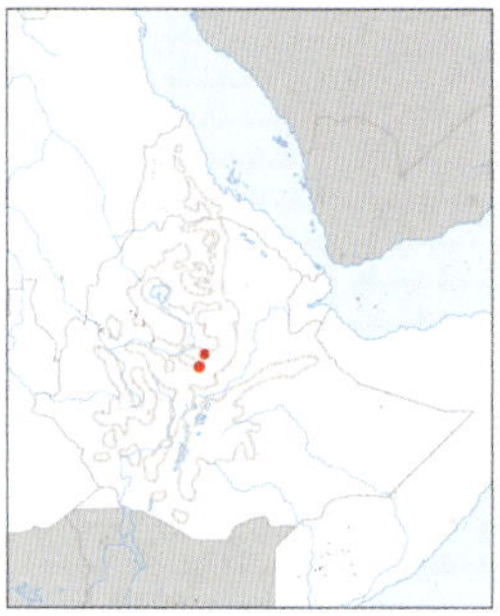

IDENTIFICATION: A relatively large, fairly slender colubrid with a pear-shaped head, distinctive greyish colour and a low-contrast blotched pattern. The broad, pear-shaped head is very distinct from neck and narrows abruptly in front of eyes. Snout rounded. Eye medium-sized, pupil vertical, iris grey; two anterior temporals. Body slender, cylindrical; tail fairly short, about 13–17 per cent of total length. Scales smooth, without apical pits, in 21 mid-body rows, ventrals 216–223, subcaudals paired, 45–62 (57–62 in males,

Pseudoboodon sandfordorum, Mulu.

Pseudoboodon sandfordorum, Mulu.

45–51 in females), cloacal scale entire. Maximum size 92cm (males smaller, up to 60cm), average 60–80cm, hatchling probably around 20cm. Colour pattern distinctive; dorsum greyish, with a series of large, rounded or oval, dark brown blotches along midline of back that show little or no tendency to fuse into a zigzag vertebral stripe, accompanied on either flank by separate row of more of less distinct dark spots. No arrow-shaped mark on top of head. Underside grey-white, with darker freckles and spots concentrated towards the tail but not arranged into any clearly defined pattern; underside of tail without a median stripe. Juveniles resemble adults.

HABITAT AND DISTRIBUTION: An Ethiopian endemic, associated with *Arundinaria* broadleaf forest, collected in montane grassland with scattered *Vachellia abyssinica* trees in intensively cultivated country with the cereals teff (*Eragrostis tef*) and barley, at 1,800–2,400m. A relatively poorly known snake, with currently just seven examples that were all obtained in the vicinity of Keriyo and the nearby Mugher River gorge, some 40km north of Addis Ababa.

CONSERVATION STATUS: Data Deficient. Probably not threatened, seems to be locally common in agricultural landscapes.

NATURAL HISTORY: Little-known. Probably nocturnal and terrestrial. One specimen was found in a pile of hay, another inside a house and a third in fork of a *Vachellia* tree, although captive animals did not climb. Captive animals were fed on lizards (*Trachylepis* spp.), frogs (*Ptychadena* spp., *Amietia nutti*) and rodents (Muridae). Cannibalism observed in captivity. It is sympatric with *Pseudoboodon lemniscatus* at Keriyo.

WOLF-SNAKES *LYCOPHIDION*

An African genus of small, harmless, secretive, flat-headed colubrid snakes, ground-dwelling or burrowing, most are speckled grey or brown. All have flattened or cylindrical bodies, smooth scales in 15–17 rows at mid-body and the usual nine large scales on top of head. The eye is small, with an unusual pupil that is sub-circular and distinctive to this genus, the iris is often silvery, shading to orange around the pupil. Wolf snakes have big recurved teeth, (hence the common name), which they use to catch sleeping lizards. They rarely try to bite when handled. Lay eggs. Often found in the rainy season, when flooded out of their holes. Twenty-three species known at present. Five species in north-east Africa, although their taxonomy is not clear-cut. Two recently described Central African species, *L. chirioi* and *L. tchadensis* have been found in Central African Republic, very close to the border with Southern Darfur, in south-west Sudan, and might occur in our area. Wolf snakes have a superficial resemblance to burrowing asps (*Atractaspis*), which are dangerous, so don't handle one unless you are totally certain of its identity.

KEY TO THE NORTH-EAST AFRICAN MEMBERS OF THE GENUS *LYCOPHIDION*

1a: Usually a white collar or blotch on neck, in dry savanna or semi-desert
Lycophidion taylori, Taylor's Wolf Snake

1b: Not usually a white collar or neck blotch, not necessarily in arid country — 2

2a: Dorsal scale rows not reduced from 17 before vent — 3

2b: Dorsal scale rows reduced from 17 to 15 before vent — 4

3a: Dorsal colour predominantly brown, only one apical pit on each dorsal scale
Lycophidion ornatum, Forest Wolf Snake

3b: Dorsal colour predominantly grey, at least two apical pits on each dorsal scale
Lycophidion irroratum, Pale Wolf Snake

4a: Ventrals 153–173 in females, throat dark like rest of belly, a broad pale band around snout, usually no other head markings, dorsal scales typically with pale stippling in apical region
Lycophidion depressirostre, Flat Snouted Wolf Snake

4b: Ventrals 164–221, throat pale, ventrum dark, pale band around snout narrow or absent, if absent top of head usually with pale vermiculation, dorsal scales typically with an undivided white apical spot or border
Lycophidion capense, Cape Wolf Snake

CAPE WOLF SNAKE

Lycophidion capense
(Smith, 1831)

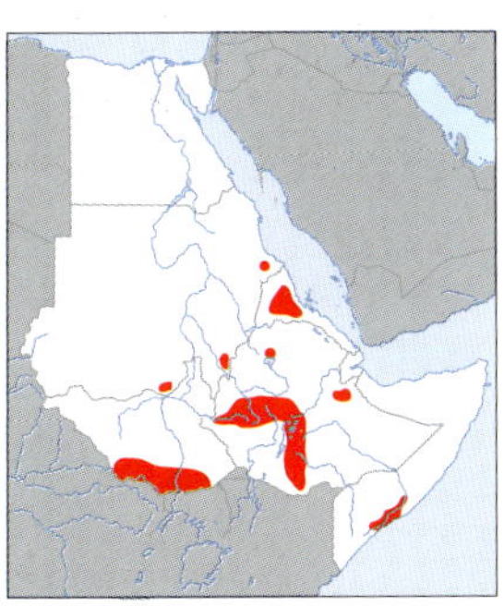

IDENTIFICATION: A small, harmless dark snake. Head very flat, slightly distinct from neck, the eye is tiny so the vertical pupil and silvery-grey iris are hard to see. Body cylindrical, tail fairly short, 8–15 per cent of total length (longer in males). Scales smooth, with a single apical pit, in 17 mid-body rows, ventrals 164–221 (higher counts in females) subcaudals paired, 26–48 (higher counts in males). Maximum size about 60cm (big specimens usually female), average 20–45cm, hatchling size 14–16cm. Colour: looks grey, brown or purplish-brown in field, occasionally black. Close up the scales are black with fine white edges. Head sometimes black on top and big adults may get very dark. Juveniles have a pale belly but the centre of each ventral darkens as they grow, so big adults have dark, light-edged belly scales. Taxonomic Notes: Several subspecies have been described, the Somali coastal animals are *L. c. loveridgei*; elsewhere in our area *L. c. jacksonii*, which some authorities regard as worthy of species status.

HABITAT AND DISTRIBUTION: High grassland, woodland, moist and fairly dry savanna, from sea level to 2,500m. Usually absent in areas with less than 250mm annual rainfall. Known from a few records around Mogadishu in Somalia, and fairly widespread in medium to high-altitude savanna of Ethiopia and central Eritrea. A cluster of records in South Sudan, in Eastern, Central and Western Equatoria; in Sudan in South Kordofan and Blue Nile State. Two curious records further north; one at Erkowit, near the Sudan Red Sea coast, and a 1904 specimen from the Faiyum in Egypt. No other Egyptian examples are known; whether its provenance is correct, and if a population still exists there, are unknown. Elsewhere, west to Central African Republic, south to eastern South Africa.

CONSERVATION STATUS: Least Concern.

Lycophidion capense, Bedelle.

NATURAL HISTORY: A slow-moving, inoffensive, nocturnal ground-dwelling snake. Spends much time investigating suitable refuges in its hunt for sleeping lizards; when inactive hides in holes, under ground cover (logs, rocks, piles of vegetable debris, building junk etc.). Most often seen in rainy season, especially if flooded out of its refuge by rising water. Hardly ever tries to bite if restrained, but will jerk and twist its body when held, and if held by the head its vigorous twisting can cause the long teeth to puncture the skin. It lays 3–8 eggs, roughly 1×2cm. A Kenyan specimen laid six eggs in September, a hatchling was taken in May in Tanzania, incubation in Botswana was 50–65 days, hatchlings emerged in middle of rainy season. Diet largely smooth-bodied skinks, occasionally small snakes. A 50cm Kenya specimen had eaten a 13cm Montane Side-striped Chameleon *Trioceros ellioti*.

FLAT-SNOUTED WOLF SNAKE

Lycophidion depressirostre
Laurent, 1968

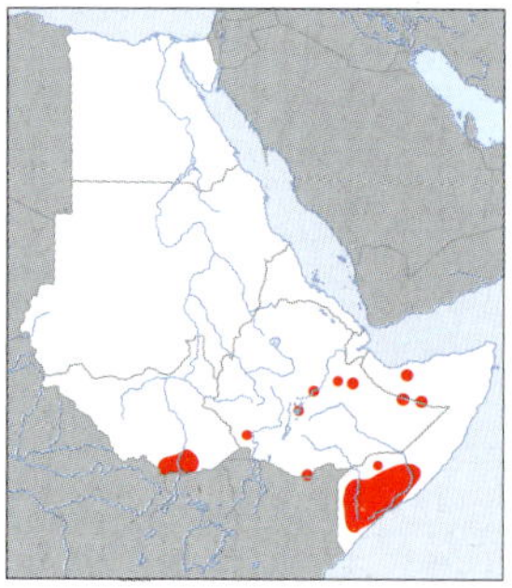

IDENTIFICATION: A small, dark grey or brown, flat-headed snake. Eye small. In good light, the pupil actually looks round, the iris is silvery-grey but the borders of the pupil are often orange. Body cylindrical, tail fairly short, 8–14 per cent of total length (longer in males). Scales smooth, with a single apical pit, in 17 mid-body rows, ventrals 153–180, subcaudals paired, 26–40 (higher counts in males). Maximum size 48cm (big specimens usually female), average 20–35cm, hatchling probably 15–17cm. Colour quite variable, may appear almost black, dark grey, light grey or grey-brown, head usually uniform but rarely stippled or spotted, dorsal scales have tiny light speckles and white edging. Occasionally a light snout band is present; sometimes snout tip is rufous. Belly dark grey, sometimes paler on chin and outer edges of ventrals.

Lycophidion depressirostre, Kenya.

HABITAT AND DISTRIBUTION: Usually in semi-desert, dry and moist savanna, a few records from lusher habitats, between sea level and 2,000m. Records somewhat sporadic. In southern Somalia, south of roughly 04°N, in the north in the Haud on the

Ethiopian/Somali border at 46°E and in Goolis Mountains. In Ethiopia, known from the vicinity of Harar and Bourka, in the Ahmar Mountains; Aware Melka, north of Awash National Park; Moyale, and Mui on the lower Omo River. In South Sudan, only in south, Juba and the vicinity, Yei and Torit. Elsewhere, south to southern Tanzania.

CONSERVATION STATUS: Least Concern.

NATURAL HISTORY: Poorly known. Nocturnal and terrestrial, emerging at dusk to hunt, when inactive hides in holes or under ground cover (log, rocks, debris, etc). Diet probably similar to other wolf snakes, i.e., lizards. They lay eggs, but no clutch details known.

Lycophidion depressirostre, Kenya.

PALE WOLF SNAKE

Lycophidion irroratum
(Leach, 1819)

IDENTIFICATION: A grey wolf snake with a pale snout band, just reaching South Sudan in our area. Body cylindrical, tail fairly short, 12–16 per cent of total length (longer in males). Scales smooth, with 2–4 apical pits, in 17 mid-body rows, ventrals 158–193, subcaudals paired, 27–54. Maximum size 45cm, average 20–35cm, hatchling size unknown. Colour: light to dark grey above, with small, poorly defined and sparely distributed dark spots; close up dorsal scales grey with irregular dark edges. A broad pinkish or light brown band curls round the snout and extends as a broad but diffuse pale line behind eyes. Ventrals dark, sometimes with pale edges.

HABITAT AND DISTRIBUTION: Woodland and moist savanna, at low altitude. In our area known from Yei, in Central Equatoria, South Sudan; a snake suspected to be this species is also known from Garamba National Park, in DR Congo, so might be in Western Equatoria. Elsewhere, west to Senegal.

CONSERVATION STATUS: Least Concern.

NATURAL HISTORY: Poorly known. Nocturnal, terrestrial and secretive, emerging at dusk to hunt lizards, when inactive hides in leaf litter, holes or under ground cover (in rotting logs, under rocks, debris, etc). Diet smooth-bodied lizards. They lay eggs.

Lycophidion irroratum, Guinea.

FOREST WOLF SNAKE

Lycophidion ornatum
Parker, 1936

IDENTIFICATION: A small, attractive forest-dwelling wolf-snake. Head flat, eye small but obvious, iris silvery, pupil edged orange. Body cylindrical, tail longer than in most wolf snakes, 12–18 per cent of total length (longer in males). Scales smooth, 17 mid-body rows, ventrals 175–212, subcaudals paired, 32–53 (usually 33–44 in Uganda, higher counts in males). Maximum size 59cm (a Kenya female), average 25–50cm, hatchlings 14–15cm. Colour: brown, red-brown, grey, pinkish or pinkish-grey, sometimes uniform, but usually has a broad white snout band, extending on side of head through eye, behind eye it splits into two, one onto temporals, the other onto upper labials. Some have a double row of dark spots on back. Belly scales black or grey, often with pale edging, so underside looks finely barred.

HABITAT AND DISTRIBUTION: Forest and deforested areas, from sea level to 2,700m but in our area known only from Gilo, at 1,800m in the Imatong Mountains in Eastern Equatoria, South Sudan.

CONSERVATION STATUS: Least Concern.

NATURAL HISTORY: Poorly known. Nocturnal and terrestrial, emerging to hunt on forest floor at dusk, when inactive hides in leaf litter, in holes or rotting log, under ground cover, etc. Clutches of 2–6 eggs recorded. Gravid females collected in Uganda in January and February, a Kenyan female laid five eggs, 2–2.6×0.6–1cm in January. Diet largely small, smooth-bodied forest lizards, which it finds by investigating holes and leaf litter, but a Congo snake climbed a tree and seized a sleeping Johnston's Chameleon *Trioceros johnstoni*.

Lycophidion ornatum, Rwanda.

TAYLOR'S WOLF SNAKE

Lycophidion taylori
Broadley & Hughes, 1993

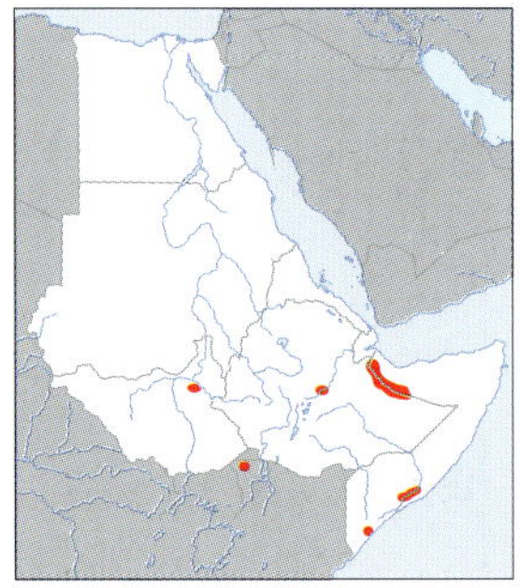

IDENTIFICATION: A rare arid-country wolf-snake, usually with a distinctive white collar. The eye is medium-sized, pupil vertical, iris brown or red-brown. Head flat, tongue pale or pinkish. Body cylindrical, tail short, 9–13 per cent of total length (longer in males). Scales smooth, with a single apical pit, in 17 mid-body rows, ventrals 161–184 (higher counts in females) subcaudals paired, 26–36 (higher counts in males). Maximum size 51cm (big specimens usually female), average 25–40cm, hatchling size unknown, a juvenile was 23cm. Colour: grey or grey-brown above, each scale with either a white dot or stipple, head grey with white stippling. Usually has a white collar or blotch on neck, may be bordered by black patches or bands, occasionally has white body markings. Belly scales uniform brown or reddish-brown, sometimes white-edged, chin stippled white.

Lycophidion taylori, Tanzania.

HABITAT AND DISTRIBUTION: Arid savanna and semi-desert at medium to low altitudes. Records sporadic in our area. Recorded near Kismayu, Balad and Afgoi in southern Somalia, a number of records along

Ethiopia/Somalia border in the Haud and around Borama. In Ethiopia reported at Awash National Park. A recent sight record from Old Fangak, Jonglei, South Sudan. Might be in Sudan. Elsewhere, south through Kenya to northern Tanzania. A curious record from Senegal.

CONSERVATION STATUS: Least Concern.

NATURAL HISTORY: Described in 1993, from museum specimen, so little known. Nocturnal. Habits presumably similar to other wolf snakes.

FOREST SNAKES *HORMONOTUS*

An African genus with a single harmless forest species; sister to the enigmatic southern African *Inyoka swazicus*, and its next nearest relative is the file snake genus *Gonionotophis*.

YELLOW FOREST SNAKE

Hormonotus modestus
(Duméril, Bibron & Duméril, 1854)

Hormonotus modestus, Gabon.

IDENTIFICATION: A ground-dwelling snake, with a distinctive pear-shaped head. Eyes big and prominent, with a pale iris and vertical pupil. Snout square, body cylindrical, tail long, 19–23 per cent of total length, longest in males. Scales smooth, 15 mid-body rows, ventrals 220–244, subcaudals paired, 78–103. Maximum size 88cm, average 40–70cm, hatchling size unknown. Colour: variable, either dull yellow, various shades of brown, yellow-brown or grey-brown. Scales on top of head often white-edged, giving a most distinctive reticulated appearance, which is a valuable aid to identification. Each lip scale usually has a dark spot. Belly yellow or cream.

HABITAT AND DISTRIBUTION: Forest or recently deforested country, between sea level and 1,300m. A single documented record from our area; Talanga Forest, at 950m, in South Sudan, but informally reported at Torit. Elsewhere, south to western Kenya and Tanzania, south-west to northern Angola, west to Guinea.

CONSERVATION STATUS: Least Concern.

NATURAL HISTORY: Little known. Nocturnal, both terrestrial and arboreal, hunting on forest floor and in trees at night. Hides by day in leaf litter or under ground cover. Presumably lays eggs. Diet said to be rodents, a DR Congo specimen had eaten a tropical house gecko. It appears to be rare.

SAVANNA FILE SNAKES *LIMAFORMOSA*

For a long time, all African file snakes, with their triangular bodies and enlarged vertebral scales, were in two genera, *Mehelya* and *Gonionotophis*. At present, they are now in four genera, three of which (*Limaformosa*, *Mehelya*, *Gracililima*) occur in our area. The genus *Limaformosa* contains six species, two occur in our area. All are large (1.3m or more), with distinctly triangular bodies and obviously enlarged vertebral scales, the skin is visible between the dorsal scales. Harmless, dark-coloured, slow-moving, secretive nocturnal snakes; they feed on a range of prey, favouring amphibians and reptiles, including highly venomous snakes. There are several superstitions about file snakes, due to their slow movement and sinister appearance; for example, that anyone seeing one that has a light-coloured vertebral ridge will go blind, another is that if one enters your house bad luck will follow, but the snake must not be molested.

KEY TO NORTH-EAST AFRICAN MEMBERS OF THE GENUS *LIMAFORMOSA*

1a. Vertebral scale row often quadricarinate; venter usually dark brown to black
Limaformosa chanleri, Chanler's File Snake

1b. Vertebral scale row bicarinate; underside white, ends of each ventral dark
Limaformosa savorgnani, Congo File Snake

CHANLER'S FILE SNAKE

Limaformosa chanleri
(Stejneger, 1893)

Limaformosa chanleri, Tanzania.

IDENTIFICATION: A fairly big, slow-moving, dark snake. Head broad and flat, eye medium-sized and dark. Tongue pink with a white tip. Body triangular in section, with an enlarged rough vertebral ridge; tail fairly long, 10–15 per cent of total length, slightly longer in males than females. Tail often mutilated and body scales and strips of skin missing, due to attacks by predators. Scales small and heavily keeled, in 15 rows at mid-body, ventrals in our area 215–243, subcaudals paired, 44–62 rows, higher counts usually males. Maximum size about 1.5m, average 70cm to 1.3m, hatchling size unknown. Colour grey, black, purplish-grey or dark brown, the interstitial skin usually paler than the scales. Ventral scales grey, with white edging. Taxonomic Notes: Known in older Somali herpetological literature as *Mehelya* (*Simocephalus*) *somaliensis* and *Mehelya* (*Simocephalus*) *fiechteri*.

HABITAT AND DISTRIBUTION: Occupies wide range of habitats including semi-desert, dry and moist savanna and woodland, but not forest, from sea level to around 1,800m. Widespread in our area but very rare. Eritrean records include Dorfu village, near Asmara, and 'eastern Eritrea' (without further data), the Haud, on Somali/Ethiopian border, at 09°N 44°E, and a little cluster of records around Mogadishu and environs, and on lower Juba River between Gelib and Kismayu. Elsewhere, south to Tanzania, south-west to Rwanda.

CONSERVATION STATUS: Least Concern.

NATURAL HISTORY: Nocturnal, largely terrestrial but will climb in search of prey. Slow-moving and secretive. Often shelter in termitaria, or in holes, under suitable ground cover, etc. They are good-natured, and rarely try to bite, but are restless in the hand, and defend themselves by discharging a foul-smelling green paste from glands near the cloaca. Lay eggs. Diet is a wide variety of ectotherms; amphibians, lizards and snakes (including a highly venomous one, they seem immune to their venom); they may take mammals. The name honours William Astor Chanler, charismatic American explorer in northern Kenya in the 1890s, who collected the first specimen on the mainland next to Manda Island on the Kenya coast.

CONGO FILE SNAKE, DE BRAZZA'S FILE SNAKE

Limaformosa savorgnani
(Mocquard, 1887)

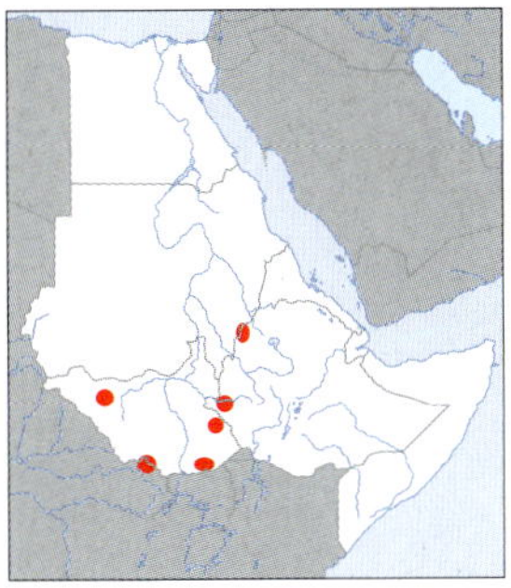

IDENTIFICATION: General appearance similar to previous species. Scales keeled, in 15 rows at mid-body. Ventrals 220–239, anal entire, subcaudals 45–62. Maximum size c.1.6m, average 90cm to 1.3m. Females larger than males. Dark brown, purplish-brown, various shades of grey or black above, dorsal scales often white-tipped, giving a speckled effect. Ventral scales white, sometimes with long edges dark, giving a barred appearance, and outer edges dark.

Limaformosa savorgnani, Uganda.

HABITAT AND DISTRIBUTION: Savanna and woodland at 300–800m; but higher and lower elsewhere. In our area, a handful of localities in South Sudan: Boma National Park, Torit, Magwe, between Wau and Chak Chak, Molongori, and known just across DR Congo border at Garamba National Park. Two Ethiopian specimens from Gambela and Kurmuk probably this species. Elsewhere, south to Uganda and Kenya, west to Cameroon.

CONSERVATION STATUS: Least Concern.

NATURAL HISTORY: As for the previous species; nocturnal, slow-moving and secretive. Ugandan specimens recorded eating snakes, lizards and amphibians.

BLACK FILE SNAKES *GRACILILIMA*

A genus with a single small species.

DWARF FILE SNAKE, BLACK FILE SNAKE

Gracililima nyassae
(Günther, 1888)

IDENTIFICATION: A slim, small dark file snake. Head not as short as other file snakes and smooth on top, eye small and dark, the vertical pupil is hard to see. Tongue pink and white. Body sub-triangular in cross-section, with a pronounced vertebral ridge. Tail fairly long, 18–27 per cent of total length, usually longer in males. Scales keeled, in 15 mid-body rows, the vertebral row enlarged and hexagonal, with a strong or weak double keel. Ventrals 164–184, anal entire, subcaudals 51–79. Maximum size about 65cm (large specimens usually females), average 35–55cm, hatchlings 20–22cm. Black or grey, the skin between the scales pink, white or grey, some scales may have fine white dots. A Zanzibar snake had a dorsolateral stripe of fine white dots. Belly may be white, cream, black or grey with pale edging.

Gracililima nyassae, Kenya.

CONSERVATION STATUS: Least Concern.

HABITAT AND DISTRIBUTION: Low savanna, coastal thicket and woodland, at sea level and higher. In our area, known only from the Juba Sugar Project, near Gelib, southern Somalia, but probably extends along the coast to Kenya, where recorded the length of the coast. It is also known in Garamba National Park, in DR Congo, so probably in south-west South Sudan, although unrecorded. Elsewhere, south to Botswana and north-east South Africa.

CONSERVATION STATUS: Least Concern.

NATURAL HISTORY: Poorly known. Nocturnal, terrestrial and secretive. Specimens in South Africa were active at night in rainy season. Shelters under ground cover and in holes. Slow-moving and inoffensive, although reported to produce a noxious cloacal secretion when handled. Between two and six eggs are laid. Hatchlings 21–23cm collected in April in southern Tanzania. Diet small lizards, particularly skinks, occasionally snakes and amphibians.

SMALL AND FOREST FILE SNAKES *MEHELYA*

Mehelya was originally used for all the African file snakes save three species of *Gonionotophis*; it is now reserved for five mostly small file snakes, three of which are known at a single locality; the other two occur in our area and are widespread in forest and savanna of West and Central Africa.

FOREST FILE SNAKE

Mehelya poensis
(Smith, 1849)

IDENTIFICATION: A long-tailed harmless forest file snake; so slim it looks like a tree snake. Head long and looks rectangular; snout noticeably square but quite pointed in profile. Eye relatively large, neck thin. Body sub-triangular in cross-section (not as pronounced as other large file snakes), with the usual enlarged row of vertebral scales. Tail very long for a file snake, 20–25 per cent of total length, generally longer in males. Dorsal scales have a single prominent keel, vertebral scales have two keels. Scales in 15 mid-body rows, ventrals 236–262, anal entire, subcaudals 75–124, male counts usually higher. Maximum size about 1.4m, average 80cm to 1.2m, hatchling size unknown. Colour olive-grey or grey, exposed skin between scales light grey or blue-grey, belly light grey to yellowish or white, darker at outer edges of ventrals.

HABITAT AND DISTRIBUTION: Forest and forest islands, from sea level to quite high altitude, but in our area reported only at Yambio, in Western Equatoria, South Sudan. Known from Garamba National Park in

Mehelya poensis, Congo.

DR Congo. Might be more widespread in southern South Sudan. Elsewhere, west through forest to Guinea, south-west to Angola.

CONSERVATION STATUS: Least Concern.

NATURAL HISTORY: Poorly known. Slow-moving, nocturnal, terrestrial and secretive. Its relatively long tail suggests that it might be arboreal to some extent. By day hides in holes, leaf litter, etc. In Ghana, a female contained eight ova, roughly 2.5 × 1 cm, in May (start of rainy season), no other breeding details known. Recorded prey includes a mouse, Red-headed Rock Agamas *Agama lionotus* and Speckle-lipped Skinks *Trachylepis maculilabris*.

SMALL-EYED FILE SNAKE

Mehelya stenophthalmus
(Mocquard, 1887)

IDENTIFICATION: A small dark file snake, not noticeably triangular in cross-section, with smooth scales. Head flat; eye small and dark, nostril curiously large and round. Body weakly sub-triangular in shape (unlike most file snakes, which are clearly triangular), tail 12–17 per cent of total length, usually longer in males. Scales smooth except vertebral row, which is weakly and sporadically keeled. Mid-body scale count 15, ventrals 192–228, subcaudals paired, 47–60, anal entire. Maximum size about 76cm, average 40–70cm, hatchling size unknown. Colour glossy black or dark grey; scales iridescent. Belly ivory-white, cream or light yellow, darkening at outer edges of ventrals.

HABITAT AND DISTRIBUTION: Forest and recently deforested areas, from sea level to 2,000m. Not formally recorded in our area, but known from Garamba National Park in DR Congo, just across border from Western Equatoria, and also in eastern Central African Republic, so almost certainly in South Sudan. Elsewhere, west to Guinea, south to Uganda.

CONSERVATION STATUS: Least Concern.

NATURAL HISTORY: A nocturnal, terrestrial file snake. Poorly known, might well spend much time burrowing or pushing through leaf litter. One Uganda specimen was found at night on forest floor. Presumably lays eggs, no details known. Diet said to be snakes, including other file snakes and an Olive Marsh Snake *Natriciteres olivacea*, but few data available, a specimen from DR Congo had eaten a Western Forest Limbless Skink *Feylinia currori*.

Mehelya stenophthalmus, Gabon.

FAMILY COLUBRIDAE COLUBRID SNAKES, OR 'ORDINARY' SNAKES

A large intercontinental family of more than 2,000 snake species in nearly 250 genera, they occur almost worldwide, although there are only a few in Australia. However, there is ongoing debate about the status of both the family and subfamilies within it. Consequently, the 'running order' here does not reflect any particular evolutionary succession; we have largely covered the genera with Palearctic affinities first and then the Afrotropical forms. As presently defined, in our area there are about 70 species of colubrid, in 25 genera. None is very large or very small, most are between 50cm and 2m long. They lack front fangs; most have obvious eyes, a loreal scale, broad belly scales and their tails taper. Most are harmless, but the group includes several rear-fanged snakes, including some that are dangerously venomous, the Boomslang and two vine or twig snakes of the genus *Thelotornis*; the big rear-fanged tree snakes of the genus *Toxicodryas* also should be treated with caution.

KEY TO THE GENERA AND SOME SPECIES OF COLUBRIDAE IN NORTH-EAST AFRICA

1a: One or more pairs of enlarged, grooved venom fangs towards rear of upper jaw	2
1b: No enlarged, grooved venom fangs towards rear of upper jaw	8
2a: Dorsal scales keeled	3
2b: Dorsal scales smooth, without keels	4
3a: Pupil distinctly keyhole-shaped, head elongate, snout long and pointed	*Thelotornis*, Vine snakes
3b: Pupil generally round but sometimes teardrop-shaped, head and snout rounded	*Dispholidus typus*, Boomslang
4a: Small, <50cm, head black, not particularly distinct from neck, body brown, mid-body scale rows 19; in our area only in northern Egypt	*Macroprotodon cucullatus*, Hooded Snake
4b: Lacks the above combination of characters; head distinctly broader than neck	5
5a: Ventral scales 141–183	*Crotaphopeltis*, Herald and White-lipped snakes
5b: Ventral scales 195–270	6
6a: Subcaudal scales >100, large, adults grow to >1m	*Toxicodryas*, Broad-headed tree snakes
6b: Subcaudal scales <100, smaller, adults rarely >1m	7
7a: Cloacal shield entire, single anterior temporal, in forest or moist savanna	*Dipsadoboa*, Marbled and black-tailed tree snakes
7b: Cloacal shield divided, two or more anterior temporals, generally in arid country	*Telescopus*, Large-eyed and cat snakes
8a: Dorsal scales strongly keeled	9
8b: Dorsal scales smooth or very weakly keeled	11
9a: Pupil of eye vertically elliptical	*Dasypeltis*, Egg-eaters
9b: Pupil of eye circular	10
10a: Mid-body scale rows 19, in our area only in northern Egypt	*Natrix*, Water snakes
10b: Mid-body scale rows 15, in tropical Africa	*Hapsidophrys lineata*, Black-lined Green Snake
11a: Black, arboreal, adults >1.4m, juveniles chequered, dorsal scales smooth or weakly keeled	*Thrasops jacksonii*, Jackson's Tree Snake
11b: Lacks the above combination of characters	12
12a: A row of scales (not just individual scales) separates eyes from upper labial scales, large, head elongate, dorsal pattern of darker brown subcircular blotches on paler background	*Spalerosophis*, Diadem snakes

12b: Lacks the above combination of characters 13

13a: Mid-body scales rows 15, part or all of dorsal surface vivid green, pupil round
Philothamnus, African green snakes

13b: Not vivid green, lacks above combination of characters 14

14a: Rostral scale large and pointed, lower jaw invisible from side, parietal scales fragmented
Scaphiophis, Hook-nosed snakes

14b: Lacks above combination of characters 15

15a: Head black, mid-body scale rows 15; north Sinai
Rhynchocalamus melanocephalus, Palestine Black-headed Snake

15b: Lacks the above combination of characters 16

16a: Pupil vertically elliptic *Aeluroglena cucullata*, Somali Snake

16b: Pupil round 17

17a: Body stout, mid-body scale rows 15–19, eye pupil round but small, large (adults to 1.3m or more), in vicinity of water *Grayia*, Central African water snakes

17b: Lacks above combination of characters 18

18a: Mid-body scale rows 15 in our area *Eirenis*, Dwarf snakes

18b: Mid-body scale rows >15 19

19a: Small, with a curious, prominent, wedge-shaped nostril; desert and near-desert
Lytorhynchus diadema, Awl-headed Snake

19b: No prominent rostral, not necessarily in desert 20

20a: Large, up to 2m, heavily built, adult usually black; in our area only in north Sinai
Dolichophis jugularis, Large Whip Snake

20b: Lacks above combination of characters 21

21a: Adult up to 1.3m or more, stocky, brown, dorsal pattern either circular blotches or crossbars but extends length of body; in our area only in north Egypt *Hemorrhois*, Whip snakes

21b: Lacks above combination of characters 22

22a: One or more small scales below and forward of eye, below preocular, separating front of orbit from upper labials *Platyceps*, Racers

22b: No small scale between preocular and upper labials 23

23a: Slim, tail long and thin, often patterned with crossbars or blotches
Meizodon, Smooth and semi-ornate snakes

23b: Stocky, tail short, pattern-less or has dorsal stripe 24

24a: Subcaudals 46, uniform brown *Aprosdoketophis andreonei*, False White-lipped Snake

24b: Subcaudals >50, usually has distinct vertebral stripe, different colour to rest of back
Natriciteres olivacea, Olive Marsh Snake

RACERS *PLATYCEPS*

A genus of mainly African and Asian, medium-sized, diurnal, fast-moving, slim, aglyphous, harmless egg-laying colubrid snakes, most are brown or grey, and often banded or spotted. They hunt by day for small vertebrates. In our area, they occupy a range of habitats, from deserts to dry savanna, at lower elevations. They can usually be identified by the presence of one or more small subocular scales, below and slightly in front of the eye, separating the preocular from the upper labial scales. Originally placed in the genus *Coluber*, about 29 species are known in southern Europe, North Africa, the Middle East and western Asia. Some 13 relatively small species occur in Africa, all of them in our area; four are endemic. Several groups of closely related and similar species in the genus are distinguished. Their taxonomy is not yet fully resolved, and several are known from only a handful of specimens. The genus *Platyceps* is most closely related to whip snakes of the genus *Hemorrhois* and diadem snakes of the genus *Spalerosophis*.

KEY TO NORTH-EAST AFRICAN MEMBERS OF THE GENUS *PLATYCEPS*

1a: Mid-body scale rows 15	*Platyceps somalicus*, Ogaden Racer
1b: Mid-body scale rows >15	2
2a: Mid-body scale rows 17–19	3
2b: Mid-body scale rows >19	8
3a: Dorsum with black crossbars, usually connected by light orange vertebral stripe from nape to tail tip	*Platyceps sinai*, Sinai Banded Racer
3b: Dorsum without black crossbars	4
4a: Fewer than 95 subcaudals, <175 ventrals in males	*Platyceps brevis*, Somali Racer
4b: More than 95 subcaudals, >175 ventrals in males	5
5a: Small, maximum size 54cm; pale body in anterior ⅔ covered by small blackish dots; mid-body scale rows 17	*Platyceps messanai*, Messana's Racer
5b: Moderate to large-sized, maximum 103–160cm; lacks previous coloration and pattern; mid-body scale rows 19	6
6a: Less slender; longitudinal nuchal stripe on nape; anterior dorsal pattern large round spots; 89–117 subcaudals; northern Egypt and Sinai	*Platyceps rogersi*, Spotted Racer
6b: Very slender; no nuchal stripe; anterior dorsal pattern crossbars or plain; 112–149 subcaudals	7
7a: 134–149 subcaudals, 213–264 ventrals; northern Sudan to Egypt including Sinai	*Platyceps saharicus*, Saharan Cliff Racer
7b: 112–132 subcaudals, 208–228 ventrals; central Eritrea south to central Somalia	*Platyceps rhodorachis subniger*, Cliff Racer
8a: Mid-body scale rows 27–29	*Platyceps scorteccii*, Scortecci's Racer
8b: Mid-body scale rows 21–23	9
9a: Very slender; uniformly very dark coloration above; >250 ventrals and 143 subcaudals in males	*Platyceps afarensis*, Afar Racer
9b: Less slender; usually lacks uniform very dark coloration above; <250 ventrals and 143 subcaudals in males	10
10a: Dorsum pale grey in adults, with black-edged scales on back creating dark reticulation; Dahlak archipelago in Eritrea	*Platyceps largeni*, Dahlak Racer
10b: Lacks coloration of previous; not known on Dahlak archipelago, Eritrea	11
11a: Ventrals >192 in males, >207 in females	*Platyceps florulentus*, Flowered Racer
11b: Ventrals not more than 189 in males or 199 in females	12
12a: Mid-body scale rows 21; head and neck with three dark bands; posterior body lacks distinct pattern in adults	*Platyceps smithi*, Smith's Racer
12b: Mid-body scale rows 21–23; large and partly confluent paravertebral spots throughout	*Platyceps taylori*, Taylor's Racer

AFAR RACER

Platyceps afarensis
Schätti & Ineich, 2004

IDENTIFICATION: A moderately sized, very slim, dark-coloured colubrid, endemic to Djibouti; known from just two specimens. Head narrow, depressed, distinct from neck, snout narrow from above, slightly prominent. Eye large with a round pupil. Body long and slim, cylindrical; tail long, very thin, 27–28 per cent of total length. Scales smooth, fairly small, in 21 mid-body rows, ventrals 251–258, subcaudals 143–144. Maximum size about 103cm, size of hatchlings unknown. Head dark above, upper labials, preocular and postoculars mostly pale; chin uniformly pale. Dorsum of larger preserved specimen is dark blackish on neck and anterior third of body, fading gradually to dark brown posteriorly; tail tip yellow-brown. The other known, smaller specimen is similar but more brownish. Underside iridescent yellowish, with small irregular dots; darker dorsal coloration extends to lateral edges of ventrals. Taxonomic Notes: Closely related to *P. rhodorachis subniger* and *P. saharicus*.

HABITAT AND DISTRIBUTION: Dry desertic steppe with sparse herbaceous vegetation, shrubs and a few *Vachellia* and *Balanites rotundifolia* trees, on a rocky highland plateau, which overlooks the coastal plains, at elevation 705m. Known only from an adult and sub-adult male, from Arta, Gulf of Tadjoura, Djibouti, about 7km from the sea. Probably more widespread in adjacent countries (southern Eritrea, eastern Ethiopia and north-west Somaliland), all suitable habitats, but under-collected.

CONSERVATION STATUS: Data Deficient.

NATURAL HISTORY: Unknown, but presumably similar to others of genus, i.e., active by day, terrestrial, fast-moving, probably lays eggs. Diet probably diurnal lizards but unknown.

Platyceps afarensis, preserved type, Djibouti.

SOMALI RACER

Platyceps brevis
(Boulenger, 1895)

IDENTIFICATION: A small, slim, harmless, barred brown or rufous snake with dorsal scales in 17–19 mid-body rows, in Somalia north of Shebelle River. Head slightly distinct from neck, snout obtuse. Eye large with a round pupil. Body cylindrical, tail long, 23–30 per cent of total length. Scales smooth, moderately sized, in 17–19 (rarely 21 in southern Somalia) mid-body rows, ventrals 158–183 (158–168 in males, 169–183 in females), subcaudals 75–95. Maximum size about 60cm, average 40–55cm, hatchling probably 20–22cm. Body light brown to sandy-yellow or greyish, usually with a crossbar from below eye across supraoculars and frontal, sometimes followed by a black spot on posterior temporal region, in juveniles this spot reaches parietals. Juveniles strongly barred, in adults fades to a greater or lesser extent. Underside white (in juveniles), yellow or cream, spotted black.

HABITAT AND DISTRIBUTION: Dry savanna and semi-desert, sea level to 1,500m. Widespread in northern Somalia, in northern Puntland, from Daarin at southern edge of Cal Miskaat Mountains south across Gardo to Callis, Las Anod, near Taleh and Halin in Nogal Valley, thence west through Haud, where observed at Bohodle. Also known from Borama in western Somaliland. The type specimen was collected at an imprecise locality in the Ogaden. Known also from the Hobyo Plains ('Afghedud' or Afguduudle, between Caaio and Andurgab) in central Somalia. Specimens with higher dorsal counts (21) of dorsals occur in the southern coastal region, at Mahaddei Uen, Bardera and Sar Uanle near Kismayu, although typical specimens with 19 dorsals were collected on Koyama Island, Bajuni archipelago, about 35km south of Kismayu.

CONSERVATION STATUS: Least Concern.

NATURAL HISTORY: Little-known, presumably similar to *P. smithi*: diurnal, terrestrial, secretive and fast-moving. Diet probably diurnal lizards, possibly snakes and mammals, but poorly known. A juvenile from Daarin was killed some hours before it was found, apparently while crossing an open area by day.

Platyceps brevis, preserved type, Ogaden.

FLOWERED RACER

Platyceps florulentus
(Geoffroy Saint-Hilaire, 1827)

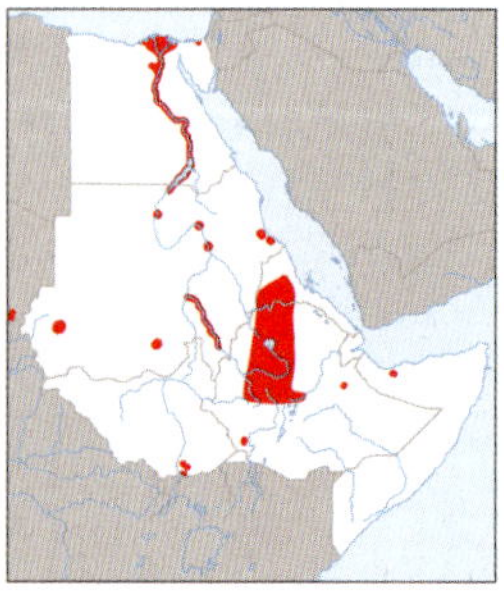

IDENTIFICATION: A moderately sized, slim, usually brownish snake with several series of small dots on anterior part of body, from Egypt to northern Somalia. Head oval, eye moderately large with a round, yellow-rimmed pupil, iris dark grey-brown. Body slender, cylindrical; tail comparatively fairly short, thin, 21–26 per cent of total length. Scales smooth, small, in 21–23 mid-body rows, ventrals 192–228 (192–214 in males, 207–228 in females), subcaudals 82–105. Maximum size about 110cm, average 60–90cm, size of hatchlings unknown. Dorsum olive-brown, grey-brown or rufous, uniform or has 1–2 dark crossbars on head and a series of crossbars or darker spots on body; markings may fade in large adults, especially towards tail. Occasional specimens are grey anteriorly and brown posteriorly. Pale preocular, postoculars and dark spot below eye usually present. Underside white, cream or orangey-yellow. Melanistic specimens known from Awash National Park in Ethiopia. Juveniles more boldly coloured. Taxonomic Notes: A polytypic species, in our area the wide-ranging nominate, *P. f. florulentus*, occurs. According to preliminary genetic data, probably a species complex.

HABITAT AND DISTRIBUTION: Semi-deserts, including lava fields, dry and moist savanna, lower levels in montane grasslands and fluvial habitats, including cultivated land, fallow fields, canal banks, wetland margins, ruins and old houses, between sea level and 2,400m. Known from Nile Valley and Delta in Egypt (with an isolated recent record at Al Arish in northern Sinai). Widespread but sporadic in Sudan; in Red Sea coastal plains and hills (Erkowit, Tokar), Kordofan (Delami) and along Nile River, where very common along Blue Nile, from Khartoum to Ar-Rusayris. Probably occurs in Darfur in west Sudan; known from eastern Chad and Kordofan, so probably in intervening country. In western lowlands and central mountains of Eritrea, but not on coastal plain or Ethiopian Afar region. Widespread on western plateau and mountains of Ethiopia approximately north of 08°30'N, at lower elevations in Omo National Park in the south-west, Awash National Park in centre and Dire Dawa in east of the country. No Djibouti records, but recorded in mountains of western and central Somaliland, east to Berbera and Goolis Mountains. In South Sudan only at Nimule, Torit and Lowai (Jabal Lowe) in Eastern Equatoria.

CONSERVATION STATUS: Least Concern.

NATURAL HISTORY: Terrestrial, but will climb into bushes and low trees. Fast-moving and diurnal, although in parts of Nile Valley they are crepuscular, moving at dawn and dusk. At night they shelter in holes and under ground cover. In Egypt and Sudan often found in cultivated land, especially gardens, and sometimes enters buildings. They lay eggs, but no clutch details unknown. Diet includes frogs, rodents, lizards and birds. A specimen from Egypt had just eaten a frog, *Ptychadena nilotica*, another contained a mouse. Racers will bite when first handled; locally reputed to be 'the fiercest snake in Egypt and Sudan'.

Platyceps florulentus, Mulu.

DAHLAK RACER

Platyceps largeni
(Schätti, 2001)

IDENTIFICATION: A small, harmless snake with a distinct reticulated pattern in adults, endemic to the Dahlak archipelago in Eritrea. Head slightly distinct from neck, snout appears narrow from above. Eye large with a round pupil. Body cylindrical, tail long, ¼ of total length. Scales smooth, moderately sized, in 21 mid-body rows, ventrals 182–197, subcaudals 87–92. Maximum size 64.2cm, average 35–55cm. Smallest known specimen less than 20cm. Two distinct colour phases, apparently related to age; one spotted/banded, one uniformly reticulate. Large specimens pale grey above, each scale black-edged to give a reticulate effect; head unpatterned. Smaller specimens brown or grey-brown, with two dark crossbars on head and a few transverse bands on anterior region of back, followed by longitudinal rows of dark spots, those along the midline sometimes fusing to form larger blotches. Belly pale with rows of dark spots on outer edges of ventral scales. Taxonomic Notes: Closely related to *P. florulentus*, with which it was confused before being described as a separate species.

Platyceps largeni, preserved, Dahlak Archipelago.

HABITAT AND DISTRIBUTION: Endemic to Eritrea, where known only from several small islands in Dahlak archipelago. Inhabits very dry, near-desert Pleistocene limestone flats of coralline origin, with sparse vegetation, 5–15m above sea level. The type specimen is from Sarad, others are from Andeber and Nocra, (all very close to Dahlak Kebir, the largest island in the archipelago) and on Museri Island, south of Dahlak Kebir. Probably more widespread, the Dahlak archipelago is poorly studied. Snakes on Dahlak archipelago are usually believed to be carpet vipers or cobras and killed, however the small islands where this snake occurs are uninhabited and thus it is not persecuted.

CONSERVATION STATUS: Data Deficient.

NATURAL HISTORY: Known from just six specimens, so little information available. One was under a boulder in very arid environs, a second was climbing the smooth surface of a cliff near an accumulation of brackish water and a third was in a bush growing within the confines of a shallow brackish well. Diurnal, terrestrial and fast-moving. Activity apparently highest in January–April and October–November, when irregular rainfall occurs and temperatures milder. Probably eats diurnal lizards, possibly other vertebrates. Presumably lays eggs.

MESSANA'S RACER

Platyceps messanai
(Schätti & Lanza, 1989)

IDENTIFICATION: A slim, harmless, pale brownish-grey snake with tiny blackish spots and a yellow neck, confined to Nogal Valley of northern Somalia. Head slightly distinct from neck. Eye large; pupil round with a reddish ring around it. Body cylindrical, tail long, 29 per cent of total length. Scales smooth, large, in 17 mid-body rows, ventrals 187 and subcaudals 100. Maximum known size 54cm, hatchling size unknown. Dorsal part of head and anterior half of body light brownish-grey, gradually changing into light cinnamon-brown on posterior parts and tail; lateral parts of head and neck yellow. A series of about a dozen small blackish irregular spots along body; more developed laterally, the last quarter of the body and tail pattern-less. Underside cream, with irregular spots in middle two quarters. Taxonomic Notes: Belongs to the *P. brevis* complex and is related to *P. brevis*, *P. smithi* and *P. somalicus*.

HABITAT AND DISTRIBUTION: A rare endemic described from a single adult male from Wadi Run, Central Nogal Valley, Puntland, north-east Somalia, at 450m. In dry stony environment in low hilly near-desertic country. Probably more widespread, but under-collected. Found in subdesert scrub; in a broad, usually dry wadi with large, deep, permanent pools along its bed, flanked by dense riparian vegetation.

CONSERVATION STATUS: Data Deficient.

NATURAL HISTORY: Unknown; probably similar to other *Platyceps*. Found in the morning crawling on a stony slope with scattered vegetation. Sympatric with *P. brevis*.

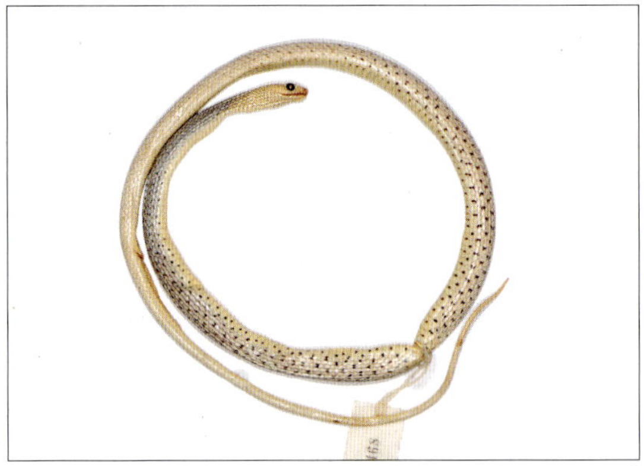

Platyceps messanai, preserved type, Nogal Valley.

SPOTTED RACER

Platyceps rogersi
(Anderson, 1893)

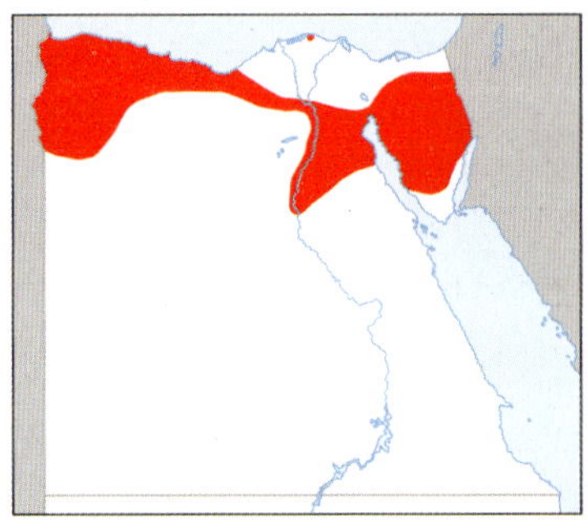

IDENTIFICATION: A moderately sized, slender, pale grey or brown snake with a pattern of large rounded spots and three vertical whitish bands on head, in northern Egypt. Eye large with a round orange-rimmed pupil. Body long and thin, cylindrical; tail relatively long, 24–36 per cent of total length. Scales smooth, matte, moderately sized, in 19 mid-body rows, ventrals 184–215 (higher counts females), subcaudals 89–117. Maximum size about 103cm, average 50–70cm, size of hatchlings unknown. Colour pale with light brown crossbars, becoming uniform towards tail. Three pale (often yellow) blotches on side of head and neck. Underside white, with dark spotting. Juveniles more highly contrasting. Taxonomic Notes: Genetic analyses indicate this is the sister species to *P. saharicus*.

HABITAT AND DISTRIBUTION: Deserts, semi-desert and arid steppes, from sea level to 1,500m in our area. Prefers open hilly country with sparse vegetation and stony ground, such as gravel plains, stony plateaux and rocky wadis. In northern Egypt, north of 28°N. Occurs throughout western Mediterranean coastal plain, Wadi El Natrun and Siwa Oasis. In the Eastern Desert found north of Wadi Araba. In Sinai widespread in the north, except dune fields. In the south less common, known mostly from St. Katherine and Wadi Watir areas. Elsewhere, northern Cyrenaica in Libya, and east to Syria, Iraq and northern Saudi Arabia. In Egypt fairly common and widespread. Common in the pet trade.

CONSERVATION STATUS: Least Concern.

NATURAL HISTORY: Mostly diurnal, in very hot months may become crepuscular or aestivate. In Siwa Oasis several shed skins found under a single palm stump, perhaps indicating communal hibernation. Terrestrial, fast-moving, rather shy. Hides in rodent burrows, in crevices, under ground cover or in bushes. If handled, it readily bites. Oviparous; a clutch of five eggs, roughly 1×4cm, reported. Diet predominantly small mammals and lizards, perhaps also frogs; juveniles favour small geckos and lacertids, occasionally invertebrates.

Platyceps rogersi, Sinai.

SAHARAN CLIFF RACER

Platyceps saharicus
(Schätti & McCarthy, 2001)

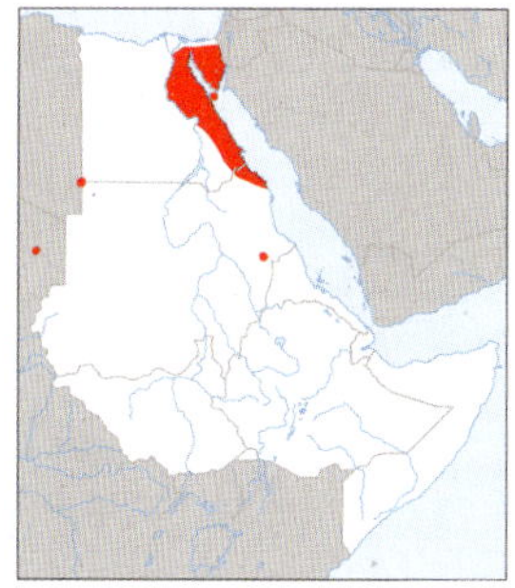

IDENTIFICATION: A large, slender desert snake, light brown with dark crossbars. Head narrow, eye large. Body long and very slim, cylindrical; tail long, very thin, 25–29 per cent of total length. Scales smooth, in 19 mid-body rows, ventrals 213–264 (slightly higher counts in females), subcaudals 134–149. Maximum size about 160cm, average size 80–100cm (males are larger), hatchlings about 35cm. Grey or brown above, with darker vertebral and lateral bars that don't coincide and fade posteriorly. Iridescent pinkish-white with dark spots below, juvenile markings more contrasting. Taxonomic Notes: Originally called *P. rhodorachis*. Recent work indicates this is a species complex; African animals now *P. saharicus*, which might be a junior synonym of *P. tessellatus*.

HABITAT AND DISTRIBUTION: Rocky and stony areas in desert, semi-desert and arid mountains; often in wadis, on cliffs, ruins, abandoned gardens, on edges of farms and in oases close to water, from sea level to 1,700m. Known from mountainous regions of Sinai and the Eastern Desert, and Gebel Uweinat in the Western Desert in Egypt. In Sudan in northernmost Nubian region (e.g. west of Marsa Oseif) and near Derudeb and Hamashkoraib in north-east. Elsewhere, west to south-central Algeria (where usually in rocky mountains, not desert), east to north-west Saudi Arabia. Small numbers in pet trade in Egypt.

CONSERVATION STATUS: Least Concern.

NATURAL HISTORY: Diurnal or occasionally crepuscular, in hottest months predominantly active early morning and shortly before sunset. Terrestrial, but climbs well on rocks. A fast and rather shy snake, when disturbed shelters between rocks, in crevices or bushes. Often found close to pools with water. Eats largely lizards (reportedly fond of fan-footed geckos), rodents, sometimes frogs. One specimen had a Schneider's Skink, *Eumeces schneiderii*, in its stomach. Prey often swallowed alive. It lays eggs.

Platyceps saharicus, Israel/Palestine.

SCORTECCI'S RACER

Platyceps scorteccii
(Lanza, 1963)

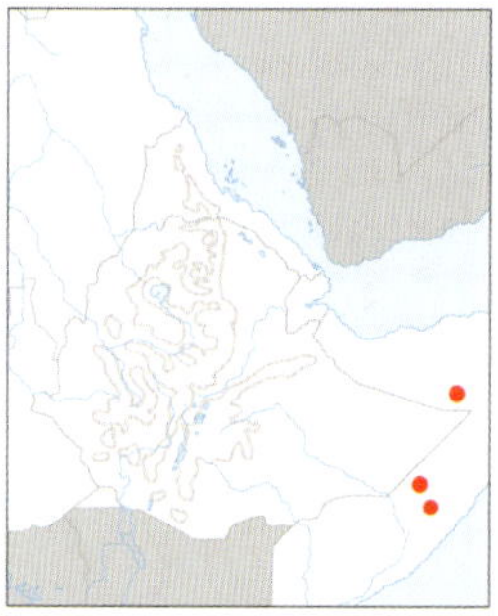

IDENTIFICATION: A moderately sized, long-headed, greyish snake with a dark head and dorsal scales in 27–29 mid-body rows, restricted to central Somalia. Head long, eye large and looks uniformly dark, with slightly vertically elliptical pupil (vertical in a freshly dead specimen). Body cylindrical, fairly slender tail long, 23–28 per cent of total length. Scales smooth, glossy and very small, in 27–29 mid-body rows, ventrals 192–198 (only females known), subcaudals 107–111. Maximum size 70cm. Coloration distinctive. Head black (although snout may be paler and black may be confined to a nape crossbar), with a dark spot below eye. Body brown (in juveniles) or grey, becoming paler posteriorly. A juvenile had irregular white spots forming part-crossbars on neck. Underside iridescent yellowish-brown. Taxonomic Notes: Not genetically analysed, the absence of subocular scales and high mid-body counts suggest it may not be a *Platyceps*.

Platyceps scorteccii, juvenile, Las Anod.

Platyceps scorteccii, Las Anod.

HABITAT AND DISTRIBUTION: Low-altitude flat xerophilous open woodland, subdesert shrub and grassland with sparse vegetation and scattered trees, in gypseous areas with reddish sandy or loamy soil, between 150 and 900m. A rare Somali endemic; long known only from the three types, at Dusa Marreb and El Bur, Galguduud region, central Somalia. Recently discovered in the Sool near Las Anod, a range extension 350km further north.

CONSERVATION STATUS: Data Deficient but probably widespread; said to be fairly common at Las Anod.

NATURAL HISTORY: Poorly known. Diurnal, terrestrial, fast-moving, shelter in holes, under ground cover and in vegetation. A Las Anod specimen was under a rock by day, two others under fencing material in a building, several were inside a wall of stacked stone, hunting Boulenger's Geckos *Hemidactylus macropholis*. Erroneously believed to be dangerous, probably due to confusion with cobras, at Las Anod. The local name is *mas* (although this a generic name elsewhere for snakes). Probably feeds largely on lacertid lizards like *Latastia* and *Pseuderemias*. Presumably lays eggs.

SINAI BANDED RACER

Platyceps sinai
(Schmidt & Marx, 1956)

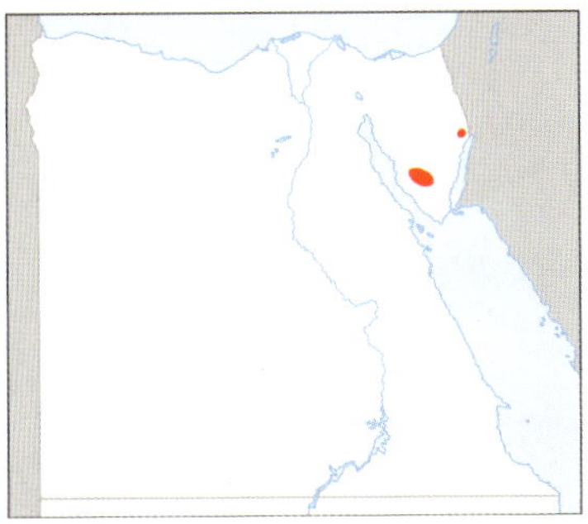

IDENTIFICATION: An elegant, vividly banded, small racer, in our area only in southern Sinai, Egypt. Eye moderately sized, with a round pupil and dark orange-brown iris. Body slender, cylindrical, tail relatively long, 25–26 per cent of total length. Scales smooth; moderately sized, in 17 mid-body rows, ventrals 184–194 (higher counts in females), subcaudals 87–100. Maximum size about 58cm, average 35–45cm, size of hatchlings unknown. Easily recognised by its conspicuous dorsal coloration of 69–81 black crossbars, which narrow posteriorly, on a beige background; and an interrupted orange vertebral stripe (not always present). White below; ventral side of tail yellowish. Taxonomic Notes: Closely related to similar small banded Arabian species (*P. elegantissimus*, *P. thomasi*, *P. variabilis*). *P. elegantissimus* appears very similar (but has fewer crossbands) and occurs in southern Wadi Araba (Eilat), close to Egyptian border.

HABITAT AND DISTRIBUTION: Mostly arid stony wadis with sand and gravel, or margins of rocky slopes, in our area at 700–1,500m. Usually found in oases and near abandoned farmland among mountains of extreme desert. Originally described from Wadi Feiran in south of Sinai Peninsula, Egypt. Additional specimens were collected in the nearby St. Catherine area and about 20km north-west of Taba, near the Israeli/Palestine border. Probably more widespread but unrecorded in the intervening country (all suitable habitats). Elsewhere, endemic to a small geographic area from southern Israel/Palestine to south-western Jordan. In Egypt rare and local. Not in pet trade.

CONSERVATION STATUS: Data Deficient.

NATURAL HISTORY: Poorly known, a secretive, shy, rare snake. Terrestrial, but climbs walls and cliffs. Mostly nocturnal or crepuscular, with an activity peak in evening and early hours of the night. Oviparous. Diet largely lizards and arthropods. A specimen from Wadi Feiran contained a half-digested adult Rüppell's Snake-eyed Skink, *Ablepharus rueppellii*. Local name at St. Katherine is *Abu Merira*, due to the colour similarity between this snake and traditional Arab head dress. In Jordan sympatric with similar *P. elegantissimus*, in Egypt with *P. saharicus*.

Platyceps sinai, Jordan.

SMITH'S RACER

Platyceps smithi
(Boulenger, 1895)

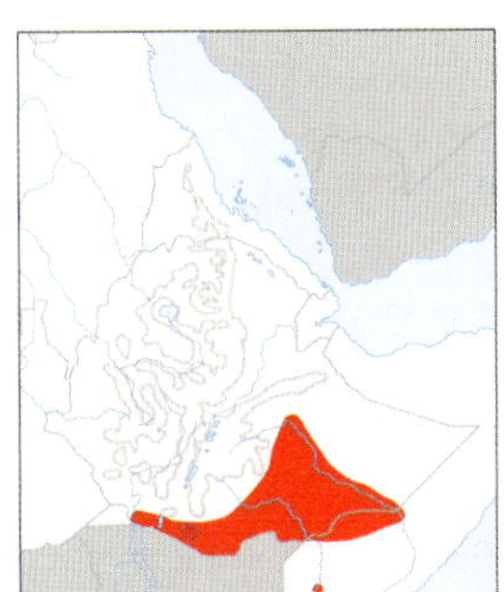

IDENTIFICATION: A small, slim, harmless, barred brown or rufous snake with dorsal scales in 21 mid-body rows. Eye large and looks uniformly dark, but has a brown iris and round pupil. Body cylindrical, tail long, 25–30 per cent of total length. Scales smooth, small, in 21 mid-body rows, ventrals 173–200 (173–189 in males, 184–200 in females), subcaudals 83–110. Maximum size about 70cm, average 40–60cm, hatchling 22–24cm. Brown, rufous or yellow-brown, usually with three dark crossbars on head and nape. Body usually strongly barred in juveniles, in adults the bars fade to rufous spots or half-bars, in some pattern completely disappears, although head bars may remain. Sometimes a dark vertebral stripe is present, or extensive fine white speckling. Belly yellow, cream or white. Taxonomic Notes: Often regarded as a subspecies of *P. brevis*, however we have treated it as a separate species.

Platyceps smithi, Kenya.

HABITAT AND DISTRIBUTION: A typical inhabitant of the Somali arid zone, in dry savanna and semi-desert, at low altitudes of 100–1,400m. Widespread in southern Ethiopia from the plains around Lakes Turkana and Chew Bahr, east to the central Ogaden, north to the surroundings of Fik, thence south along the Shebelle River to northern parts (Rahanweyn country) of inter-riverine southern Somalia. Absent from Benadir coastal region. Elsewhere, widespread in dry Kenya, entering northern Tanzania just south of Mt Kilimanjaro.

CONSERVATION STATUS: Not Assessed.

NATURAL HISTORY: Little-known. Diurnal and terrestrial, but will climb into low bushes. It is secretive and fast-moving. Lays eggs, one Kenya specimen laid three elongate eggs in March. Diet probably diurnal lizards, possibly snakes and mammals, but poorly known. A captive specimen at Nairobi Snake Park fed readily on small lacertid lizards.

OGADEN RACER

Platyceps somalicus
(Boulenger, 1896)

IDENTIFICATION: A small, slim snake with large dorsal scales, pale coloration and wide-spaced black crossbars; known from a single Ethiopian specimen with a unique mid-body scale count of 15 rows. Head slightly distinct from neck, snout obtuse. Eye large; pupil round. Body cylindrical, tail tip missing in only known specimen. Scales smooth, large, in 15 mid-body rows, ventrals 172, subcaudals more than 54 (tail truncated). Type 44cm long. Dorsum of preserved specimen pale grey-brown, with widely spaced (by four scales) black crossbars along length of back, a series of dark spots on sides, a blackish transverse bar between eyes and a large dark blotch on each temple. Underside white. Taxonomic Notes: Tentatively placed in the *P. brevis* complex, probably related to *P. brevis*, *P. messanai* and *P. smithi*.

HABITAT AND DISTRIBUTION: A presumably rare Ethiopian endemic, known from a single female, collected at an imprecise location in the Audo Mountains, Bale Province, somewhere between 750 and 1,450m. The habitat is dry savanna with abundant herbaceous vegetation and scattered low scrub, on fine-grained red sand, almost clay, in low hilly country with rich seasonal streams. Probably more widespread in surrounding country (all suitable habitat), but under-collected.

CONSERVATION STATUS: Data Deficient.

NATURAL HISTORY: Nothing known.

Platyceps somalicus, preserved type.

CLIFF RACER

Platyceps rhodorachis
(Jan, 1863)

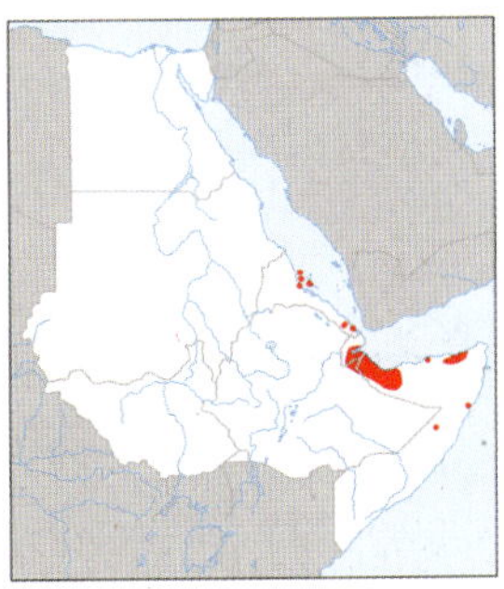

IDENTIFICATION: A moderately to large-sized, very slender snake with anterior dorsal pattern of numerous crossbars; side of head and neck yellow. Eye large with round yellow-rimmed pupil, iris brown. Body long and very slim, cylindrical; tail long, very thin, 24–30 per cent of total length. Scales smooth, moderately sized, in 19 mid-body rows, ventrals 208–228 (slightly higher counts in females), subcaudals 112–132. Maximum size about 130cm, average 80–100cm (males are larger), hatchlings about 35cm. Dorsum pale grey to pale brown or olive, typically with dark transverse bars either side of midline that either alternate or fuse mid-dorsally, usually present at least on anterior portion of body; large adults tend to become paler posteriorly and crossbars may completely disappear towards tail. Occasional individuals are black or dark grey. Top of head, snout, temporal region darker; preocular, postoculars and upper labials paler, dark spot below eye. Underside iridescent white, cream or pinkish. A juvenile from Bosaso in Puntland was uniformly pale brown above, belly yellow-cream, head as described above. Our notes refer to *P. r. subniger*, the African subspecies of this widespread snake.

HABITAT AND DISTRIBUTION: Desert and semi-desert, typically in hot rocky or sandy places with little vegetation, between sea level and 1,500m. Occurs from Eritrea to central Somalia, typically on coasts (although inland records exist). Eritrean records include several Dahlak Islands (Dahlak Kebir, Difnein, Harat), Massawa on adjacent mainland (and nearby Sheikh Said Island), Gaarre and Assab. Widespread in Djibouti, including Maskali and Musha islets, and the Guban strip along the coast to Puntland's Bosaso plains, Candala, south to Eyl. Some sporadic inland Somali records (south of Toyo Plains and Hargeisa), with southernmost record at Galkayo in central Somalia. The melanistic type was from 'Ogaden' (presumably in Ethiopia, but without details). At higher elevations, Boettiger's Racer occurs at inland localities in Somaliland and Cal Miskaat Mountains in Puntland. Almost certainly more widespread in Horn of Africa, but under-collected. Elsewhere; the nominate subspecies occurs north-west across the Middle East, to Tibet. Widespread and locally common.

Platyceps rhodorachis, Dahlak Archipelago.

CONSERVATION STATUS: Least Concern.

NATURAL HISTORY: Terrestrial, diurnal and very fast-moving. Specimens from north-west Somaliland were collected throughout the dry season and the early *Gu* rains (April–June), found only in stony places in arid localities with little vegetation. In Djibouti found in rocky areas, and in or near mangroves. In

Bosaso town juveniles are often encountered in buildings. An individual at Candala was collected in a stony canyon; a freshly dead specimen at Rugay, Sanaag region, Somaliland, was also near a stony canyon with sparse shrubs, *Commiphora* and *Vachellia* vegetation. Diet mostly lizards, rodents and frogs. A specimen from Randa, in Djibouti, had eaten a sand frog *Tomopterna kachowskii*. Lays eggs; a gravid female had three eggs in the oviduct.

TAYLOR'S RACER

Platyceps taylori
(Parker, 1949)

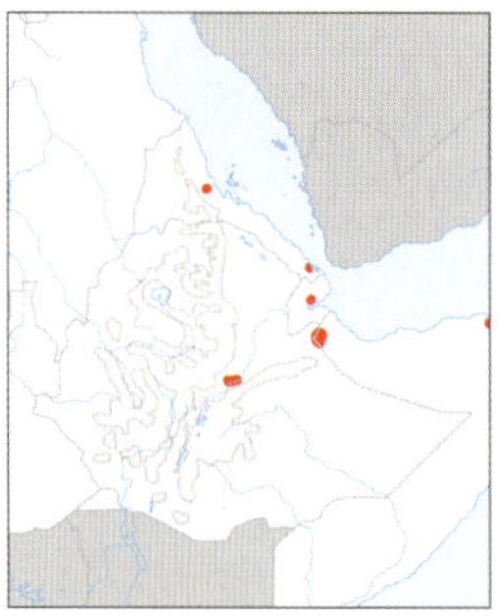

IDENTIFICATION: A moderately sized, stocky snake with large and partly confluent paravertebral spots, of eastern Ethiopia and Eritrea to Puntland. Eye moderately large with a round yellow-rimmed pupil. Body stocky, cylindrical; tail long, very thin, 24–29 per cent of total length. Scales smooth, small, in 21 or 23 midbody rows, ventrals 179–198 (179–186 in males, 191–199 in females), subcaudals 81–96. Maximum size about 70cm, average 40–55cm, hatchlings about 20–23cm. Coloration distinctive; dorsum brownish-grey with two or more wavy-edged crossbars on head that have dark margins and paler centres, followed by 5–8 longitudinal rows of prominent dark spots on body, median series largest and may form large roundish blotches, an undulating dorsal pattern or zigzag bands; dorsal markings separated by narrow, paler areas. Underside white with scattered black dots on outer edges of ventral scales. Taxonomic Notes: Closely related to *P. florulentus*.

HABITAT AND DISTRIBUTION: Dry savanna, lava fields and other desertic environments, between sea level and about 1,200m. In Eritrea, recently observed at Gahtielay on central coastal plains, and Assab. A lone Djibouti record (Bankouale) but probably more widespread there. A population occurs in Awash National Park in Ethiopia, where it seems to be relatively common. The type was described from Borama District west of Harirad, in desertic low country of north-west Somaliland. There is an isolated record from Galgalo Oasis, in Puntland. Probably occurs in the intervening country between these six isolated populations (all suitable habitat) but no records.

Platyceps taylori, Fantalle.

Platyceps taylori, juvenile, Fantalle.

CONSERVATION STATUS: Least Concern.

NATURAL HISTORY: Terrestrial, fast-moving and diurnal. Has been found in thorn scrub on open grassy plains, but also in rocky areas almost devoid of vegetation. Diet predominantly lizards, but occasionally rodents, small snakes and invertebrates. Specimens from Somaliland contained the remains of *Hemidactylus* geckos and the lacertid *Latastia boscai*. Captive animals favoured lizards but would take small mice. Two animals from Awash in Ethiopia were found hidden by day under lava rocks. Taylor's Racer has the same range as North-east African Carpet or

Saw-Scaled Viper *Echis pyramidum*, in the Horn of Africa they occur syntopically in many places. The coloration of this species resembles that of vipers and we hypothesise that *P. taylori* mimics these dangerous snakes.

DWARF SNAKES *EIRENIS*

A primarily Middle Eastern genus of small, slender, aglyphous, harmless snakes with a small head slightly distinct from neck, small eyes with round pupils, nostril in an undivided nasal, smooth dorsals in 15–19 rows at mid-body, and cloacal scale undivided. Terrestrial and secretive. Some species diurnal, others nocturnal. Dwarf snakes are oviparous and reported to feed mainly on insects and other invertebrates, such as scorpions and spiders, sometimes supplemented with small lizards. At present 23 species are known, two occur in Africa, both in our area. Taxonomy not yet totally clarified; several new species described recently. Nearest relatives are Whip snakes *Dolichophis* from southern Europe and the Middle East.

KEY TO NORTH-EAST AFRICAN MEMBERS OF THE GENUS *EIRENIS*

1a: Dorsum with spots, subcaudals in 69–78 rows, not in Egypt	*Eirenis africanus*, African Dwarf Snake
1b: Dorsum with transverse dark bands, subcaudals in 48–67 rows, in Egypt	*Eirenis coronella*, Crowned Dwarf Snake

AFRICAN DWARF SNAKE

Eirenis africanus
(Boulenger, 1914)

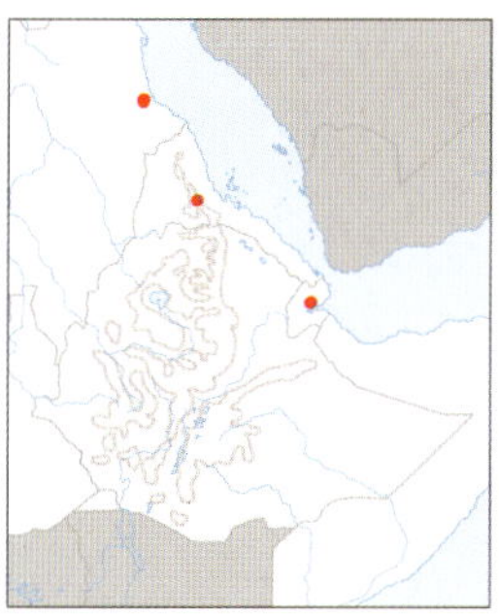

IDENTIFICATION: A fairly small, rare, generally yellowish-brown snake with a dark crossbar on neck and spotted anterior body, confined to hills of the Red Sea coast from Sudan to Djibouti. Eye moderate, pupil round. Body slender, cylindrical, tail about 25 per cent of total length. Scales smooth, in 15 mid-body rows, ventrals 146–162 (146 in male, 160–162 in female) with a divided cloacal scale, subcaudals paired, 69–78 rows. Maximum size 45cm, hatchling size unknown. Dorsum grey-brown or yellowish-grey with a broad, dark brown crossbar on neck and four longitudinal rows of dark spots from neck to mid-body, decreasing in size towards tail. Top of head darker than sides. Underside either white or yellowish-grey, slightly paler than back.

Eirenis africanus, Djibouti.

HABITAT AND DISTRIBUTION: Semi-evergreen bushland with scattered trees and bushes, at 1,000–1,200m. A rare snake, known from just three examples. A recent Djibouti specimen was found at Dittilou, in a mountainous landscape on a small plateau, in a steep, rocky dry area covered with shrubs and *Vachellia* bushes. The type was from Erkowit in north-east Sudan. Also known from Asmara in Eritrea, at 2,300m, but the specimen might have been obtained at lower elevation. Probably widespread in the intervening country in suitable habitats, but no records.

CONSERVATION STATUS: Data Deficient.

NATURAL HISTORY: Very poorly known. Secretive, diurnal, terrestrial and fairly fast-moving. The Djibouti snake was active by day in April and fled under a large boulder. Presumably lays eggs and feeds mainly on invertebrates and small lizards.

CROWNED DWARF SNAKE

Eirenis coronella
(Schlegel, 1837)

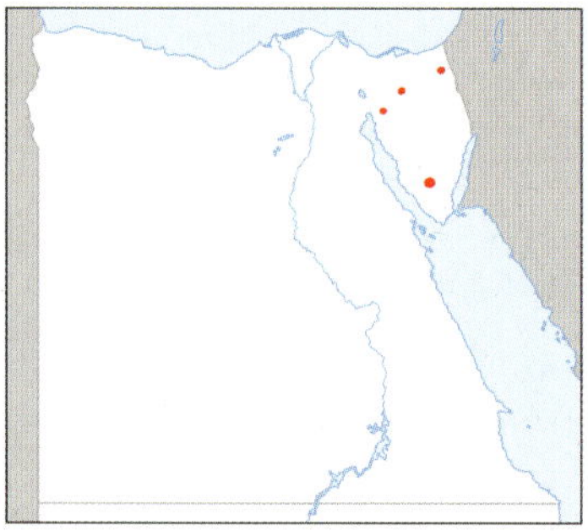

IDENTIFICATION: A small, sandy or grey snake with a dark crossbar on neck and banded body, confined to Sinai in Egypt. Eye moderate in size, pupil round. Body relatively stout, cylindrical, tail about 25–35 per cent of total length. Scales smooth, in 15 mid-body rows, ventrals 141–162 (141–145 in males, 156–162 in females) with a divided cloacal scale, subcaudals paired, 48–67 rows (61–67 in males, 48–63 in females). Maximum size about 33cm, hatchling size unknown. Dorsum sandy or grey, with 32–65 light brownish crossbars; a broad dark-brownish collar on neck. Underside white, sometimes with small dark spots on anterior body. Taxonomic Notes: Three subspecies, the endemic *E. c. ibrahimi* in our area.

HABITAT AND DISTRIBUTION: Semi-arid mountain and hilly country. Found on lightly vegetated stony hillsides, up to at least 1,600m. Very tolerant of high temperatures and aridity. Well known from St. Catherine area in Sinai. Recorded also at several localities in north Sinai, notably Gebel Dalfa, Metla Pass and probably also Bir Gilgafa. Probably more widespread than known. Local and fairly rare, currently not in pet trade.

Eirenis coronella, Israel/Palestine.

CONSERVATION STATUS: Least Concern, although in Egypt it is Near Threatened.

NATURAL HISTORY: Poorly known. A diurnal species with a secretive lifestyle, active early in morning and before sunset. Feeds on insects and other arthropods, including scorpions and centipedes, one had eaten a large spider. Usually found under stones or in rock cracks. Oviparous, a female of the nominate subspecies held five elongate eggs. Mating occurs in spring, during March–April, but possibly also in autumn, followed by ovulation in spring.

LARGE WHIP SNAKES *DOLICHOPHIS*

A genus of four large, fast-moving, harmless diurnal snakes of south-west Europe and the Middle East long placed in the genera *Coluber* and *Hierophis*; recent molecular data indicate *Dolichophis* may not be monophyletic. One species just reaches our area, in Sinai.

LARGE WHIP SNAKE

Dolichophis jugularis
(Linnaeus, 1758)

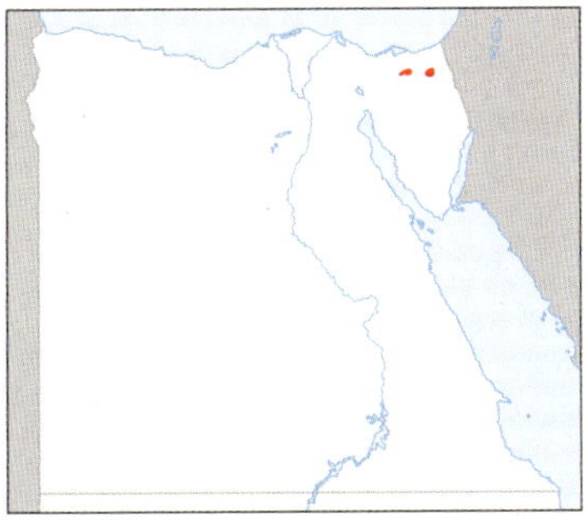

IDENTIFICATION: A huge, alert, fast-moving black or brown snake of Sinai. Head relatively small, just broader than neck, eye medium-sized with a round pupil. Body cylindrical and muscular, tail fairly long, 20–22 per cent of total length, slightly longer in males than females. Scales large and smooth, in 19 rows at midbody, ventrals 189–220, subcaudals paired, 98–128. Maximum size about 2.5m, but this is unusual, average 80cm to 1.4m, hatchlings 18–20cm. Adults shiny black above, although lips are barred black and white; underside grey, throat spotted. Juveniles very different; grey or grey-brown above, with rows of fine black spots that often coalesce into irregular crossbars, interspersed by fine white spotting; venter creamy-grey, with pale grey rounded spots. Change from juvenile to adult colour occurs at c.1m long.

HABITAT AND DISTRIBUTION: Usually in open Mediterranean countryside, with light vegetation, often common around farmland. Often on rocky hillsides with some vegetation in Sinai below 200m, but higher elsewhere. In our area, known from Jebel El Halal and slightly further west in northern Sinai. Elsewhere, east to Iran, north to Turkey.

Dolichophis jugularis, Turkey.

Dolichophis jugularis, Jordan.

CONSERVATION STATUS: Least Concern.

NATURAL HISTORY: Diurnal but may be crepuscular (or even nocturnal) during hot periods. Largely terrestrial, but will climb bushes and rocks. Shelters in holes, under ground cover and in rock fissures. Fast-moving and nervous, may bask for long periods but moves away swiftly if disturbed; if cornered may elevate fore body, and if handled will bite. May aestivate in winter, for various periods dependant on elevation and temperature. Females lay up to 18 eggs, usually in summer; incubation around two months. Eat any small vertebrate they can catch, including their own species.

MEDITERRANEAN WHIP SNAKES *HEMORRHOIS*

A genus of four fairly large, slim or sturdy, fast-moving, harmless diurnal snakes of lands around Mediterranean and Near East. Two species just reach our area. Originally placed in genus *Coluber*, molecular analysis indicates they are closely related to snakes in the genera *Platyceps* and *Spalerosophis*.

KEY TO NORTH-EAST AFRICAN MEMBERS OF THE GENUS *HEMORRHOIS*

1a: Dorsum with fewer than 70 large rounded spots, dorsal scales weakly keeled, in Egypt east from Cairo, <95 subcaudals
Hemorrhois nummifer, Coin-marked Snake

1b: Dorsum with >80 dark bars, dorsal scales smooth, in Egypt west of Nile, >83 subcaudals
Hemorrhois algirus, Algerian Whip Snake

COIN-MARKED SNAKE

Hemorrhois nummifer
(Reuss, 1834)

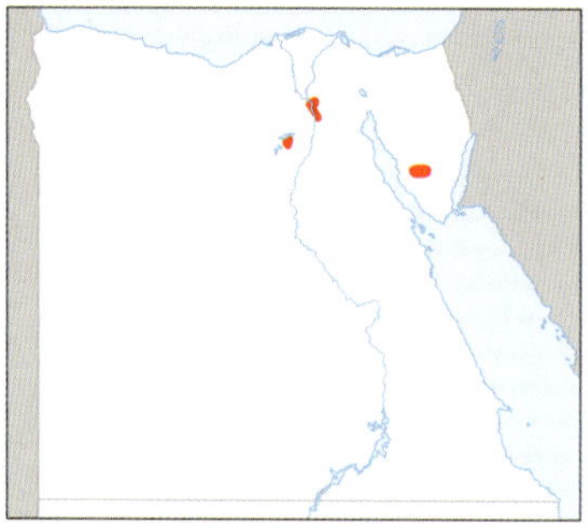

IDENTIFICATION: A large, fast-moving snake of north-east Egypt, with a clear pattern of circular blotches. Eye medium-sized with a round pupil and golden iris. Body cylindrical and muscular, tail fairly long, 18–24 per cent of total length, slightly longer in males than females. Scales large and weakly keeled or smooth, in 23–25 rows at mid-body, ventrals 193–230, subcaudals paired, 79–94. Maximum size about 1.5m, but this is unusual, average 80cm to 1.1m, hatchlings 20–21cm. Colour pattern distinctive: grey to grey-brown, with a series of large, near-circular rich brown blotches on spine, these may be edged by yellow scales; they coalesce into a narrow line on tail. A series of smaller brown spots extend along flanks. There is usually a darker chevron on head. Underside cream with dark spots on margin of every third or fourth ventral. Young animals have higher-contrast markings.

HABITAT AND DISTRIBUTION: Occupies a range of open habitats, provided there is some vegetation, favours rocky terrain in hilly and mountainous near-desert country, with sparse vegetation, usually in or near rather mesic areas such as oases, cultivated fields and gardens, along agricultural drainage canals; also ruins, old houses, stone walls and similar places at low elevation in Egypt (but higher elsewhere). In our area, known from Cairo and Helwan, the Faiyum, and southern Sinai near Wadi Feiran. Unconfirmed records from Nile Delta and nearby Alexandria (not shown on map). Probably more widespread. Elsewhere, on shores of eastern Mediterranean, to Turkey, north-east to Kazakhstan. Rare and local, sometimes in the pet trade.

Hemorrhois nummifer, Israel/Palestine.

CONSERVATION STATUS: Least Concern, but Vulnerable in Egypt.

NATURAL HISTORY: Active by day, occasionally crepuscular or nocturnal. Terrestrial, but climbs well, sometimes ascends several floors up in buildings. Shelters in holes, under ground cover and in rock fissures; snakes in Cairo found in houses and stables. Fairly fast-moving. Often basks and remains still if approached, relying on camouflage. If cornered hisses loudly, flattens the head into a subtriangular shape, elevates the fore body, and if handled will bite. Although non-venomous, local swelling following a bite has been recorded. Sometimes active in winter. Mates in early summer, females lay 3–12 eggs, roughly 2×4cm, usually in June–July; incubation *c*.2 months. They eat any small vertebrate they can catch, including venomous snakes; a Cairo animal had a young pigeon in its stomach. Often confused with local vipers.

ALGERIAN WHIP SNAKE

Hemorrhois algirus
(Jan, 1863)

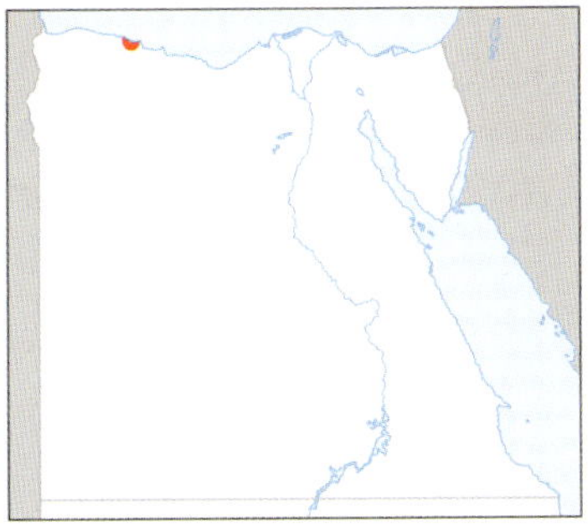

IDENTIFICATION: A medium-sized, slender snake, head usually dark, body paler with well-spaced dark transverse bars, on Mediterranean coast of Egypt west of Nile Delta. Eye large with a round pupil. Body long, slender and cylindrical; tail long, c.18–30 per cent of total length. Body scales smooth, shiny and small, in 23 or 25 mid-body rows, ventrals 205–240, subcaudals 83–117, cloacal scale may be divided but usually not. Maximum size about 140cm, average 70–100cm (females larger). Quite variable in colour; dorsum pale grey, brown or even yellow with about 90 well-spaced dark transverse bars (occasionally rounded spots), additionally 1–2 alternating smaller round spots on flanks. Top of head, or at least its posterior part, usually distinctly darker, greyish or blackish; sides of head around eyes and upper labials paler, a darker oblique spot below eye usually present. Iris golden or greyish-yellow. Belly cream-coloured, with darker margins. Taxonomic Notes: Two subspecies recognised – *H. a. algirus* occurs in our area and as far west as north-west Algeria.

HABITAT AND DISTRIBUTION: Poorly known. In our area known from only a single specimen taken among rubble in a ruin atop a sparsely vegetated limestone ridge, south-west of Marsa Matruh in Mediterranean coastal subdesert in north-west Egypt, at 110m. Might extend east to Alexandria. Elsewhere, west along coast to Mauritania; also in Hoggar Range in southern Algeria. Rare and uncommon.

CONSERVATION STATUS: Least Concern, but Vulnerable in Egypt.

NATURAL HISTORY: Dry stony or rocky areas, frequently with sparse vegetation; also oases, wadis, farmlands and wetlands; will enter houses. Diurnal, but often active after dusk (crepuscular) during hot months. In morning often basks in sun, then retreats to shelter, becoming active again in afternoon or after sunset. Feeds mostly on lizards (e.g., *Tarentola* or outside our area *Agama*), rodents (jirds or gerbils) and birds; a very fast hunter. Mostly ground-dwelling, but sometimes climbs on bushes or rocks when searching for prey. Breeding starts early in spring; females then lay several soft-shelled elongate eggs.

Hemorrhois algirus, Tunisia.

Hemorrhois algirus, Tunisia.

FALSE SMOOTH SNAKES *MACROPROTODON*

A mainly North African genus of small to medium-sized, rather slender, crepuscular or nocturnal, slow-moving, rear-fanged (but essentially harmless) egg-laying snakes. They have a distinct coloration consisting of usually darker head (all black or with markings) and paler body. They have venom glands and grooved rear fangs but are small and not considered dangerous. They inhabit dry country and feeds mostly on lizards and small snakes. Four species known, one in our area. Taxonomy not yet fully resolved.

EASTERN FALSE SMOOTH SNAKE

Macroprotodon cucullatus
(Geoffroy Saint-Hilaire, 1827)

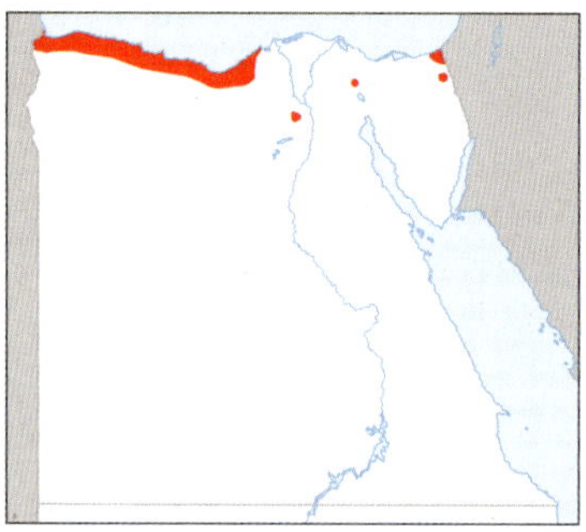

IDENTIFICATION: A small, fairly slender, short-tailed snake, head usually darker than body, with faint dorsal markings, in northern Egypt. Eye rather small and angled slightly upwards, the pupil not perfectly round but vertically elliptical to a small extent. Body moderately long, fairly slender, cylindrical; tail short, about 15–18 per cent of total length. Body scales smooth, shiny and moderately large, in 19 mid-body rows, ventrals 153–182 (153–163 in males, 167–182 in females), subcaudals divided 43–55, cloacal scale divided. Maximum size about 47cm, average size 35–40cm, hatchlings 12–17cm. From a distance appears reddish-brown or beige, the back a mixture of low contrast darker markings and light speckling. Head may be jet-black above or have numerous slightly darker, paired irregular blotches, which may form a V-shape. An oblique streak runs from eye towards angle of mouth. Sometimes there is a fragmented collar. Belly yellowish, usually uniformly so.

HABITAT AND DISTRIBUTION: Dry, rocky and sandy country, in our area coastal near- or semi-desert, from sea level to about 310m. Also in altered habitats and near human settlements. Known from Mediterranean coastal zone of northern Egypt and Sinai, from Libyan border east to just west of Nile Delta; a somewhat dubious record from Abu Rawash west of Cairo. Known also from Ismailia and north-east Sinai, notably El Quseima and Sheikh Zowayid. Elsewhere; east to Israel/Palestine, west across Libya to central Tunisia.

CONSERVATION STATUS: Least Concern, although endangered, uncommon and declining in Egypt.

NATURAL HISTORY: Essentially crepuscular and nocturnal. Ground-dwelling and slow-moving. A relatively rare and secretive snake. Unaggressive, but sometimes bites, or balls up with head hidden among body coils. Shelters by day under ground cover or in holes. After sunset hunts lizards (especially geckos) and small snakes; cannibalism recorded. Immobilises prey with its venom, larger prey is held within body coils. Alexandria specimens were found under rocks near the coast, one under an old kerosene tin on a grassy spot in an otherwise sandy place, one in a barley field on sandy ground. Females lay 2–6 eggs, probably only once every two years. Suspected to hibernate briefly in winter.

Macroprotodon cucullatus, Tunisia.

WATER AND GRASS SNAKES *NATRIX*

A Western Palearctic genus of medium-sized, semi-aquatic, largely diurnal harmless snakes, with round pupils and strongly keeled dorsal scales. They are non-aggressive but in defence they exude a very offensive-smelling substance from their cloaca. Excellent swimmers and divers, they are found at or near waterbodies. They prey mainly on frogs, but fishes and other prey are also taken. Originally, the genus contained many species from North America, Europe, North Africa and Asia, but these are now reallocated to other genera; five species remain, one in our area. This genus and African Marsh Snakes *Natriciteres* (described later with other Afrotropical genera) are sometimes placed in a subfamily, Natricinae.

DICE SNAKE

Natrix tessellata
(Laurenti, 1768)

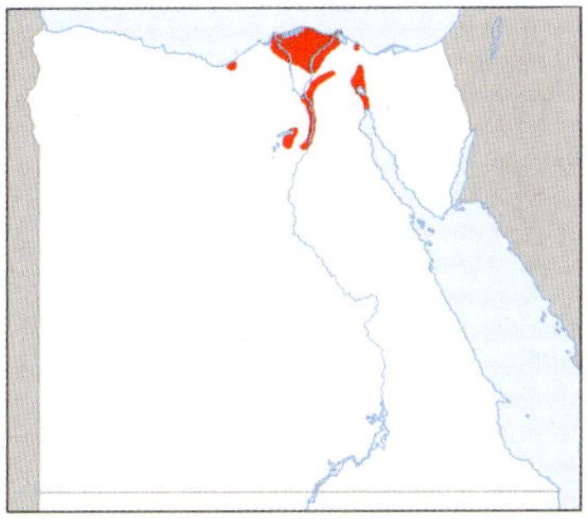

IDENTIFICATION: A fairly large drab-looking snake, brown or olive-green, associated with water sources in northern Egypt. Eye medium-sized with a round pupil. Body cylindrical and muscular, tail relatively short, 17–18 per cent of total length, slightly longer in males than females. Scales large and strongly keeled, in 19 rows at mid-body, ventrals 160–197, subcaudals paired, 48–86. Maximum size about 90cm (occasionally, old females reach 1.3m), average 60–80cm, hatchlings 15–24cm. Adults dark olive-green, brown to grey-brown above, irregularly but densely spotted with black and white, forming vague lines, or sometimes monocoloured, or have subdued darker blotches. Lips may have fine dark barring. Underside yellow or dirty grey-white anteriorly, becoming chequered black and yellow towards tail. Taxonomic Notes: Recently published genetic data suggest there may be cryptic species concealed in this taxon.

HABITAT AND DISTRIBUTION: Usually near freshwater sources, e.g., river and canal banks, swampy areas, dams and lakes or garden ponds. In our area at low altitude, usually below 100m. Occurs along most of lower Nile Valley, north throughout Nile Delta. Also the Faiyum, Suez Canal zone, Port Said and at El Hamman just west of Alexandria. Elsewhere, east to western China, west to Germany, a huge range. In Egypt common and widespread, commonly taken for pet trade. Individual populations may be threatened by habitat loss, killed on roads and by farmers, who believe it is venomous.

CONSERVATION STATUS: Least Concern.

NATURAL HISTORY: May be nocturnal (particularly in hot months), crepuscular or diurnal. Terrestrial, but essentially semi-aquatic, hunting for prey in reedbeds, at water's edge or in water, swimming and often diving for long periods. May bask on vegetation overhanging water. Shy and inoffensive, moves fairly quickly to escape. Mating starts in spring, usually early April, females lay 5–25 eggs, occasionally more (37 recorded), usually in June; in wet organic material at water's edge. Incubation time at 30°C was c.5 weeks. They eat fish, frogs and tadpoles in our area, also newts elsewhere; rodents and birds occasionally taken. They may feign death if threatened, and show auto-bleeding behaviour. Often occur at high density in suitable areas. Longevity at least 14 years in wild.

Natrix tessellata, Czech Republic.

DIADEM SNAKES *SPALEROSOPHIS*

A Saharo-Sindian genus of medium to large, fairly slender, aglyphous, harmless snakes with rather short tails. Head elongate and distinct from neck, moderate eyes with round pupils, separated from upper labials by row of subocular scales, prefrontals usually broken into three or more scales, temporals into numerous small scales. They occur in northern Africa and south-west Asia, from Senegal and Morocco to Somalia, northern India and Kazakhstan. Six species are known, two in our area, one endemic. The systematics of this genus are in need of revision; molecular data show they are closely related to racers of the genus *Platyceps*.

KEY TO NORTH-EAST AFRICAN MEMBERS OF THE GENUS *SPALEROSOPHIS*

1a: Mid-body scales in 36–39 rows, 6–10 temporals in first row, frontal and parietal scales fragmented, known only from Galgalo Oasis in Puntland — *Spalerosophis josephscorteccii*, Scortecci's Diadem Snake

1b: Mid-body scales in 25–33 rows, 3–6 temporals in first row, frontal and parietal scales not fragmented, range Egypt to central Sudan — *Spalerosophis diadema*, Diadem Snake

DIADEM SNAKE

Spalerosophis diadema
(Schlegel, 1837)

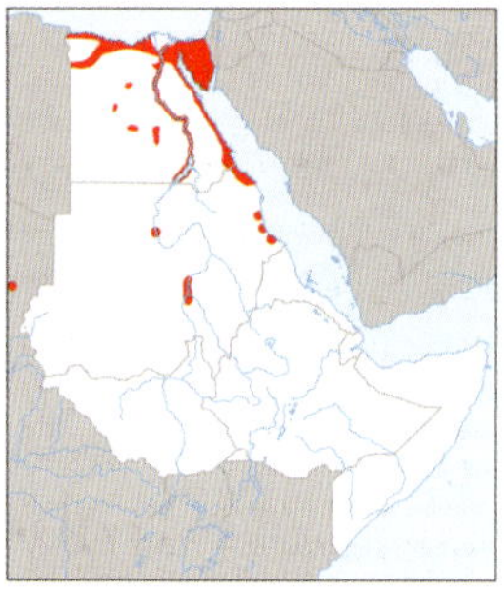

IDENTIFICATION: A fairly large snake with numerous fragmented temporals and a dorsal pattern of large, rhomboid blotches, in arid country in Sudan and Egypt. Eye moderate with a round pupil and golden iris, separated from upper labials by series of suboculars. Prefrontals broken up into several scales (usually 4–6); and they have small numerous temporals, 3–6 in first row. Body slender, cylindrical, tail about 16–22 per cent of total length. Dorsal scales obtusely keeled, in 25–33 mid-body rows, ventrals 210–248, with an undivided cloacal scale, subcaudals paired, 62–93 rows. Maximum size about 1.6m, average around 1m, hatchling 36–42cm. Colour variable. Dorsum reddish-buff, greyish or yellowish with a pattern of large, rhomboid or oval, brown blotches with pale edges, alternating with dark sub-rectangular spots on flanks that extend onto tail. Dorsal surface of head has fine light and dark marbling and a dark bar between eyes, a dark lateral stripe extends from eyes to mouth angle. Two short stripes on side of neck. Underside whitish or yellowish. Taxonomic Notes: Three subspecies recognised, in our area the widespread *S. d. cliffordii* occurs.

HABITAT AND DISTRIBUTION: Inhabits a wide range of habitats in desert, semi-desert and marginal agricultural lands, as well as near human settlements and gardens, from sea level to 1,500m. Diadem Snakes are distributed in north-east Africa in Egypt and Sudan, but not in true desert. Widespread on western Mediterranean coastal plain, known from all major oases of the Western Desert and the southern Nile Delta. Extends along the Nile to the Sudan border, south along the Red Sea coast, and throughout Sinai. Sudanese records sporadic: at Durrur, Suakin, Tokar and Red Sea coastal plain, and Dongola and Khartoum in Nile Valley. Probably more widespread in semi-arid Sudan, but under-collected; reported from Kassala, Blue Nile, Kordofan and probably Darfur States (not shown on map). Elsewhere, this subspecies occurs east to Iran, west right across the Sahara to Mauritania. In Egypt it is popular in pet trade, with large numbers being collected.

CONSERVATION STATUS: Least Concern.

NATURAL HISTORY: Terrestrial and fast-moving; diurnal in cold months, crepuscular and nocturnal in hot ones. Feeds predominantly on small rodents and birds, juveniles take mostly lizards; all killed by powerful constriction. Quick to bite if handled. If cornered, convolutes its neck into an S-shape, flattens and triangulates its head and hisses, producing a noise similar to the undulatory friction of carpet vipers (*Echis*) and strikes, hence often confused with dangerous snakes. Mate in spring; females lay 3–15 elongate eggs, roughly 2.5×6cm, about 40–50 days after copulation; they hatch 60–80 days later.

Spalerosophis diadema, Tunisia.

SCORTECCI'S DIADEM SNAKE

Spalerosophis josephscorteccii
Lanza, 1964

IDENTIFICATION: A large snake, dimorphic in pattern with fragmented frontal and parietal scales, confined to Galgalo Oasis in Puntland, known from just four specimens. Head sub-triangular, resembling that of a rock python, and sharply distinct from neck. Snout wide; eye moderate with round pupil and grey iris, separated from upper labials by series of suboculars; prefrontals broken into several scales (usually 12–16); frontals fragmented as well as parietals, which are always separated by additional scales; small numerous temporals, 7–10 in first row. Body slender, cylindrical, tail 19–22 per cent of total length. Dorsal scales smooth, in 36–39 mid-body rows, ventrals 228–237, with a divided cloacal scale, subcaudals paired, 98–105 rows. Maximum known size 1.2m, hatchling size unknown. Two colour phases, spotted and uniform, known. Three uniform specimens are very light pearl-grey in colour with a faint pattern; the smallest has a dark patch under eye, a dark lateral stripe from eyes to mouth angle, weak pigmentation on neck just behind head and faint longitudinal vertebral stripe. The single spotted phase specimen, a juvenile, has an elegant series of 43 brown transversally elongate blotches along spine, alternating with smaller subcircular flank blotches, the head pattern as in uniform phase, but more pronounced. Underside yellowish.

HABITAT AND DISTRIBUTION: Endemic, rare and essentially unknown. All four specimens are from Galgalo (or Galgala), in foothills at eastern end of Cal Madow Mountains in extreme north-west Puntland, northern Somalia. Galgalo is a mountain oasis with springs, settlements and gardens, at an altitude of c.875m, and rich in life due to the rain-fed hillside run-off, unlike its surroundings. Several other unusual species occur there, including the poorly known and highly endangered palm *Livistona carinensis*, attaining 30m. Surrounding landscape is dry rocky hilly country with lightly vegetated, succulent rich, stony steep hillsides.

Spalerosophis josephscorteccii, preserved, Galgalo Oasis.

Spalerosophis josephscorteccii, preserved.

CONSERVATION STATUS: Data Deficient.

NATURAL HISTORY: Poorly known. The types were collected by Giuseppe Scortecci and local helpers at the edge of the oasis in a thicket of *Salvadora persica*. An uncommon and secretive species, a prolonged later expedition by Benedetto Lanza yielded just two further juveniles, from local collectors, without field data. It is probably nocturnal and feeds on small vertebrates. The specific name honours Giuseppe Scortecci (1898–1973), Italian herpetologist, explorer of Somalia, and an authority on African and desert herpetofauna, who was curator at the Museum of Natural History in Milan, rising to become Professor of Zoology and later Dean of the University of Genoa, Italy.

DWARF, KUKRI OR BLACK-HEADED SNAKES *RHYNCHOCALAMUS*

A genus of five species of small, slim, harmless burrowing snakes, confined to southwest Asia and Turkey, one species just reaches our area, in the Sinai. There is no agreement on a common name for the genus.

PALESTINE BLACK-HEADED DWARF SNAKE

Rhynchocalamus melanocephalus
(Jan, 1862)

IDENTIFICATION: Unmistakable in our area, a small, slim, brown, black-headed snake of Sinai. Head almost indistinct from the neck, flat and bullet-shaped, eye small, the pupil round. Body long, slim and cylindrical, tail about 26–27 per cent of total length. Scales smooth, in 15 mid-body rows, ventrals 179–231, with a divided cloacal scale, subcaudals paired, 53–70. Maximum size about 55cm, average 30–45cm, hatchling size unknown. Dorsum uniformly coloured, usually shades of brown, rufous or chestnut-brown, but occasionally grey-brown or yellow. Top of head and nape black, lips, chin, throat are white or pinkish white.

HABITAT AND DISTRIBUTION: In Sinai known from gravel plains at altitudes above 1,000m, but elsewhere in a range of dry, often mountainous, rocky habitats, down to sea level. Sinai animals, the only records from our area, were from the Wadi El Sheikh and north-east of St. Katherine. Elsewhere, north to Syria and Jordan, records further north are now a separate species.

CONSERVATION STATUS: Least Concern.

NATURAL HISTORY: Poorly known. Terrestrial and nocturnal, sometimes active by night, and specimens found under ground cover by day. Diurnal and crepuscular activity has been reported. Feeds on insects and other arthropods, including centipedes, and small lizards. A captive animal refused to eat until given ant larvae, but then took grasshoppers and woodlice. They lay eggs, two snakes contained three and five oviductal eggs in July, male spermiogenesis noted in May–August.

Rhynchocalamus melanocephalus, Jordan.

AWL-HEADED OR LEAF-NOSED SNAKES *LYTORHYNCHUS*

A genus of seven species of small, biscuit-coloured, blotched or barred snakes, largely sand-dwelling, of south-west Asia and North Africa. They have an enlarged protruding rostral scales for pushing through sand. One species occurs in our area.

AWL-HEADED SNAKE

Lytorhynchus diadema
(Duméril, Bibron & Duméril, 1854)

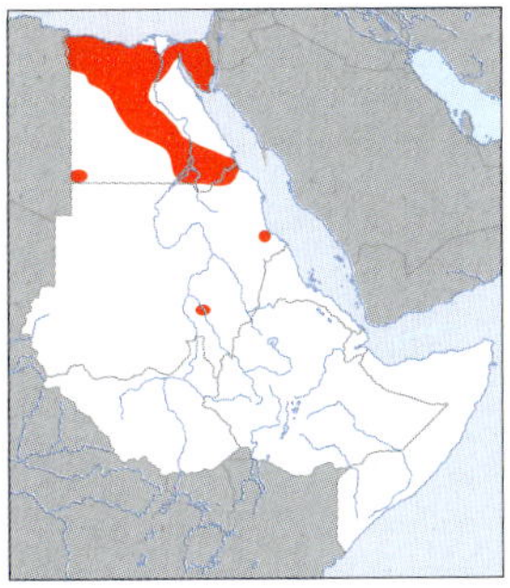

IDENTIFICATION: A small, quite stocky desert snake with a distinctly pointed rostral (nasal) scale. Eye medium-sized, the pupil round. Body stocky and cylindrical, tail about 16–19 per cent of total length. Scales smooth, in 19 mid-body rows, ventrals 152–195, with a divided cloacal scale, subcaudals paired, 33–49. Maximum known size about 46cm, average 25–40cm, hatchling size unknown. Dorsum usually shades of brown, sometimes rufous, with a series of darker brown or black subcircular blotches along the back, these may be white-edged; flanks speckled with smaller blotches to a greater or lesser extent. Spot on nape often extends forwards to top of head, broadens sideways at level of eyes and encloses a light spot, and a dark bar often extends down behind eye. Underside white or cream.

HABITAT AND DISTRIBUTION: Semi-desert, sandy desert, dunes and sand patches (even small ones) in gravel plains or sheet rock areas, largely at low elevations, but up to 1,500m in Sinai. Widespread in Egypt, save agricultural areas of Nile Delta, but records very sporadic. Few Sudanese records (Erkowit, Wad Madani); but records from Jebel Awaynat, on Sudan/Egypt/Libya border and along Egypt's southern border indicate it is probably widespread across desert areas of Sudan. Elsewhere, east to western Iran and Oman, west to Mauritania and northern Senegal.

CONSERVATION STATUS: Least Concern.

NATURAL HISTORY: A terrestrial nocturnal snake that burrows in soft sand, and by day shelters in sand, in holes, under ground cover. Emerges at night and crawls about on sand; near Ismailia observed near midnight in dunes. Feeds largely on lizards, particularly lacertids which it finds sleeping in their refuges, but may take arthropods. Mating noted April–May, 3–5 elongate eggs (roughly 1×2.5cm) are laid in June–July. If threatened, they usually coil up and hide the head under the coils, but occasional individuals known to rear up like a cobra and slightly flatten the neck. Has a distinctive track, as it crawls across the sand it probes left and right with its nose, leaving small impressions in the sand either side of the slide mark.

Lytorhynchus diadema, Tunisia.

Lytorhynchus diadema, Ismailia.

SMOOTH AND SEMI-ORNATE SNAKES *MEIZODON*

A sub-Saharan African genus of small, harmless colubrid snakes, mostly brown or grey, with thin bodies, smooth scales and eyes with round pupils. Diurnal and secretive; they lay eggs; and feed mostly on lizards and frogs. Five species known, three of which occur in north-east Africa. The Kenyan endemic Tana Delta Smooth Snake *Meizodon krameri*, at present known only from Tana Delta, might well extend into southern Somalia. Smooth snakes have not been included in any recent molecular analysis, so their relationships with other colubrids is uncertain, although it is suggested they are closely related to the racers *Platyceps*.

KEY TO NORTH-EAST AFRICAN MEMBERS OF THE GENUS *MEIZODON*

1a: Scales in 19 rows at mid-body	*Meizodon regularis*, Eastern Crowned Smooth Snake
1b: Scales in 21 rows at mid-body	2
2a: Ventrals 159–206, usually barred, grey to brown, sometimes black	*Meizodon semiornatus*, Semi-ornate Snake
2b: Ventrals 201–235, no barring, red or pink with scattered dark dots on body and black head	*Meizodon plumbiceps*, Black-headed Smooth Snake

BLACK-HEADED SMOOTH SNAKE

Meizodon plumbiceps
(Boettger, 1893)

IDENTIFICATION: A slim, small, pink or red, black-spotted smooth snake with a dark head. Eye fairly large, pupil round or slightly elliptic. Body slender, cylindrical, tail about 25 per cent of total length. Scales smooth, in 21 mid-body rows, ventrals 201–235 (201–211 in males, 215–235 in females), subcaudals paired, 76–90 rows. Maximum size about 55cm, average 30–45cm, hatchling size unknown. Coloration is distinct and an excellent aid to identification. Red or pink, with four or fewer, irregular rows of widely scattered black dots, varying in size, on back. A broad black collar extends onto top of the head but not snout. Chin and throat speckled and blotched black, in shape of an arrowhead. Belly scales cream or pale brown. Originally regarded as a subspecies of Semi-ornate Smooth Snake *M. semiornatus*, it was elevated to a full species in 1982, following a study of Juba River specimens.

HABITAT AND DISTRIBUTION: Coastal thicket, riverine woodland, dry savanna and semi-desert, from near sea level to about 1,500m. The type was from an unspecified locality in the Ogaden. Apparently confined to south-east Ethiopia, southern Somalia, and a single northern Kenyan example from Malka Murri. The only precise Ethiopian locality is Sheikh Hussein. In southern Somalia it occurs widely between the Juba and Webi Shebelle Rivers, known from Mareri near Gelib, Labadaah/Bulo Yaag (south of Gelib), Afmadu, Bardera, Baidoa, Dinsor, Afgoi and Giohar. An improbable record from south-west Arabia.

Meizodon plumbiceps, preserved, Baidoa.

Meizodon plumbiceps, preserved, Baidoa.

CONSERVATION STATUS: Least Concern.

NATURAL HISTORY: Little known, but probably similar to other snakes in the genus, i.e., terrestrial, diurnal and secretive, lays eggs, eats lizards. Some Juba River specimens were found in termitaria. In southern Somalia, syntopic with *M. semiornatus*.

EASTERN CROWNED SMOOTH SNAKE

Meizodon regularis
Fischer, 1856

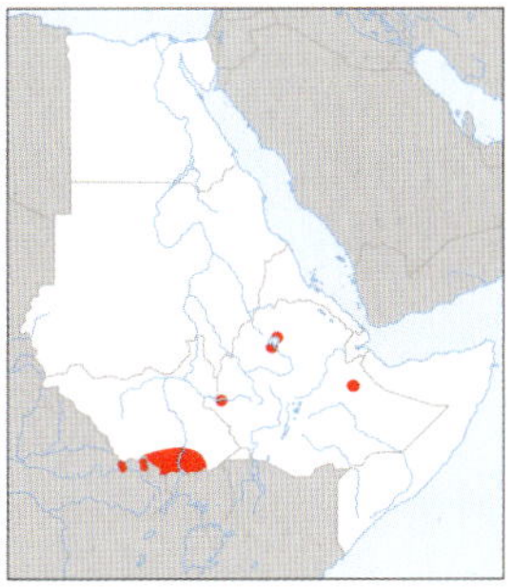

IDENTIFICATION: A slim, harmless smooth snake, unlike other *Meizodon* has only 19 scales at mid-body. Eye medium-sized, pupil round. Body slender, cylindrical, tail about 20 per cent of total length. Scales smooth, in 19 rows at mid-body, ventrals 175–205 (175–192 in males, 190–205 in females), cloacal scale divided, subcaudals 60–79 in East Africa. Maximum size about 73cm, average 35–50cm, hatchling size unknown. Juveniles grey, the head and neck black with four or five distinct, white or yellow crossbars. Eye finely bordered with black, lips and chin white, throat white or yellow, the ventrals grey. Adults in two colour phases; uniform grey, or dark grey with regular light, broad dirty grey crossbars.

HABITAT AND DISTRIBUTION: A typical species of Sudanian savanna and Congolian forest-savanna mosaic ecoregions, in moist savanna, forest edges, relatively dry savanna-woodland and lower levels of montane grassland, at medium altitudes (mostly 500–2,000m). Known from eastern (20km east of Harar), northern (Zegie and 40km south of Gondar) and western (Gambela) Ethiopia. In South Sudan in Eastern and Central Equatoria, and also probably West Equatoria as it has been collected in the Garamba region of north-east DR Congo. Notable localities in South Sudan: Juba, Katire, Lotti Forest, Nimule and Torit. Elsewhere, south to western Kenya and Uganda, west to Guinea-Bissau.

CONSERVATION STATUS: Least Concern.

NATURAL HISTORY: Poorly known. Fast-moving, diurnal and terrestrial, often in waterside vegetation. Hides under ground cover or in holes at night. Lays eggs, a Nigerian specimen had four eggs in oviducts in June, roughly 3.5×0.6cm. Diet unknown, probably mostly lizards, possibly small amphibians; the closely related (possibly conspecific) West African *M. coronatus* eats only lizards.

Meizodon regularis, West Africa.

Meizodon regularis, Uganda.

SEMI-ORNATE SMOOTH SNAKE

Meizodon semiornatus
(Peters, 1854)

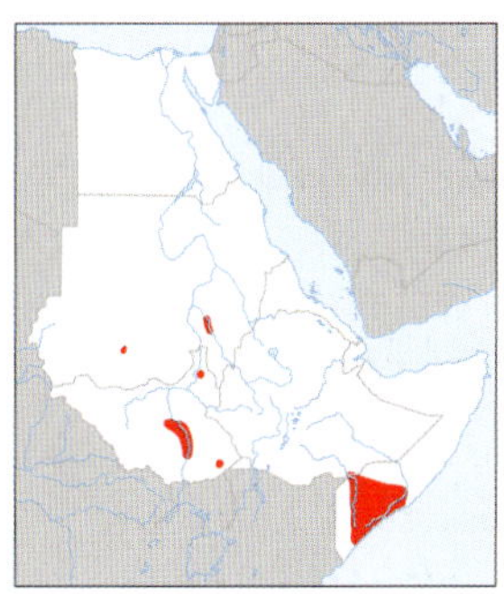

Meizodon semiornatus, Kenya.

Meizodon semiornatus, Heglig.

IDENTIFICATION: A small, thin, harmless colubrid snake, juveniles are conspicuously barred on front of half of body; hence the name. Eye fairly large, pupil round. Tail about 25 per cent of total length. Scales smooth, 21 mid-body rows, ventrals 159–206 (higher counts in females), subcaudals paired, 74–88, cloacal scale divided. Maximum size about 80cm, average 40–70cm, hatchlings 18–20cm. Colour very variable, adults usually grey, brown or olive, occasionally black or rufous-black; sometimes a yellow wash on underside of head and neck, the posterior flanks may be speckled grey, with a dark brown dorsal tail stripe. Juveniles paler, light grey, yellow-brown or blue-grey, with a series of fine or broad dark crossbars or half-bars on front ⅔ of body, which fade wholly or partially, from top down as they grow. Ventrals usually grey, grey-white or yellow, the chin and throat white. Top of head black in some juveniles. Taxonomic Notes: Two subspecies recognised, a southern nominate and *M. s. tchadensis* in the north. They differ as follows: *M. s. semiornatus*; ventrals in males 169–199, in females 176–207, two anterior temporals, long tail (usually 30–38 per cent of snout–vent length); *M. s. tchadensis*: ventrals in males 200–206, in females 211–225, one anterior temporal, short tail (23–27 per cent of snout–vent length).

HABITAT AND DISTRIBUTION: Widespread in a range of habitats; coastal thicket and woodland, semi-desert, moist and dry savanna and high grassland, from sea level to 2,200m, but most common at lower elevations. They prefer thick vegetation or wooded areas, and are also found near water (grass, reeds). In our area, *M. s. semiornatus* is found at several localities in southern Ethiopia and Somalia, from Dolo south along the Juba River to Benadir coast, then north along the coast to Giohar. No records in northern Kenya or southern Ethiopia except Dolo, but found at Kapoeta in East Equatoria in South Sudan. Elsewhere, south from Kenya to Botswana and South Africa. *M. s. tchadensis* is sporadically recorded in South Sudan (White Nile basin around Mongalla, Malek, Bor, Paloich) and south-east and south-central Sudan (Wad Madani, Sennar, Heglig). Few records west of the Nile in Sudan, but almost certainly there as widespread in Chad. Elsewhere, to northern Cameroon (but not known from Central African Republic).

CONSERVATION STATUS: Least Concern.

NATURAL HISTORY: Terrestrial, fast-moving, usually diurnal but occasionally active at dusk. When inactive, hides under ground cover, in holes, etc., small aggregations of 2–3 individuals found under loose tree bark in Botswana. Semi-ornate Smooth Snakes are inoffensive and rarely try to bite when handled. They lay 2–3 eggs, roughly 3.5 × 1cm. Male combat has been observed, males neck-wrestling. Diet lizards, especially skinks and geckos; small frogs and rodents also recorded. In southern Somalia one had the hind legs of a *Ptychadena* in the stomach. Said to be the most common snake species in parts of its range, e.g., at Belet Amin on Juba River in southern Somalia.

SOMALI SNAKE *AELUROGLENA*

A genus with a single, small, harmless species known from just five specimens, all from northern Somalia.

SOMALI SNAKE

Aeluroglena cucullata
Boulenger, 1898

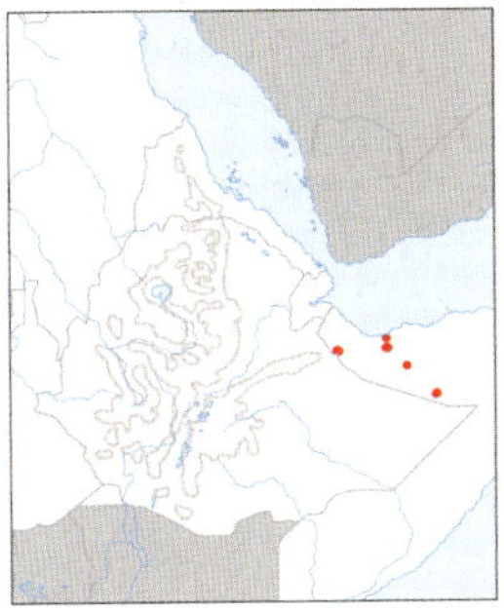

IDENTIFICATION: A small, slim, buff snake with a black head, from northern Somalia. Eye medium-sized, with a vertical pupil. Body slender, cylindrical, tail about 20–30 per cent of total length. Scales smooth, in 21 mid-body rows, ventrals 188–223, with a divided cloacal scale, subcaudals paired, 68–75 rows. Maximum size 38cm, hatchling size unknown. Dorsum pale grey-brown, with top of head dark brown or black, a broad transverse bar of similar colour across neck and small irregularly scattered dark spots on anterior part of body. Anterior upper labials white, snout pale brown or yellowish. Underside cream, except throat, which is mottled grey-brown and may have a dark median stripe. Taxonomic Notes: Resembles *Meizodon*, but differs in shape of pupil and number of maxillary teeth.

HABITAT AND DISTRIBUTION: A north Somali endemic of arid rocky coastal plains, dry medium to high savanna and *Vachellia–Commiphora* dry lowland savanna, at elevations of 150–1,400m. A rare species, the five known specimens were collected in diverse and widely separated environments, so probably widespread but under-collected. Originally described from Goolis Mountains in Somaliland at high elevation, then reported from a similar environment in Borama in extreme west Somaliland. Other specimens were collected at lower elevations near Berbera and the Haud (near Bohodle, just north of Ethiopia border).

CONSERVATION STATUS: Data Deficient.

NATURAL HISTORY: Little known. Terrestrial, diurnal, fast-moving and secretive. Diet unknown. The Haud animal was discovered by day on sandy ground, in an area clothed with trees and bushes; local people showed great fear of this snake, insisting that it was a juvenile of a very deadly black species (probably a burrowing asp).

Aeluroglena cucullata, preserved, Bohodle to Berbera.

Aeluroglena cucullata, preserved, Bohodle to Berbera.

FALSE WHITE-LIPPED SNAKE *APROSDOKETOPHIS*

A curious genus, erected in 2010 to accommodate a single small harmless snake, collected 90 years earlier in Somalia, lacking field data. It somewhat resembles a White-lipped Snake *Crotaphopeltis hotamboeia*, but it has no fangs, has circular pupils and is stockier.

FALSE WHITE-LIPPED SNAKE

Aprosdoketophis andreonei
Wallach, Lanza & Nistri, 2010

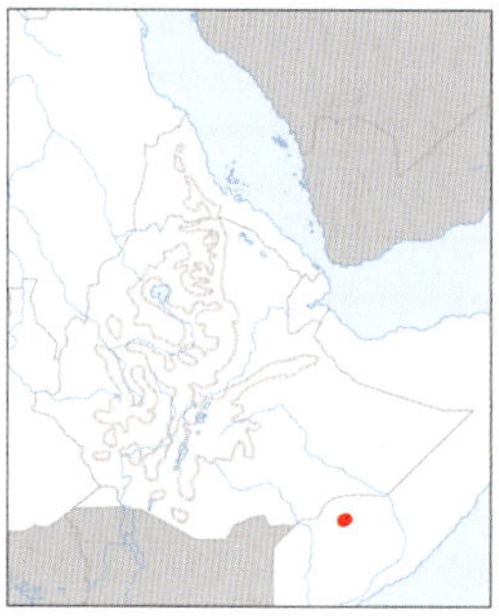

IDENTIFICATION: A small snake, the head is oval, weakly distinct from neck, pupils round, nasal scale divided with an interior flap or valve, eight upper labials. Body short, robust and cylindrical; tail short, 17 per cent of total length. Dorsal scales smooth, lacking apical pits, in 19 rows at mid-body. Ventrals 146, with a divided cloacal scale, subcaudals paired, 46 rows. It is 39cm long. In preservative the top of head and sides are brown, the lips paler. Back purplish-brown with a faint reticulate pattern; individual dorsal scales speckled reddish-brown with a light posterior margin. Lowest dorsal scale row creamy-white, as is underside, although brown stippling is present on outer edges of each ventral, and every third to fourth ventral has light brown peppering extending to midline. Subcaudals immaculate cream.

HABITAT AND DISTRIBUTION: Habitat at Oddur, Bakool Region, in Rahanuin country of southern Somalia, where the solitary specimen is said to have been collected, is semi-arid scrub or broken xerophytic woodland in the Somali arid zone, typically with red sandy soil, at altitude of 500m. There are no natural barriers save rivers in that area, so it may be more widespread.

CONSERVATION STATUS: Data Deficient.

NATURAL HISTORY: Unknown. Stocky, so may be terrestrial, the round pupils suggest it is diurnal. Reduced dentition and lack of fangs might suggest a specialised diet (perhaps lizards), the short tail might indicate secretive or subfossorial habits. Loss of apical scale pits may be a reaction to arenicolous habits; as might valve flaps in the nostril, preventing wind-blown sand from entering. The specimen is an adult female, with 19 ova and another 26 developing follicles, suggesting fecundity.

Aprosdoketophis andreonei, preserved type.

Aprosdoketophis andreonei, preserved type.

CAT AND TIGER SNAKES *TELESCOPUS*

A genus of small to medium-sized, rear-fanged, slender, large-eyed snakes, with broad heads, wide snouts, thin necks and large protruding eyes with a vertical pupil. Although venomous, they are not dangerous to humans. All are nocturnal, some are partially arboreal, they feed on a range of prey, favouring lizards. They lay eggs. Cat snakes are found in dry habitats such as savanna or semi-desert. At least 14 species are known, in Africa, the Middle East, Asia and southern Europe. Ten occur in Africa, six are found in north-east Africa, one is endemic. Cat snakes originated in Africa, and are most closely related to the similar-looking genera *Crotaphopeltis* and *Dipsadoboa*. They are bad-tempered snakes, some species if handled will inflict slow deliberate bites on the handler. The common name 'Cat snakes' refers to their vertically elliptical pupils, similar to those of a cat. When threatened, they form nested C-shaped coils, and appear to mimic highly dangerous Carpet vipers *Echis*.

KEY TO THE NORTH-EAST AFRICAN SPECIES OF *TELESCOPUS*

1a: Loreal enters eye; in Sinai, Egypt	*Telescopus hoogstraali*, Hoogstraal's Cat Snake
1b: Loreal separated from eye by a preocular; not in Sinai, Egypt (except *T. dhara*)	2
2a: Scales in 17–18 rows at mid-body, one anterior temporal	*Telescopus pulcher*, Beautiful Cat Snake
2b: Scales in 19–25 rows at mid-body, two anterior temporals	3
3a: Subcaudals 49–52; dorsum pale brown or orangey pink, with 28–37 sub-rectangular dark brown dorsal blotches	*Telescopus gezirae*, Blue Nile Tiger Snake
3b: Subcaudals 55–97; lacks such coloration	4
4a: Two upper labial scales in contact with eye; head almost always dark, body paler	*Telescopus tripolitanus*, North African Cat Snake
4b: Three upper labial scales in contact with eye; head not darker than body	5
5a: Dorsum usually reddish-buff to brick-red; no dark band from snout, through eye to corner of mouth; snout square, somewhat downwardly directed; cloacal scale usually entire; in Sinai, Egypt	*Telescopus dhara*, Arabian Large-eyed Cat Snake
5b: Dorsum usually brownish to black; dark band from snout, through eye to corner of mouth in some populations; snout sub-rectangular or rounded, not downwardly directed; cloacal scale entire (subspecies *somalicus*) or divided (*obtusus*); not in Sinai, Egypt	*Telescopus obtusus*, African Large-eyed Snake

ARABIAN LARGE-EYED CAT SNAKE

Telescopus dhara
(Forskål, 1775)

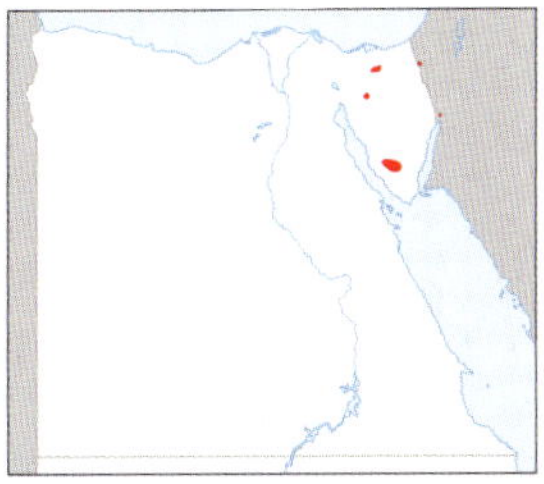

IDENTIFICATION: A medium-sized, slender and reddish snake, restricted to Sinai, Egypt. Pupil vertical, iris orangey-yellow. Neck long and very thin; body slender, triangular or sub-triangular in section; tail short, about 13–16 per cent of total length. Scales smooth, in 21 (rarely 19 or 23) rows at mid-body, ventrals 228–274, cloacal scale usually entire, subcaudals paired, 56–81 rows. Maximum size up to 132cm, average 50–85cm (males much smaller), hatchling 16–20cm. Dorsum pale reddish-buff to brick-red, sometimes sandy-beige, usually with a series of more than 50 faint large darker blotches along mid-dorsal line, separated by contrasting narrow pale transverse lines, or plain (less often); head notably reddish, with a short, dark, indistinct nuchal streak; upper labials have darker posterior edges. Underside uniformly opalescent-white. Taxonomic Notes: Recent molecular analysis confirms that this species and *T. obtusus*, long confused and conflated, are separate species; and further indicates a deep divergence between 'true' *T. dhara* of Oman and Yemen and those from North Arabia and Sinai, so a name change is 'on the cards'.

HABITAT AND DISTRIBUTION: Rocky environments in mountain and hilly regions, usually with vegetation and trees, often in or near oases, wadis, cultivation and human settlements; in our area to 1,500m. Known from south Sinai in Egypt (Wadi Feiran, Wadi El Sheikh, and near St. Catherine airport). Probably also in Gebel Maghara and Gebel El Giddi in the north. Elsewhere, Arabia, north to Jordan. In Egypt uncommon and local, although more common and widespread elsewhere.

CONSERVATION STATUS: Least Concern.

NATURAL HISTORY: Nocturnal or crepuscular, depending upon season, and quite tolerant of low temperatures. A slow-moving and generally terrestrial snake. However, it is an agile climber, and females readily ascend bushes, searching and waiting in ambush for prey. Diet mostly lizards (including chameleons), birds and their eggs, occasionally bats. When threatened, it elevates the head and forms nested C-shaped coils, mimicking dangerous Carpet vipers *Echis*. Females lay up to ten elongate eggs in June–October, some laying two clutches per season.

Telescopus dhara, Oman.

BLUE NILE TIGER SNAKE

Telescopus gezirae
Broadley, 1994

IDENTIFICATION: A medium-sized, slender, banded snake, known from two specimens taken in the Sennar of Sudan. Body slender, triangular or sub-triangular in section; tail short, about 13–16 per cent of total length. Scales smooth, in 21 rows at mid-body, ventrals 195–207, cloacal scale divided, subcaudals paired, 49–52 rows. Size of two known specimens 50 and 72cm. Coloration very distinct; dorsum pale brown or orangey-pink, with 28–37 sub-rectangular dark brown dorsal blotches, each separated by 2–6 scale rows; irregular small dark lateral blotches extend onto ventrals, which are otherwise yellowish-white. Head pattern-less, the same colour as body. Strongly resembles the well-known Tiger Snake *T. semiannulatus* of south-east Africa.

HABITAT AND DISTRIBUTION: A Sudanese endemic; the two known specimens were from Wad Madani and about 30km north of Sennar, on banks of Blue Nile, at about 400m. The region is now under agriculture, but originally was semi-arid savanna. Recent expeditions have failed to find further specimens, so it must be rare.

CONSERVATION STATUS: Data Deficient.

NATURAL HISTORY: Never knowingly seen alive, but presumably similar to other *Telescopus*.

Telescopus gezirae, preserved type.

HOOGSTRAAL'S CAT SNAKE

Telescopus hoogstraali
Schmidt & Marx, 1956

IDENTIFICATION: A medium-sized, dark-headed snake with vertical pupils, the loreal scale enters the eye. Unusually for a *Telescopus*, eye small and less prominent. The neck is thin; body slender, triangular or sub-triangular in section; tail short, about 9–15 per cent of total length. Scales smooth, in 19 rows at mid-body, ventrals 214–254, cloacal scale divided, subcaudals paired, 46–59 rows. Maximum size about 100cm, average 60–80cm, hatchling size unknown. Dorsum pale greyish-brown, the pattern consists of 28–56 weak, somewhat darker, thin bands. Head grey, vermiculated black, the neck is dorsally black; upper and lower labials tending to have pale anterior edges. Underside grey, densely marked with fine black spots, paler laterally, darker medially; several first ventrals have black edges. Juveniles have a grey ground coloration.

HABITAT AND DISTRIBUTION: Arid or semi-arid sparsely vegetated rocky, hilly and mountainous regions, but also on sandy soils. Known from Sinai in Egypt, notably the St. Katherine area and nearby Wadi El Sheikh in the south, and Gebel Maghara in the north, at elevations of 300–2,280m. Probably widespread in hilly central Sinai, but no records. Elsewhere, to western Jordan. In Egypt rare and local.

CONSERVATION STATUS: Endangered.

NATURAL HISTORY: Poorly known, few specimens. A ground-dwelling, slow-moving and strictly nocturnal species. Found under rocks by day. Diet includes small vertebrates such as lizards, snakes, rodents and even birds. Egg-laying, a clutch of 20 was reported in July, which hatched in September. When disturbed, coils up with the neck and fore body in a S-shape, widens the head into a triangular shape, hisses and strikes.

Telescopus hoogstraali, Israel/Palestine.

AFRICAN LARGE-EYED CAT SNAKE

Telescopus obtusus
(Reuss, 1834)

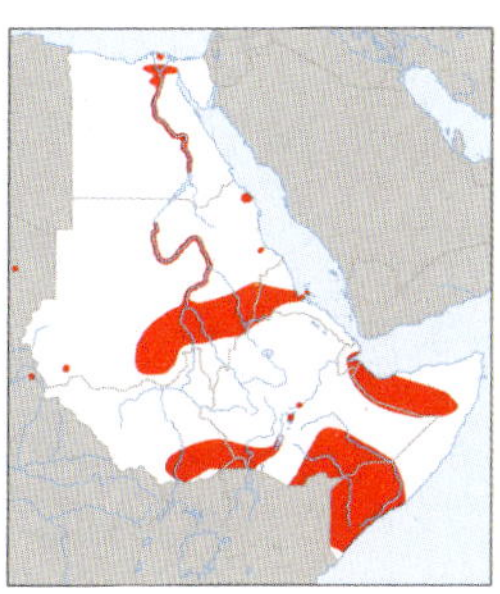

IDENTIFICATION: A medium-sized, slender, broad-headed, brownish snake. Iris yellow. Neck long and very thin; body slender, triangular or sub-triangular in section; tail long, about 15–20 per cent of total length. Scales smooth, in 19–25 (usually 20–23) rows at mid-body, ventrals 200–278, cloacal scale entire or divided, subcaudals paired, 56–97 rows. Maximum known size 187cm, average 50–80cm (males much smaller), hatchlings 17–21cm. Colour very variable: black, slate-grey, shades of brown or golden-yellow, monocoloured, usually plain, or has a series of faint, large darker square blotches along mid-dorsal line, occasionally has narrow, paler crossbars, flanks uniform or in blotched phase with two rows of dark spots. Freshly sloughed specimens often quite iridescent. Lighter below, usually cream, pearly or pink. Taxonomic Notes: As presently defined, two subspecies; *T. o. somalicus* in south-east Ethiopia and southern Somalia, and *T. o. obtusus* everywhere else. However, the species' morphology varies substantially, Ethiopian and Somali animals are genetically very different from Eritrean specimens; and molecular studies indicate that it is probably a species complex. Typical animals corresponding to types from Egypt are known only in Egypt and Sudan, and characterised by head darker on top, the presence of a dark band from snout through eye to corner of mouth, pale unspotted upper labials (an excellent aid to identification), high ventral counts (230–278) and a divided cloacal scale. *T. o. somalicus* usually uniformly dark, lacks band on a less depressed head, snout rounded and short, a smaller number of ventrals (200–244) and usually an entire cloacal scale.

HABITAT AND DISTRIBUTION: Semi-desert, dry savanna and coastal thicket, often in rocky areas, from sea level to about 1,500m. In the north, largely along Nile Valley, with a few scattered records elsewhere. Occurs in a broad belt across southern Sudan, into

Telescopus obtusus, Borama.

Eritrea, over most of Somaliland, in the Danakil region of Ethiopia and coastal Djibouti; but absent from the highlands and forested south-west of Ethiopia. Reappears in southern South Sudan and extreme south-west Ethiopia. The southern subspecies is confined to southern Somalia, approximately from the Shebelle River across Rahanweyn country to the Benadir coast and the Juba River, thence into Kenya and northern Uganda. Elsewhere, to Chad and isolated records in Algeria.

CONSERVATION STATUS: Least Concern.

NATURAL HISTORY: Nocturnal and semi-arboreal, hunting at night, crawling across ground, climbing bushes and trees, and investigating holes, by day shelters under ground cover and in holes, but may sometimes rest coiled up in trees or bushes, or in bird nests. In Sudan and South Sudan often in houses, attracted by prey such as mice and geckos. Tolerant of dry conditions and low night-time temperatures. Slow-moving and ponderous, if threatened it will raise the head and fore body in a loop, hiss loudly and strike, and if restrained it turns and bites slowly and deliberately. Large-eyed snakes lay clutches of 5–20 eggs, average size 1 × 2.5cm; at Mareri, on Juba River in Somalia, a female was gravid in November and held seven eggs. They eat a range of prey, including lizards, birds, bats and rodents. They take chameleons, a Mareri hatchling had eaten a *Lygodactylus*, a Hudun (Nogal Valley in Somaliland) animal an *Agama rueppelli*, in Khartoum in Sudan a nestling sparrow (*Passer*) was found in a stomach. Greatly feared in Somalia, where it is called *Mas Gadut* ('red snake'), but possibly confused with Red Spitting Cobra *Naja pallida*.

BEAUTIFUL CAT SNAKE

Telescopus pulcher
(Scortecci, 1935)

Telescopus pulcher, Shanshacade.

IDENTIFICATION: A small, very rare, slender snake of Somaliland, with a bullet-shaped snout, distinct coloration and carpet-viper like pattern. Head oval, fairly narrow, depressed; snout wide, bullet-shaped; eye moderate in size, prominent; pupil vertical, iris orange; tongue pink; loreal separated from orbit by preocular; usually a single anterior temporal. Body slender, triangular or sub-triangular in section; tail 16–18 per cent of total length. Scales smooth, in 17–18 rows at mid-body; ventrals 160–180; cloacal scale entire; subcaudals paired, 50–57 rows. Maximum size about 45cm, hatchling 16–18cm. Coloration very distinct: dorsum orange or red-brown, with dark whitish markings forming a series of 25–30 large, irregular or X-shaped blotches along vertebral midline, on flanks is a separate row of distinct dark spots or bars on a whitish background. Top of head orange to red-brown with an irregular but fairly symmetric pattern of white and black. Underside uniformly whitish. Looks like the dangerous Carpet or Saw-scaled Vipers *Echis* spp. Taxonomic Notes: This very distinct species was originally described in genus *Migiurtinophis*, but a recent molecular analysis revealed it is undoubtedly a *Telescopus*, most closely related to *T. dhara* of Arabia and Sinai.

HABITAT AND DISTRIBUTION: Endemic to northern Somalia. Most specimens from Somali arid zone, a region with soft red sandy ground in semi-arid scrub or broken xerophytic woodland and dry open *Vachellia–Commiphora* savanna, at 790–1,340m. Originally described from an unspecified locality in 'Migiurtinia' (= Puntland), all subsequent specimens in Somaliland. Found in Davegoriale in the Haud on border with Ethiopia and also at Shanshacade, then north to Burao as far as foothills of the Goolis and Wagar

Mountains, where recorded from a locality 15km south-east of Sheikh. Also the environs of Hargeisa (Toon village in south, Dhamal plain and Bacaado in north). Probably more widespread, especially in the Ethiopian Haud and Ogaden, but under-collected. Seems widespread in Somaliland and locally fairly common, although it is persecuted due to its resemblance to Carpet vipers.

CONSERVATION STATUS: Least Concern.

NATURAL HISTORY: Poorly known. Nocturnal and terrestrial, hunting at night on ground. By day, shelters underground, in holes or under stones. Diet unknown, but probably eats lizards (especially *Pseuderemias*, *Latastia*, *Pristurus*, *Hemidactylus*). Presumably lays eggs, like others of genus. Recently observed adults and juveniles were active at night; a juvenile was found under a large boulder by day in September. Two juveniles with umbilical scars, 18–20cm long, collected in August and September, had probably hatched during hot midsummer rainy season. Sympatric with African Large-eyed Cat Snake *T. obtusus*.

NORTH AFRICAN CAT SNAKE

Telescopus tripolitanus
(Werner, 1909)

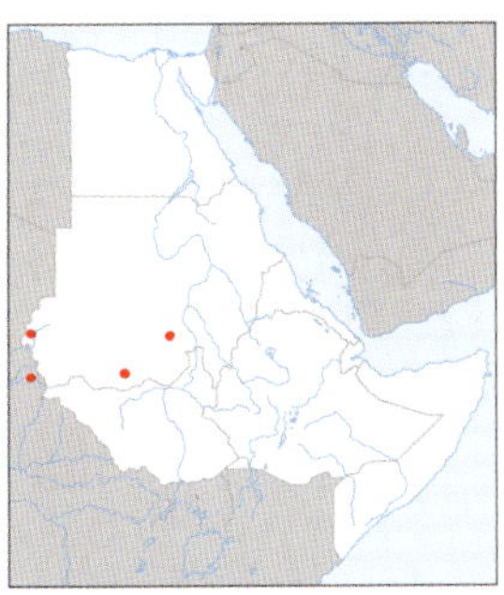

IDENTIFICATION: A medium-sized, slender, black-headed snake with vertical pupils, restricted to Sudan. Iris greyish to dark brown. Neck very thin; body slender, triangular or sub-triangular in section; tail short, about 13–18 per cent of total length. Scales smooth, in 21 (occasionally 22–23) rows at mid-body, ventrals 207–248, cloacal scale divided, subcaudals paired, 55–83 rows. Maximum size 97cm, average 40–60cm, hatchling size unknown. Coloration very distinct: dorsum grey-brown, with 28–50 dark blotches separated by 1–3 scale rows; blotches sometimes poorly defined, with indistinct edges, and disappear with age. Head usually all black or much darker than body, or in some populations almost same colour as body. Underside uniformly pale except sometimes a grey band under tail. Juveniles pale brown to orange with very contrasting black head and become darker with age. Many populations retain this juvenile coloration in adults. Taxonomic Notes: Specific status well supported by recently published molecular data. Most closely related to *T. obtusus*.

HABITAT AND DISTRIBUTION: Rocky semi-arid regions with bushes or solitary trees, margins of oases and wadis at edge of Sahara, often near human settlements and gardens, at low to medium altitudes. A typical species of the Sahel. Has an almost circum-Saharan distribution from Libya west across Tunisia, Algeria, Morocco and Western Sahara to Mauritania and Senegal, then east from Mali to Chad and northern Central African Republic. In our area known only at Umm Ruwaba and Neem in Kordofan, and Geneina, West Darfur, at 460–800m; its occurrence at Birao in Central African Republic suggests it may be more widespread in the intervening country (all suitable habitat), but no records.

CONSERVATION STATUS: Least Concern.

NATURAL HISTORY: Slow-moving, secretive, crepuscular and nocturnal. Terrestrial, but can climb in search of food. Like others of its genus, probably oviparous and may feed primarily on small reptiles. Adults active on the ground after sunset and at night, a juvenile found under rocks by day. A *Tarentola* gecko was found in the stomach of a Mali specimen, a Morocco juvenile was found swallowing a juvenile *Agama bibronii*.

Telescopus tripolitanus, Nigeria.

MARBLED AND GREEN TREE-SNAKES *DIPSADOBOA*

An African genus of slim-bodied, broad-headed, large-eyed elongate tree-snakes. They are rear-fanged, but not known to be dangerous to humans. Ten species in the genus, in West, Central and southern Africa, mostly in forest but also moist savanna; two species occur in our area. Secretive and nocturnal, and little is known of their habits. The genus *Crotaphopeltis* is their closest relatives.

KEY TO NORTH-EAST AFRICAN MEMBERS OF THE GENUS *DIPSADOBOA*

1a: Subcaudals single, in our area only in South Sudan	*Dipsadoboa weileri*, Black-tailed Tree-snake
1b: Subcaudals paired, in our area only in Somalia	*Dipsadoboa flavida*, Cross-barred Tree-snake

CROSS-BARRED TREE-SNAKE

Dipsadoboa flavida
(Broadley & Stevens, 1971)

IDENTIFICATION: A small, slender tree-snake with a broad head and prominent yellow eyes. Two subspecies known; the notes here apply to *D. f. broadleyi*. Dorsal scales smooth, in rows of 17 at mid-body. Cloacal scale entire, ventrals in males 177–197, in females 170–194. Subcaudals paired, 90–100 in males, 79–100 in females. Tip of tail usually rounded. Largest male about 63cm, largest female 61cm, hatchlings about 20cm. Colour rich brown or red-brown, marbled yellow, particularly on front half of body, becoming brown with narrow yellow crossbars towards tail. Head mottled yellow-brown, often with a brown eye-stripe. Those from Somalia often very yellow. Underside pinkish, white or cream. Tongue white towards tip with a black base, or black band near fork.

Dipsadoboa flavida, Kenya.

HABITAT AND DISTRIBUTION: Coastal thicket, woodland and moist savanna, from sea level to around 1,200m. In our area, only known from Juba Sugar Project, near Gelib, on Juba River, but occurs right along Kenya coast, so probably extends south to Kenya border. Elsewhere, south to northern South Africa.

CONSERVATION STATUS: Least Concern.

NATURAL HISTORY: Nocturnal and partially arboreal; usually found under bark and in crevices. Feeds on geckos and frogs, particularly tree frogs; and specimens collected in Kenyan coastal forests at night were mostly on bushes around a swamp, chasing tree frogs. When angry, they raise the fore body off the ground, flatten the head and strike. They lay eggs.

BLACK-TAILED TREE-SNAKE

Dipsadoboa weileri
(Lindholm, 1905)

IDENTIFICATION: A moderate-sized, broad headed, elongate tree-snake of West and Central African

Dipsadoboa weileri, Congo.

forests. Mid-body dorsal scale rows 17 (rarely 19) ventrals, 181–205, subcaudals single, 56–73. Largest male 96cm, largest female 88cm, average 60–80cm, hatchling size unknown. Adults grey or black dorsally, yellow below, except tail which is grey below.

HABITAT AND DISTRIBUTION: One record in our area, but same specimen listed from both Watoka, near Yei, Central Equatoria, at 800m, and Imatong Mountains in Eastern Equatoria, both South Sudan. Yei seems more probable. Elsewhere, west to Nigeria and across Dahomey Gap to Ghana, south-west to Angola. Widespread and unlikely to be threatened by human activity.

CONSERVATION STATUS: Least Concern.

NATURAL HISTORY: Virtually unknown. Probably arboreal. Moderate-sized eyes with vertical pupils, indicating a nocturnal way of life. Largely limited to equatorial forest and grassy fields, where it feeds mostly on forest frogs and toads. Pregnant females vary from 59 to 82cm and may contain 4–8 eggs.

HERALD AND WHITE-LIPPED SNAKES *CROTAPHOPELTIS*

A tropical African genus of small (<1 m), dark, rear-fanged snakes with large eyes and vertical pupils, scales at mid-body in 17–21 rows. They have blade-like rear-fangs, but their venom is not dangerous to humans. Terrestrial and nocturnal, most eat amphibians. Six species known, three occur in north-east Africa, none is endemic. Although effectively harmless, they are fierce little snakes, with a distinctive threat display; they flatten the head into a triangle, flare the lips, hiss and strike readily if threatened.

KEY TO NORTH-EAST AFRICAN MEMBERS OF THE GENUS *CROTAPHOPELTIS*

1a: Lower dorsal scale rows and belly orange or yellow	*Crotaphopeltis degeni*, Yellow-flanked Snake
1b: Lower dorsal scale rows and belly not yellow or orange	2
2a: Usually white speckling on body, males with short hemipenes, not extending beyond 13th subcaudal	*Crotaphopeltis hotamboeia*, White-lipped Snake
2b: No white speckling on body, males with long hemipenes, extending beyond 16th subcaudal	*Crotaphopeltis braestrupi*, Tana Herald Snake

TANA HERALD SNAKE

Crotaphopeltis braestrupi
Rasmussen, 1985

IDENTIFICATION: A small dark snake, head sub-triangular, broader than neck, pupils vertical, iris dark. Tail slender, 20 per cent of total length. Dorsal scales mostly smooth, faintly keeled towards tail, in 19 rows at mid-body, ventrals 156–174, subcaudals paired, 47–67 (high counts in males), cloacal scale entire. Maximum size 67cm, average 30–55cm, hatchling probably around 15–18cm. Colour quite variable; usually black, grey or grey-brown, sometimes with a dark horseshoe on nape; snout sometimes yellow. Juveniles often red-brown. Ventrals cream or yellow. Closely resembles White-lipped Snake, but it is never white-speckled (White-lipped Snake usually is); see also key.

Crotaphopeltis braestrupi, Kenya.

CONSERVATION STATUS: Least Concern.

HABITAT AND DISTRIBUTION: In our area confined to southern Somalia, and known from the lower Juba and Webi Shebelle Rivers and environs, the only inland record is at Oddur in Bakool. Elsewhere, on the north Kenyan coast.

CONSERVATION STATUS: Least Concern.

NATURAL HISTORY: Poorly known. Nocturnal and terrestrial. Juba River specimens were found along irrigation canals, in swamps and gardens. Probably has same threat display as White-lipped Snake. The holotype was caught in a swamp, hunting tree frogs (*Hyperolius pusillus*), the diet is largely amphibians.

YELLOW-FLANKED SNAKE, DEGEN'S WATER SNAKE

Crotaphopeltis degeni
(Boulenger, 1906)

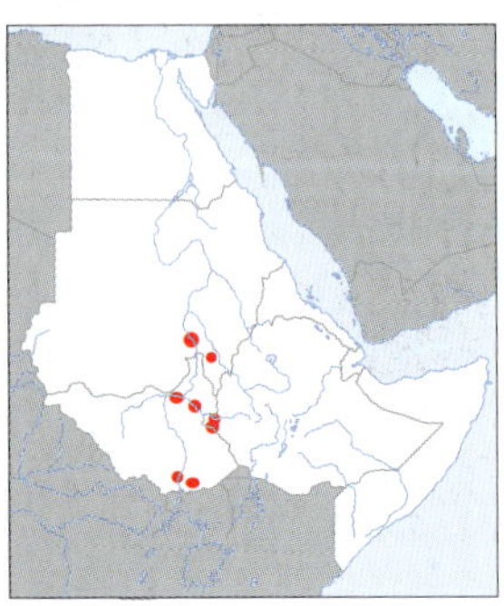

IDENTIFICATION: A small, dark, broad-headed and rear-fanged snake. Eye large, pupil vertical. Body cylindrical, tail short, 10 per cent of total length. Scales smooth, in 19 rows at mid-body, ventrals 168–183, subcaudals paired, 28–40, cloacal scale entire. Maximum size about 60cm, average 30–50cm, hatchling size unknown. Black or blue-black above, ventrals vivid orange or yellow , this colour extending onto dorsal scales to a greater or lesser extent, and above region of uniform yellow the scales may be yellow-tipped. Chin and lips yellow, orange, white or dark. Tongue dark with a white tip.

HABITAT AND DISTRIBUTION: Moist savanna, in swamps and on floodplains, usually near water, at low elevations in our area, 700m or below, but ranges much higher in East Africa. Essentially associated with Nile Basin; and probably very widespread, but records sporadic: at Gambela in Ethiopia, Paloich, Lankien, Malakal and Juba in South Sudan; in Sudan recorded at Kosti and 12°N 34°E, near Ad Damazin on Blue Nile. Elsewhere, south to Tanzania, west to Cameroon.

CONSERVATION STATUS: Least Concern.

NATURAL HISTORY: Terrestrial and nocturnal (occasional specimens seen moving by day), active in damp areas; hides by day under ground cover, in waterside vegetation or holes. Swims well and will hunt in water. Fairly slow-moving on land. Has a similar but sometimes more dramatic threat display to the next species; inflates the body, hissing, flattening head to a triangle, flaring the lips, raising the fore body in a sinusoidal curve and striking with a loud hiss. A clutch of six eggs recorded. Diet amphibians; might eat small fish.

Crotaphopeltis degeni, Gambela.

WHITE-LIPPED SNAKE, WHITE-LIP

Crotaphopeltis hotamboeia
(Laurenti, 1768)

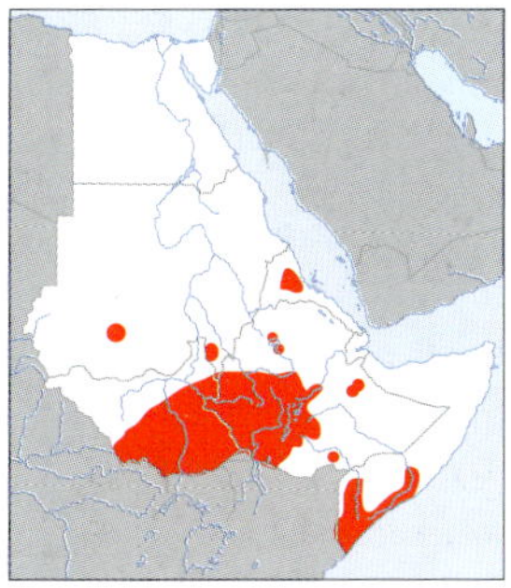

IDENTIFICATION: One of tropical Africa's commonest snakes, widespread in savanna, a small, dark, broad-headed and rear-fanged snake. Body cylindrical, tail short, 11–15 per cent of total length. Dorsal scales mostly smooth (keeled posteriorly), in 19 (occasionally 21) rows at mid-body, ventrals 139–181, subcaudals paired, 29–57, cloacal scale entire. Maximum size about 80cm, possibly larger, big specimens usually females, average 40–70cm, hatchlings 14–18cm, Colour: black, grey or olive-green, lips usually white, but sometimes with dark blotches or head colour encroaches on lip scales; head scales iridescent. Nearly always has a series of crossbars, consisting of fine white dots, particularly visible when snake inflates itself. Olive specimens often have a black patch on each temple. Ventrals cream or white, snout sometimes yellowish. Tongue has a purple base, then darkens, and tip white.

HABITAT AND DISTRIBUTION: Moist savannas, woodland and dry savanna in presence of water, from sea level to 2,600m, an often common and adaptable nocturnal snake, readily takes to suburbia and farmland so long as frogs and ground cover persist. Occurs on lower Webi Shebelle and Juba Rivers in southern Somalia, and near Kenyan border, widespread in Ethiopia in highlands, central Rift Valley and western lowlands; sporadic records in south-east (Filtu, Siddam-Bale Bridge). A handful of records from central Eritrea (Mendefera, Saganeiti, Maldi). Presumably widespread in Nile Basin of South Sudan, but records sporadic, at Boma, Torit, Imatong Mountains, Juba and its vicinity, Kapoeta, Maridi, Li Rangu, Katire, Yei. Records in south-east Chad indicate it is probably present in Southern Darfur in Sudan; known from Holi in Northern Kordofan. Elsewhere, south to southern South Africa, west to Senegal.

CONSERVATION STATUS: Least Concern.

NATURAL HISTORY: Nocturnal and terrestrial, prowling at night, hunting for frogs. Hides by day under ground cover, in holes, etc. Slow-moving, but has a fierce and distinctive threat display; flattens the head into a triangular shape and flares the lips, lifts the fore body off the ground, hisses and puffs, and if the aggressor gets close will strike and hiss loudly at same time; if it connects it will sometimes bite viciously, but it may strike to miss. Looks dangerous and is thus often needlessly killed. Has big, blade-like fangs, and can cause frightful lacerations to frogs, but the venom has no effect on humans. White-lips lay 6–19 eggs, roughly 3 × 1.5cm, which take 50–80 days to hatch, depending on temperature. Hatchlings collected in March, April and September in Kenya; a Kenyan female laid seven eggs in December. Diet largely amphibians but takes other small vertebrates.

Crotaphopeltis hotamboeia, Kenya.

Crotaphopeltis hotamboeia, Jinka.

BOOMSLANG *DISPHOLIDUS*

Large, rear-fanged tree snakes. For a long time the genus was monotypic; several proposed subspecies and species received little acceptance. However, a recently described form on Pemba Island does seem valid. The mainland form, found virtually throughout sub-Saharan savanna will probably also prove to represent a species complex. Unlike most rear-fanged snakes, it has a deadly venom. However, it poses little risk to the ordinary person, as it is arboreal and unaggressive.

BOOMSLANG

Dispholidus typus
(Smith, 1828)

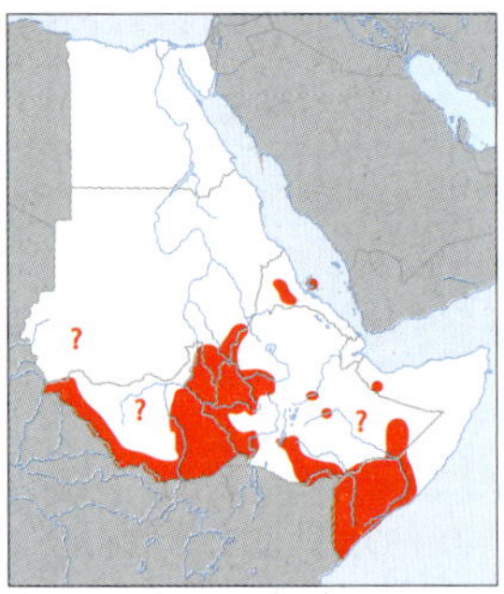

IDENTIFICATION: A big, highly venomous, back-fanged tree snake, usually green, grey or brown, but may be several other colours, with a huge eye, short egg-shaped head and keeled dorsal scales. Pupil almost round but elongate at front (sometimes dumb-bell-shaped). Iris yellow, black-veined in adults, light green in juveniles. Body laterally compressed, tail long, 25–30 per cent of total length in males, 22–28 per cent in females. Dorsal scales strongly keeled, in 17–21 (usually 19) rows at mid-body, ventrals 161–201, with double keels, cloacal scale divided, subcaudals 91–142. Maximum size about 2.1m, average 1.2–1.5m, hatchlings 30–40cm. Colour very variable: males usually shades of green, females brown or grey. Other colour phases include green or yellow with scales black-edged, uniform black, black or brown with yellow or cream spots, brown with a rufous head, striped black and grey-white, light yellow-brown with white spots, and blue-grey. Juveniles have a different colour and pattern to adults; they have a grey or brown, blue-spotted vertebral stripe, flanks light grey or pinkish, speckled black, top of head brown, olive or grey, chin and throat light, often with a big yellow blotch on side of neck; eye vivid apple-green. They gain adult colouration at lengths of 60–90cm; large adults sometimes darken. Taxonomic Notes: See generic introduction.

HABITAT AND DISTRIBUTION: Coastal thicket, woodland, moist and dry savanna and semi-desert, around forest edges and clearings but not usually in dense forest or high grassland; from sea level to 2,400m. Probably occurs throughout our region,

Dispholidus typus, Nigeria.

Dispholidus typus, Kenya.

Dispholidus typus, juvenile.

Dispholidus typus, mating, Kenya.

apart from areas at high altitude and true and near-desert areas of Sudan, but few records, particularly in Sudan, although it extends to 15°N further west. Not recorded in most of northern Somalia (save Hargeisa), no Djibouti records and virtually none above 2,000m in Ethiopia. Elsewhere, west to Senegal, south to South Africa.

CONSERVATION STATUS: Least Concern.

NATURAL HISTORY: A diurnal snake, largely arboreal. Fast-moving, climbs effortlessly. Appears to have binocular vision and can spot prey before it moves, unusually for a snake. Non-aggressive and moves away if approached, but if cornered inflates the neck enormously and flicks the tongue up and down. If restrained it will strike. Boomslangs lay 8–27 eggs, roughly 2×4cm, in deep moist holes, damp tree hollows or vegetation heaps, they hatch in 3–4 months. Prey includes arboreal lizards (especially chameleons), birds, fledglings and eggs, frogs and rodents. Often mobbed by birds. Boomslangs have a deadly venom, but are diurnal, unaggressive, active and alert, with good eyesight and live in trees; they avoid confrontation. A specific monovalent anti-venom is available from South African Vaccine Producers.

VINE OR TWIG SNAKES *THELOTORNIS*

An African genus of medium-sized, largely diurnal tree snakes, with a curious dumbbell or keyhole-shaped eye pupil. They are slim with a long tail. Rear-fanged, gentle and non-aggressive. Four species known, two just reach our area. The slim body and pointed head have led to a legend in some parts of Africa that they can shoot through humans like an arrow. Vine snakes have a deadly venom that causes non-clotting blood but, like Boomslang, the threat posed by vine snakes to humanity is negligible; they are diurnal, secretive, unaggressive, active and alert, with good eyesight and live in trees. They either remain motionless, or move quickly away from confrontation.

KEY TO NORTH-EAST AFRICAN MEMBERS OF THE GENUS *THELOTORNIS*

1a : In our area confined to South Sudan	*Thelotornis kirtlandii*, Forest Vine Snake
1b: In our area confined to southern Somalia	*Thelotornis mossambicanus*, Eastern Vine Snake

FOREST VINE SNAKE, FOREST TWIG SNAKE

Thelotornis kirtlandii
(Hallowell, 1844)

IDENTIFICATION: A long, very thin tree snake, with a green head, of South Sudan. Head long and arrow-shaped, with a large pale eye and a horizontal, keyhole-shaped pupil, which is diagnostic of vine snakes; no other African snake has a pupil so shaped. Iris green and yellow. A narrow groove runs from each eye to tip of snout, giving the snake binocular vision. Tongue bright red with a black tip. Body cylindrical, tail very long and thin, 33–42 per cent of total length. Scales feebly keeled, in 19 rows at mid-body (very occasionally 17); ventrals 162–189, subcaudals 132–172. Maximum size about 1.6m, average 90cm to 1.4m, hatchling 25–32cm. Speckled grey-brown, head green, chin and lips white, sometimes spotted black.

HABITAT AND DISTRIBUTION: In thicket, forest, forest patches and woodland. In our area only in southern South Sudan (Imatong Mountains, Talanga Forest, Torit, 50km south of Torit, Katire). Elsewhere, west to Sierra Leone, south-west to northern Angola.

CONSERVATION STATUS: Least Concern.

NATURAL HISTORY: Diurnal and arboreal, in trees, bushes, thickets and reedbeds. Usually quite low down. Fairly quick-moving but relies on camouflage for defence. Usually avoids confrontation, but if molested will inflate the neck. They lay eggs, clutches of 4–12, size 35×15mm; an Ivory Coast snake laid two pairs of eggs in an arboreal ant nest, which may provide protection against predators; the eggs hatched in January. They eat mostly arboreal lizards, especially chameleons, but may take other snakes and amphibians. Although totally unaggressive, they have a deadly venom.

Thelotornis kirtlandii, Congo.

EASTERN VINE SNAKE

Thelotornis mossambicanus
(Bocage, 1895)

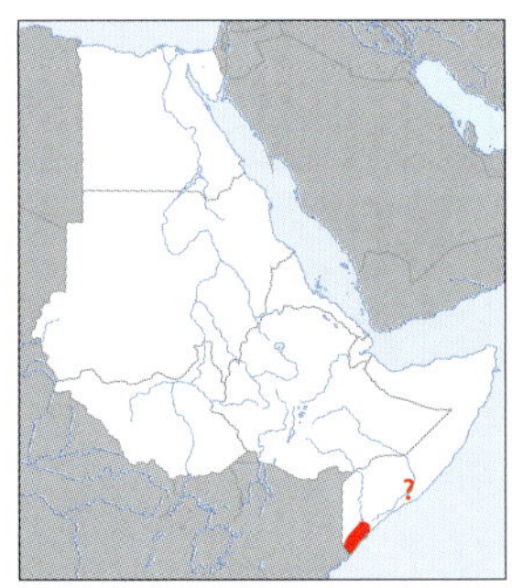

IDENTIFICATION: A very thin, medium-sized grey tree snake on south Somali coast. Head long, snout pointed. Eye large, pupil keyhole-shaped, iris yellow. Tongue red with a black tip. Body thin, tail long, 35–40 per cent of total length. Body scales keeled, in 19 rows at mid-body (occasionally 17 or 21), ventrals 144–169, subcaudals paired, 123–158, cloacal scale divided. Maximum size about 1.4m, average 90cm to 1.3m, hatchlings 23–30cm. From a distance, appears silvery-grey or grey-brown, close-up pale grey or whitish with intense dark speckling; head green or brown, sometimes temporal region behind eye is brown, heavily speckled black. Below, slightly paler but also speckled black. Taxonomic Notes: The status of vine snakes of coastal East Africa is debated. Kenyan coastal specimens assigned to Usambara Vine Snake *T. usambaricus*; but it seems unlikely that Somali animals are a different species.

Thelotornis mossambicanus, Tanzania.

HABITAT AND DISTRIBUTION: Savanna, woodland and coastal thicket, between sea level and 1,700m, but in our area confined to a few localities just above sea level in extreme southern coastal Somalia, from Gelib south to Kenyan border (and reported near Mogadishu). Elsewhere, to Mozambique and eastern Zimbabwe.

CONSERVATION STATUS: Least Concern.

NATURAL HISTORY: Diurnal, arboreal and secretive; similar to previous species. Inflates neck if molested. In the breeding season, male combat observed, with rivals trying to force their opponent's heads down. They lay 4–16 eggs, roughly 1.5×3.5cm. Diet mostly lizards, including chameleons, sometimes frogs and other snakes. Like the previous species, they are inoffensive but have a deadly venom.

BLACK TREE SNAKES *THRASOPS*

A tropical African genus of large, forest-dwelling, harmless black tree snakes with keeled or smooth scales. They inflate the neck in a threat display. Four species known, one in our area. They look like black Boomslangs, and recent molecular analysis reveals they are actually very close to them.

JACKSON'S TREE SNAKE

Thrasops jacksonii
Günther, 1895

IDENTIFICATION: A big black tree snake with a large dark eye and short head. Smells strongly of liquorice, especially when freshly sloughed. Pupil round but iris black, so whole eye just looks black. Body laterally compressed, tail long, ⅓ of total length. Body scales usually unkeeled in our area, in 19 rows at mid-body (sometimes 17 or 21), ventrals 187–214, subcaudals 129–155, cloacal scale divided. Maximum size about 2.3m, average 1.4–1.8m, hatchlings 32–35cm. Adults uniform glossy black above and below, the throat may be grey or white. Juveniles checked black and orange or yellow, above and below, tail spotted yellow, the juvenile colour changes when the snake is 40–60cm long.

HABITAT AND DISTRIBUTION: Forest, forest islands, woodland and riverine forest, from 600 to above 2,400m. In our area, known only in the Imatong Mountains of South Sudan; localities include between Katire and Talanga, and Gilo. Elsewhere, extends south through western East Africa to north-west Zambia.

Thrasops jacksonii, Angola.

CONSERVATION STATUS: Least Concern.

NATURAL HISTORY: Diurnal and arboreal, a superb climber, ascending high into trees, to 30m or more. If approached or threatened in a tree and unable to slide away, it will launch itself off into space and its long light body enables it to move sideways while falling; when it hits the ground it quickly slides away. Has been seen jumping from trees into water. If cornered, will inflate its neck and anterior body like a Boomslang, move sideways and make huge, lunging strikes, looking very large and threatening. Lays 7–12 eggs, roughly 1.5×3cm. A generalist feeder, prey includes arboreal lizards (especially chameleons), mammals including bats, it raids nests for eggs, nestlings and adult birds, and has been seen to drop out of a tree to catch a frog.

BROAD-HEADED TREE SNAKES *TOXICODRYAS* (FORMERLY *BOIGA*)

An African genus of broad-headed, big-eyed, rear-fanged tree snakes, widespread across forested Africa, originally placed in Australasian genus *Boiga*. Interestingly, molecular analysis indicates their closes relatives are the African Egg-eaters *Dasypeltis*, and then *Boiga*. For a long time, only two species were known but recent systematic work has split both species into eastern and western forms. Two species just reach our area. Their venoms contain neurotoxins and are produced in relatively large quantities, so they might represent a danger to humans; however bites by tree snakes are very rare. No anti-venom is available.

KEY TO THE NORTH-EAST AFRICAN MEMBERS OF THE GENUS *TOXICODRYAS*

1a: Mid-body scale rows 21–25, large, up to 2.8m long	*Toxicodryas vexator* Eastern Blanding's Tree Snake
1b: Mid-body scale rows 19, small, never more than 1.25m long	*Toxicodryas adamanteus*, Eastern Powdered Tree Snake

EASTERN BLANDING'S TREE SNAKE

Toxicodryas vexator
Greenbaum *et al.*, 2021

IDENTIFICATION: A very big, stocky tree snake, with a thin neck, a short, broad, flattened head and prominent eyes set well forward, with vertical pupils. Nostrils large. Body triangular in section and rubbery in texture; head also rubbery and flexible. Tail long and thin, 20–26 per cent of total length. Scales large and velvety, in 21–25 (21–23 in East Africa) rows at mid-body, ventrals 240–289, subcaudals 120–141. Maximum size about 2.8m, average 1.4–2m, hatchling 28–35cm. Two main colour phases, males glossy black above, yellow below, the yellow may extend up sides or be confined to narrow stripe on mid-belly, lips yellow, bordered with black. Eye dark. Females usually brown, grey or yellow-brown, yellow-brown below, with faint or clear darker crossbars or diamonds on flanks. Skin between scales bluish-grey, obvious when snake inflates body. Head brown, lips may be yellow-brown, eye yellowish or brown with a vertical pupil. Juveniles all brown (both sexes) with clear irregular black bars that are roughly diamond-shaped, sometimes with pale edges and/or paler centres. Black adults startlingly similar to Gold's Tree Cobra *Pseudohaje goldii*, but have more velvety skin. Taxonomic Notes: Formerly Blanding's Tree Snake *T. blandingii*, but that name now reserved for specimens from the west.

HABITAT AND DISTRIBUTION: Forest, woodland, forest-savanna mosaic and riverine woodland across its range, but in our area only in Imatong Mountains and at Torit, South Sudan (although the Torit specimen might have originated in the Imatongs). Elsewhere, west to the Congo River, south-west to northern Angola.

Toxicodryas vexator, female, Congo.

Toxicodryas vexator, male, Congo.

CONSERVATION STATUS: Not Assessed.

NATURAL HISTORY: Unusually, this is a nocturnal tree snake; a superb climber that actively hunts prey. By day it rests in leaf clumps, hollows in trees, etc. When threatened, elevates and inflates front of body, flattens head, opens mouth and makes a huge, lunging strike. Quite tolerant of urbanisation, and enters caves and buildings in search of bats. Lays 3–14 eggs, approximately 20×40mm. Eats a range of prey; birds, birds' eggs, arboreal lizards, frogs, arboreal rodents and bats; juveniles eat largely lizards; adults eat mostly birds and mammals.

EASTERN POWDERED TREE SNAKE

Toxicodryas adamanteus
Greenbaum *et al.*, 2021

IDENTIFICATION: A fairly big, broad-headed, rear-fanged tree snake, usually pinkish or red-brown. Head broad, eye large and prominent, pupil vertical. Body laterally compressed, with a vertebral ridge. Tail 20–25 per cent of total length. Scales smooth, 19 rows at mid-body, ventrals 236–276, subcaudals 96–132. Maximum size about 1.25m, average 80cm to 1.2m, hatchling size unknown. Colour pinkish to brown, red-brown or pinkish-grey, with darker crossbars alternatively narrow and uniform, or broadening to subtriangular, enclosing a pale spot. The back is usually very finely dusted with brown or black specks (hence the common name), as is the belly, which is pale pink with dashed dark lines on side of each ventral scale. Tongue pink with a white tip. Taxonomic Notes: Formerly

Toxicodryas adamanteus, Congo.

Powdered Tree Snake *T. pulverulenta*, but that name is now reserved for specimens from the west.

HABITAT AND DISTRIBUTION: Forest, woodland and forest-savanna mosaic, between sea level and 2,000m, possibly higher, but in our area known only at Talanga Forest in South Sudan, at 950m. Elsewhere, west to Cameroon, south-west across DR Congo to north-west Angola.

CONSERVATION STATUS: Not Assessed.

Toxicodryas adamanteus, Congo.

NATURAL HISTORY: Arboreal, it is an elegant, careful climber, and nocturnal, which is unusual for a big tree snake. During the day shelters in holes and cracks in trees, creeper and epiphyte tangles, disused plant pods and bird nests, and sleeps curled up, will also sleep in a fork or on a branch. When threatened, elevates and inflates the fore body, flicks its red tongue, hissing loudly and makes lunging strikes. Lays eggs, a DR Congo specimen contained eggs in June. Diet apparently rodents and arboreal lizards.

GREEN SNAKES *PHILOTHAMNUS*

A sub-Saharan African genus of 24 species of harmless, thin, diurnal, fast-moving agile tree snakes with aglyph dentition. Most are green, but other colour variants exist, black (melanistic) varieties occur in some species. Taxonomy under review, with cryptic species probably concealed in some taxa, and identification is difficult with some animals. Most *Philothamnus* green snakes live near water sources and eat amphibians. Few grow larger than 1m. They are often mistaken for Green Mambas, *Dendroaspis* sp., or Boomslangs, *Dispholidus typus*, and subsequently killed. They bite if restrained and can cause deep lacerations; their saliva appears to be toxic to frogs, causing paralysis. Some have a curious, currently unexplained display, holding the head and neck horizontally and send a shimmy of transverse waves down the neck. If threatened, some inflate their necks, exposing startling hidden colours. Nine species in our area.

KEY TO NORTH-EAST AFRICAN MEMBERS OF THE GENUS *PHILOTHAMNUS*

1a: Cloacal scale entire	*Philothamnus heterodermus*, Forest Green Snake
1b: Cloacal scale divided	2
2a: Broad bronze stripe extends length of back	*Philothamnus hughesi*, Hughes' Green Snake
2b: No broad bronze dorsal stripe	3
3a: Subcaudals rounded or angular, not keeled	4
3b: Subcaudals sharply angular and keeled like ventrals	8
4a: Body extremely slender, ventrals 168–194, no hidden white spots on dorsals	*Philothamnus heterolepidotus*, Slender Green Snake
4b: Body moderate, ventrals 138–179, hidden white spots on dorsal scales	5
5a: Two upper labials entering orbit	6
5b: Three upper labials entering orbit	7
6a: Subcaudals 60–104, Imatong Mountains, snout often yellow	*Philothamnus hoplogaster*, South-eastern Green Snake
6b: Subcaudals 88–128, Ethiopia and Somalia, snout not yellow	*Philothamnus battersbyi*, Battersby's Green Snake

7a: Ventrals keeled, maximum size 90cm	*Philothamnus bequaerti*, Uganda Green Snake
7b: Ventrals not keeled, maximum size 1.2m	*Philothamnus angolensis*, Angola Green Snake
8a: Usually two upper labials entering orbit, ventrals 157–188, hidden white spots on dorsal scales lacking or of very low contrast	*Philothamnus punctatus*, Speckled Green Snake
8b: Usually three upper labials entering orbit, ventrals 170–209, hidden white spots on dorsal scales	*Philothamnus semivariegatus*, Spotted Bush Snake

ANGOLAN GREEN SNAKE

Philothamnus angolensis
Bocage, 1882

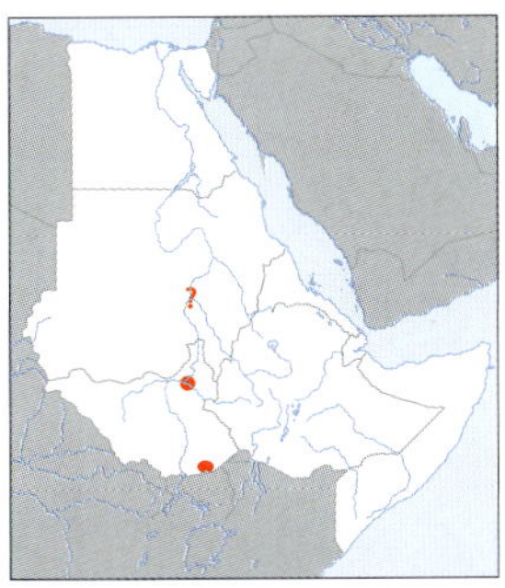

IDENTIFICATION: A long thin green tree snake with a big eye and golden iris, round pupil, the supraocular scale above eye slightly raised. Body cylindrical or sub-triangular in section. Tail long, 27–35 per cent of total length; longer in males. Scales smooth, in 15 rows at mid-body, ventrals 143–167 in males, 148–184 in females, subcaudals 90–134 in males, 90–122 in females. Maximum size about 1.2m, average 70cm to 1m, hatchling 22–26cm. Colour usually emerald-green, sometimes with a blue or yellowish tinge; occasionally black. Some have black flecks or irregular blotches on body. A white spot on lower edge of each dorsal scale is visible when snake inflates body in defence; skin between scales black. Ventrals white, greenish-white or greenish-yellow.

HABITAT AND DISTRIBUTION: Moist savanna and woodland, from sea level to nearly 2,000m, usually in vegetation near water. In our area, at Hellet Nuer (south of Malakal) and Katire in South Sudan, with a curious Sudanese record at Khartoum (possibly transported on a boat). Elsewhere, west to Cameroon, south to South Africa.

CONSERVATION STATUS: Least Concern.

NATURAL HISTORY: Arboreal and diurnal, spends day hunting in waterside vegetation. A fast, secretive and graceful climber, going high, up to 8–10m in trees. Sleeps at night coiled in high vegetation. Swims readily. When angry inflates the body as described. Hisses and makes lunging strikes with open mouth, it has sharp teeth and draws blood when it bites. Lays 4–16 eggs, roughly 2×3cm; communal nests containing up to 85 eggs in Uganda, with embryos on point of hatching in October and February. Incubation takes c.2 months. Diet mostly frogs, but known to take nestlings and birds caught on nest, might eat lizards.

Philothamnus angolensis, Tanzania.

BATTERSBY'S GREEN SNAKE

Philothamnus battersbyi
Loveridge, 1951

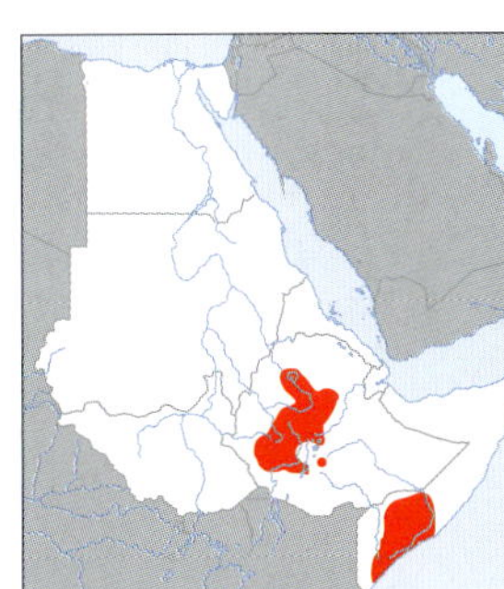

IDENTIFICATION: A fairly long thin green tree snake with a big eye, golden-brown iris and round pupil. Body cylindrical or sub-triangular in section. Tail long, 26–32 per cent of total length, longer in males. Scales smooth, in 15 rows at mid-body, ventrals 152–176, subcaudals 88–129, higher counts usually males. Maximum size about 90cm, average 50–80cm, hatchling probably 15–17cm. Vivid emerald-green or quite dull grey-green, with a bluish or white spot on front lower edge of each dorsal scale, visible when inflates body in defence; skin between scales black. Ventrals greenish-white, light green or greenish-yellow.

HABITAT AND DISTRIBUTION: In moist savanna and woodland, often near rivers, lakes and dams from sea level to almost 3,000m. Sporadically recorded in southern Somalia (including Belet Uen, Kismayu, Gelib and Mogadishu). In Ethiopia widespread on plateau west of Rift Valley, from forests of south-west north to Gondar; a couple of records from valley floors. Elsewhere, south to northern Tanzania and the Harenna Forest.

CONSERVATION STATUS: Least Concern.

NATURAL HISTORY: Arboreal and diurnal, spends day in waterside trees and bushes, either waiting in a prominent position for prey to pass or actively hunting, investigating suitable spots for frogs. When motionless, often shows the curious 'shimmy' described in the generic introduction, sending sinuous waves along body. A fast, secretive and graceful climber, but descends to ground to hunt or to move across country. Sleeps at night coiled on outer branches of waterside trees and big bushes, may also hide in hollow logs, or under vegetation or other ground cover. Emerges to bask in early morning and once warm will hunt; may also bask near sundown. Swims readily and will hunt fish and frogs underwater. Though it usually disappears swiftly and silently if disturbed, when cornered and angry it inflates body to expose bright spots on scales, and may hiss and make lunging strikes with open mouth; has sharp teeth and draws blood when it bites, jerking head side to side and causing lacerations. Lays 3–11 eggs, roughly 2×3cm; communal nests containing >100 eggs recorded in Kenya. During egg-laying, concentrations of up to 40 snakes (gravid and non-gravid females, and males) have been seen in waterside vegetation in Kenya. Diet mostly frogs and fish, hunted from trees and riverbanks but also in water; other prey includes chameleons and skinks. Often called a 'green mamba' and killed. Very tolerant or urbanisation, so long as hedges, streams and frogs remain; still present in centre of Addis Ababa along quite polluted streams.

Philothamnus battersbyi, Bahar Dar.

UGANDA GREEN SNAKE

Philothamnus bequaerti
(Schmidt, 1923)

IDENTIFICATION: A slender green tree snake, eye fairly large, pupil round, body cylindrical, tail 28–30 per cent of total length. Scales smooth, in 15 rows at mid-body, ventrals keeled, 155–179 (higher counts usually females), subcaudals paired, 93–123, higher counts usually males. Cloacal scale divided, except in a type specimen from Niangara in DR Congo, which has an entire cloacal scale. Maximum size about 90cm, average 40–70cm, hatchling size unknown. Uniform green above, paler below, dorsal scales have a concealed white spot on lower edge.

HABITAT AND DISTRIBUTION: Sporadically recorded in low moist savanna and forest, between 300 and 1,200m, in southern South Sudan and south-west Ethiopia. Ethiopian localities include Didessa River

Bridge and Gambela area; South Sudan localities mostly in Jonglei Western, Central and Eastern Equatoria, including Imatong Mountains, Torit, Boma Hills, Li Yubu, Li Rangu, Juba, Yei and Mongalla. Elsewhere, south-west to Uganda, west across north edge of forest to Cameroon.

CONSERVATION STATUS: Least Concern.

NATURAL HISTORY: Poorly known but presumably similar to other green snakes, i.e. diurnal, fast-moving, largely arboreal, climbs and swims expertly, in waterside vegetation, lays eggs, eats amphibians.

Philothamnus bequaerti, Rwanda.

FOREST GREEN SNAKE

Philothamnus heterodermus
(Hallowell, 1857)

IDENTIFICATION: A slim green snake from south-west South Sudan. Snout slightly upturned, pupil round. Tail 24–27 per cent of total length in females, 26–29 per cent in males. Scales smooth, in 15 midbody rows in East Africa, ventrals faintly keeled, 141–166, subcaudals 71–100, males with higher counts, cloacal scale entire. Maximum size c.76cm, average 45–70cm, hatchling size unknown but a Nigerian juvenile, caught in February, was 25cm. Colour varies but in East Africa usually green, body scales with concealed white spot, pale green or yellow below.

HABITAT AND DISTRIBUTION: Forest and moist savanna. In our area, single record at Bangangai Game Reserve, west of Yambio in South Sudan, at 650m. Elsewhere, west to Guinea.

CONSERVATION STATUS: Least Concern.

NATURAL HISTORY: Poorly known. Diurnal. A survey in Ghana found this species spent as much time in trees as on ground. When inactive, coils up in branches or hides in leaf litter. Presumably inflates body to expose white spots when threatened. Ghana and Nigeria specimens contained 1–4 eggs, roughly 1 × 3cm. Diet presumably frogs, but toads (*Bufo* species) were ignored by captive specimens.

Philothamnus heterodermus, Uganda.

SLENDER GREEN SNAKE

Philothamnus heterolepidotus
(Günther, 1863)

IDENTIFICATION: A large, very slender green snake, sometimes with bronzy sheen. Eye quite large, pupil round, iris yellow, golden-brown or red-brown. Body cylindrical, tail long, 33–38 per cent of total length in males, 28–33 per cent in females. Scales smooth, in 15 mid-body rows, ventrals always unkeeled, 164–194, subcaudals paired, 101–144, male counts highest. Cloacal scale divided. Maximum size just over 1m, average 60–80cm, hatchling size unknown. Colour quite variable; immaculate vivid green, greenish-white or yellowish-white below, outer edges of ventrals sometimes finely black-edged, some specimens have a bronzy sheen. Interstitial skin black, chin and throat white or bluish-white.

HABITAT AND DISTRIBUTION: Forest, woodland and moist savanna, at 400–2,000m, elsewhere to sea level. Occurs in south-west Ethiopia, (Gambela) and sporadically recorded in South Sudan; localities include Malakal, the Sudd (Adok, Fangak), Boma, Torit, and Garamba National Park in DR Congo. A curious Sudanese record from Khartoum. Elsewhere, west to Sierra Leone, south-west to Angola.

CONSERVATION STATUS: Least Concern.

NATURAL HISTORY: Probably similar to other green snakes, i.e., partially arboreal, diurnal, climbs and swims well, lives near water sources. In Bugesera, Rwanda, several collected asleep at night in tall grass. Absence of white warning spots on dorsal scales may mean it does not inflate body in defence. Lays eggs, three ova recorded in a Uganda female in March. Diet frogs, fond of tree frogs (*Hyperolius*).

Philothamnus heterolepidotus, Rwanda.

SOUTH-EASTERN GREEN SNAKE

Philothamnus hoplogaster
(Günther, 1863)

IDENTIFICATION: A fairly large green snake with a short head. Pupil round, iris golden. Body cylindrical, tail 27–33 per cent of total length; longer in males. Scales smooth, in 15 mid-body scale rows (occasionally 13 or 11, but nearly always 15 on neck), ventrals 138–167, subcaudals paired, 60–118, higher counts males, cloacal scale divided. Maximum size about 96cm, average 40–70cm, hatchling 15–20cm. Colour vivid green, coastal specimens sometimes bright blue on lower flanks; snout may be orange, yellow or greenish-yellow. Occasionally a series of black or rufous crossbars or blotches on neck and front third of body. Interstitial skin black. A black specimen from western Kenya.

Philothamnus hoplogaster, Kenya.

HABITAT AND DISTRIBUTION: A single record from our area, at Gilo, Imatong Mountains, South Sudan, at 1,900m. Elsewhere, in forest and savanna, west to Cameroon, south to South Africa.

CONSERVATION STATUS: Least Concern.

NATURAL HISTORY: Diurnal, active on round and in trees, sleeps at outer edges of vegetation at night. Climbs and swims well. Lays clutches of 3–8 eggs, roughly 1×3cm. Male combat observed in Zambia. Diet mostly frogs but will take fish and will hunt in water; known to take lizards, and juveniles recorded eating insects. Doesn't appear to inflate its neck, unlike most *Philothamnus*, but if restrained will hold neck and head out rigidly, away from hand, and then strike.

HUGHES' GREEN SNAKE

Philothamnus hughesi
Trape & Roux-Esteve, 1990

IDENTIFICATION: A slim, fairly large green snake. Eye fairly large, tail c.30 per cent of total length. Scales smooth, in 15 rows (occasionally 13) at mid-body, ventrals 152–163, unkeeled, subcaudals 93–105, unkeeled. Maximum size 93.3cm, average 60–80cm, hatchling probably around 15cm. Head and neck vivid green, neck scales heavily blotched blue. A broad brownish-bronze stripe extends length of back, edged by a lighter bronze stripe; front half looks heavily speckled on account of white-and-blue scale margins and black interstitial skin. Underside light yellow or bluish-green.

Philothamnus hughesi, Central African Republic.

HABITAT AND DISTRIBUTION: In our area, only at Gilo, Imatong Mountains, South Sudan, at 1,900m, but also known from northern Uganda. Elsewhere, extends west to Cameroon, south to Tanzania.

CONSERVATION STATUS: Least Concern.

NATURAL HISTORY: Unknown but presumably similar to other *Philothamnus*, i.e. diurnal, partially arboreal, climbs and swims well, lays eggs, eats frogs.

SPECKLED GREEN SNAKE

Philothamnus punctatus
Peters, 1866

IDENTIFICATION: A big, slim green (sometimes turquoise or blue) snake with a raised 'eyebrow' (supraocular scale) and prominent eyes, usually with black speckles but not crossbars. Snout upturned, pupil round, iris golden or pale yellow. Body cylindrical, tail long, 33–37 per cent of total length in males, 28–33 per cent in females. Scales smooth, in 15 rows at mid-body, ventrals strongly keeled, 157–188, subcaudals paired, 126–170 rows, keeled. Maximum size about 1.2m, average 75cm to 1.1m, hatchling 21–24cm. Ground colour bright green or yellow-green, usually finely speckled black. Occasional blue or turquoise specimens in coastal areas. Ventrals cream, white, bluish-green or pale green.

HABITAT AND DISTRIBUTION: Coastal forest, woodland and thicket, moist and dry savanna and semi-desert, from sea level to about 1,200m. Unlike most green snakes, not tied to water (but often found in waterside vegetation) and ranges widely into dry and open country, provided there are trees or large bushes. Recorded in Somaliland, at Borama, Sheikh, Goolis and Wagar Ranges, and a handful of localities on Ethiopian/Somali border, south-east as far as

Bohodle. Seemingly absent from the Ogaden and central Somalia, but widespread in southern Somalia south of 05°N. Elsewhere, south to Mozambique.

CONSERVATION STATUS: Least Concern.

NATURAL HISTORY: A diurnal, fast-moving, agile snake, largely arboreal, climbs high in trees and coastal thickets, using its keeled belly scales to ascend vertical trees. Sleeps at night high on outer branches of vegetation. When threatened and unable to escape inflates its neck and makes lunging strikes. Lays 3–6 eggs, two Somali females had five eggs each in oviducts in June and October, size 1×3cm. Diet mostly lizards, unusually for a *Philothamnus*, major prey items include dwarf geckos (*Lygodactylus*), other geckos and chameleons, will also take frogs and reputed to eat nestling birds. Often mistaken on East African coast for a Green Mamba, *Dendroaspis angusticeps,* and needlessly killed.

Philothamnus punctatus, Kenya.

SPOTTED BUSH SNAKE, SPOTTED WOOD SNAKE

Philothamnus semivariegatus
(Smith, 1847)

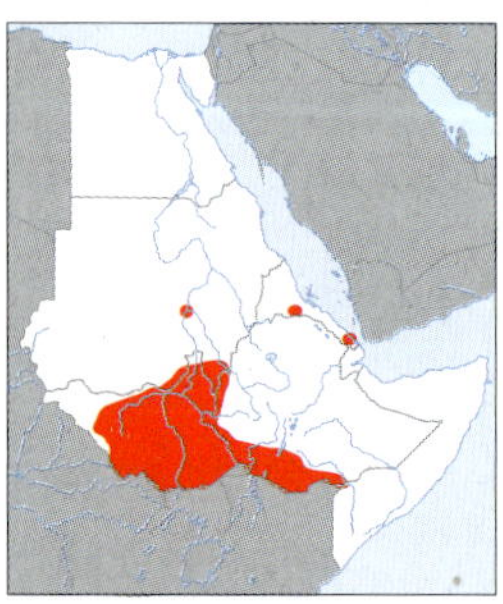

IDENTIFICATION: A big, slim green snake, usually with black crossbars. Eye large, pupil round, iris golden or orange. Tongue bright blue with a black tip. Supraocular raised, giving a 'raised eyebrow' look. Body cylindrical, tail 27–31 per cent of total length in females and 32–36 per cent in males. Scales smooth, in 15 rows at mid-body, ventrals 170–209, keeled, subcaudals 98–166, paired and keeled, high counts usually males, cloacal scale divided. Maximum size about 1.3m; the biggest form in genus, average 70cm to 1.1m, hatchlings 22–25cm. Colour various shades of green, yellow-green or turquoise, usually with distinct, broad or narrow black crossbars along back, flanks spotted black, sometimes has black spots on head. Occasionally has no black markings at all. Ventrals yellowish or light green. Some have a turquoise or blue head, and in parts of range the tail has no spots and is bronzy-green.

Philothamnus semivariegatus, Kenya.

Philothamnus semivariegatus, Tanzania.

HABITAT AND DISTRIBUTION: Widely distributed, although often uncommon, in coastal bush, woodland, moist and dry savanna, and semi-desert, will move into forest clearings and edges, from sea level to

2,000m, possibly higher. Records somewhat sporadic in our area. Known from Gura at 2,000m in Eritrea (and possibly from Assab), Ethiopian records include Nechisar National Park, the Siddam-Bale bridge, the Gambela area and Murri on Kenya border. In Sudan, known from Khartoum and Ar-Rusayris, South Sudan records mostly in south, including Boma Hills, Mongalla, Torit, Yei, Talanga Forest, Juba, Mt Lofi, Mt Imurok. Known from Garamba National Park in DR Congo. Probably much more widespread but secretive in an under-collected area. Elsewhere, south to eastern South Africa, west to Senegal.

CONSERVATION STATUS: Least Concern.

NATURAL HISTORY: A diurnal, fast-moving, agile snake, largely arboreal, but will descend to cross open ground, climbs high in trees, using its keeled belly scales to ascend vertical tree trunks; it is a superb climber, able to use the smallest projections. Sleeps at night high up on outer branches of vegetation. Unusually, ranges widely in open country, and is not tied to water sources. Alert and secretive, its camouflage is superb, being hard to spot even in smallest tree, and if approached moves swiftly and silently away. When threatened and unable to escape it inflates its neck, exposing the concealed blue-white lower edges of its scales, and makes lunging strikes. When resting in a tree, will exhibit the curious 'shimmy' shown by other *Philothamnus*; it will do this on the ground as well, even while moving. Captives, unlike congenerics, seems to remain irascible for a long time. Clutches of 3–12 eggs, roughly 3–4×1cm, are laid. Diet mostly arboreal lizards, but will also take frogs.

KEELED GREEN FOREST SNAKES *HAPSIDOPHRYS*

A genus of slim green forest tree snakes, three species are known, one from the island of Príncipe, the other two in Central African forests; both mainland species just reach our area. Both are harmless. Similar to some *Philothamnus*, but differ in having strongly keeled dorsal scales. Emerald Snake was originally placed in the genus *Gastropyxis*.

KEY TO MEMBERS OF THE GENUS *HAPSIDOPHRYS*

1a: Subcaudals 93–115, tail <35 per cent of total length	*Hapsidophrys lineatus*, Black-lined Green Snake
1b: Subcaudals 119–172, tail >35 per cent of total length	*Hapsidophrys smaragdinus*, Emerald Snake

BLACK-LINED GREEN SNAKE

Hapsidophrys lineatus
Fischer, 1856

IDENTIFICATION: A slim, very large-eyed tree snake, green with fine black stripes. Head quite rounded, tongue light blue with a black tip, pupil of eye round, iris golden or orange. Body cylindrical, tail 27–33 per cent of total length. Dorsal scales strongly keeled, in 15 mid-body rows, ventrals keeled, 150–175, subcaudals paired, 93–115. Cloacal scale entire. Maximum size about 1.1m, average 70–90cm, hatchling size unknown. Emerald- or dark green, each dorsal scale edged black on upper and lower margins, thus appears finely striped green and black. Ventral scales pale or yellow-green, sometimes a fine black or blue-black median line under tail. Head scales may be dark-edged. Bears some resemblance to black-and-green form of Boomslang, so should be treated with caution.

HABITAT AND DISTRIBUTION: Forest and woodland. A single South Sudan record, at Kajiko North (03.47°N, 30.35°E), south of Yei, altitude 1,000m. Elsewhere, south-west to northern Angola, west to Guinea, in forest.

Hapsidophrys lineatus, Congo.

CONSERVATION STATUS: Least Concern.

NATURAL HISTORY: Poorly known. Presumably diurnal, although there are anecdotal reports of night activity. Arboreal, a fast-moving agile climber, its keeled ventrals are a climbing aid. Reputed to be timid, but Kenyan specimens settled down quite quickly in captivity. Lays eggs. Diet largely lizards and amphibians.

EMERALD SNAKE

Hapsidophrys smaragdinus
(Schlegel, 1837)

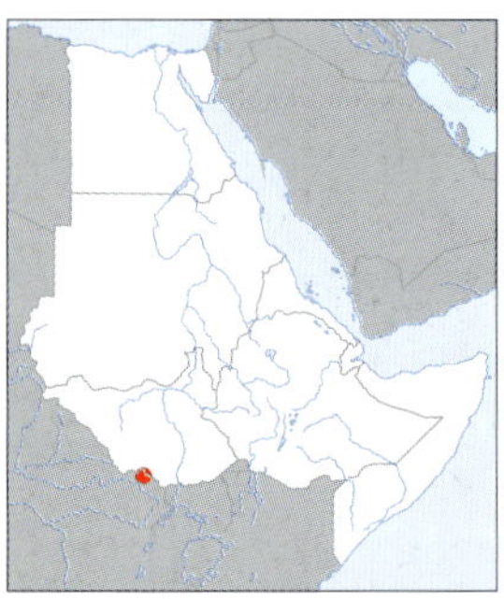

IDENTIFICATION: A slim, green tree snake with a bright yellow belly and distinct dark stripe through eye. Head long, snout quite pointed, eyes large and prominent, pupil round, iris yellow or golden-brown. Body cylindrical, tail very long, 33–42 per cent of total length. Dorsal scales heavily keeled, giving body a distinctly ridged appearance, 15 mid-body scale rows, ventrals 150–174, heavily keeled; belly square and box-like (originally placed in genus *Gastropyxis*, meaning 'box-belly'), subcaudals paired, 129–172, cloacal scale divided. Maximum size about 1.1m average 70cm to 1m, hatchling size unknown. Colour: light to dark green or turquoise, outer edge of ventrals also green, centrally yellow, chin, lips and throat yellow. A dark line through eye, tapering away from orbit.

HABITAT AND DISTRIBUTION: In woodland, forest and deforested areas. Not formally recorded from our area but recently observed in Garamba National Park, at 800m, near South Sudan border. Elsewhere; west to Guinea-Bissau, south-west to northern Angola.

CONSERVATION STATUS: Least Concern.

NATURAL HISTORY: Diurnal and largely arboreal, a fast-moving agile climber, able to ascend vertical tree trunks using its keeled ventral scales. Basks in forest clearings. Sleeps at night on a branch, often near water. Not known to inflate its throat when threatened. Lays 2–6 eggs. Diet largely arboreal lizards and tree frogs. If seized by tail struggles violently and tail may break off, it is not regenerated.

Hapsidophrys smaragdinus, Liberia.

HOOK-NOSED SNAKES *SCAPHIOPHIS*

A curious African genus of burrowing colubrids; they are fairly big, robust, harmless terrestrial snakes, living mostly in burrows or soft soil and sand. They are well adapted to a subterranean life, having slit-shaped recessed nostrils, a huge tough projecting rostral scale and a mouth opening under the head, when the mouth closes it fits with valve-like precision into the flanged upper lip. Body stout and muscular. They have small eyes, with round pupils. The mouth is black inside, the teeth strangely tiny. They lay eggs. Two species are known, both occur in north-east Africa.

KEY TO NORTH-EAST AFRICAN MEMBERS OF THE GENUS *SCAPHIOPHIS*

1a: Mid-body scales 19–25, ventrals 170–201 in males, 189–221 in females, dorsal scales broad, 1.5 times as long as wide *Scaphiophis albopunctatus*, Hook-nosed Snake

1b: Mid-body scales 27–31, ventrals 204–216 in males, 225–243 in females, dorsal scales narrow, more than twice as long as wide *Scaphiophis raffreyi*, Ethiopian Hook-nosed Snake

HOOK-NOSED SNAKE

Scaphiophis albopunctatus
Peters, 1870

IDENTIFICATION: A fairly big, thick-bodied, harmless snake. Head short and blunt, with a large projecting rostral scale. Last upper labial huge. Eye small, pupil round, iris golden or golden-brown. Tongue black. Tail fairly short, 14–18 per cent of total length. Body scales smooth, details in key. Subcaudals 49–76. Maximum size about 1.6m, average 90cm to 1.3m, hatchling size unknown but a Kenyan juvenile was 34cm. Colour quite variable; may be shades of grey or pinkish-grey, uniform black, brown or orange with broad grey vertebral stripe; often with a sprinkling of black or dark brown spots on head and back; ventrals orange, pink, cream or ivory-white. Juveniles grey or blue-grey with white spots, in some specimens head greenish, this colour, initially to pinkish-grey and then to adult colour when around 60–80cm long.

HABITAT AND DISTRIBUTION: Semi-desert, dry and moist savanna and woodland, between 300 to about 1,500m. In our area known only from southern South Sudan; localities include Lerwa, Juba, Nimule, Torit, Ohilang and Kagelu. Elsewhere, west to Ghana, south around the southern edge of Central African forest to Gabon.

CONSERVATION STATUS: Least Concern.

NATURAL HISTORY: Poorly known. Lives in holes, but may move about by day. If threatened, these snakes (especially juveniles) have a spectacular threat display; elevating the body, opening the black mouth

Scaphiophis albopunctatus, Kenya.

Scaphiophis albopunctatus, threat display, Kenya.

Scaphiophis albopunctatus, Kenya.

wide and making a huge strike, designed to miss. If this fails, it jams its head under its coils. They lay eggs, up to 48 recorded. Diet largely mammals, caught in holes and slammed against the wall.

ETHIOPIAN HOOK-NOSED SNAKE

Scaphiophis raffreyi
Bocourt, 1875

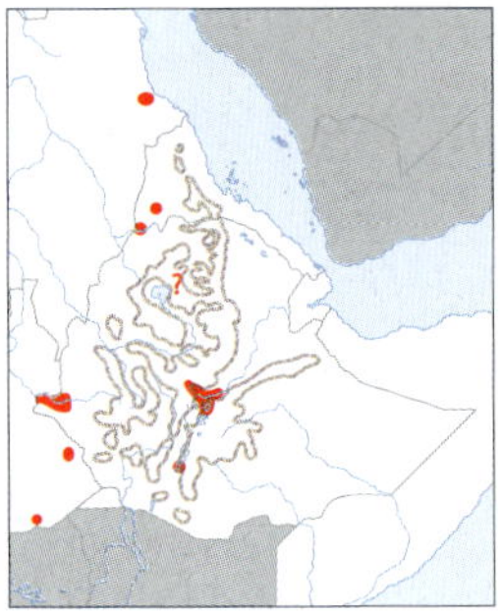

IDENTIFICATION: A fairly big, thick-bodied, harmless snake. Head short and blunt, with a large projecting rostral scale. Last upper labial huge. Eye small, pupil round. Tongue black. Tail fairly short, 15–20 per cent of total length. Body scales smooth and elongate, details above in key. Subcaudals 55–68 in females, 72–79 in males. Maximum size about 1.5m, average 90cm to 1.3m, hatchling size unknown. Colour dull grey, brown or red-brown, often with a sprinkling of black spots on head and back; ventrals orange, cream or white. Juveniles grey or blue-grey with white spots, changing to adult colour when 60–80cm long. Compare Hook-nosed Snake.

HABITAT AND DISTRIBUTION: Moist and dry savanna and semi-desert, at low to mid altitudes. In Ethiopia, a population exists in high northern Rift Valley and foothills west of there: between Debre Zeit and Welenchiti, south to Meki and Lake Zwai. Also at Arba Minch, Gambela area and vicinity of Debra Tabor. Two records in Eritrea: Barentu and Cunana, two from South Sudan (Boma National Park, Ikwoto) and one in Sudan; Khor Okwat in Red Sea hills of north-east; Sudanese specimens informally reported at Kassala and Kordofan as *S. albopunctatus* might be this species. Might be much more widespread but under-collected due to its secretive habits. Elsewhere, south to Kenya and Uganda.

CONSERVATION STATUS: Least Concern.

NATURAL HISTORY: Poorly known but probably similar to previous species. An Ethiopian juvenile was caught during ploughing. A large adult attempted to cross a road around mid-morning, so probably diurnal.

Scaphiophis raffreyi, Kenya.

Scaphiophis raffreyi, Meki.

EGG-EATING SNAKES *DASYPELTIS*

An African genus (one species reaches south-west Arabia) of small to medium-sized harmless snakes, with blunt bullet-shaped heads and small eyes, vertical pupils, and strongly keeled scales (a useful field character). They closely resemble some dangerous snakes, particularly carpet vipers (*Echis*) and night adders (*Causus*); so should be treated cautiously until confidently identified .They have rudimentary teeth but are highly adapted for eating birds' eggs; vertebral processes protrude into gullet; acting like saw blades, they cut through the eggshell, which is then regurgitated and the egg contents swallowed. They vary a lot in colour but scale counts do not; this has created taxonomic problems. At present there are 17 species in the genus (several described recently); nine in our area.

KEY TO NORTH-EAST AFRICAN MEMBERS OF THE GENUS *DASYPELTIS*

1a: A distinct, not faint, rhombic pattern, or clear dark crossbars, extends length of body	2
1b: No distinct rhombic pattern extending length of body	9
2a: In our area only Egypt and possibly extreme north-east Sudan	*Dasypeltis bazi*, Egyptian Egg-eater
2b: In our area, not in Egypt and extreme north-east Sudan	3
3a: Confined to Eritrea; dorsal pattern of pale oblique crosses, in a darker surround	*Dasypeltis crucifera*, Cross-marked Egg-eater
3b: Not confined to Eritrea, dorsal pattern lacks pale crosses	4
4a: Dorsal pattern of large, square or rectangular saddles, interspersed by very narrow, straight white crossbars, only in lowlands of Djibouti, eastern Ethiopia and Somaliland	*Dasypeltis taylori*, Taylor's Egg-eater
4b: Light and dark crossbars of similar size, or saddles, but if present latter not usually square, more elongate or rounded and crossbars between them not always conspicuously narrow	5
5a: In our area confined to far western Sudan	*Dasypeltis sahelensis*, Sahel Egg-eater
5b: Not in western Sudan	6
6a: Dorsal crossbars and lips distinctly yellow or orange, confined to northern Ethiopian and Eritrean plateau	*Dasypeltis abyssina*, Ethiopian Egg-eater (part)
6b: Dorsal crossbars and lips not distinctly yellow or orange, not confined to northern Ethiopian and Eritrean plateau	7
7a: Vertical dark bars on flanks parallel with dark blotches on spine	*Dasypeltis confusa*, Confusing Egg-eater
7b: Vertical dark bars on flanks not parallel with pale blotches on spine	8
8a: Fewer than 80 pattern cycles between nape and base of tail, usually below 2,000m	*Dasypeltis scabra*, Common Egg-eater (part)
8b: More than 85 pattern cycles between nape and base of tail, usually above 2,000m	*Dasypeltis atra*, Montane Egg-eater (part)
9a: Bright orange or red, with or without markings	10
9b: Not bright orange or red	11
10a: Usually above 1,500m	*Dasypeltis atra*, Montane Egg-eater (red form only) (part)
10b: Below 1,500m	*Dasypeltis medici lamuensis*, Rufous Red Egg-eater (Lamu subspecies)
11a: Black in colour	*Dasypeltis atra*, Montane Egg-eater (part)
11b: Not black in colour	12
12a: Brown, with or without markings	13
12b: Not brown	14
13a: High altitudes, only on northern Ethiopian and Eritrean plateau	*Dasypeltis abyssina*, Ethiopian Egg-eater (part)
13b: Not at high altitude	*Dasypeltis scabra*, Common Egg-eater (part)
14a: One or two postoculars, usually above 1,500m	*Dasypeltis atra*, Montane Egg-eater (part)
14b: Two postoculars, at altitudes below 1,500m	*Dasypeltis scabra*, Common Egg-eater (part)

ETHIOPIAN EGG-EATER

Dasypeltis abyssina
(Duméril, Bibron & Duméril, 1854)

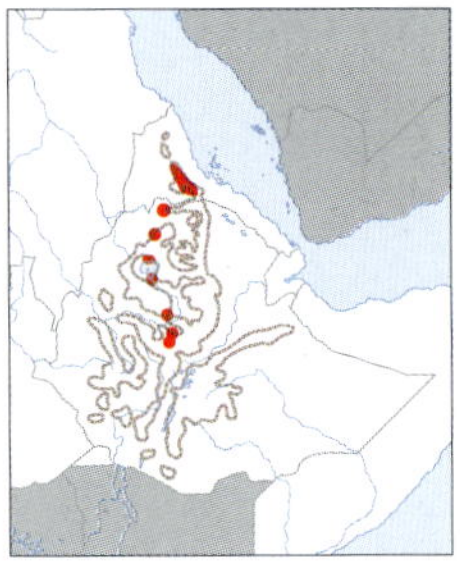

IDENTIFICATION: A large egg-eater of northern Ethiopia plateau, either distinctly black and yellow, or uniform brown. Head bullet-shaped, eye quite large and prominent, pupil vertical, iris dark yellow. Body cylindrical, tail 10–17 per cent of total length (longest in males). Scales strongly keeled and serrated, in 23–26 rows at mid-body, ventrals 226–271, subcaudals paired, 49–68. Maximum size about 90cm, average 40–70cm, hatchlings c.20cm. Two colour phases. One brown, with a series (usually 70–100) of black sub-rectangular blotches on spine, interspersed by short, narrow yellow crossbars, matched by series of vertical black flank bars. Upper labials usually yellow with black edging; underside cream, heavily marbled black. The other colour phase is uniform brown or rufous-brown, lighter brown below. Taxonomic Notes: Recently resurrected from synonymy of *D. scabra*.

HABITAT AND DISTRIBUTION: Mid- to high-altitude savanna, light woodland, grassland and pastures with ground-nesting birds, between 1,800 and 2,700m. Known only from north-west Ethiopian and Eritrean plateau, from Sululta Plain just north of Addis Ababa, north through highlands into Eritrea, as far as Gheleb.

Dasypeltis abyssina, Mulu.

CONSERVATION STATUS: Least Concern.

NATURAL HISTORY: Similar to other egg-eaters, i.e., nocturnal, largely terrestrial, eats birds eggs, lays eggs; see Montane Egg-eater *D. atra*. Gravid females found December–January at Mulu, Ethiopia. Most active May–June, when ground-nesting birds lay their eggs.

MONTANE EGG-EATER

Dasypeltis atra
Sternfeld, 1912

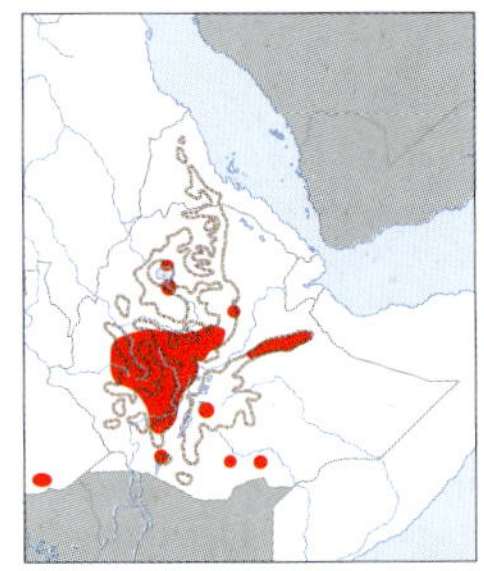

IDENTIFICATION: A medium-sized snake with a bullet-shaped head, eyes small but prominent, pupil vertical, iris golden, flecked black, tongue black. Body cylindrical, tail 12–18 per cent of total length (longest in males). Scales strongly keeled and serrated, in 22–27 rows at mid-body, ventrals 202–237, subcaudals paired, 49–72. Maximum size about 1.1m, average 50–80cm, hatchlings 22–25cm. Several colour forms thus; (a) uniform black, above and below (sometimes throat is grey), (b) uniform red-brown, red or pink, paler below, (c) uniform brown, (d) brown with a row of many darker blotches on centre of back, often white between blotches, faint dark lateral bars may be present, pale brown or yellow-brown below. Taxonomic Notes: A recent molecular analysis indicates *D. atra* is polyphyletic; it might be just a melanistic or dark morph of *D. scabra*, rather than a discrete species.

HABITAT AND DISTRIBUTION: High-altitude savanna, grassland, woodland and forest, at 1,000–2,800m, but usually above 1,500m. In our area, widespread on Ethiopian plateau, both sides of Rift Valley, from El Dire north-east to Jijiga, north-west to Gondar (also at Negele and Filtu). Known from Imatong Mountains in South Sudan. Elsewhere, south to northern Tanzania, south-east to high Albertine Rift.

Dasypeltis atra, Addis Ababa.

Dasypeltis atra, Mulu.

CONSERVATION STATUS: Least Concern.

NATURAL HISTORY: Terrestrial and semi-arboreal, hunting on ground but also climbing trees and bushes to find birds' nests. Nocturnal, sheltering by day in abandoned birds' nests, tree holes and cracks, under ground cover or in holes. Diet birds' eggs. If threatened, has a remarkable threat display (although not all individuals show it), forming body into C-shaped coils, which are rotated against each other, the friction between the keeled scales producing a remarkable hissing or crackling noise, like water falling on a red-hot plate. At same time, the snake opens its mouth very wide to expose the black interior, lowering and broadening the lower jaw, if further molested they hiss loudly and simultaneously strike; a tremendous threat display, but it is all bluff. Their teeth are so tiny that they can do no damage; consequently they usually strike to miss. Montane Egg-eaters lay 7–14 eggs, roughly 3×1.5cm. Sympatric with *D. scabra* at several localities, e.g., Negele and Filtu.

EGYPTIAN EGG-EATER

Dasypeltis bazi
Saleh & Sarhan, 2016

IDENTIFICATION: A medium-sized rhombic patterned egg-eater from Egypt. Head bullet-shaped, eye quite large and prominent, pupil vertical, iris yellowish-grey. Body cylindrical, tail 12–16 per cent of total length (longest in males). Scales strongly keeled and serrated, in 23–24 rows at mid-body, ventrals 213–234, subcaudals paired, 50–70. Maximum size about 70cm, average 40–60cm, hatchling size unknown. Colour quite variable but usually brown or brownish-grey, with brown or rufous oval or rhombic shapes on centre of back (may be elongate), interspersed by light crossbars. Flanks usually have fine irregular, oblique darker bars. Ventrals white or cream, either uniform or blotched to greater or lesser extent. Almost identical in colour and appearance to dangerous North-east African Carpet Viper *Echis pyramidum*.

HABITAT AND DISTRIBUTION: Known with certainty only in Faiyum Depression, a basin dominated by agriculture some 80km south-west of Cairo, at 30–50m below sea level. Reported from Cairo and Abu Rawash, without unequivocal evidence; a specimen from Erkowit in north-east Sudan might also be this species.

Dasypeltis bazi, Fayum.

CONSERVATION STATUS: Not Assessed.

NATURAL HISTORY: Presumably similar to other egg-eaters, i.e. nocturnal, largely terrestrial, eats birds' eggs, lay eggs; see Montane Egg-eater *D. atra*.

CONFUSING EGG-EATER

Dasypeltis confusa
Trape & Mané 2006

IDENTIFICATION: A small to medium-sized snake with a bullet-shaped head; eye quite large and prominent, pupil vertical, iris golden or silvery yellow. Body cylindrical, tail 12–17 per cent of total length (longest in males). Scales strongly keeled and serrated, in 23–26 rows at mid-body, ventrals 213–242, subcaudals paired, 53–73. Maximum size about 97cm, average 50–80cm, hatchlings around 23cm. Usually brown or grey, along spine is a chain of around 40–60 sub-rectangular dark markings, in some specimens these may be linked by lighter patches, and on flanks are irregular dark vertical or oblique bars, but top of these always parallel with dark spinal markings, never the pale ones or gaps between dark markings. There are often narrow dark crossbars on head, and a narrow V-shape on nape followed by a broad V-shape. Underside beige, unmarked or with black edges.

Dasypeltis confusa, Chad.

HABITAT AND DISTRIBUTION: Widespread across savannas of Central and West Africa, but in our area known only from West, Central and Eastern Equatoria in southern South Sudan; localities include Torit, Nimule and Juba. A doubtful record from Addis Ababa is probably a misidentification.

CONSERVATION STATUS: Least Concern.

NATURAL HISTORY: Similar to other egg-eaters, i.e. nocturnal, largely terrestrial, eats birds' eggs, lay eggs; see Montane Egg-eater *D. atra*.

CROSS-MARKED EGG-EATER

Dasypeltis crucifera
Bates, 2018

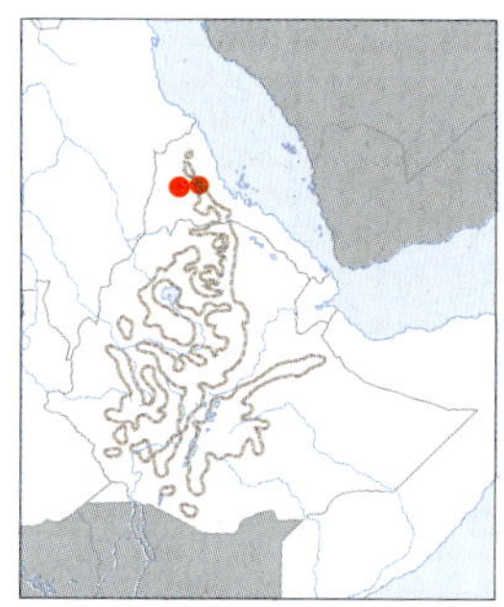

IDENTIFICATION: An Eritrean endemic, a small slim egg-eater with distinctive pale crosses on spine. Has a bullet-shaped head, eye quite large and prominent, pupil vertical. Body cylindrical, tail 11–16 per cent of total length (longest in males). Scales strongly keeled and serrated, in 23–26 rows at mid-body, ventrals 226–247, subcaudals paired, 49–61. Largest specimen just over 50cm total length. Colour pattern unusual; dark brown oval saddles and dark flank bars enclose pale cross-markings; between 51 and 62 pattern cycles. Underside largely immaculate yellow-brown (in preservative), with a few dark markings at edges.

HABITAT AND DISTRIBUTION: Recently described and known from just four specimens, from Agordat (600m) and vicinity of Keren (1,400m) in north-west Eritrea, in dry savanna.

CONSERVATION STATUS: Data Deficient.

Dasypeltis crucifera, preserved.

NATURAL HISTORY: Never knowingly seen alive but presumably similar to other egg-eaters, i.e., nocturnal, largely terrestrial, eats birds' eggs, lays eggs; see Montane Egg-eater *D. atra*.

LAMU RED EGG-EATER

Dasypeltis medici lamuensis
Gans, 1957

IDENTIFICATION: A big monocoloured egg-eater of the southern Somali coast. A bullet-shaped head, eyes quite large and prominent, pupil vertical, iris orange or grey, depending on colour pattern, tongue black. Body cylindrical, tail 15–21 per cent of total length (longest in males). Scales strongly keeled and serrated, in 23–27 rows at mid-body. Notes here refer to the northern subspecies (Lamu Red Egg-eater *D. m. lamuensis*) which is the only one in our area: ventrals 226–237, subcaudals paired, 71–109. Maximum size about 1.1m, average 50–80cm, hatchling 23–27cm. Colour quite variable; but Somali specimens usually red, vivid pink, orange, grey or blue-grey, sometimes with clusters of fine white dot on spine but no other markings. Close examination shows that each scale is finely stippled, often with white, a useful field character.

Dasypeltis medici lamuensis, Kenya.

HABITAT AND DISTRIBUTION: Dry savanna and coastal thicket, at low altitudes. In our area, only in extreme southern Somalia, from lower Juba River south to Kenya border. This subspecies extends into Kenya and northern Tanzania, the nominate subspecies to north-east South Africa.

CONSERVATION STATUS: Least Concern.

NATURAL HISTORY: Similar to other egg-eaters, see notes on Montane Egg-eater *D. atra*. Lays 6–28 eggs, roughly 3×1.5cm, a juvenile was taken in May at Vipingo on Kenya coast.

SAHEL EGG-EATER

Dasypeltis sahelensis
Trape & Mané, 2006

IDENTIFICATION: A small egg-eater of the Sahel. Has a bullet-shaped head; eye quite large and prominent, pupil vertical, iris golden or orange. Body cylindrical, tail 11–15 per cent of total length (longest in males). Scales strongly keeled and serrated, in 21–23 rows at mid-body, ventrals 207–237, subcaudals paired, 45–67. Maximum size about 65cm, average 40–60cm, hatchling unknown but probably around 20cm.

Dasypeltis sahelensis, Morocco.

Usually brown or grey, a chain of sub-rectangular or oval dark markings on spine, interspersed by white bars or blotches, and dark irregular vertical flanks bars align with white blotches. Often narrow brown partial or total V-shapes on head and nape, followed by a broad V-shape. Underside white with darker marbling.

HABITAT AND DISTRIBUTION: A species of Sahel, a single record from our area; at Al Junaynah (= Geneina), 800m, in *Vachellia* savanna in Western Darfur, Sudan. Elsewhere, west to Senegal, Mauritania and Morocco.

CONSERVATION STATUS: Least Concern.

NATURAL HISTORY: Similar to other egg-eaters, i.e. nocturnal, largely terrestrial, eats birds' eggs, lay eggs; see Montane Egg-eater *D. atra*. Probably most active in April–May, when passerine birds start nesting.

COMMON EGG-EATER, RHOMBIC EGG-EATER

Dasypeltis scabra
(Linnaeus, 1758)

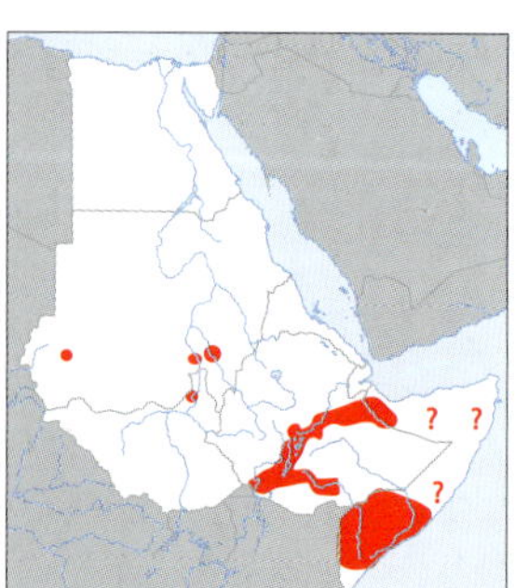

Dasypeltis scabra, Awash National Park.

Dasypeltis scabra, Filtu.

IDENTIFICATION: A small to medium-sized snake with a bullet-shaped head; eyes small but prominent, pupil vertical, iris golden flecked black, tongue black. Body cylindrical, tail 12–18 per cent of total length (longest in males). Scales strongly keeled and serrated, in 20–27 rows at mid-body, ventrals 180–263, subcaudals paired, 38–94. Maximum size about 1.1m, average 50–80cm, hatchlings 20–26cm. Colour quite variable but commonest is 'rhombic' form, which is brown, rufous or grey, with a series of darker, oval or rhombic shapes on centre of back, often interspersed by paler blotches, often with one or two V-shapes on neck, flanks usually marked with fine irregular, oblique darker bars. Ventrals may be white, cream, either uniform or blotched to a greater or lesser extent. Specimens with faint patterns, or unicoloured animals, which may be reddish or any shade of brown or grey, also occur. Note; in parts of its range closely resembles both carpet vipers and Rhombic Night Adder, so great care should be taken with identification; the C-shaped coil threat display of this species is also shown by North-east African Carpet Viper.

HABITAT AND DISTRIBUTION: Occupies a very wide range of habitats, from coastal forest and thicket to moist and dry savanna, semi-desert, near-desert, high grassland and woodland, usually absent only in desert, areas above 2,400m and closed forest. However, many older records from our area have been moved to recently described or resurrected forms, complicating the picture; for example, most Somaliland examples assigned to *D. taylori*, those from South Sudan to *D. confusa*, and those from highland Ethiopia and Eritrea to *D. abyssina* (and *D. crucifera*). Consequently, as presently defined, this species seems to be confined in our area to southern Somalia, parts of south-east Ethiopia and the Rift Valley, just entering western Somaliland, north-east South Sudan and south-east Sudan. Also recorded at Kutum in Darfur. Elsewhere, south to South Africa (populations further west probably a different species, as presently defined).

CONSERVATION STATUS: Least Concern.

NATURAL HISTORY: As other egg-eaters, see Montane Egg-eater *D. atra*. This species is nearly always willing to use the C-coil threat display with its accompanying sizzling noise; do not confuse it with a carpet viper. Common Egg-eaters lay 6–25 eggs, roughly 3.5 × 1.5cm. A specimen lived 31 years in captivity. They can tolerate long periods of starvation, a captive went 12 months without food before it commenced feeding; this is presumably an adaptation since in some areas of Africa birds have a short breeding season, and in between times the snake will be unable to find food.

Dasypeltis scabra, threat display, Rwanda.

Dasypeltis scabra, Borama.

TAYLOR'S EGG-EATER

Dasypeltis taylori
Bates & Broadley, 2018

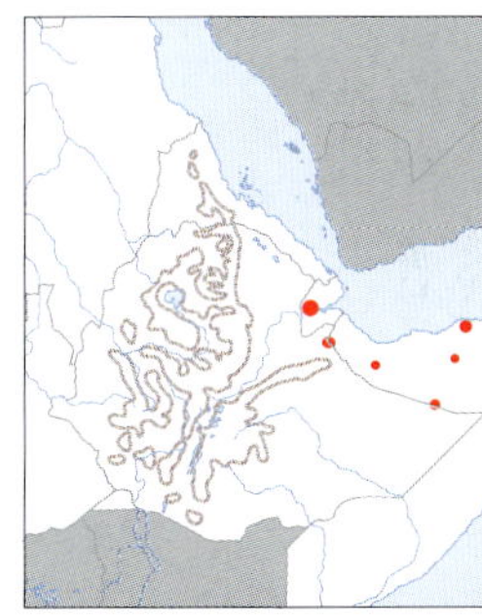

IDENTIFICATION: A rhombic-patterned egg-eater with narrow white crossbars, of dry country associated with the Afar Triangle. Has a bullet-shaped head; eye quite large and prominent, pupil vertical. Body cylindrical, tail 10–17 per cent of total length (longest in males). Scales strongly keeled and serrated, in 23–25 rows at mid-body, ventrals 196–222, subcaudals paired, 44–68. Largest known specimen just less than 55cm total length. Dorsal colour some shade of brown; pattern consists of rectangular, often elongate dark brown saddles, interspersed with white, often narrow, crossbars and brown, irregular flank bars,

Dasypeltis taylori, Djibouti.

underside cream with large grey spots laterally and smaller grey spots elsewhere.

HABITAT AND DISTRIBUTION: Recently described from a handful of specimens, in semi-desert north and east of Afar Triangle, between sea level and 1,400m; in Djibouti, Ogo (north-west of Borama but just in Ethiopia) and Hargeisa, Garadag, Daallo Forest (north of Erigavo) and the Haud in Somaliland.

CONSERVATION STATUS: Data Deficient.

NATURAL HISTORY: Presumably similar to other egg-eaters, i.e. nocturnal, largely terrestrial, eats birds' eggs, lay eggs; see Montane Egg-eater *D. atra*.

WATER SNAKES *GRAYIA*

An African genus of four species of harmless, large, heavy-looking Central African water snake, with small round pupils. Three species occur in north-east Africa. Some authorities have placed the genus in its own subfamily; the Grayiinae.

KEY TO THE NORTH-EAST AFRICAN MEMBERS OF THE GENUS *GRAYIA*

1a: Mid-body scale count 15, tail long, >40 per cent of total length	*Grayia tholloni*, Thollon's Water Snake
1b: Mid-body scale count 17–20, tail shorter, <36 per cent of total length	2
2a: Seven upper labials, at least 84 subcaudal scales	*Grayia smythii*, Smyth's Water Snake
2b: 8–9 upper labials, more than 88 subcaudal scales	*Grayia ornata*, Ornate Water Snake

SMYTH'S WATER SNAKE

Grayia smythii
(Leach, 1818)

IDENTIFICATION: A huge stocky water snake, conspicuously banded when young. Head scales dark-edged. Eye quite small, with a tiny round pupil (good field character), iris yellow or brown. Tail fairly long, >30 per cent of total length, but often truncated, presumably by predators. Males longer tailed. Dorsal scales smooth, in 17 rows at mid-body; ventrals 149–167, subcaudals 90–102. Maximum size about 1.7m, although larger animals, up to 2.5m, reported in West Africa. Average size 1–1.5m, hatchling size unknown. Ground colour varies from yellowish-brown to rufous, olive or black. Chain of broad dark blotches along spine makes juveniles appear banded, as they become adult bands may change to white, black-edged crossbars or fade, leaving a mottled brown or olive snake, with vague dark crossbars and blotches, often darkening towards tail. Head usually light brown or pinkish-brown, chin and lips yellow, head scales black-edged. Ventrals orange, yellow or cream, usually spotted, the spots forming thin longitudinal lines. Taxonomic Notes: Much debate about spelling of specific name, *smythii*, as it was named after Norwegian botanist Christen Smith, so *smithii* or *smithi* would appear more appropriate. The debate continues.

HABITAT AND DISTRIBUTION: In waterside vegetation or in water, in savanna and forest. In our area, known from Nimule, Torit, Mongalla and Juba, at 550–600m, in South Sudan, and at Koji, 1,266m, south-west of Mizan Teferi, in south-west Ethiopia. Also known from Garamba National Park in DR Congo; these records indicate it is probably widespread across South Sudan and south-west Ethiopia. Elsewhere, west to Senegal, south-west to northern Angola.

CONSERVATION STATUS: Least Concern.

Grayia smythii, Nigeria.

NATURAL HISTORY: Diurnal and aquatic, spends time in water but shelters in thick waterside vegetation, will also live in stone jetties and other man-made waterside constructions, if there are places to hide. Moves slowly and ponderously on land, but swims well. Often found in fish traps, it eats adult frogs (particularly *Xenopus*, clawed frogs), fish and tadpoles. Placid in temperament but will hiss if threatened and may threaten with open mouth. Females become fertile when *c.*75cm long. Lays a clutch of 8–20 eggs, roughly 5×2.5cm, in batches, 3–4 eggs at a site. Enemies include herons, Nile Monitors and other snakes.

THOLLON'S WATER SNAKE

Grayia tholloni
Mocquard, 1897

IDENTIFICATION: A fairly large, brown, banded water snake of Central African wetlands. Eye quite large, pupil round, iris yellow centrally. Tail very long, >40 per cent of total length. Scales smooth, in 15 rows at mid-body, ventrals 133–147, subcaudals 110–129. Maximum size about 1.1m, average 70–90cm, hatchling size unknown. Colour shades of brown or grey-brown, becoming paler on flanks, body with indistinct bands of light and dark scales, prominent in juveniles, they may fade in adults but do not disappear. Like many aquatic or semi-aquatic snakes, lip scales cream, yellow or orange with distinct black sutures. Belly yellow or cream, with black spots on ends of ventral scales.

Grayia tholloni, Uganda.

Grayia tholloni, Lake Langano.

HABITAT AND DISTRIBUTION: Lakes, rivers and streams in moist savanna and woodland, between 500 and 1,500m in our area. Less tied to big waterbodies than Smyth's Water Snake, possibly because it is smaller, and may be found in quite small streams. In Ethiopia, known from some Rift Valley lakes (Langano, Chamo and Abaya) and Gambela area, also Torit and Talanga Forest in South Sudan. Probably widespread but under-collected in south-west Ethiopia and South Sudan; known from Garamba National Park in DR Congo. Elsewhere, south-west to Angola, west to Nigeria.

CONSERVATION STATUS: Least Concern.

NATURAL HISTORY: Not well known, but presumably similar to Smyth's Water Snake, i.e. diurnal, aquatic and terrestrial, sheltering in waterside vegetation. Presumably lays eggs, no details known, it might give live birth. It eats fish, frogs and tadpoles.

ORNATE WATER SNAKE

Grayia ornata
(Bocage, 1866)

Grayia ornata, Congo.

IDENTIFICATION: A big, brown, black-banded water snake. Eye quite large, pupil round. Tail fairly long, 25–32 per cent of total length. Scales smooth, in 17–19 rows at mid-body, ventrals 144–161, subcaudals 73–88. Maximum size about 1.5m, average 80cm to 1.3m, hatchling size unknown. Usually some shade of brown, paler on flanks, with distinct black bands that may bifurcate low on flanks. Like many aquatic or semi-aquatic snakes, lip scales often pale with distinct black sutures. Belly yellow or cream, darkening towards tail. A longitudinally striped form has been documented from the Congo.

HABITAT AND DISTRIBUTION: Lakes, rivers and streams in moist savanna and woodland, at low altitudes. Not formally recorded from our area, but taken at Bagbele, 800m, in Garamba National Park, on border of Western Equatoria, South Sudan, so probably widespread in that area. Elsewhere, south-west to Angola, west to Nigeria.

CONSERVATION STATUS: Least Concern.

NATURAL HISTORY: Not well known, but likely to be similar to other *Grayia*, i.e. diurnal, aquatic and terrestrial, sheltering in waterside vegetation. Presumably lays eggs, no details known, it might give live birth. It eats fish, frogs and tadpoles.

MARSH SNAKES *NATRICITERES*

A sub-Saharan African genus of small harmless snakes, associated with water sources; their greatest diversity is in moist savannas of south-east Africa. They have small heads with round pupils and fairly short tails, which they can shed unusually easily, but are not regenerated, so many specimens have damaged or truncated tails. Six species; one in our area. Placed in the Natricinae subfamily, their closest relatives are swamp snakes, *Limnophis*, of southern Africa.

OLIVE MARSH SNAKE

Natriciteres olivacea
(Peters, 1854)

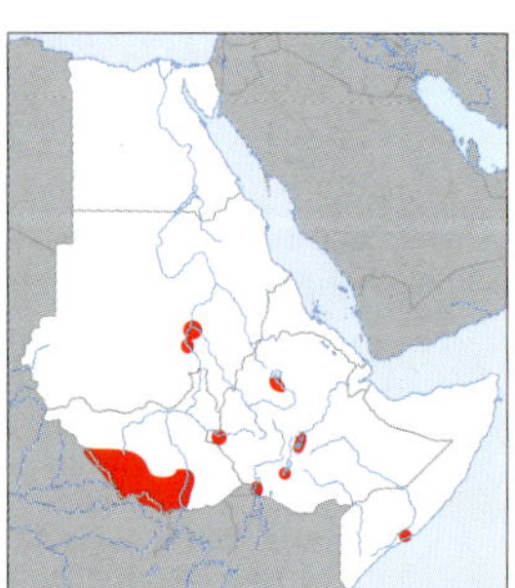

IDENTIFICATION: A medium-sized slender marsh snake, with distinctive black-edged lip scales, of variable coloration, but usually has a broad dorsal stripe. Stripe usually dull green and sometimes hard to see but may be another bright colour. Body cylindrical, tail short, about 25 per cent of total length. dorsal scales smooth, in 19 rows at mid-body (rarely 17). Ventrals 120–153. Cloacal scale divided. Subcaudals 51–87. Maximum size about 55cm, average 25–35cm, hatchling size unknown. Dorsum various shades of brown, olive-green, grey or blue-black; all-black (melanistic) specimens recorded in Congo. Usually a vertebral band five scales wide. In our area this is usually darker than the dorsal colour, but may be maroon. Dorsal stripe may also be bordered by a series of tiny white dots. Ventral scales usually yellow, the ends bordered olive, grey or red; in other areas mauve or pale blue.

Upper labials usually yellow with distinctive black edging. Chin and throat white.

HABITAT AND DISTRIBUTION: Coastal bush, moist savanna and dry savanna around watercourses, from sea level to above 2,200m. A solitary old Somali record at Mogadishu. A handful of Ethiopian records (Lakes Tana, Awassa, Arba Minch and Gambela). Widespread but sporadic in South Sudan; known from Juba, Mongalla, Bahar El Ghazal, Wau, Bo River near Southern National Park, and Upper Nile State; in Sudan recorded at Duem, Khartoum and further south along White Nile, but surprisingly few records; particularly considering that >130 individuals were collected in Garamba National Park, DR Congo. Elsewhere, west to Guinea, south-west to Angola, south to Zimbabwe.

CONSERVATION STATUS: Least Concern.

NATURAL HISTORY: A diurnal species (sometimes crepuscular) which lives at edge of streams and other water sources; it swims well. Slow-moving and inoffensive, although said to release a nauseous odour when handled. Lays a clutch of 3–11 eggs, roughly 2×1.5cm. Diet amphibians and fish. Members of this genus often seem to be predated by birds such as Hamerkops, *Scopus umbretta*. Tips of tails easily 'dropped' (autotomised) and many animals are found in nature with the tail tip missing.

Natriciteres olivacea, Bahar Dar.

FAMILY PSEUDOXYRHOPHIIDAE

A subfamily of about 90 species and 22 genera, mostly in Madagascar, but a few on other African offshore islands. Two genera occur on mainland Africa. One species, Slug-eater *Duberria lutrix*, occurs in our area.

SLUG-EATING SNAKES *DUBERRIA*

A south-east and East African genus of small, harmless slug-eating snakes; they have short blunt heads, large eyes, the body is cylindrical and the tail very short. They give live birth. Four species are known, one reaches our area.

SLUG-EATER

Duberria lutrix
(Linnaeus, 1758)

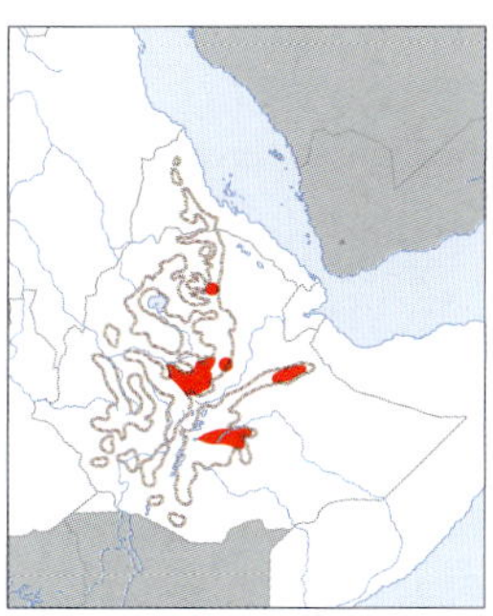

IDENTIFICATION: A stout, harmless little brown snake with a short head, often in mid to high-altitude grassland. Eye large and dark with a round pupil, but hard to see. Body cylindrical, tail short, 14–22 per cent of total length (males have longer tails). Scales smooth, 15 mid-body rows, ventrals 107–129 in Ethiopia, subcaudals paired, 20–34, cloacal scale entire. Maximum size about 45cm (big specimens usually females), but in East African averages 20–30cm, hatchling 8–11cm. Usually some shade of brown, often a fine black vertebral line is present; flanks dark.

Duberria lutrix, Mulu.

Duberria lutrix, Mulu.

Sides of head extensively marbled dark and light. Ventral scales light, with dark blotching. All-black (melanistic) specimens known. Taxonomic Notes: Several subspecies described, which genetics suggest may merit specific status; *D. l. abyssina* in our area.

HABITAT AND DISTRIBUTION: High grassland and moist savanna. In our area confined to Ethiopian highlands, from about 1,800m to 3,200m, possibly higher. Occurs on both sides of Rift Valley, in high-altitude grassland. Records sporadic but widespread; west of Rift Valley from Lake Hashenge south to Sululta Plain, west of Addis Ababa, and Ankober. On east side it occurs widely in Bale Mountains and surrounding high country, north and east in Ahmar Mountains. Elsewhere, south to South Africa.

CONSERVATION STATUS: Least Concern.

NATURAL HISTORY: Terrestrial, diurnal and secretive, foraging in grass and vegetation, when inactive hides in grass tufts, under vegetation heaps, logs, rocks, in soil cracks or down holes. Gentle and good-natured, it never tries to bite, but if handled it may squirm and defecate. Slug-eaters give live birth, up to 12 young recorded in our area; clutch size related to size of mother. Sometimes common; on an irrigated strawberry farm at Mulu, Slug-eaters comprised >90 per cent of snake population, in surrounding non-irrigated grassland <5 per cent.

FAMILY PROSYMNIDAE

An African family containing only the small unusual snakes in the genus *Prosymna*, shovel-snouts.

SHOVEL-SNOUTS *PROSYMNA*

An African genus of small harmless burrowing snakes, none larger than 45cm, most are grey or brown, often with speckling, although some species are spectacularly coloured and striped or banded. Most have smooth scales, in 15–21 rows at mid-body, and very short tails. They have a sharp-edged, broad rostral scale for pushing through soil or sand, short heads not distinct from the neck and a relatively large eye. They are rarely encountered except after rainstorms, when they may move about on the surface or are flooded out of their holes. About 18 species are known, five in our area, one a northern Somali near-endemic. Several species have a curious defence mechanism, they snap the body into coils, looking like the flat spring of a watch, and if prodded they will jerk violently, coil and uncoil. They lay eggs. Unusually, they eat nothing but reptile eggs.

KEY TO NORTH-EAST AFRICAN MEMBERS OF THE GENUS *PROSYMNA*

1a: Internasal scales paired	*Prosymna somalica*, Somali Shovel-Snout
1b: Internasal scales single	2
2a: Upper labials 5	*Prosymna greigerti*, Southern Sahel Speckled Shoved-Snout
2b: Upper labials 6	3
3a: Dorsal scales with paired apical pits, one postocular scale	*Prosymna ruspolii*, Prince Ruspoli's Shovel-Snout
3b: Dorsal scales with single apical pit, two postocular scales	4
4a: Body stocky, rostral very angular in profile, body lacks white dots	*Prosymna ambigua*, Angolan Shovel-Snout
4b: Body slender, rostral rounded in profile, back often has white dots	*Prosymna stuhlmanni*, East African Shovel-Snout

ANGOLAN SHOVEL-SNOUT, MIOMBO SHOVEL-SNOUT

Prosymna ambigua
Bocage, 1873

IDENTIFICATION: A small grey shovel-snout with a single internasal, in extreme south-west of our area. Head scarcely distinct from neck; rostral sharply pointed; eye fairly smaller, pupil round; internasal and prefrontal single; six (rarely four, five or seven) upper labials, the third and fourth border eye; two postoculars. Body cylindrical and stocky; tail short, 7–16 per cent of total length. Scales smooth, with single apical pit, reducing from 17 to 15 rows between ventrals 19–66; ventrals 138–150 in males, 158–171 in females; subcaudals paired, 27–33 in males, 17–20 in females. Largest male 30cm, but female 40cm; average 20–30cm. Usually appears grey, but varies from dark purple-brown to blue-grey, dorsal scales often dark-edged. Chin white or has some brown. Ventrals with a blackish half-moon marking proximally, each has its free edge coloured white. Subcaudal scales patterned similarly to ventrals. Taxonomic Notes: Two subspecies recognised; details here apply to *P. a. bocagii*.

Prosymna ambigua, Tanzania.

HABITAT AND DISTRIBUTION: Moist woodland and savanna-forest mosaic around equatorial lowland forest block, in our area at 600–800m. Known only from southern South Sudan, localities notably Torit and Molongori in East Equatoria and Kagelu, Mt Korobi and Kajiko in Central Equatoria. Recorded further west at Bagbele and Garamba in northern DR Congo, near border with South Sudan. Elsewhere, widespread in northern Uganda, west to Cameroon, north-west to Central African Republic and Chad; the nominate subspecies occurs south of the Congo Basin.

CONSERVATION STATUS: Least Concern.

NATURAL HISTORY: Poorly known but probably similar to others of genus, i.e. nocturnal, burrowing, spending its time in holes or buried in soft soil or sand, most active in rainy season, eats snake and lizard eggs, lays a clutch of small eggs. A female at Molongori held six eggs measuring about 16–23×7mm, at start of July. Has very long and thin hemipenes, exceeding length of tail. Displays the 'watch-spring' behaviour described in the generic introduction. Sympatric with *P. greigerti greigerti* in Torit and Molongori in South Sudan.

SOUTHERN SAHEL SPECKLED SHOVEL-SNOUT

Prosymna greigerti
Mocquard, 1906

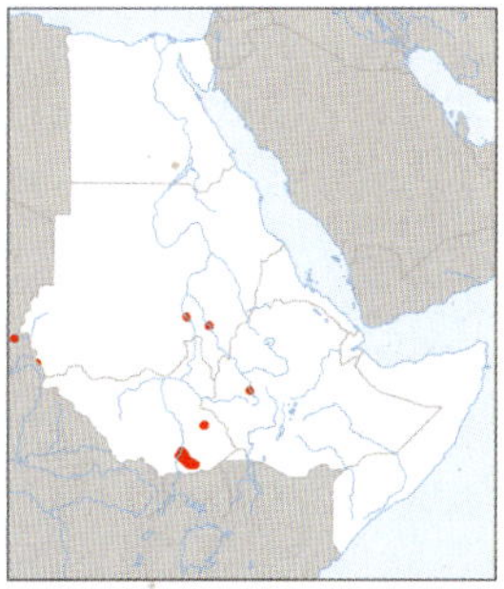

IDENTIFICATION: A small, slender snake with a shovel-shaped snout, from western Ethiopia and Sudan/South Sudan. Head slightly distinct from neck; snout pointed; eye large, pupil round; internasal single; five upper labials, the second and third border eye; one postocular. Body cylindrical and slim, tail short, 7.3–12 per cent of total length. Scales smooth, with paired

Prosymna greigerti, Neem.

apical pits, reducing from 17 to 15 rows between ventrals 17–29; ventrals 144–165 in males, 159–190 in females; subcaudals paired, 28–41 in males, 19–27 in females. Largest male 32cm, largest female 37.2cm. At a distance appears speckled grey. Back dark grey or purple; each dorsal scale has white edge; subspecies *collaris* distinct in having black nuchal band with a pair of white spots anterior, and one spot or collar posterior to it. Lower edges of rostral and upper labials white, as is ventrum; pale underside sometimes extends to lower flanks. Taxonomic Notes: Two subspecies (sometimes considered full species) occur sympatrically at some localities in Chad; the southern nominate and *P. g. collaris* further north. Differ slightly in number of ventral scales but distinguished by presence or absence of dark nuchal band and two adjoining series of white spots.

HABITAT AND DISTRIBUTION: Dry and moist savanna and woodland, at low to mid-altitudes, 380–1,200m. A typical inhabitant of dry Sahelo-Sudanian zone, the nominate form occurs in more mesic habitats (the Sudanese zone) in southern Sudan, notably at Neem, Singa in Sennar and on border with Central African Republic at Am Dafok in South Darfur, probably widespread in the intervening country, but no records. In South Sudan, at Boma, widespread in Central and East Equatoria, where localities include Mongalla, Juba, Torit, Molongori, Katire and Magwe. A single, western, Ethiopian record; at Didessa River bridge, west of Nekemte. There is an imprecise record ('upper Nile Valley', probably somewhere south of Khartoum) of subspecies *collaris*, which favours drier habitats (the Sahel); nearest record from Bahar in eastern Chad. Probably more widespread in Sudan (both forms), but overlooked due to small size and secretive lifestyle. Elsewhere, west to Senegal.

CONSERVATION STATUS: Least Concern.

NATURAL HISTORY: Poorly known but probably similar to others of genus. South Sudan specimens were under garden rubbish or hoed up during cultivation. West African snakes laid clutches of 2-4 eggs in early March. This snake shows the watch-spring behaviour described in the generic introduction. Very secretive; they do not try to bite.

PRINCE RUSPOLI'S SHOVEL-SNOUT

Prosymna ruspolii
(Boulenger, 1896)

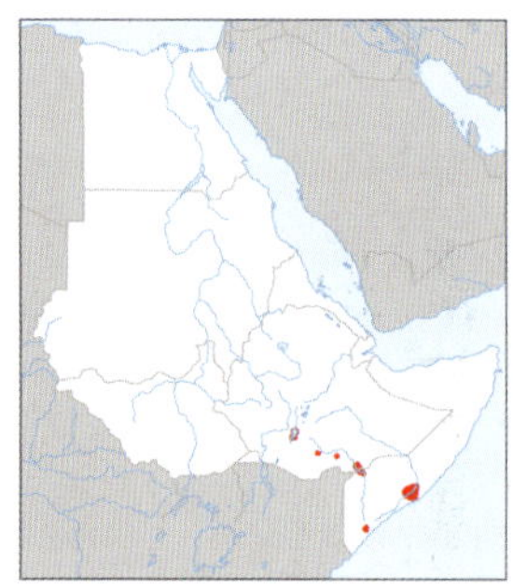

IDENTIFICATION: A small snake with a shovel-shaped snout and single internasal, from southern Ethiopia and Somalia. Head slightly distinct from neck; rostral with an acutely angular horizontal edge; eye large, pupil round; internasal and prefrontal single; six upper

Prosymna ruspolii, Negele.

Prosymna ruspolii.

Prosymna ruspolii, after eating egg, Negele.

labials, the third and fourth border eye; one postocular. Body cylindrical and fairly stocky, tail short, 11–16 per cent of total length. Scales smooth, with paired apical pits, reducing from 17 to 15 rows between ventrals 25–44; ventrals 130–138 in males, 138–152 in females; subcaudals paired, 31–36 in males, 23–28 in females. Largest male 23cm, largest female 30cm. Dorsum dark grey, almost blackish, or brown, but often appears prominently speckled because each scale has a pale margin; upper lip and ventrum greyish-white; snout paler, pinkish-brown. Tongue pink with a white tip. Taxonomic Notes: Two subspecies recognised, nominate *P. r. ruspolii* in our area. Based on molecular data, *P. ruspolii* together with closely related *P. greigerti* and extralimital *P. meleagris* belong to the *meleagris* group in *Prosymna*.

HABITAT AND DISTRIBUTION: Semi-desert, subdesert or dry wooded *Vachellia–Commiphora* steppe, dry grassland and coastal thicket, from sea level to about 1,600m. Widespread; scattered localities in southern Ethiopia and Somalia. Known at Lake Abaya, a mesic habitat in southern Rift Valley in Ethiopia, and further east and south at Negele and Filtu Plains. In Somalia, at 'Magala Umberto I', a historical village and island on Ganale River, north of Dolo. In southern Somalia reported at Belet Amin on Juba River, Duduble, Giohar, Afgoi, Jonderma and Mogadishu near or on Shebelle River on Benadir coast. Probably more widespread, but due to its secretive lifestyle fairly under-collected. Elsewhere, south to Kenya and northern Tanzania (subspecies *keniensis*).

CONSERVATION STATUS: Least Concern.

NATURAL HISTORY: Probably similar to others of genus. An adult was recently found by day under a large flat stone on a dry grassland plain south-east of Negele. A large female at Belet Amin laid four eggs, each measuring 27.5×7mm. Two long-term captive adults from Filtu fed readily on the eggs (*c*.10–11mm in diameter) of Ethiopian East African House Gecko *Hemidactylus angulatus*, a common gecko (for part of year) in southern Ethiopia.

SOMALI SHOVEL-SNOUT

Prosymna somalica
Parker, 1930

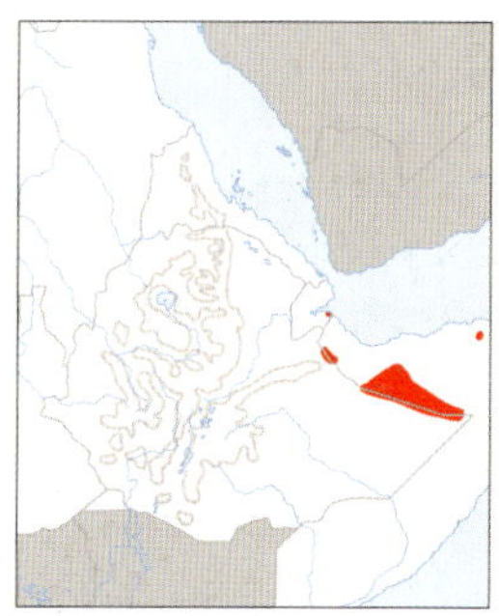

IDENTIFICATION: A small snake with a shovel-shaped snout and paired internasals. Head slightly distinct from neck; rostral sharply pointed; eye fairly large, pupil round; internasals paired, in broad contact (an excellent aid to identification); six upper labials, the third and fourth border eye. Body cylindrical and stocky; tail fairly short, 12–20 per cent of total length. Scales smooth, usually with paired apical pits, reducing from 17 to 15 rows between ventrals 25–41; ventrals

116–125 in males, 131–143 in females; subcaudals paired, 30–38 in males, 22–29 in females. Maximum size 26cm, average 18–23cm, hatchling size unknown. Dorsum red-brown to dark purple-brown, but appears prominently speckled because each scale has a pale margin. Head occasionally black, at least in juveniles. Underside white, sometimes with a brown or dark grey patch on throat and first ventrals.

HABITAT AND DISTRIBUTION: A north Somali endemic of dry environments, from semi-desert with lava fields to medium and high savanna, also in *Vachellia–Commiphora* dry lowland savanna, between near sea level and 1,900m (usually 600–1,500m). A handful of widely scattered localities, suggesting it is rare but widespread. The type was from 21km west of Qoriley in the Haud in Somaliland, subsequently collected at several localities situated very precisely along border with Ethiopia and Somaliland as far north as Harirad in Borama District. Inland Somaliland records from Burao and several kilometres north of Erigavo; a recent isolated specimen from Djibouti. Probably more widespread in adjacent parts of Ethiopia and Puntland.

CONSERVATION STATUS: Least Concern.

NATURAL HISTORY: Probably similar to others of genus. Mostly collected in sandy and stony country with patches of grass and *Vachellia–Euphorbia* bush. An adult male, recently found in early September in high country (1,880m) north of Erigavo, was under a rock by day following heavy rain, on a hillside with low bushes, *Aloes* and other succulents; at Shanshacade one was crawling between bushes on red sandy soil following a short rainstorm at night.

Prosymna somalica, Shanshacade.

EAST AFRICAN SHOVEL-SNOUT

Prosymna stuhlmanni
(Pfeffer, 1893)

IDENTIFICATION: A small slender snake with a single internasal and distinctly pale brownish yellow snout, from extreme southern Somalia. Head comparatively long and small, slightly distinct from neck; snout more rounded than in other species. Eye large, pupil round; internasal and prefrontal single; six upper labials, the third and fourth border eye; usually two postoculars. Body cylindrical, long and slim; tail short, 11–15 per cent of total length. Scales smooth, with single apical pit, reducing from 17–15 rows between ventrals 20–61; ventrals 124–153 in males, 138–164 in females; subcaudals paired, 27–39 in males, 17–30 in females. Largest male 27cm, largest female 30cm. Colour quite variable, red-brown, blue-grey or black, each scale with a paler centre, but most north-east African specimens light to dark grey, the scales are shiny. Some have a single or double row of white spots on centre of back. Snout distinctly brownish-yellow or almost greenish-yellow, ventrum white, pearly or cream. Tongue pink with a white tip. Taxonomic Notes: Together with closely related *P. ambigua* and extralimital *P. janii*, it belongs to the *ambigua* group in *Prosymna*.

HABITAT AND DISTRIBUTION: Semi-desert, dry and moist savanna and coastal thicket, at low to mid-altitudes, in our area close to sea level. In north-east Africa recorded only at Belet Amin, a village on Juba River, about 65km north-east of Kismayu, in southern Somalia. Probably occurs in coastal zone south of Belet Amin, towards Kenya border. Elsewhere, widespread in coastal and south-east Kenya, south to South Africa.

CONSERVATION STATUS: Least Concern, but poorly known in Somalia.

Prosymna stuhlmanni, Kenya.

Prosymna stuhlmanni, Tanzania.

NATURAL HISTORY: Probably similar to others of genus. Shy, secretive and never bites. Specimens in Kenya found under sisal stumps or inside rotten logs. Clutches of 3–4 eggs recorded, one 1.9×0.8cm, the other 3×0.7cm; gravid females collected in January–February in southern Tanzania. This snake does not appear to show 'watch-spring' behaviour described in generic introduction but has an unusual threat display, some individuals lifting up front half of body and flattening neck like a cobra, while flicking the tongue slowly in and out in deliberately; if further teased it tips the head right back, while continuing to flick the tongue. In the single known locality in our area it is sympatric with *P. ruspolii*.

FAMILY PSAMMOPHIIDAE SAND SNAKES AND ALLIES

A family of rear-fanged snakes, with about 55 species in eight genera. They are mostly African, with a limited penetration into Europe and Asia, and two Malagasy species. Six genera and 22 species occur in our area. Many are longitudinally striped, relatively large, fast-moving and diurnal.

KEY TO THE NORTH-EAST AFRICAN GENERA OF PSAMMOPHIIDAE

1a: Upper surface of snout longitudinally concave, or profile convex, with distinct raised supraocular scales ('eyebrows'), large, in northern Egypt or desert and semi-desert	*Malpolon*, Montpellier and Moila snakes
1b: Upper surface of snout flat, or profile not markedly convex, most species not in Egypt or desert	2
2a: Rostral large, usually projecting, snout hooked in profile	3
2b: Rostral normal and snout not hooked in profile	4
3a: Longitudinally striped	*Kladirostratus togoensis*, Northern Sharp-nosed Beaked Snake
3b: Without longitudinal stripes	*Rhamphiophis*, Beaked snakes
4a: Nostril pierced in an entire nasal shield	*Hemirhagerrhis*, Bark snakes
4b: Nostril pierced between two or more shields	5
5a Maxillary teeth interrupted below anterior border of eye by two much-enlarged, fang-like teeth	*Psammophis*, Sand snakes (not including *P. lineatus*)
5b: Maxillary teeth subequal in size and continuous up to interspace separating them from posterior pair of enlarged fangs	6

6a: Nostril pierced between two nasal shields only, tail long with >80 subcaudals
Psammophis lineatus, Striped Olympic Sand Snake

6b: Nostril pierced between two nasals and an internasal shield, tail moderate, <80 subcaudals
Psammophylax multisquamis, Kenyan Striped Skaapsteker

MONTPELLIER AND MOILA SNAKES *MALPOLON*

A genus of medium to large, robust, impressive snakes. Primarily diurnal, fast-moving, mildly venomous, opisthoglyphous, egg-laying, and active hunters with a wide dietary range. Two large species with elongate heads and distinct forehead furrows, distributed mainly around Mediterranean in Europe and North Africa, a third species with a distinctly short head and pointed snout occurs across North Africa and Arabia. They have large eyes with a round pupil, smooth dorsal scales (sometimes weakly grooved), in 17–19 rows at mid-body and a transverse row of small scales on posterior margin of cloacal opening. Inhabit dry, open, mostly sunny sites with some vegetation. Two species in our area. Genus *Malpolon* is most closely related to members of the genus *Rhamphiophis*.

MOILA SNAKE, FALSE COBRA

Malpolon moilensis
(Reuss, 1834)

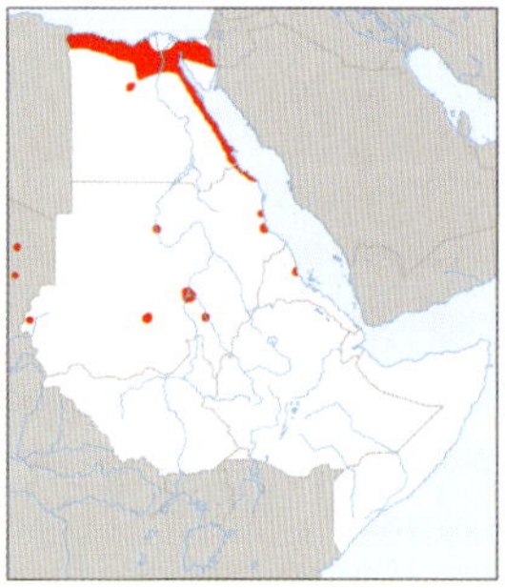

IDENTIFICATION: A medium-sized, muscular, sandy-coloured snake with pointed snout and sharp *canthus rostralis*. Body slim in younger and fairly stout in large individuals, cylindrical; tail moderately long, about 19–26 per cent of total length. Body scales smooth, matte and fairly large, in 17 (rarely 19) mid-body rows, ventrals 152–180, subcaudals divided 48–73, cloacal scale divided. Maximum size 130–150cm (although snakes >110cm are rare), average 70–95cm, hatchling 15–20cm. Dorsum sandy, sometimes more reddish-yellow, with regularly spaced, round, diffuse, brown spots (more contrasting in juveniles); a subocular patch below each eye, one or two dark bars in posterior temporal region of head and adjacent neck on each side very distinctive in this species. Iris orangey-golden, with a darker periphery. Belly uniform white or cream-colored. Taxonomic Notes: In recent years placed in genus *Rhagerhis*, but molecular data reinstate it in *Malpolon*.

HABITAT AND DISTRIBUTION: Semi-desert, desert edges or sparsely vegetated areas in desert, but will venture into quite stony desert, generally at low altitudes in our area, although up to 1,800m in Morocco. Most Egyptian records in north (northern Sinai, Mediterranean coast), a cluster of records from vicinity of Cairo, Wadi Natrun and Faiyum, sporadic records further south (Asyut, Wadi El Gemal). Sudanese records equally sporadic: Suakin, Haiya, Khartoum, Dongola and Darfur. Um Ruwaba, Wadi El Awateib and 'Bara' (=Barah) seem to be southernmost Sudanese records. A single Eritrean record, Hasmet, on northern coastal plain. Range probably much more extensive in Sudan. Elsewhere, west to Mauritania, east throughout Arabia.

Malpolon moilensis, Tunisia.

CONSERVATION STATUS: Least Concern.

NATURAL HISTORY: Essentially diurnal and crepuscular, even nocturnal (especially in hottest periods). Rapid and fairly shy, ground-dwelling. Periods of inactivity spent hidden in rodent burrows or beneath logs and boulders. Has been observed 'sand-shovelling' using its head (the large head scales, sharp *canthus rostralis* and pointed snout are good digging tools) for unknown reasons, perhaps to uncover prey. If threatened flattens and flares its neck, forming a defensive hood and has been mistaken for a cobra. If handled it tries to bite. Rear-fanged, some local swelling following a bite has been noted. Diet largely small mammals, but also takes birds, lizards and other snakes; juveniles may eat large insects. Females lay 4–18 eggs, about 9–10×24–29mm in size, in moist substrates. Ritual fighting between males recorded.

EASTERN MONTPELLIER SNAKE

Malpolon insignitus
(Geoffroy Saint-Hilaire, 1827)

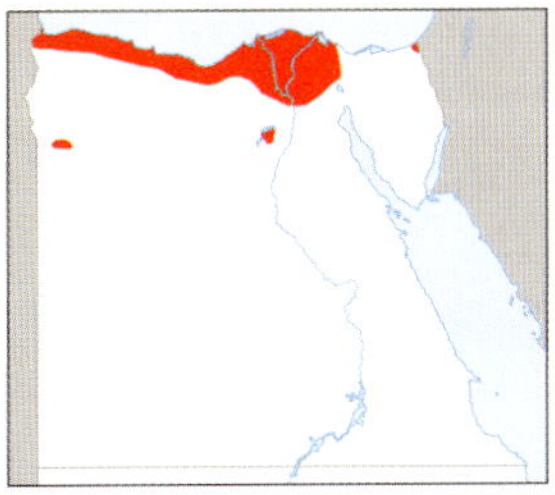

IDENTIFICATION: A large, robust and impressive snake with a deep distinctive furrow on snout, known from northern Egypt and north-east Sinai. Snout round at tip with a distinctly concave furrow longitudinally (an excellent aid to identification), eye large with round pupil and prominent supraocular scales (said to have a 'penetrating' impression). Body slim but muscular, cylindrical; tail relatively long, *c.*25–28 per cent of total length. Body scales smooth, large and weakly grooved, giving a somewhat coarse appearance, in 19 mid-body rows, ventrals 156–182, subcaudals divided 75–102, cloacal scale divided. Maximum size 182cm, average size 110–150cm (males much larger), hatchling 20–33cm. Dorsum usually more or less uniform grey, greenish-grey, olive, brown to blackish, sometimes with whitish spots (especially in females); head and neck sometimes reddish (especially in males). Iris golden, with darker periphery. Belly cream-colored or variegated in orange. Colour and pattern more contrasting in young, often with darker head. Taxonomic Notes: Two subspecies recognised, in our area only nominate *M. i. insignitus* is present.

HABITAT AND DISTRIBUTION: Favours Mediterranean habitats and adjacent semi-desert, in our area from sea level to about 200m. In lightly vegetated steppe-like habitats, marginal cultivated areas, wetlands, saltmarshes, littoral scrub, in rocky but also sandy areas. Not in true desert. Known from northern Egypt on Mediterranean coast, mostly within 50km of coast, in Nile Delta and Valley, in more mesic regions as far south as Cairo and Faiyum. An isolated record at Siwa Oasis. Specimens from Rafah are only known Sinai records. Elsewhere, west to Morocco, east to Iran, north to southern Russia. In Egypt common but rather local. Collected for pet trade.

CONSERVATION STATUS: Least Concern.

NATURAL HISTORY: Primarily diurnal, but in hotter periods also crepuscular. A shy, fast-moving terrestrial species, but climbs well. Shelters in rodent burrows or under rocks. Mildly venomous but confident in defence, bites have led to severe local reactions. If cornered, it hisses and strikes. An active hunter; feeds mainly on lizards, small mammals, also birds and other snakes; juveniles also eat large arthropods. Prey is killed by venom or swallowed alive. They lay 4–12 (occasionally 18) eggs, about 12–25×40–55mm in size.

Malpolon insignitus, Egypt.

BARK SNAKES *HEMIRHAGERRHIS*

An African genus of slim, little, rear-fanged snakes. Diurnal and partially arboreal, in savanna and semi-desert. Most eat lizards. Their small size and non-aggressive habits mean they are harmless to humans. Recently one of the two known species, Bark Snake *H. nototaenia*, was divided into three species. Thus four species are now known, three in north-east Africa.

KEY TO NORTH-EAST AFRICAN MEMBERS OF THE GENUS *HEMIRHAGERRHIS*

1a: Brown or tan, with a straight-edged, dark or pale vertebral stripe, or no stripe at all, ventrals 140–159 — *Hemirhagerrhis kelleri*, Striped Bark Snake

1b: Grey or brown, usually with an irregular black vertebral stripe or line of dots or crossbars, ventrals 152–187 — 2

2a: Temporal formula usually 1+2, lower labials usually nine, in our area often with distinct crossbars, underside heavily stippled grey-brown — *Hemirhagerrhis nototaenia*, South-eastern Bark Snake

2b: Temporal formula usually 2+3, lower labials usually ten or more, in our area usually has zigzag vertebral stripe, underside pale with grey streaks — *Hemirhagerrhis hildebrandtii*, Kenyan Bark Snake

STRIPED BARK SNAKE

Hemirhagerrhis kelleri
Boettger, 1893

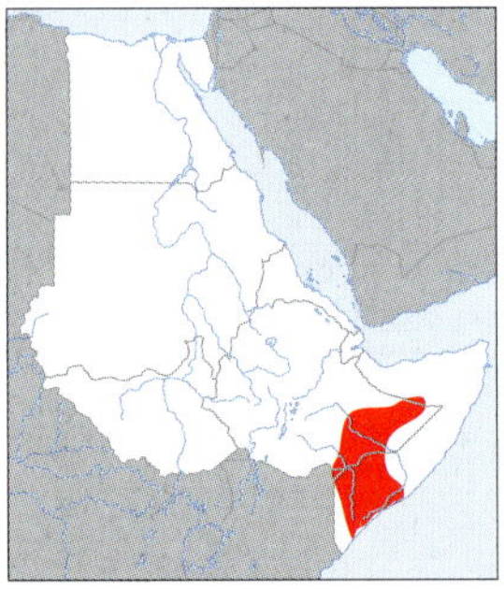

IDENTIFICATION: A slim brown striped snake with a small eye. Pupil round or slightly oval vertically, iris golden-brown, but dark head stripe runs through it. Supraocular slightly raised. Body cylindrical, tail c.21–27 per cent of total length. Scales smooth, in 17 rows at mid-body, ventrals 137–159, subcaudals 61–78. Maximum size about 40cm, average 25–35cm, hatchling size unknown. Usually has a dark brown vertebral stripe (sometimes with a light yellow stripe in centre), a fawn dorsolateral stripe and pale brown flank stripe. Vertebral stripe occasionally very pale or absent, such individuals may have vertebral and lateral lines of small brown spots. Dorsal scales often have a fine black dot at posterior end. Head has fine brown and tan reticulations. Ventrals white or cream with three dashed brown or rufous, white-centred stripes.

HABITAT AND DISTRIBUTION: Dry savanna, coastal thicket and semi-desert, from sea level to about 1,500m, but commonest below 1,200m. Far eastern Ethiopia and widespread in southern Somalia from south of Belet Uen, along the Webi Shebelle, to north of Mogadishu, thence to Kenya. A single record in the Somali Haud (west of Odenleh) on Ethiopian border. A couple of South Sudanese records, at Talanga and Torit, and known from Kidepo, just across border in Uganda. Elsewhere, south to Tanzania.

CONSERVATION STATUS: Least Concern.

NATURAL HISTORY: Poorly known. Diurnal and partially arboreal, climbing into small trees and bushes. Fairly quick-moving, at night it shelters under ground

Hemirhagerrhis keller, Kenya.

cover or in holes. Lays eggs, a clutch of three recorded. Captives fed readily on lizards, especially day geckos (*Lygodactylus*). A population in southern Somalia fed almost exclusively on gecko eggs, a Tanzanian snake had eaten a Somali-Maasai Clawed Gecko *Holodactylus africanus*.

KENYAN BARK SNAKE

Hemirhagerrhis hildebrandtii
(Peters, 1878)

IDENTIFICATION: A little grey snake with a zigzag vertebral line. Very similar to previous species, eyes prominent but small, pupil round, iris silvery-brown, and facial stripe through eye. Tail 24–34 per cent of total length. Scales smooth, in 17 rows at mid-body, ventrals 157–177, subcaudals 79–105, cloacal scale divided. Maximum size about 48cm (possibly larger), average 25–40cm, hatchling size unknown. Colour grey, most have a pattern of broad black subtriangular blotches connected by a dark vertebral stripe, but some also have narrow crossbars; flanks mottled grey, sometimes with black blotching. Sides of head and neck often paler than rest of body, especially in inland specimens. Tail tip may be orange or rufous. Below pale grey or buff, lightly stippled black or brown.

HABITAT AND DISTRIBUTION: Widespread in semi-desert, dry and moist savanna and coastal thicket, from sea level to about 1,200m. A couple of records in Ethiopian Ogaden (Harradigit, Ado) and a cluster of records at junction of Kenyan, Ethiopian and Somali borders (Murri, Mandera, Dolo, Rahanuin Region, Dinsoor) south to Kismayu and Kenya border. Informally reported in Galguduud Region. Known from Lokitaung near South Sudan border in north-west Kenya. Elsewhere, south to northern Tanzania.

CONSERVATION STATUS: Least Concern.

Hemirhagerrhis hildebrandtii, Kenya.

NATURAL HISTORY: As previous species; fast-moving, diurnal and semi-arboreal. Active during even in hottest hours of day. Climbs big trees by working its way up the bark, its camouflage is superb. May dart out to seize lizards from ambush, often feeds on *Lygodactylus* dwarf geckos, but also recorded preying on eggs of Tree Gecko *Hemidactylus platycephalus* at Kismayu in Somalia. A Kenyan female laid four eggs in early September.

SOUTH-EASTERN BARK SNAKE

Hemirhagerrhis nototaenia
(Günther, 1864)

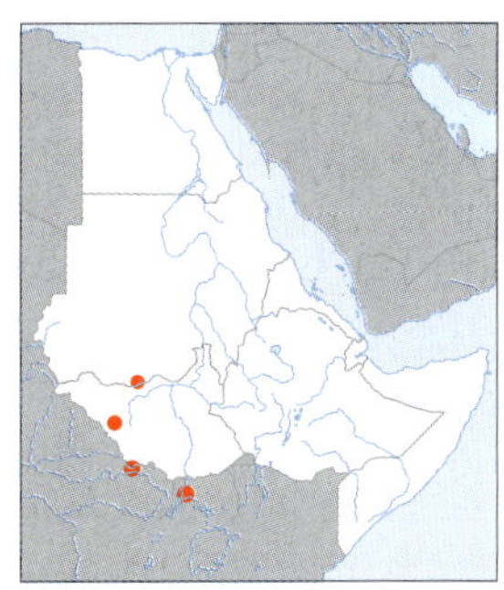

IDENTIFICATION: A little brown-grey snake with a broad vertebral stripe edged by irregular dark spots, in South Sudan. Eyes prominent but small, pupil round, iris yellow, tongue red. Supraocular raised, giving a 'raised eyebrow' impression. Tail 22–30 per cent of total length. Scales smooth, in 17 rows at mid-body, ventrals 151–187, subcaudals 58–90, cloacal scale divided. Maximum size about 44cm, average 25–40cm, hatchling size unknown but an 18cm juvenile in Kenya. Most specimens north of equator seem to be brown, with a dark, broad or narrow vertebral stripe, edged with a series of fine darker spots; flanks and underside striped light and dark grey. Taxonomic Notes: Sudanese and West African specimens very different from the typical south-east African form; their status would bear investigation.

HABITAT AND DISTRIBUTION: Dry and moist savanna and woodland, at low altitudes in our area, 500–800m. Known from Yambio, Western Equatoria, and Bahar El Ghazal State (without details) in South Sudan, and Abyei, Southern Kordofan, Sudan. Informally reported at Torit and 50km north of Nimule, South Sudan. Elsewhere west across the Sahel to Burkina Faso, south to Botswana and South Africa.

CONSERVATION STATUS: Least Concern.

NATURAL HISTORY: Diurnal and semi-arboreal, often found on rough bark of trees (hence the name), looking for geckos and their eggs. Non-aggressive and doesn't usually bite if picked up. When approached it often freezes, relying on its camouflage. If detected in open it slides rapidly away. Shelters under bark, ground cover, in tree cracks, etc. Lays 2–8 elongate eggs, roughly 0.6×2.5cm, gravid females in southern Tanzania in October and May. Diet mostly lizards, especially tree-dwelling geckos, gecko eggs, skinks and small agamas, and occasionally amphibians.

Hemirhagerrhis nototaenia, Uganda.

SAND SNAKES *PSAMMOPHIS*

A Saharo-Sindian and African genus of diurnal, fast-moving rear-fanged snakes, with long heads and big eyes with round pupils, slim elongate bodies and long tails; evolved to move fast through lightly vegetated or grassy habitats. Most are striped above and below. Dorsal scales smooth, in 11–19 rows at mid-body. They have very small, smooth hemipenes, and this may be connected with size and diurnal habits, enabling them to separate rapidly and flee if disturbed while copulating. Sand snakes have relatively large, grooved fangs, some bites to humans have resulted in severe local symptoms (pain, swelling, itching, etc.). When undisturbed, sand snakes show a curious behaviour known as 'self-rubbing', smearing a secretion from a gland just inside the nose along the belly and lower dorsal scales; the purpose is uncertain. About 33 species are known, most in Africa, a few reach the Middle East and western Asia, in a range of habitats (save true desert and closed forest). Thirteen species occur in our area (and a 14th, *P. praeornatus*, is reported from the Central African Republic, not far west of the Sudanese border).

Taxonomy of some species is confused, despite intense research spanning many years; complicated by fact that several are polymorphic, some hybridise in certain places but not in others, and some change colour and shape as they grow. The relatively big savanna and semi-desert snakes in the so-called 'Hissing Sand Snake *P. sibilans* complex' are a case in point. Names used for members of this group include *sibilans*, *mossambicus*, *orientalis*, *afroccidentalis*, *phillipsii*, *sudanensis*, *subtaeniatus* and *rukwae*. They are not always easy to separate; experts have argued acrimoniously over the identity of crucial specimens; the key and DNA may give different identities for the same animal. Bear this in mind if you use our key; some forms may not fit, or appear to be intermediates.

KEY TO NORTH-EAST AFRICAN MEMBERS OF THE GENUS *PSAMMOPHIS*

1a: In Egypt and northern Sudan, underside always plain, without striping, nine upper labials — *Psammophis aegyptius*, Saharan Sand Snake

1b: Lacks above combination of characters — 2

2a: Transverse dashes on outer edges of neck ventrals, no large fang-like teeth in front of fangs, which are below level of back of eye — *Psammophis lineatus*, Striped Olympic Sand Snake

2b: No transverse dashes on outer edges of neck ventrals, enlarged fang-like teeth below front margin of eye, enlarged fangs below back margin of eye — 3

3a: Dorsal scales in 17 rows at mid-body — 4

3b: Dorsal scales in <17 rows at mid-body — 10

4a: Upper labials usually nine — 5

4b: Upper labials usually eight — 6

5a: Dorsum with a clear straight-edged black, yellow-edged, vertebral stripe — *Psammophis punctulatus*, Speckled Sand Snake

5b: Dorsum lacks black vertebral stripe — *Psammophis schokari*, Schokari Sand Snake

6a: Usually first five lower labials in contact with anterior sublinguals — 7

6b: Usually the first four lower labials in contact with anterior sublinguals — 8

7a: Top of head often with a pale median stripe on the snout which forks when reaching the frontal and then borders the frontal, usually less than 100 subcaudals — *Psammophis rukwae*, Lake Rukwa Sand Snake

7b: Top of head often with a pale median stripe on the snout which is interrupted before the frontal or forks only after the middle of the frontal, usually more than 100 subcaudals — *Psammophis sibilans*, Hissing Sand Snake

8a: Ventrum uniform or with ill-defined brown or black line or dashes — *Psammophis mossambicus*, Olive Sand Snake

8b: Ventrum usually with a pair of well-defined black lines, yellow between them, ends of ventrals white — 9

9a: Top of head without a pale median stripe, usually lacks body stripes — *Psammophis orientalis*, Eastern Stripe-belled Sand Snake

9b: Top of head with a yellow median stripe on the snout and frontal, body conspicuously striped — *Psammophis sudanensis*, Northern Stripe Bellied Sand Snake

10a: Dorsal scales in 13–15 rows at mid-body — 11

10b: Dorsal scales in 11 rows at mid-body — *Psammophis angolensis*, Dwarf Sand Snake

11a: Dorsal scales in 13 rows at mid-body — *Psammophis pulcher*, Beautiful Sand Snake

11b: Dorsal scales in 15 rows at mid-body — 12

12a: Two upper labials, usually fifth and sixth, enter orbit, head usually uniform brown, bordered below by a dark stripe through eye — *Psammophis biseriatus*, Link-marked Sand Snake

12b: Three upper labials, usually 4–6th entering orbit, head with dark-bordered tan blotches on a paler background and dark-bordered pale longitudinal stripe at junction of intranasal and prefrontals — *Psammophis tanganicus*, Tanganyika Sand Snake

SAHARAN SAND SNAKE

Psammophis aegyptius
Marx, 1958

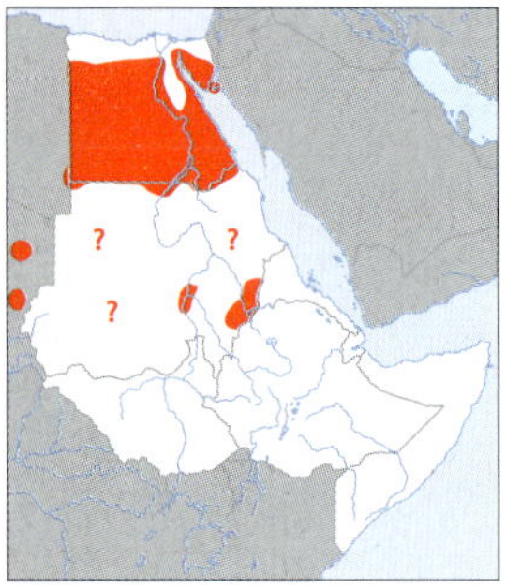

IDENTIFICATION: A fairly big, fast-moving, faintly striped or unicoloured sand snake of true desert. Head long, snout pointed, supraocular slightly raised, eyes large, pupil round, iris yellow-brown, loreal very elongate with pronounced longitudinal groove in most specimens, 12–14 lower labials. Body cylindrical and muscular, tail about $\frac{1}{3}$ of total length. Scales smooth, in 17–19 rows at mid-body, ventrals 183–203, subcaudals 110–122, paired, cloacal scale divided. Maximum size about 1.5m, average 80cm to 1.2m, hatchlings 25–30cm. Dorsal colour brown, grey or yellowy-brown, underside yellowish in juveniles, becoming covered with brown fine spots as they grow; occasional individuals salmon-red below. Usually a dark line through eye to back of head, and lips often white, sometimes speckled brown. Taxonomic Notes: Most closely related to very widespread Schokari Sand Snake *P. schokari*, they split about 10 Mya.

HABITAT AND DISTRIBUTION: Sandy and rocky desert, sand seas and arid ravines, even very open and arid areas, usually at low elevation but in Sinai ascends to above 1,500m. Widely but sporadically recorded

Psammophis aegyptius, Hamata.

Psammophis aegyptius, Hamata.

Psammophis aegyptius, White Nile State.

almost throughout Egypt south of 30°N, wherever there is true desert, known from Tiran Island, South Sinai. In Sudan reported at Jebel Awaynat, at border with Libya, Sudan and Egypt, Wadi Halfa, and Khartoum and environs (Hafir and Ab-Dareesh). Khartoum animals have high ventral counts (around 190) and are plain above, underside unpatterned or delicately spotted, loreal strangely long, but no doubt they are this species. Specimens from Sennar, Kordofan and those from northern Nubian desert, west and north of Dongola have not been evaluated recently but we assign them to this species; following recent research in Chad that indicates all specimens from the northern $\frac{2}{3}$ of that country belong to this species. Some recently sequenced Somaliland specimens with plain dorsal coloration and pattern-less belly, from Tokhoshi, Rugay and Las Anod (not shown on map) nest within this species, although traditionally assigned to *P. schokari*. Elsewhere, just reaches southern Israel/Palestine in east, west to Algeria and Niger.

CONSERVATION STATUS: Least Concern.

NATURAL HISTORY: An astonishing lifestyle. Terrestrial but fast-moving, agile and climbs well, and is active by day, dusk and sometimes at night. Although it takes lizards (particularly when young) and desert mammals, adults favour migrating birds, and they shelter near large boulders, in rock cracks and in trees and bushes, waiting for such migrants to settle. In Egypt, most isolated hills harbour a few of these snakes. They lay eggs, and when handled or threatened will inflate the throat and bite.

SCHOKARI SAND SNAKE

Psammophis schokari
(Forskål, 1775)

IDENTIFICATION: A fairly large, slender sand snake of near-desert. Head long, eye large, pupil round, iris red-brown, loreal as long as nasals without a pronounced longitudinal groove in most specimens, 10–12 lower labials. Body slim and cylindrical, tail long, about ⅓ of total length. Scales smooth, in 17 midbody scale rows, ventrals 156–179 (up to 189 in Ethiopia and Somalia), cloacal scale divided, subcaudals 109–121 (in Egypt; 77–115 in Sudan to northern Somalia). Maximum size about 1.5m, average 80cm to 1.2m, hatchling size unknown. Colour quite variable: some uniform brown, others have a broad darker brown, black-speckled vertebral stripe, and a lighter flank stripe (and irregular dark head markings), others have a narrow yellow vertebral stripe and two yellow, black-edged dorsolateral stripes; this phase is sometimes associated with higher rainfall regions. A dark stripe through eye extends onto neck; a useful field character. Lips white, speckled brown to a greater or lesser extent. Cream, pale grey or reddish below, with a dark grey mid-ventral band and rows of dark streaks towards outer edges of ventrals. Taxonomic Notes: Records from Sudan, Eritrea, Djibouti and northern Somalia are assigned tentatively to this species; an as-yet unpublished molecular study indicates this and the preceding species are extremely diverse, and their boundaries may need redrawing.

HABITAT AND DISTRIBUTION: Near-desert, semi-desert and dry savanna, usually in areas with some permanent vegetation, however sparse, from sea level to about 1,500m. Fairly widespread in northern Egypt and Sinai, extending south to about 29°N, a few records in the Nile Valley. Probably occurs right along Red Sea coast in Sudan (known from Tokar, Suakin, Durrur). Eritrean records include coast at Massawa, Gulbub, Saati and Ghinda, as well as inland at Keren and Tesenay. Known from Djibouti and across northern Somalia (most of arid Somaliland and Puntland). Elsewhere, a huge range west to Senegal, east to Pakistan.

Psammophis schokari, Berbera.

Psammophis schokari, Las Anod.

Psammophis schokari, Tunisia.

CONSERVATION STATUS: Least Concern.

NATURAL HISTORY: A fast-moving, agile, diurnal snake, active during even during heat of day, and partially arboreal. At Berbera (Somaliland) one was found lying on a low *Vachellia* bush at night at a height of about 50cm. Females lay 2–6 eggs (occasionally up to ten). An active hunter, moving with head raised. Diet mostly lizards, but larger individuals may take rodents and birds; a Sinai snake had eaten a chameleon. A Djibouti specimen from Djalelo ate a Common Long-tailed Lizard, *Latastia longicaudata longicaudata*.

DWARF SAND SNAKE

Psammophis angolensis
(Bocage, 1872)

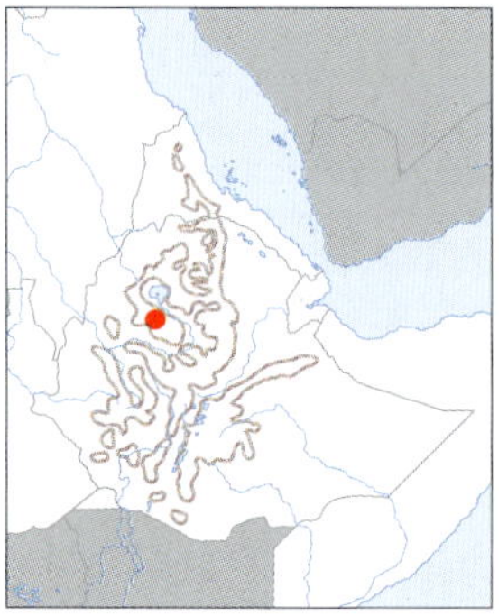

IDENTIFICATION: A small striped sand snake with fine pale crossbars on head; in our area known from just two specimens in north-west Ethiopian highlands. Eye large, pupil round, iris red-brown, tongue black. Body cylindrical, tail 27–30 per cent of total length. Scales smooth, in 11 rows at mid-body, ventrals 135–156, subcaudals 57–82, cloacal scale divided. Maximum size about 50cm, average 30–40cm, hatchling 14–15cm. Distinctly marked; top of head black or deep brown with three fine yellow or white crossbars between eye and nape, behind third is a broad dark collar, followed by 1–3 dark crossbars, which may coalesce. A broad dark brown, finely yellow-edged vertebral stripe, flanks light or yellow-brown, sometimes with fine dark hairlines on lower flanks. Ventrals white or cream, as are lips, chin and throat. Taxonomic Notes: Lack of any further records in Ethiopia (and total absence from Kenya) suggests that the two Ethiopia specimens might not be *P. angolensis* but something new, although mid-body count of 11 places them in that species.

HABITAT AND DISTRIBUTION: In our area, known only by two specimens from Dangila, at 1,200m in Ethiopia's north-west highlands, in high-altitude savanna and grassland that is now almost entirely under agriculture. Elsewhere, in savanna from Tanzania south to northern South Africa.

CONSERVATION STATUS: Least Concern.

NATURAL HISTORY: Southern African specimens are diurnal, terrestrial and secretive, usually living in well-vegetated areas. Quick-moving, they don't usually bite if picked up but struggle. They lay 2–5 elongate eggs, roughly 0.5 × 1.5cm; females gravid in September and 15cm hatchlings collected in November and January in Tanzania. Diet lizards, especially skinks and small geckos, skink eggs and frogs. Rare and nowhere common, the only other sand snake of similar size is Beautiful Sand Snake *P. pulcher*.

Psammophis angolensis, South Africa.

BEAUTIFUL SAND SNAKE

Psammophis pulcher
Boulenger, 1895

IDENTIFICATION: A small striped snake, with a rounded snout, big eye and round pupil. Body cylindrical, tail long, 36–39 per cent of total length. Scales smooth, in 13 rows at mid-body, ventrals 140–147, subcaudals paired, 97–108. Maximum size 43cm. Hatchling size unknown. A fine brown, orange or maroon, black-edged vertebral line bordered by two uniform grey or brown dorsolateral stripes, darkening downwards to become a black flank stripe that extends onto side of head but not the lips, which are cream. The lower 1.5 dorsal scale rows are white. Belly yellow centrally, bordered by fine red lines, outer edges of ventrals white.

HABITAT AND DISTRIBUTION: One of Africa's rarest snakes, prior to 2010 known from just four specimens. The holotype was collected on the Webi Shebelle River in western Ogaden, Ethiopia, by Donaldson Smith in 1894, at 500m. Second and third specimens were found in Kenya, at Voi and Mwingi, and a fourth is known from 'Somaliland', without further data (might refer to the Ogaden). In August 2011, however, a Kenyan safari guide, Sean Flatt, collected several specimens at Bisanadi National Reserve, east of Meru National Park in Kenya. Apparently inhabits dry savanna at low altitudes. Probably quite widespread in the Ogaden, but very rare.

CONSERVATION STATUS: Least Concern.

NATURAL HISTORY: Poorly known but probably similar to other small *Psammophis*. Flatt noted that they were active by day, a male and a female were caught outside same hole on successive days and captives fed on dwarf geckos *Lygodactylus*. Most were on the ground. A male introduced into a cage with a female in August wrapped himself around her and rubbed his nose along her flank, she turned and rubbed noses, face to face, while they mated.

Psammophis pulcher, Kenya.

STRIPED OLYMPIC SAND SNAKE

Psammophis lineatus
(Brandt, 1838)

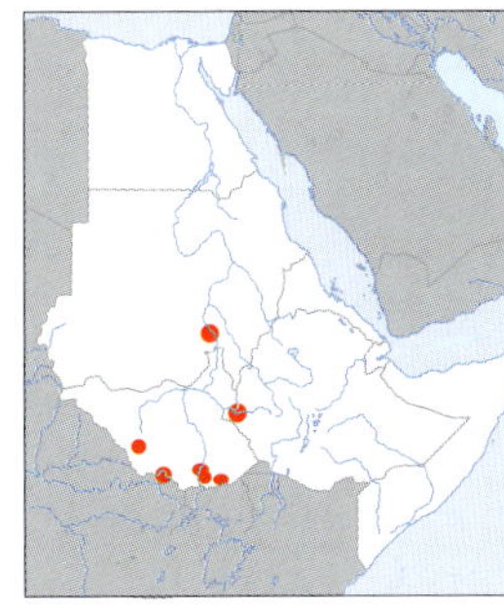

IDENTIFICATION: A medium-sized sand snake, usually in marshy areas, with short crossbars on the neck's ventral scales. Head long, pupil round, iris brown. Body muscular and cylindrical, tail fairly long, about ⅓ of total length but often mutilated. Scales smooth, in 17 rows at mid-body, ventrals 138–167, subcaudals paired, 73–105. Maximum size about 1.2m, average 70cm to 1m, hatchling size unknown but a Ghana juvenile with an umbilical scar was 24cm. Colour essentially brown or olive. Some uniform above, others striped; juveniles usually striped. May have any combination of a broad brown dorsal stripe, a fine yellow vertebral line, a fine yellow dorsolateral stripe and grey-brown lateral stripe, black-edged on lower margin. Head sometimes green. Usually 2–3 pale yellow neck crossbars, more common in juveniles. Lip scales sometimes black-edged; preocular scale often distinctly yellow, which can be seen in field. Belly yellow, greenish-yellow or cream, with distinct transverse black marks at outer edges, particularly on neck ventrals; these are a valuable clue to identification. Taxonomic Notes: Long placed in genus *Dromophis*, Olympic snakes, but molecular analysis proved they were *Psammophis*.

HABITAT AND DISTRIBUTION: Moist savannas, especially on floodplains, around lakes and marshes, and in damp grassland, but will move some distance from water. Occurs at altitudes of 300–600m in our area, elsewhere down to sea level and also higher. Known from Gambela area in Ethiopia. Probably occurs almost throughout Nile Basin in South Sudan and southern Sudan, due to presence of suitable habitat, but very few records: South Sudan records include Torit, Mongalla, Juba, Bahr El Ghazal; also Garamba

National Park just across border in DR Congo; Sudan records include Al Jabalain near Kosti. Elsewhere, south to northern Botswana, west to Senegal.

CONSERVATION STATUS: Least Concern.

NATURAL HISTORY: A diurnal, terrestrial snake, fast-moving, it swims well and will hunt in water. If approached darts away quickly; if seized some seem reluctant to bite but others do so freely. Takes cover in waterside vegetation, in holes or under ground cover at night, a Ghana specimen was under a clump of mud. Clutches of 6–9 eggs, roughly 1.5×3cm, recorded. In wild feeds largely on amphibians but a captive took a range of lizards.

Psammophis lineatus, Nigeria.

OLIVE SAND SNAKE, HISSING SAND SNAKE

Psammophis mossambicus
Peters, 1882

IDENTIFICATION: A big brown snake, common in many parts of north-east Africa. Has a rounded deep snout ('roman nose'), overhanging supraocular and preocular scales (said to have a 'penetrating expression'), a big eye with a round pupil, the iris may be yellow or brown. Body cylindrical and muscular, tail long, 27–33 per cent of total length. Scales smooth, in 17 rows at mid-body, ventrals 151–183, cloacal scale divided, subcaudals paired, 82–110. Maximum size about 1.8m (possibly larger), average 1–1.5m, hatchling 28–33cm. Quite variable, usually olive-brown above, but may be any shade of brown or grey. All dorsal scales or just central ones may be black-edged, giving a finely striped appearance. Labial scales, chin and sides of neck may be yellow, orange or dull red, the lips black-speckled, as may the lower flanks. Belly usually yellow, cream or white (white anteriorly can shade to yellow posteriorly), sometimes with a faint or clear dark line or series of dashes on each side, but never a pair of clear black lines.

HABITAT AND DISTRIBUTION: Widespread in a range of habitats, typically in savanna with >400mm annual rainfall, so not in desert and semi-desert, and absent from areas above 2,500m and closed forest. A 2019 revision of the complex by Jean-François Trape and his team indicates that within our area, this species (at least, as currently defined) occurs in South Sudan and Sudan, but not Ethiopia or further east. Probably quite widespread in South Sudan, and Sudan south of about 13°N; a more precise distribution will depend upon re-examination of museum specimens assigned to other species; previously, this species wasn't listed as occurring in West or Central Africa.

Psammophis mossambicus, Kenya.

Psammophis mossambicus, Heglig.

Psammophis mossambicus, Heglig.

Elsewhere; west to Nigeria, south to eastern South Africa.

CONSERVATION STATUS: Least Concern.

NATURAL HISTORY: Diurnal, partially arboreal. Fast-moving and alert, if threatened dashes away and hides. Often mistaken for more dangerous species due to its size and colour. Emerges mid-morning to bask and then hunts. If restrained will bite vigorously. From 8–30 eggs, roughly 1.5×3cm, are laid in a suitable moist hole or damp tree crack, or in leaf litter, and take about two months to hatch. Hatchlings collected November–December in Tanzania, and eggs were laid in September there. A wide range of prey recorded, including snakes (one ate a young Black Mamba), lizards, rodents, frogs and even birds; juveniles eat mostly lizards. Often eaten by birds of prey, caught as they bask on tops of bushes. Some local symptoms of envenomation recorded following bites to snake-handlers, including pain, which may be severe, swelling, discoloration, nausea and intense local itching.

LAKE RUKWA SAND SNAKE

Psammophis rukwae
Broadley, 1966

IDENTIFICATION: A large and rather rare sand snake. Head fairly long, eye large, pupil round, iris red-brown. Body slim and cylindrical, tail long, about ⅓ of total length. Scales smooth, in 17 mid-body scale rows, ventrals 148–183, cloacal scale divided, subcaudals 82–122. Maximum size about 1.3m, average 80cm to 1.2m, hatchling size unknown. Usually striped; there is fine pale vertebral line astride a broad brown or rufous dorsal stripe, flanks yellow or pale brown, ventrals yellow or cream with a very fine black line each side; sometimes yellow between lines and white on outer ventrals. From other striped sand snakes by combination of dorsal stripes (often subdued) and fine (not thick) black ventral lines.

HABITAT AND DISTRIBUTION: Largely in moist savanna, between sea level and 1,500m. The 2019 Trape *et al*. revision of the *P. sibilans* group indicates this species occurs in south and west Ethiopia, Sudan and South Sudan; a recent analysis confirms a specimen from Eritrea. Probably widespread; but pending re-examination of museum material, definite records are scarce; they include 'south of Asosa' and Aykel in Ethiopia, Asmara (Eritrea), Torit and Thar Jath oilfield in South Sudan. Elsewhere west to Cameroon, south to Lake Rukwa in Tanzania.

CONSERVATION STATUS: Least Concern.

NATURAL HISTORY: Not well known, but probably similar to other members of complex, i.e. diurnal and partially arboreal, fast-moving, hunts by sight, lays eggs; diet small vertebrates, particularly lizards and other snakes.

Psammophis rukwae, Asmara.

NORTHERN STRIPE-BELLIED SAND SNAKE

Psammophis sudanensis
Werner 1919

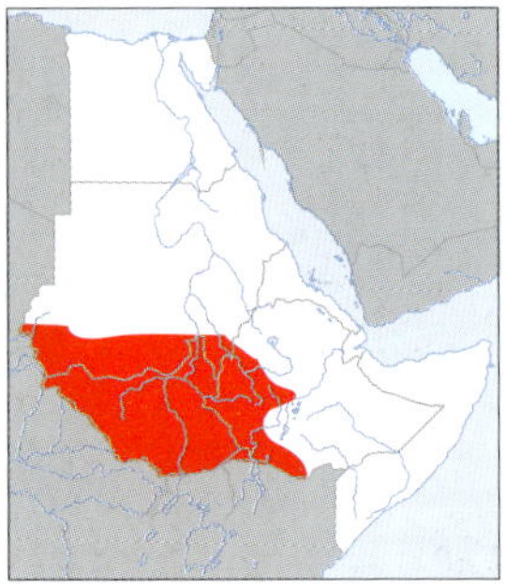

IDENTIFICATION: A fairly big striped sand snake with a long head and pointed snout. Eye large, pupil round, yellow or brown. Body slim and muscular, tail long, about ⅓ of total length. Scales smooth, in 17 rows at mid-body, ventrals 146–180, subcaudals 82–129, cloacal scale divided. Maximum size about 1.2m, average 70cm to 1m, hatchling 26–28cm. Conspicuously striped, the broad brown vertebral stripe usually has a yellow hairline at its centre. On each side is a fine yellow dorsolateral stripe, often finely edged black, and a broad brown flank stripe. Lowest dorsal scale rows cream or white, as are outer edges of ventrals. There is a clear black line each side of the ventrals, with yellow between the lines. Head distinctly marked, with three yellow stripes on snout, top of head is brown with pale cross- and angled bars. Neck barred or dark-blotched. Lips usually light, speckled brown.

HABITAT AND DISTRIBUTION: Semi-desert, dry savanna and moist savanna and grassland, from sea level to 2,700m. The type was from Kadugli, Southern Kordofan, Sudan, also recorded at Torit and environs, Juba, Yei, Malek and Murikuren, all in South Sudan, and although the 2019 revision does not list this species from Ethiopia, specimens fitting its description occur on the western plateau and the south, at Mulu, Yabello, Weyto and Konso. Probably occurs right across South Sudan and southern Sudan, it is widespread and common in Chad. Elsewhere, south to Tanzania, west to Senegal.

CONSERVATION STATUS: Least Concern.

NATURAL HISTORY: Diurnal, swift-moving and partially arboreal. Climbs well, often shelters in thickets, bushes and hedges. At night may sleep on branches, or in holes, hollow logs or under ground cover. Hunts by sight, chasing its prey, which it seizes and chews, holding the victim in its coils. If threatened, moves off swiftly, one jumped into a river to escape, juveniles are described as actually jumping off the ground in their efforts to escape. Lays 4–10 eggs, Kenyan specimens were gravid in January and February, and hatchlings collected in May. Diet mostly lizards and other snakes. A favoured prey of snake-eagles, *Circaetus*.

Psammophis sudanensis, Yabello.

Psammophis sudanensis, Mulu.

Psammophis sudanensis, Chad.

EASTERN STRIPE-BELLIED SAND SNAKE

Psammophis orientalis
Broadley, 1977

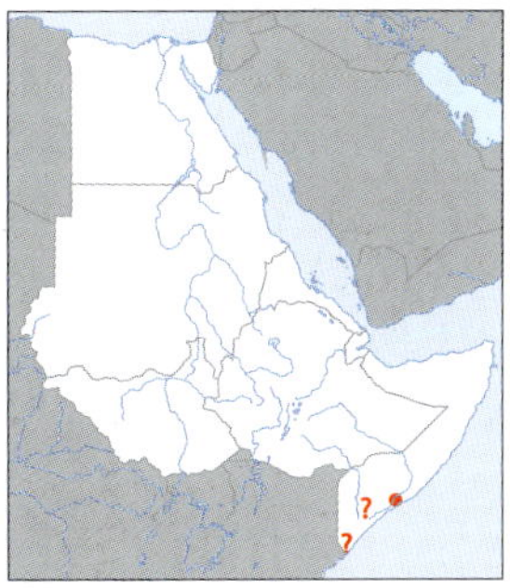

IDENTIFICATION: A fairly big, fast-moving, faintly striped or unicoloured sand snake. Head long, snout pointed, supraocular slightly raised, eyes large, pupil round, iris yellow-brown. Body cylindrical and muscular, tail about ⅓ of total length. Scales smooth, in 17 rows at mid-body, ventrals 148–170, subcaudals 94–116, paired, cloacal scale divided. Maximum size about 1.25m, average 80cm to 1.1m, hatchlings 27–32cm. All shades of brown, from pale yellow-brown to deep brown, unicoloured or with paler dorsolateral stripes. Belly has two broad black or dark brown stripes, with yellow between them and yellow or white on outer edges. Upper labials may be yellow, cream or bluish-green, immaculate or blotched brown. Usually distinguished from similar sand snakes by combination of obvious black ventral lines and faint or absent dorsal stripes.

HABITAT AND DISTRIBUTION: Coastal thicket and moist savanna, from sea level to 1,300m. In our area, a single Mogadishu specimen, tentatively identified as this species, but its extension along Somali coast from Kenya border would not be unexpected. Elsewhere, south along East African coast and inland to Malawi, Mozambique and eastern Zimbabwe.

CONSERVATION STATUS: Least Concern.

NATURAL HISTORY: Fast-moving, diurnal, hunting mostly on ground but will ascend low trees and bushes, or rocks. By night shelters in holes, under ground cover, in hollow logs or sleeps in a thicket on branches. Hunts by sight, chasing moving prey. If restrained will bite fiercely and thrashes around, rotating vigorously, liable to break its tail if held too far back. Two to ten elongate eggs, roughly 1 × 3cm are laid, a Kenya specimen laid nine eggs in mid-February; hatchlings collected in November and December in southern Tanzania. Diet mostly lizards and other snakes but takes rodents, frogs and birds. A common snake on East African coast, often seen around habitation and in gardens, and tolerant of urbanisation, where rock walls and ornamental trees provide habitat for the lizards it eats.

Psammophis orientalis, Kenya.

HISSING SAND SNAKE, AFRICAN BEAUTY SNAKE

Psammophis sibilans
(Linnaeus, 1758)

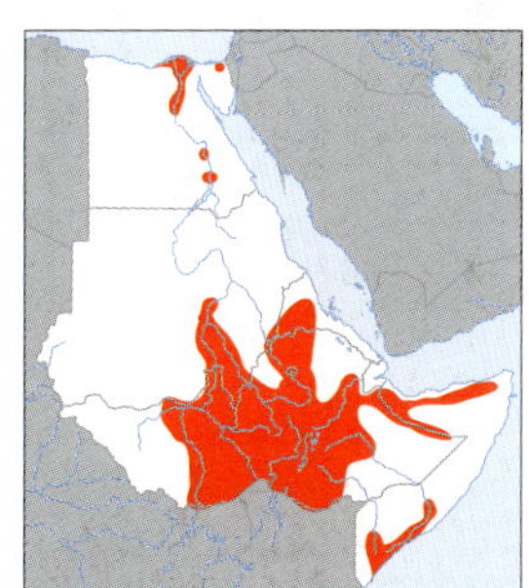

IDENTIFICATION: A fairly large, usually striped sand snake of Egypt, Sudan, Ethiopia and Eritrea. Head fairly long, eye large, pupil round, iris red-brown. Body slim and cylindrical, tail long, about ⅓ of total length. Scales smooth, in 17 mid-body scale rows, ventrals 154–178, cloacal scale divided, subcaudals 98–112. Maximum size about 1.5m, average 80cm to 1.3m, hatchling size unknown. Often has a pale median stripe on snout and poorly defined crossbars on nape. Lips immaculate yellow or have a few dark spots. Usually striped; a fine yellow vertebral line bordered by stripes of brown, black-edged scales, often a yellow dorsolateral stripe and darker flank stripe. Underside yellow, sometimes with faint or broken black hairlines. Striping may fade in large adults; some are uniform brown. Juveniles usually more contrasting than adults. Taxonomic Notes: A recent molecular analysis indicates this species is a complex of taxa, some still undescribed.

HABITAT AND DISTRIBUTION: Moist and dry savanna, often near rivers, and along watercourses in desert, and often quite tolerant of urbanisation. Between sea level and 2,600m. Endemic to north-east Africa. In Egypt confined to Nile Delta and Valley, south to Sudan and South Sudan, where most records in vicinity of Nile and its tributaries (localities include Shendi, Khartoum, Wad Madani, near Kosti, and Malakal). Records further south (Yei, Torit, Nimule, Juba) might be this species or other forms. Doesn't appear to extend west of about 28°E. Widespread in savanna of highland Ethiopia, and central Eritrea, and extends into Somaliland, across mountain ranges of north to about 50°E; recently recorded at Ufeyn in southern foothills of Cal Miskaat in Puntland. Some southern Somali records (Mareri, Gelib, Mordinle, Belet Amin, Kismayu) might be this form or another species. Common in Ethiopian and Eritrean highlands.

CONSERVATION STATUS: Least Concern.

NATURAL HISTORY: Diurnal and partially arboreal, fast-moving, hunts by sight. Lays 6–25 eggs. Feeds on any small vertebrate it can catch, particularly lizards and other snakes. Bites fiercely if seized and some painful local symptoms have been recorded following a bite.

Psammophis sibilans, Debre Zeit.

SPECKLED SAND SNAKE

Psammophis punctulatus
Duméril, Bibron & Duméril, 1854

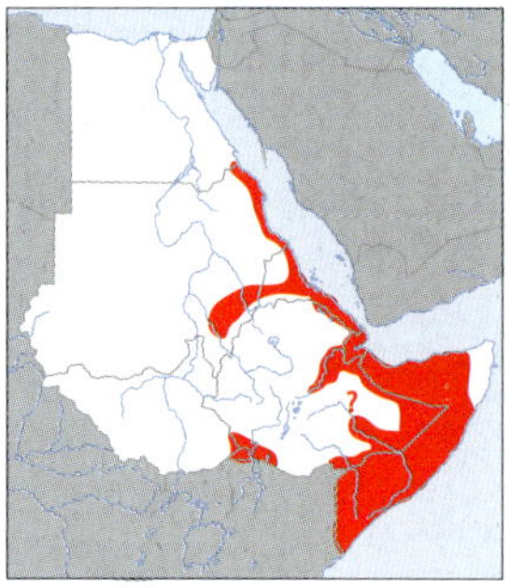

IDENTIFICATION: A big, attractive, thin, striped and speckled snake with a long head. Pupil round, iris golden-yellow. Body slim and cylindrical, tail very long and thin, ⅓ of total length. Scales smooth, in 17 rows at mid-body, ventrals 170–198, subcaudals 130–178, cloacal scale divided. Maximum size about 1.9m, average 1–1.4m, hatchling size unknown. Two subspecies in our area. *P. p. trivirgatus* has body yellow (grey in juveniles) with three black longitudinal stripes, head orange or dull red above (very occasionally grey), lips white. Nape usually grey. Flanks and belly grey or white, heavily speckled black. The nominate differs by its grey head, flanks may be pinkish, and lacks dorsolateral black stripes, having a single black vertebral stripe and higher subcaudal counts.

HABITAT AND DISTRIBUTION: Dry savanna and semi-desert, from sea level to about 1,400m. Nominate *P. p. punctulatus* occurs from Awash National Park north-east up Awash River Valley, to Djibouti (and just into extreme north-west Somaliland), through lowland Eritrea, into Sudan, west to Kassala and Blue Nile State, north along Red Sea coast to Gebel Elba in Egypt. *P. p. trivirgatus* occurs across the Ogaden, and is widespread in Somalia (rare east of 48°E); it just reaches South Sudan in Ilemi Triangle. Elsewhere; south to northern Tanzania.

CONSERVATION STATUS: Least Concern.

NATURAL HISTORY: Diurnal and partially arboreal. Probably the fastest-moving snake in Africa, its long

Psammophis punctulatus punctulatus, Gewane.

Psammophis punctulatus trivirgatus, Kenya.

powerful body enabling it to move at great speed, a useful adaptation for a big diurnal snake in a very hot, sparsely vegetated habitat. By day may hunt actively, and investigate holes, but also waits in ambush in trees and bushes. When prey spotted, it descends swiftly, seizing and holding the victim in its coils, chewing vigorously until the victim succumbs. By night, usually shelters in a hole or under ground cover but may also sleep in a tree, on a branch or in a recess. If restrained, lashes vigorously, gyrating and biting as well. Mating observed in Kenya in March and June, the two snakes closely watched by other males that had presumably followed the female's scent trail; a male and female were found in same hole in July in Voi, Kenya. Copulation usually in dry season; appearance of hatchlings coincides with rainy season. Some 3–12 eggs are laid, roughly 1×3cm, oviposition occurred in July–August; in northern Kenya a juvenile with an umbilical scar was captured in July. Diet includes lizards; known prey includes lacertids such as *Latastia* and *Heliobolus*, also agamas and skinks, and snakes, it is big enough to take small vertebrates as well. Can be found in city of Hargeisa, in gardens and sometimes enters houses. A bite from a large specimen caused intense local itching, slight swelling and pain; bites are best avoided.

LINK-MARKED SAND SNAKE

Psammophis biseriatus
Peters, 1881

IDENTIFICATION: A medium-sized, slim grey sand snake. Head long and pointed, eye large, pupil round, iris yellow-brown. Body thin and cylindrical, tail long, 35–40 per cent of total length. Scales smooth, in 15 rows at mid-body, ventrals 138–168, subcaudals 97–134. Maximum size about 1m, average 50–80cm, hatchling size unknown, a Kenya juvenile was 32cm. Has a broad olive or brown vertebral stripe, edged with evenly spaced black subtriangular blotches, scales on vertebral row light yellow-brown; resembling a chain, hence common name. Flanks and ventrals light grey with fine black speckling. Upper lips white, sometimes speckled, bordered above by dark line; top of head olive or brown, uniform or with irregular but symmetrical dark-edged pale blotching. Taxonomic Notes: This and next species (*P. tanganicus*) long regarded as subspecies. Separated by colour and lip scales (see key) but intermediates exist; which can be tentatively assigned on basis of head colour.

HABITAT AND DISTRIBUTION: Dry savanna, semi-desert and coastal thicket, from sea level to about 1,300m. Widespread in southern and central Somalia, north to Nogal Valley and the Haud; further north largely replaced by *P. tanganicus*. In Ethiopia, mostly in low country on southern border with Kenya; it might just enter south-east South Sudan in Ilemi Triangle, but no records. Elsewhere, south to Tanzania.

Psammophis biseriatus, Kenya.

Psammophis biseriatus, Kenya.

CONSERVATION STATUS: Least Concern.

NATURAL HISTORY: Diurnal and largely arboreal, in dry thornbush country where excellent camouflage makes it hard to see; looks like a dry twig. Spends much time motionless in branches, waiting for prey to pass, then advances slowly and carefully before making a final quick rush to seize its victim. At night may sleep in branches or hide in a hole or under ground cover. Fast-moving but tends to rely on camouflage, when approached in a bush it may freeze, or move surreptitiously away. If seized will try to bite, lashes wildly and rotates, breaking the tail if held near tip. Clutches of 2–4 elongate eggs laid in July and September in north-east Kenya; hatchlings collected in September in Kenya. Diet mostly lizards including dry-country geckos, lacertids, skinks and chameleons. A prolonged bite from one of these snakes caused local pain and haemorrhaging, swelling and lymphadenitis.

TANGANYIKA SAND SNAKE

Psammophis tanganicus
Loveridge, 1940

IDENTIFICATION: A medium-sized, slim grey sand snake, often like a brighter version of previous species. Head long, snout pointed but slightly bulbous, eye large, pupil round, iris yellow-brown. Body thin and cylindrical, tail long, 35–40 per cent of total length. Scales smooth, in 15 rows at mid-body, ventrals 143–169, subcaudals 81–123. Maximum size c.1m, average 50–80cm, hatchling size unknown. Colour pattern variable; broad vertebral stripe (sometimes with a fine rufous or yellow spinal stripe) grey, fawn or brown, edged with evenly spaced dark or rufous blotches which may form crossbars, flanks and

Psammophis tanganicus, Berbera.

Psammophis tanganicus, Awash National Park.

ventrals may be any mixture of grey, rufous, fawn or brown, side of neck often heavily mottled orange-brown. Head may be grey or brown above, usually with a clear, but sometimes faded dark double-crescent mark on nape and a cluster of regular dark blotches on top of head. Upper lips white or yellow, immaculate, or edged with a mix of maroon, brown and orange.

HABITAT AND DISTRIBUTION: Semi-desert and dry savanna, from sea level to about 1,300m. A single Eritrean record near Assab, known from northern Djibouti, and much of low eastern Ethiopia, south to border on Daua River, thence into Kenya. Widespread in northern Somalia, largely north of Nogal Valley but does extend into the Haud, where it meets *P. biseriatus*. The situation beyond there is somewhat confused. Reported from south and south-west Libya, the Sudan (Kordofan, En Nahud, Red Sea coast, Erkowit), but apparently without supporting specimens; so not shown on map. Elsewhere, south to Tanzania.

CONSERVATION STATUS: Least Concern.

NATURAL HISTORY: Poorly known but probably similar to Link-marked Sand Snake, i.e. diurnal and largely arboreal, living in dry thornbush country, looks like a dry twig, largely hunts by ambush. Seems to be more prone to hunting on ground. A gravid female was captured in December in Awash National Park. Diet largely lizards, but may take other small vertebrates. Lays eggs.

Psammophis tanganicus, Dire Dawa.

SHARP-NOSED BEAKED SNAKES, SHARP-NOSED SKAAPSTEKERS
KLADIROSTRATUS

A genus erected in 2019 to accommodate two small, diurnal, striped, savanna-dwelling rear-fanged snakes with distinctly pointed noses; previously placed in *Rhamphiophis* (beaked snakes), then *Psammophylax* (skaapstekers). The name honours the eminent South African-based herpetologist Bill Branch, 1947–2018 (the Greek word *klados* means branch).

NORTHERN SHARP-NOSED BEAKED SNAKE

Kladirostratus togoensis
(Matschie, 1893)

IDENTIFICATION: A small striped snake with a big eye, sharply pointed snout and conspicuous black ventral lines; in south-west South Sudan. Head short and deep, pupil round, iris brown above but black flank stripe runs through it. Body cylindrical, tail 17–25 per cent of total length. Body scales smooth, 17 rows. Ventrals 165–188, subcaudals 57–76. Maximum size about 85cm, average 50–70cm, hatchling size unknown. Conspicuously striped, with a fine brown vertebral line in centre of broad fawn stripe; a broad dark brown flank stripe starts on snout and runs through eye, lower dorsal yellow or tan. Belly yellow centrally, bordered either side by black line, which is fine or broad but always unbroken. Outer edges of ventrals white.

Kladirostratus togoensis, Nigeria.

HABITAT AND DISTRIBUTION: Moist savanna, from sea level to above 1,600m. In our area only at Yei, at 800m in Central Equatoria, in wooded savanna of South Sudan. Also known from Murchison Falls in north-west Uganda and Garamba National Park in north-east DR Congo. Elsewhere, west across savanna to Guinea.

CONSERVATION STATUS: Least Concern.

NATURAL HISTORY: Poorly known. Diurnal and terrestrial, often in holes. Said to live 'in wet lowlands'. Lays eggs. Diet includes amphibians and rodents. One was found in burnt grassland with its body held vertically, supposedly pretending to be a branch.

SKAAPSTEKERS *PSAMMOPHYLAX*

An African genus of small to medium, usually striped, fast-moving, rear-fanged snakes. They are terrestrial and diurnal, inhabiting savanna and grassland. Venom fairly toxic, but little is injected when they bite humans, and no serious symptoms have been recorded. Unusually, male skaapstekers are larger than females. Six species; one in our area. Skaapsteker means 'sheep-stabber' in Afrikaans, as they were mistakenly believed to kill sheep. There have been moves to change this pejorative name, but it is well known in East Africa, so we retain it.

KENYAN STRIPED SKAAPSTEKER

Psammophylax multisquamis (Loveridge, 1932)

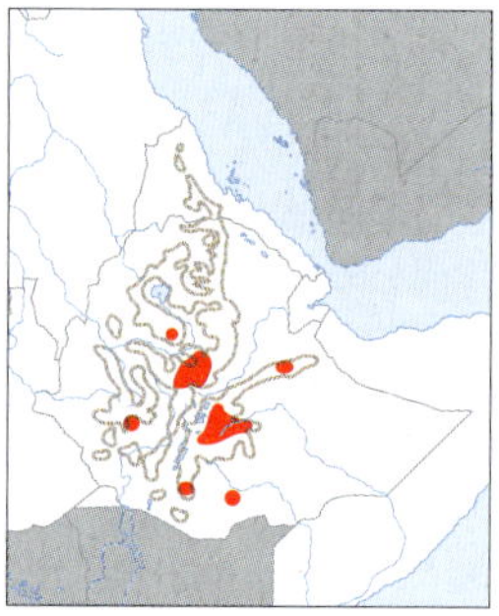

Psammophylax multisquamis, faded adult, Mulu.

Psammophylax multisquamis, Mulu.

IDENTIFICATION: A big striped snake. Head quite short, eye medium to small, pupil round. Body cylindrical, tail quite short, 18–20 per cent of total length. Dorsal scales smooth, in 17 rows at mid-body, ventrals 160–184, subcaudals 51–66. Maximum size about 1.4m (larger individuals always males), average 60–90cm, hatchling size unknown. A striped snake, the central dorsal stripe fawn, light brown or grey, often with a fine yellow vertebral line, bordered by darker brown or grey lateral stripe. Ventrals cream, often with a bright yellow median stripe and line of orange or rufous dashes on each outer edge. Colours may fade in large adults, leaving stripes still visible but not contrasting much.

HABITAT AND DISTRIBUTION: Tall grassland and moist savanna at medium to high altitudes, from about 1,900 to above 3,000m. Widespread in highland Ethiopia on both sides of Rift Valley. A record from Jimma in south-west forest (not shown on map) is probably erroneous. Reported from Torit, South Sudan, but without a specimen. Elsewhere, in highland Kenya.

CONSERVATION STATUS: Not Assessed, but very common in high grasslands of Ethiopia.

NATURAL HISTORY: Diurnal and terrestrial, shelters under ground cover, in rotting logs, vegetation clumps or holes. Seems to have a curious pattern of activity (there is speculation that it might be nocturnal to some extent), as it is often under cover during day. Often common in high grassland. Sometimes basks in a strange, kinked fashion. Quite fast-moving, slides away if threatened and able to escape, but if cornered or seized, it sometimes shams death, turning its head and neck over, opening its mouth and lolling lifeless. Lays eggs, clutches of 4–16 found, and females coil around them, presumably protectively. Diet rodents, lizards and amphibians, but may also take small snakes.

BEAKED SNAKES *RHAMPHIOPHIS*

An African genus of relatively large, back-fanged, diurnal snakes with hooked or pointed snouts; they have a short skull and the rostral bone is braced by the nasal, an adaptation for shoving through soil or sand. They have strong muscular bodies. Big adults sometimes bite vigorously but their venom does not appear to be particularly toxic to humans. Three species are known, in semi-desert, dry and moist savanna; all occur in our area. They are under-represented in most museums, due to their secretive burrowing way of life. Before the building of the Kariba Dam on Zambezi River, Rufous-beaked Snake *R. rostratus* was represented in Zimbabwe museums by a handful of specimens but after the water rose more than 30 were collected, flooded out of their holes. Beaked snakes are fast-moving, some species can spread a curious flattened hood.

KEY TO NORTH-EAST AFRICAN MEMBERS OF THE GENUS *RHAMPHIOPHIS*

1a: Mid-body scale rows 19, head and neck red or rufous in adults, body grey
Rhamphiophis rubropunctatus, Red-spotted Beaked Snake

1b: Mid-body scales rows 17, medium-sized to small, head and neck not red in adults — 2

2a: Dark stripe through eye, dorsal scales often dark-edged, giving a latticed effect, mostly in east of our area
Rhamphiophis rostratus, Rufous Beaked Snake

2b: No dark stripe through eye in our area, dorsal scales not dark-edged, mostly in far west of our area
Rhamphiophis oxyrhynchus, Western Beaked Snake

WESTERN BEAKED SNAKE

Rhamphiophis oxyrhynchus
(Reinhardt 1843)

IDENTIFICATION: A fairly large, muscular snake, with a big eye; in South Sudan. Head short, snout pointed, eye large with a round pupil and a golden or red-brown iris, hard to see in normal light. Tongue pink with a white tip. Body cylindrical, tail 27–32 per cent of total length. Scales smooth, in 17 rows at mid-body, ventrals 170–196, subcaudals paired, 88–106. Maximum size about 1.5m, average 70cm to 1.2m, hatchling 28–35cm. Colour quite variable: grey, pink, brown, yellow-brown or orange, in large specimens dorsal scales horizontally darker in centre, giving body a finely striped appearance. Ventrals immaculate white, cream or yellow, sometimes throat is yellow and belly white. Juveniles have a series of small rufous flank blotches. Taxonomic Notes: This species and Rufous Beaked Snake *R. rostratus* were long thought to be subspecies. The boundary between the them lies across South Sudan.

HABITAT AND DISTRIBUTION: Moist savanna, from about 500m to 1,300m, elsewhere to sea level. Essentially a West and Central African snake, in our area only in southern South Sudan, at Wau, Kediba

Rhamphiophis oxyrhynchus, Ghana.

and Juba. Slightly further south-east, at Torit, *R. rostratus* occurs. Elsewhere, south into northern Uganda, west to Senegal.

CONSERVATION STATUS: Least Concern.

NATURAL HISTORY: Diurnal and terrestrial (but will climb bushes), spends much time in holes looking for prey. Quick-moving and alert, when moving through the bush it often pauses, head up. When it sees prey it may jerk head side to side, targeting its prey. Copulation in captivity occurred in open, taking between 20 minutes and three hours, in February and April. A Ugandan female was gravid in October, which could mean that hatchlings appear in rainy season, March–April. Hatchlings in Ghana were captured in April–May, at start of rainy season. Multiple clutches recorded in captivity, up to four per year, but if this occurs in the wild is not known. Clutches of 6–18 eggs, roughly 2×4.5cm, recorded. Eats a wide variety of prey; rodents, lizards, frogs and snakes.

RUFOUS BEAKED SNAKE

Rhamphiophis rostratus
Peters, 1854

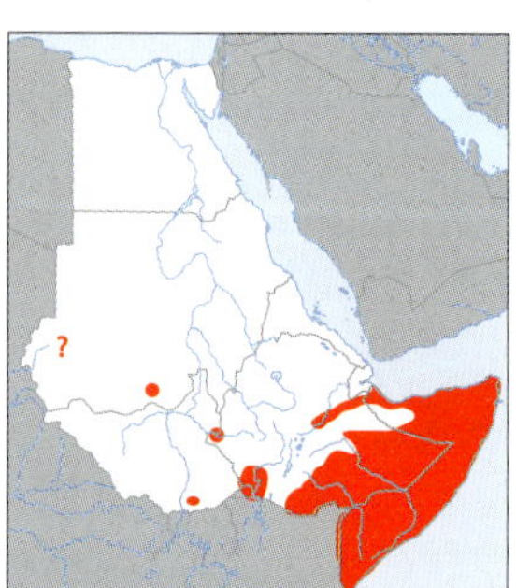

IDENTIFICATION: A fairly large, muscular, rear-fanged snake. Head short, snout pointed, eye large with a broad dark line through it. Pupil round, iris golden or red-brown, and hard to see in normal light. Body cylindrical and muscular, tail 25–30 per cent of total length. Scales smooth, in 17 rows at mid-body, ventrals 148–192, subcaudals paired, 90–125, cloacal scale divided. Maximum size about 1.6m, average 80cm to 1.2m, hatchling 25–32cm. Colour quite variable: field impression of a white snake with dark speckling; may be grey, pink, brown, orange, white, in darker specimens scales towards tail have pale centres. Ventrals immaculate white, cream or yellow, sometimes finely dark-edged. Juveniles heavily speckled, irregular red-brown, more concentrated on lower flanks, fading by time they reach 60–70cm long.

HABITAT AND DISTRIBUTION: Semi-desert, dry and moist savanna, coastal thicket and woodland, from sea level up to about 1,500m. From southern Djibouti border, virtually throughout Somalia and low eastern and southern Ethiopia, below 1,400m. Distribution elsewhere in our region disjunct and complicated. Recorded at Gambela in south-west Ethiopia; Torit in South Sudan, Darfur and Gebel Moro (Southern Kordofan), Sudan, with a surprising record from Bitea, eastern Chad. Elsewhere, south to Botswana.

CONSERVATION STATUS: Least Concern.

NATURAL HISTORY: Secretive, diurnal and terrestrial (but will climb bushes), spends much time in holes looking for prey. When moving it often pauses, head up, and jerks the head side to side like a chicken; this behaviour is also shown in a more pronounced and

Rhamphiophis rostratus, Kenya.

continuous way by aroused males demonstrating to a possibly receptive female. Active all hours of day, even at hottest times, when inactive hides in holes, squirrel warrens, etc, often in abandoned termitaria; it seems to occupy a suitable refuge for a length of time, foraging in the environs. Fast-moving and alert, it moves quickly away from danger, but if threatened and unable to escape flattens neck into a small hood and raises the fore body. If picked up, will hiss, jerk convulsively and may bite. It can dig holes with its pointed snout, breaking the soil with the rostral and turning the head sideways to scoop soil out. Clutches of 4–12 eggs laid, roughly 3.5×2cm, a Tanzanian specimen laid 15 eggs over a five-day period. Newly hatched juveniles collected in December in Tanzania. Eats a range of prey; rodents (including naked mole rats), lizards, frogs and snakes, and reported eating beetles. Two captive specimens competing for a mouse showed agonistic behaviour, neck-wrestling with each other. Often killed on roads due to being active by day.

RED-SPOTTED BEAKED SNAKE

Rhamphiophis rubropunctatus
(Fischer, 1884)

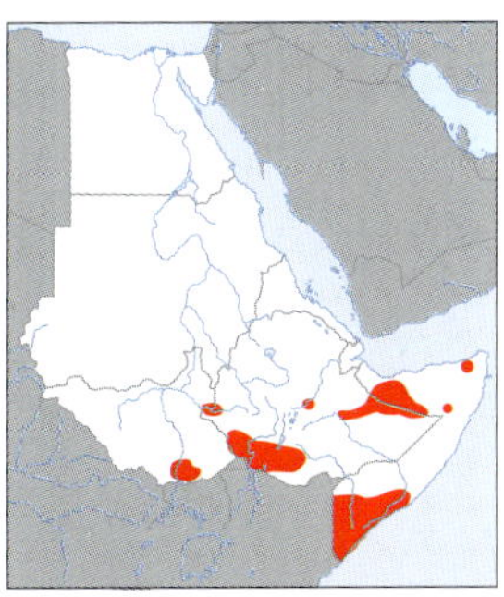

IDENTIFICATION: A big, grey or grey-brown, orange-headed, back-fanged snake. Common name inappropriate, as only juveniles are red-spotted. Head short, snout weakly pointed, tongue black, eye large and dark, pupil round, iris brown. Body elongate and cylindrical, tail about ⅓ of total length. Scales smooth, in 19 rows at mid-body, ventrals 207–245, subcaudals paired, 130–160, cloacal scale divided. A big snake, reaching 2.5m, average 1.4–2m, hatchling size unknown. Usually grey, with an orange head, sometimes anterior body is orange or yellow-orange. Ventrals cream, light brown or light orange. Juveniles cream or white, heavily marked with red or red-brown sub-triangular spots, and a grey dorsal stripe; fades to adult colour at a length of 50–60cm. Interestingly, a colour phase of Boomslang is identical, grey with a red head, and occurs in same area.

HABITAT AND DISTRIBUTION: Coastal thicket and woodland, semi-desert and near-desert, moist and dry savanna, from sea level to about 1,200m. Widespread but sporadic in southern Somalia, south of 03°N, a handful of records from southern Ethiopia south and west of Arba Minch in low country below the forest in Negele area, and north to Gambela; equally sporadic in South Sudan, at Torit, Lohutok, Juba and Mongalla. A cluster of records in Somaliland and the northern Ogaden, including Ogo Mountains and Hargeisa, Adadley, Lafaruug, the Haud, Metaker and Garowe. Distribution probably much more continuous, but despite its huge size this is a rare and secretive snake, hence under-reported. Elsewhere, south to Tanzania.

CONSERVATION STATUS: Least Concern.

NATURAL HISTORY: Poorly known but probably similar to Rufous Beaked Snake. Secretive, diurnal and mostly terrestrial, will climb low trees and bushes. Often hunts prey in holes, when inactive shelters there; in squirrel warrens, abandoned termitaria, hollow logs, etc. A specimen in northern Kenya was active in a barren lava field during heat of day. Lays eggs. Diet includes rodents and squirrels; a captive juvenile took lacertids. One of our biggest rear-fanged snakes, but little known, and often uncommon even in prime habitat (although around Hargeisa is quite common on sandy soils). A big snake, so bites should be avoided.

Rhamphiophis rubropunctatus, juvenile, Hargeisa.

FAMILY ATRACTASPIDIDAE AFRICAN BURROWING SNAKES

A family of 12 genera and c.70 species of mostly small African burrowing snakes; a few extend into or occur in the Middle East. Most are secretive, rear-fanged and essentially harmless. However, the family includes, in their own subfamily, 22 species of burrowing asp, *Atractaspis*, which are highly evolved, venomous, front-fanged snakes that cause much pain and suffering.

KEY TO THE NORTH-EAST AFRICAN SUBFAMILIES OF ATRACTASPIDIDAE

1a: Large movable fangs at front of upper jaw	Subfamily Atractaspidinae, Burrowing asps
1b: No enlarged teeth at front of upper jaw	Subfamily Aparallactinae, African burrowing snakes

SUBFAMILY APARALLACTINAE

A subfamily of small burrowing snakes, with around 45 species in nine genera. Mostly African. Ten species, in six genera, in our area. Most are small, dark-coloured burrowing snakes, with small, short heads, small eyes, no loreal scale and grooved rear-fangs, but not dangerous to humans.

KEY TO THE NORTH-EAST AFRICAN GENERA OF APARALLACTINAE

1a: Subcaudals in a single row	2
1b: Subcaudals in a double row	3
2a: Slim, subcaudals more than 30	*Aparallactus*, Centipede eaters
2b: Stocky, subcaudals less than 16	*Brachyophis*, Somali short snakes
3a: No preocular	4
3b: Preocular present	5
4a: Anterior temporal present	*Micrelaps vaillanti*, Guineafowl Snake
4b: Anterior temporal absent	*Amblyodipsas*, Purple-glossed snakes
5a: Two pairs of shields between rostral and frontal	*Polemon christyi*, Christy's Snake-eater
5b: Single shield between rostral and frontal	*Chilorhinophis butleri*, Butler's Black and Yellow Burrowing Snake

CENTIPEDE EATERS *APARALLACTUS*

A sub-Saharan genus of small slim dark snakes with a tiny head and eye with round pupil. Unusually, they have single subcaudals. The two local species have relatively large rear fangs. They live underground, in holes or soft soil, although they may emerge, especially on damp nights following rain, and slide around looking for prey or a mate. They eat centipedes. Most lay eggs but Jackson's Centipede Eater *A. jacksonii* gives live birth. Eleven species are known, with the majority in south-east Africa; two species occur in our area.

KEY TO NORTH-EAST AFRICAN MEMBERS OF THE GENUS *APARALLACTUS*

1a: Postoculars 1, separated from anterior temporal	*Aparallactus lunulatus*, Plumbeous Centipede Eater
1b: Postoculars 2, rarely 1, in contact with anterior temporals	*Aparallactus jacksonii*, Jackson's Centipede Eater

JACKSON'S CENTIPEDE EATER

Aparallactus jacksonii
(Günther, 1888)

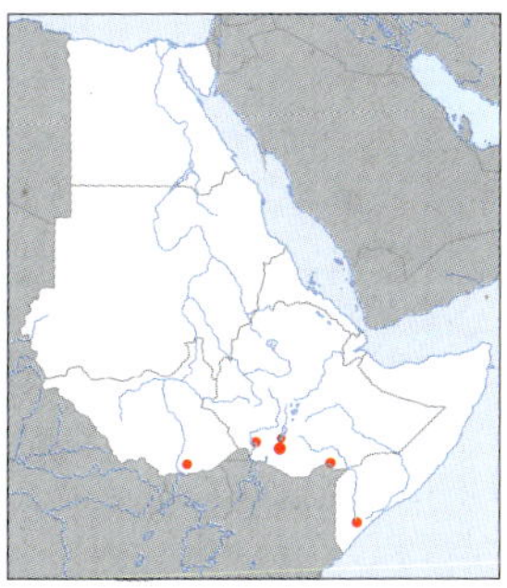

IDENTIFICATION: A small, slim brown snake with a black head, eye small and dark, pupil round. Body cylindrical, tail 13–20 per cent of total length, longest in females. Scales smooth in 15 rows at mid-body, ventrals 134–166, cloacal scale entire, subcaudals single, 33–52. Maximum size about 28cm, average 20–26cm, hatchling 10–11cm. Dorsum pinkish, reddish or nut-brown, head black above, usually a thin yellow crossbar on nape, becoming a yellow blotch at angle of jaw, with broad black collar behind that. There are sometimes irregular black dots on anterior third of body, back sometimes with fine black longitudinal lines. Yellow, cream or pinkish below. Taxonomic Notes: Two subspecies. *A. j. jacksonii* has low ventral and subcaudal counts (ventrals 134–157, subcaudals 33–40); *A. j. oweni* has higher ventral counts 151–166, subcaudal counts 41–52.

HABITAT AND DISTRIBUTION: Coastal bush, moist and dry savanna, from sea level to above 2,000m, although largely at lower altitudes. In our area, northern *A. j. oweni* is known from a few isolated records (Torit in South Sudan, Murri, Arba Minch and near Lake Turkana in Ethiopia); the southern nominate at Juba Sugar Project near Gelib, southern Somalia. Elsewhere; south through Kenya to northern Tanzania.

CONSERVATION STATUS: Least Concern.

NATURAL HISTORY: Lives in holes, but sometimes on surface at night. Unlike other centipede eaters, they give live birth, 2–3 offspring around 10cm long. Diet centipedes.

Aparallactus jacksonii, Kenya.

PLUMBEOUS CENTIPEDE EATER

Aparallactus lunulatus
(Peters, 1854)

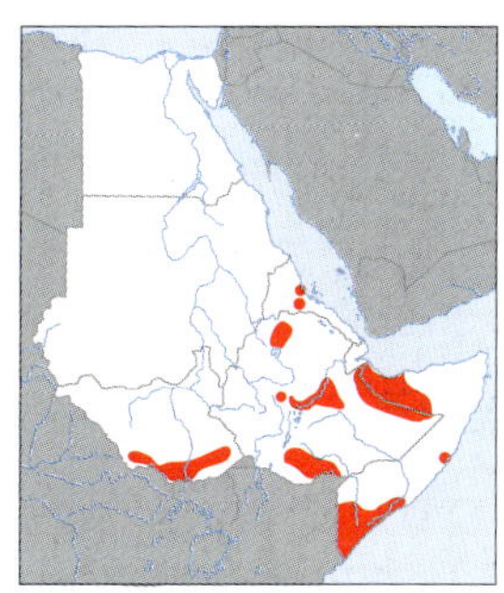

IDENTIFICATION: A small, slim dark snake; colour very variable. Head short, eye small and dark, pupil round. Body cylindrical, tail 20–28 per cent of total length, longest in females. Scales smooth in 15 rows at mid-body, ventrals 140–176, cloacal scale entire, subcaudals single, 43–70. Maximum size about 52cm, average 30–40cm, hatchlings 9–10cm. Three main colour phases in north-east Africa: (a) grey, each scale white-edged, head yellow, with broad black collar, (b) uniform grey, black, grey-brown or plumbeous, paler below, sometimes with darker head, (c) brown or yellow, dorsal scales sometimes dark-edged, a dark collar may be present. Most Benadir coast specimens uniform brown. Juveniles may have dark, posteriorly narrowing crossbars. Taxonomic Notes: Several subspecies described, none clear-cut, but remarkable colour and genetic variation suggest cryptic species are concealed in this taxon.

HABITAT AND DISTRIBUTION: Coastal bush, dry and moist savanna and semi-desert, from sea level to 2,600m. In Eritrea, at Massawa and Saganeiti. In Ethiopia, a handful of widely scattered records west of Rift Valley (Nazret, Mulu, near Gondar), widespread in low south-east and north-east, extending into

Aparallactus lunulatus, Kenya.

Aparallactus lunulatus, Borama.

Aparallactus lunulatus, Kenya.

Aparallactus lunulatus, Borama.

southern Djibouti, northern Somalia, east beyond Borama, Hargeisa and Berbera to Bohodle and Erigavo. Largely absent from the Ogaden and central Somalia (single record from Hobyo area) but quite widespread in southern Somalia south of 03°N. In southern South Sudan, scattered records from Jonglei, and Eastern, Central and Western Equatoria, at Li Rangu, Nimule, Juba, Torit and Boma. Elsewhere, west to Ghana, south to northern South Africa.

CONSERVATION STATUS: Least Concern.

NATURAL HISTORY: Lives in holes, sometimes on surface at night. Eats centipedes; one ate a scorpion. Sometimes bites fiercely if handled. Clutches of 2–4 eggs recorded, roughly 3×0.5cm. Most recent Somaliland and northern Ethiopian snakes were under large stones in grassy savanna during wet season.

SOMALI SHORT SNAKES *BRACHYOPHIS*

An African genus of small (20–30cm), very stout, rear-fanged and harmless snakes. They have a short head with wedge-shaped snout, very small eyes with round pupils, an indistinct neck, very short tapering blunt-tipped tail, large, rounded, smooth scales in 15 rows at mid-body, and a large heart-shaped occipital scale on top of head. No loreal or temporal scales. They are highly specialised burrowers, living in soft sand or sandy soil in arid or semi-arid environments. Secretive, crepuscular or nocturnal, feeding on other fossorial reptiles, particularly skinks, worm lizards, worm and blind snakes, and probably lay eggs. Three species occur in our area, all endemic to coastal southern and central Somalia. Rare and poorly known. Formerly, one species with three subspecies was recognised, but recent work by Van Wallach suggests all three are worthy of specific status. Somali short snakes have been traditionally placed in the Aparallactinae, and morphological analyses suggest they are closely related to *Micrelaps*, but to date no tissue has been available to run a molecular analysis to confirm their placement.

KEY TO MEMBERS OF THE GENUS *BRACHYOPHIS*

1a: Ventral and subcaudal scales yellowish with darker anterior margins, lateral edges of scales pale; body pattern not reduced; dark bar below eye (on fourth upper labial) present; second lower labial in contact with first inframaxillar scale (chin-shield) *Brachyophis revoili*, Révoil's Somali Short Snake

1b: Ventral and subcaudal scales uniform black or very dark red-brown, only very lateral edges of scales paler or entirely dark; body pattern usually strongly reduced; dark bar below eye (on fourth upper labial) absent; second lower labial not in contact with first inframaxillar scale (chin-shield) 2

2a: Ventral and subcaudal scales uniform black or very dark red-brown, only with very lateral edges of scales paler; white collar always present on nape, scales without dark edges; two pairs of inframaxillar scales (chin-shields); top of head more or less entirely dark *Brachyophis cornii*, Corni's Somali Short Snake

2b: Ventral and subcaudal scales uniform black or very dark red-brown, often including lowest row of dorsals; white collar on nape absent, scales always have dark edges; three pairs of inframaxillar scales (chin-shields); top of head pale, only dark spot/s on frontal, parietal and occipital scales present *Brachyophis krameri*, Krameri's Somali Short Snake

CORNI'S SOMALI SHORT SNAKE

Brachyophis cornii
Scortecci, 1932

IDENTIFICATION: A small, stocky snake with very pointed snout, short tail, dark head and dark collar on neck. Head indistinct, short, somewhat compressed and spatulate, as broad as neck. Snout wedge-shaped, eyes very small, with round pupil. Second lower labial not in contact with anterior inframaxillar scale (chin-shield), two pairs of inframaxillaries. Body thick, short, cylindrical, tail very short, strongly tapers, blunt, 6.5–7.7 per cent of total length. Scales smooth, shiny, large, terminally rounded, in 15 rows at mid-body; ventrals 105–108 (105–106 in males, 108 in females), obtusely keeled laterally; cloacal scale divided; subcaudals entire, 13–14 in males, ten in females. Transversally enlarged dorsal scales cover tip or whole tail in some. Maximum length 17.7cm, average 15–17cm, hatchling size unknown. Dorsal colour white, uniform or scales slightly edged black, giving a fine reticulate appearance, or with narrow irregular transverse lines (>50). Top of head dark, upper labials completely pale or only posterior one or two dark, dark spot or bar below eye absent, pale (completely white scales) and dark crossbar on nape always present. Ventral and subcaudal scales uniform black or very dark red-brown, only very lateral edges of scales paler.

HABITAT AND DISTRIBUTION: Coastal plains with white sand dunes and sparse vegetation composed of small bushes in extremely arid conditions. Inland it favours more arid shrublands and grasslands with yellowish or orangey soft sand, between sea level and c.120m. A poorly known species confined to coastal central Somalia. Recorded at three localities, around Hobyo and between 'Durgale' (Xero Dhagaxleey) and 'Magghiole' (about 35km west of Xero Dhagaxleey) in north and Meregh in south. Probably more widespread in the intervening area (all suitable habitat), but no records. Unlikely to be threatened by human activity; the area is underpopulated.

CONSERVATION STATUS: Least Concern.

NATURAL HISTORY: Known from fewer than five specimens. Habits probably similar to *B. revoili*.

Brachyophis cornii, preserved type, Meregh.

KRAMERI'S SOMALI SHORT SNAKE

Brachyophis krameri
Lanza, 1966

Brachyophis krameri, preserved type, Geriban.

IDENTIFICATION: A small, stocky snake with very pointed snout, short tail, pale coloration and reticulate pattern, confined to Geriban in central Somalia. Head indistinct, short, compressed, as broad as neck. Snout long, wedge-shaped; eye very small with a round pupil. Second lower labial not in contact with anterior inframaxillar scale (chin-shield), three pairs of inframaxillaries. Body thick, short, cylindrical, tail very short, strongly tapers, blunt, 5.8–6.9 per cent of total length. Scales smooth, shiny, large, terminally rounded, in 15 rows at mid-body; ventrals 105–108 (109–112 in males, 112–118 in females), obtusely keeled laterally; cloacal scale divided; subcaudals entire, 11–12 in males, 11 in females. Maximum size 18.5cm, all other five known specimens were 13.1–18cm. hatchling size unknown. Coloration very distinctive, dorsals ashy-grey, all scales uniformly edged black or brown, giving a reticulate effect, no neck- and body-bands or cross-bars. Top of head pale, a blackish spot on frontal, parietals and occipital always present, dark spot or bar below eye absent. Ventral and subcaudal scales uniform black, first row of dorsals adjacent to ventrals, also usually dark-coloured or entirely black or brown. Chin and throat paler, marbled by black or brown.

HABITAT AND DISTRIBUTION: Sandy areas in very arid steppes and grasslands with scattered low (20–40cm high) shrubs and almost no trees, at elevations around 190m. Known only from the type locality in flat country with orangey soft sand at Geriban (also known as Jariiban or Xamur), Mudug region, central Somalia. Occurrence in Nogal region in extreme southern Puntland and further south towards Galkayo is expected. Dune fields typical of the habitats of the other two Somali short snake species are absent here.

CONSERVATION STATUS: Not assessed, but its habitat is underpopulated.

NATURAL HISTORY: Unknown, probably similar to *B. revoili*. The six known specimens were collected in sand by local helpers.

RÉVOIL'S SOMALI SHORT SNAKE

Brachyophis revoili
Mocquard, 1888

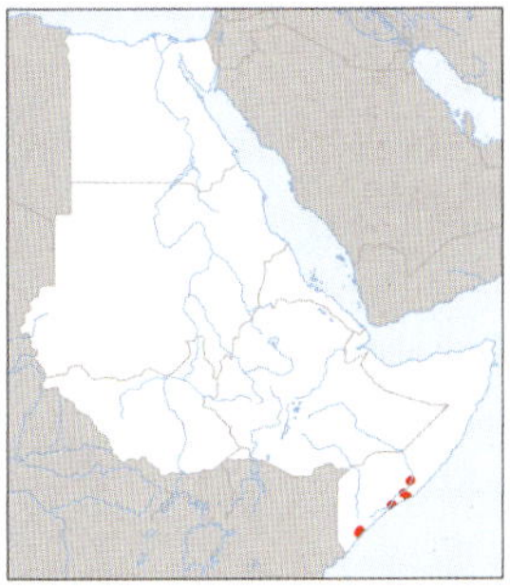

IDENTIFICATION: A small, stocky snake with very pointed snout, short tail and dark spot below eye. Head indistinct, moderately long, somewhat compressed and spatulate, slightly broader than neck. Snout wedge-shaped, eye very small with round pupil. Second lower labial in contact with anterior inframaxillar scale (chin-shield), three pairs of inframaxillaries. Body thick, short, cylindrical, tail very short, strongly tapers, blunt, 5–8 per cent of total length. Scales smooth, shiny, large, terminally rounded, in 15 rows at mid-body; ventrals 103–115 (103–107 in males, 112–115 in females), obtusely keeled laterally; cloacal scale divided; subcaudals entire, 11–13 in males, 8–11 in females. Transversally enlarged dorsal scales cover tip or whole tail in some specimens. Maximum length 33.5cm, average 20–26cm, hatchling size unknown, smallest specimen 10.5cm. Dorsal colour yellowish-cream or whitish. In some specimens anterior half of dorsal and lateral scales blackish-brown or reddish-brown, creating a reticulate appearance; in others a series of 30–50 irregular reddish-brown transversal bands, which may be so large as to reduce background colour to simple transverse thin stripe, each formed by a single scale row. Posterior half of top and sides of head dark, snout paler; a distinct dark spot or bar below eye (on fourth upper labial) and a pale crossbar on nape (of all-white scales) is always present. Ventral and subcaudal scales yellowish with darker anterior margins, the lateral edges of scales pale.

HABITAT AND DISTRIBUTION: Usually in fixed sand dunes, with mixed bushland and scattered trees, along coast between Kismayo and Mogadishu, between sea level and 115m. Commonly recorded in city limits. Endemic to inter-riverine region along Benadir coast in southern Somalia. Known from Kismayu and Jumbo in south, then Merca, Afgoi and Mogadishu in north. Only known inland record from Mahadday Weyne c.65km north of Mogadishu on Shebelle River. Old records from Kenya and Yemen erroneous. Locally fairly common. Unlikely to be threatened by human activities.

CONSERVATION STATUS: Least Concern.

NATURAL HISTORY: Poorly known. Burrows in soft sand, secretive and shy. Calm and doesn't try to bite. Slow-moving and crepuscular, hides in sand or under

Brachyophis revoili, Mogadishu.

Brachyophis revoili, Mogadishu.

ground cover by day, emerges to prowl in late afternoon and early night on surface. Mostly active during wet period, in dry season rarely seen on ground. Diet unknown but probably feeds on fossorial lizards and snakes – skinks (*Mochlus*), worm lizards (*Agamodon*), worm and blind snakes. A captive snake from Mogadishu took *Mochlus* skinks, biting until the venom immobilised the prey, which was then swallowed without constriction. It has a large Duvernoy's venom gland and enlarged rear fangs; it seems that venom plays an important role in obtaining food. A Mogadishu specimen was dug up from a small dune covered by patches of grass and *Vachellia* scrubs, together with *Agamodon anguliceps*. Another was observed active at night in an old abandoned house, among windblown sand and pieces of concrete. Presumably lays eggs, but no details. The specific name honours Georges Emmanuel Joseph Révoil (1852–1894), French explorer, photographer and diplomat in Africa. He was the first to travel extensively in Somalia, and made significant collections of reptiles and other zoological and botanical material. He collected the types of *Brachyophis revoili*, *Xenagama batillifera*, *Agama rueppelli*, *Latastia longicaudata revoili*, *Hemidactylus tropidolepis* and *Agamodon compressus*.

TWO-COLOURED AND BLACK-HEADED SNAKES *MICRELAPS*

A genus of small burrowing snakes with blunt heads and round pupils, usually boldly marked. Thin, with fairly short tails. They have large grooved rear fangs but are small and not regarded as dangerous. The dorsal scales are smooth, in 15 rows at mid-body. They live in holes or in soft sand or soil. Little is known of their diet and breeding details. Four species are known, one in our area. They are provisionally included here in the Aparallactinae, although molecular analyses have provided confusing data about their placement.

GUINEA-FOWL SNAKE, DESERT BLACK-HEADED SNAKE

Micrelaps vaillanti
(Mocquard, 1888)

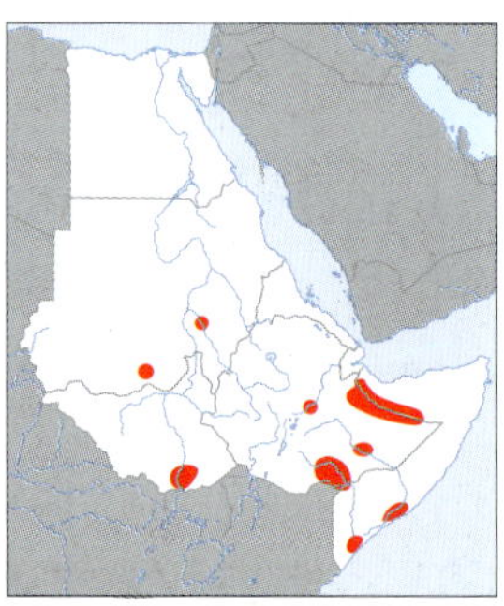

IDENTIFICATION: A little burrowing snake, sometimes uniformly dark, or grey with a black head. Head slightly broader than neck, snout rounded. Body cylindrical, tail 5–10 per cent of total length in females, 9–15 per cent in males. Scales smooth, in 15 rows at mid-body, ventrals 170–253, cloacal scale divided, subcaudals paired, 17–30. Maximum size about 50cm, but average 20–35cm, hatchling 12–13cm. From a distance grey animals look speckled, often with a black head, close up each grey dorsal scale has a white spot at free end. Underside white, pinkish or cream. Dark animals may be black, deep brown or purply grey, these seem to be associated with higher and wetter elevations.

Micrelaps vaillanti, Kenya.

Micrelaps vaillanti, Negele.

HABITAT AND DISTRIBUTION: Semi-desert, dry and moist savanna, from sea level up to 1,700m altitude. Sporadically recorded in lowland east Ethiopia, in Awash National Park, and in western Somaliland east to the Haud. Quite widespread in southern Somalia, south of 04°N. In South Sudan mostly associated with southern White Nile, two records from Sudan (Kadugli; Wad Madani). Probably occurs in suitable intervening areas, but under-collected. Elsewhere, south to northern Tanzania.

CONSERVATION STATUS: Least Concern.

NATURAL HISTORY: Poorly known. Burrows in soft soil and sand, secretive, active at night, hides in holes and under ground cover by day. Sudanese specimens were dug up in a field. A Negele snake was under a large flat rock in grassland. A female from lower Juba River in Somalia had six eggs in her oviducts in October; another at same locality had eaten a blind snake *Afrotyphlops*. A captive specimen ate a skink, *Panaspis*.

SNAKE-EATERS *POLEMON*

A genus of poorly known, fairly small, usually dark-coloured snakes, some of which were originally placed in the genus *Miodon*. They have a very short, stubby tail, which is a good field character (bear in mind that the venomous burrowing asps, *Atractaspis*, also have short tails). They have grooved rear fangs and eat other snakes, but are not dangerous to humans. Snake-eaters have smooth scales, in 15 rows at mid-body. They live mostly underground, in leaf litter and in holes. Twelve species are known, but just one occurs in our area.

CHRISTY'S SNAKE EATER

Polemon christyi
(Boulenger, 1903)

IDENTIFICATION: A small dark snake with a very short, stubby tail. Head slightly broader than neck, eye very small, pupil round. Body cylindrical, tail short and broad, tapering to short spike, 3–5 per cent of total length in females, 6–9 per cent in males. Scales smooth, in 15 rows at mid-body, ventrals 199–250, subcaudals paired, 15–24. Maximum length about 84cm, average 45–70cm, hatchling size unknown. Iridescent blue-black, lead-black or blue-grey above, ventrals and subcaudals uniform black or black with white or silver edges. Juveniles grey-brown above, whitish below. Rather similar to a burrowing asp, so should not be handled.

HABITAT AND DISTRIBUTION: Usually in forest, thick woodland, well-wooded savanna and recently deforested areas. A single record in our area; from Katire, Eastern Equatoria, South Sudan, at 1,000m. Elsewhere, through western East Africa to northern and north-west Zambia.

CONSERVATION STATUS: Least Concern.

NATURAL HISTORY: Poorly known. Lives mostly under surface, in holes and leaf litter, emerges to prowl at night on surface in rainy season. Slow-moving and inoffensive. No breeding details known. Diet a range of other snakes (including, apparently, its own species).

Polemon christyi, Kenya.

BLACK-AND-YELLOW SNAKES *CHILORHINOPHIS*

An African genus of rarely observed, small, very elongate, cylindrical-bodied snakes. They are striped black and yellow and rear-fanged. Although their fangs are relatively large, they are so small as to be quite harmless to humans. They have little short heads with round or vertically elliptic pupils. Two species are known, one in our region.

BUTLER'S BLACK-AND-YELLOW BURROWING-SNAKE

Chilorhinophis butleri
Werner, 1908

IDENTIFICATION: A very slender, black-and-yellow striped snake, known from a single specimen from South Sudan. Head short and blunt, eye small, pupil round. Body cylindrical, tail very short and blunt, 9.6 per cent of total length. Scales smooth, in 15 rows at mid-body, ventrals 256, subcaudals paired, 33. Head black above, body yellow with three longitudinal black stripes, ventrals yellow, tail has a black crossbar, below which ventrals white-spotted. Taxonomic Notes: Two other subspecies are known, in Tanzania and Mozambique.

HABITAT AND DISTRIBUTION: Known from a single specimen, found dead at Mongalla on White Nile in Central Equatoria, South Sudan, in savanna, at 440m.

CONSERVATION STATUS: Least Concern.

NATURAL HISTORY: The other subspecies are secretive, burrowing snakes, living in leaf litter, holes and soft soil; if threatened they raise the head and tail, and move back and forth, presumably confusing a predator as to which end to attack; if further molested they hide their heads in their coils and wave their tail in the air.

Chilorhinophis butleri, preserved type.

PURPLE-GLOSSED SNAKES *AMBLYODIPSAS*

An African genus of small to medium, usually dark-coloured snakes, with a distinctive purple iridescence on the scales, which may be useful for identification in the field. Scales smooth, in 15–21 rows at mid-body. Mostly burrowers, living underground, but will hunt on surface on wet nights. They eat mostly other snakes, worm lizards and ground-dwelling lizards. Nine species known, two in our area. They have long grooved rear fangs and may bite hard if restrained. Liable to be confused with burrowing asps, but lack a spike at the tail tip; however differences are subtle, so do not handle.

KEY TO NORTH-EAST AFRICAN MEMBERS OF THE GENUS *AMBLYODIPSAS*

1a: Dorsal scale rows usually 15 on neck and 17 mid-body; southern Somalia
Amblyodipsas polylepis, Common Purple-glossed Snake (East African subspecies)

1b: Dorsal scale rows usually 17 on neck, 17–19 at mid-body, range South Sudan in our area
Amblyodipsas unicolor, Western Purple-glossed Snake

COMMON PURPLE-GLOSSED SNAKE

Amblyodipsas polylepis
(Bocage, 1873)

IDENTIFICATION: Notes refer to northern subspecies, *A. p. hildebrandtii*. The nominate, *A. p. polylepis*, occurs further south. A small dark stocky snake with a blunt tail. Head short and rounded, not distinct from neck, eye very small, pupil round but hard to see. Body cylindrical, tail short and blunt, 5–10 per cent of total length, longest in males. Scales smooth, in 17–19 rows at mid-body, ventrals 159–203, cloacal scale divided, subcaudals paired, 17–29. Maximum length about 68cm. Females larger than males; the nominate southern subspecies grows up to 1.1m. Average 40–60cm, hatchling size unknown. Black, grey, purple-brown to pinkish-brown, sometimes blackish-brown above and below, usually with a distinctive purple gloss on scales, especially if freshly sloughed; individuals about to slough become very grey. Dorsal and ventral scales sometimes pale-edged.

HABITAT AND DISTRIBUTION: Coastal woodland and thicket, moist savanna and evergreen hills forest, from sea level to about 1,500m, but in our area only known from Juba Sugar Project, near Gelib, southern Somalia. Elsewhere, south to South Africa.

Amblyodipsas polylepis, Kenya.

CONSERVATION STATUS: Least Concern.

NATURAL HISTORY: Poorly known. Lives in holes and leaf litter, shelters under ground cover, may prowl on surface on damp nights. Nervous and quick moving, will bite if held. If molested, may hide its head under coils and raise tail tip, or elevate and flatten the tail, curling the tip down, possibly mimicking a cobra. Lays eggs. Diet: snakes, smooth-bodied skinks, worm lizards and caecilians.

WESTERN PURPLE-GLOSSED SNAKE

Amblyodipsas unicolor
(Reinhardt, 1843)

IDENTIFICATION: A medium-sized dark stocky snake with a blunt tail. Head short, eye very small, pupil round but hard to see, snout rounded and protruding. Body cylindrical, tail short and tapering, 9–13 per cent of total length in males, 7–9 per cent females. Scales smooth, in 17 rows at mid-body (sometimes 15), ventrals 165–179 in males, 192–207 in females, cloacal scale divided, subcaudals paired, 30–41 in males, 21–29 in females. Maximum length about 1.14m. Females larger than males. Average 60–90cm, hatchling size unknown. Dark grey, purple-grey or black, above and below, often with a distinctive purple gloss on scales,

Amblyodipsas unicolor, Ghana.

especially if freshly sloughed; West African specimens may be metallic blue-grey.

HABITAT AND DISTRIBUTION: Dry and moist savanna, woodland and forest, in our area known only from Li Rangu, 760m, in savanna of Western Equatoria, South Sudan. Elsewhere, west to Senegal through the Guinea savanna.

CONSERVATION STATUS: Least Concern.

NATURAL HISTORY: Lives mostly below surface, in holes and leaf litter or under ground cover, in hollow logs etc., emerges to prowl at night on surface in rainy season. Nervous and quite quick-moving, if restrained will turn and bite furiously and tenaciously. If restrained above ground, it also shows a modified form of head-pointing behaviour as in Burrowing asps, arching the neck and pushing the nose into the ground. Lays eggs but no clutch details known. Diet similar to other Purple-glossed snakes, including small snakes, worm lizards and burrowing skinks.

BURROWING ASPS AND HARLEQUIN SNAKES

SUBFAMILY ATRACTASPIDINAE

A subfamily of 24 species in two genera. The two species of Harlequin snake occur in South Africa.

BURROWING ASPS *ATRACTASPIS*

An African genus (with a limited penetration into the Middle East) of 22 species of similar-looking, small (usually <70cm), nocturnal, dark, blunt-headed snakes, with cylindrical bodies and a tail ending in a short spike. Eleven species occur in our area. They have hollow, hinged front fangs but are not vipers, rather specialised members of the family Atractaspididae. They have been called Burrowing vipers, Burrowing asps, Mole vipers, Stiletto snakes, and Side-stabbing snakes; we have opted for Burrowing asp, to avoid confusion with vipers. Their taxonomy is incompletely resolved.

Burrowing asps largely live underground, feeding on small vertebrates. They lay eggs, and closely resemble several innocuous small snakes, particularly Purple glossed-snakes, *Amblyodipsas,* some Snake-eaters (*Polemon*) and to a lesser extent Wolf snakes (*Lycophidion*) and Shovel-snouts (*Prosymna*), so care is needed. Identification to species is not easy; you need a dead specimen and a binocular microscope to count scales, although the locality may give some clues. Live animals have a distinctive display, arching the neck and pointing the head at the ground, looking like a croquet hoop or an inverted U; they also release a strong-smelling chemical from the cloaca if threatened. If further molested, they may convulse, jerking and lashing about. They cannot be held safely by hand, due to their short heads, long fangs and ability to slip a fang out without opening the mouth. If held by the tail they can jerk upwards and bite the hand. The fangs are used for stabbing; they may enter a rodent nest and stab all the babies before commencing feeding. Apart from their fangs, they are virtually toothless. Several species (including large ones of the *A. microlepidota* complex) have extremely long venom glands.

They are not aggressive, avoiding humans, even if teased they do not open their mouths and lunge forwards. But they are medically significant; over their huge range they cause a lot of bites, particularly in Africa's tropical lowlands. They are active on the ground at night, they may enter buildings and if trodden on they turn and bite immediately. By day, people farming or gardening with short tools are also at risk. Although almost all bites result in local symptoms, a few species are implicated in human deaths (although the clinical picture is usually incomplete). No anti-venom is available, a bite must be treated symptomatically.

KEY TO THE NORTH-EAST AFRICAN MEMBERS OF THE GENUS *ATRACTASPIS*, BURROWING ASPS

1a: In our area, known only from Sinai	*Atractaspis engaddensis*, Palestine Burrowing Asp
1b: Not in Sinai, elsewhere in Africa	2

2a: White markings on head, always in Horn of Africa	3
2b No white markings on head	4
3a: Ventrals 215–230, no white vertebral stripe	*Atractaspis scorteccii*, Somali Burrowing Asp
3b: Ventrals 230–243, has a white vertebral stripe	*Atractaspis leucomelas*, Ogaden Burrowing Asp
4a: Cloacal scale divided	5
4b: Cloacal scale entire	6
5a: In southern Somalia and north-east Kenya, mid-body scales 19	*Atractaspis engdahli*, Engdahl's Burrowing Asp
5b: Not in southern Somalia or north-east Kenya, mid-body scale rows 21–27	*Atractaspis irregularis*, Variable Burrowing Asp
6a: One anterior temporal scale, often slim, usually in moist savanna or forest	7
6b: Two anterior temporal scales, often stocky, in savanna and semi-desert	8
7a: First upper labial in contact with posterior nasal	*Atractaspis aterrima*, Slender Burrowing Asp
7b: First upper labial not in contact with posterior nasal	*Atractaspis bibronii*, Bibron's Burrowing Asp
8a: Found west of 25°E	*Atractaspis watsoni*, Watson's Burrowing Asp
8b: Found east of 25°E	9
9a: Mid-body scale rows 25–27 (rarely 29), largely north of 15°N	*Atractaspis magrettii*, Magretti's Burrowing Asp
9b: Mid-body scale rows 27–37, largely south of 15°N	10
10a: Ventral counts strongly sexually dimorphic, males (246–252) much higher than females (211–238); Sudan, South Sudan and low western Ethiopia	*Atractaspis phillipsi*, Sudan Burrowing Asp
10b: Ventral counts not strongly sexually dimorphic, males (227–246) little different from females (229–257); South Sudan, eastern Ethiopia, Somalia	*Atractaspis fallax*, Eastern Small-scaled Burrowing Asp

SLENDER BURROWING ASP

Atractaspis aterrima
Günther, 1863

IDENTIFICATION: A small, slim, fast-moving burrowing asp. Head blunt, eyes tiny. Tail short and blunt, 3–5 per cent of total length, ending in a pointed cone. Scales in 19–21 (rarely 23) rows at mid-body, ventrals 239–300, cloacal scale entire, subcaudals single, 17–26. Maximum size about 80cm, average 30–50cm. Hatchling size unknown; a Garamba juvenile was 20cm. Usually black or blackish-grey (occasionally blackish-brown or blackish-purple), above and below; juveniles can be brown. Some West African specimens have a white dot near tail tip.

HABITAT AND DISTRIBUTION: An unusually wide choice of habitat, in dry and moist savanna, woodland

Atractaspis aterrima, Tanzania.

and forest, from sea level to 2,000m. In our area, known only from Katire, Eastern Equatoria, South Sudan, at 1,000m, but also recorded at Garamba National Park in DR Congo, so probably more widespread in southern South Sudan. Elsewhere; west to Senegal, south to Tanzania.

CONSERVATION STATUS: Least Concern.

NATURAL HISTORY: Poorly known. Lives underground, but may emerge at night (especially during or after rain) and move around on ground. If molested they show the distinctive responses mentioned in the generic introduction (pointing the head, releasing a chemical, jerking and lashing). Diet mostly rodents and smooth-bodied reptiles, in Tanzania recorded eating caecilians, some captives will eat invertebrates. Lays eggs but details unknown.

BIBRON'S BURROWING ASP

Atractaspis bibronii
Smith, 1849

IDENTIFICATION: A small, fairly slim dark snake, with a prominent snout. Tongue white. Body cylindrical, tail short and blunt, ending in a pointed cone, 5–7 per cent of total length. Scales in 19–25 rows at mid-body, ventrals 213–262, cloacal scale entire, subcaudals single, 18–28. Maximum size about 70cm, average 30–50cm. hatchling 20cm. May be black, brown, pinkish or purplish-brown (especially East African coastal animals) or grey, often with a purplish sheen on scales; belly usually pinkish in eastern coastal Africa. Taxonomic Notes: Probably a species complex; northern animals (including Somalia), originally regarded as a subspecies, *A. b. rostrata*, form a separate clade and will probably be split off in the future.

HABITAT AND DISTRIBUTION: A range of habitats, generally in lowlands; in our area recorded only at Afgoi, just north-west of Mogadishu, in southern Somalia, near sea level. Elsewhere, south-west to DR Congo, south to South Africa.

CONSERVATION STATUS: Least Concern.

NATURAL HISTORY: Nocturnal and terrestrial, spending much time in holes, rotting logs or under ground cover but active on ground on wet nights. Slow-moving. Clutches of 3–8 eggs recorded, roughly 3.5 × 1.5cm. Known prey items include rodents, shrews, burrowing skinks, skink eggs, snakes and worm lizards. Like previous species, when molested they show the distinctive response, arching the neck, pointing the snout into the ground, jerking convulsively and releasing a distinctive chemical smell. This species is a major cause of snakebite in low and coastal south-east Africa, although no deaths have been recorded.

Atractaspis bibronii, Kenya.

Atractaspis bibronii, Congo.

ENGDAHL'S BURROWING ASP

Atractaspis engdahli
Lönnberg & Andersson, 1913

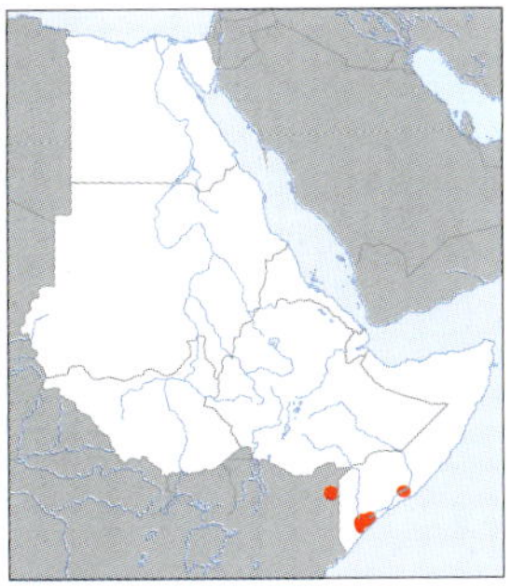

IDENTIFICATION: A small burrowing asp of dry country in southern Somalia. Snout appears rounded from above but shark-like in profile. Body cylindrical, tail short, 5–6 per cent of total length. Scales in 19 rows at mid-body, ventrals 219–232, cloacal scale divided, subcaudals paired, 19–23. Maximum size about 45cm, average 25–40cm, hatchling probably about 15cm. Black, blue-black or red-brown (juveniles can be quite light brown), paler below.

HABITAT AND DISTRIBUTION: Endemic to Horn of Africa. Coastal plain, thicket, grassland, dry savanna and semi-desert, from sea level to 250m. Known from a handful of specimens in southern Somalia (Belet Amin, Kismayu, Gelib, Afgoi) and also at Wajir Bor, in north-east Kenya; distribution probably continuous between southern Somali coast and north-east Kenya but few specimens.

CONSERVATION STATUS: Least Concern.

NATURAL HISTORY: Unknown, presumably similar to other *Atractaspis*. Specimens captured in termitaria, in holes and prowling semi-desert at night, the Kenya specimen was in a dry well.

Atractaspis engdahli, preserved.

VARIABLE BURROWING ASP

Atractaspis irregularis
(Reinhardt, 1843)

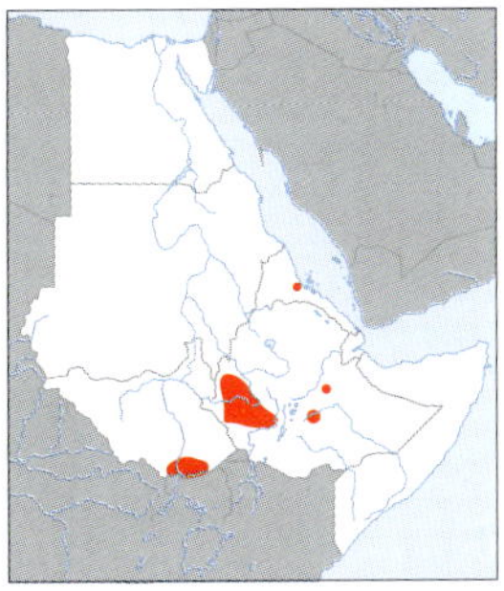

IDENTIFICATION: A small burrowing asp of the forest and outliers to the east. Body cylindrical, tail short, 5–8 per cent of total length, ending in a pointed cone. Scales smooth, 23–27 rows at mid-body (occasionally 21, usually 23 in east, 25–27 in west), ventrals 213–263, cloacal scale divided, subcaudals paired, 20–32, higher counts males. Maximum size about 66cm (possibly larger), average 30–50cm. Hatchling size unknown. Usually shiny black or blackish-grey, scales iridescent. Ventrals dark grey, black or rufous, sometimes with white edges. Taxonomic Notes: A highly variable species, with several described subspecies.

HABITAT AND DISTRIBUTION: Moist savanna, woodland, forest-savanna mosaic and forest, from sea level to 1,800m. A single Eritrean record, at Debre Bizen near Nefasit. Ethiopian records include Batanissa, in Ahmar Mountains, Sheikh Hussein, and a handful from the west and south-west. In South Sudan, at Kajiko North, south of Yei, Central Equatoria; Torit and environs, with a cluster of records in and around Imatong Mountains in Eastern Equatoria. Elsewhere, west to Liberia, south to north-west Angola.

Atractaspis irregularis, Liberia.

CONSERVATION STATUS: Least Concern.

NATURAL HISTORY: As other burrowing asps. A female contained six eggs. Eats rodents as well as small reptiles.

OGADEN BURROWING ASP

Atractaspis leucomelas
Boulenger, 1895

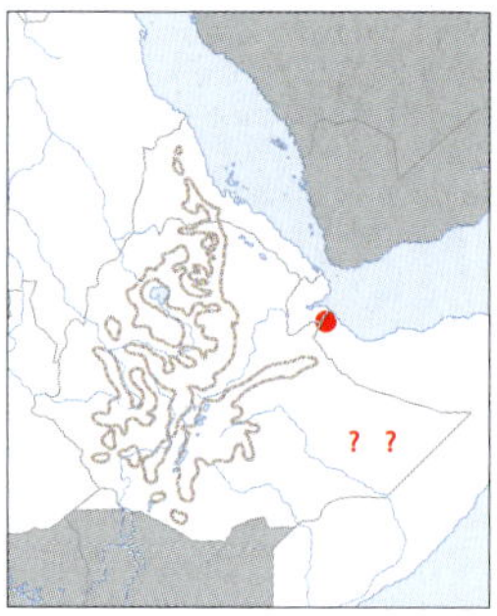

IDENTIFICATION: A small slim burrowing asp of the northern Horn of Africa. Distinctly marked with a white head and white vertebral stripe. Scales in 23 rows at mid-body; ventrals 230–243, cloacal scale entire, subcaudals single, 27–29. Maximum size uncertain but probably around 50cm, average 20–30cm. Distinctly marked, with a white or pale head, and dark crown; back dark with a fine yellow-white vertebral stripe. Four lowest scale rows and ventrals white; with a broad black throat-band.

HABITAT AND DISTRIBUTION: Known from just two specimens, in dry savanna and grassland; one from the Ogaden in eastern Ethiopia (precise locality unknown); the second on Somali/Djibouti border, in a region of sandy soil with grass and stunted thornbush at 300m. A possible sight record from Djibouti.

Atractaspis leucomelas, preserved type.

CONSERVATION STATUS: Data Deficient.

NATURAL HISTORY: Unknown. One specimen found by day but without further details.

SOMALI BURROWING ASP

Atractaspis scorteccii
Parker, 1949

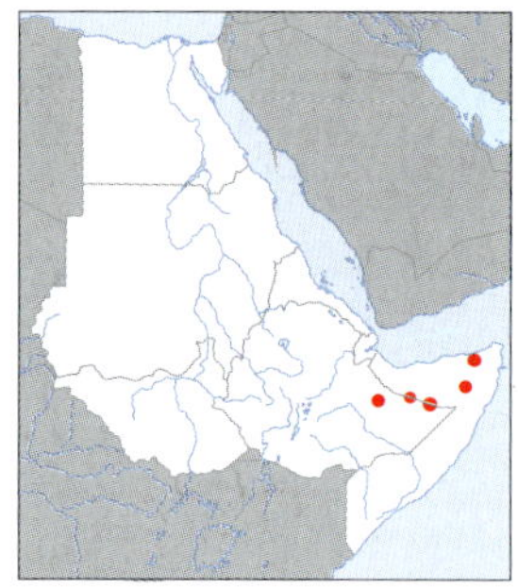

IDENTIFICATION: A large, stout burrowing asp with a white head, in dry country of far north-east Africa. Scales in 23–25 rows at mid-body; ventrals 215–230, cloacal scale entire, subcaudals single, 30–34. Maximum size about 85cm, average 50–75cm. Hatchling size unknown. Usually black or dark brown, with a broad white ring right around neck. Top of head mottled dark brown. Lips, chin and throat white, often with a dark patch behind and slightly below eye. Belly white, mottled dark brown, and outer edges of belly scales dark. Taxonomic Notes: Elevated out of the synonymy of *A. leucomelas*.

HABITAT AND DISTRIBUTION: Known from only a handful of specimens in sandy and stony country in dry grassland and semi-desert of eastern Ethiopia (Degeh Bur), Somaliland (Haud and the Somali/Ethiopian border at 45–47°E) and in Puntland (at

Atractaspis scorteccii, preserved type.

Gardo and Wadi Hamug Jero, 22km south of Candala), at 600–1,500m. These disjunct records suggest it is widespread but rare. A living specimen from Gardo was recently photographed; it was black, the neck and back of head white.

CONSERVATION STATUS: Data Deficient.

NATURAL HISTORY: Essentially unknown; the Wadi Hamug Jero specimen was active at night near a termite mound. The white neck-band may have some significance, a snake with a similar pattern (Central African Snake-eating Snake *Miodon collaris graueri*) eats snakes.

WATSON'S BURROWING ASP

Atractaspis watsoni
Boulenger, 1908

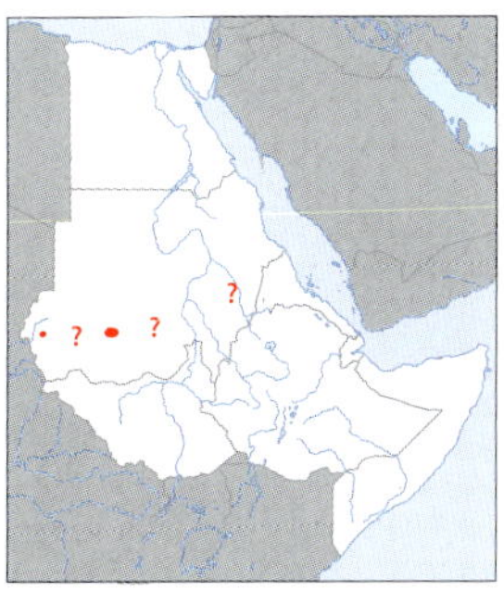

IDENTIFICATION: A fairly big black burrowing asp of savanna and Sahel, just reaches west of our area. A stocky snake with a short broad head. Body cylindrical, tail short, ends in a spike, less than 10 per cent of total length. Mid-body scales 29–31 (occasionally 27). Ventrals 213–242, cloacal scale entire, subcaudals single, 21–30. Maximum length 72cm, average 30–65cm, hatchling size unknown. Black or brownish-black above, slightly paler below. Taxonomic Notes: This snake and the following four species are members of the small-scaled burrowing asp, *A. microlepidota*, complex, a group of relatively large, stocky, broad-headed burrowing asps of the Sahel, north-east Africa and Middle East. All were long regarded as subspecies (or just variable individuals) of *A. microlepidota*. Recent work suggests that at least six taxa are involved, although their relative distributions are very poorly known.

Atractaspis watsoni, Chad.

HABITAT AND DISTRIBUTION: Sudan savanna and Sahel of West Africa, between sea level and 800m. Not known with absolute certainty from our area (see Taxonomic Notes) but examples from Suga Al Gamal (Kordofan) and Zalingai (Darfur) in Sudan, and Tesenay in Eritrea, are probably this species. Elsewhere, west to Senegal and Mauritania.

CONSERVATION STATUS: Least Concern.

NATURAL HISTORY: Poorly known. Nocturnal and terrestrial; usually underground but active at night on ground. If molested they show the usual distinctive *Atractaspis* response, arching neck and pointing head at the ground, looking like an inverted U. They may also release a distinctive-smelling chemical from the cloaca. Presumably lays eggs. Diet mostly reptiles; might take other small vertebrates.

PALESTINE BURROWING ASP

Atractaspis engaddensis
Haas, 1950

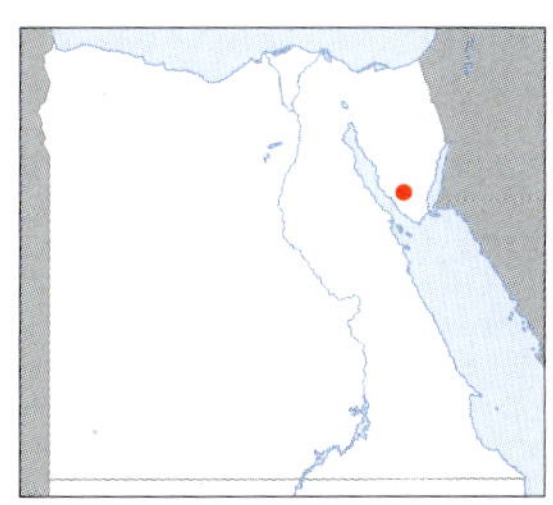

IDENTIFICATION: A relatively large burrowing asp, the only one in Egypt. Body cylindrical, tail short, 7 per cent of total length. Scales in 25–29 rows at mid-body, ventrals 263–284 (usually 275–282 in Egypt), cloacal scale entire, subcaudals single, 32–40 (usually 34–36 in Egypt). Maximum size about 90cm, largest Egyptian snake measured 68.5cm, average 40–60cm, hatchling probably about 15cm. Shiny black above and below.

Atractaspis engaddensis, Israel/Palestine.

Atractaspis magrettii, preserved.

HABITAT AND DISTRIBUTION: Semi-desert, often in large wadis (dry riverbeds) around oases. Not in mainland Africa, but recorded at Wadi Feiran and Mt Sinai area in Sinai, at 200–300m. Elsewhere, east to Jordan and western Saudi Arabia.

CONSERVATION STATUS: Least Concern.

NATURAL HISTORY: Terrestrial and nocturnal, active by night in warmer months; hides under ground cover and in holes bye day. If threatened, shows the nose-pointing behaviour, then hides its head under its coils and may elevate and wave the tail. Three to four eggs are laid, September–November, incubation time three months. Feeds mostly on smooth-bodied lizards and snakes, but will take rodents. At least one fatal bite case is attributed to this species.

MAGRETTI'S BURROWING ASP

Atractaspis magrettii
Scortecci, 1929

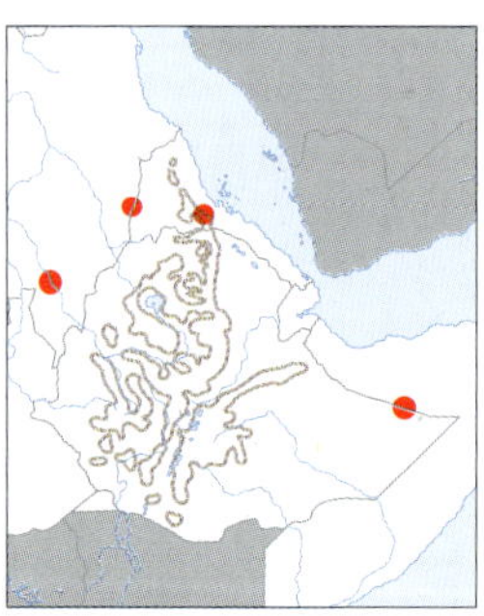

IDENTIFICATION: A rare burrowing asp from Eritrea and nearby Sudan, possibly Somalia and Ethiopia. A stocky snake with a short broad head. Body cylindrical, tail short, ends in a spike, <10 per cent of total length. Mid-body scales 25–27 (one had 29). Ventrals 228–248, cloacal scale entire, subcaudals single, 31–33. Maximum recorded length 75cm, average 30–60cm, hatchling size unknown. Uniformly black above, slightly paler below.

HABITAT AND DISTRIBUTION: Semi-desert, dry savanna and semi-arid high plateau, between 400 and 2,400m. Recorded at a handful of localities in Eritrea, and north-east Sudan at Kassala and Sennar. The type came from Mandafena, near Adi Keyh, at 2,400m in mountains of central Eritrea, a curiously high-altitude location for a big burrowing asp. A handful of specimens from the border between the northern Ogaden in Ethiopia and Somalia are tentatively also assigned to this species.

CONSERVATION STATUS: Least Concern.

NATURAL HISTORY: Essentially unknown, presumably similar to other burrowing asps. A Kassala specimen was reported to have killed an adult human in six hours.

SUDAN BURROWING ASP

Atractaspis phillipsi
Barbour, 1913

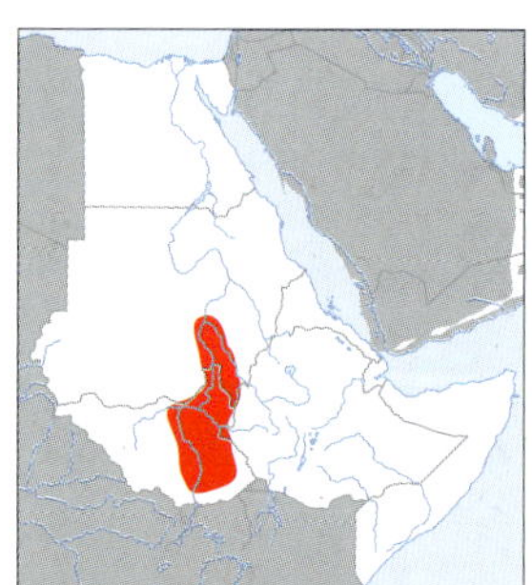

IDENTIFICATION: A fairly big steel-grey or black burrowing asp of central Sudan, South Sudan and western Ethiopia. Stocky with a short broad head. Body cylindrical, tail short, ends in a spike, <10 per cent of total length. Mid-body scales 29–31 (33 in some southern animals). Ventrals 222–258 (possibly as few as 211); cloacal scale entire, subcaudals single, 23–33. Maximum recorded length about 90cm, average 50–70cm, hatchling size unknown. Uniform black, steel-grey or grey-black above and below.

HABITAT AND DISTRIBUTION: Dry savanna, grassland and swamp, at altitudes between 400 and 600m on plains of Sudanese and South Sudanese Nile. Occurs south from Khartoum, along the Blue Nile to Singa, and the White Nile to Juba, east to Gambela and western Benishangul-Gumuz (Kutaworke) in Ethiopia.

CONSERVATION STATUS: Data Deficient.

NATURAL HISTORY: Essentially unknown, presumably similar to other burrowing asps.

Atractaspis phillipsi, near Kutaworke.

EASTERN SMALL-SCALED BURROWING ASP

Atractaspis fallax
Peters, 1867

IDENTIFICATION: A really big burrowing asp of the Horn of Africa and northern East Africa. Often stout, with a broad blunt head. Body cylindrical, tail short, 7–9 per cent of total length. Scales smooth, in 27–37 rows at mid-body, ventrals 227–257, cloacal entire, subcaudals single (occasionally paired, or a mix of both) 20–39. One of the biggest burrowing asps, reaching 1.1m in Kenya, possibly larger. Average 60–90cm, hatchling size unknown. Colour quite variable; black, grey, brown, often distinctly purple-brown, scales may be iridescent. Coastal specimens often light purple-brown, darkening towards head and neck, which black. Juveniles often light brown. Belly usually dark.

HABITAT AND DISTRIBUTION: Coastal bush and thicket, dry and moist savanna, grassland and semi-desert, from sea level to about 1,800m, but commonest in dry savanna below 1,000m. In Ethiopia occurs downstream from Awash National Park to Tendaho. Widespread in southern Somalia and the eastern Ogaden, but largely absent from northern Somalia, save records in the Haud, Las Anod and Gardo. Elsewhere, into Kenya and northern Tanzania.

CONSERVATION STATUS: Least Concern.

NATURAL HISTORY: Nocturnal and terrestrial; usually underground, but active at night on ground in rainy season. Sometimes active by day. If molested shows the usual distinctive *Atractaspis* response, arching neck and pointing head at ground, looking like an inverted U, then writhing and hiding the head. They may also release a distinctive-smelling chemical from the cloaca. Captive s have been photographed elevating the head and flaring the neck into a moderate hood, like a cobra. A Kenyan female laid eight eggs. Diet snakes, smooth-bodied lizards and occasional rodents. Said

Atractaspis fallax, Awash National Park.

to be relatively common in parts of its range. In Somalia it is known as *Jilbris* (or *Jilbis*), 'the snake of seven steps' (i.e., if you are bitten you take seven steps and die) or 'father of ten minutes' (for obvious reasons). A very difficult snake to catch, big adults are strong, active and bite furiously if restrained. The large size of some adults means that bites may require effective symptomatic treatment.

FAMILY ELAPIDAE ELAPIDS (COBRAS, MAMBAS AND RELATIVES)

A family of dangerous snakes, with short immovable poison fangs at front of upper jaw. More than 390 species and 50 genera occur worldwide in tropical regions; 16 species in five genera occur in our area.

KEY TO THE NORTH-EAST AFRICAN ELAPID SUBFAMILIES

1a: Tail oar-shaped, lives in the sea or stranded on foreshore	Subfamily Hydrophiinae, Sea snakes
1b: Tail not oar-shaped, does not live in sea	Subfamily Elapinae, Land elapids

SUBFAMILY ELAPINAE

There are 15 species of Elapinae in four genera in our area. All have round pupils, lack a loreal scale and lay eggs, and most are large (apart from the small garter snakes). With care, elapids are usually identifiable by eye, using a mix of size, location, appearance and behaviour. They are dangerous and most have caused fatalities (apart from garter snakes).

KEY TO THE NORTH-EAST AFRICAN ELAPID GENERA

1a: Black, does not spread a hood, in Egypt, mid-body scales 23	*Walterinnesia aegyptia*, Desert Black Snake
1b: Does not have above combination of characters	2
2a: Mid-body scales in 13 rows, adults small, < 80cm, juveniles virtually always have broad bands	*Elapsoidea*, Garter snakes
2b: Mid-body scales in 15 or more rows, big, adults usually >80cm, juveniles not banded or have narrow bands	3
3a: Three preoculars, head coffin-shaped, long and narrow, hood present but narrow	*Dendroaspis*, Mambas
3b: One to two preoculars, head not coffin-shaped, squarer, hood broad	*Naja*, True cobras

MAMBAS *DENDROASPIS*

Mambas occur in sub-Saharan Africa. All are large (>2m), agile, slender, diurnal elapid snakes, with long heads. Four species are known; three in our area, the Black Mamba is widespread, the other two are marginal. Mambas are greatly feared, but their reputation is largely unjustified and based on legend; they are not aggressive and avoid confrontation. However, all mambas have dangerous neurotoxic venoms.

KEY TO THE GENUS *DENDROASPIS*, MAMBAS

1a: Olive, grey or brown, mouth lining dark, mid-body scales in 21–25 rows, widespread in savanna	*Dendroaspis polylepis*, Black Mamba
1b: Mostly or all green, mouth lining pink, mid-body scales in 15–19 rows, usually in forest and woodland	2

2a: Uniform light green, lip scales immaculate, mid-body scales 17–19, in southern Somalia
Dendroaspis angusticeps, Green Mamba

2b: Darker green, lip scales finely black-edged, mid-body scales 15–17, in South Sudan
Dendroaspis jamesoni, Jameson's Mamba

GREEN MAMBA, EASTERN GREEN MAMBA

Dendroaspis angusticeps
(Smith, 1849)

Dendroaspis angusticeps, Kenya.

IDENTIFICATION: A distinctive, vivid green mamba of low-altitude forests and woodlands, might just reach southern Somalia. A big, slender, tree snake with a long head, relatively small eye with a round pupil, and yellow iris. Tail long and thin, 20–23 per cent of total length. Scales smooth, in 17–19 (sometimes 21) rows at mid-body, ventrals 201–232, subcaudals 99–126. Maximum size 2.3m (possibly more), average 1.5–2m, hatchlings 30–40cm. Bright uniform green above, sometimes with a sprinkling of yellow scales, pale green below.

HABITAT AND DISTRIBUTION: Coastal bush, thicket and forest, woodland and plantations with good tree cover, at low altitudes. In our area, might occur in extreme southern Somalia, as it is known from just across Kenya border. Elsewhere, south to South Africa.

CONSERVATION STATUS: Least Concern.

NATURAL HISTORY: A fast-moving, diurnal, secretive tree snake, climbs expertly, sometimes high, but will descend to ground. Not aggressive, if threatened tries to escape, never tries to bite. Sleeps at night in branches, not seeking a hole to sleep in but coiled up in a thick patch. May emerge to sunbathe in early morning. Sometimes shelters under thatched roofs. An even-natured snake, doesn't spread a hood or threaten with open mouth. Male combat has been observed. Lays up to 17 eggs, roughly 6×3cm, which take 2–3 months to hatch. Hatchlings found on north Kenya coast in June–July. Feeds mostly on birds, their nestlings, rodents and bats; occasionally lizards. They hunt by ambush and foraging. Sometimes relatively common. A dangerous snake, with a potentially deadly neurotoxic venom, but being alert, tree-dwelling, diurnal and secretive, it rarely comes into contact with humans.

JAMESON'S MAMBA

Dendroaspis jamesoni
(Traill, 1843)

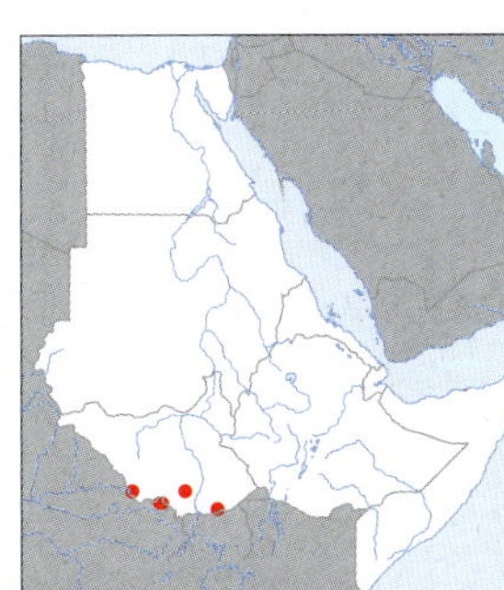

Dendroaspis jamesoni, Kenya.

IDENTIFICATION: A big, slender, mamba of Central African forests, head scales finely black-edged. A long head and small eye with a round pupil. Tail long and thin, 20–25 per cent of total length. Scales smooth, in 15–17 rows at mid-body, ventrals 202–227, subcaudals 94–113. Maximum size 2.64m (possibly more, big specimens usually males), average 1.5–2.2m, hatchling probably around 30cm. Dull green above, mottled black and yellow-green, pale green below, neck and throat yellow. In some (particularly juveniles) the black-and-yellow dorsal scales form narrow bands. Scales on head and body narrowly edged black. Taxonomic Notes: Sudanese specimens are of eastern *D. j. kaimosae*, with a black tail.

HABITAT AND DISTRIBUTION: Forest, woodland, swamp-forest, forest-savanna mosaic, thicket and deforested areas, even in and around towns where tree cover remains. In our area, known only in South Sudan: at Yambio and Aza Forest, near Bingo, Western Equatoria, the Imatong Mountains and Torit, Eastern Equatoria, but probably widespread in denser woodlands of south; known from just across DR Congo border in Garamba National Park. Elsewhere, west to Nigeria and southern Benin, south-west to Angola.

CONSERVATION STATUS: Least Concern.

NATURAL HISTORY: A fast-moving, diurnal tree snake, climbs expertly, but will descend to ground. Not aggressive, if threatened tries to escape but occasionally may flatten neck to a narrow hood. Reluctantly strikes if provoked. Shelters in hollow trees, cracks, tangles of vegetation. Male-to-male combat and mating observed in Nigeria in dry season, December to January; mating in Kenya in September. Lays eggs, clutches of 5–15 in Cameroon, in March, 7–16 in Nigeria, mostly in April–June. Diet in Kenya includes rodents, birds and arboreal lizards; juveniles also take frogs and lizards. Research indicates it doesn't move great distances. A dangerous snake, with a potentially deadly neurotoxic venom, it is also cardiotoxic. However, being alert, tree-dwelling, diurnal and secretive, it rarely comes into contact with humans.

BLACK MAMBA

Dendroaspis polylepis
Günther, 1864

IDENTIFICATION: A long, slender, fast-moving snake, with a long, narrow, 'coffin-shaped' head, with a fairly pronounced brow ridge and medium-sized eye with a round pupil. Inside of mouth bluish-black. Body cylindrical, tail long and thin, 17–25 per cent of total length. Scales in 23–25 (rarely 21) rows at mid-body. Ventrals 239–281, 142–182, subcaudals 105–132. Maximum size probably about 3.5m (possibly larger), adults average 2.2–2.7m, hatchlings 30–60cm.

Dendroaspis polylepis, South Africa.

Dendroaspis polylepis, Tanzania.

Juveniles grow rapidly and may reach 1.5–2m in a year. Olive, brownish, yellow-brown or grey, sometimes khaki or olive-green, but almost never black as name suggests. Juveniles greenish-grey. Scales smooth and have distinct purplish bloom in some adults. Belly cream, ivory or pale green. The back half is often distinctly speckled black on flanks, especially in animals from dryer areas. Some have rows of lighter and darker scales towards the tail, giving the impression of oblique lateral bars of grey and yellow.

HABITAT AND DISTRIBUTION: In coastal bush, woodland, moist and dry savanna and occasionally semi-desert, from sea level to about 1,600m, very rarely above this. Most common in well-wooded savanna or riverine forest, especially where there are rocky hills and big trees. A single old Eritrean record; from Anseba River, and a tentative sight record in Djibouti City. A few records west of Ethiopian Rift Valley; known from 30km north-west of Lake Tana, Mulu Gorge, and Giaimo near lower Omo River. Might be widespread, in deeply dissected river gorges running west off western plateau. Occurs in Awash National Park, thence down Awash River east to near Dire Dawa, and into Somaliland, at Borama and Hargeisa and east to Wagar Range. A lone record in southern Somalia, at Juba Sugar Project near Gelib. In eastern Ethiopia at Degeh Bur, Sof Omar caves, along Kenya border and around Lakes Chamo, Abaya and Chew Bahir; probably widespread along lower Omo River. In South Sudan, in Central and Eastern Equatoria (localities include Torit, Lohutok, Murikuren, Juba, Nimule and Lainya). Informally reported in Nuba Mountains, Kordofan, and known from Mboki, in Central African Republic, 150km west of Sudan border town of Ri Yubu, Western Equatoria. This is a large but secretive snake, and its distribution is probably more or less continuous across southern Ethiopia and southern South Sudan.

CONSERVATION STATUS: Least Concern.

NATURAL HISTORY: A big, fast-moving, alert, agile diurnal snake, equally at home on ground or climbing rocks or trees, often with head and neck raised high. Usually moves away if approached, but if threatened may rear up, flatten neck into a narrow hood, hiss loudly and open mouth to display the black interior, while shaking head from side to side. If further antagonised, may respond with a long upwards strike. Black Mambas shelter, often for several months, in holes, termitaria, rock fissures and tree cracks, Hamerkop, *Scopus umbretta*, nests, and even in tree beehives and roof spaces. Often basks before hunting and in late afternoon. They often share their refuges with other snakes, including cobras and pythons. Lays 6–17 eggs (12–18 more usual) measuring *c*.65×30mm, which usually hatch after 2–3 months, at start of rainy season. They take a variety of prey, mammals (mice, squirrels, bats, bushbabies, mongooses and hyrax), birds and snakes, one was observed eating winged termites as they left a nest. Males indulge in combat, neck-wrestling with bodies intertwined; this has been misidentified as courtship. They are often mobbed by birds and by squirrels and savanna monkeys. Although unaggressive, this is a very dangerous snake with powerful, fast-acting neurotoxic venom, it is confident in defence and may persist in suburbia and farmland. A bite is a medical emergency.

Dendroaspis polylepis, Kenya.

AFRICAN GARTER SNAKES *ELAPSOIDEA*

A genus of small (<70cm), ground-dwelling and burrowing venomous elapids (not to be confused with harmless American garter snakes, genus *Thamnophis*). Ten species occur in sub-Saharan Africa, five in our area. Taxonomy not finalised; oddly, they are not closely related to any other African elapid. Some Central African specimens hard to identify, particularly *E. loveridgei*, *E. laticincta* and *E. semiannulata moebiusi*, where characters overlap. Juveniles vividly banded, the banding usually (but not always) fades in adults; it may serve to stop adults eating juveniles. Garter snakes have short front fangs. They are usually reluctant to bite. When disturbed, they often inflate and flatten the body, making the bands prominent, elevate the body (either head and neck, or middle) and hiss; sometimes they flatten and roll the tail into a vertical plane. Usually active by night in rainy season. Although elapids, they are not of medical significance, bites usually cause only local pain and swelling.

KEY TO NORTH-EAST AFRICAN MEMBERS OF THE GENUS *ELAPSOIDEA*

1a:	Known only from Somalia, unbanded	2
1b:	Known elsewhere in our area, usually banded	3
2a:	Ventrals 136–141	*Elapsoidea chelazziorum*, Southern Somali Garter Snake
2b:	Ventrals 165–180	*Elapsoidea broadleyi*, Broadley's Garter Snake
3a:	West and Central African savannas, largely north of 05°N, bands usually fade centrally in adults	4
3b:	Largely south of 05°N, bands retained for life	*Elapsoidea loveridgei*, East African Garter Snake
4a:	8–17 body bands, upper labial in contact with prefrontal, widespread in South Sudan and southern Sudan	*Elapsoidea laticincta*, Central African Garter Snake
4b:	10–21 body bands, no contact between prefrontal and upper labial, might just reach south-west South Sudan	*Elapsoidea semiannulata moebiusi*, Northern Half-banded Garter Snake

BROADLEY'S GARTER SNAKE

Elapsoidea broadleyi
Jakobsen, 1997

IDENTIFICATION: A small, glossy snake, the head short, slightly broader than neck, eyes set well forward, with round pupils. Body cylindrical, tail very short, 6–7 per cent of total length. Scales smooth, in 13 rows at mid-body, ventrals 165–180, subcaudals 22–23 in males, 18 in only known female. Maximum size 76.6cm, average 40–70cm, hatchling 12–14cm. Known solely from three preserved specimens and a sight record; in preservative brown above, white below, without any trace of banding; sight record was uniform blackish-brown.

HABITAT AND DISTRIBUTION: Recorded at Gelib and Afmadu, and a specimen photographed near Kismayu, in dry low-altitude savanna of southern Somalia.

CONSERVATION STATUS: Data Deficient.

NATURAL HISTORY: Unknown, but presumably nocturnal and terrestrial, eats reptiles and amphibians.

Elapsoidea broadleyi, Kismayu.

SOUTHERN SOMALI GARTER SNAKE

Elapsoidea chelazziorum
Lanza, 1979

IDENTIFICATION: A small, glossy snake, known from only two specimens collected in Afgoi, near Mogadishu, Somalia, in late 1970s. Head short, slightly broader than neck, eyes set well forward, with round pupils. Body cylindrical, tail very short, 7–8 per cent of total length. Scales smooth, in 13 rows at mid-body, ventrals 136–141, subcaudals 17–21. The two known specimens were18 and 41.5cm total length. Adult dark greyish-brown, front of head, upper lips and throat off-white, with brownish spots, belly dark grey. Juvenile black above, head and throat off-white above, with scattered tiny brown dots, belly pale grey. Neither individual was banded, unusually.

HABITAT AND DISTRIBUTION: Taken in an area of sandy soil, in dry open woodland typical of consolidated coastal dunes, at Afgoi, in valley of Shebelle River, in dry low-altitude savanna of southern Somalia.

CONSERVATION STATUS: Data Deficient.

Elapsoidea chelazziorum, preserved.

NATURAL HISTORY: Unknown. Presumably similar to other garter snakes. One specimen was active at night after first rainstorm of the season.

CENTRAL AFRICAN GARTER SNAKE

Elapsoidea laticincta
(Werner, 1919)

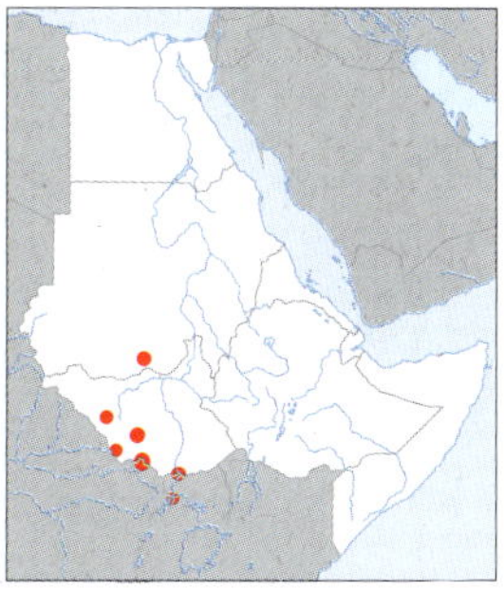

IDENTIFICATION: A small, glossy snake with a short head; fairy large eyes, round pupils. Body cylindrical, tail fairly short, 6–9 per cent of total length. Scales smooth, in 13 rows at mid-body, ventrals 139–151, subcaudals 13–25 (20 or more in males). Maximum size about 56cm, average 25–40cm, hatchling probably about 12cm. Young have 8–19 pale brown or reddish-brown bands on a black body, the bands about half as wide as black spaces between. Belly brownish-yellow or dull orange. As it grows the bands darken in the centre, but scales at outer edges of bands remain pale, so that adults appear black, with a series of fine white double rings along body; occasionally lose all their markings.

Elapsoidea laticincta, captive.

HABITAT AND DISTRIBUTION: In moist and dry savanna, woodland and forest-savanna mosaic, between 300 and above 1,000m. In our area, known from Sudan (Kadugli), quite widespread in South Sudan, records at Bengangai Game Reserve, Yambio, Nimule, Rumbek; and Bahr El Ghazal State. Elsewhere south into Uganda, west to Central African Republic.

CONSERVATION STATUS: Least Concern.

NATURAL HISTORY: Slow-moving, nocturnal and terrestrial. Usually active in rainy season. Inoffensive, some will let themselves be handled freely, making no attempt to bite, but if teased or molested may flatten and inflate body, showing the bands prominently, and may lift front half of body off the ground and jerk from side to side. Lays eggs, but no information on size or number. Diet includes snakes, smooth-bodied lizards and frogs.

Elapsoidea laticincta, juvenile, Uganda.

EAST AFRICAN GARTER SNAKE

Elapsoidea loveridgei
Parker, 1949

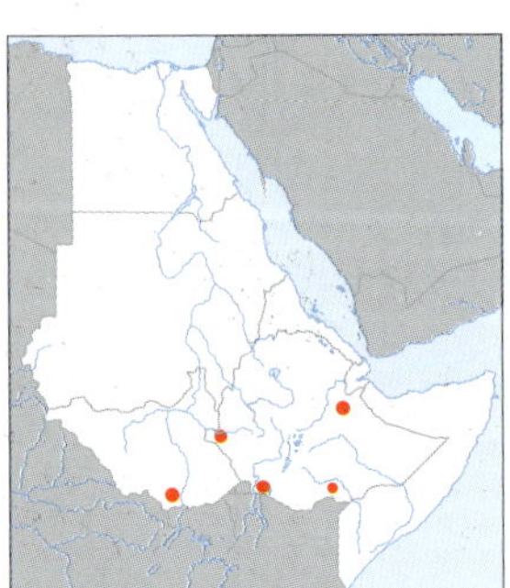

IDENTIFICATION: A small, glossy garter snake, in highlands of Ethiopia and South Sudan; adults retain bands. Head short, eyes set well forward, with round pupils. Body cylindrical, tail fairly short, 6–9 per cent of total length. Scales smooth, in 13 rows at mid-body. Three subspecies, with those in our area tentatively assigned to the many-banded *E. l. multicincta*. In our area has 20–24 bands or pairs of white transverse lines, 143–157 ventrals and 19–24 subcaudals. Maximum size about 65cm, average 30–55cm, hatchling probably about 12–15cm. Juveniles vividly marked with 19–36 narrow white, red, pinkish or white-edged grey-brown bands on grey or black body. Unlike many other garter snakes, bands usually persist in adults and their colour varies; pink, white or yellow, sometimes the band darkens centrally, leaving two fine white bands, or fades totally. Taxonomic Notes: True identity of garter snakes in Ethiopia, and parts of southern Sudan, complicated. In their review, Largen & Rasmussen tentatively allocated all Ethiopian examples to this species, and the South Sudanese example from Watoka might also be this form.

HABITAT AND DISTRIBUTION: Mid-altitude woodland, moist savanna and grassland, at 300–2,400m. Four widely scattered Ethiopian records (Kounhi, Filtu, Caschei and Gambela), and Watoka in Central Equatoria, South Sudan. Elsewhere, south to Tanzania, south-west to eastern DR Congo.

CONSERVATION STATUS: Least Concern.

NATURAL HISTORY: Terrestrial. Fairly slow-moving. Nocturnal, emerging at dusk, hides in holes or under ground cover by day. Believed to spend much of its activity in holes, but will prowl on surface at night, especially when damp, after rain, which acts as a trigger. Inoffensive, many will not attempt to bite when handled gently, but if molested may flatten and inflate body, showing the bands prominently, and may lift front half of body off the ground and jerk from side to side; if restrained may try to bite. Lays 2–6 eggs; a Ugandan female was gravid in September, hatchlings collected in Kenya in April, May and July. Diet includes small snakes, lizards and frogs, reptile eggs and rodents; a beetle larva and a centipede also recorded. A captive specimen bitten by a large Black-necked Spitting Cobra, *Naja nigricollis*, survived despite internal bleeding.

Elapsoidea loveridgei.

HALF-BANDED GARTER SNAKE

Elapsoidea semiannulata
Bocage, 1882

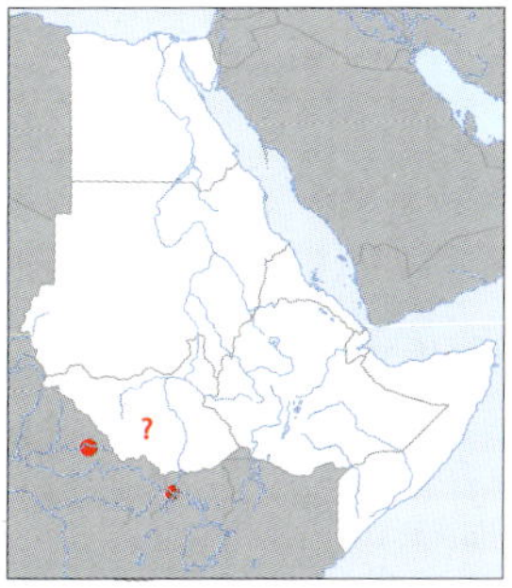

A white-banded garter snake, which might just reach our area. Small and glossy, neither fat nor thin, with a short head and eyes set well forward, with round pupils. Body cylindrical, tail fairly short. Scales smooth, in 13 rows at mid-body. Two subspecies; northern *E. s. moebiusi* might occur in our area; ventrals number 148–167, subcaudals 15–28. Maximum size about 70cm in northern form, average 30–50cm, hatchling 14–16cm. Juveniles vividly marked with 8–24 narrow white or yellow bands on grey or black body. White bands half or less the thickness of black interspaces. When it reaches a length of 20cm or so the bands become grey in centre and then darken, until all that is left is pairs of fine white lines, which may disappear in large adults. Underside white.

HABITAT AND DISTRIBUTION: Semi-desert, moist and dry savanna, forest clearings and edges. Not formally recorded in our area, but records at Mahagi Port, DR Congo, and Mboki, far south-east Central African Republic, indicate it is probably present in south-west South Sudan. Elsewhere, west to Senegal, south-west to Angola.

CONSERVATION STATUS: Least Concern.

NATURAL HISTORY: As other garter snakes. Lays 2–8 eggs.

Elapsoidea semiannulata, Ivory Coast.

TYPICAL COBRAS *NAJA*

Famous for their ability to spread a hood, worshipped and feared in ancient kingdoms of Egypt, *Naja* is a genus of more than 30 species of highly venomous and medically significant snakes, some of them very large, in Africa and Asia. Taxonomic research has dramatically increased the number of African cobras, from four to 21. Six occur in our area. They are a major snakebite hazard; at least one species, often more, occurs virtually everywhere in Africa and due to their secretive behaviour and nocturnal habits, they often persist in urban areas. Four species of spitting cobra occur in our region, they have a cytotoxic venom and defend themselves by spitting venom which, if it lands in the eyes, is horribly painful, and if untreated can lead to long-term damage and blindness. The non-spitting cobras of our area (Egyptian and Eastern Forest Cobras) have potent neurotoxic venom.

KEY TO THE NORTH-EAST AFRICAN MEMBERS OF THE GENUS *NAJA*

1a: Upper labials separated from eye by subocular scales — *Naja haje*, Egyptian Cobra

1b: At least one upper labial in contact with eye — 2

2a: Sixth upper labial largest and in contact with postoculars, one anterior temporal, do not spit venom — *Naja subfulva*, Eastern Forest Cobra

2b: Sixth upper labial not largest and not in contact with postoculars, 2–3 anterior temporals, all capable of spitting venom — 3

3a: Adults usually <1.5m, entire dorsal surface usually brown, maroon, rufous, red, orange or yellow, unmarked dorsally save 1–3 dark rings on neck — 4

3b: Adults not necessarily smaller than 1.5m, dorsal surface usually black, brown, olive or grey, throat bars do not usually form rings — 5

4a: Dorsal colour brown, maroon, grey or olive, with 2–3 bands on neck, mostly in north-east Africa north of 11°N — *Naja nubiae*, Nubian Spitting Cobra

4b: Dorsal colour usually orange, pink or red, sometimes pinkish-grey, red-brown, yellow, olive-green or steel-grey, with 1–2 bands on neck, in east of our area, south of 11°N — *Naja pallida*, Red Spitting Cobra

5a: Never black, underside yellow with single broad, deep brown throat bar, never pink on neck, reaches 2.7m long — *Naja ashei*, Large Brown Spitting Cobra

5b: Dorsal colour brown, grey, or black, never has a single brown throat bar, often pink/orange on neck, usually <2m long — *Naja nigricollis*, Black-necked Spitting Cobra

EGYPTIAN COBRA

Naja haje
(Linnaeus, 1758)

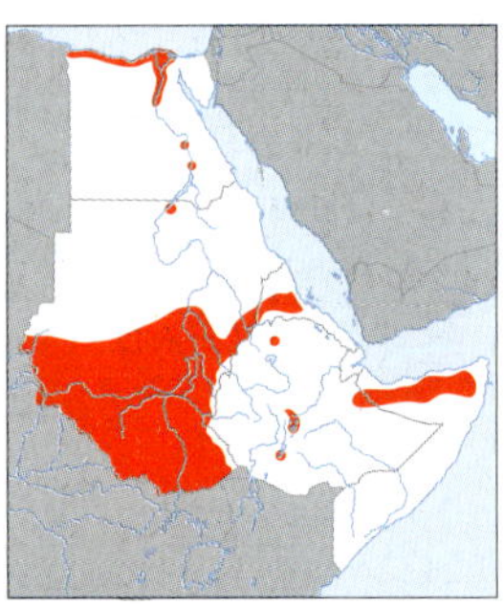

IDENTIFICATION: A big thick-bodied cobra, with a broad head, distinct from neck, eye fairly large, pupil round. Egyptian Cobra has a subocular scale row, separating eye from upper labials. Body cylindrical, tail fairly long, 15–18 per cent of total length. Scales smooth, in 21 (very occasionally 19) rows at midbody, ventrals 191–222, subcaudals paired, 53–68. Maximum size 2.5m (possibly larger), average 1.3–1.8m, hatchling 25–30cm. Colour astonishingly variable, but in our area usually brown, tan, yellow or rufous above. Light brown, yellow or cream below, often with a broad grey or brown throat bar, 8–20 scales deep, visible when hood spread. Often has irregular brown or yellow blotches and speckles on neck, back and underside. Some brown specimens are patterned with yellow scales on back, and pale ventral colour may extend onto flanks as irregular blotches; some Ethiopian snakes have light bands on posterior body, in Somaliland yellow, often with blackish head and neck. Juveniles often yellow or orange, with distinctive dark ring on neck, or a black head and neck, sometimes with a cream or yellow collar, often with fine banding.

HABITAT AND DISTRIBUTION: Mostly in semi-desert, moist and dry savanna, woodland (never forest), grassland, active and degraded agricultural lands, in Egypt not in true desert but vegetated wadis,

Naja haje, Kassala.

Naja haje juvenile, Lake Langano.

Naja haje, Egypt.

agriculture, etc. From sea level to around 2,000m. In Egypt, along western Mediterranean coast to Nile Delta, no further east, but known from Faiyum, extends south along Nile. Widespread in Sudan but very patchy; localities include Wadi Halfa, Khartoum, Kassala, Dinder National, Geneina, En Nahud and Kadugli, and known from northern Central African Republic, just west of Darfur. South Sudan records include Yei, Boma, Aweil, Palotaka near Torit, Lohutok, Atet and Juba, and east to Gambela in Ethiopia at base of western escarpment. Widespread in a broad belt across central Eritrea. A curiously disjunct distribution in Ethiopia; one specimen from 90km north-east of Gondar, a cluster of records in western foothills and floor of northern Rift Valley, from Debre Zeit south-east to Koka Dam, thence south to Arba Minch. Range in Somaliland east from Borama along northern mountain chain to Cal Miskaat and Carcar Mountains. Not recorded in coastal areas and the Haud. Often uncommon, or curiously absent from pockets in its range, possibly displaced from some areas by more successful snakes, but might be present and simply overlooked; in other restricted areas can be surprisingly common.

Naja haje, Keren.

CONSERVATION STATUS: Least Concern.

NATURAL HISTORY: Mostly terrestrial, but occasionally climbs. Clumsy but quick-moving. Active day and night, often basks in day, but will hunt at any time. Lives in termite holes, rock fissures, hollow trees or holes; known to share its refuge with Black Mambas. Rears up if threatened, spreads a hood, and if very stressed may rush towards an aggressor. Sometimes shams death. Lays up to 23 eggs, averaging 50–60×30–35mm. Eats a wide range of prey, fond of toads but will take eggs, small mammals and other snakes. This is a big, highly dangerous snake with potent neurotoxic venom; responsible for most fatal snakebites in Somaliland and Puntland. A bite is a medical emergency.

EASTERN FOREST COBRA, BROWN FOREST COBRA

Naja subfulva
Laurent, 1955

IDENTIFICATION: A big black or brown cobra of riverine woodland and forest. A fairly thick-bodied snake, it has a big head and large dark eye with a round pupil. Body cylindrical, tail fairly long and thin, 15–20 per cent of total length. Scales smooth and glossy, in 19 (rarely 17 or 21) rows at mid-body, ventrals 196–226, subcaudals 55–71. Maximum size about 2.7m, average 1.4–2.2m, hatchling 25–40cm. Several colour forms exist. Those in Ethiopian forest are glossy black, the juveniles sometimes white-spotted and this may persist in some adults. The chin, throat and anterior part belly are white, cream or pale yellow, with broad black crossbars and blotches, although in some most of underside is black. Sides of head strikingly marked in black and white, giving impression of vertical black and white bars on lips. Those in South Sudan and on Somali coast usually brownish or blackish-brown above, sometimes spotted darker brown, with a darker or black tail, but also have black lip edging; a good field character. Underside paler; belly yellow or cream, heavily speckled brown or black. Taxonomic Notes: Recently the Forest Cobra *N. melanoleuca* complex was split into five species; this is the easternmost. Banded Forest Cobra *N. savannula* occurs in savanna much further west, but some indeterminate, partially banded, specimens occur in north-west Uganda, so might occur in our area.

HABITAT AND DISTRIBUTION: Forest, woodland, coastal thicket, moist savanna and grassland, from sea level to above 2,000m. In Somalia, known from the Juba Sugar Project on lower Juba River and lower Juba Region, south to Kenya border. In Ethiopia, recorded from the south-west, in savanna and forest, localities include Bedelle, Godare Forest, Gambela and south-east of Bonga. South Sudan records include Nzara (Yambio), the Imatong Mountains (Katire and Gilo), Torit, Juba, Mongalla and Terekeka. Also recorded in Garamba National Park, DR Congo, so probably occurs across most of southern South Sudan, and north through Jonglei to the Gambela pedicle in Ethiopia, but unrecorded. A live specimen from just north of the border on the White Nile (not on map) is the only Sudan record. Elsewhere, west to Cameroon, south to South Africa.

CONSERVATION STATUS: Least Concern.

NATURAL HISTORY: Terrestrial, but is a fast graceful climber, known to ascend trees to a height of 10m or more. Quick-moving and alert. Swims well, almost semi-aquatic in some areas. Active both day and night, mostly by day in uninhabited areas and by night in urban areas. When not active, hides in holes, moribund termitaria, rock fissures, under ground cover, etc. If molested rears up to a considerable height and spreads a long narrow hood; if further stressed and

Naja subfulva, Kenya.

Naja subfulva, Tanzania.

Naja subfulva, Kenya.

unable to retreat may rush forward and bite. Male-on-male combat has been observed. Lays 15–26 eggs, roughly 30×60mm, which take around two months to hatch. Eats a wide variety of prey; fond of mammals (squirrels, mice, shrews, elephant shrews) and amphibians, will also take fish (including mudskippers), other snakes , monitor lizards and other lizards, birds and their eggs. Captives can seem almost cunning in their timing of escape bids, and have been reported as trying to attack rather than escape. When restrained it will jab the sharp tail tip into the handler. This is a very dangerous snake, with potent neurotoxic venom; a bite is a medical emergency.

BLACK-NECKED SPITTING COBRA

Naja nigricollis
Reinhardt, 1843

IDENTIFICATION: A big savanna spitting cobra, usually with black and pink or orange neck bars, in lowlands of west of our area. A broad head, cylindrical body and smooth scales; tail 15–20 per cent of total length. Mid-body scale count 17–23, ventrals 175–212, subcaudals 50–70. Maximum size seems to vary, in our area about 1.8m. Colour variable. Adults usually black, sometimes grey, or dark brown; juveniles grey. Upper neck black above, throat usually barred black and red, orange or pink (occasionally yellow, pale grey or white).

HABITAT AND DISTRIBUTION: Moist and dry savanna and woodland, but rarely semi-desert or forest, but will colonise forest clearings and fringes, and follow rivers into dry country; often in suburbia if there are suitable refuges and prey. Sea level to above 1,800m. Records in our area complicated by allocation of many to *N. ashei*, originally concealed in this taxon. A couple of records in western Ethiopia (10km west of Mabil in Blue Nile Gorge, south of Debre Markos, and a somewhat dubious record from Bedelle). In Sudan, apparently widespread in most southern states, north to Khartoum, but records lacking. In South Sudan probably very widespread, known from following states: Unity, North and West Bahr El Ghazal, Warrab, Lakes, Jonglei, and Upper Nile. Elsewhere, west to Mauritania, south to northern Botswana.

CONSERVATION STATUS: Least Concern.

NATURAL HISTORY: Terrestrial, but climbs well, may escape by climbing. Quick-moving and alert. Large adults mostly active by night, but will also hunt or bask during day, especially in uninhabited areas. Juveniles often active by day. When not active, takes cover in termitaria, holes, hollow trees, old logs, brush piles, etc. When threatened, rears and spread hood and will spit if further molested, but may also spit without spreading hood. Big adults can spit 2m or more. Male-on-male combat observed. Five to 20 eggs laid, roughly 25×40mm, often in termitaria; the larger the

Naja nigricollis, Ghana.

female the more eggs she lays. They eat a wide variety of food, including amphibians, various mammals, lizards, snakes, even fish; will raid chicken runs and take eggs and chicks. Quick to exploit suitable habitats; aided by their secretive nocturnal habits. This is a dangerous snake with a potent cytotoxic venom, it bites and spits. A bite is a medical emergency; venom in the eye should be gently washed out with copious amounts of water.

Naja nigricollis, Heglig.

Naja nigricollis, Nigeria.

ASHE'S SPITTING COBRA, LARGE BROWN SPITTING COBRA

Naja ashei
Wüster & Broadley, 2007

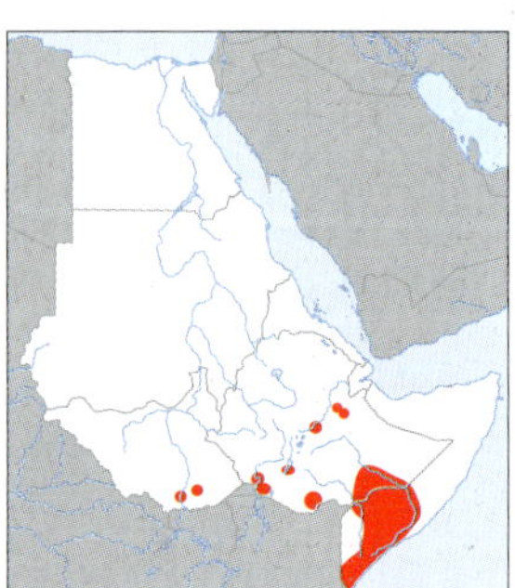

IDENTIFICATION: A recently described, huge brown spitting cobra of coasts and dry country in eastern Africa. The eye has a round pupil and orange-brown iris. Body cylindrical, tail fairly long, 15–18 per cent of total length. Scales smooth, in 20–23 rows at midbody, ventrals 192–207, subcaudals 55–65. Maximum size *c.*2.7m (maybe larger), average 1.2–1.8m, hatchling 35–38cm. Various shades of brown or olive, sometimes grey, dorsally; some can be quite yellow-brown, or uniform, others have sprinkling of yellow scales, which may form narrow yellow bars. Underside dull light yellow, each ventral scale edged pale brown; usually a deep, dark brown bar on underside of neck, but may fragment into several bars with yellow scales between. Inland juveniles light grey-brown, throat bar dark brown or black, head deep brown.

HABITAT AND DISTRIBUTION: Dry savanna, semi-desert and coastal thicket, from sea level to about 1,300m. Occurs in southern Somalia south of 04°N. Sporadic records in Ethiopia include Awash River north and north-east of Awash National Park, Gode in the Ogaden, Arba Minch and lower Omo River south of Omo National Park. Just enters South Sudan in the south-east, at Lohutok and Juba. Elsewhere, south to northern Tanzania.

CONSERVATION STATUS: Least Concern.

NATURAL HISTORY: Terrestrial, but may climb bushes and low trees. Quick-moving and alert. Active day and night. Shelters in holes (particularly termite hills), old logs or under ground cover. When disturbed rears up, spreads a broad hood and can spit venom up to 3m. Lays eggs. Eats any small vertebrate it can kill and

Naja ashei, juvenile, Kenya.

Naja ashei, Kenya.

Naja pallida, Kenya.

swallow. Known to take carrion. Like the previous species this is a dangerous snake, with potent cytotoxic venom, it bites and spits. A bite is a medical emergency; venom in the eye should be gently washed out with copious amounts of water.

RED SPITTING COBRA

Naja pallida
Boulenger, 1896

IDENTIFICATION: A relatively small, brown or rufous cobra of dry country in the Horn of Africa. A smallish head but large eye, with a round pupil. Body cylindrical, tail fairly long, 15–19 per cent of total length. Scales smooth, in 21–27 rows at mid-body, ventrals 192–228, subcaudals 56–81. Maximum size c.1.5m (possibly larger), most adults 70cm to 1.2m, hatchling about 16–20cm. Colour very variable, often related to soil colour, may be orange, rufous, brown, olive-green, yellow, pink or steel-grey. Usually has broad black throat ring (sometimes 2–3) several scales deep, and often a black 'suture' on scales below eye, like a teardrop. Underside usually a subdued or paler version of dorsal colour. In large adults, this band may fade or disappear. Colour may become dull with increasing size.

Naja pallida, Fantalle.

Naja pallida, Kenya.

HABITAT AND DISTRIBUTION: Dry savanna and semi-desert, from sea level to about 1,200m. From Djibouti and northern Somalia south throughout Somalia and northern Ogaden. In Ethiopia, extends upstream on Awash River to Awash National Park. No records south of Webi Shebelle River in the Ogaden but probably present, occurs west along Ethiopia's southern border to Lake Chew Bahir, north end of Lake Turkana and lower Omo River. Almost certainly in extreme south-east South Sudan, as it occurs at Lokitaung (just on Kenya side). Elsewhere, south to northern Tanzania.

CONSERVATION STATUS: Least Concern.

NATURAL HISTORY: Terrestrial, but may climb bushes and low trees. Quick moving and alert. Adults mostly nocturnal, hiding in holes, termite hills and under ground cover. Juveniles often active by day. When disturbed rears up, spreads a hood and may spit venom. Lays 6–15 eggs, size about 50×25mm. Juveniles taken in Kenya in July and October. Diet mostly amphibians, but will take other small vertebrates. May visit house looking for water or frogs. This a dangerous species with potent cytotoxic venom, it bites and spits. A bite is a medical emergency; venom in the eye should be gently washed out with copious amounts of water. In Puntland and Somaliland, it is responsible for most of the serious snakebites.

NUBIAN SPITTING COBRA

Naja nubiae
Wüster & Broadley, 2003

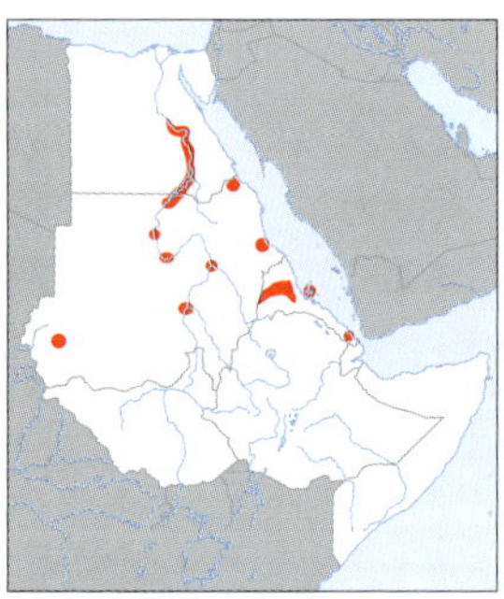

IDENTIFICATION: A relatively small, usually brown, grey or rufous cobra of Nile Valley and north-east Africa. A smallish head and large eye, with a round pupil. Body cylindrical, tail fairly long, 15–19 per cent of total length. Scales smooth, in 23–27 rows at midbody, ventrals 207–226, cloacal scale entire, subcaudals 58–72. Maximum size about 1.5m (possibly larger), most adults 70cm to 1.2m, hatchlings about 16–20cm. Variable; warm brown, maroon, grey or olive, juveniles usually paler; scales sometimes black-edged giving a latticed effect. Usually has a dark band on neck, which may cross underside of hood, and two dark bands always cross the throat below this. A black 'teardrop' created by black edges to labial scales below and behind eye. Underside white, cream or brown. Taxonomic Notes: Described relatively recently, previously most specimens were assigned to *N. nigricollis* or *N. pallida*.

HABITAT AND DISTRIBUTION: Dry savanna, semi-desert, vegetated rocky hills and riverside vegetation, from sea level to about 1,500m. Occurs from Asyut in Egypt, on Nile, south in Nile Valley to Khartoum; possibly further. Sporadic records on Red Sea coast (Jebel Elba in Egypt, Erkowit in Sudan, Assab in Eritrea), and in central Eritrea (including Keren,

Naja nubiae, north of Dongola.

Naja nubiae, Aswan.

Naja nubiae, captive.

Ghinda, Agordat and Dahlak Kebir Island). An apparently isolated populations at Jebel Marra in western Sudan. Elsewhere, recorded in Ennedi Massif in north-east Chad and Aïr Mountains in northern Niger.

CONSERVATION STATUS: Least Concern.

NATURAL HISTORY: Terrestrial, but may climb bushes and low trees. Quick-moving and alert. Adults mostly nocturnal, hiding in holes, rock fissures, or under ground cover during day, but juveniles often active by day. When disturbed rears up and spreads hood, and may spit twin jets of venom. Apparently swims well. Lays eggs. Diet probably similar to Red Spitting Cobra, although captives feed readily on mice. Often used by Egyptian 'snake charmers', despite fact that it can spit and thus represents a hazard to all. This a dangerous species with potent cytotoxic venom, it bites and spits. A bite is a medical emergency. Venom in the eye should be gently washed out with copious amounts of water.

DESERT BLACK SNAKES, BLACK DESERT COBRAS *WALTERINNESIA*

Black desert elapids; two species occur in dry country and deserts of Egypt and the Middle East; one in our area. Oddly, their nearest relatives are Shield cobras (*Aspidelaps*) of southern Africa.

DESERT BLACK SNAKE

Walterinnesia aegyptia
Lataste, 1887

IDENTIFICATION: A big black snake of Egyptian desert, usually active at night. Fairly broad head, slightly distinct from neck and smallish, dark eye with a round pupil (hard to see). Body cylindrical, tail long and thin, 12–15 per cent of total length, ending in a spike. Scales smooth and glossy, in 23 rows at mid-body; ventral 186–198, subcaudals 43–53; largely single but some often in pairs, unusually. Maximum size about 1.3m, average 80cm to 1.2m, hatchling 20–35cm. Black or dark grey, often with a bluish sheen; paler below, whitish-grey, sometimes yellow-brown; becomes grey before sloughing. Sides of head and lip scales often paler in Egypt, becoming light grey or yellow-brown on snout. Juveniles uniform black.

HABITAT AND DISTRIBUTION: Rocky desert, gravel or sandy plains, vegetated wadis and hilly and mountainous desert, but also near agriculture and irrigated areas, and near villages. Absent from extensive areas of soft sand and dunes. Sea level to 1,200m. In Egypt, east and north-east from Cairo to Sinai (few records from interior of Sinai), then east to Saudi Arabia. Informally reported from Egypt's Eastern Desert.

CONSERVATION STATUS: Least Concern.

NATURAL HISTORY: Secretive, terrestrial and largely nocturnal; lives in holes, rock fissures and under ground cover. Emerges at night to hunt, and can be active on relatively cold nights. Quick-moving and alert; if confronted, hiss loudly and raise body, preparatory to strike, but in an angled S-shaped coil, not vertical like a cobra. If teased they may also vibrate the tail vigorously, then hide the head and neck under their coils, defecate and release a foul-smelling fluid from the cloaca. However, they seem reluctant (at last in captivity) to actually strike; tending to bang the aggressor with mouth closed. They lay 2–20 eggs; captives laid in May and the eggs hatched 74–77 days later; this fits with the idea of newborn being able to put on weight before winter. Feeds on a wide range

Walterinnesia aegyptia, Israel/Palestine.

of vertebrates; fond of toads around oases and in desert feeds on Dabb or Spiny-tailed lizard (*Uromastyx* spp.), and often found in areas with these lizards. Sometimes used by 'snake charmers' (including one who thought it was harmless). Although rare and reluctant to bite, this is a dangerous snake; a bite must be treated as a medical emergency.

SUBFAMILY HYDROPHIINAE

A family of front-fanged, venomous snakes, living on land and in sea; the terrestrial species look quite different from marine snakes. There are around 50 species in the genus *Hydrophis*, sea snakes. One, Yellow-bellied Sea Snake *Hydrophis platurus*, occurs in the western Indian Ocean and is reported from Djibouti and Somalia.

YELLOW-BELLIED SEA SNAKE

Hydrophis platurus
(Linnaeus, 1766)

Hydrophis platurus, Kenya.

IDENTIFICATION: A medium-sized, black-and-yellow sea snake. Head long and tapering, eye small with a round pupil, nostrils on top of head. Body laterally compressed; tail short, vertically flattened and oar-like, c.10 per cent of total length. Scales smooth, 49–76 mid-body rows, ventrals 264–406, maximum size about 90cm, average 60–85cm, hatchling 24–25cm. Bright warning colours, the broad dorsal stripe is usually black, greenish-black or very dark brown, belly yellow, tail spotted black and cream or yellow. Dorsal stripe may be straight-edged (most African specimens this colour), wavy, break up into black saddles or totally disappear.

HABITAT AND DISTRIBUTION: Widespread across Indian and Pacific Oceans; in all tropical seas save the Atlantic. Not recorded in the Red Sea, but stranded specimens recorded in Djibouti, at Obock, and Hobyo in Somalia.

CONSERVATION STATUS: Least Concern.

NATURAL HISTORY: Lives mostly in open water, not around reefs, but usually within 300km of land. Usually only washed ashore when dying or in storms; it can only wriggle in an ungainly fashion on land, but is a superb swimmer. Three to eight young are born alive. Concentrations of up to ten have been seen in East African waters. Eats fish. A dangerous snake, but bites are almost unknown.

FAMILY VIPERIDAE VIPERS

A family of dangerous snakes, with long, perfectly tubular, folding venom fangs, and of international medical significance. Vipers or adders (the names are interchangeable, although it is suggested that adder be restricted to those that give live birth) are split into three subfamilies; the Viperinae, Old World vipers; the Crotalinae, pit vipers of America and Asia; and the Azemiophinae, two species, in one genus, of curious slim Asian vipers.

SUBFAMILY VIPERINAE

A subfamily of 13 genera and just over 100 species, in Asia, Europe and Africa, five genera are confined to Africa and six occur in both Africa and Asia. At present some six genera and 17 species occur in our area. All are venomous.

KEY TO THE NORTH-EAST AFRICAN GENERA AND SOME SPECIES OF VIPER

1a: Pupil round, nine large scales on top of head	*Causus*, Night adders
1b: Pupil vertical, many small scales on top of head	2
2a: Subcaudal scales single	3
2b: Subcaudal scales paired	4
3a: Head pear-shaped, usually on ground in dry country and semi-desert of northern Africa, usually shades of brown or grey	*Echis*, African carpet vipers
3b: Head not pear-shaped, usually in trees and bushes, green and yellow	*Atheris squamigera*, Green Bush Viper
4a: In sub-Saharan Africa, all large, adults >1m long, with geometric patterns	*Bitis*, African vipers
4b: In Sahel or North Africa, all small, <80cm, without geometric patterns	5
5a: Upper dorsals with knobbed (lumpy) keels, usually in or near Sahara	*Cerastes*, North African Desert Vipers
5b: Upper dorsals with narrow keels, in Sinai	*Pseudocerastes persicus*, Western False Horned Viper

BUSH VIPERS *ATHERIS*

A tropical African genus of small, attractive, largely green, broad-headed vipers, most are arboreal and inhabit forest and dense woodland. About 18 species are known (their taxonomy is confused), and a single species just reaches our area, in South Sudan.

GREEN BUSH VIPER

Atheris squamigera
Hallowell, 1854

IDENTIFICATION: A large bush viper, usually green but sometimes other colours, often with fine yellow crossbars. Head broad and triangular, covered in small strongly-keeled scales. The small eye, with a vertical pupil, is set far forward. Tail long and prehensile, 15–17 per cent of total length in females, 17–20 per cent in males. Scales keeled, in 15–25 rows at mid-body, ventrals 133–175, subcaudals 45–67 (higher counts in males). Maximum size c.80cm, average 40–65cm, hatchling 16–22cm. Usually green or yellow-green above, sometimes with fine yellow crossbars, greenish-blue below. Hatchling more strongly banded than adult, but has a bright yellow or pink tail tip, which may be used as a lure.

Atheris squamigera, Kenya.

HABITAT AND DISTRIBUTION: All types of forest and well-wooded savanna. A single record in our area, at Watoka (or Iwatoka), south-west of Yei in Central Equatoria, South Sudan, at 1,000m. Elsewhere, west to Nigeria, south-west to Angola.

CONSERVATION STATUS: Least Concern.

NATURAL HISTORY: Largely arboreal, slow-moving snakes, but will descend to ground at night. Nocturnal, but may bask. When threatened, a C-shaped coil threat display similar to Carpet vipers may result. Diet usually small mammals but takes other vertebrates. Seven to nine young are born. This snake is dangerously venomous, and a bite is a medical emergency.

AFRICAN VIPERS *BITIS*

A genus of stout-bodied, broad-headed nocturnal African vipers, all beautifully marked and fairly easily identified in the field. Eighteen species are known; five in our area, two Ethiopian endemics. All are large, slow moving and dangerous.

KEY TO THE NORTH-EAST AFRICAN MEMBERS OF THE GENUS *BITIS*

1a: Prominent horns (not just raised scales) on nose	*Bitis nasicornis*, Rhinoceros Viper
1b: No prominent horns on nose	2
2a: Black with fine yellow vermiculations, in high montane forest in south-east Ethiopia	*Bitis harenna*, Bale Mountains Viper
2b: Not black with fine yellow vermiculations, not just high montane forest in south-east Ethiopia	3
3a: Pale line between eyes, dorsal pattern of V-shapes pointing towards tail	*Bitis arietans*, Puff Adder
3b: No pale line between eyes, dorsal pattern not consisting of V-shapes	4
4a: Dorsal pattern of yellow butterfly shapes and dark hexagons, in highland forest in Ethiopia	*Bitis parviocula*, Ethiopian Mountain Viper
4b: Dorsal pattern does not consist of yellow butterfly shapes and dark hexagons, not in Ethiopia	*Bitis gabonica*, Eastern Gaboon Viper

PUFF ADDER

Bitis arietans
(Merrem, 1820)

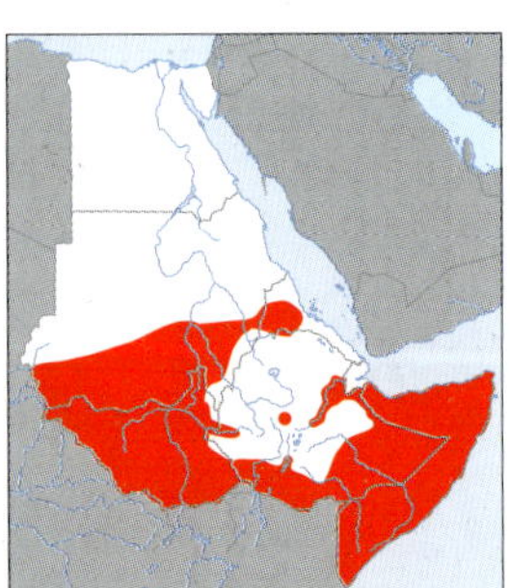

IDENTIFICATION: Africa's biggest viper occurs throughout savanna; a fat snake, with a broad, flat, triangular head, pale line between eyes and V-shapes on back. Head covered in small overlapping strongly keeled scales. Small eye with a vertical pupil, set far forward, nostrils upturned. Neck thin, body fat and depressed, tail fairly short, 10–15 per cent of total length in males, 6–9 per cent in females. Dorsal scales keeled, in 27–41 rows at mid-body, ventrals 123–147, subcaudals 14–39 (usually >25 in males). Maximum size depends on locality, in most of Africa rarely >1.2m, average 60cm to 1m. Much larger examples, up to 1.8m or possibly more, occur in dry savanna of north and east Kenya, Somalia and east Ethiopia. Hatchlings 15–23cm. Colour may be related to local soil colour, grey, brown, rufous or yellow, usually with a series of V-shapes on back that may morph into crossbars. Underside yellow or white, interspersed by short black dashes. Usually a pale line between eyes.

Bitis arietans, Mulu.

Taxonomic Notes: Two subspecies recognised, both in our area; *B. a. somalica* has distal subcaudals keeled.

HABITAT AND DISTRIBUTION: Throughout African savannas, also woodland, semi-desert and near desert, from sea level to 2,200m, sometimes higher. In our area, throughout Somalia and much of Ethiopia below 2,400m (but largely absent from closed forest of south-west). Handful of Eritrea records, largely in centre; not reported in Djibouti (but probably present). Virtually throughout South Sudan and Sudan, except true desert, in Sudan as far north as Khartoum, also at Kassala and Jebel Marra. The *somalica* subspecies occurs from Ethiopian Rift Valley east through Awash region and the Ogaden to northern Somalia (where it is called *Abguri*).

CONSERVATION STATUS: Least Concern.

NATURAL HISTORY: Terrestrial (may climb low bushes or trees). Usually nocturnal, sometimes active by day in rainy season. Occasionally basks. Hides under ground cover but also, perfectly concealed, among

Bitis arietans somalica, Dire Dawa.

vegetation, in leaf drifts, under bushes, etc. Hunts largely by ambush. Crawls in straight line if unhurried ('caterpillar crawl') but can also adopt usual serpentine motion. When threatened, inflates body, hisses loudly – hence common name – and raises and draws back front third of body. If a target is in range, a rapid strike follows. Captives often bad-tempered and noisy. Males indulge in neck-wrestling combat. Females produce a pheromone to attract males. They give live birth; usually 10–35, but an all-time record was 156 young. Diet mostly mammals, from rats to hyrax or dik-dik, but takes other vertebrates, including snakes and small tortoises. Large prey are struck, released and followed; small prey may be simply seized and swallowed. Enemies include monitor lizards, large cobras, predatory birds and small carnivorous mammals. A very dangerous snake with a potent cytotoxic venom; its relative abundance, good camouflage and disinclination to move if approached means many people are bitten. A bite is a medical emergency.

Bitis arietans, Unity State.

RHINOCEROS VIPER, RHINO-HORNED VIPER

Bitis nasicornis
(Shaw & Nodder, 1792)

IDENTIFICATION: Unmistakable if seen clearly; a large stout viper, with a narrow, flat, triangular head, covered in small strongly keeled scales, on end of nose is a cluster of 2–3 pairs of horn-like scales. Small eye set well forward; pupil vertical, iris green or gold, with black flecks. Neck thin, body triangular in section, tail short, 13–18 per cent of total length in males, 7–10 per cent in females. Body scales rough and heavily keeled. Scales in 30–43 rows at mid-body, ventrals 117–140, cloacal scale entire, subcaudals paired, 12–32 (higher counts in males). Maximum size about 1.4m, average 80cm to 1.1m, hatchling 18–25cm. Pattern consists of blue-green, yellow-centred vertebral oblongs, on flanks are upward-pointing crimson triangles, bordered blue and green. Head blue or green above with a black arrow mark. Underside grey-white or green, with dark marbling.

HABITAT AND DISTRIBUTION: Forest, woodland and forest-savanna mosaic. In our area, known only from suitable habitat in Imatong Mountains, South Sudan, at Gilo, in forest above Katire, and Lerua, between 1,120 and 2,100m. Also reported from just across border on Ugandan side, near Loturut. Elsewhere, west to Sierra Leone, south-west to Angola.

CONSERVATION STATUS: Vulnerable.

NATURAL HISTORY: Terrestrial, but regularly climbs into thickets and trees, up to 2.5m above ground. Slow-moving, but strikes quickly. Swims, and has been found in shallow pools. Nocturnal, tends to hide by day among leaf litter, around fallen trees, in holes or among root tangles of big forest trees, will also climb into clumps of leaves or cracks in trees. Hunts by ambush. Often bad-tempered, with a very loud hiss. Diet largely mammals and amphibians. From 6–38 young recorded. This is a highly venomous viper; a bite is a medical emergency.

Bitis nasicornis, Imatong Mountains.

Bitis nasicornis, Congo.

EASTERN GABOON VIPER, GABOON ADDER

Bitis gabonica
(Duméril, Bibron & Duméril, 1854)

IDENTIFICATION: Unmistakable, a fat viper of Central and southern African forests, with a broad,

Bitis gabonica, Imatong Mountains.

white, flat, triangular head with a remarkable geometric-patterned body. Between the raised nostrils is a pair of tiny horns. Little eyes, with a vertical pupil and cream, yellow-white or orange iris, set far forward. Tongue black with red tip. Body very stout and depressed, tail short, 9–12 per cent of total length in males, 5–8 per cent in females. Scales heavily keeled, 35–46 rows at mid-body, ventrals 124–140, subcaudals 17–33 (higher counts in males). Maximum size about 1.75m, possibly larger. Average 90cm to 1.5m, female usually larger than male, hatchling 25–37cm. Head white or cream with fine dark central line and black spots on rear corners, and dark blue-black triangle behind and below each eye. On centre of back a series of pale, sub-rectangular blotches, interspaced with dark, yellow-edged, hourglass markings, on flanks a series of fawn or brown rhomboidal shapes, with pale vertical central bars. Belly pale, with irregular black or brown blotches.

HABITAT AND DISTRIBUTION: Forest, woodland and well-wooded savanna. In our area only in southern South Sudan, at 600–1,800m. Several records in and around Imatong Mountains (Gilo, Katire, Lerwa, Laboni–Pajok Road, near Magwe). Also at Li Yubu (Yubo), Western Equatoria. Recorded in Garamba National Park, just across DR Congo border. Unlike Rhinoceros Viper, not strictly a forest snake, and might prove more widespread in southern woodlands of South Sudan.

CONSERVATION STATUS: Vulnerable.

NATURAL HISTORY: A slow-moving, usually placid, nocturnal viper, hides in leaf litter or in vegetation, in clearings, sunlit thickets and forest margins. Adults hunt by ambush, waiting (often for days) for prey; juveniles actively forage. Males more active than females. Sometimes bask. Gives live birth; usually 11–34 young, but up to 60 recorded. Diet largely mammals but also takes birds and other vertebrates. Males fight in mating season. This is a very dangerous snake and, although somewhat good-natured and reluctant to bite, it has a powerful cytotoxic venom; a bite is a medical emergency.

ETHIOPIAN MOUNTAIN VIPER

Bitis parviocula
Böhme, 1976

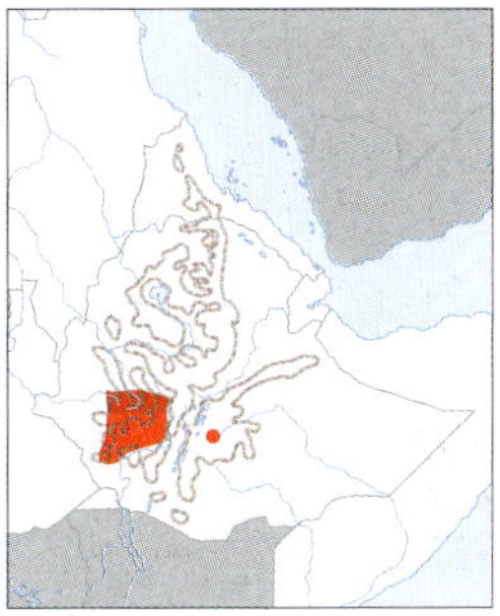

IDENTIFICATION: A fat viper endemic to forests of southern Ethiopia. The small eyes, with a vertical pupil and yellow iris, are set fairly far forward; there is usually a dark triangle set in centre of broad yellow head. Tongue deep blue and black. Body very stout and depressed, tail short. Scales heavily keeled, 37–39 rows at mid-body, ventrals 141–146, subcaudals 20–21. Maximum size about 1.3m, possibly larger. Average 70cm to 1m, hatchling 17–24cm. Head yellow with a dark triangle between eyes and a vague hammer shape behind this. Ground colour yellowish or greenish-grey, often darkening posteriorly. On centre of back is a series of black hexagons, interspersed by yellow butterfly shapes, with black, white-spotted subtriangular shapes below. On flanks is a series of dark, pale-edged upward-pointing triangles. Underside greenish or yellow, with irregular black or brown blotches. Some captive-bred animals are striped.

HABITAT AND DISTRIBUTION: Broadleaf forest, coffee plantations or deforested areas of south-west Ethiopia. West of Rift Valley probably fairly widespread in area between Bedelle, Metu, Bonga and Mizan Teferi; where forest remains. Maybe more widespread, known from Sheka Biosphere Reserve north of Mizan Teferi, and Mt Bibita, 60km south-west of Mizan Teferi. Probably occurs in extensive forest west of

Bitis parviocula, captive.

Bitis parviocula, south-west Ethiopia.

Metu–Mizan Teferi road. East of Rift Valley, a single enigmatic record from Dodolla, at 2,400m on north side of Bale Mountains; the town itself is in high grassland but suitable forest exists not far south of there. Unverified sight records at Wondo Genet, much further west, and Harenna Forest.

CONSERVATION STATUS: Endangered. Habitat is being felled for firewood and agriculture; this area is densely populated. Also in demand in the pet trade. However, it can survive in coffee plantations.

NATURAL HISTORY: Poorly known but presumably similar to other large forest *Bitis*, i.e. slow-moving, nocturnal, usually motionless, hidden in leaf litter, eats mostly mammals and birds. Captives seem fairly placid, and resemble Gaboon Vipers in temperament, although a restrained juvenile hissed and struggled furiously. Farmers in forest east of Metu indicate that it is commonest in valley bottoms and often shelters in cracks and recesses in large trees. Most captive specimens are from coffee plantations; in Ethiopia coffee is grown as an understorey tree in forest. A male and female were killed near dusk just west of Bonga on a road in September; they may have been mating. Captives gave birth to 11–16 offspring in May, near start of main rainy season at Bonga. Smaller and larger clutches, 6–21, informally reported in August–September. Captives feed readily on rodents and chicks. A highly dangerous snake, and a bite would be a medical emergency.

BALE MOUNTAIN VIPER

Bitis harenna

Gower, Wade, Spawls, Böhme, Buechley, Sykes & Colston, 2016

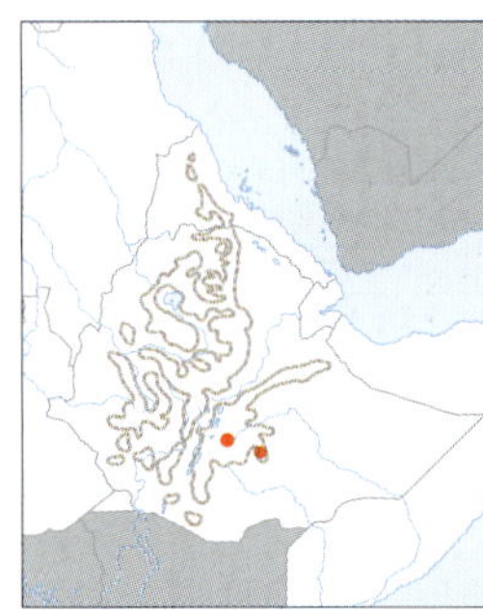

IDENTIFICATION: A fat black-and-yellow viper in a small area of south-central Ethiopia, east of Rift Valley. Described in 2016; only a handful of specimens in museums. The small eyes, with a vertical pupil and dark iris, are set fairly far forward. Body stout and depressed, tail short. Scales heavily keeled, 38–39 rows at mid-body, ventrals 145, subcaudals 20. One museum specimen is 66cm long, but individuals of *c.*1m have been observed; hatchling probably around 15–20cm. Head black or blue-black, edged with a yellow arrowhead shape and yellow line across head behind eyes. Body black, with a fine reticulate network of yellow or cream lines; Underside cream or yellow, with sub-rectangular dark blotches.

HABITAT AND DISTRIBUTION: Several live specimens have been observed at edges of coniferous and broadleaf forest; so presumably favours clearings in forest. Type specimen labelled from Dodolla, a town in high grasslands in northern foothills of Bale Mountains, Ethiopia, at 2,400m, but may have been transported there. Live animals have been observed in Bale Mountains National Park, along road that descends from Sanetti Plateau south towards Dolo Mena, within forest at 2,400 and 2,100m. Probably occurs throughout forest that extends 10–30km either side of this road, perhaps further in forest patches to west. Unverified sight records at Wondo Genet, 100km to west on a forested hill.

CONSERVATION STATUS: Not Assessed. However, known from only a small area, so needs to be surveyed.

NATURAL HISTORY: Largely unknown but presumably similar to other large forest *Bitis*, i.e. slow-moving, usually motionless, hidden in leaf litter, eats mostly mammals and birds, gives live birth. Observers have remarked on its physical similarity to puff adders of the same size. Its habitat is wet and relatively cold, and for several months of year receives little sunlight, so it may be cold-adapted, or perhaps is quick to utilise opportunities to bask. May be selectively diurnal; specimens have been observed active by day. Presumably a bite would be a medical emergency, as it is a big viper, but no cases are documented.

Bitis harenna, Harenna Forest.

NIGHT ADDERS *CAUSUS*

A genus of distinct, small, stout, frog-eating vipers, rarely larger than 75cm or so. Found in sub-Saharan Africa in savanna and forest, occasionally in semi-desert along watercourses. Seven species are known, four in our area. Unusually for vipers, they have a round pupil, nine large scales on top of head and lay eggs. Despite their common name, they are active by day and night. Their fangs are relatively short. Several species have exceptionally long venom glands that extend well down neck. Often common, and quick to bite, in some areas they are responsible for a high proportion of snakebite cases. However, their venom is weak, no documented human fatalities are known and no anti-venom is available. When angry, they often hiss and puff, elevate the body and flatten the neck like a little cobra.

KEY TO NORTH-EAST AFRICAN MEMBERS OF THE GENUS *CAUSUS*

1a: Usually vivid green when adult	2
1b: Brown, grey or dull olive-green when adult	3
2a: Subcaudals single, eyes large, in forest, rostral not upturned	*Causus lichtensteinii*, Forest Night Adder
2b: Subcaudals paired, eyes moderate, not in forest, rostral slightly upturned	*Causus resimus*, Velvety-green Night Adder
3a: Upper labials not dark-edged, ventrals 118–150, no dark bar between eyes, dark dorsal markings very rarely edged with paler scales	*Causus maculatus*, West African Night Adder
3b: Upper labials dark-edged, ventrals 134–166, usually a dark bar between eyes, dark dorsal markings usually edged by paler scales	*Causus rhombeatus*, Rhombic Night Adder

FOREST NIGHT ADDER

Causus lichtensteini
(Jan, 1859)

IDENTIFICATION: A small, relatively slim green forest night adder, in southern South Sudan. Head slightly more rounded than usual in night adders, distinct from neck. Body cylindrical or slightly depressed, and (especially in juveniles) slimmer than other night adders. Tail short and blunt, 6–8 per cent of total length in females, 8–11 per cent in males. Unusually, subcaudal scales single; number 14–23, with higher counts in males. Scales soft, velvety and feebly keeled, in 15 rows at mid-body, ventrals 134–156, higher counts in females. Maximum size about 70cm but this is exceptional, average 30–55cm, hatchling c.15cm. Back usually shades of green, sometimes brown, or has a series of vague, pale-centred rhombic or chevron markings; sometimes 2–3 orange tail bars. Belly scales yellowish, cream or pearly. Juveniles very variable; deep green, brown or turquoise, uniform or with dark crossbars; lips white. Hatchlings have white V on neck, which fades in adults. Lips and throat yellow. Tongue blue at base, then black with an orange band.

HABITAT AND DISTRIBUTION: In forest and woodland, forested wetland and deforested areas. In and around Imatong Mountains in Eastern Equatoria, South Sudan, above 1,000m: localities include Lotti Forest, Talanga Forest, Katire, and between them. Elsewhere, west to Sierra Leone, south across DR Congo to northern Zambia.

CONSERVATION STATUS: Least Concern.

NATURAL HISTORY: Terrestrial and secretive. Fairly slow-moving, but strikes quite quickly. Despite its name, seems most active by day among shade and leaf litter of forest floor. Swims well. When inactive, hides in holes or under ground cover. If angered, it hisses, puffs and inflates itself, coils up and then strikes. Eats frogs and toads. Lays 4–8 eggs.

Causus lichtensteini, DR Congo.

WEST AFRICAN NIGHT ADDER

Causus maculatus
(Hallowell, 1842)

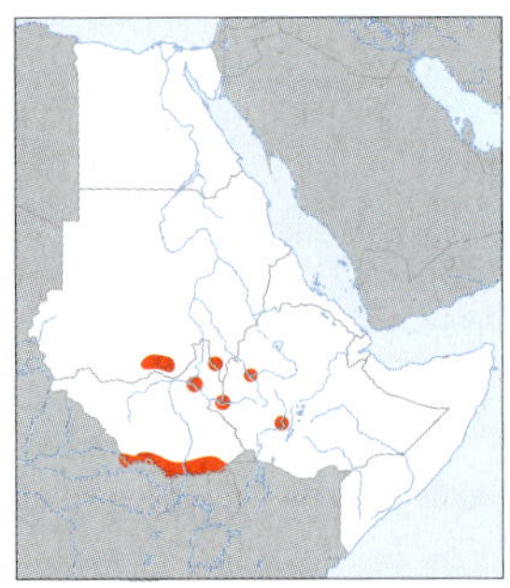

IDENTIFICATION: A small, stout brown night adder, in a wide range of habitats in Central and West Africa. It has a rounded snout. Body cylindrical or slightly depressed, Tail short, 7–9 per cent of total length in females, 9–11 per cent in males. Scales soft and feebly keeled, in 17–22 rows at mid-body, ventrals 118–154, subcaudals 15–26, higher counts usually males. Maximum size c.70cm (possibly slightly larger), average 30–60cm, hatchling 13–16cm. Grey or brown above, with dark crossbars or rhomboids, and a V on neck. Some have no markings. Underside white, cream or pinkish-grey, the ventrals sometimes black-edged.

HABITAT AND DISTRIBUTION: Occurs in wide

Causus maculatus, Uganda.

Causus maculatus, DR Congo.

range of habitats; forest, woodland, savanna and semi-desert; in our area from 300 to about 1,000m. A handful of records from south- and west-draining rivers in Ethiopia (lower Blue Nile Gorge, Gambela, Omo/Gibe confluence). Most South Sudan records from south, in Eastern Equatoria (Torit, Talanga Forest, Molongori, Ikotos) or Central Equatoria (Yei and environs, Kagelu), and Western Equatoria (Li Rangu). Seemingly absent from northern South Sudan, but reappears in South Kordofan, Sudan (Kadugli, Talodi). Elsewhere, west to Mauritania and Senegal, south to southern DR Congo.

CONSERVATION STATUS: Least Concern.

NATURAL HISTORY: Terrestrial, but will climb bushes in pursuit of frogs. Active by day, night and at twilight, but usually only active in rainy season. Hisses, puffs, strikes and flattens neck if molested. Eats frogs and toads. Lays 6–20 eggs, roughly 26×16mm. Common in some areas.

VELVETY-GREEN NIGHT ADDER

Causus resimus
(Peters, 1862)

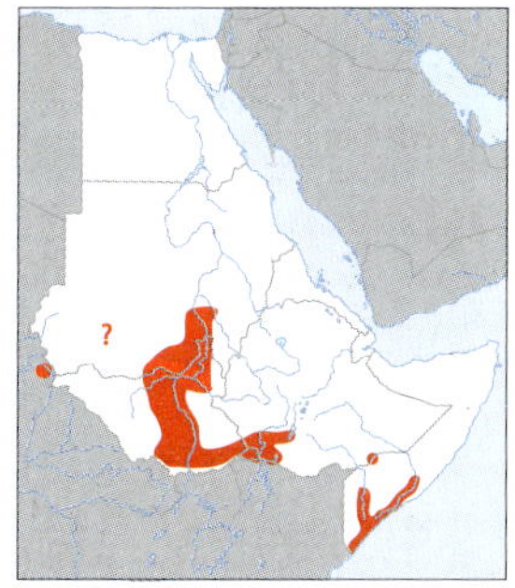

IDENTIFICATION: A small, stout viper with a fairly short head, tip of snout slightly upturned. Body cylindrical or slightly depressed; Tail short, 7–9 per cent of total length in females, 8–10 per cent in males. Scales soft, velvety and feebly keeled, in 19–21 rows at midbody, ventrals 131–152, subcaudals 15–25, males usually >20, females usually <20. Maximum size about 75cm, average 30–60cm, hatchling 12–15cm. A beautiful snake, the back is usually vivid green, (sometimes brown) of various shades, plain or densely spotted. Chin and throat yellow, belly scales yellowish, cream or pearly. Often black scales form a V-shaped outline on head (especially in juveniles) and scattered black scales may form indistinct rhomboid or V-shapes on back and oblique dark bars on flanks. Tongue pale blue and black. Hidden margin of scales often vivid blue, which is visible when the snake inflates its body in anger.

HABITAT AND DISTRIBUTION: Coastal savanna and thicket, dry and moist savanna and woodland, from sea level to 1,800m. Widespread in southern Somalia, south of 04°N. Ethiopian records include Dolo, on

Causus resimus, Arba Minch.

Somali border and a cluster from low savanna of south-west (Arba Minch, Mui, Mago and Omo National Parks, Lake Chew Bahir). Sporadically recorded in South Sudan, at a number of localities in Eastern and Central Equatoria (Lafon, Kapoeta, Terangore, Juba, Mongalla). Other South Sudan records from Maridi (Western Equatoria), Malakal and environs (Upper Nile State), Paloich and Rumbek (Lakes State). Probably fairly widespread in Nile Basin. In Sudan, known from Sennar (type locality, Jebel Ghule), Heglig, Diffra Oilfields near Muglag, and at Kosti. A specimen from Am Dafok in Central African Republic is right on border with Darfur. Elsewhere, west to Nigeria, south to Tanzania, with an isolated population in Angola.

CONSERVATION STATUS: Least Concern.

Causus resimus, Omo National Park.

Causus resimus, West Kordofan.

NATURAL HISTORY: Terrestrial, but will climb low vegetation in pursuit of frogs. Active by day and night, and will bask. Fairly slow-moving but strikes quite quickly. Swims well. When inactive, hides in holes, bush piles, under ground cover, etc. If angered, it responds like other night adders, hissing, puffing and striking. Diet amphibians. Lays 4–12 eggs; Somali females laid 4–11 eggs, in July.

RHOMBIC NIGHT ADDER

Causus rhombeatus
(Lichtenstein, 1823)

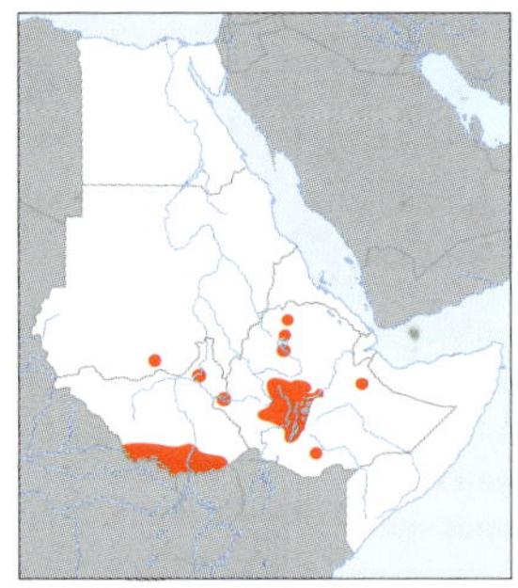

IDENTIFICATION: A small, stout viper with a fairly short head slightly distinct from neck, a rounded snout and medium-sized eye with round pupil, top of head covered by nine large scales (unlike most vipers). Body cylindrical or slightly depressed, Tail short, 9–12 per cent of total length. Scales soft and feebly keeled, in 15–23 rows at mid-body, ventrals 134–166, subcaudals 21–35, higher counts in males. Maximum size about 95cm (possibly slightly larger), average 30–60cm, hatchling 13–16cm. Usually brown, with 20–30 dark, pale-edged rhombic blotches along back; flanks have oblique dark bars or scales. Sometimes pattern-less. Usually a black, or black-edged, V-shape on head. Cream to pinkish-grey below, the colour sometimes graduated so belly looks barred.

HABITAT AND DISTRIBUTION: Moist savanna, grassland and woodland, from sea level to 2,400m, nearly always in vicinity of water. Widespread but sporadic in Ethiopian highlands both sides of Rift Valley, from north of Simien Mountains south via Gondar and Lake Tana to Sululta Plain just west of Addis Ababa, south along central Rift Valley from Koka Dam to Arba Minch, sporadic records in south-west (Gambela, Bonga) and eastern highlands (Harar, and below loop of Webi Shebelle River). A few records in South

Causus rhombeatus, Mulu.

Causus rhombeatus, Mena.

Sudan (Li Rangu, Mt Korobe near Yei, Mongalla, Torit, Kapoeta, and between Katire and Talanga Forest), and 'Upper Nile' (without specimens). In Sudan reported at Kadugli, Kordofan, records from 'Dongola' and 'Merowe Dam' in northern desert seem unlikely. Elsewhere, south to South Africa, west to Nigeria.

CONSERVATION STATUS: Least Concern.

Causus rhombeatus, Yambio.

NATURAL HISTORY: Terrestrial, but will climb low vegetation in pursuit of frogs. Active by day and night, and will bask. Fairly slow-moving, but strikes quite quickly. Swims well. When inactive, hides in holes, bush piles, under ground cover, etc. If angered, responds like other night adders, hissing, puffing and striking. Diet amphibians. May also lift front part of body off ground, flatten neck and move forward with tongue extended, looking like a small cobra. Eats frogs and toads. Lays 7–26 eggs, roughly 26–37 × 16–20mm, in southern Africa these take about 2.5 months to hatch.

CARPET OR SAW-SCALED VIPERS *ECHIS*

A Saharo-Sindian genus of small to medium-sized, fairly slender, dangerous vipers, with pear-shaped heads, a very distinct neck, short snout, moderate to large eyes and vertical pupils. Dorsal scales strongly keeled, lateral dorsal scales with coarsely serrated keels, cloacal scale entire, subcaudals single. Unique among snakes in possessing an external aperture corresponding to an epidermoid supralabial gland on last upper labial (lip) scale; its function is unknown. Carpet vipers are ground-dwellers in arid habitats, nocturnal, egg-laying (some Asian species viviparous) and feed largely on lizards, rodents, birds, frogs and invertebrates (scorpions, solfugids, centipedes). Widely distributed across northern Africa and the Middle East, to India; often very abundant. Although small (none much longer than c.90cm, usually 25–50cm), their venom is potent and also varies remarkably both among and within species, apparently dictated by diet. They are responsible for the bulk of venomous snake bites, many fatal, across a huge swathe of Sub-Saharan Africa, from Senegal east to Somalia. Eleven species recognised, seven in Africa, four in our area, one endemic. Taxonomy not yet fully resolved. *Echis* is most closely related to North African desert vipers, *Cerastes*, and originated in the Oligocene, probably in Arabia.

KEY TO NORTH-EAST AFRICAN MEMBERS OF THE GENUS *ECHIS*

1a: Eye separated from upper labials by 3–4 scales, nasal separated from rostral, enlarged gular scales in midline not present *Echis coloratus*, Burton's Carpet Viper

1b: Eye separated from upper labials by 1–2 scales, nasal in contact with rostral, enlarged gular scales in midline 2

2a: Distinctive white blotches ('eye-spots') on flanks; strongly keeled and overlapping scales on top of head, regularly arranged in one direction *Echis romani*, Roman's Carpet Viper

2b: No distinctive white blotches ('eye-spots') on flanks; scales on top of head weakly keeled, not overlapping and arranged in different directions, especially on snout and between eyes 3

3a: Large scales on head and body, 9–10 between eyes, mid-body scale rows 23–25, ventrals <150 *Echis hughesi*, Somali Carpet Viper

3b: Small scales on head and body, 10–13 between eyes, mid-body scale rows 25–32, ventrals >155 *Echis pyramidum*, North-east African Carpet Viper

BURTON'S CARPET VIPER

Echis coloratus
Günther, 1878

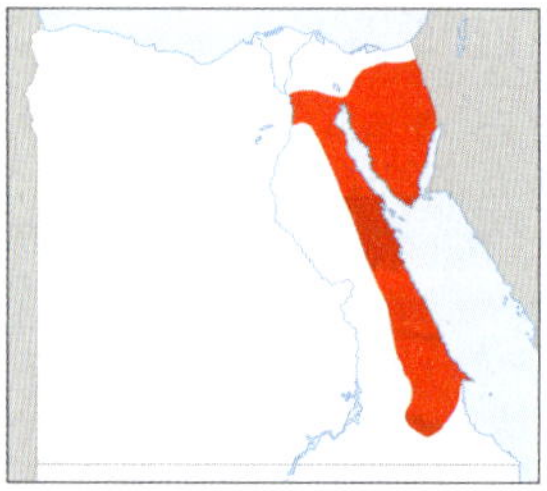

IDENTIFICATION: A medium-sized, robust, beautifully coloured viper with a pear-shaped head and very thin neck, in eastern Egypt and Sinai. Snout rather long and broad; eye moderate, pupil vertical, iris sandy-grey; scales on top of head small, weakly keeled, not strongly overlapping and arranged regularly in one direction, 12–15 between eyes; 15–22 around eye; 12–17 upper labials; 14–16 lower labials; eye separated from upper labials by 3–4 scales. Body stocky, cylindrical, tail short, 8–12 per cent of total length. Scales rough and heavily keeled, in 31–35 rows at mid-body. Ventrals 174–205, subcaudals single, 40–56, cloacal scale entire. Maximum size about 80cm (possibly slightly larger), average 40–60cm, hatchling 20–23cm. Coloration distinctive, dorsum pale grey, brownish or biscuit-coloured, with series of 40–50 large pale grey or whitish saddles on back, rufous-brown patches between saddles, and dark spots on flanks. Usually a diffuse dark stripe extends back from eye, dorsal side of head uniformly coloured. Underside white, sometimes with tiny spots. Taxonomic Notes: Two subspecies; in our area *E. c. coloratus* occurs.

Echis coloratus, Sinai.

HABITAT AND DISTRIBUTION: Usually in rocky desert, with fewer than 150mm of annual rainfall. Fond of dry stone hills, rock pavement, dry wadis and ravines, often in areas with some vegetation. Seems to avoid areas of loose sand, but sometimes on steep slopes and often near water. Sea level to quite high altitudes (2,000m) in our area, only in eastern Egypt and Sinai. A handful of specimens from hills and rocky wadis just east and south-east of Cairo, extending east to Suez, across Sinai, and south along west side of Gulf of Suez, just into extreme north-east Sudan. In Sinai fairly common, but absent from dune fields in north. Elsewhere, to Saudi Arabia. In Egypt widespread, but uncommon. Occasionally collected for pet trade.

CONSERVATION STATUS: Least Concern.

Echis coloratus, Sinai.

NATURAL HISTORY: Terrestrial. Nocturnal, active from sunset onwards, particularly during early hours of night. By day hides in rock fissures, under big rocks, rock piles and ground cover; researchers found that these snakes often chose ambush positions near water, both under cover and on elevated objects (rocks, logs etc), and occasionally climbed vegetation. Appears more good-natured than other carpet vipers but if severely provoked will form a series of C-shaped coils, like others, these coils are shifted against each other in opposite directions, and the friction between scales produces a sound like water falling on a very hot plate, at same time as moving backwards or forwards. If further agitated will strike. Lays 4–10 eggs, approximately 2×3.5cm, glued to hard surfaces; an adaptation to living in a rocky environment. Preferred diet includes rodents, birds and frogs, but lizards and invertebrates also taken, indicating it hunts by day and night; both by ambush and prowling. This is a highly venomous snake with a small and inhospitable range, but the venom is cytotoxic and hemorrhagic, a bite needs hospital treatment.

SOMALI CARPET VIPER

Echis hughesi
Cherlin, 1990

IDENTIFICATION: A small pale carpet viper with a faint pattern, endemic to north-east Somalia. Pear-shaped head with short snout, pale yellowish and prominent eyes with vertical pupils set near front of head. Compared to other species has large head scales: 9–10 between eyes, 14–17 around eye, 9–11 (usually ten) upper labials, 9–10 lower labials, eye separated from upper labials by 1–2 scales. Scales on top of head only weakly keeled. Neck thin, body cylindrical or sub-triangular in section, tail short, *c.*10–14 per cent of total length. Scales large, rough and heavily keeled, in 23–25 rows at mid-body, ventrals 135–149, subcaudals single, 28–32, cloacal scale entire. Maximum size about 35cm, average 26–32cm, hatchling size unknown. Ground colour never described in living animals, but probably sandy or yellowish-brown; usually has 21–32 oblique pale crossbars on back, with darker blotches, often with even darker edging. Flanks show irregular darker blotches of varying size, the pale crescent spots are missing or faint. Unlike, other species with similar coloration, the faint pattern in Somali Carpet Viper is somewhat diffuse. Underside white, unspotted.

HABITAT AND DISTRIBUTION: Near-desert, semi-desert and dry savanna, usually in rocky environments with sparse vegetation, at 84–996m. Eastern Somaliland, Puntland and central Somalia, from Fararali and Meleden in Daror Valley south of Cal Madow and Cal Miskaat Mountains in Sanaag and Bari Regions, south through Sol Haud to Nogal Valley, where known from Wadi Run, Eyl, Garowe and between Hamur and Gombeya. In Mudug of central Somalia reported from Hobyo and environs, then extending *c.*80km north-west. Sympatric with North-east African Carpet Viper *E. pyramidum* in Meleden, the latter species then occurs across Karin Gap and Galgala to Bosaso Plain, and along coast further west.

Echis hughesi, preserved, Wadi Run.

Probably widespread and locally common.

CONSERVATION STATUS: Least Concern.

NATURAL HISTORY: Unknown, but presumably similar to others of genus. Venom probably similar to that of *E. pyramidum*. Specific name honours Barry Hughes (1935–), English herpetologist with a deep interest in snakes of sub-Saharan Africa. He recognised the distinction of this form (in 1976), long before it was formally described.

NORTH-EAST AFRICAN CARPET VIPER, SAW-SCALED VIPER

Echis pyramidum
(Geoffroy Saint-Hilaire, 1827)

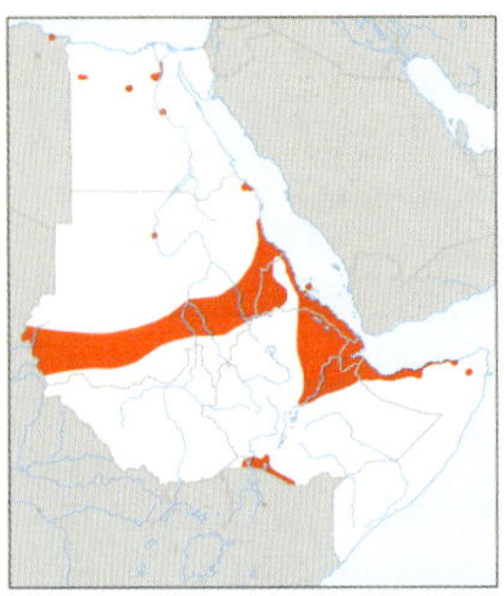

IDENTIFICATION: A medium-sized, fairly stocky snake, usually brownish, with a pear-shaped head and thin neck. Snout short, rounded; pale uniformly yellowish prominent eyes with vertical pupils, tongue reddish. Compared to other *Echis* has moderate head scales: 10–13 between eyes, 16–20 around eye, 10–12 upper labials, 10–13 lower labials, eye separated from upper labials by 1–2 scales. Scales on top of head only weakly keeled, not strongly overlapping and arranged regularly in one direction. Body cylindrical or sub-triangular in section, tail short, 8–12 per cent of total length. Scales rough and heavily keeled, in 25–32 rows at mid-body; ventrals 155–188; subcaudals single, 21–45, higher counts usually males; cloacal scale entire. Maximum size about 70cm (possibly slightly larger), average 30–50cm, hatchling 10–13cm. Usually brownish, but can be yellowish or grey. Typically has a series of 22–40 oblique pale blotches on back, with intervening dark spaces, and on each side usually a row of triangular, sub-triangular or circular dark markings, with distinct large pale crescent spots. Specimens with very faded or almost invisible markings known. Top of head marked usually with pale elongated spot bifurcated at front and posteriorly with transverse arms. Belly whitish, unspotted or finely spotted brown, depending on locality. Taxonomic Notes: Currently two subspecies recognised, see below. Ongoing research suggests that the nominate occurs only in Egypt and Sudan in our area; other populations represent different species.

HABITAT AND DISTRIBUTION: Near-desert, semi-desert and dry savanna, with <500mm of annual rainfall, between sea level and 1,500m, but most frequently below 1,000m. As presently defined, *E. p. pyramidum* is known from scattered localities in northern Egypt, where found mostly in mesic microhabitats of Siwa, Bahariya and Faiyum oases and along Nile Valley, known from Cairo, El Wasta and Asyut. A probable continuous distribution starts from Gebel Elba in extreme south-east Egypt and continues across Red Sea Hills and plains to Sudan and Eritrea, thence west through dry Sahel savanna to central and

Echis pyramidum, Awash National Park.

Echis pyramidum, Berbera.

western Sudan (North Kordofan and Darfur States), including Kassala, Gadarif, Sennar, Khartoum, around Rabak on White Nile, Er Rahad, Fertangul, Muglad, Nyala and Geneina. Also probably present along Nile Valley of Northern State (old record at Dongola). Known from Eritrean lowlands and some of Dahlak Islands (Dahlak Kebir, Dissei, Nokra) south to Danakil region, as far south as Awash National Park, thence east to Djibouti and the Guban coastal strip in Somaliland (not south of mountains) as far east as Bosaso, Karin Gap and Meleden in Puntland. Elsewhere, isolated records in Libya and Central African Republic. Southern *E. p. leakeyi* known in our area only around Lakes Chew Bahr and Turkana in south-west Ethiopia, and Ilemi Triangle in south-east South Sudan. Differs from nominate by smaller size (not exceeding 60cm), overall lower counts of head and body scales, larger and more prominent eyes with an orangey iris, enlarged supraoculars, and darker or rufous coloration with dorsal pale blotches usually wider than they are long. Widespread and locally very common, especially in south of range. In Egypt uncommon and very local, often in pet trade.

CONSERVATION STATUS: Least Concern.

NATURAL HISTORY: Terrestrial, although it occasionally climbs low bushes to avoid hot or wet surfaces. Moves relatively quickly. Nocturnal, active from twilight onwards. During day hides in holes, under or in logs, under rocks or brush piles, may partially bury itself in sand or coil up in or around grass tufts. In sandy coastal habitats of Berbera in Somaliland, these vipers always shelter in holes shaded from high daytime temperatures by crowns of the evergreen tree *Balanites rotundifolia*. In dry and cold months activity is low; these vipers are most abundant as rain starts and temperatures increase. A spirited snake, when angry forms a series of C-shaped coils, which are shifted against each other in opposite directions, and friction between them produces a sound like water falling on a very hot plate, at same time the snake may move backwards or forwards. If further agitated will strike continuously and vigorously, to point of overbalancing. Can strike over a long distance and may move towards a perceived threat; also, sidewinds at considerable speed. Lays 4–25 eggs, *c.*1.7×3cm in size; hatchlings in July–August in Kenya and Somaliland. In captivity up to three clutches per year. Eats a huge variety of prey; some populations eat mostly invertebrates (scorpions, solfugids, centipedes, grasshoppers), but most small vertebrates are also taken, especially lizards, plus rodents and snakes, even its own kind. Ability to feed on a wide range of prey may explain why it is abundant in places; in Moille area of northern Kenya, over 6,500km2, nearly 7,000 were collected in four months. In Somaliland specimens were observed at night in an ambush position on ground, waiting for prey (*Pristurus*, *Hemidactylus*, *Tarentola* lizards, and scorpions) about 50–100cm from trunks of *Vachellia* trees. In right locality and season in Somaliland these snakes are abundant; on a night walk one can be seen roughly every 10m. Similar behaviour observed on Dissei Island, the vipers waiting for prey (*Tarentola* and *Hemidactylus* geckos) at base of walls. A dangerous snake, in medical terms the most significant of our area, on account of its cytotoxic and hemorrhagic venom, abundance, willingness to bite, and it often entering houses. Many people are bitten. Somali name is *Abeeso*, in Sudan *Um Genaib* or *Washasha*, in Egypt *Afa Ghariba*.

Echis pyramidum, Dissei Island.

Echis pyramidum leakeyi, Weyto.

ROMAN'S CARPET VIPER

Echis romani
Trape, 2018

IDENTIFICATION: A distinctly marked carpet viper, relatively large, usually with pale eye-spots on flanks. Head rather short and wide; eye large, pupil vertical, iris bicoloured, lower half dark brown, upper half pale sandy-grey; strongly keeled overlapping scales on top of head arranged regularly in one direction, 12–13 (occasionally 10–11) between eyes, 13–17 scales around eye, ten upper labials, ten lower labials, eye separated from upper labials by 1–2 scales. Body fairly stout, cylindrical or sub-triangular in cross-section, tail short, 9–16 per cent of total length. Scales small, rough and heavily keeled, in 29–33 rows at mid-body. Ventrals 146–168, subcaudals single, 18–29 (25–29 in males, 18–23 in females), cloacal scale entire. Maximum size about 72cm, average 30–50cm, hatchling size unknown but a 13cm specimen was collected in May. Ground colour usually rufous, brown or grey, with a series of 28–32 grey or brown, more or less square blotches on spine, and whitish eye-spots on flanks; animals on iron-rich red sand of Nhem (South Sudan) vivid rufous. On top of head are two pale spots, the anterior small, posterior elongate. Belly pale, and usually covered with bold brown or reddish spots. Taxonomic Notes: Recently elevated from synonymy of *E. ocellatus*, originally regarded as eastern population of that species. Often mentioned in older Sudanese herpetological literature but under names *E. carinatus* or *E. pyramidum*.

HABITAT AND DISTRIBUTION: Dry and moist savanna, even forest edges, with 600–800mm of annual rainfall, at 200–1,200m. In our area at just two localities, both in Sudanian savanna and Sahel *Vachellia* wooded grassland, but probably more widespread. Old records at Kadugli in Nuba Mountains in South Kordofan, Sudan, describe it frequenting grass, scrub and rocks at base of hills between Kadugli and Semma, at around 500m. Recently discovered west of Kadugli, between 430 and 500m, in vicinity of Kulshabi, Neem and Muglad. Probably occurs in Southern Darfur. Elsewhere, from south-west Chad and extreme north-west Central African Republic, to northern Cameroon, west-central Nigeria and extreme southern Niger.

Echis romani, Neem.

Echis romani, Neem.

CONSERVATION STATUS: Least Concern.

NATURAL HISTORY: Like other carpet vipers, terrestrial, but may climb low bushes to avoid hot (or wet) surfaces. Moves relatively quickly for a viper. Nocturnal, active from twilight onwards, most active during first few hours of night. By day hides in holes, under or in logs, rocks, brush piles, etc. When disturbed, readily forms C-shaped coils, stridulates and strikes vigorously over a long distance; moving back or forward, it may strike so enthusiastically that it overbalances, and may even move towards an aggressor, most unusual behaviour in a snake. Lays eggs, Nigerian females laid 10–22 eggs, usually in March–April; these hatch after two months to coincide with rainy season. Eats a wide range of prey, invertebrates (especially scorpions and centipedes) and vertebrates; including amphibians, lizards, other snakes and mammals. This is a dangerous snake, for the reasons detailed above, and in South Kordofan many people are bitten every year.

Around Kadugli the most commonly encountered snake, often entering dwellings, with 11 snakes once found in a single hut; an old man was bitten and died six hours later. Local names include *Um Shideigat* (Kordofan), *Alwashasha* (central and north Sudan) and *Kokor* (Red Sea Hills).

NORTH AFRICAN DESERT VIPERS *CERASTES*

A genus of broad-headed, brown- or biscuit-coloured desert vipers, rarely larger than 70cm. Some have a thin spiny horn above eye, all have strongly keeled dorsal scales, the lateral ones with finely serrated keels, obliquely positioned; when these are rubbed together they produce a defensive hissing signal, like that of carpet vipers. Four species; two in our area.

KEY TO THE NORTH-EAST AFRICAN MEMBERS OF THE GENUS *CERASTES*

1a: Small, <50cm, never has horns, eyes on junction of side and top of head, ventrals always <126
Cerastes vipera, Sahara Sand Viper

1b: Large, up to 75cm or more, often has horns, eyes on side of head, ventrals always >129
Cerastes cerastes, Sahara Horned Viper

SAHARA HORNED VIPER

Cerastes cerastes
(Linnaeus, 1758)

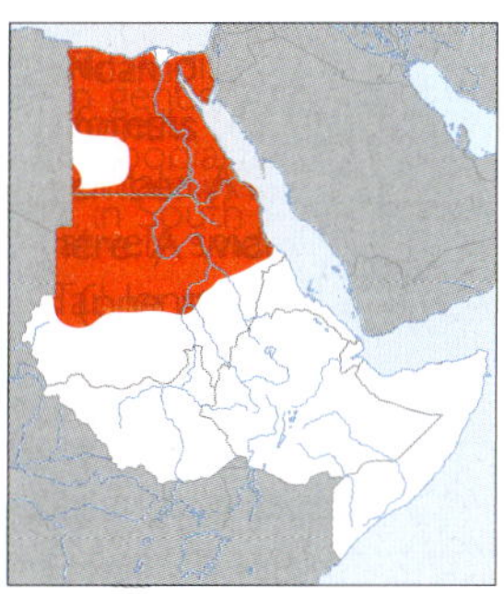

IDENTIFICATION: A short, stout viper, usually with prominent horns, in Sahara and environs. Head broad, flat and triangular, covered with small scales, eye prominent, on side of head, with a vertical pupil in bright light. Neck thin. Often a long horn above each eye; this consists of a single scale, not a group of scales. Individuals without horns often have prominent brow ridges. Body broad and flattened, tail short, 7–12 per cent of total length. Scales heavily keeled, in 23–39 rows at mid-body, ventrals 130–165, cloacal scale entire, subcaudals 23–45. Maximum size in wild about 75cm, but captives up to 90cm, most 35–60cm, hatchling 12–15cm. Colour yellowish, biscuit-coloured, reddish or greyish-brown, with a series of almost rectangular brown blotches on back. A dark line extends back from each eye. Tail tip may be black in some. Taxonomic Notes: Two subspecies recognised; nominate in our area.

HABITAT AND DISTRIBUTION: Desert and semi-desert. A versatile snake, usually in wadis with vegetation and sandy soil but also on gravel plains, in mountainous country, even very rocky areas provided some sand or soil patches exist. Not usually in dunes, but found on edges. Sea level to over 1,200m, possibly higher. Virtually throughout Egypt, save Nile Delta, cultivated areas, and dune fields of northern Sinai. In Sudan probably throughout deserts, but records patchy; Khartoum and Omdurman, Burj At Tuyur, Suakin, Port Sudan, Wadi Halfa, Atbara and Dongola; southernmost appears to be Jebel Teiga in South Darfur. Elsewhere, west to Mauritania. Common and widespread.

CONSERVATION STATUS: Least Concern.

NATURAL HISTORY: Terrestrial and slow-moving, but strikes quickly and can sidewind. Usually nocturnal, but may bask in cold months. Shelters in rodent burrows or other holes, under ground cover (wood, rocks, etc.) or hidden in bush or among grass tufts, often buries (or partially buries) itself in soft sand, sometimes in lee of a bush or rock, but may simply coil up in a sheltered place out of sun and wind, where can be surprisingly visible. Fairly placid, but strikes quickly, and rubs its coils in a C-shaped display when angry, like Carpet vipers, making a crackling sound.

Cerastes cerastes, captive.

Cerastes cerastes, Egypt.

May move long distances at night in search of prey. Lays 10–23 eggs, which hatch after 45–80 days. Diet rodents, lizards and birds, the fact that it eats birds indicates how quickly it strikes. A captive in Cairo Zoo took 6–8 birds a year and lived 14 years. One of the more common hieroglyphic characters in Egypt is a Sahara Horned Viper. In some areas it is greatly feared, and is believed to be able to spring a huge distance in air to bite, or even to be able to fly! Unsolicited reports of horned desert vipers in parts of Somalia invite investigation. This is a venomous species; no fatalities are known, but bites need medical attention.

SAHARA SAND VIPER

Cerastes vipera
(Linnaeus, 1758)

IDENTIFICATION: A small, fat, flattened stout viper with strangely angled eyes on top of its head, in sand dunes of the Sahara and environs. Head broad, flat, triangular and covered by small scales, eye prominent, on side of head and has vertical pupil in bright light. Neck thin. Body broad and flattened, tail short, 7–14 per cent of total length. Scales heavily keeled, in 23–29 rows at mid-body, ventrals 102–125, cloacal entire, subcaudals 18–26. Maximum size c.50cm, but mostly 20–30cm (females larger), hatchling 12–15cm. Ground colour yellowish, biscuit-coloured, reddish or greyish-brown, with series of almost rectangular brown blotches on back. A dark line extends back from each eye. Tail tip black in females.

HABITAT AND DISTRIBUTION: A desert snake, usually found in dunes or other areas of soft sand. From sea level to above 1,000m, possibly higher. In our area, probably almost throughout Egypt where there are dunes, although few records in south-west. Only two Sudanese records, Wadi Ibib on northern Red Sea coast and El Ga'ab Depression, west of Dongola in Northern State, but probably fairly widespread in

Cerastes vipera, Israel/Palestine.

Cerastes vipera, Giza.

northern desert; as occurs at Fada in Chad at 17°N. Elsewhere, west across Sahara to Mauritania.

CONSERVATION STATUS: Least Concern.

NATURAL HISTORY: Terrestrial, slow-moving, but strikes quickly and can sidewind. Usually nocturnal but basks in cooler months, sometimes hibernates. Usually buries itself in soft sand, sometimes with eyes exposed, but may hide in holes or under ground cover. Fairly placid, but shows C-shaped coil display if threatened. Mating occurs April–May, and 3–6 young are born alive in latter half of August. Diet mostly lizards, captive snakes took rodents. A venomous viper, bites usually cause only local symptoms, but may need medical attention.

FALSE HORNED AND SPIDER-TAILED VIPERS *PSEUDOCERASTES*

A genus of broad-headed, medium to fairly large vipers in eastern Egypt and the Middle East; reaching 80–90cm. Three species are known; one just reaches our area.

WESTERN FALSE HORNED VIPER

Pseudocerastes fieldi
Schmidt, 1930

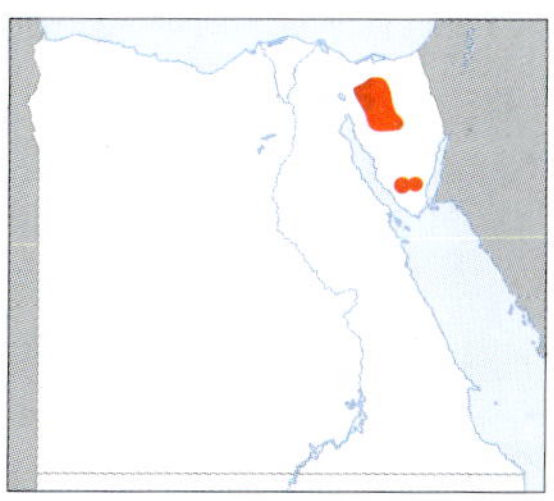

IDENTIFICATION: A short, stout viper, with prominent horns, in Sinai. Head broad, flat, triangular and covered by small scales, eye prominent, on side of head, with a vertical pupil in bright light. Neck thin. A stout backward-curving horn above each eye, consisting of a cluster of scales. Body broad and slightly flattened, tail fairly short, 11–16 per cent of total length. Oddly for a viper, flank scales show no or only very slight keeling, mid-dorsal scales keeled. Scales in 21–23 rows at mid-body, ventrals 127–142, cloacal entire, subcaudals 35–38. Maximum size about 90cm, mostly 40–70cm, hatchling 15–19cm. Pale brown, yellowish, pinkish or mottled grey, with darker (usually brown or reddish) blotches or crossbars, often very straight-edged, and a row of blotches or dots on sides. These may be very faded, and in some the markings are so faint the snake just looks uniform. A dark line through and behind eye. Tail tip often black. Belly surface white.

HABITAT AND DISTRIBUTION: Semi-desert and desert country, in sandy, gravel and rocky regions, but prefers areas with some vegetation. Rarely in dunes or other areas of soft sand, or on very steep slopes, but occurs on scree rubble. Tends to avoid habitation. From sea level to over 2,200m. In our area, it occurs in Sinai (although not dune fields in north), thence east to western Iraq.

CONSERVATION STATUS: Least Concern.

NATURAL HISTORY: Terrestrial, and slow-moving, but can sidewind fairly rapidly. Usually nocturnal, but sometimes active by day or at dusk, especially in spring. Shelters in rodent burrows or other holes, rock cracks or under bushes. Fairly placid, but can strike quickly and over some distance. They mate in May or June, and lay 10–21 eggs in August, which contain large embryos that hatch after about a month of incubation. Diet includes mammals and birds, juveniles eat lizards. A venomous snake and bites need medical attention.

Pseudocerastes fieldi, captive.

Gazetteer

We have listed here (hopefully) all the localities mentioned in the text, and alternative names where both are used. They are listed as follows: name, country, decimal latitude and longitude. To convert decimal to traditional (degree, minutes, seconds) latitude and longitude, multiply for both values the two-figure value after the decimal point by 0.6; alternatively converters can be found °N the internet.

Ab-Dareesh, Sudan 15.00°N 32.46°E
Abal Mt, Puntland (Somalia) 11.28°N 49.63°E
Abay Valley, Ethiopia c.09.96°N 37.60°E
Abaya Lake, Ethiopia 06.33°N 37.83°E
Abeche, Central African Republic 13.85°N 20.85°E
Abelti (bridge), Ethiopia 08.23°N 37.58°E
Abiyu, Ethiopia 08.20°N 35.33°E
Abri, Sudan 20.80°N 30.35°E
Abu Hamad, Sudan 19.53°N 33.32°E
Abu Idres, see Ab-Dareesh
Abu Likri, Sudan 10.98°N 28.81°E
Abu Rawash, Egypt 30.05°N 31.10°E
Abu Rudeis, Egypt 28.90°N 33.19°E
Abu Sari, Sudan 20.28°N 30.58°E
Abu Simbel, Egypt 22.35°N 31.62°E
Abu Tabari (in Wadi Howar), Sudan 17.58°N 28.52°E
Abu Zor, Sudan 12.39°N 34.35°E
Abyei, Sudan 09.60°N 28.43°E
Ad-dabba, see Debba
Ad Damazin, Sudan 11.79°N 34.36°E
Adadley, Somaliland (Somalia) 09.77°N 44.69°E
Addis Ababa, Ethiopia 08.94°N 38.79°E
Adi Caie, Eritrea 14.84°N 39.73°E
Adi Keyh, see Adi Caie
Adi Teklezan, Eritrea 15.57°N 38.78°E
Adi Ugri, see Mendefera
Adis Abeba, see Adis Abbaba
Ado, Ethiopia 07.33°N 45.25°E
Adok, South Sudan 08.18°N 30.32°E
Adoshebai Valley, Ethiopia 05.75°N 37.25°E
Adua, Ethiopia 14.16°N 38.90°E
Adwa, see Adua
Af Abed, see Afabet 16.19°N 38.69°E
Afabet, Eritrea 16.19°N 38.69°E
Afar (Triangle, region), Ethiopia 13.00°N 41.00°E
Afghedud, Somalia 05.76°N 48.03°E
Afgoi, Somalia 02.14°N 45.11°E
Afguduudle, see Afghedud
Afmadow, see Afmadu
Afmadu, Somalia 00.52°N 42.07°E
Afrera, Ethiopia 13.21°N 40.86°E
Agabar, Somaliland (Somalia) 09.92°N 43.92°E
Agordat, Eritrea 15.54°N 37.88°E
Ahmar Mountains, Ethiopia 09.41°N 41.44°E
Ailet, Eritrea 15.57°N 39.15°E
Ain Sukhna, Egypt 29.60°N 32.32°E
Akaki, Ethiopia 08.87°N 38.78°E
Akasha, Sudan 21.08°N 30.72°E
Akik, Sudan (Hillat Agig) 18.23°N 38.19°E
Al Arish, Egypt 16.97°N 42.83°E
Al Fashir, Sudan 13.63°N 25.35°E
Al Jabalain, Sudan 12.59°N 32.80°E
Al Junaynah, see Geneina
Al Qutaynah, Sudan 14.86°N 32.37°E
Al Sunut Forest, Sudan 15.59°N 32.50°E
Alaua, Ethiopia 10.07°N 42.98°E
Alemaya Lake, Ethiopia 09.40°N 42.01°E
Alessandra Island, Somalia 00.50°N 42.77°E
Alexandria, Egypt 31.19°N 29.91°E
Ali, Ethiopia 07.25°N 39.93°E
Ali Sabieh (Region), Djibouti 11.14°N 42.73°E
Am Dafok, Central African Republic 10.46°N 23.29°E
Amarr, Ethiopia 05.33°N 38.00°E
Amba Giorgis, Ethiopia 12.77°N 37.62°E
Ambo, Ethiopia 08.98°N 37.85°E
Ambukol, Sudan 21.27°N 30.92°E
Amhara (Region), Ethiopia c.11.50°N 38.50°E
Amibarra, Ethiopia 09.40°N 40.18°E
Amle, Sudan 19.53°N 30.18°E
Amoud University, Somaliland (Somalia) 09.94°N 43.22°E
Andeber Island, Eritrea 15.70°N 39.91°E
Andurgab, Somalia c.05.33°N 48.50°E
Ankober, Ethiopia 09.60°N 39.73°E
Anseba (Valley, River), Eritrea c.15.87°N 38.47°E
Aqaba, Jordan 29.53°N 35.00°E
Aqaba, Gulf of c.28.78°N 34.72°E
Arabsiyo, Somaliland (Somalia) 09.67°N 43.75°E
Arava Valley, Israel/Palestine and Jordan c.30.42°N 35.16°E
Ar-Rusayris, Sudan 11.87°N 34.39°E
Arba Minch, Ethiopia 06.03°N 37.55°E
Arbascio, Eritrea 15.50°N 38.73°E
Arbegona, Ethiopia 06.68°N 38.72°E
Archers Post, Kenya 00.64°N 37.67°E
Arero, Ethiopia 04.75°N 38.82°E
Argo Island, Sudan 19.55°N, 30.44°E
Aroma, Sudan 15.85°N 36.13°E
Aroweina, Somaliland (Somalia) 10.06°N 44.68°E
Arrusi Mountains, Ethiopia 07.75°N 38.75°E
Arsi (Region), Ethiopia 07.50°N 39.50°E
Arta, Djibouti 11.53°N 42.85°E
Artu, Somaliland 10 46°N 42 55°E
Asayita, Ethiopia 11.57°N 41.44°E
Asalaya, Sudan 13.25°N 32.73°E
Asela, Ethiopia 07.95°N 39.13°E
Asendabo, Ethiopia 07.77°N 37.23°E
Asmara, Eritrea 15.34°N 38.93°E
Asosa, Ethiopia 10.07°N 34.53°E
Assab, Eritrea 13.00°N 42.74°E
Asuan, see Aswan
Aswan, Egypt 24.09°N 32.90°E
Asyut, Egypt 27.18°N 31.18°E
Atbara, Sudan 17.70°N 33.99°E
Atbara River, Sudan c.17.68°N 33.97°E
Atet, South Sudan 06.26°N 31.73°E
Audo Mountains, Ethiopia 06.17°N 41.75°E
Aware, Ethiopia 08.26°N 44.16°E
Aware Melka, Ethiopia 09.13°N 39.97°E
Awash National Park, Ethiopia 08.92°N 40.02°E
Awash River, Ethiopia 08.87°N 40.09°E
Awassa Lake, Ethiopia 07.05°N 38.43°E
Awdal (Region), Somaliland (Somalia) c.10.50°N 43.50°E
Aweil, South Sudan 08.76°N 27.39°E

Buulobarde, see Bulo Barde
Caaio, Somalia c.05.33°N 48.50°E
Cairo, Egypt 30.06°N 31.25°E
Cal Madow Mountains, Somaliland (Somalia) c.10.97°N 48.66°E
Cal Miskaat Mountains, Puntland (Somalia) c.10.96°N 49.99°E
Callis, Puntland (Somalia) 08.43°N 49.05°E
Candala, Puntland (Somalia) 11.47°N 49.87°E
Candhodhexe, Somaliland (Somalia) 08.43°N 45.73°E
Carcar Mountains, Puntland (Somalia) 09.91°N 49.39°E
Caschei, Ethiopia 04.80°N 36.45°E
Ceel Cillan, see El Ellan
Ch'efe Donsa, see Ciaffedenza
Chagni, Ethiopia 10.95°N 36.50°E
Chak Chak, South Sudan 08.67°N 26.85°E
Chancho, Ethiopia 09.31°N 38.75°E
Chinahasen, Ethiopia 09.51°N 42.61°E
Choke Mt, Ethiopia 10.72°N 37.85°E
Chamo Lake, Ethiopia 05.83°N 37.55°E
Chelago, Ethiopia 04.18°N 40.03°E
Chenafena, Eritrea 14.79°N 39.02°E
Chencha, Ethiopia 06.25°N 37.57°E
Cheren, see Keren
Chew Bahir Lake, Ethiopia 04.63°N 36.83°E
Chew Bahr, see Chew Bahir Lake
Chira, Ethiopia 08.43°N 36.13°E
Chirim, Ethiopia 05.93°N 35.33°E
Chisimaio, see Kismayu
Chovaye Island, Somalia 00.87 S 42.16°E
Ciaffedenza, Ethiopia 08.97°N 39.12°E
Coiama Island, Somalia 00.64 S 42.34°E
Cunama (Region), Eritrea c.14.33°N 36.75°E
Cyrenaica (Region), Libya c.31.00°N 22.50°E
Daallo Mt (Forest), Somaliland (Somalia) 10.76°N 47.30°E
Daarbuduq, Somaliland (Somalia) 09.86°N 44.54°E
Daarin, Puntland (Somalia) 10.75°N 49.73°E
Dabus River, Ethiopia 10.62°N 35.15°E
Dadar, Puntland (Somalia) 11.12°N 49.87°E
Dagah Shabell, Somaliland (Somalia) 10.17°N 45.20°E
Dahab, Egypt 28.48°N 34.50°E
Dahlak Archipelago, Eritrea c.15.83°N 40.20°E
Dahlak Kebir (island), Eritrea 15.74°N 40.09°E
Dakhata Valley, Ethiopia 09.23°N 42.56°E
Dakhla (Oasis), Egypt 25.50°N 29.17°E
Dallami, see Delami
Dalol, Ethiopia 14.24°N 40.19°E
Danakil Depression, Ethiopia c.14.19°N 40.56°E
Dangila, Ethiopia 11.27°N 36.83°E
Darfur (Region), Sudan c.13.00°N 25.00°E
Daror, Ethiopia 08.22°N 44.69°E
Daror Valley, Puntland and Somaliland (Somalia) c.10.34°N 50.42°E
Darror, see Daror Valley
Daua River, Ethiopia c.04.61°N 40.62°E
Davegoriale, Somaliland (Somalia) 08.73°N 44.82°E
Day Forest National Park, Djibouti 11.77°N 42.66°E
Debark, see Debarq
Debarq, Ethiopia 13.16°N 37.90°E
Debba, Sudan (Al Dabba) 18.05°N 30.95°E
Debre Berhan, Ethiopia 09.68°N 39.53°E
Debre Birhan, see Debre Berhan
Debre Libanos, Ethiopia 09.72°N 38.87°E
Debre Marcos, Ethiopia 10.35°N 37.73°E
Debre Tabor, Ethiopia 11.85°N 38.02°E
Debre Zeit, Ethiopia 08.75°N 38.98°E
Degeh Bur, Ethiopia 08.22°N 43.56°E
Dehel Island, Eritrea 15.91°N 39.64°E
Dejen, Ethiopia 10.17°N 38.13°E
Dekemhare, Eritrea 15.07°N 39.05°E
Delami, Sudan 11.86°N 30.46°E
Delgo, see Dulgo
Delimbe, Central African Republic 09.79°N 22.62°E
Delo, Ethiopia 07.25°N 40.47°E
Dembi, Ethiopia 08.08°N 36.46°E
Derudeb, Sudan 17.53°N 36.08°E
Dessie, Ethiopia 11.13°N 39.63°E
Didinga Hills, South Sudan 04.33°N 33.57°E
Dhamal Plain, Somaliland (Somalia) 09.84°N 44.15°E
Dharkayn Geenyo, Ethiopia 08.00°N 47.00°E
Dhubaato, Somaliland (Somalia) 09.74°N 44.45°E
Didessa River Bridge, Ethiopia 09.03°N 36.15°E
Didessa River Valley, Ethiopia c.09.77°N 36.02°E
Diffra, Sudan 10.04°N 28.40°E
Difnein Island, Eritrea 16.64°N 39.32°E
Dilla, Somaliland (Somalia) 09.78°N 43.35°E
Dilling, also Dalang, Sudan 12.05°N 29.65°E
Dinder National Park, Sudan 12.29°N 35.00°E
Dinsor, Somalia 02.41 N 42.98°E
Dinsoor, see Dinsor
Dinshu, Ethiopia 07.10°N 39.78°E
Dire Dawa, Ethiopia 09.58°N 41.87°E
Dittilou, Djibouti 11.78°N 42.69°E
Dissei Island, Eritrea 15.46°N 39.74°E
Djalelo, Djibouti 11.35°N 42.79°E
Djibouti City, Djibouti 11.59°N 43.15°E
Dodola, see Dodolla,
Dodolla, Ethiopia 06.98°N 39.18°E
Dohul, see Dehel Island
Dolo Mena, see Mena
Dolobscio, Somalia 05.57°N 48.08°E
Dongola, Sudan 19.18°N 30.48°E
Doolow, see Dollo
Dollo Somalia 04.16°N 42.07°E
Doriyanle, Somalia 02.57°N 42.61°E
Dorze, Ethiopia 06.20°N 37.58°E
Duem, Sudan (Ed-Dueim) 14.00°N 32.32°E
Dulgo, Sudan 12.51°N 23.26°E
Durdeb, see Derudeb
Dur Ghella, see Durghella
Durgale, Somalia 05.58°N 48.27°E
Durghella Island, Eritrea 15.77°N 39.79°E
Durro River, Ethiopia c.07.03°N 41.67°E
Durru, see Durrur
Durrur, Sudan 19.84°N 37.25°E
Durruqsi, Somaliland (Somalia) 08.52°N 45.47°E
Dusa Marreb, Somalia 05.54°N 46.39°E
Ed Daein, Sudan 11.46°N 26.13°E
Ed Damazin, Sudan 11.79°N 34.36°E
Ed Debba, Sudan, see Debba
Edfu, Egypt 24.98°N 32.88°E
Eggi, Somalia 03.18°N 45.54°E
Eil, see Eyl
Eilat, Israel/Palestine 29.56°N 34.95°E
Eilat Mountains, Israel/Palestine 29.58°N 34.92°E
El Afweyn, Somaliland (Somalia) 09.93°N 47.22°E
El Bur, Somalia 04.68°N 46.61°E
El Dere, Somalia 03.85°N 47.18°E
El Dire, Ethiopia 04.98°N 37.12°E

El Donfar, Puntland (Somalia) ... 10.63°N 49.05°E
El Ellan, Somalia ... 02.50°N 42.52°E
El Fasher, Sudan ... 13.63°N 25.35°E
El Figra, Puntland (Somalia) ... 10.60°N 50.53°E
El Fira, *see* El Figra
El Fula, Sudan ... 11.73°N 28.36°E
El Gaal, Puntland (Somalia) ... 11.38°N 50.40°E
El Ga'ab (Depression), Sudan ... *c.* 19.22°N 30.19°E
El Hammam, Egypt ... 30.85°N 29.39°E
El Kadada, Sudan ... 16.75°N 33.57°E
El Magdhaba, Egypt ... 30.90°N 34.02°E
El Matamma, Sudan ... 16.71°N 33.36°E
El Mocoile, Somalia ... 03.82°N 45.22°E
El Mur, Puntland (Somalia) ... 09.53°N 49.14°E
El Obeid, Sudan ... 13.18°N 30.22°E
El Omayed (Protected Area), Egypt ... 30.75°N 29.20°E
El Quseima, Egypt ... 30.67°N 34.37°E
El Sod, Ethiopia ... 04.20°N 38.40°E
El Tarfa, Egypt ... 28.70°N 33.95°E
El Ure, Somalia ... 03.83°N 43.10°E
El Wasta, Egypt ... 29.33°N 31.20°E
Ela Beridi, *see* Elaberet
Elaberet, Eritrea ... 15.71°N 38.63°E
En Nahud, Sudan ... 12.70°N 28.43°E
Enjebara, *see* Injibara
Ennedi Massif, Chad ... 17.53°N 22.27°E
Entedebir Island, *see* Andeber,
Entoto hills, Ethiopia ... 09.07°N 38.80°E
Equatoria, Central (State), South Sudan ... *c.* 04.75°N 31.00°E
Equatoria, Eastern (State), South Sudan ... *c.* 04.90°N 33.80°E
Equatoria, Western (State), South Sudan ... *c.* 05.40°N 28.40°E
Er Rahad, Sudan ... 12.72°N 30.65°E
Erigavo, Somaliland (Somalia) ... 10.61°N 47.37°E
Erkowit, Sudan ... 18.77°N 37.10°E
Es Suweigat, Sudan ... 18.92°N 32.30°E
Eyl, Puntland (Somalia) ... 07.98°N 49.82°E
Fada, Chad ... 17.19°N 21.58°E
Fadhi-Gaab, Somaliland (Somalia) ... 09.69°N 47.01°E
Fakoki, Sudan ... 12.45°N 27.51°E
Fangak, South Sudan ... 09.07°N 30.88°E
Fantalle (Volcano), Ethiopia ... 08.98°N 39.93°E
Farafra (Oasis), Egypt ... 27.06°N 27.97 E
Farasan Kebir Island, Saudi Arabia ... 16.73°N 41.91°E
Fayum (Oasis), Egypt ... 29.31°N 30.84°E
Felege Neway, *see* Felenguai
Felenguai, Ethiopia ... 06.30°N 36.88°E
Ferdusa Ruins, Somaliland (Somalia) ... 09.94°N 45.19°E
Fertangul, Sudan ... 12.93°N 30.08°E
Filfil, *see* Fil-fil,
Fil-fil, Eritrea ... 15.62°N 38.94°E
Filtu, Ethiopia ... 04.97°N 40.38°E
Fik, Ethiopia ... 08.14°N 42.29°E
Fiq, *see* Fik
Forêt du Day, *see* Day Forest National Park
Fung (province), Sudan ... *c.* 12.15°N 34.34°E
Gaan Libaah, *see* Gan Libah
Gaarre, Eritrea ... 13.20°N 42.03°E
Gadarif, Sudan ... 14.03°N 35.39°E
Gadat, Ethiopia ... 06.33°N 36.83°E
Gahtielay, Eritrea ... 15.52°N 39.15°E
Galadi, Ethiopia ... 06.98°N 46.41°E
Galgalo (Oasis), Puntland (Somalia) ... 10.99°N 49.06°E
Galguduud (Region), Somalia ... *c.* 05.00°N 47.00°E
Galkayo, Puntland (Somalia) ... 06.77°N 47.43°E
Gambela, Ethiopia ... 07.83°N 34.50°E
Gan Libah, Somaliland (Somalia) ... 09.86°N 44.82°E
Garadag, Somaliland (Somalia) ... 09.49°N 46.87°E
Garamba (National Park), DR Congo ... 04.20°N 29.48°E
Garbaqabat, Somaliland (Somalia) ... 08.88°N 47.40°E
Gardo, Puntland (Somalia) ... 09.50°N 49.08°E
Garissa, Kenya ... 00.46 S 39.64°E
Garoe, *see* Garowe
Garowe, Puntland (Somalia) ... 08.40°N 48.49°E
Garoowe, *see* Garowe
Garsaale, *see* Ghersale
Gash River, Eritrea ... *c.* 14.78°N 37.47°E
Gaza (region), Israel/Palestine ... *c.* 31.50°N 34.47°E
Gebel Dalfa, Egypt ... 30.78°N 34.19°E
Gebel El Giddi, Egypt ... 30.16°N 33.15°E
Gebel Elba, Egypt ... 22.18°N 36.35°E
Gebel Maghara, Egypt ... 30.68°N 33.36°E
Gebel Marra, *see* Jebel Marra
Gebel Moro, Sudan ... 10.88°N 30.12°E
Gebel Urayf an Naqah, Egypt ... 30.36°N 34.43°E
Gedarif, *see* Gadarif
Geedaley, Somalia ... 05.27°N 47.92°E
Geerissa, Somaliland (Somalia) ... 10.61°N 43.43°E
Gelib, Somalia ... 00.49°N 42.79°E
Gemeri Lake, Ethiopia ... 11.55°N 41.66°E
Gemmaiza, South Sudan ... 05.69°N 31.80°E
Genale, Somalia ... 01.81°N 44.70°E
Genale River, Ethiopia ... *c.* 04.32°N 41.94°E
Gendoa River, Ethiopia ... *c.* 12.50°N 36.50°E
Geneina, Sudan ... 13.45°N 22.45°E
Geriban, Puntland (Somalia) ... 07.21°N 48.85°E
Gerisa, *see* Geerissa
Gestro River, Ethiopia ... *c.* 04.33°N 42.05°E
Gewane, Ethiopia ... 10.17°N 40.65°E
Gheleb, Eritrea ... 15.82°N 38.74°E
Ghersale, Somalia ... 02.65°N 45.38°E
Ghimbi, Ethiopia ... 09.17°N 35.83°E
Ghinda, Eritrea ... 15.45°N 39.09°E
Giarcosa, Ethiopia ... 04.17°N 39.83°E
Giamo, Ethiopia ... 06.63°N 35.80°E
Gibe River, upper Ethiopia ... 08.00°N 37.40°E
Gid Daa (wells), Puntland (Somalia) ... 11.20°N 49.87°E
Gildessa, Ethiopia ... 09.71°N 42.12°E
Gilf El Kebir Plateau, Egypt ... 23.68°N 25.94°E
Gilo, South Sudan ... 04.04°N 32.84°E
Ginir, Ethiopia ... 07.14°N 40.71°E
Giohar, Somalia ... 02.78°N 45.50°E
Giovai Island, Somalia ... 00.87 S 42.16°E
Giumbo, Somalia ... 00.25 S 42.62°E
Giza, Egypt ... 30.06°N 30.99°E
Goba, Ethiopia ... 07.02°N 39.98°E
Godare, Ethiopia ... 07.43°N 35.00°E
God Heeli, Somaliland (Somalia) ... 09.04°N 46.86°E
Gode, Ethiopia ... 05.95°N 43.55°E
Godolefassi, Eritrea ... 14.87°N 38.85°E
Gojjam (highlands, province), Ethiopia ... *c.* 10.73°N 37.50°E
Golgia, Somalia ... 00.72°N 41.72°E
Gombeya, *see* Kombeyo
Gondar, Ethiopia ... 12.60°N 37.47°E
Gondokoro, South Sudan ... 04.90°N 31.66°E
Geneina, Sudan ... 13.45°N 22.45°E
Gongo, Central African Republic ... 04.87°N 21.34°E
Goolis Mountains, Somaliland (Somalia) ... 09.97°N 45.02°E
Go'oo (village), Somaliland (Somalia) ... 09.79°N 44.91°E

Gore, Ethiopia....08.13°N 35.52°E
Goscia (zone), Somalia....c.01.19°N 42.57°E
Goulf, Ethiopia....06.83°N 41.42°E
Goz Abu Goma, Sudan....13.14°N 32.73°E
Green Island, Eritrea....15.59°N 39.48°E
Guban (Region), Somaliland (Somalia)....c.10.31°N 44.37°E
Guder, Ethiopia....08.97°N 37.78°E
Guge Mountains, *see* Gughe Mountains
Gughe Mountains, Ethiopia....06.17°N 37.42°E
Gura, Eritrea....15.02°N 39.07°E
Guriasamo, Puntland (Somalia)....11.27°N 49.80°E
Haaway, *see* Havai
Habeno Mt, Puntland (Somalia)....11.09°N 49.44°E
Habo, Puntland (Somalia)....11.79°N 50.52°E
Hadaden, Somaliland (Somalia)....08.29°N 46.14°E
Hagere Maryam, Ethiopia....05.63°N 38.24°E
Hafun (Peninsula), Puntland (Somalia)....10.42°N 51.25°E
Haiya, Sudan....18.34°N 36.39°E
Halin, Somaliland (Somalia)....09.09°N 48.62°E
Hamashkoraib, Sudan....17.12°N 36.74°E
Hamata, Egypt....24.28°N 35.38°E
Hamur, Somaliland (Somalia)....08.87°N 48.63°E
Harar, Ethiopia....09.31°N 42.12°E
Harat Island, Eritrea....16.11°N 39.46°E
Harenna Forest, Ethiopia....c.06.60°N 39.73°E
Hargeisa, Somaliland (Somalia)....09.56°N 44.07°E
Hargeysa, *see* Hargeisa
Harirad, Somaliland (Somalia)....10.33°N 42.84°E
Harmil Island, Eritrea....16.52°N 40.15°E
Hàro, Somalia....05.83°N 48.18°E
Harradigit, Ethiopia....07.75°N 45.50°E
Hashenge Lake, Ethiopia....12.58°N 39.50°E
Hasmet, Eritrea....17.73°N 38.68°E
Haud (Region), Somaliland (Somalia) and Ethiopia
....c.08.46°N 45.55°E
Havai, Somalia....01.17°N 43.71°E
Hawd, *see* Haud
Hay Booqan, *see* Heibogan
Heglig, Sudan....11.98°N 27.87°E
Heibogan (Cave), Puntland (Somalia)....09.55°N 49.11°E
Heis, Somaliland (Somalia)....10.89°N 46.92°E
Hellet Nuer (= Adok), South Sudan....08.18°N 30.32°E
Helwan, Egypt....29.84°N 31.30°E
Hertale Lake, Ethiopia....09.91°N 40.39°E
Hiraan, *see* Hiran
Hiran (Region), Somalia....c.04.00°N 45.50°E
Hobyo, Somalia....05.35°N 48.53°E
Hoja Wajeer, *see* Ola Uager
Holi, Sudan....12.92°N 27.58°E
Homa Bay, Kenya....00.53 S 34.46°E
Hordio, Puntland (Somalia)....10.56°N 51.13°E
Hosanna, *see* Hoseana
Hoseana, Ethiopia....07.55°N 37.85°E
Hudet, Ethiopia....04.75°N 39.23°E
Hudisso, Somaliland (Somalia)....10.00°N 45.16°E
Hudun, Somaliland (Somalia)....09.15°N 47.48°E
Huluul, Somaliland (Somalia)....09.97°N 46.70°E
Hurghada, Egypt....27.26°N 33.81°E
Huri Hills, Kenya....03.52°N 37.78°E
Idali, South Sudan....04.71°N 32.74°E
Ikotos, South Sudan....04.08°N 33.11°E
Ikwoto, South Sudan....04.08°N 33.11°E
Ilala Sala Plains, Ethiopia....08.91°N 40.08°E
Ilemi Triangle, South Sudan....05.18°N 35.34°E
Illubabor (Province), Ethiopia....08.28°N 35.76°E
Imatong Mountains, South Sudan....04.04 N, 32.85°E
Imi, Ethiopia....06.47°N 42.18°E
Imurok Mt, South Sudan....04.30°N 32.40°E
Inaafmadow, Somaliland (Somalia)....09.15°N 45.94°E
Inimvua River, South Sudan....c.04.63°N 29.71°E
Injibara, Ethiopia....11.00°N 36.92°E
Iskushuban, Puntland (Somalia)....10.28°N 50.23°E
Ismailia, Egypt....30.60°N 32.27°E
Isratu Island, Eritrea....16.33°N 39.88°E
Iwatoka, *see* Watoka
Jabal Karari, Sudan....15.78°N 32.51°E
Jabal Lowe, *see* Lowai
Jamaame, Somalia....00.07°N 42.74°E
Jannaale, *see* Genale
Jebel Al Dair (Biosphere Reserve), Sudan....12.45°N 30.7°E
Jebel Aulia, *see* Jebel Awlia
Jebel Awaynat, Egypt....22.02°N 25.06°E
Jebel Awlia, Sudan....15.39°N 32.61°E
Jebel Barkal, Sudan....18.53°N 31.83°E
Jebel El Halal, Egypt....30.65°N 34.03°E
Jebel El Shayeb, Egypt....26.98°N 33.48°E
Jebel Lado, South Sudan....05.09°N 31.58°E
Jebel Marra, Sudan....13.75°N 24.50°E
Jebel Meidob, Sudan....15.23°N 26.50°E
Jebel Teiga, Sudan....12.70°N 24.82°E
Jebel Uweinat, *see* Jebel Awaynat
Jesomma, Somalia....04.04°N 45.73°E
Jibal Hamashrbab, Sudan....18.80°N 37.04°E
Jidhi, Somaliland (Somalia)....10.59°N 43.04°E
Jijiga, Ethiopia....09.35°N 42.80°E
Jilib, *see* Gelib
Jimma, Ethiopia....07.67°N 36.83°E
Jinka, Ethiopia....05.65°N 36.65°E
Jofay, *see* Giovai Island
Jonglei (State), South Sudan....07.40°N 32.40°E
Jowhar, *see* Giohar
Juba, South Sudan....04.85°N 31.58°E
Juba River, Somalia....c.02.06°N 42.21°E
Judean Desert, Israel/Palestine....c.31.50°N 35.30°E
Jur River, South Sudan....c.08.67°N 29.16°E
Jurile, Puntland (Somalia)....10.38°N 49.07°E
Kadugli, Sudan....11.01°N 29.72°E
Kafta Sherano National Park, Ethiopia....14.25°N 37.26°E
Kagelu, South Sudan....04.05°N 30.61°E
Kajiko, South Sudan....03.85°N 30.60°E
Kakuma, Kenya....03.71°N 34.86°E
Kamba, Ethiopia....06.15°N 37.20°E
Kapoeta, South Sudan....04.77°N 33.59°E
Karima, Sudan....18.55°N 31.85°E
Karin, Puntland (Somalia)....10.98°N 49.22°E
Karin Gap, Puntland (Somalia)....11.02°N 49.20°E
Kasigau Mt, Kenya....03.83°S 38.66°E
Kassala, Sudan....15.47°N 36.43°E
Katire, South Sudan....04.04°N 32.80°E
Kebre Mengist, *see* Kebri Mangest
Kebri Beyah, Ethiopia....09.10°N 43.17°E
Kebri Mangest, Ethiopia....05.88°N 38.98°E
Kediba, South Sudan....05.51°N 30.72°E
Keffa (zone), Ethiopia....07.18°N 36.05°E
Keilak, Sudan....10.83°N 29.35°E
Kelafo, Ethiopia....05.59°N 44.20°E
Keren, Eritrea....15.78°N 38.45°E
Kereyu Lodge (in Awash), Ethiopia....08.88°N 40.10°E

Keriyo, Ethiopia........09.04°N 38.65°E
Kerma, Sudan........19.59°N 30.41°E
Kessa, Ethiopia........10.95°N 36.97°E
Key Afer, Ethiopia........05.52°N 36.73°E
Khanka Dunes, Egypt........30.18°N 31.47°E
Kharga (Oasis), Egypt........25.31°N 30.58°E
Khartoum, Sudan........15.55°N 32.53°E
Khor Arbaat, Sudan........19.75°N 37.20°E
Khor Atar, South Sudan........09.34°N 31.42°E
Khor Attar, *see* Khor Atar
Khor Okwat, Sudan........19.38°N 37.17°E
Kibre Mengist, *see* Kebri Mangest
Kidepo, Uganda........03.85°N 33.76°E
Kismayu, Somalia........00.36°S 42.55°E
Kinyeti Mt, South Sudan........03.95°N 32.91°E
Kiridh, Somaliland (Somalia)........08.97°N 46.13°E
Kodok, South Sudan........09.89°N 32.11°E
Kofele, *see* Koffelle
Koffelle, Ethiopia........07.07°N 38.78°E
Kohaito (Plateau), Eritrea........14.88°N 39.42°E
Koji, Ethiopia........06.81°N 35.25°E
Koka (Dam), Ethiopia........08.47°N 39.15°E
Kokuro, South Sudan........04.58°N 35.72°E
Kom Ombo, Egypt........24.48°N 32.95°E
Kom Oshim, Egypt........29.55°N 30.88°E
Kombeyo, Somaliland (Somalia)........08.92°N 48.58°E
Konso, Ethiopia........05.25°N 37.48°E
Koobi Fora, Kenya........03.95°N 36.21°E
Kordofan, North (State), Sudan........*c.*14.68°N 29.93°E
Kordofan, South (State), Sudan........*c.*11.27°N 30.83°E,
Korobi Mt, South Sudan........03.58°N 30.52°E
Kosha, Sudan........20.83°N 30.54°E
Kosti, Sudan........13.16°N 32.66°E
Kounhi, *see* Kuni
Koyama Island, *see* Coiama
Kulet El-Dob, Egypt........24.23°N 35.25°E
Kuni (Valley), Ethiopia........09.00°N 40.87°E
Kuraymah, *see* Karima
Kurmuk, Sudan........10.55°N 34.28°E
Kutaworke, Ethiopia........10.55°N 34.50°E
Kutum, Sudan........14.20°N 24.67°E
Laas Dhuure, Somaliland (Somalia)........10.18°N 45.98°E
Laascaanood, *see* Las Anod
Laasqoray, *see* Las Korey
Labadaah/Bulo Yaag, Somalia........*c.*00.50°N 42.77°E
Laboni, South Sudan........03.81°N 32.76°E
Lac Badana, Somalia........00.76°S 42.0°E
Lafaruug, Somaliland (Somalia)........10.01°N 44.78°E
Lafon, South Sudan........05.03°N 32.47°E
Laikipia, Kenya........00.33°N 36.77°E
Lainya, South Sudan........04.34°N 31.07°E
Laisamis, Kenya........01.59°N 37.81°E
Lake Maghra, Egypt........30.27°N 28.93°E
Lakes (State), South Sudan........*c.*06.81°N 29.68°E
Lalibela, Ethiopia........12.03°N 39.05°E
Langano Lake, Ethiopia........07.60°N 38.72°E
Lankien, South Sudan........08.50°N 32.06°E
Las Anod, Puntland (Somalia)........08.48°N 47.36°E
Las Geel, Somaliland (Somalia)........09.78°N 44.45°E
Las Khoreh, *see* Las Korey
Las Korey, Somaliland (Somalia)........11.16°N 48.20°E
Leer, South Sudan........08.30°N 30.14°E
Lerua, South Sudan........03.99°N 32.57°E
Lerwa, *see* Lerua

Let Marefia, Ethiopia........09.63°N 39.75°E
Li Rangu, South Sudan........04.73°N 28.36°E
Li Yubu, South Sudan........05.39°N 27.25°E
Liria, South Sudan........04.64°N 32.08°E
Lofi Mt, South Sudan........04.24°N 32.99°E
Logiya, Ethiopia........11.73°N 40.98°E
Lohutok, South Sudan........04.66°N 32.86°E
Loka, South Sudan........04.22°N 30.96°E
Lokichoggio, Kenya........04.21°N 34.35°E
Lokomarinyang, South Sudan........05.02°N 35.59°E
Logwi, South Sudan........04.11°N 32.71°E
Lokwi, *see* Logwi
Lomoriti, South Sudan........03.90°N 32.72°E
Lote, Ethiopia........06.21°N 36.80°E
Lotti (lowland forest), South Sudan........04.04°N 32.54°E
Lowai, South Sudan........04.50°N 32.13°E
Loyengalani, Kenya........02.76°N 36.72°E
Lugh, Somalia........03.81°N 42.55°E
Lui, South Sudan........05.34°N 30.45°E
Lulubsho, *see* Dolobscio
Luxor, Egypt........25.70°N 32.64°E
Mabil, Ethiopia........10.33°N 36.82°E
Madash°N District, Puntland (Somalia)........11.02°N 48.97°E
Mader Moge, Somaliland (Somalia)........10.80°N 47.31°E
Magajebe, *see* Magangib
Magala Umberto I (historical village), Somalia
........04.30°N 42.05°E
Magangib, Somalia........05.20°N 48.42°E
Magghiole, Somaia........05.61°N 47.93°E
Mago National Park, Ethiopia........05.50°N 36.25°E
Magwe, South Sudan........04.14°N 32.29°E
Mahaddei Uen, *see* Mahadday Weyne
Mahadday Weyne, Somalia........02.97°N 45.53°E
Mahagi Port, DR Congo........02.14°N 31.24°E
Mahas, Somalia........04.39°N 46.09°E
Mait, *see* Maydh
Maji, Ethiopia........06.18°N 35.58°E
Malakal, South Sudan........09.53°N 31.66°E
Maldi, Eritrea........15.65°N 38.81°E
Malek, South Sudan........06.07°N 31.60°E
Malka Mari, *see* Malka Murri
Malka Murri, Ethiopia........04.27°N 40.78°E
Mana Moofi, *see* Mofi
Mandafena, Eritrea........14.91°N 39.32°E
Mandah, Somaliland (Somalia)........11.00°N 43.18°E
Mandara, Sudan........14.43°N 33.45°E
Mandera, Kenya........03.93°N 41.86°E
Mandheera, Somaliland (Somalia)........09.92°N 44.71°E
Manyate, Ethiopia........06.49°N 39.75°E
Manzala Lake, Egypt........31.28°N 32.17°E
Mareb River, Eritrea........*c.*15.03°N 37.17°E
Mareri, Somalia........00.43°N 42.69°E
Margherita, *see* Jamaame
Marial Bay, South Sudan........08.67°N 28.15°E
Maridi, South Sudan........04.92°N 29.47°E
Marsa Alam, Egypt........25.06°N 34.89°E
Marsa Halaib, Egypt........22.22°N 36.63°E
Marsa Matruh, Egypt........31.35°N 27.24°E
Marsa Oseif, Sudan........21.77°N 36.87°E
Marsabit Mountains, Kenya........02.26°N 37.93°E
Mashraar Ragg, *see* Meschra El Rek
Maskali Island, Djibouti........11.71°N 43.16°E
Massawa, Eritrea........15.61°N 39.44°E
Maxaas, *see* Mahas

Maydh, Somaliland (Somalia) 11.00°N 47.11°E
Mboki, Central African Republic 05.32°N 25.96°E
Medpi Saul, Eritrea 15.46°N 38.87°E
Meel Daaq, *see* Bur Meldac
Mega, Ethiopia 04.05°N 38.30°E
Mekele, Ethiopia 13.50°N 39.48°E
Meki, Ethiopia 08.15°N 38.82°E
Mena, Ethiopia 06.42°N 39.85°E
Mendefera, Eritrea 14.89°N 38.82°E
Mendi, Ethiopia 09.80°N 35.10°E
Merca, Somalia 01.72°N 44.77°E
Mereb River, Eritrea c.14.90°N 38.99°E
Meregh, Somalia 03.77°N 47.30°E
Merga (Nukheila), Sudan 19.05°N 26.32°E
Meroe, Sudan 18.47°N 31.81°E
Merowe, *see* Mer°E
Meschra El Rek, Sudan 08.42°N 29.27°E
Metaker, Ethiopia 08.78°N 42.18°E
Metahara, Ethiopia 08.91°N 39.91°E
Metema, *see* Metemma
Metemma, Ethiopia 12.95°N 36.16°E
Metla Pass, Egypt 30.01°N 32.89°E
Mettu, Ethiopia 08.30°N 35.58°E
Metu, *see* Mettu
Mieso, Ethiopia 09.23°N 40.76°E
Mile, Ethiopia 11.42°N 40.76°E
Milmil, Ethiopia 08.17°N 43.67°E
Misfaq, Egypt 31.03°N 33.17°E
Misherat, Sudan 12.28°N 34.13°E
Missarole (springs), Somalia 03.43°N 45.06°E
Mizan Teferi, Ethiopia 07.00°N 35.59°E
Modjo, Ethiopia 08.59°N 39.12°E
Mofi, Somalia 00.09°N 42.71°E
Mogadishu, Somalia 02.04°N 45.34°E
Mogatta, Sudan 14.67°N 35.85°E
Moggio River, Ethiopia c.07.87°N 42.02°E
Molongori, South Sudan 04.17°N 32.84°E
Mongalla, South Sudan 05.20°N 31.77°E
Moucha, *see* Musha Island
Moullou River, Ethiopia c.09.25°N 40.78°E
Moyale, Ethiopia 03.53°N 39.05°E
Mudug (Region), Puntland (Somalia) and Somalia c.07.00°N 48.00°E
Muger River, Ethiopia c.09.49°N 38.58°E
Mugher, *see* Muger River
Muglad, Sudan 11.03°N 27.73°E
Muhammad Gol, *see* Muhammad Qol
Muhammad Qol, Sudan 20.90°N 37.16°E
Mui, Ethiopia 05.83°N 35.75°E
Mukayle, *see* El Mocoile
Mulu, Ethiopia 09.40°N 38.65°E
Munfuudhi, Somalia 00.81°N 42.62°E
Murchins°N Falls, Uganda 02.28°N 31.69°E
Murikuren, South Sudan 04.30°N 33.16°E
Murle, Ethiopia 05.15°N 36.22°E
Murri, Ethiopia 04.23°N 40.70°E
Murukurren, *see* Murikuren
Museri Island, Eritrea 15.49°N 40.36°E
Musha Island, Djibouti 11.72°N 43.21°E
Mustahil, Ethiopia 05.24°N 44.73°E
Muqatta, *see* Mogatta
Mwingi, Kenya 00.94°S 38.06°E
Naaso Hablood, Somaliland (Somalia) 09.59°N 44.12°E
Nabq, Egypt 28.04°N 34.43°E
Nagishot, South Sudan 04.27°N 33.57°E
Najd (Plateau), Saudi Arabia c.25.00°N 44.50°E
Nasser Lake, Egypt 23.16°N 32.79°E
Nazareth, Ethiopia 08.55°N 39.27°E
Nazret, *see* Nazareth
Nasir, South Sudan 08.62°N 33.06°E
Nechisar National Park, Ethiopia 05.94°N 37.68°E
Neem, Sudan 10.56°N 28.38°E
Nefasit, Eritrea 15.33°N 39.06°E
Negele, Ethiopia 05.32°N 39.58°E
Negev (Desert), Israel/Palestine c.30.50°N 34.92°E
Neghele, *see* Negele
Nekemte, Ethiopia 09.08°N 36.55°E
Nekhel, Egypt 29.97°N 33.74°E
New Cairo City, Egypt 30.03°N 31.47°E
Ngangala, South Sudan 04.71°N 31.92°E
Ngatana, Kenya 02.18°S 40.18°E
Ngumi Island, Somalia 00.73°S 42.29°E
Nhem, *see* Neem
Nimule, South Sudan 03.60°N 32.06°E
No Lake, South Sudan 09.49°N 30.45°E
Nocra, *see* Nokra
Nogal (River), Puntland (Somalia) c.08.01°N 49.78°E
Nogal Valley, Puntland and Somaliland (Somalia) c.08.58°N 48.58°E
Nokra Island, Eritrea 15.71°N 39.94°E
Nuba Mountains, Sudan 12.00°N 30.75°E
Nugaal (Region), Puntland (Somalia) c.08.25°N 49.25°E
Nuweiba, Egypt 29.05°N 34.66°E
Nyala, Sudan 12.05°N 24.88°E
Nzara, South Sudan 04.64°N 28.25°E
Obbia, *see* Hobyo
Obele, Ethiopia 08.78°N 43.45°E
Obock, Djibouti 11.97°N 43.29°E
Odamuda, Ethiopia 07.83°N 41.17°E
Oddur, Somalia 04.12°N 43.89°E
Odenleh, Somaliland (Somalia) 08.32°N 46.04°E
Ogaden (Region), Ethiopia c.08.00°N 44.00°E
Ogo Mountains, Somaliland (Somalia) c.09.95°N 43.54°E
Ohilang, South Sudan 04.68°N 32.78°E
Oisame, Puntland (Somalia) 09.00°N 49.27°E
Ola Uager, Somalia 01.24°S 41.54°E
Old Fangak, South Sudan 09.07°N 30.88°E
Ololokwe Mt, Kenya 00.85°N 37.53°E
Om Hajer, Eritrea 14.32°N 36.65°E
Omdurman, Sudan 15.64°N 32.48°E
Omo River, Ethiopia c.05.18°N 36.20°E
Oodweyne, Somaliland (Somalia) 09.41°N 45.06°E
Oog, Somaliland (Somalia) 08.93°N 46.62°E
Opari, South Sudan 03.93°N 32.06°E
Oroflillo (Well), Somalia 02.22°N 42.58°E
Oromia (Region), Ethiopia c.08.00°N 39.00°E
Ouanda-Djalle, Central African Republic 08.90°N 22.80°E
Pajok, South Sudan 03.87°N 32.48°E
Paloich South Sudan 10.47°N 32.49°E
Palotaka, South Sudan 04.02°N 32.47°E
Port Said, Egypt 31.27°N 32.30°E
Port Sudan, Sudan 19.62°N 37.21°E
Qarabulli, Libya 32.75°N 13.71°E
Qardho, *see* Gardo
Qattara Depression, Egypt c.30.23°N 27.77°E
Qohaito, *see* Kohaito Plateau
Qool-Cadday, Somaliland (Somalia) 09.21°N 44.16°E

Qoow, Puntland (Somalia) 11.23°N 49.00°E
Qoriley, Somaliland (Somalia) 08.00°N 47.57°E
Qoryale, Somaliland (Somalia) 09.08°N 45.98°E
Qued Wada Weyn, Djibouti 11.07°N 42.78°E
Qulan Islands, Egypt 24.36°N 35.40°E
Quljeet, Somaliland (Somalia) 10.09°N 43.01°E
Quseir, Egypt 26.11°N 34.28°E
Quezan, Sudan 10.83°N 34.80°E
Rabak, Sudan 13.18°N 32.74°E
Rafah, Egypt 31.28°N 34.24°E
Rahanweyn country (region), Somalia c.04.75°N 43.75°E
Rahanuin country, *see* Rahanweyn
Rahib (Wells), Sudan 17.55°N 27.07°E
Ramu, Kenya 03.93°N 41.22°E
Randa, Djibouti 11.85°N 42.66°E
Ras El Hekma, Egypt 31.25°N 27.85°E
Ras Kamboni, Somalia 01.64°S 41.59°E
Ras Mohamed, Egypt 27.73°N 34.25°E
Rauabi, Egypt 30.00°N 34.43°E
Rawabi, *see* Rauabi
Redjaf, South Sudan 04.75°N 31.59°E
Rejaf, *see* Redjaf
Renk, South Sudan 11.74°N 32.80°E
Riyadh, Saudi Arabia 24.69°N 46.72°E
Robe, Ethiopia 07.12°N 40.00°E
Rocca Littorio, *see* Galkayo
Romani, Egypt 31.01°N 32.66°E
Romia Island, Eritrea 16.53°N 40.05°E
Roseires Dam, Sudan 11.87°N 34.39°E
Rugay, Somaliland (Somalia) 10.85°N 47.31°E
Rugdeia Sogheira, Ethiopia 09.65°N 41.01°E
Rugno, Somalia 04.10°N 46.95°E
Rumbek, South Sudan 06.81°N 29.68°E
Ruunyooc, *see* Rugno
Ruwenzori Mountains, Uganda 00.33°N 29.92°E
Saati, Eritrea 15.58°N 39.26°E
Sabderat, Eritrea 15.45°N 36.65°E
Safaga, Egypt 26.75°N 33.94°E
Saganeiti, Eritrea 15.05°N 39.19°E
Sahil (Region), Somaliland (Somalia) c.10.24°N 45.33°E
Sai Island, Sudan 20.71°N 30.32°E
Saja Forest, Ethiopia 07.51°N 36.12°E
Salahle, Somaliland (Somalia) 09.03°N 44.21°E
Salole, Ethiopia 04.52°N 39.57°E
Samaysa Dheer, Puntland (Somalia) 09.86°N 49.09°E
Sanaag (Region), Somaliland (Somalia) c.10.50°N 48.00°E
Sanafir Island, Saudi Arabia 27.93°N 34.71°E
Sancurar, Ethiopia 04.03°N 40.13°E
Sanetti Plateau, Ethiopia 06.85°N 39.86°E
Santaalo (hill), Somaliland (Somalia) 10.21°N 44.87°E
Sar Uanle, Somalia 00.49°S 42.42°E
Saraar Plains, Somaliland (Somalia) 09.48°N 46.25°E
Sarad Island, Eritrea 15.83°N 39.90°E
Sassabane, Ethiopia 07.87°N 43.67°E
Saylac, *see* Zeila
Sciotel, Eritrea 15.58°N 38.33°E
Sebeta, Ethiopia 08.92°N 38.62°E
Semma, Sudan 10.95°N 29.70°E
Senafe, Eritrea 14.70°N 39.42°E
Sennar, Sudan 13.57°N 33.57°E
Serdo, Ethiopia 11.97°N 41.30°E
Setit River, Ethiopia 14.25°N 36.57°E
Sguss, Somalia 02.42°N 45.40°E
Shafartak (bridge), Ethiopia 10.10°N 38.28°E
Shanshacade, Somaliland (Somalia) 08.66°N 45.93°E
Sharm-El-Sheikh, Egypt 27.92°N 34.33°E
Shashamene, Ethiopia 00.70°N 38.60°E
Shayata (Oasis) Egypt 29.33°N 25.16°E
Shebelle River, *see* Webi Shebelli River
Sheikh Awad, Egypt 28.65°N 33.90°E
Sheikh Hussein, Ethiopia 07.75°N 40.7°E
Sheik Mahomed, Ethiopia 07.33°N 40.50°E
Sheikh, Somaliland (Somalia) 09.93°N 45.19°E
Sheikh Pass, Somaliland (Somalia) 09.98°N 45.17°E
Sheikh Zowayid, Egypt 31.22°N 34.08°E
Sheik-Said Island, *see* Green Island
Sheka (zone), Ethiopia 07.56°N 35.40°E
Sheka (Biosphere Reserve), Ethiopia 07.56°N 35.40°E
Shendi, Sudan 16.69°N 33.44°E
Shimala River, Sudan 10.75°N 34.81°E
Shimbiris Mt, Somaliland (Somalia) 10.74°N 47.25°E
Shire, Ethiopia 14.14°N 38.40°E
Siddam-Bale Bridge, Ethiopia 05.68°N 39.53°E
Simien Mountains, Ethiopia 13.17°N 37.97°E
Sinai Mt, Egypt 28.54°N 33.97°E
Singa, Sudan 13.15°N 33.93°E
Sinkat, Sudan 18.84°N 36.84°E
Sissib (wells), Somalia 05.40°N 48.48°E
Siwa (Oasis), Egypt 29.21°N 25.55°E
Soba, Sudan 15.50°N 32.68°E
Sobat River, South Sudan c.09.16°N 32.17°E
Soddo, Ethiopia 08.58°N 35.82°E
Sodere, Ethiopia 08.40°N 39.38°E
Sof Omar (caves), Ethiopia 06.91°N 40.85°E
Sol Haud (plateau), Somaliland (Somalia) c.09.67°N 48.00°E
Sool (region), Somaliland (Somalia) 08.75°N 47.55°E
St. Katherine, Egypt 28.56°N 33.95°E
Suakin, Sudan 19.10°N 37.33°E
Sudd (swamps), South Sudan 08.00°N 31.00°E
Sue River, South Sudan c.07.40°N 28.11°E,
Suez (Canal), Egypt 29.93°N 32.56°E
Suga al Gamal, Sudan 12.80°N 27.70°E
Sukorre, Puntland (Somalia) 10.98°N 49.36°E
Sululta (Plain), Ethiopia 09.18°N 38.75°E
Sungikai, Sudan 12.33°N 29.77°E
Taba, Egypt 29.49°N 34.90°E
Tabaghbagh, Egypt 29.10°N 26.39°E
Tabo, Sudan 19.35°N 30.46°E
Tadjoura, Gulf of, Djibouti 11.79°N 42.88°E
Talanga (Forest), South Sudan 04.01°N 32.75°E
Taleh, Somaliland (Somalia) 09.15°N 48.42°E
Talla-Talla Kabir Island, Sudan 18.83°N 38.17°E
Talodi, Sudan 10.63°N 30.38°E
Tana (River) Delta, Kenya 02.35°S 40.23°E
Tana Lake, Ethiopia 12.00°N 37.33°E
Tanguru, Sudan 10.23°N 30.69°E
Taufikia, Sudan 09.43°N 31.63°E
Tekeze Valley, Ethiopia c.13.75°N 38.01°E
Tendaho, Ethiopia 11.70°N 40.95°E
Tepi, Ethiopia 07.20°N 35.45°E
Teppi, *see* Tepi
Terangore, South Sudan 04.48°N 32.78°E
Terekeka, South Sudan 05.44°N 31.75°E
Tesenay, Eritrea 15.11°N 36.66°E
Thar Jath (oil field), South Sudan 08.47°N 30.32°E
Tiran Island, Egypt 27.95°N 34.55°E
Tissisat Falls, Ethiopia 11.49°N 37.59°E
Tizi-Fongoro, Central African Republic 10.93°N 22.84°E

Tog Wajaale, Somaliland (Somalia) .... 09.60°N 43.34°E
Togdheer (Region), Somaliland (Somalia) .... c.09.33°N 45.42°E
Tokar, Sudan .... 18.43°N 37.73°E
Tokhoshi, Somaliland (Somalia) .... 11.35°N 43.41°E
Tonga, South Sudan .... 09.47°N 31.05°E
Tonj, South Sudan .... 07.27°N 28.68°E
Tonsa River, Eritrea .... c.13.42°N 39.92°E
Toon, Somaliland (Somalia) .... 09.39°N 44.12°E
Toyo Plains, Somaliland (Somalia) .... 09.34°N 44.62°E
Tungaru, *see* Tanguru
Turfa River, Ethiopia .... 07.53°N 42.33°E
Torit, South Sudan .... 04.41°N 32.57°E
Tsazega, Eritrea .... 15.35°N 38.80°E
Turkana Lake, Ethiopia .... 04.49°N 36.08°E
Turmi, Ethiopia .... 04.97°N 36.49°E
Uarandi, Somalia .... 05.78°N 47.72°E
Uarscèk, *see* Warshiikh
Uargalo, Somalia .... 06.26°N 47.52°E
Uegit, Somalia .... 03.81°N 43.25°E
Ufeyn, Puntland (Somalia) .... 10.65°N 49.75°E
Um Elsheikh Island, Sudan .... 21.08°N 37.15°E
Umfudu, *see* Munfuudhi
Umm Ruweba, Sudan, also Um Ruwaba .... 12.91°N 31.22°E
Unity (State), South Sudan .... c.08.65°N 29.85°E
Upper Nile (State), South Sudan .... c.10.00°N 32.70°E
Villaggio Duca degli Abruzzi, *see* Giohar
Villaggio Sguss, *see* Sguss
Voi, Kenya .... 03.40°S 38.56°E
Wachile, Ethiopia .... 04.55°N 39.07°E
Wad Banda, Sudan .... 13.10°N 27.95°E
Wad Bena Naga, Sudan .... 16.51°N 33.11°E
Wad Madani, Sudan .... 14.40°N 33.52°E
Wad Medani, *see* Wad Madani
Wadi Abbad, Egypt .... 25.01°N 33.02°E
Wadi Abd El Malek, Egypt .... 23.83°N 25.30°E
Wadi Aideb, Egypt .... 22.18°N 36.35°E
Wadi Araba, Egypt .... 28.98°N 31.18°E
Wadi Balah, Egypt .... 29.06°N 33.44°E
Wadi Degla, Egypt .... 29.97°N 31.30°E
Wadi Dudo, Puntland (Somalia) .... 09.24°N 50.66°E
Wadi El Awateib, Sudan .... 16.33°N 33.08°E
Wadi El Firaq, Egypt .... 22.86°N 25.76°E
Wadi El Gedeirat, Egypt .... 30.64°N 34.41°E
Wadi El Gemal, Egypt .... 24.66°N 35.09°E
Wadi El Natrun, Egypt .... 30.44°N 30.19°E
Wadi El Rada, Egypt .... 24.25°N 35.23°E
Wadi El Sheikh, Egypt .... 28.70°N 33.68°E
Wadi Faradeero, Somaliland (Somalia) .... 10.24°N 45.08°E
Wadi Feiran, Egypt .... 28.64°N 33.26°E
Wadi Gebal, Egypt .... 28.52°N 33.90°E
Wadi Gharandal, Egypt .... 29.33°N 33.00°E
Wadi Gindali, Egypt .... 30.04°N 31.85°E
Wadi Goddalola, Puntland (Somalia) .... 11.33°N 49.55°E
Wadi Halfa, Sudan .... 21.80°N 31.37°E
Wadi Howar, Sudan .... c.17.50°N 27.13°E
Wadi Ibib, Sudan .... 22.83°N 35.77°E
Wadi Kid, Egypt .... 28.16°N 34.38°E
Wadi Natrun, Egypt .... 30.44°N 30.19°E
Wadi Qena, Egypt .... c.26.21°N 32.74°E
Wadi Run, Puntland (Somalia) .... 08.80°N 48.89°E
Wadi Sartout, Egypt .... 24.32°N 35.02°E
Wadi Watir, Egypt .... 29.00°N 34.69°E
Wadi Wazarah, Egypt .... 28.45°N 34.04°E
Wagar (Range), Somaliland (Somalia) .... 10.00°N 45.39°E
Wajir, Kenya .... 01.75°N 40.02°E
Watoka, South Sudan .... 03.75°N 30.63°E
Waqadariya, Puntland (Somalia) .... 11.12°N 47.80°E
Warder, Ethiopia .... 06.98°N 45.33°E
Wargalo, *see* Uargalo
Warmadow, Puntland (Somalia) .... 11.33°N 49.75°E
Warrab (State), South Sudan .... c.08.00°N 28.85°E
Warshiikh, Somalia .... 02.30°N 45.80°E
Wau, South Sudan .... 07.70°N 27.99°E
Webi Shebelli River, Somalia .... c.03.13°N 45.54°E
Weeraar, Somaliland (Somalia) .... 10.26°N 43.25°E
Welenchiti, Ethiopia .... 08.67°N 39.43°E
Welkite, Ethiopia .... 08.28°N 37.78°E
Weyt'o, Ethiopia .... 05.22°N 37.00°E
Wisil, Somalia .... 05.43°N 48.12°E
Wonchi Crater Lake, Ethiopia .... 08.80°N 37.90°E
Wondo Genet, Ethiopia .... 07.08°N 38.62°E
Wonji, Ethiopia .... 08.45°N 39.28°E
Woqooyi Galbeed (region), Somaliland (Somalia) .... c.10.00°N 44.50°E
Wush-Wush, Ethiopia .... 07.23°N 35.97°E
Xaaro, *see* Hàro
Xagal, Somaliland (Somalia) .... 10.25°N 45.74°E
Xero Dhagaxleey, Somalia .... 05.58°N 48.24°E
Xidda, Puntland (Somalia) .... 09.96°N 49.05°E
Xuddur, *see* Oddur
Xudun, *see* Hudun
Yabello, Ethiopia .... 04.88°N 38.08°E
Yagoori, Somaliland (Somalia) .... 08.75°N 46.96°E
Yaka, Puntland (Somalia) .... 09.22°N 49.12°E
Yambio, South Sudan .... 04.57°N 28.39°E
Yei, South Sudan .... 04.09°N 30.68°E
Yirga Chefe, Ethiopia .... 06.16°N 38.20°E
Yuffleh, Somaliland (Somalia) .... 10.38°N 47.20°E
Yufle, *see* Yuffleh
Zakouma (National Park), Chad .... 10.84°N 19.66°E
Zalingai, Sudan .... 12.91°N 23.47°E
Zara Gold, Eritrea .... 16.48°N 37.92°E
Zaranik (Protected Area), Egypt .... 31.12°N 33.48°E
Zege, *see* Zegie
Zegie, Ethiopia .... 11.70°N 37.35°E
Zeila, Somaliland (Somalia) .... 11.35°N 43.47°E
Zeylac, *see* Zeila
Zwai Lake, Ethiopia .... 08.00°N 38.83°E

Glossary

Aestivation: when an animal becomes torpid during periods of high temperatures or drought.

Allopatric: occurring in separate regions, not overlapping.

Anterior: toward the front of the animal.

Annuli: (singular, annulus): in worm lizards, a ring of rectangular, non-overlapping scales.

Arboreal: living in trees or bushes

Autotomy: the loss of a part of the body (for example the tail in lizards) as a defence mechanism by an animal under threat.

Axil: the angle between the leaf and the upper part of the stem.

Bicuspid: having two points or peaks.

Brille: the transparent scale that covers a reptile's eye.

Buccal: relating to the mouth.

Caniculate: having longitudinal channels.

Canthal: any scale that lies along the canthus rostralis.

Canthi rostrales: (singular, canthus rostralis): in chameleons and agamas, the ridges or crests that connect the ridge above the eye to the snout.

Carapace: the upper surface of a chelonian's shell.

Cardiotoxic: a toxin that affects the heart.

Carnivorous: feeding on other animals

Casque: the raised helmet-like structure on the back of the head of some chameleons.

Chin shields: (snakes, *Hemidactylus*) large scales behind the infralabial scales on the lower jaw of a reptile

Chromosome: a long DNA molecule usually in cell nuclei, containing part or all of the genetic material of an organism

Cloaca: in vertebrates, common chamber and outlet into which the intestinal, urinary, and genital tracts open.

Cloacal spurs: (or post-cloacal spurs, anal spurs or post-cloacal tubercles) sexually dimorphic tubercles (or enlarged scales) located on the ventrolateral border of the tail base, typically found in most of geckos.

Colubrid: a 'typical' snake of the family Colubridae (a family now much reduced); in its older sense a snake without front fangs and with broad ventral scales that span the width of the belly.

Concave: curved inwards.

Convex: curved outwards.

Costals: plates or shields on the carapace of a chelonian, between the marginals and vertebrals (see Figure 1).

Crepuscular: active at twilight.

Cusp: cone-shaped prominence on the surface of beak.

Cycloid: a scale (usually on a skink) with an evenly curved, free border.

Cytotoxic: a toxin that affects cells.

Diapsid: animals that developed two holes in each side of their skulls, although some have now lost them; the group includes all reptiles.

Dimorphic: existing in two distinct forms.

Diurnal: active by day.

Dorsal: relating to the back, or upper surface between the neck and tail.

Dorsals: scales on the back.

Dorsolateral: between the side and the back.

Endemic: restricted to a certain place; in herpetology, often refers to a country, a group of countries, or an isolated habitat.

External vocal sacs: a sound-resonating extensible sac, possessed by frogs, that extends greatly outside the body when inflated.

Fossorial: a species that lives beneath the soil surface, usually by burrowing, or living in holes.

Frontonasal: scale at the front of a lizard's snout (see Figure 7).

Frontoparietals: paired scales in the upper centre of a lizard's head (see Figure 7).

Gregarious: forage in flocks.

Gular: relating to the throat.

Hemipenis: one of two intromittent organs of male squamates, usually concealed within the body and everted for reproduction.

Herbivorous: eating plant matter.

Heterogenous: diverse in character or content.

Hemipenial pockets: pouches that the hemipenes fit into when not everted (usually in geckos).

Holotype: a single type specimen that the description of a new species is based upon.

Homogenous: similar or unvaried.

Imbricate: overlapping.

Immaculate: in terms of appearance, uniformly coloured, without any spots or other markings.

Infra-tympanic gland: a discontinuous row of small glands or a continuous ridge, extending below the tympanum to the base of the forearm.

Inner metatarsal tubercle: pad on the hind limbs of frogs, often enlarged and keratinised for digging.

Inguinal: relating to the area just in front of the point where the hind limbs connect to the body, the groin.

Intergular: a small shield between the gular shields at the front of a turtle plastron.

Internasal: the scales between the nostrils of a snake (see Figure 22).

Interparietal: a scale at the back of a reptile head (see Figure 7).

Interstitial: between the scales (usually refers to the thin skin between a reptile's scales).

Juxtaposed: (granules, scales); next to one another but not overlapping.

Keel: length-wise ridge on the scales of some snakes, geckos and other lizards and the shells of some turtles.

Lamellae: any thin plate-like scale or scale structure, usually refers to the small scales across the underside of lizards' digits.

Loreal: a scale between the eye and the nostril (but not touching either) of a snake; its presence or absence is a useful diagnostic tool.

Melanistic: a dark or black individual, usually as a result of excess of black pigment, melanin.

Mental: a scale right at the front of the lower jaw of a reptile (see Figure 8).

Monophyletic: of a group of organisms that are descended from a single common ancestor, not shared with any other group.

Mucronate: terminating in a sharp point.

Mya: million years ago.

Nasal: either a scale that borders the nostril or a bone in the nose of some chelonians.

Neurotoxin: toxin that affects the nerves.

Nocturnal: active by night

Nominate: the first defined species, and if subspecies exist, the specific name is the same as the subspecific name, e.g. *Bitis arietans arietans.*

Nuchal: relating to the nape of the neck; in snakes the nuchals scales are small scales at the back end of the enlarged head scales, in a tortoise the term refers to the small scale at the front centre of a carapace, some lizards have nuchal crests.

Occipital: relating to the back of the head or the skull.

Occiput: the back part of the skull or the head, where it joins the vertebral column.

Ocelli: eye-like spots or markings.

Omnivorous: feeding on a variety of food of both plant and animal origin.

Opisthoglyphous: see rear-fanged.

Osteoderms: very small bones within the scales of some reptiles, toughening them.

Ovoviviparous: reproduction by eggs which are hatched within the body of the parent (Snakes).

Parapatric: of two species, occurring immediately adjacent to one another but their ranges do not overlap.

Paravertebral: usually refers to scales on either side of the vertebral column.

Parietals: relates to the crown of the head; in squamate reptiles it usually refers to the large scales just behind the eye on top of the head (see Figures 7 and 22).

Parotid gland: (more correctly parotoid) is an external skin gland on the back, neck, and should er of toads, it can secrete a number of milky alkaloid substances, often toxic.

Pectoral: scales on the plastron of a turtle (see Figures 2 and 4).

Pentadactyl: having limbs with five digits.

Pineal eye: a rudimentary third eye, on the top of the head in some lizards and amphibians, often covered with skin.

Plantar: to do with the sole or underside of the (hind) feet.

Plastron: the lower surface of a chelonian's shell.

Pleural scales (Chelonians): scales one or both sides, with the typical symmetric pattern.

Polygonal bones: the ribs and backbones of turtles and tortoises.

Pores (femoral): small deep openings on the underside of some lizard thighs.

Posterior: towards the rear.

Precloacal: in front of the cloaca.

Preoculars: scales in front of the eye.

Quadricarinate: having four ridges or keels.

Rear-fanged: applies to snakes with enlarged teeth, in the upper jaw at or near the level of the eye, grooved to carry venom.

Rostral: related to the nose, also reptile scale at the front of the head (see Figures 6, 7, 16, 17 and 19).

Rugose: wrinkled, ridges or corrugated.

Rupicolous: living on or around rocks.

Scansorial: adapted to or specialised for climbing.

Scansors: specialised pad on the toe tips (usually of a gecko), consisting of minute hairs, setae, that by their gripping properties enable the animal to climb vertical surfaces.

Scutes: scales.

Skin ridges: elevated elongate skin folds, usually on the back of frogs.

Solifugids: order of relatively large, hard biting but non-venomous arachnids.

Spectacle: see brille.

Spiracles: small external opening on amphibian larvae that leads to the gill chambers.

Subcaudals: scales under the tail, usually refers to such relatively broad scales in snakes.

Subimbricate: weakly overlapping.

Subocular: scale beneath the eye.

Subocular tentacle: filamentous tentacle below the eye of some frogs.

Sulcus: a furrow or faintly depressed linear groove.

Supra-orbital: above the eye.

Supraciliaries: relating to the region above the eye, scales above the eye (see Figure 7); usually small and present in certain lizards.

Supranasals: scales above the nose.

Supraocular: scales above the eye (see Figures 7, 15, 19 and 22).

Sympatric: of two or more species, occurring in the same geographical areas or regions.

Syntopy: of two species, occurring in the same place at the same time.

Tectiform: roof-shaped; sloping downwards on two sides from a raised central ridge.

Temporals: scales behind the eye but behind the postocular scales (see Figures 6, 21 and 22).

Temporo-lateral: the rear side of the head.

Terrestrial: lives on the ground (as opposed to arboreal).

Thanatosis: pretending to be dead.

Tricarinate: of scales, having three keels or ridges.

Tridactyle: having three digits.

Trihedral: having three sides or faces.

Tubercles: small, rounded protuberances or projections, usually refers to enlarged button-like scales in geckoes.

Tympanum: the eardrum.

Type: the specimen(s) on which the original description of a species is based.

Unicuspid: having a single cusp.

Venter: refers to the underside.

Vermiculation: markings that look like worms or worm tracks.

Vertical: a group of structures arranged in a circle, usually refers to a cluster of small scales.

Viviparous: reproduction by giving birth to live young that develops in the mother's body.

Vestigial: a rudimentary organ or part of a body that is no longer useful but still exists.

Vomerine teeth: short conic projections on the vomer, in the palate of amphibians.

Warts: small grainy lumps, often hemispherical, on the skin of amphibians.

Whorls: a ring of similar-sized scales on the tails of agamid lizards, usually separated by a row of larger scales.

Line Drawings

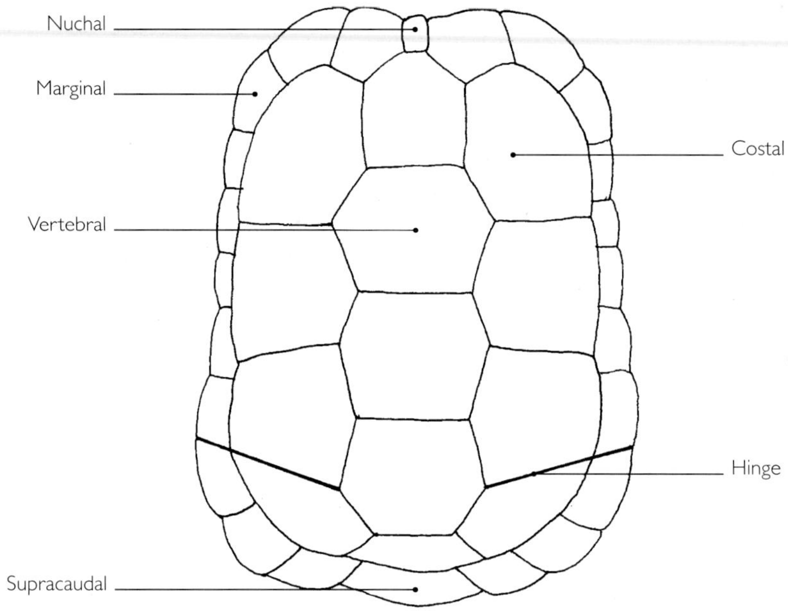

Figure 1: The upper shield (carapace) of a tortoise shell (after Branch)

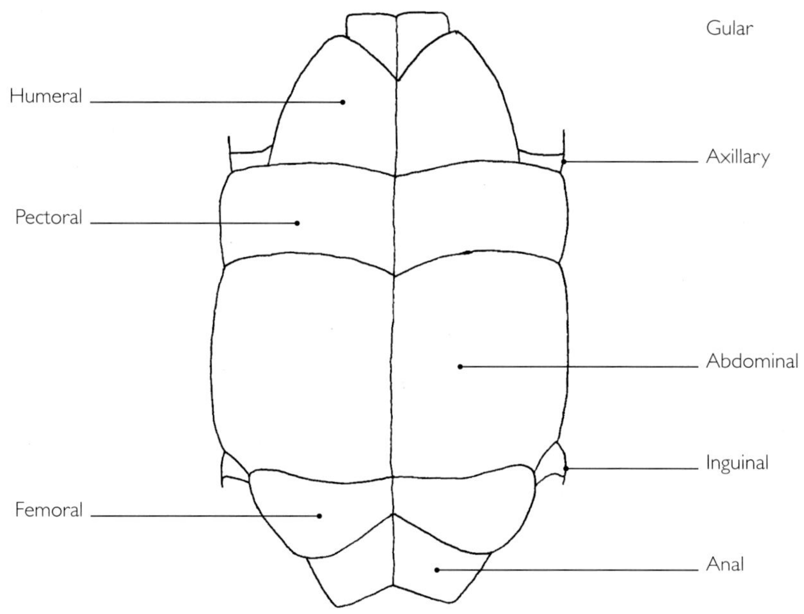

Figure 2: The under shield (plastron) of a tortoise shell

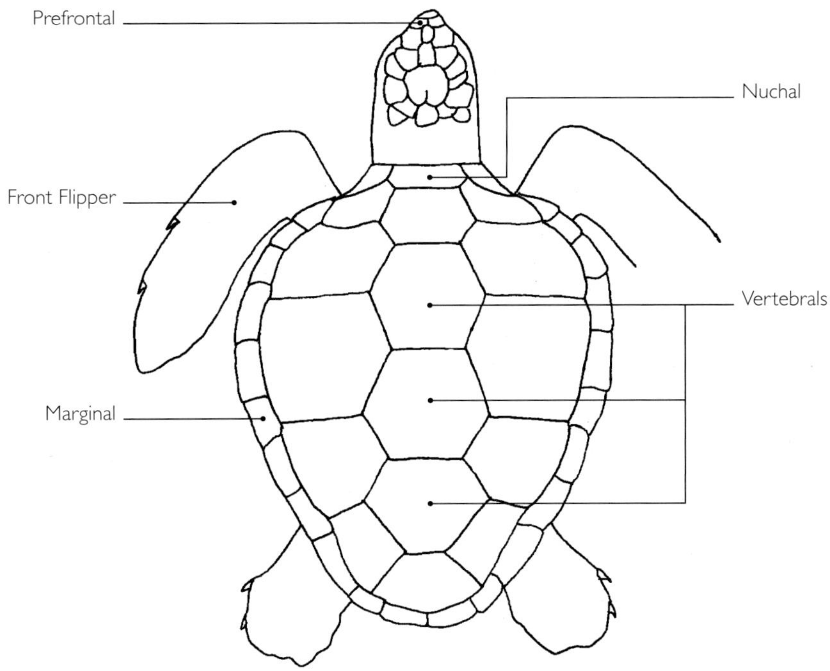

Figure 3: The upper shield (carapace) of a turtle shell

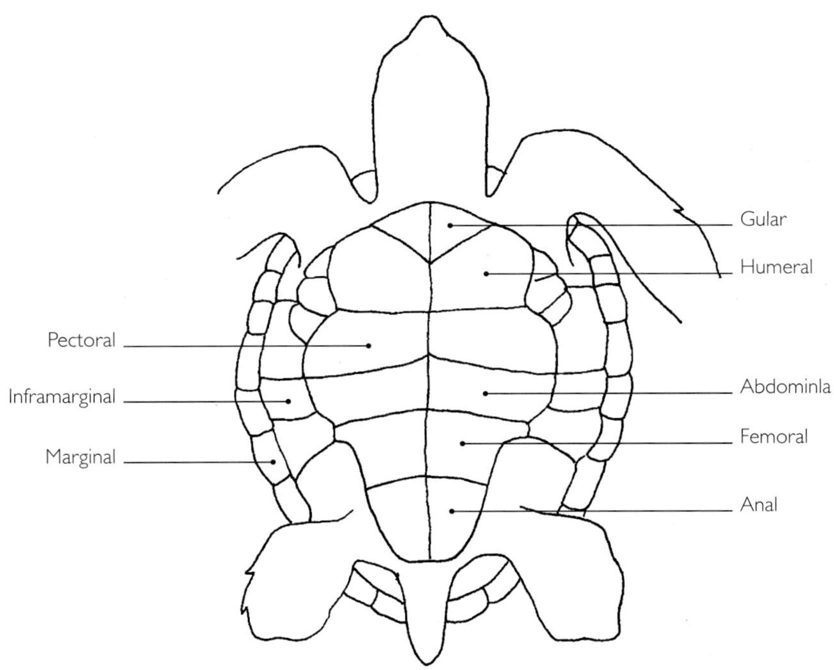

Figure 4: The under shield (plastron) of a turtle shell (after Branch)

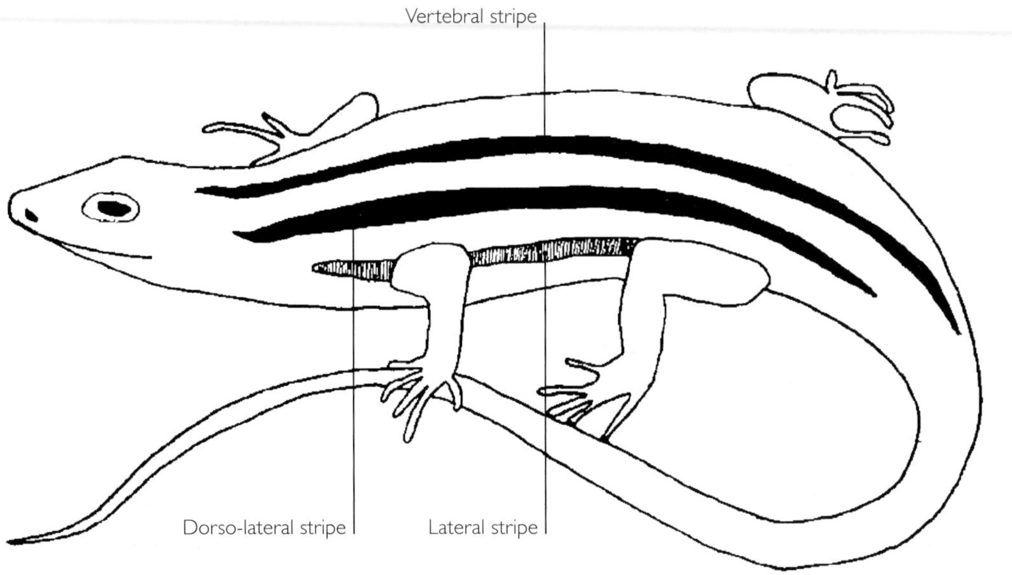

Figure 5: Lizard stripes

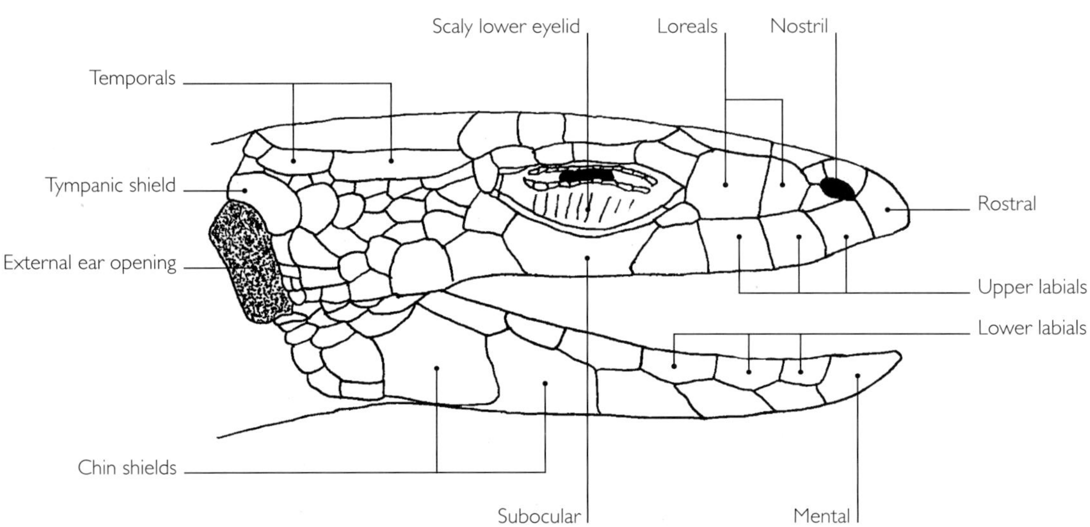

Figure 6: Lizard head scales – side view

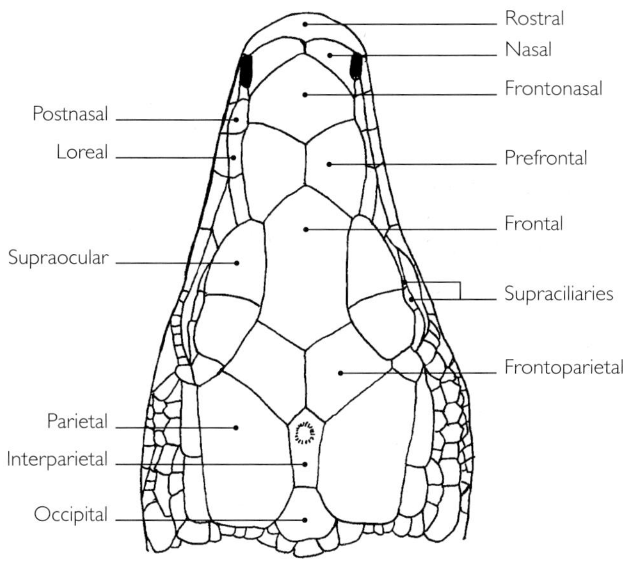

Figure 7: Lizard head scales - from above

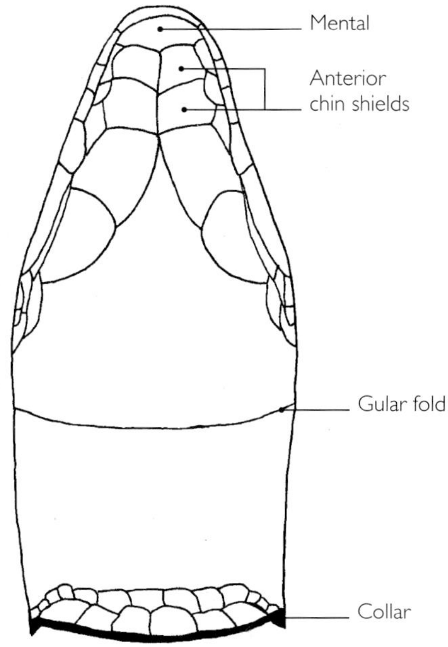

Figure 8: Lizard head scales – from below

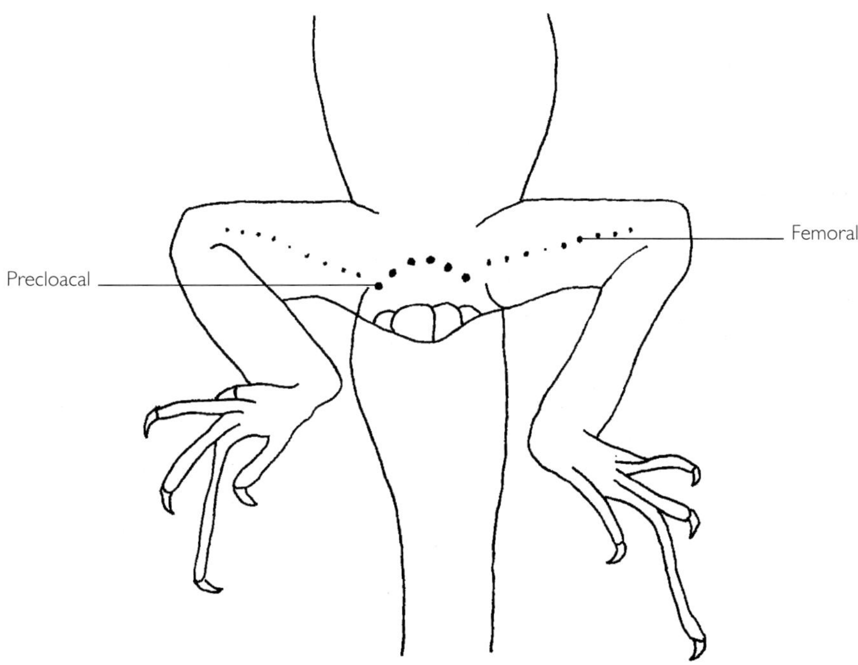

Figure 9: Lizard pores (all after Branch)

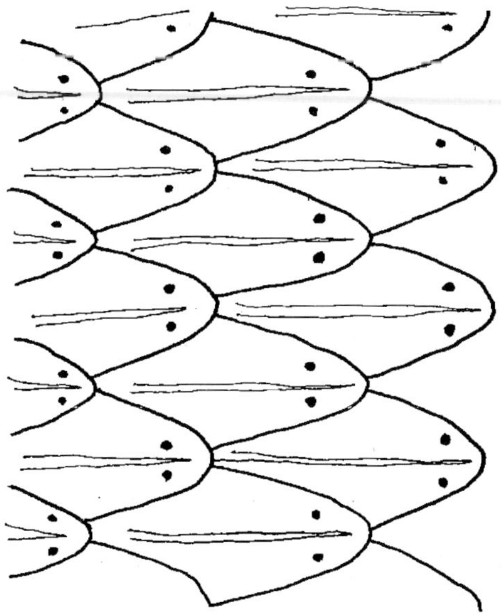

Figure 10: Keeled imbricate lizards scales with pores

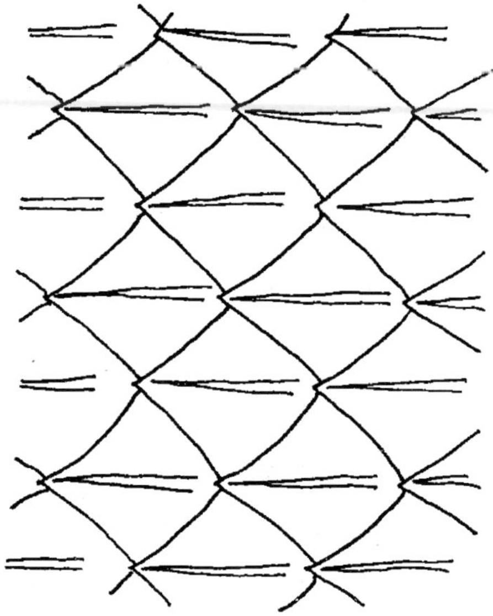

Figure 11: Mucronate scales

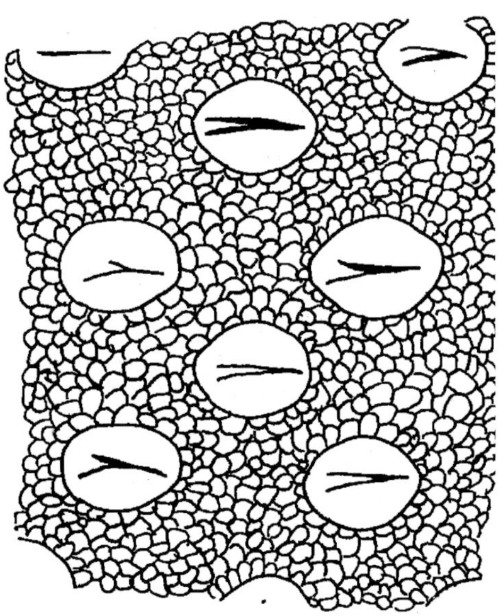

Figure 12: Granular scales with enlarged tubercular scales

Figure 13: Cycloid scales, tricarinate and smooth (all after Branch)

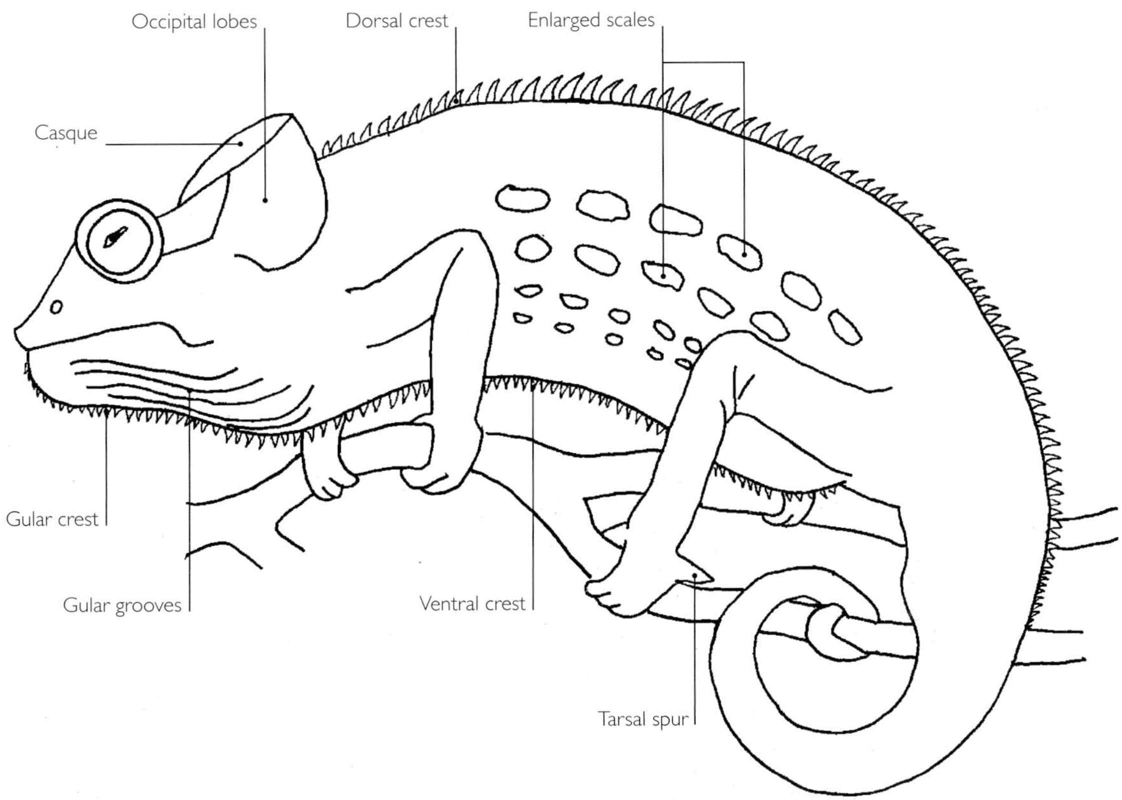

Figure 14 : Chameleon anatomy (after de Witte)

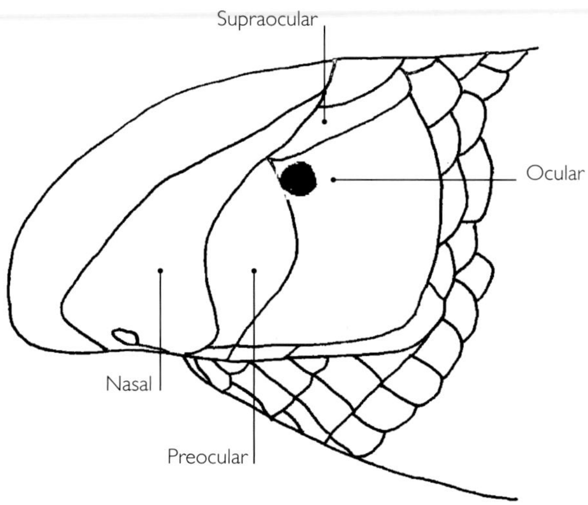

Figure 15: Head scales of a blind snake, side view

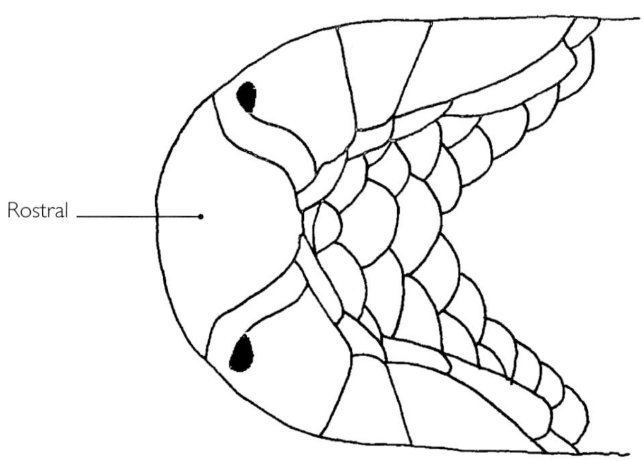

Figure 16: Head scales of a blind snake, from below

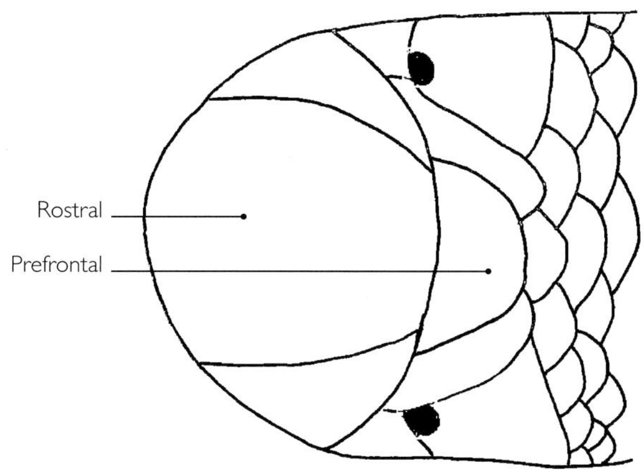

Figure 17: Head scales of a blind snake, from above

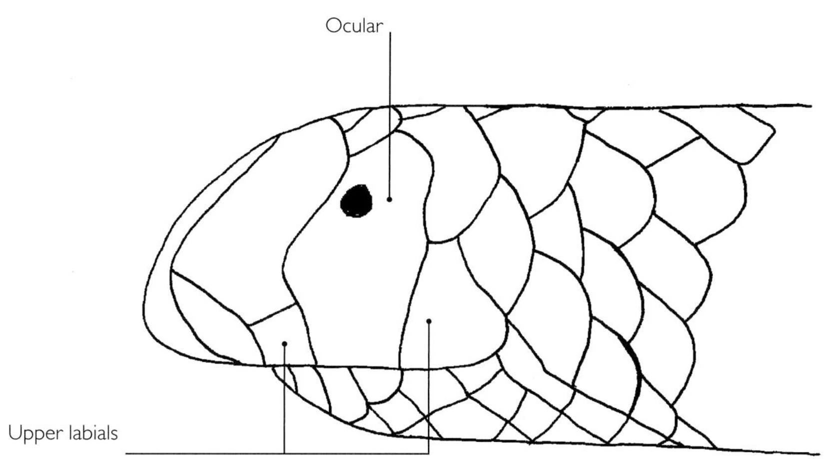

Figure 18: Head scales of a worm snake, side view

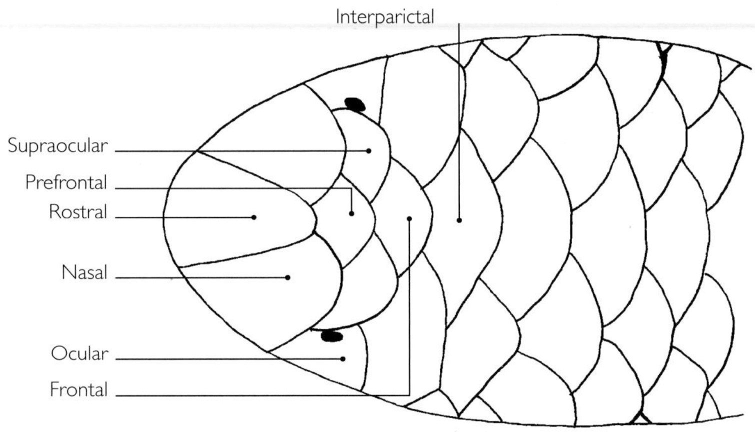

Figure 19: Head scales of a worm snake, from above

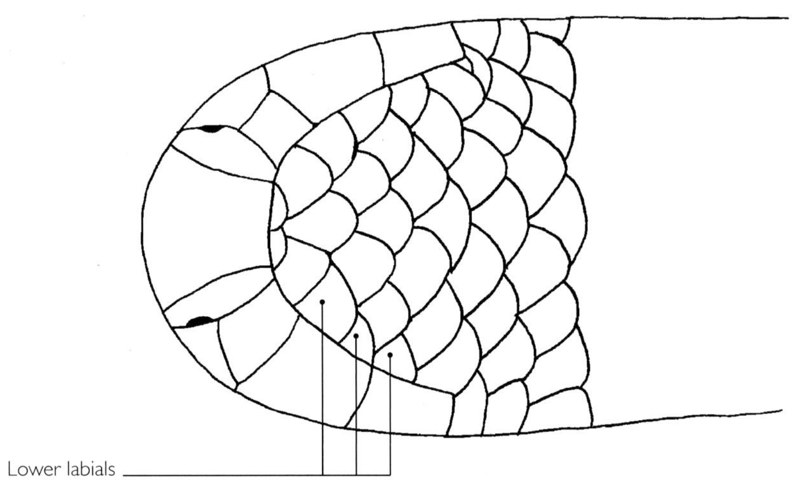

Figure 20: Head scales of a worm snake, from below

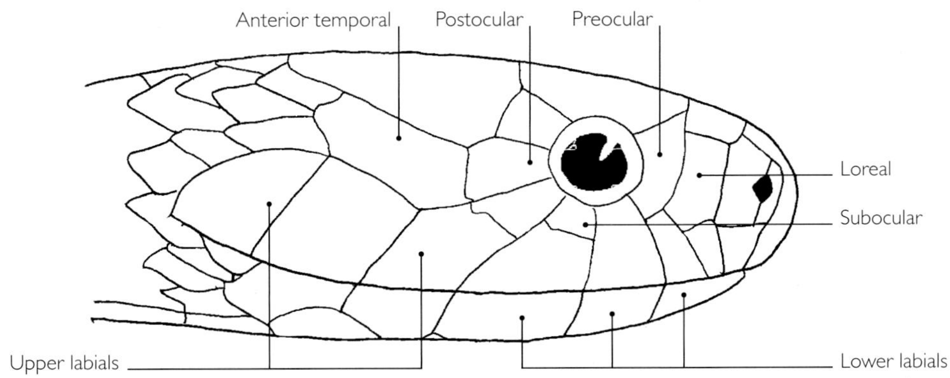

Figure 21: Head scales of a colubrid snake, side view

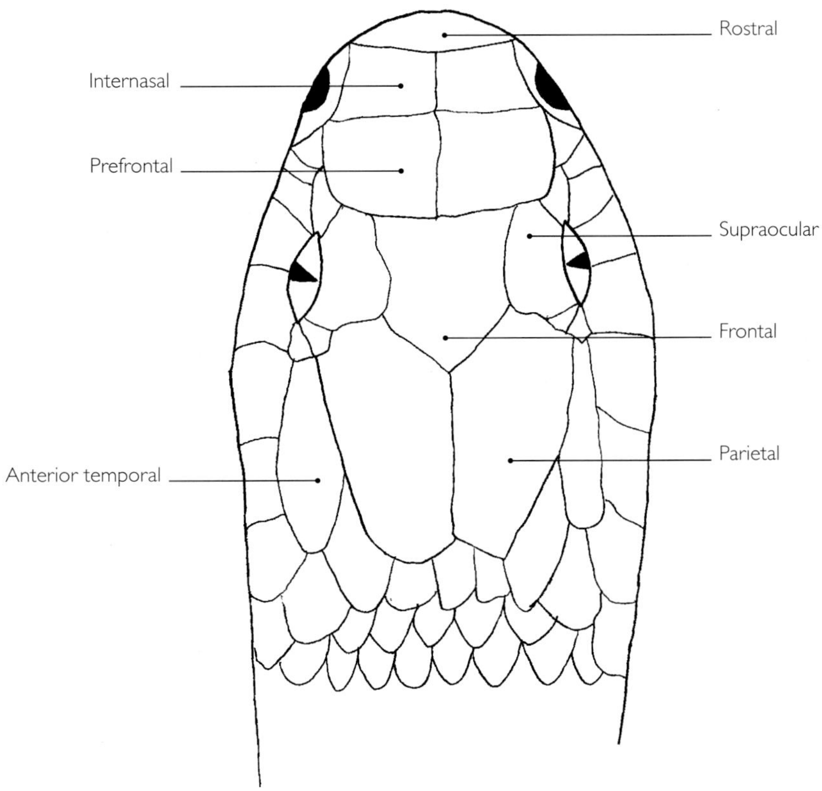

Figure 22: Head scales of a colubrid snake, from above

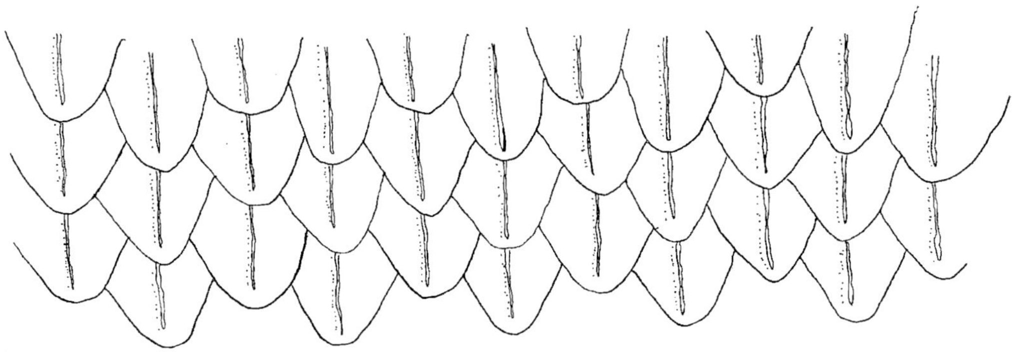

Figure 23: Keeled dorsal scales

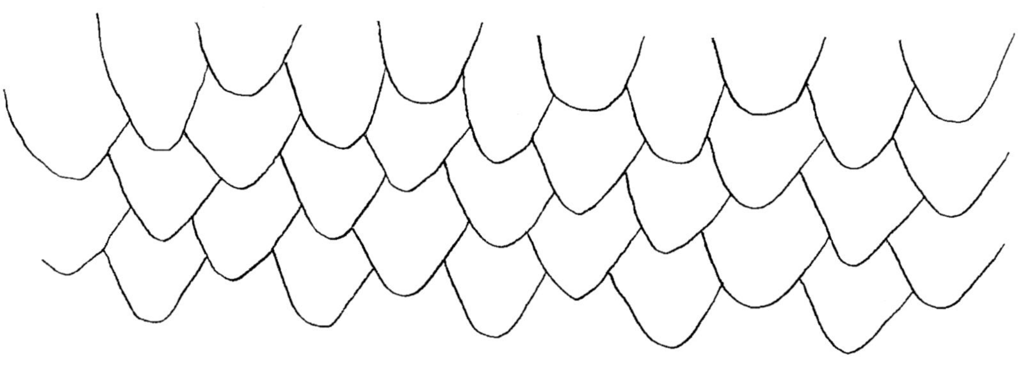

Figure 24: Unkeeled dorsal scales

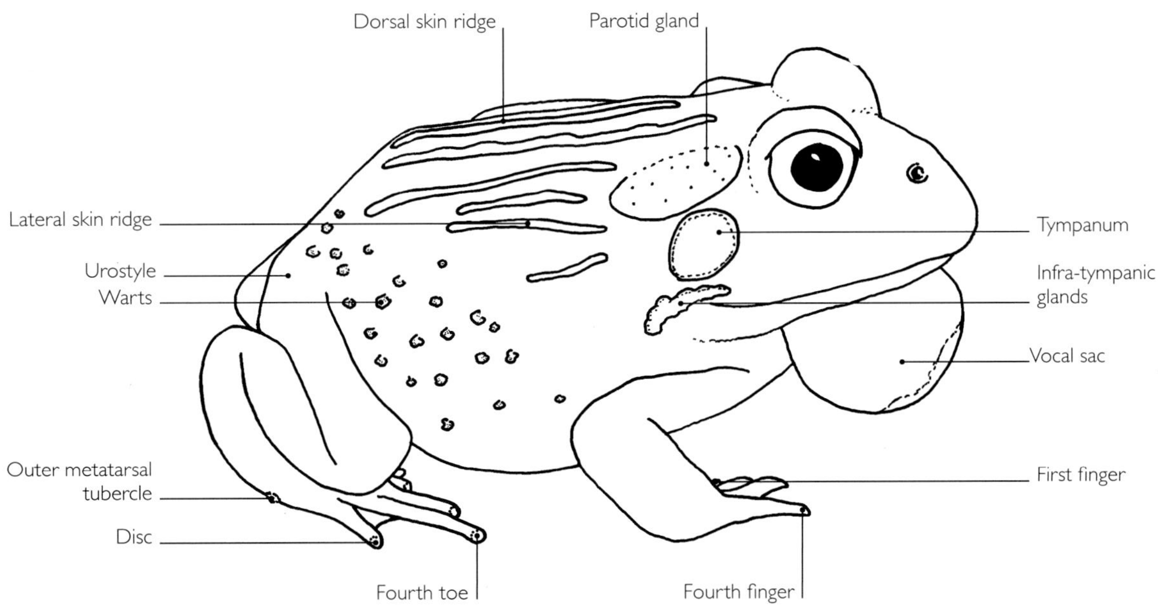

Figure 25: Features of a frog

Image Credits

Bloomsbury Publishing would like to thank the following for providing photographs and permission to use copyright material.

Key: T = top; M = middle; B = bottom; L = left; R = right; TL = top left; TR = top right; TM = top middle; ML = middle left; MR = middle right; BL = bottom left; BR = bottom right; BM = bottom middle.

Abeje Kassie: 32B, 82T, 115T, 115BR, 238; **Abubakar S Ringim:** 437T, 493; **Abubakr Mohammad:** Cover (back), 9T, 14T, 15, 16B, 19B, 37BR, 43, 45M, 116R, 127BL, 135TR, 135TM, 135BM, 146, 172M, 176TL, 259M, 295T, 295B, 296M, 296B, 308TR, 308BR, 321TM, 339B, 342R, 345T, 386T, 386M, 410, 486R, 528T, 538M, 542B, 543T, 580, 584T, 586T, 591B, 598B, 604T, 604B; **Alan Channing:** 30, 42T, 59M, 70, 80T, 80B, 99BL, 599B; **Alex MacKay:** 55, 69M; **Andre van Hecke:** 78B, 485B; **Annamaria Nistri:** 194, 346B; **Anthony Childs:** 470, 486L, 514T, 555, 584B; **Arthur Tiutenko:** 113BL; **Aviad Bar:** 290, 320, 322, 323T, 364B, 425B, 468BL; **Barry Hughes:** 139T, 583B; **Bill Branch:** 251B, 418T, 440T, 455, 459B, 498BR, 501B, 512B, 527, 540, 567R, 577R; **Bill Roland Schroeder:** 496T; **Bio-Ken Archive:** 141, 250BL, 424L, 439B, 556ML, 574T; **Breda Zimkus:** 91R; **Colin Tilbury:** 396T, 396B, 566B; **Daniel Hollands:** 36T, 313BR; **David Blackburn:** 52MR; **David Hegner:** 298B, 341B; **Davide dal Lago:** 536, 578; **Dietmar Emmrich:** 439T, 514B; **Drawing Guiseppe Scortecci:** 126B; **Edoardo Razzetti:** 156B, 224; **Eli Greenbaum:** 495; **Felix Amat (courtesy C Carranza):** 264T; **Frank Tillack:** 519T; **G D Robinson:** 12M; **George Dannhauser:** 407T; **Gerald Dunger:** 249T, 302T, 302B, 340T, 340B, 418M, 418B, 437M, 498T, 523TL, 542T, 549B, 584M; **Harald Hinkel:** 343B, 507L, 521T; **Harvard University:** 101T; **Ivan Ineich:** 381B, 444R, 473; **Jaromír Chvástek:** 94B; **Jaroslav Karhánek:** 330, 334T, 334B, 467, 474; **Jean-Francois Trape:** 177B, 252R, 274, 281, 282T, 282B, 331, 341T, 363, 404R, 444L, 485T, 517B, 518, 569; **Jean-Luc Perret:** 69B, 72T, 72B, 79, 127BR; **Jesse Erens:** 32T, 65B, 115BL, 183T, 183B; **Jiří Moravec:** 358T; **Jiří Šmíd:** 225, 241, 321TL, 435T; **Johan Marais:** 574B; **Jonas Arvidsson:** 595; **Julian Baker:** 10T, 12T, 13T; **Justin Mathews:** 544BR; **Kate Jackson:** 137, 453R; **Konrad Mebert:** 459T, 503T, 503M, 503B, 504, 512T, 523TR, 524, 596, 597M; **Laurent Chirio:** 157, 509; **Lorenzo Vinciguerra:** 454R, 531R; **Luca Luiselli:** 149; **Malcolm Largen:** 26, 41, 45T, 52BR, 62B, 68T, 78T, 82M, 82B, 92R, 107, 113T, 127T, 471B; **Marc Vaillant:** 236, 254B, 289; **Margaret Bartkowiak:** 264B; **Marius Burger:** 498BL, 571T; **Mark-Oliver Rödel:** 31, 52ML, 64R, 100, 148, 579B; **Massimo Delfino:** 360; **Max Dehling:** 508L; **Michele Menegon:** 8L, 47, 56, 61, 62T, 66B, 83TL, 83TR, 83BR, 84T, 84B, 90, 91L, 92L, 93, 114R, 288, 399, 454L, 500, 501T, 505, 565; **Mike McLaren:** 428B, 521BR, 576; **Mike Poole:** 408, 559L, 559R; **Mike Roberts:** 369; **MNHN, Antoine Fraysse** (courtesy Ivan Ineich): 463; **Mohammed Salah:** 123T, 155; **Nik Borrow:** 301T, 301B; **Nuh Ismail Awed Shire:** 468TR, 468B; **Paul Freed:** 607; **Pavel Novák:** 73, 496B, 585M; **Petr Nečas:** 66T, 284T, 394B, 397B, 398T, 402, 442T, 443T, 452T, 597B; **Philipp Wagner:** 343T; **Philippe Geniez:** 40L, 263, 371L, 371R, 475T, 475B; **Pierre-Andre Crochet:** 175, 240T, 240B, 244, 245T, 245B, 338T, 338B, 490R; **Quentin Luke:** 12B, 14B, 20, 21, 205, 287B, 592T; **Robert C Drewes:** 53, 59T, 64L, 69T, 71R, 88L, 88R, 128, 198T, 507R; **Robert Sindaco:** 118TR, 178B, 278MR, 321MR, 321BR, 323B, 361, 469, 482; **Robin James Backhouse:** 561; **Ron Gutberlet:** 445; **Royjan Taylor:** 588; **Salvador Carranza:** 167; **Sandra Goutte:** 81T, 81B, 89B, 108TL, 108TR, 108B, 112T, 113BR, 114L; **Saulo Bambi (courtesy A Nistri):** 269; **Scott Weinstein:** 570R; **Sebastian Kirchhof:** 585B; **Sherif Baha El Din:** 178T, 216, 357, 362, 364T; **Stephen Spawls:** Cover (front, BL), 8R, 11, 13M, 13B, 16T, 17T, 17B, 19T, 40R, 44T, 44B, 48T, 48M, 48B, 56, 57, 59, 76, 99BR, 102T, 106T, 109, 110L, 110R, 111M, 111B, 118TL, 118B, 119, 120B, 125T, 129T, 129B, 139B, 140L, 140R, 147T, 158L, 158R, 174, 176B, 177T, 182T, 204, 213, 250BR, 252L, 256R, 257T, 257B, 258, 259T, 260T, 262R, 266, 267, 268, 272, 277B, 278BR, 280T, 284B, 285T, 294M, 294B, 296T, 313T, 313BL, 323M, 324, 345B, 346TR, 381T, 385, 388B, 389B, 390T, 390B, 393T, 393M, 394M, 395B, 398M, 404L, 419, 421R, 423, 427, 430R, 435B, 440B, 442B, 446T, 447T, 447B, 448B, 449T, 449B, 450T, 450B, 452B, 456, 458, 465T, 465B, 483B, 494, 497L, 499, 506, 508R, 510TL, 510TR, 510B, 513T, 513BR, 516, 517T, 517M, 519B, 520TL, 520TR, 523B, 525T, 525B, 526, 531L, 534, 535, 541, 542M, 544BL, 545, 546, 547B, 548TL, 548ML, 548BR, 550T, 550B, 552T, 552B, 560T, 563L, 563R, 571B, 573, 575B, 579T, 581TL, 582L, 582R, 583T, 585TL, 585TR, 586BL, 591TL, 593, 594T, 598T, 599TL, 601T, 602L, 606TL, 606BR; **Steve Russell:** 138M, 391, 397T, 457, 597T; **Stuart Nielsen:** 60, 209T; **Thomas Hakansson:** 566T; **Tim Spawls:** Cover (spine), 71L, 77, 83BL, 111T, 395T, 411T, 513BL, 599TR; **Tomáš Mazuch:** Cover (front: main, BM left, BM right, BR) 2, 6, 9B, 10B, 22, 35, 36B, 37TL, 37TR, 37ML, 39T, 39B, 45B, 46T, 46B, 50, 65T, 68B, 87T, 87B, 95L, 95R, 99T, 99M, 101BL, 101BR, 105, 112B, 116L, 121, 122T, 122B, 123B, 124, 125B, 126T, 130T, 130B, 133L, 133R, 134L, 134R, 135BR, 138T, 151T, 151B, 152T, 152B, 154T, 154B, 156TL, 156TR, 161T, 161B, 163T, 163M, 163B, 165L, 165R, 166T, 166M, 166B, 168T, 168B, 169, 170T, 170M, 170B, 172T, 172B, 173T, 173B, 176TR, 181T, 181M, 181B, 182B, 184L, 184R, 185, 186, 191TL, 191TR, 191B, 192TL, 192ML, 192MR, 192BR, 193T, 193M, 193B, 195T, 195M, 195B, 196T, 196M, 196BL, 196BR, 198M, 198B, 200T, 200B, 201T, 201B, 202T, 202B, 203, 206, 207T, 207M, 207B, 208T, 208B, 209M, 209B, 210T, 210B, 211T, 211M, 211B, 212T, 212B, 214T, 214B, 215, 217, 218T, 218B, 219T, 219B, 220L, 220R, 221T, 221B, 222, 223L, 223R, 226T, 226B, 227L, 227R, 228T, 228B, 229, 230T, 230B, 231L, 231R, 233, 234TL, 234TR, 234B, 235L 235R, 237T, 237B, 242T, 242B, 243TL, 243ML, 243MR, 243BR, 249B, 250TL, 250TR, 251T, 253T, 253B, 254T, 255T, 255M, 255B, 259B, 260B, 261, 262L, 270T, 270B, 271T, 271B, 274T, 275T, 275B, 276, 277T, 278TL, 278TLM, 278TLB, 280B, 285B, 286, 287T, 293, 294T, 297, 298T, 299T, 299B, 303TL, 303TR, 303B, 304TL, 304BL, 304R, 305, 306T, 306B, 307T, 307B, 308TL, 308BL, 310L, 310R, 311, 312T, 312B, 314T, 314M, 314B, 315, 316, 317L, 317R, 318, 325T, 325M, 325B, 326T, 326BL, 326BR, 327, 328T, 328M, 328B, 329, 332T, 332M, 332B, 333T, 333B, 335T, 335B, 336, 339T, 342L, 346TL, 348TL, 348ML, 348MR, 348B4, 349T, 349B, 350, 351, 352T, 352M, 352B, 353T, 353BL, 353BR, 354, 355, 358B, 359T, 359B, 365L, 365R, 367T, 367B, 368L, 368R, 370T, 370B, 372L, 372R, 373T, 373B, 374T, 374B, 375, 376T, 376B, 377T, 377B, 378TL, 378TR, 378B, 379T, 379BL, 379BR, 382, 384, 386B, 387, 388T, 389T, 393B, 394T, 398B, 401L, 401R, 403T, 403B, 405, 407B, 409, 411B, 415T, 415B, 416B, 417, 420, 421L, 422T, 422B, 424R, 425T, 428T, 428M, 429R, 430L, 431, 432L, 432R, 434T, 434B, 435M, 436, 437B, 443B, 448T, 453L, 464, 466, 471T, 472T, 472B, 477L, 477R, 478, 479, 481T, 481MR, 481BR, 482, 483T, 484T, 484B, 487T, 487B, 490L, 491R, 492, 497R, 520B, 521BL, 528B, 529T, 529B, 530, 532, 533, 538T, 538B, 539T, 539M, 539B, 543B, 544T, 547T, 548MR, 549T, 553, 556TR, 556MR, 556BL, 558T, 558B, 562, 567L, 568L, 568R, 575T, 577L, 581TR, 586BR, 589, 591TR, 592B, 594B, 600, 601B, 602R, 603L, 603R, 606ML; **Tony Kamphorst:** 581B; **Václav Gvoždík:** 33, 38, 42B, 49, 58, 74M, 74B, 85, 89T, 102B, 103, 104T, 106B, 120TL, 120TR; **Van Wallach:** 488T, 488B; **Victor Wasonga:** 416T; **Vladimír Trailin:** 52TL, 54, 74T, 94T, 104B, 138B, 147B, 153T, 153B, 256L, 300T, 300M, 300B, 429L, 446B, 560B; **Watamu Turtle Watch:** 142, 143, 144, 145; **Yannick Francioli:** 570L, 587, 606MR.

Index

Bold page numbers refer to species accounts.